The Association gratefully acknowledges
publication support in the form of a loan from the
Michel T. Halbouty Fund of the AAPG Foundation.

CIRCUM-PACIFIC ENERGY AND MINERAL RESOURCES

Memoir 25

CIRCUM-PACIFIC ENERGY AND MINERAL RESOURCES

Papers from the Circum-Pacific Energy and Mineral Resources Conference held August 26-30, 1974, in Honolulu, Hawaii.

Edited by
MICHEL T. HALBOUTY,
JOHN C. MAHER,
and HAROLD M. LIAN

Published by The American Association of Petroleum Geologists
Tulsa, Oklahoma, U.S.A., 1976

Published October 1976

Library of Congress Catalog Card No. 76-41597

ISBN 0-89181-301-2

The AAPG staff responsible for editing and production of this book was:

Peggy Rice, Book Editor
Deborah L. Zikmund, Assistant Editor
assisted by
E. M. Tidwell, Assistant Editor
Nancy Greeson, Keyboarder
John D. Brainard, Cover Design

Printed by Banta West, Inc., Sparks, Nevada

DEDICATION

John Charles Hazzard
(1903-1973)

This volume is dedicated to the memory of John Charles Hazzard in recognition of his outstanding scientific research and contributions in all aspects of petroleum geology. His scientific stature and vision enhanced petroleum exploration, as well as a cooperative understanding of the earth sciences, in the Circum-Pacific region. He was one of the small group of devoted scientists who aided in the organization of the Conference and, before his untimely death, he laid the groundwork for the Technical Program from which the papers are published in this Memoir.

CONTENTS

HYDROGEOLOGY

FOREWORD

Scientific conferences are numerous, but rarely does one become a landmark in new fields and areas of endeavor. The first Circum-Pacific Energy and Mineral Resources Conference, held in August 1974 in Honolulu, was one of these rare occasions. It marked the beginning of new international cooperation on a continuing basis among government, industry, and academia in about half the world.

In February of 1973, the Executive Committee of The American Association of Petroleum Geologists (AAPG) approved a recommendation from many of its members that it sponsor a conference on the energy and mineral resources of the vast Circum-Pacific region. The Committee for Coordination of Joint Prospecting for Mineral Resources in Asian Offshore Areas (CCOP) and the Pacific Science Association (PSA) later joined as cosponsors, although AAPG retained full fiscal responsibility. The Conference was supported by a $30,000 grant from the National Science Foundation.

The tremendous interest in the meeting is illustrated by the fact that 46 other cooperating scientific organizations representing approximately 275,000 scientists supported the Conference. This unique broad support was instrumental in the tremendous success of the Conference. Large conferences usually require a lead time of at least 4 years but, in the case of the first Circum-Pacific Conference, plans were effectively implemented within the short time of 18 months through the dedicated efforts of many scientists of 26 nations bordering the Pacific Ocean.

The results of the Conference were far-reaching not only in the knowledge disseminated, ideas exchanged, and the international contacts gained, but also in the creation of a permanent Circum-Pacific Council on Energy and Mineral Resources to promote, in varied ways, geological and geophysical exploration, protection of the natural environment, and goodwill and understanding of energy and mineral problems among Pacific nations. The Council maintains and supports (1) a Conference Committee to repeat the Conference each 4 years, at different locations if desired, (2) a Geoscience Training Committee to organize and manage traveling workshops, and (3) a Map Committee to prepare a new map series of the Pacific region. Members of these committees are distributed around the Pacific.

The Map Committee is preparing a series of colored maps which include a geographic map, a geologic map, a tectonic map, a mineral-resources map, and an energy-resources map for each of four quadrants of the Pacific region and several maps of the Antarctic region showing geology and resources. These colored maps, at a 1:10,000,000 scale, will show geologic relationships from oceanic to adjacent continental areas with minimum distortion. In addition, a Pacific Basin map being prepared at a 1:20,000,000 scale is designed to integrate the major geographic, oceanographic, and geologic features that will appear on the Circum-Pacific quadrant and Antarctica maps. This project represents one of the largest cooperative map projects ever undertaken, and this map series alone would be worth all the effort expended on the Conference. The geographic maps will provide base maps for many individual projects by geologists of many nations for a long time in the future. The geologic, tectonic, and resources maps will be instrumental in establishing new targets for exploration for the energy and mineral resources of the whole Pacific Ocean and borderlands. Some maps will be completed and shown at the second Circum-Pacific Energy and Mineral Resources Conference to be held in Honolulu in August 1978.

Michel T. Halbouty
Houston, Texas 77056

John C. Maher
Menlo Park, California 94025

Harold M. Lian
Los Angeles, California 90051

SPECIAL EDITORS
August 3, 1976

INTRODUCTION

The first Circum-Pacific Energy and Mineral Resources Conference was a landmark in international scientific cooperation between governments and industry. There were 813 delegates from 31 nations, along with 423 members of their families, for a total attendance of 1,240 persons. They represented industries, universities, and governmental agencies from 31 countries, namely: Australia, Canada, Chile, Colombia, Ecuador, England, Fiji, France, Germany, Guam, Indonesia, Italy, Japan, Korea, Kuwait, Malaysia, Mexico, Netherlands, New Caledonia, Norway, New Zealand, Peru, Philippines, Saudi Arabia, Singapore, Taiwan, Thailand, Union of Soviet Socialist Republics, United Kingdom, United States of America, and Venezuela. The majority of the delegates were from the petroleum and mining industries, as might be expected, but many representatives came from other industries dependent on energy and minerals for their products, and from financial institutions, research institutions, universities, and government. One of the surprising and significant statistics from registration information was that 46 percent of the registrants were identified as upper-echelon decision-makers—presidents, vice-presidents, directors, chairmen, chief geologists, chief geophysicists, exploration managers, branch chiefs, agency heads, or other officers in charge of organization or programs.

The principal objective of the Conference was to provide a forum for presenting and discussing papers on the potential energy and mineral resources of the Circum-Pacific region emphasizing the exploration and development of energy and mineral resources in a manner compatible with the environment. The papers were designed (1) to focus attention on present and future availability of energy and mineral resources in land and offshore areas bordering the Pacific Ocean; (2) to consider where and how exploration efforts may be most effectively directed; (3) to enlist as speakers and authors those scientists with experience and special know-how in the Circum-Pacific region; and (4) to encourage the free exchange of pertinent ideas and information among Circum-Pacific countries so as to solve common problems of exploration and development of energy and mineral resources and protection of the environment. We can state that the objectives of the Conference were fully achieved.

The papers included subjects in geology and geophysical research; exploration and development of oil, gas, and coal and of geothermal, solar, and nuclear energy; exploration and development of ferrous and nonferrous metallic minerals and nonmetallic minerals; and the pertinent economic and environmental problems. One hundred twenty-six papers presented by authors from 16 countries are published in whole or in part in this *Memoir*. These consist of 17 general papers, 7 papers on coal, 15 papers on geothermal energy, 28 papers on hydrocarbons, 13 papers on hydrogeology, and 46 papers on minerals. This volume should prove to be an indispensable reference for scientific, technical, and economic development among all Pacific nations.

Many chairmen and committee members working in concert exerted the tremendous effort necessary to the success of this landmark Conference. All deserve the heartfelt thanks of the international earth-science community. Lists of the Conference committees and cooperating organizations which assisted in sponsoring the Conference follow.

It is anticipated that the voluminous amount of data contained in the papers published in this *Memoir* will prove to be keys to the discovery of new mineral and energy resources in the Circum-Pacific region.

Michel T. Halbouty
Houston, Texas 77056

John C. Maher
Menlo Park, California 94025

Harold M. Lian
Los Angeles, California 90051

SPECIAL EDITORS

CIRCUM-PACIFIC ENERGY AND MINERAL RESOURCES CONFERENCE
HONOLULU, HAWAII
August 26-30, 1974

Convention Headquarters: Mid-Pacific Conference Center, Hilton Hawaiian Village

CONFERENCE ORGANIZATION

MICHEL T. HALBOUTY
General Chairman

JOHN C. MAHER
General Vice-Chairman, Operations

EXECUTIVE COMMITTEE

J. C. MAHER, *Chairman*
M. T. HALBOUTY
K. H. CRANDALL
E. W. ELLSWORTH
D. A. HENRIKSEN
H. M. LIAN

DAVID McGARRY
General Vice-Chairman, SW Pacific

C. Y. LI
General Vice-Chairman, NW Pacific

F. B. CONSELMAN
General Vice-Chairman, Americas

E. W. ELLSWORTH
Conference Manager

KATHY WATSON
AAPG Convention Manager

COORDINATING COMMITTEE

MICHEL T. HALBOUTY, *Chairman*
T. D. BARBER
F. B. CONSELMAN
K. H. CRANDALL
W. H. CURRY
E. W. ELLSWORTH
R. W. FORCE
D. A. HENRIKSEN
C. Y. LI
J. C. MAHER
DAVID McGARRY
J. C. MOODY

ADVISORY COMMITTEE

K. H. CRANDALL, *Chairman*
J. M. BROWNING
R. A. BURKE
IAN CAMPBELL
W. H. CURRY
E. M. DAVIN
L. H. FUNKHOUSER
M. W. HAAS
H. D. HEDBERG
R. W. HEGGLAND
D. A. HENRIKSEN
M. J. HILL
M. L. HILL
LINN HOOVER
K. M. HORLER
J. C. MAHER
J. C. MAXWELL
R. H. NANZ, JR.
R. B. PALMER
R. P. SHELDON
L. W. STACH
L. G. WEEKS
R. J. WEIMER
J. D. WEIR
S. A. WENGERD

A. E. ROBERTS
Technical Program Coordinator

E. I. RICH
Program Editor

S. T. WATSON
Slide Editor

J. H. SAWYER
Operations Coordinator

TECHNICAL PROGRAM COMMITTEE

H. M. LIAN, *Chairman*

R. D. STEWART, *Vice-Chairman*

J. F. MASON, *Deputy Chairman*

A. E. MORRIS, *Deputy Chairman*

KASPAR ARBENZ
Hydrocarbons Coordinator

S. J. MUESSIG
Minerals Coordinator

L. T. GROSE
Geothermal Coordinator

J. H. FETH
Hydrogeology Coordinator

ISAMU KOBAYASHI
Vice-Chairman, NW Pacific
MASAMI HAYAKAWA
YASUFUMI ISHIWADA
SHUN-ICHI SANO
YOSHIHIKO SHIMAZAKI
TATSUO TATSUMI
SHIGEMOTO TOKUNAGA
SOKI YAMAMOTO

R. J. FOSTER
Vice-Chairman, SW Pacific
N. O. JONES
DAVID KEAR
J. F. LOVERING
J. C. NIXON
K. A. RICHARDS
B. P. WEBB
L. W. WILLIAMS

D. A. MABRA
Vice-Chairman, NE Pacific
D. A. CORDOBA
GABRIEL DENGO
J. R. DUNCAN
HUGH GABRIELSE
GEORGE GRYC
J. D. MOODY
T. L. WRIGHT

M. CHURKIN, JR.
Vice-Chairman, Russia

J. E. MACK, JR.
Vice-Chairman, SE Pacific
ROGER BRISTOW
REYNALDO CHARRIER
D. A. CORDOBA
GABRIEL DENGO
HERMAN DUQUE (CARO)
J. B. KENNERLY
WILLIAM PACE
CARLOS DEL SOLAR
S. R. VALENZUELA

CIRCUM-PACIFIC MAP COMMITTEE

J. A. REINEMUND, *Chairman*
D. M. KINNEY
Vice-Chairman, Operations
J. C. MAHER
Deputy Vice-Chairman, AAPG
CHIKAO NISHIWAKI
Vice-Chairman, NW Pacific
DAVID KEAR
Vice-Chairman, SW Pacific
C. A. BURK
Vice-Chairman, NE Pacific
JOSE CORVALAN
Vice-Chairman, SE Pacific
F. A. WADE
Vice-Chairman, Antarctica
M. T. HALBOUTY
Advisory Committee Chairman
A. L. CLARK
Computer Committee Chairman

PUBLICATIONS COMMITTEE

F. E. KOTTLOWSKI, *Chairman*
G. H. FRIEDMAN
A. A. MEYERHOFF
AAPG Associate Editors for Pacific Region

AAPG MEMBER RELATIONS COMMITTEE

A. O. SPAULDING, *Chairman*
MEMBERS—PRESIDENTS of all AAPG Sections and Societies, and those of the AAPG International Associates in Colombia, Ecuador, England, Mexico, and Venezuela

AWARDS COMMITTEE

F. B. CONSELMAN, *Co-Chairman*
M. L. HILL, *Co-Chairman*

BUDGET AND FINANCE COMMITTEE

D. A. HENRIKSEN, *Chairman*
O. VAN BEVEREN

PRINTING COMMITTEE

KATHY WATSON, *Chairman*
W. B. BEATTY

PUBLICITY COMMITTEE

E. E. BRABB, *Chairman*
J. A. CLARK
J. R. JACKSON, JR.
O. F. KOTICK
W. H. LeROY
J. C. McCASLIN
J. L. MUELLER
C. B. PAMENTER
F. P. SONNENBERG
V. E. STEPHENS
K. A. YENNE

REGISTRATION COMMITTEE

D. H. PFEIFFER, *Chairman*
M. A. MEYLAN, *Vice-Chairman*

TRANSPORTATION COMMITTEE

BOB GRIEDER, *Chairman*
P. W. GESTER

HONOLULU COMMITTEES

J. E. ANDREWS, *Local Chairman*
S. V. MARGOLIS, *Local Vice-Chairman*

EDUCATIONAL EXHIBITS COMMITTEE

W. M. ADAMS, *Chairman*
E. BERG, *Vice-Chairman*
J. H. FETH
KUMIJI IIDA
W. A. MINOZA
G. L. PICKARD

ENTERTAINMENT COMMITTEE

P. M. KROPNICK, *Chairman*

FIELD TRIP COMMITTEE

A. T. ABBOTT, *Chairman*
D. A. DAVIS
L. I. KNOWLES
DANIEL LUM
G. A. McDONALD
D. W. PETERSON

HOTELS AND HOUSING COMMITTEE

DANIEL LUM, *Chairman*
R. H. DALE
K. J. TAKASAKI

INFORMATION COMMITTEE

JOHANNA RESIG, *Chairman*
PAMELA MULLER

RECEPTION COMMITTEE

C. D. FEIN, *Chairman*
ALICE LUM

TECHNICAL SERVICES COMMITTEE

CHESTER LAO, *Chairman*
STANLEY SUMIDA

CIRCUM-PACIFIC ENERGY AND MINERAL RESOURCES CONFERENCE
HONOLULU, HAWAII

August 26-30, 1974

Sponsored Jointly By

The American Association of Petroleum Geologists
The Committee for Coordination of Joint Prospecting for Mineral Resources in Asian Offshore Areas
The Pacific Science Association

in cooperation with

American Association for the Advancement of Science
American Geophysical Union
American Institute of Mining, Metallurgical, and Petroleum Engineers
American Institute of Professional Geologists
American Society of Photogrammetry
Association of Earth Science Editors
Association of Engineering Geologists
Australasian Institute of Mining and Metallurgy
Australian Petroleum Exploration Association
Canadian Society of Petroleum Geologists
Colombian Society of Petroleum Geologists and Geophysicists
Committee for Coordination of Joint Prospecting for Mineral Resources in Southwest Pacific Areas
Division of Earth Sciences, National Research Council, National Academy of Sciences
Earth Sciences Section of the National Science Foundation of the United States
Geochemical Society
Geological Association of Canada
Geological and Mineral Institute of Korea
Geological Society of America
Geological Society of Malaysia
Geological Society of the Philippines
Geological Society of Thailand
Geological Survey of Papua New Guinea
Geoscience Information Society
Geothermal Resources Council
Hawaii Department of Land and Natural Resources
Honolulu Board of Water Supply
Institute of Petroleum (London)
International Decade of Ocean Exploration of the National Science Foundation of the United States
International Union of Geological Sciences
Japan Geothermal Energy Association
Mexican Geological Society
Mineralogical Society of America
National Association of Geology Teachers
Pan American Institute of Geography and History, Commission of Geophysics
Petroleum Exploration Society of Great Britain
Royal Netherlands Geological and Mining Society
Scripps Institution of Oceanography
Seismological Society of America
Servicio Geológico de Bolivia
Society of Economic Geologists, Inc.
Society of Economic Paleontologists and Mineralogists
Society of Exploration Geophysicists
Society of Mining Engineers of AIME
Society of Petroleum Engineers of AIME
Southeast Asia Petroleum Exploration Society
U.S. - Japan Cooperative Program in Natural Resources

OFFICERS OF SPONSORING ORGANIZATIONS AT TIME OF CONFERENCE

THE AMERICAN ASSOCIATION OF PETROLEUM GEOLOGISTS

Merrill W. Haas, *President*
Duncan A. McNaughton, *Vice-President*
John E. Kilkenny, *President-Elect*
Bernold M. Hanson, *Secretary*
George C. Grow, Jr., *Treasurer*
Frank E. Kottlowski, *Editor*
Hugh N. Frenzel, *Chairman, House of Delegates*

THE COMMITTEE FOR COORDINATION OF JOINT PROSPECTING FOR MINERAL RESOURCES IN ASIAN OFFSHORE AREAS

C. Y. Li, *Secretary*
Member countries and their permanent representatives to CCOP are as follows:
Indonesia: Ismet Akil, *Chief, Geological Dept., Pertamina*
Japan: Isamu Kobayashi, *Director of the Geological Survey of Japan*
Khmer Republic: Sean Pengae, *Minister of Industry*
Republic of Korea: Joung Hwan Lee, *Director of the Geological and Mineral Institute of Korea*
Malaysia: Chung Sooi Keong, *Director General of the Geological Survey of Malaysia*
Philippines: Juanito Fernandez, *Director of the Bureau of Mines*
Republic of Vietnam: Vo An Tuan, *Deputy Director of the Office of the National Petroleum Board*
Singapore: Tan Suan Bing, *Head of the Geological Department*
Thailand: Saman Buravas, *Director General of the Department of Mineral Resources*

THE PACIFIC SCIENCE ASSOCIATION

Ian McT. Cowan, *President and Chairman of Pacific Science Council*
Roland W. Force, *Chairman, Executive Committee of Council*
Yoshio Hiyama, *Chairman, Finance Committee*
W. S. Hoar, *Chairman, Membership Committee*
J. A. R. Miles, *Chairman, Policy Committee*

GENERAL

Importance of Circum-Pacific Energy and Mineral Resources to World[1]

ROGERS C. B. MORTON[2]

It is a pleasure to add my own warm welcome to those of Acting Governor Ariyoshi and Mayor Fasi. I am pleased beyond measure by the magnificent collaboration among the governments and scientists of the Circum-Pacific countries in planning this conference and in jointly addressing themselves to examining the energy and mineral resources of this great region.

I am told that nearly 67 nations are participating in this conference, the first ever called to consider the total mineral-resource potential of an entire region. The published transactions of this meeting will lay a foundation for the continuing effort to locate and assess the mineral riches of this enormous sector of the world.

I want especially to commend the ongoing effort to map the Pacific Ocean area. Although it will be several years before the Circum-Pacific Map Project will be completed, every increment that is published will have immediate usefulness to those who have some responsibility for providing the vital raw materials on which our countries must depend.

A major function of the U.S. Department of the Interior is that of mapping the features of the United States, both natural and man-made. On the personal side, I have an aviator's respect for maps and a firsthand knowledge of their critical importance to the user, and I am wholeheartedly in support of the effort to provide these vital guides to explorationists who will be called upon to help us navigate efficiently to the mineral supplies of tomorrow. I wish success to all who are connected with this most important undertaking.

The United States has had close and enduring ties with the Pacific region since early in the last century: from the China trade carried on by the great Clipper ships, to Commodore Perry's mission to Japan, to the accession of Guam and Hawaii in 1898, to the present responsibilities the United States holds under the United Nations for the welfare of the Micronesian peoples residing in the Trust Territory of the Pacific islands, and to friendly and mutually beneficial relations with Circum-Pacific nations. More than 7,500 mi (12,000 km) of United States coastline borders the Pacific, and for more than 30 years the United States has been involved in close technical and scientific cooperation with other Pacific countries.

Scientists from my own country in the Department of the Interior have worked in most of the countries represented at this conference, and have worked jointly with many of their scientists. Contrary to the implications of its name, the United States Department of the Interior has helped establish or strengthen earth-resources agencies through major assistance programs in more than 20 Pacific countries.

The Interior Department and other U.S. agencies are involved in many major scientific research efforts in or adjacent to the Pacific region, including Antarctic mapping and research, the International Geological Correlation Program, and ocean exploration and drilling. The Pacific region is not only a major area to explore for resources; it is also the world laboratory for the study of active processes of geotectonics as they relate to the origin of energy and mineral resources.

By any measurement, the Pacific region is enormous. A billion people live along its shores. The Ring of Fire, in a giant arc from New Zealand to Patagonia, numbers more active volcanoes than all the rest of the world combined. The entire land area of the earth could easily be contained in the Pacific basin. To identify and assess the mineral wealth of such a region will be a formidable task indeed.

Precisely because of its gigantic proportions, the Pacific region must be explored for its ability to contribute to the rising needs of the world's people. The problems posed by these needs are themselves of giant proportions. There are at least

[1]Manuscript received, August 29, 1974.

[2]Secretary, U.S. Department of the Interior; now Secretary, Department of Commerce, Washington, D.C. 20230. Secretary Morton's speech was read by Vincent E. McKelvey, Director, U.S. Geological Survey.

3.8 billion of us now, and the experts expect the world population to be 7 billion by the turn of the century. Not only will there be more people, but we can be sure they will not be content with today's level of living. Understandably and justifiably, they will want something better.

"Something better" for 7 billion people means enormous new demands for materials of all kinds, especially minerals. This is not a conjecture on my part; it is an observation of trends already evident.

Twenty-five years ago, for example, the United States produced approximately half the world's minerals, by value, and consumed about the same share. Today it accounts for no more than 30% of mineral consumption and 25% of production. The growth in demand for minerals by other countries has far exceeded that of the United States, and our share of consumption has diminished accordingly.

The result of this large sudden increase in demand for materials of all kinds was to break the stable mold of long-term real prices of mineral commodities. Since 1963 these prices have been advancing—at first slowly, but lately very rapidly. In 1973 the price of a pound of copper on the London Metal Exchange was $0.60; in 1974 it was $0.86. Lead went from $0.16 to $0.24 per pound, zinc from $0.21 to $0.35, and tin from $2.18 to $4.49 during the same interval. The price of steel scrap in the United States almost tripled from 1973 to 1974, and is currently about $145 a ton. The point here is not the prices themselves, but what they say about the outlook for future supply.

Let me cite the simple but critical example of food. The people of the highly industrialized nations eat well, and one of the primary features of their diet is its high content of animal proteins. But it is costly to "eat high on the hog," if I may use an American rural phrase, for this high-protein diet requires for its production about five times as much primary energy in the form of sunlight as does the average diet in Asia or Africa. To achieve a better quality diet for all peoples will require that a vast amount of additional crop and pasture land be put into use. The additional acreage will require tremendous amounts of fertilizer, water, pesticides, machinery, and all other concomitants of efficient agricultural operation. The provision of these supplies will demand, in turn, enormous amounts of minerals and energy.

In regard to fertilizer, one of the most severe problems to come out of the skyrocketing prices of petroleum since 1973 has been their impact on the production of chemical fertilizers—which are critically important to the cultivation of the high-yield strains of wheat and rice that are the foundation of the "Green Revolution."

Meanwhile, the population bomb ticks relentlessly on: 225 babies born every minute, 93 deaths every minute, 70 million people added each year—a rate which would double the world's population by the year 2010.

I do not pretend to know what the world's demands for minerals will be as we enter the next century, except that they seem likely to be multiples of what they are today. The exact numbers are not important. What is important is that people everywhere come to realize that the provision of mineral supply, including the mineral fuels, will be the most important problem they have to deal with in the last quarter of this century. Everything else depends on it, including the production of food and fiber. Mineral supply is now everybody's business, everywhere in the world!

There has been much argument in recent years over the question as to whether or not the earth's resources will soon be exhausted by ever-increasing demands. Some people talk about the future glut of energy and minerals as producers respond to higher prices, and some see an urgent need to achieve zero growth in population and resource consumption. The earth's carrying capacity, of course, will not long sustain a doubling of the human population every 35 years, which is the implication of the present rate of increase. At some time, we shall indeed reach a point of equilibrium between population and resources—if not by design, then later of necessity.

There is a world of difference between (1) conceding and preparing for the ultimate arrival of a steady state between population and resources and (2) supinely resigning ourselves to some inevitable doomsday soon to come. The fact is that we have the capacity to create additional resources by discovery of deposits whose existence and location are presently unknown to us. And we can increase our ability to use the resources we have more efficiently. These are the processes by which men have satisfied their increasing needs throughout human history.

Our demands now are much greater than ever before—but so are our capabilities for meeting them if we are willing to make the effort and apply ourselves to the problem with diligence and imagination.

Everything that we can infer about the presence of minerals in the earth's crust leads us to believe that vast resources remain to be discovered and developed. We have many recent reminders of this: the huge iron and bauxite deposits of Australia, discovered within the last 25 years in a continent that had been explored for more than 200 years; new discoveries of tin deposits in the British Isles, where tin had been mined for 4,000 years; and newly discovered deposits in the lead belt of my own country where lead had been mined extensively for 2 centuries. Here in Hawaii we are in the midst of a region that contains enormous deposits of metal-rich nodules on the seafloor for which the recovery technology is now within our grasp.

It is true that we are now facing shortages in some energy and mineral commodities, but I believe we still face a largely unexplored horizon, rich in opportunities to provide the resources the world needs. More of a problem than the adequacy of our potential is the problem of our attitude about what it takes to meet future needs. The specter of scarcity has risen and been put to rest so many times in the past that there is a tendency to regard the present warnings as no more than passing annoyances.

But we can ignore these signals only at our peril. New sources of yet-to-be-discovered minerals do not materialize overnight. There must be years of exhaustive exploratory and development effort, billions of dollars of investment, and a vast amount of patient, cooperative effort among nations to solve the political, technological, and economic problems that bar access to the new supplies. It takes time also to develop the technologies to use resources that are now too costly to produce, and to increase significantly the efficiency of resource use to make them go further and last longer.

The essential ingredient—as yet largely missing—is a reservoir of popular support for the costly and difficult actions that must be taken by both government and private enterprise to assure that timely discoveries of new supplies are made and that conservation science and technology advance in a timely manner. At best, there is a long interval between the time when remedial actions are begun and the time when their effect begins to be evident in terms of increased supplies. Unless these actions are begun early, before urgency sets in, the likelihood is that their results will come too late to be of any real benefit. I am saying simply that the time to begin these actions is *now*.

This conference points in the true direction we should be traveling. It is a bold venture—to join together to examine the total mineral-resource potential of half of the world's area. It is a recognition of the essential interdependence of all the world's peoples and the importance of working together to help solve world resource problems.

An adequate attack on these problems, including research to improve technology, requires joint government and private-company efforts; the cooperation of national and international agencies; and mechanisms for articulating and implementing these cooperative actions. The Circum-Pacific Conference and Map Project are prime examples of the kind of mechanisms so critically needed to develop new concepts and turn them into specific results.

To the sponsoring organizations, and to Michel T. Halbouty in particular, I take this opportunity to express my gratitude for assuming the burden and responsibility for arranging this vitally needed convocation at the geographic crossroads of the Pacific.

May this be the beginning of a long and fruitful effort by the Pacific countries to develop the resources of this great region for the benefit of mankind.

The Macroproblem of Modernization[1]

HARLAN CLEVELAND[2]

Abstract Mankind has demonstrated an enormous capacity to solve the extremely difficult and complex technical problems of modern civilization—but this is not enough. The challenge which must now be met is to recognize the fundamental interdependence of all nations and to devise a global strategy which will ensure that our miraculous technologies will lead to our fulfillment rather than our destruction.

INTRODUCTION

Each of us here at this luncheon now has partaken of enough protein and calories to last one of the world's disadvantaged people for 2 days or more. So we should be ready to face the world. It is high time we do just that—face the world, not part of it, not one or another function or commodity or industry, but all of it at once.

I make no apology for speaking in the down-to-earth language of general theory. As practical men and women, we can do nothing more practical today than think comprehensively, inclusively, about the situation as a whole.

We humans are operators—we are very good at getting rapidly and efficiently to where we do not want to be when we get there.

Ten years ago, a conference such as this would have focused on the *exploitation* of resources and the *production* of energy. But the moral climate does not permit so narrow a focus today. The key questions about Pacific energy and mineral resources are no longer just "How?" and "When?" and "How much?"—they also include "Why?" and "For whom?"

To build systems for the discovery and development of energy and minerals, in new forms and places including the deep ocean, will present a technical challenge of enormous complexity, at least as complex as the challenge of outer space and advanced-weapons systems. At the University of Hawaii, we are launched on a major effort to find and package geothermal energy from volcanoes, and our legislature has authorized a natural energy laboratory which will work to convert to human use the flood of energy from our year-round sunlight. Because I am largely ignorant of the problems which the scientists and engineers among you will have to solve in developments like these, I am serenely confident that you will solve them.

But my profession is to conceive and build the institutions to divide the wealth which you produce, train the people whom you need, analyze the alternative choices that your technical prowess makes possible, and formulate the values which your energy and minerals will serve. In this area about which I *do* know something, it is more difficult to be confident that Man will learn in time how to harness his runaway power to a restraining purpose.

One of America's most thoughtful historians, the one who first dared to include the future as part of history, recently has produced a powerful and pessimistic prognosis. "Is there hope for man?" Robert Heilbroner (1974) asks, and gloomily answers his own question.

We are entering, says Heilbroner, an era ". . . in which rapid population growth, the presence of obliterative weapons, and dwindling resources will bring international tensions to dangerous levels for an extended period. . . . " Human nature is quite unable to cope with the changes human beings have set in train. "Thus in all likelihood we must brace ourselves for the consequences. . .the risk of 'wars of redistribution' or of 'preemptive seizure,' the rise of social tensions in the industrialized nations over the division of an ever more slow-growing or even diminishing product, and the prospect of a far more coercive exercise of national power as the means by which we will attempt to bring these disruptive processes under control. . . . If then, by the question 'Is there hope for man?' we ask whether it is possible to meet the challenges of the future without the payment of a fearful price, the answer must be: No, there is no such hope."

We have on our agenda more urgent tactical issues and more fascinating technical puzzles. But we have no more important mission at this conference than to consider how, in the great laboratory of the Pacific, we are going to prove Heilbroner wrong. And the agony of the historian's writing shows how badly Heilbroner would like to be proved wrong.

[1]Luncheon address, Circum-Pacific Energy and Mineral Resources Conference. Manuscript received, September 16, 1974.

[2]President, University of Hawaii, 1969–74; now Director, Program in International Affairs, Aspen Institute for Humanistic Studies, P. O. Box 2820, Princeton, New Jersey 08540.

Ten Parts of the Macroproblem of Modernization

A good place to start is to look honestly at the changing assumptions, aspirations, and attitudes that make the world's peoples so much more interdependent than their institutions—and their leaders' rhetoric—yet express. Each of the elements mentioned is familiar in the abstract. But it helps to look at them all at once, as 10 parts of the macroproblem of modernization.

1. Values once taken as basic are now in question. Long-time defenders of democracy are beginning to wonder, with Heilbroner, whether authoritarian government is not, after all, the only way out. In the "post-industrial" countries, quality of life is challenging quantitative growth as the central goal of national development. Charity turns out to be too fragile and unjust a basis for aid from the rich to the poor. Doctrines such as freedom of the seas and national control of communications are questioned as ethically and technologically obsolete. Those steadying institutions, the extended family and religious traditions, are enormously weakened—whether beyond repair is part of the macroproblem.

2. Concepts of the world as a biosphere, of different peoples as common residents of a global village, of future generations as possibly having no future if we do not act now, are taken seriously by many scholars and writers, if not yet by most statesmen and political leaders.

3. For the first time in history, barriers of prudence are seen as limiting what people can safely do to nature, or to other people. Even if it cannot yet be defined, we sense now that there is an outer limit beyond which Man cannot exploit resources without self-destruction; in the mastery of our environment, the costs are all too obviously catching up with the benefits. Nuclear weapons are likewise setting an upper limit to warfare; *guerre à outrance*, war to the uttermost, has so far been rejected in practice, if not always in theory, by every leader who acquaints himself with the potentials of nuclear conflict.

4. The "population bomb" is still ticking, and is no less dangerous now that the simultaneous equation of births and deaths is a staple of paperback literature. The least alarmist experts think the world's population will double by the year 2000 or shortly thereafter. In the last quarter of this century, the requirements for shelter and facilities for all these people represent an enormous quantum of effort, equivalent (says the Club of Rome) to a thousand years of past construction. We can probably grow enough food for a world population of 6 or 7 billion, but we have not yet demonstrated that we can get it past the institutional obstacles into malnourished and starving households. And climatologists are now warning that weather changes in the decades ahead may also be adverse to food production.

5. The "soft" technologies of the biological revolution seem likely to do even more to change the world than the industrial/nuclear/electronics revolutions we have just been experiencing. Machines and drugs and procedures that can inspire or tranquilize people, alter human personality, raise or lower intelligence, enhance or impair memory and learning, make births more various or uniform, and extend the very frontiers of death, present agonizing new choices for individuals, for families, and for public policy.

6. Each technological breakthrough thus adds to mankind's moral burden, by opening doors that used to be nailed shut and making it possible to avert tragedies that previously could only be deplored. Now that many diseases are preventable and food for starving people can physically be produced and delivered, the act of *not* preventing, producing, and delivering becomes suddenly outrageous. When weather can be modified at human command, the ethics of changing other people's weather will raise a whole new set of reasons for international cooperation.

7. The revolution in communications and transportation connects peoples, and their leaders, with each other in ways that are inherently revolutionary. No violence is "local"; no disruption is "one of a kind"; and no fad or fashion is limited by geography or culture if the mass media are looking on. No place need be remote that has a jet runway, direct-dial telephones, satellite radio and television, and terminals connected to a major computer; satellite exchange of data packages will make the world of information rounder than the world of coal and steel could ever be. The technologies that shorten time and distance are not necessarily "progress"; moreover, they make society's web more vulnerable to individual desperadoes or small groups of terrorists.

8. A proliferation of serious trade disputes, the breakdown of the dollar as the world's "key currency," and a new transnational pattern of "inflation with unemployment" threaten to produce ". . . a vicious cycle of competition, autarchy, rivalry and depression such as led to the collapse of world order in the Thirties" (quoted from speech by Secretary of State Henry Kissinger at 1974 Washington Energy Conference). More workable international rules are urgently needed, both for commerce and for money.

9. The rich nations and the poor nations are overdue for a new, unpatronizing relationship. The coming struggles over resources, including energy, food, and metals, will require ways of thinking and institutional arrangements which get beyond the postwar concepts of national development planning aimed at economic growth and supported by aid as charity. Again the moral issues are breaking through the surface of events. For example, "Why should the poor in the rich countries help the rich in the poor countries?" (Camps, 1974).

10. In the years just ahead, minimum human needs threaten to outdistance dramatically the world's capacity to produce and deliver them. (Again the moral dimension is paramount: do the rich let enough of the poor die to relieve the population pressure in the Malthusian way, or do we rescue them whatever it costs? And if the starving are rescued, do we face what Heilbroner [1974] gloomily calls ". . . the danger that the Malthusian check will be offset by a large increase in food production, which will enable additional hundreds of millions to reach childbearing age"?)

Need for Interdependence Among Nations

The interdependence of nations has long been a familiar litany. It must have been 20 years ago when the nationalistic American was first described as sitting in his Danish chair, wearing his Hong Kong suit, drinking Brazilian coffee out of an English china cup, writing a letter on Irish linen paper with a Japanese ball-point pen. The letter is to his Congressman, and in it he complains that too much gold is going overseas.

Yet the nation-state has proved to be very durable indeed, as a focus of loyalty and a channel for political ambition. And in the United States at least, each international crisis strengthens in their resolve those who would limit our foreign entanglements.

The 1973–74 energy crisis was no exception. The rhetoric of Project Independence soon went beyond prudent conservation and active research on alternatives to fossil fuels, to put an overriding value on cutting off imports of energy by 1980. That goal is not only unrealistic—my impression is that by spending one third of a trillion dollars we could only reduce our dependence from 18% to about 10% by 1985—but almost certainly undesirable as well. If America strives for autarchy at enormous costs, weaker and more vulnerable nations will strive even harder to protect themselves. The resulting rivalries and the ugly scramble for scarce resources could bring the collapse of world order which Secretary of State Kissinger warned against early in 1974.

Despite an occasional flourish such as Project Independence, the internationalization of the world economy is now so far advanced as to be an element in each nation's domestic politics. National political leaders live on the presumption that they can affect the destinies of the people they lead. Yet no thoughtful national leader can honestly claim that his government has under control the global trends that determine whether his constituents prosper or are impoverished, develop or stagnate, get sick or stay well, live or die. The determining factors are no longer "domestic affairs."

The very protection of "national sovereignty," not to mention the fixing of oil prices, requires collective action with other sovereignties.

Energy, pollution, human rights, racial tensions, education, science, technology, business, labor, food, transportation, population, culture, communication, terrorism, revolution, arms, drugs, religion, ideology—name a category and its international aspects are likely to outweigh its "domestic" aspects, now or soon.

No nation now controls even that central symbol of national independence—the value of its money; inflation and recession are partly transnational. They certainly are not just domestic diseases for which all-American committees should be established to seek a cure. Inflation and unemployment are, of course, part of a global epidemic. The shortage of capital for investment in new resources, or new forms of energy, is worldwide. The status of many new resources—minerals from the seabed, protein from ocean waters—is uncertain, and can only be determined by governments working together. (Whether to divide the oceanic spoils or to place them under an international authority is the piece of our macroproblem which many nations fruitlessly debated at Caracas during the summer of 1974.)

So, Who Speaks for Man?

For those of us who would bet on interdependence, there is a modicum of good news—some developments may help mankind enhance control of its destiny.

For one thing, systems thinking has created new ways to help an individual encompass in a single mind "the situation as a whole." The fast computer and the complex simulations it makes possible become key tools for analyzing conflict, forecasting economic processes, assessing the future impact of present technologies, studying events as the interplay of many conditions, constructing decision models, and applying mathematical probability to social and political guesswork. These tools enable groups of human beings, drawn from different cultures and polities, to develop new comparative approaches to policy-choice analysis and make complex decisions in the more mindful knowledge of alternative futures.

Do not misunderstand me. There are dangers in excessive dedication to systems analysis. Part of the tragic body count in Vietnam may be traceable to trying to quantify the tactics of war while neglecting the strategies of peace. Still, it is useful for the decision-maker to be able to analyze what can be analyzed—as long as he remembers only to guess at what can only be guessed at.

Do we not have a galaxy of public international organizations—regional alliances, common markets, and the U.N. family—to reflect and express the interdependence of nations? We do, but they are, of course, the subject of global dismay and disenchantment—and with some reason.

The United Nations Charter and other compacts of 25 or 30 years ago did not predict or provide for what we would now regard as "gut" issues in "harmonizing the actions of nations": the distribution of energy, the pricing and distribution of food and other world resources, the management of inflation, the globalization of pollution, the explosion of population, the exploitation of the oceans and the seabeds, the limitation of strategic weapons, the spread of nuclear technology, the control of traffic in arms and in drugs, the modification of weather, the control of outer space. Some of these omissions have been partly filled in by some effective programs of development aid; a well-managed World Bank group; more than a dozen local peacekeeping operations; conventions to ban nuclear spread and bombs in orbit; bilateral talks about strategic weapons; and conferences for international sensitivity training about population, pollution, and poverty. But the range and rhetoric of international cooperation have been simply no match for the pace of events.

Meanwhile, remarkably effective nongovernmental vehicles for transnational operations have been developing alongside the public international agencies, and in some degree beyond the reach of effective national regulation. In business, especially, multinational enterprises (MNE) have solved many problems of communication, accountability, and staffing which organizations tied to national governments still find baffling. The world of interdependence they serve and reflect has rewarded them with an extraordinary rate of growth: MNEs now do nearly half a trillion dollars' worth of business, more than one seventh of the gross world product, and are growing at about 10% per year. It is estimated seriously that multinational corporations will account for half of the gross world product by the year 2000. There is obviously much to be learned about international cooperation from their successes, and excesses; the important task ahead is to reconcile their private purposes with their public responsibility.

Beyond the technical tools for the management of interdependence, our world environment of unglued values and threatened disasters defines an urgent requirement for world political leadership. But who are the contemporary leaders who presume—as Churchill, Nehru, Betancourt, Paul-Henri Spaak, and Hammarskjöld did—to address the human condition in its glorious and desperate entirety? Who in America is speaking, and is listened to, as Woodrow Wilson and Franklin Roosevelt and Adlai Stevenson were in their time? Try coming up with 10 names. Three names. One name. The available political leaders—our President and the prime ministers, chairmen, monarchs, and military politicians around the world—are talking to and for fragmented national constituencies. Who speaks for Man? Nobody?

The answer cannot be "Nobody"; so it has to be "Ourselves." We are gathered here because in one way or another we are, or want to be, leaders in thinking and acting about the future of the Circum-Pacific area. If each of us merely says what he came to say and departs, we shall miss a great opportunity. To match the macroproblem of modernization, we need to build a Pacific-wide consortium of the concerned, a community of continuous consultation about the purposes and values which our miraculous technologies are going to serve.

If we miss this chance, now or soon, we who presume to leadership and are called "President" or "Chairman" or "Director," in our own communities will deserve that devastatingly snide comment of Girardoux: "The privilege of the great is to watch catastrophe from a terrace."

References Cited

Camps, Miriam, 1974, The management of interdependence—a preliminary view: Council on Foreign Relations, Inc., 58 East 68th Street, New York City, New York 10021, p. 79.

Heilbroner, R. L., 1974, An inquiry into the human prospect: New York, W. W. Norton and Co., p. 35, 127, 135.

Boundaries of Seabed Jurisdiction off Pacific Coast of Asia[1]

NORTHCUTT ELY and ROBERT F. PIETROWSKI, JR.[2]

Abstract In the Yellow Sea, East China Sea, South China Sea, and Gulf of Thailand, 13 states are involved in seabed-jurisdiction problems. Of these, only two have entered into bilateral boundary agreements and only four have become parties to the Convention on the Continental Shelf. Numerous problems remain to be solved. In general, we conclude that (1) the Asian coastal states have exclusive competence to license exploration for and exploitation of the seabed resources of the marginal seas surrounded by these states, and (2) the application of equidistance methods, with due regard to special circumstances, should be the guiding principle in determining seabed jurisdiction between opposite states.

INTRODUCTION

This discussion touches two major problems: (1) the geographic extent of the coastal state's jurisdiction over the resources of the seabed and (2) the determination of seabed boundaries among states having interests in the same seabed area.

Under existing international law, as developed by the practice of states and articulated in the 1958 Geneva Convention on the Continental Shelf, a coastal state's jurisdiction extends beyond its territorial sea to the natural resources of the seabed and subsoil adjacent to its coast. For convenience, we call it "seabed jurisdiction." This is not full territorial jurisdiction, such as that which the coastal state exercises over its land territory and territorial sea. It relates only to the real estate, and not to the water column or the air space above. The first problem is to determine how far into the sea this criterion of "adjacency" carries that seabed jurisdiction.

The second problem, that of delimiting seabed boundaries among states, arises when a particular seabed area is adjacent to the coasts of more than one state or, in the terms employed in the North Sea judgment, is the common prolongation of the landmasses of two or more states. Valid seabed boundaries may be created by treaties between two (or more) states, or by the unilateral act of a single state, provided that the act conforms to established principles of international law. Where two or more states enter into a boundary treaty, that treaty determines the boundary between the particular states, regardless of the boundary that would result from the application of general rules of international law. However, a treaty which establishes a boundary is not binding on a nation which is not a party to the treaty. Conversely, a nation which is not a party to a treaty cannot enforce the boundary created thereby against the parties. Of the world's 126 coastal nations, 29 have entered into treaties delimiting specific seabed areas beyond their territorial seas, and 49 have ratified the 1958 Geneva Convention on the Continental Shelf, which articulates general principles that control the delimitation of boundaries among parties to the Convention in the event that they cannot agree on a boundary.

Because most of the world's coastal states have neither ratified the Convention on the Continental Shelf nor entered into bilateral boundary treaties, customary international law (i.e., the generalization of state practice which is felt to be obligatory) plays a major role in determining seabed boundaries. A customary rule of law is generally binding on all states except those which have consistently refused to recognize it. Therefore, if a state establishes a boundary by a unilateral act, it can enforce that boundary against neighboring states if the boundary is in accord with customary law.

This discussion of seabed boundaries focuses on areas beneath the Yellow Sea, East China Sea, South China Sea, and Gulf of Thailand, all of which are of vital importance in terms of hydrocarbon potential. The particular boundaries selected for discussion illustrate the application of certain abstract legal criteria to several different factual situations. Only two bilateral boundary agreements have been concluded in these areas. Of the 13 states which border these seas, only four have ratified the Convention on the Continental Shelf. Thus, most of the boundaries which we discuss here must be ascertained from an examination of customary law.

LEGAL CRITERIA FOR DELIMITING SEABED BOUNDARIES

Area of "Competence"

The rationale justifying the exercise of exclusive jurisdiction by a coastal state over seabed resources seaward of its territorial sea has been expressed in various ways. Four are of particular significance:

[1]Manuscript received, October 21, 1974.

[2]Law Offices of Northcutt Ely, Washington, D.C. 20037.

1. In 1945, President Truman, in his Continental Shelf Proclamation regarding United States exclusive jurisdiction, gave four reasons to justify exclusive jurisdiction by a coastal state over seabed resources. We have, in the following quotation, inserted numbers to correspond to the four reasons given by President Truman.

> [T]he exercise of jurisdiction over the natural resources of the subsoil and sea bed of the continental shelf by the contiguous nation is reasonable and just, [1] since the effectiveness of measures to utilize or conserve these resources would be contingent upon cooperation and protection from the shore, [2] since the continental shelf may be regarded as an extension of the land-mass of the coastal nation and thus naturally appurtenant to it, [3] since these resources frequently form a seaward extension of a pool or deposit lying within the territory, and [4] since self-protection compels the coastal nation to keep close watch over activities off its shores which are of the nature necessary for utilization of these resources.

2. In 1958, the Convention on the Continental Shelf, in Article 1, provided: "For the purpose of these articles, the term 'continental shelf' is used as referring (a) to the seabed and subsoil of the submarine areas adjacent to the coast but outside the area of the territorial sea, to a depth of 200 metres or, beyond that limit, to where the depth of the superjacent waters admits of the exploitation of the natural resources of the said areas; (b) to the seabed and subsoil of similar submarine areas adjacent to the coasts of islands."

Article 2 of the Convention goes on to note that the coastal state exercises exclusive "sovereign rights" to the exploration and exploitation of these resources, that these rights are not dependent on occupation, actual or "notional," and that, if the coastal state does not exercise them, no one can. The phrase ". . . to a depth of 200 metres or, beyond that limit, to where the depth of the superjacent waters admits of the exploitation of the natural resources. . . " has caused considerable controversy. Yet the legislative history of the Convention makes clear that the drafters did not intend to limit coastal-state jurisdiction to the geologic continental shelf. The *travaux preparatoires*, the practice of states, the judgment of the International Court of Justice in the North Sea Cases (*infra*), and the writings of experts suggest that exclusive coastal-state jurisdiction embraces the entire continental margin, in the sense that the coastal state alone can grant rights to explore and exploit the resources of the seabed and subsoil of this whole adjacent area. Therefore, the fact that a particular seabed area is not *geologic* continental shelf does not mean that it is not subject to coastal-state jurisdiction.

The other side of the coin is that, under the Convention, the geographic extent of the coastal state's seabed jurisdiction is limited by the criterion of "adjacency"; that is, jurisdiction does not extend to the midocean, but is limited to the continental margin plus, perhaps, in special circumstances, limited areas of the seafloor contiguous with the continental slope.

3. In 1969, the International Court of Justice, in the *North Sea Continental Shelf Cases*, heard a dispute between Germany, on the one hand, and Denmark and the Netherlands, on the other hand, as to the lateral boundaries of Germany's seabed jurisdiction. The Court characterized the Truman Proclamation as having particular significance, being the foundation of the continental-shelf doctrine. It described Articles 1 and 2 of the Convention as "enshrining" customary international law as well as stating conventional law. The Court's rationale, several times restated, was that the exclusive jurisdiction that a coastal state exercises over natural resources of the seabed beyond its territorial sea is based on the fact that the area subject to such jurisdiction is the physical prolongation of the state's land territory into and under the sea. Thus:

> What confers the *ipso jure* title which international law attributes to the coastal State in respect of its continental shelf, is the fact that the submarine areas concerned may be deemed to be actually part of the territory over which the coastal State already has dominion,—in the sense that, although covered with water, they are a prolongation or continuation of that territory, an extension of it under the sea.

The judgment is, to date, the Court's only opinion dealing with seabed boundaries beyond the territorial sea. The Court did not prescribe specific boundaries in the judgment; instead, it articulated the legal principles which govern seabed boundaries, leaving the actual delimitation to the parties.

The rationale of this case is not related to the water depth. The result, on the Court's reasoning, should have been the same if the North Sea were 10 times as deep as it is.

4. A fourth jurisdictional concept seems to be emerging: the contention that the coastal state should have an "economic resource zone" of, perhaps, 200 n. mi (370 km), measured from its coast, in which it should have exclusive jurisdiction over the resources of the seabed and, in addition, either exclusive or preferential rights in fisheries and other resources of the water column. The coastal state's rights in this zone would be

distinguishable from its rights in the territorial sea in that free overflight and free transit of vessels (something greater, perhaps, than the right of innocent passage) would be available to foreigners in the larger area. This concept has not yet become a part of either customary or conventional law, but has received such general acquiescence in the current United Nations Law of the Sea negotiations as to indicate that it may well be the wave of the future. It is particularly significant because it breaks away from the "prolongation" rationale of the North Sea judgment and is more nearly related to the idea of "proximity" found in the Truman Proclamation. This difference could be important, as we shall see.

To this point, we have discussed the seaward extent of the coastal state's competence to govern the utilization of seabed resources. We turn now to the criteria which relate to seabed boundaries between states in an area in which two or more states may claim to possess such competence, for example, where the area in dispute is the prolongation of the land territories of more than one state (in the Court's phrase). The substitution of a spatial concept of jurisdiction—for example, the 200-n. mi (370 km) economic resource zone—would eliminate the problem of examining whether a particular area is the underwater prolongation of the territory of one state or the other, and replace it with the more simple problem of measuring the surface of the sea separating the states. However, on either hypothesis, if one arrives at the conclusion that the area in dispute is within the zone of competence of two or more states, one must examine the criteria available for demarcation of boundaries in such an area of common interest.

Interstate Boundary Criteria: Equidistance Concept

The most widely used yardstick for determining continental-shelf boundaries (and, for that matter, surface boundaries of the territorial seas) is the median line or equidistance principle. It has been summarized well by Robert D. Hodgson, geographer of the State Department, as follows:

> A median line (at times called "lateral line") has proved to be the best solution for delineating water areas between sovereignties. In both theory and practice the geometrical principle involved in determining the median line is the most satisfactory which has so far been devised, lending itself admirably to the construction of equitable boundaries between states. It depends upon precise measurement rather than subjective factors. Without delving into its technical characteristics, a median line is defined as a line, or boundary, every point of which is equidistant from the nearest points on the lines from which it is measured. Oddly enough, the technique upon which the construction of such lines depends is purely trial and error, that is, establishment of points contingent upon being so placed that they be no farther from one than from the other fixed point representing the two sovereignties.
>
> The spirit of the articles on median lines is to provide a means whereby boundary agreements between states may be facilitated. But since median-line boundaries are objective they can frequently be used at least as a point of departure in the reaching of agreement. Site of known or potential resources, location of a navigation channel, or traditional offshore practices of a state are among special circumstances which may give rise to modifying or even disregarding completely a median line in affixing a boundary. For example, a boundary in the territorial sea may only roughly approximate a median line, compensating for loss of an area in one place by gain in another. Despite such departures from a formula the actual precisely constructed median line stands as a potential means of establishing fair and lasting offshore boundaries.

Boundaries must be determined among countries whose coasts face one another (frontal boundaries), and among countries occupying adjacent areas of the same coast (lateral boundaries). Where equidistance principles are employed in determining a frontal boundary, it usually is referred to as a median line. In some situations, such as areas at the head of a gulf, states may occupy positions that are both opposite and adjacent.

The two basic situations are recognized in Article 6 of the Geneva Convention on the Continental Shelf, which provides, as to frontal boundaries:

> 1. Where the same continental shelf is adjacent to the territories of two or more States whose coasts are opposite each other, the boundary of the continental shelf appertaining to such States shall be determined by agreement between them. In the absence of agreement, and unless another boundary line is justified by special circumstances, the boundary is the median line, every point of which is equidistant from the nearest points of the baselines from which the breadth of the territorial sea of each State is measured.

And as to lateral boundaries:

> 2. Where the same continental shelf is adjacent to the territories of two adjacent States, the boundary of the continental shelf shall be determined by agreement between them. In the absence of agreement, and unless another boundary line is justified by special circumstances, the boundary shall be determined by application of the principle of equidistance from the nearest points of the baselines from which the breadth of the territorial sea of each State is measured.

The baseline from which the territorial sea is measured generally is the low-water line along the coast as indicated on charts officially recognized by the coastal state. In places where the coastline is deeply indented or fringed by islands, or where certain types of bays exist, appropriate points may be joined to form "straight baselines."

Although Article 6 conditions the applicability of the median or equidistance line in the absence of agreement (and states, of course, can agree on any boundary line that they choose), it is natural that the negotiators, in appraising the bargaining room available to each side, would wish to know the parameters that would be controlling in the absence of agreement.

Acceptability of Equidistance Concept in Customary Law

It must not be thought that the equidistance principle announced in Article 6 is a panacea or that it is universally applicable as a matter of customary as well as conventional law.

In regard to the decision of the International Court of Justice in the *North Sea Continental Shelf Cases*, the problem was one of determining lateral boundaries between Germany and Denmark, and between Germany and the Netherlands. Because of the concave character of the German coast, strict application of the equidistance principle, as proposed by her two neighbors, would have pinched off Germany's share of the seabed far short of the median line between the continent and the United Kingdom, giving the other two states areas of the seabed lying in front of Germany. The Court rejected this contention, and also rejected Germany's counter-contention that the controlling principle should be the determination of "fair shares" of a resource not yet belonging to anyone. Instead, the Court held that the problem was one of articulating criteria for the division of an area already belonging to the contending states, in common, and belonging to them alone, because the area was the common prolongation of their land territories.

As to the equidistance principle, the Court said:

> The continental shelf area off, and dividing, opposite States, can be claimed by each of them to be a natural prolongation of its territory. These prolongations meet and overlap, and can therefore only be delimited by means of a median line; and, ignoring the presence of islets, rocks and minor coastal projections, the disproportionally distorting effect of which can be eliminated by other means, such a line must effect an equal division of the particular area involved. If there is a third State on one of the coasts concerned, the area of mutual natural prolongation with that of the same or another opposite State will be a separate and distinct one, to be treated in the same way. This type of case is therefore different from that of laterally adjacent States on the same coast with no immediately opposite coast in front of it, and does not give rise to the same kind of problem—a conclusion which also finds some confirmation in the difference of language to be observed in the two paragraphs of Article 6 of the Geneva Convention. . . as respects recourse in the one case to median lines and in the other to lateral equidistance lines, in the event of absence of agreement.
>
> 58. If on the other hand, contrary to the view expressed in the preceding paragraph, it were correct to say that there is no essential difference in the process of delimiting the continental shelf areas between opposite States and that of delimitations between adjacent States, then the results ought in principle to be the same or at least comparable. But in fact, whereas a median line divides equally between the two opposite countries areas that can be regarded as being the natural prolongation of the territory of each of them, a lateral equidistance line often leaves to one of the States concerned areas that are a natural prolongation of the territory of the other.

The inference seems clear that the Court, in making this distinction, does not reject the median-line concept as between opposite coasts.

With respect to adjacent states, the Court held that boundaries must be delimited ". . . in accordance with equitable principles, and taking account of all relevant circumstances." The relevant circumstances which the Court identified as factors to be taken into account included the following:

> "(1) the general configuration of the coasts of the Parties, as well as the presence of any special or unusual features;
>
> "(2) so far as known or readily ascertainable, the physical and geological structure, and natural resources, of the continental shelf areas involved;
>
> "(3) the element of a reasonable degree of proportionality, which a delimitation carried out in accordance with equitable principles ought to bring about between the extent of the continental shelf areas appertaining to the coastal State and the length of its coast measured in the general direction of the coastline, account being taken for this purpose of the effects, actual or prospective, of any other continental shelf delimitations between adjacent States in the same region."

These factors would also appear to be relevant to a delimitation governed by Article 6 of the Convention on the Continental Shelf, because the "special circumstances" exception to the equidistance principle was included in Article 6 to insure that delimitation would be effected on equitable principles. Thus, these factors should also be considered in any determination of whether special circumstances exist.

The judgment in the North Sea Cases also indicates that in some cases the equidistance principle as applied to adjacent states may satisfy the "equitable principles" and "relevant circumstances" criteria articulated by the Court.

"Special Circumstances"

We noted the exception of "special circumstances" in Article 6 of the Convention, emphasized also in the North Sea judgment's articulation of customary law.

The equidistance principle, although simple in theory, usually is complicated in application to either frontal or lateral boundaries by the presence of islands in the vicinity of the seabed area to be delimited. Here we encounter the significance of the "special circumstances" exception.

International law has long recognized that islands have territorial seas. Furthermore, it is agreed that at least some islands generate sovereign rights to the natural resources of the seabed that is a natural prolongation of such islands. Because the territorial sea of an island is part of the territorial sea of the coastal state that owns the island (although it need not be connected to the "main" territorial sea thereof), a rigorous application of the equidistance principle would require that the baselines of *all* islands owned by the states involved in the delimitation be used in construction of the lateral or median lines to determine boundaries of seabed jurisdiction. It is immediately evident, however, that such a delimitation might be grossly inequitable. Suppose, for example, that two coastal states, A and B, are situated opposite each other and that the seabed between them is the common prolongation of both. Located halfway between A and B is an island owned by A. If the seabed boundary between A and B is constructed such that every point on that boundary is "equidistant from the nearest points of the baselines from which the breadth of the territorial sea of each state is measured," three fourths of the seabed area would belong to A, because "the baselines from which the breadth of the territorial sea of each state is measured" include the baselines of the island owned by A. Obviously, if the island belonging to A were close enough to B's coast, essentially the entire seabed area between the two states would belong to A under the equidistance method. If the island were not present, half of the seabed area would belong to A and half to B.

Application of the equidistance method is complicated further by the legal definition of the term "island." As in the case of the continental shelf, the legal definition is quite different from the geologic definition. For legal purposes, an island is any natural formation which is permanently above water. There is no criterion of size, habitability, or usefulness. Thus, any islet, rock, or reef is an "island," and may be offered as a candidate for determination of a base point for purposes of applying the equidistance principle. This fact makes even greater the possibility that inequity will result from a rigorous application of equidistance methods because, in such a case, vast areas of seabed may be affected by ownership of an isolated wet rock pile. Consequently, as we have noted, Article 6 provides for departures from the equidistance concept in the event of "special circumstances." It does not define "special circumstances," and no clear rules of law have yet evolved in this area. However, a study of the 29 boundary agreements referred to previously, and of 10 other boundary agreements relating to territorial seas (to which the equidistance principle and the "special circumstances" exception apply by virtue of Article 12 of the Convention on the Territorial Sea and the Contiguous Zone), suggests the following treatment of "islands," so as to avoid inequitable distortion of the boundary:

1. Any island, regardless of size, is to be given recognition in delimiting equidistance boundaries if any part of that island is within 24 n. mi (44.5 km) of its sovereign's mainland or major island. This result is suggested by the fact that 12 n. mi (22.2 km) generally is recognized as the maximum territorial sea breadth permissible under existing law.

2. However, an island, no part of which is within 24 n. mi (44.5 km) of its sovereign's mainland or other major land territory, and which is uninhabited or is inhabited only by caretakers or other tokens of sovereignty, falls into the disqualifying status of "special circumstances." If an island is too insignificant to attract its sovereign's population or investment, it cannot be used fairly as a baseline from which to expand the areal extent of its sovereign's seabed jurisdiction.

3. Location is obviously a factor. An island owned by a distant state, located in proximity to the mainland of another, poses a potential inequity, unless a formula is found to restrict its effect to a zone, something like an enclave, surrounding the island. This has been done in several cases.

4. An island's size, in and of itself, is not necessarily a special circumstance justifying departure from the equidistance principle, but the size of an island is an unavoidable factor in determining the seabed area that it should command. We have discovered no settled formulas.

5. Even if an island is denied recognition in the construction of the equidistance line because of its small size and distance from its sovereign's major land territory, that island, no matter how small, should generate the exclusive jurisdiction of its sovereign over the natural resources of the seabed and subsoil within 12 n. mi (22.2 km) of its baseline, even if the sovereign claims less than a 12-n. mi territorial sea. This result is suggested by Article 1(b) of the Convention on the Continental Shelf (which attributes legal continental shelves to islands), by the legal definition of the term "island" in Article 10 of the Convention on the Territorial Sea and the Contiguous Zone, and by the fact that any state may, under existing law, claim a 12-n. mi territorial sea and thereby exercise full territorial sovereignty over the seabed areas within 12 n. mi of such islands.

6. Where it is determined that special circumstances exist, the island in question need not be entirely ignored in the construction of the equidistance line. Instead, it may be given partial effect.

The presence of islands can cause problems other than boundary problems between states. For example, if small midocean atolls or similar "flyspecks" are permitted to command seabed areas disproportionate to their importance as land territories—whether by literal application of the "prolongation" concept or the 200-n. mi (370 km) economic-zone concept—vast areas of the seabed will be appropriated. There are several thousand islets in this category. Limitations obviously are called for which will relate fairly the area of commanded seabed to the area of the islet. The community of nations has not faced up to this problem.

Application of Legal Criteria to Specific Geography

Having identified the criteria which seem to be the guidelines for determining seabed boundaries under customary and conventional law, we proceed, with considerable caution, to attempt to apply these criteria to the seabed areas off the Pacific coast of Asia.

Yellow Sea

In the Yellow Sea, the entire seabed area is geologic continental shelf, and thus is, by hypothesis, the common prolongation of the land territories of the littoral states. Those states are South Korea, North Korea, and the People's Republic of China (PRC). Because none of these states is a party to the Convention on the Continental Shelf and because none has entered into a boundary agreement with any of the others, customary law alone determines boundaries in this area. South Korea and the PRC are opposite states, and therefore the boundary between them should be the median line. This median line should recognize all islands, regardless of size, which are within 24 n. mi (44.5 km) of major land territory of Korea or the PRC. It should also recognize the Korean island of Cheju, because of its size, population, commercial importance, and relationship to the South Korea mainland.

Islands complicate the Yellow Sea boundary problem in several ways. A string of islands lies along the parallel which, onshore, is the boundary between North Korea and South Korea. They are occupied by United Nations forces. The median line between the Korean peninsula and the mainland of China would be seriously affected, as against China, if these islands were used as baselines. Several small inhabited islets belonging to South Korea lie 50 n. mi (92.6 km) or more offshore. Similarly, some small islets owned by the PRC are more than 24 n. mi (44.5 km) off the coast of China. The recognition of, or denial of recognition to, this large inventory of islands is a matter of judgment on which opinions may well diverge. It is possible, however, to construct a median line which applies roughly the same elements of judgment to all islets, no matter which nation owns them.

This Yellow Sea problem, peculiarly enough, carries over into the determination of seabed boundaries in the northern part of the East China Sea. This is true because the Yellow Sea median line, where it intersects the East China Sea median line (the next problem), is a trisection point which affects the seabed boundary of the mainland of China in the East China Sea.

East China Sea

The East China Sea presents two difficult seabed-boundary problems: (1) determining the effect of an ocean trough on interstate boundaries, and (2) determining the effect to be given small isolated islands.

The littoral states here are South Korea, Japan, the Republic of China (Taiwan), and the PRC. Most of the seabed in this area is geologic continental shelf, but this shelf is interrupted by the Okinawa Trough, which parallels Japan's Ryukyu Islands and separates them from the geologic shelf of the mainland. If the effect of this trough is to interrupt the continuity of the *legal* continental shelf, then Japan's seabed jurisdiction within the Ryukyus terminates somewhere in the vicinity of the Okinawa Trough; if, on the other hand, the trough does not interrupt the prolongation of the continental landmass, the seabed between the

Ryukyus and the mainland is the common prolongation of both, and Japan's jurisdiction extends past the Okinawa Trough to the median line, as against China. Furthermore, if the Asian continental margin ends in the deep-ocean trench system seaward from the Ryukyus, the Ryukyus themselves are part of the prolongation of the mainland, and the intervening seabed qualifies as legal continental shelf regardless of depth or configuration. However, the seaward extent of the Asian landmass is still a matter for debate. The literature examined suggests the following three possibilities:

1. The mainland and the continental shelf, the latter bounded by the Okinawa Trough, are to be considered as one geologic province, differentiated from a second province which includes the Okinawa Trough, the Ryukyu Islands, and the Ryukyu Trench. The inference would seem to follow that the west slope of the Okinawa Trough must be deemed to be the outer boundary of the continental margin, at least if this term is to be equated with the concept of a prolongation of the land territories of adjacent coastal states.

2. There are about 25 marginal seas in the world enclosed by island arcs. The East China Sea is one of these. Characteristically, the enclosed sea contains a deep area, as well as great amounts of sediments, whereas beyond the island arc is a deep trench which separates the whole complex from the abyssal ocean floor. Geologically, it is not possible to make a generalized statement that the beds of these marginal seas and the enclosing island arcs are to be identified with the continent, as geologic extensions of it, or identified with the oceanic crust. Perhaps they are to be placed in a third category, intermediate between that of the continental landmass and that of the abyssal ocean floor.

3. On the basis of the evidence referred to in the preceding paragraph, inasmuch as the beds of the marginal seas and their enclosing island arcs certainly cannot be identified with the abyssal ocean floor, perhaps they should be viewed as part of the complex which constitutes the continental margin, irrespective of the origin of the islands and the enclosed seabeds. The widespread phenomenon of deep trenches seaward from the Asian island arcs probably is associated with tectonic spreading of oceanic plates which, it is believed, are in effect sliding under the Asian continent via the deep oceanic trenches such as the Ryukyu and Mariana Trenches.

At present the consensus appears to favor either the second or the third of these hypotheses, but no definite conclusions can be drawn.

If one accepts the imminence of boundary problems in seabed areas throughout the world, it is clear that the validity of seabed boundaries cannot be absolutely conditioned on the geologic nature of the seabed in question. The diplomats simply are not going to wait until the geology is precisely identified before they start drawing boundaries. Moreover, it would be unrealistic to condition the jurisdiction of a coastal state on a matter which is very likely to be ascertained only after a party has paid for the right to drill. Security of tenure, a *sine qua non* to a mineral concession, requires that the competence of the grantor state be known before, not after, the concessionaire has drilled. Therefore, if "prolongation" is indeed the criterion, the question, for legal purposes, is not whether a seabed area constitutes the geologic prolongation of a landmass, but whether it is a "physical" or "geomorphic" prolongation. Of course, geologic continuity is evidence of physical prolongation. However, given the practical problems of delimitation, the lawyer's primary concern in establishing seabed boundaries should be with the topography of the seabed, not with the underlying geology. Some scientists point out that the seabed between the Ryukyu Islands and the Asian mainland, if viewed as a relief feature of the earth's surface, can reasonably be considered a physical prolongation of both the mainland and the Ryukyus. The Okinawa Trough, considered in relation to the Ryukyu Trench, is a depression in this prolongation, but does not terminate it; the prolongation is terminated by the Ryukyu Trench. On the basis of this hypothesis, the competence of the coastal states encompasses the seabed from the mainland, past the Ryukyus, out to the Ryukyu Trench.

There remains the question of whether the Okinawa Trough constitutes a special circumstance which justifies departure from the median-line principle in the determination of seabed boundaries among the states which share the offshore element of the Asian landmass, defined in the expansive manner that we have suggested. As to the effect of troughs on seabed boundaries, we had hoped to find examples in the practice of states, but the examples are very few. The states which have agreed on seabed boundaries in the vicinity of ocean troughs (e.g., the Norwegian Trough in the North Sea) apparently have not treated them as special circumstances, and have drawn median lines between land territories. The answer, as to the Okinawa Trough, is important not only in the southern part of the East China Sea, where it affects Japan's seabed boundary as against the two Chinas, but also in the north, where it affects

Japan's boundary as against Korea. Indeed, Korea at one time claimed almost the whole intervening seabed on the assumption that this trough separated it from Japan.

In any event, certain islands in the East China Sea do appear to be in the category of "special circumstances." These are the Senkaku Shotō near Taiwan and the Danjo Guntō in the vicinity of Kyushu. The Senkakus, which are claimed by Japan, the Republic of China (Taiwan), and the PRC, are uninhabited islets and rocks which lie about 85 n. mi (157.4 km) from Taiwan and 80 n. mi (148.2 km) from the Ryukyus; their recognition would cause a major distortion of the median line in the southern part of the sea. Japan's Danjo islets, in the northern sector, are inhabited only by lighthouse and radio-station attendants and are relatively isolated from other Japanese territory, lying 35 n. mi (64.8 km) from the Gotō Rettō and 75 n. mi (138.9 km) from Kyushu. Their recognition would distort the median line between Japan and Korea.

Moreover, it will be recalled that the median line in the East China Sea between Japan and Korea, however determined, intersects the median line in the Yellow Sea between Korea and China, however determined, establishing a trisection point which controls seabed boundaries between the PRC, Korea, and Japan. Thus the determination of the seabed boundary between Korea and Japan, which might be thought to be only their business, affects the PRC. Uniformity of treatment of islands in both the Yellow Sea and East China Sea becomes an important matter for all states bordering both seas.

South China Sea

A similar, although even more complex, situation exists in the South China Sea. Here, the sovereign littoral states include the Republic of China (Taiwan), the Philippine Republic, Malaysia, Indonesia, Singapore, South Vietnam, North Vietnam, and the PRC. The Republic of China and Malaysia are parties to the Convention on the Continental Shelf, but do not have a common seabed boundary. The only continental-shelf boundary agreement in this area is between Indonesia and Malaysia. Thus customary law determines most seabed rights in the South China Sea. Here, as in the case of the East China Sea, there is a question as to the status of the seabed as a geologic extension of the Asian landmass. Much of the seabed in this area is clearly not continental shelf as defined geologically. The China Basin, which underlies much of the South China Sea, reaches depths in excess of 4,000 m, and the Palawan Trough is almost 3,000 m deep. However, these depths are not nearly as great as those of the deep ocean trenches seaward from this area. The Philippine Trench is approximately 10,500 m deep, and the Mariana Trench is more than 11,000 m deep. Termination of the topographic prolongation—and, hence, legal prolongation—of the Asian continent thus would appear to occur in the trench system seaward from the island arcs bordering the South China Sea, irrespective of any question of geologic continuity. If so, the natural resources of the seabed and subsoil are subject to the exclusive jurisdiction of the littoral states by extrapolation of the rationale of the North Sea judgment (prolongation). This result also is suggested by practical considerations. If the China Basin and the Palawan Trough terminate the legal prolongation of the Asian continent, large areas of the seabed are removed from the exclusive jurisdiction of the surrounding states. These states have a very special interest in this area, an interest which was articulated by the Truman Proclamation almost 30 years ago. These considerations—support and protection, appurtenance, continuity of deposits, and national security—suggest that the entire seabed of any marginal sea, not least the complex South China Sea, should be subject to the exclusive jurisdiction of the littoral states, and that this jurisdiction should not depend on geologists' opinions as to "prolongation," which, at present, would have to be based on fragmentary data.

Regarding interstate boundaries, most of the boundaries in the South China Sea are between opposite states, and, as in the Yellow and East China Seas, customary law appears to require that these boundaries be determined by equidistance methods, with allowance for special circumstances as discussed. Thus, islands located within 24 n. mi (44.5 km) of major land territory of any littoral state should be used in the construction of median lines, and small islands beyond 24 n. mi should be considered special circumstances and disregarded. Included in this latter group are hundreds of small midsea islets, including the Spratleys and the Paracels, whose ownership in many cases is disputed. Their recognition as base points would hopelessly and inequitably distort boundaries otherwise based on median lines between major landmasses.

A problem of local but particular interest in the South China Sea is the delimitation of lateral boundaries between Malaysia and the British-protected state of Brunei. Without considering the question of whether Brunei is bound by the United Kingdom's ratification of the Convention on the Continental Shelf, we observe that Brunei is situated on a concave coast such that application of the equidistance principle would pinch off her share of the continental shelf, just as it would have pinched off Germany's share in the North Sea. The question thus arises whether the concavity of Brunei's coast justifies a departure from the equidistance principle. Given the disparity between the relative lengths of the coastlines involved and the shelf areas that would accrue to the two states under the equidistance principle, it appears that geographic circumstances justify a departure from the lateral equidistance line.

Gulf of Thailand

The seabed-boundary problems in the northern part of the Gulf of Thailand are between Cambodia and Thailand; in the south, they involve Thailand, Cambodia, and South Vietnam.

The entire gulf may be considered, on geologic evidence, to be the prolongation of the land territories of the littoral states.

In the north, Cambodia and Thailand are both adjacent and opposite states. Both coasts are fringed by islands. Both countries have proclaimed straight baselines, and both own islands at a distance of more than 24 n. mi (44.5 km) from their coasts. Each has granted petroleum concessions bounded by median or equidistance lines, but there is a substantial overlap. The disparity results entirely from the treatment accorded islands in establishment of base points. It is entirely possible to construct a median line which gives identical treatment to islands similarly situated, accepting those within 24 n. mi (44.5 km) of the mainland and excluding those seaward of that distance. No islands seaward of the 24-n. mi line are important enough to justify the distortion that their recognition would create.

In the south, the situation is more complicated. Not only are the three gulf states—Cambodia, South Vietnam, and Thailand—involved, but also Malaysia shares with Thailand the Malay Peninsula, so is a party to the determination of the line running from Thailand to South Vietnam which closes the Gulf of Thailand. This line presents the unusual problem of determination of an equidistance line as among four nations—two on the Cambodia–South Vietnam side and two on the Thailand-Malaysia side, each pair confronting the other pair as an opposite coast, and at the same time constituting adjacent states on the same coast. The problem is further complicated by a dispute between Cambodia and South Vietnam as to their land boundaries and ownership of the important island of Phu Quoc, close to the coast. The seabed boundaries in the southern part of the Gulf of Thailand, where it merges into the South China Sea, cannot be completely determined until the territorial dispute between Cambodia and South Vietnam is decided.

Conclusion

We have attempted to state, subject to a caveat against any dogmatic conclusions, our views on two related subjects: (1) the geographic extent of the competence of the Asian states, mainland and island, taken as a group, to license the development of the seabed resources of the areas offshore from the Pacific coast of Asia; and (2) the principles which should determine seabed boundaries among these states.

As to the first of these, we concluded that, taken as a group, the Asian coastal states have exclusive competence to license the development of the seabed resources of the marginal seas that they respectively enclose, notwithstanding the great distances and, in some cases, the great water depths that are involved.

We reached this conclusion by application of the principles of customary law announced in the Truman Proclamation, which embodies notions of proximity and of predominant interests of the littoral states, and in the International Court of Justice's North Sea Continental Shelf decision, which embodies primarily the concept of a submarine prolongation of land territories of the littoral states. Logically, the Court's rationale is not limited by depth or distance. Our discussion illustrates the lawyer's uneasiness when called upon to render a boundary opinion which, in the present state of the law, must be premised on geologists' opinions as to whether a given submarine area is the prolongation of the land territories of this state or that. By contrast, there is great appeal in the logic of viewing all marginal or semienclosed seas as a jurisdictional problem *sui generis*, one to be solved by application of the "self-protection" or special-concern rationale of the Truman Proclamation, rather than the "prolongation" or "extension of the landmass" rationale of that proclamation and of the North Sea judgment. The littoral states have a special, and predominant, interest in the resources of the sea that they surround. Their collective seabed jurisdiction should be recognized to the exclusion of all others, without regard to the character or source of the rocks of the seabed.

Regarding the second subject, the determination of seabed boundaries, perhaps the application of equidistance principles to the surface of the sea should be enough to ask—and this is difficult, as we have shown. Part of the merit of the 200-n. mi (370 km) economic-resource zone may lie in this simplification of seabed-boundary problems, although this factor has not been highlighted. Customary law appears to require application of the equidistance principle in the case of opposite states which fail to agree on a boundary, but not in the case of adjacent states. With respect to opposite states, special circumstances may warrant a departure from application of the equidistance line. The practice of states recognizes as "special circumstances" small islands, especially uninhabited islands, situated more than 24 n. mi (44.5 km) from the sovereign's major land territory, and islands whose location in proximity to the territory of another state would seriously distort the median line. State practice does not compel the treatment of ocean troughs as special circumstances. With respect to adjacent states, customary law requires that boundaries be in accord with "equitable principles" and take into account the "relevant circumstances" articulated by the Court in the North Sea Cases, namely, configuration of the coasts, special or unusual features, physical and geological structures, and relative lengths of the coasts. These considerations would appear to be relevant also to modification of median lines as between opposite states.

In those few examples where the Convention on the Continental Shelf is applicable (i.e., where both states are parties to the Convention), the boundary is to be determined by equidistance methods regardless of whether the states are opposite or adjacent, but departures from the equidistance line are justified by the existence of "special circumstances." Here we enter a landscape of islands. Not only do islands, as a class, suggest a category of "special circumstances," but each island, isle, islet, or atoll itself constitutes a "special circumstance."

Selected References

Dietz, R. S., and J. C. Holden, 1970, Reconstruction of Pangaea—breakup and dispersion of continents, Permian to present: Jour. Geophys. Research, v. 75, p. 4939–4955.

Ely, Northcutt, 1971, United States seabed minerals policy: Natural Resources Lawyer, v. 4, p. 597–621.

——— 1972, Seabed boundaries between coastal states: the effect to be given islets as "special circumstances": Internat. Lawyer, v. 6, p. 219–236.

Emery, K. O., and Z. Ben-Avraham, 1972, Structure and stratigraphy of China Basin: AAPG Bull., v. 56, p. 839–859.

——— et al, 1969, Geological structure and some water characteristics of the East China Sea and the Yellow Sea: Bangkok, Committee for Coordination of Joint Prospecting for Mineral Resources in Asian Offshore Areas, ECAFE Tech. Bull., v. 2, p. 3–43.

Gutteridge, J. A. C., 1959, The 1958 Geneva Convention on the Continental Shelf: British Yearbook Internat. Law, v. 25, p. 102–123.

Hedberg, H. D., 1970, Continental margins from viewpoint of the petroleum geologist: AAPG Bull., v. 54, p. 3–43.

International Law Commission, 1957a, Yearbook 1951, v. 1: U.N. Doc. A/CN. 4/SER. A/1951, p. 267–301, 400–417.

——— 1957b, Yearbook 1956, v. 2: U.N. Doc. A/CN. 4/SER. A/1956/Add. 1, p. 795–801.

——— 1959, Yearbook 1953, v. 1: U.N. Doc. A/CN. 4/SER. A/1953, p. 72–169.

Jennings, R. Y., 1971, Jurisdictional adventures at sea—who has jurisdiction over the natural resources of the seabed?: Natural Resources Lawyer, v. 4, p. 829–840.

Karig, D. E., 1971a, Origin and development of marginal basins in the western Pacific: Jour. Geophys. Research, v. 76, p. 2542–2561.

——— 1971b, Structural history of the Mariana Arc system: Geol. Soc. America Bull., v. 82, p. 323–344.

Menard, H. W., 1967, Transitional types of crust under small ocean basins: Jour. Geophys. Research, v. 72, p. 3061–3073.

National Petroleum Council, 1969, Petroleum resources under the ocean floor: Washington, D.C., 107 p.

——— 1974, Ocean petroleum resources: Washington, D.C., 39 p.

Oda, Shigeru, 1968, Proposals for revising the Convention on the Continental Shelf: Columbia [Univ.] Jour. Internat. Law, v. 7, p. 1–31.

——— 1973, The delimitation of the continental shelf in southeast Asia and the Far East: Ocean Management, v. 1, p. 327–346.

Park, Choon-ho, 1972, Continental shelf issues in the Yellow Sea and the East China Sea: Law of the Sea Inst. Occas. Paper No. 15, September, 64 p.

——— 1973, Oil under troubled waters: the northeast Asia sea-bed controversy: Harvard Jour. Internat. Law, v. 14, p. 212–260.

Parke, M. L., et al, 1971, Structural framework of continental margin in South China Sea: AAPG Bull., v. 55, p. 723–751.

Parr, C. J., 1974, Non-fuel minerals—another dimension to the problem of resource availability: Rocky Mountain Mineral Law Inst., v. 20, in press.

Pearcy, G. E., 1959, Geographical aspects of the law of the sea: Assoc. Am. Geographers Annals, v. 49, p. 1–23.

Special Subcommittee on the Outer Continental Shelf, 1970, Report to the Committee on Interior and Insular Affairs of the United States Senate, December 21: Washington, D.C., U.S. Govt. Printing Office, 222 p.

Tao Cheng, 1969, Communist China and the Law of the Sea: Am. Jour. Internat. Law, v. 63, p. 47–73.

——— 1974, The Sino-Japanese dispute over the Tiao-Yu-Tai (Senkaku) Islands and the Law of Territorial Acquisition: Virginia Jour. Internat. Law, v. 14, p. 221–266.

Mineral Trade in Circum-Pacific Region—Summary[1]

THOMAS A. HENRIE[2]

Introduction

Few nations have within their boundaries all the mineral resources needed to support vigorous economic growth. Domestic resources must be augmented by foreign mineral supplies. The need for an increasing volume and variety of foreign mineral resources has been reflected in the growth of world trade in mineral commodities.

The so-called developed countries in the world have required a large and diversified input of mineral raw materials. The Circum-Pacific region has been important in the overall world mineral-supply picture because it is endowed with large resources of many important minerals. Several countries in this region traditionally have been large exporters of nickel, tin, and copper, as well as lead and zinc. More recently, exports of bauxite, iron ore, uranium, coal, and petroleum have added a new dimension to regional mineral exports.

World Mineral Trade

The growth of world mineral trade during the decade 1961–71 kept pace closely with increases in total world trade (Table 1). During the period 1963–71, the trade in major mineral commodities, which include mineral fuels and related materials, iron and steel, nonferrous metals, crude fertilizer materials, and mineral and metalliferous ores and concentrates, consistently accounted for roughly one fifth of the total world trade. Although figures are not yet available, the same rising trend in mineral trade has continued since 1971.

Of the billions of dollars in major mineral commodities traded in 1971, mineral fuels accounted for 48.4%; iron and steel, 24.1%; nonferrous metals, 14.1%; metalliferous ores, concentrates, and scrap, 10.0%; and crude nonmetallic minerals, 3.4%.

The major portion of all mineral imports, or 75%, was received by countries with developed market economies (Table 2). The so-called developing nations received about 14% of the total. A little over 9% went to countries that had centrally planned economies.

Of the $55.4 billion of major mineral commodities imported by the developed countries in that year, 52% came from other developed countries and 42% from developing countries. Of the total commodities valued at $29.7 billion exported by the developing countries, 79% went to the developed countries and 18% went to other developing nations. Shipments of mineral commodities from the countries with developing market economies were nearly on par with the value of shipments from the countries with developed market economies—about $30 billion compared with $36 billion.

The magnitude and direction of mineral-trade flow were determined not only by the obvious and fundamental factors of supply and demand, but also by such indirect factors as investment climate and national policies and priorities. On the demand side, we can observe the global rise in national income and industrial output. Increases in the production of capital goods, consumer durables, and other mineral-intensive products have spurred the demand for more mineral raw materials.

World consumption of most major mineral commodities grew rapidly during the last 2 decades. The world demand for mineral raw materials will continue to rise in the future, although probably not at such high rates as in the last 2 decades. According to forecasts made by the Federal Bureau of Mines, world requirements for most major nonfuel mineral commodities in the year 2000 will be two to five times the 1972 consumption, or more. In millions of metric tons, requirements for aluminum will increase from 11.8 to 68; copper, from 6.5 to 21.6; iron, from 435 to 849; lead, from 3.7 to 5.8, and zinc, from 5.5 to 11.2, if it is assumed that a growth trend of 5.4% will continue to 2000. Similar increases can be expected for energy requirements, but the pattern of consumption probably will shift significantly as the world's resources of petroleum and natural gas become more and more depleted.

Minerals were deposited without regard to national boundaries, and most nations are short of domestic resources or minable reserves of one or more critical minerals. Consequently, these coun-

[1]Manuscript received, November 11, 1974.

[2]Associate Director, Mineral and Materials Research and Development, Bureau of Mines, U.S. Department of the Interior, Washington, D. C. 20241.

Table 1. Total World Trade in All Commodities and World Trade in Major Mineral Commodities,[1] 1963 - 71 (billion dollars)[2]

	1963	*1964*	*1965*	*1966*	*1967*	*1968*	*1969*	*1970*	*1971*
Major mineral commodity trade	31	36	41	43	46	52	58	68	74
Total trade	154	172	186	203	214	239	272	311	348
Mineral trade as a percentage of total trade	20	21	22	21	21	22	21	22	21

[1] Data presented were for selected major commodity groups of the Standard International Trade Classification–Revised (SITC-R), and as such exclude some mineral commodities classified in that data array together with other (nonmineral) commodities. SITC-R categories included are: ores, concentrates, and scrap–SITC Division 28; iron and steel–SITC Division 67; nonferrous metals–SITC Division 68; nonmetals (crude only)–SITC Division 27; mineral fuels–SITC Section 3. Major items not included were the metals, metalloids, and metal oxides of SITC Division 51; manufactured fertilizer of SITC Division 56; and nonmetallic mineral manufactures of SITC Groups 661, 663, and 667.

[2] Source: U.S. Bureau of Mines, *Minerals Yearbook*, Area Reports: International, various issues.

Table 2. Distribution of Major Mineral Commodity Trade by Developed, Developing, and Centrally Planned Economies, 1971 (billion dollars)[1]

	Sources of Exports[2]			
	Market-Economy Countries		*Centrally Planned Economies*	*Total Imports*
	Developed	*Developing*		
Market-economy countries:				
Developed	29.0	23.5	2.9	55.4
Developing	4.4	5.2	0.6	10.2
Centrally-planned-economy countries	1.8	0.5	4.6	6.9
Undistributed	0.6	0.5	0.2	1.3
Total Exports	35.8	29.7	8.3	73.8

[1] Source: Statistical Office of the United Nations, *Monthly Bulletin of Statistics*, April 1973 and July 1973.

[2] Sources and destinations grouped according to United Nations' practice; developed-economy countries are Australia, Austria, Belgium, Canada, Denmark, Finland, France, West Germany, Greece, Iceland, Ireland, Italy, Japan, Luxembourg, Netherlands, New Zealand, Norway, Portugal, Republic of South Africa, Spain, Sweden, Switzerland, Turkey, United Kingdom, United States, and Yugoslavia; centrally-planned-economy countries are Albania, Bulgaria, People's Republic of China, Czechoslovakia, East Germany, Hungary, North Korea, Mongolia, Poland, Romania, U.S.S.R., and North Vietnam; less-developed market-economy countries include all countries not specifically listed.

tries have relied on imports to fill these requirements. For the United States, this has been the case for tin, nickel, bauxite, chromite, asbestos, industrial diamonds, and manganese, just to name a few. For Japan, this has been true for many important minerals, including iron, coking coal, petroleum, copper, bauxite, and nickel.

There are also minerals that can be produced domestically, but only at a significantly higher cost than imported commodities. However, most nations, with the notable exception of the Soviet Union and the People's Republic of China, have relied on imports rather than higher cost domestic production. Therefore, the volume and pattern of international mineral trade have been determined by factors of supply and demand as well as by national policies and priorities.

Mineral Production and Trade in Circum-Pacific Region

The Circum-Pacific region comprises more than 40 countries, including those with developed market economies—i.e., Australia, Canada, Japan, New Zealand, and the United States—as well as many Latin American countries, the Soviet Union, the People's Republic of China, and the nations of Southeast Asia and Oceania.

The region occupies roughly 72 million km^2 of land area, or 54% of the world's total land area of approximately 132 million km^2; it accommodates about half of the world's population. The per-capita consumption up to 1974 has been low in many parts of this region, but should grow in the future.

Because of the mineral resources, the region probably occupies a much greater importance than its land area and population would indicate. The value of its crude mineral production in 1968 accounted for over 60% of the world's total output (Callot, 1968). The region probably has at least maintained this position in recent years because of significant increases in iron ore, copper, and bauxite production, as well as general increases in mineral production in the Soviet Union and the People's Republic of China.

The long-range role of the region as a mineral supplier is also promising because of large areas favorable for minerals that are yet to be fully explored. With improved relations between the People's Republic of China and the market-economy countries, there are signs that China may increase significantly in importance as a supplier of mineral raw materials, particularly to the Japanese market. There are, of course, signs that the Soviet Union is also showing a special interest in increased trade with Japan.

In this Circum-Pacific region are included some mineral resources of special importance such as iron ore, copper, tin, and nickel. The region traditionally has produced by far the larger share of the total world output of these commodities.

In iron ore, the region produced about 460 million metric tons in 1972, roughly 60% of the total world output. In copper, its mine production of 4.5 million tons of contained copper in 1972 was about two thirds of the total world output. The Circum-Pacific region has consistently produced between 80 and 90% of the total world output of tin. The same situation has been true for nickel. As a result of vast discoveries of bauxite in Australia, the region is rapidly increasing in importance as a supplier of this mineral.

The Circum-Pacific region accounted for about a third of the total annual mineral trade of the world in 1971; it had $24 billion in exports and $25.6 billion in imports (Table 3), compared with the total world mineral trade of about $74 billion in that year. About half of both the mineral exports and the mineral imports were with other countries within the region. The five countries in the region with developed market economies accounted for the major share of the region's mineral trade—$13.7 billion, or 57%, of the exports, and $20.5 billion, or 80%, of the imports (Table 3).

These five countries sustained a mineral-trade deficit of nearly $7 billion in 1971. In contrast, the developing countries maintained a trade surplus in mineral commodities, and this surplus amounted to $1.5 billion in 1971. The countries with centrally planned economies, primarily the Soviet Union and the People's Republic of China, have always sought ways to minimize dependence on foreign sources of supplies. This policy has resulted in a fairly low level of imports and a surplus in overall mineral trade. Table 3 shows that in 1971 imports of mineral commodities by these countries totaled $2.0 billion and exports, $5.7 billion. The low levels of regional mineral trading of the countries with centrally planned economies were, of course, due to their relations with other countries in Europe with centrally planned economies and to the geographic proximity of the Soviet Union to its European neighbors.

Inasmuch as the Circum-Pacific region is one of the world's most important producers of iron ore, copper, tin, and nickel, the region also should play an extremely important role in the foreign trade of these mineral commodities.

Of iron ore and concentrates produced in the region, about a third of the output has gone into foreign trade. In 1971, for example, 144 million

Table 3. Mineral Trade in the Circum-Pacific Region—1971 (billion dollars)[1]

	Total Mineral Trade		*Regional Mineral Trade*	
	Exports	*Imports*	*Exports*	*Imports*
Developed Countries				
Canada	4.7	2.0	3.3	1.0
United States	3.7	9.3	2.0	4.7
Japan	3.8	8.4	2.6	4.2
Australia	1.5	.5	1.1	.2
New Zealand	.02	.3	.01	.1
Subtotal	13.7	20.5	9.0	10.2
Developing countries	4.6	3.1	2.8	1.5
Centrally planned countries	5.7	2.0	.3	.5
Total, all nations	24.0	25.6	12.1[2]	12.2[2]

[1] Source: Various United Nations publications including *World Trade Annual, Yearbook of International Trade Statistics and Commodity Trade Statistics*, and the official trade statistics of various nations.

[2] In theory, total regional exports and imports should sum to the same figure. Statistical omissions and gaps have resulted in this apparent discrepancy.

tons went into foreign trade, representing 47% of the world's total production for this commodity. The major exporters were Australia, the Soviet Union, and Canada. About two thirds of all regional exports of iron ore and concentrates remained within the region. Japan received about half of the region's total exports in 1971 and the United States, about 17%. Most of the remaining exports were shipped to Western Europe.

Both copper concentrates and metal production entered into world trade. Japan received most of the region's exports of copper concentrates, which amounted to 450,000 metric tons of contained copper in 1971 out of the total regional shipments of 576,000 tons during that year. Regional trade in copper metal was much more significant. Nearly 1.5 million tons were shipped in 1971, almost half of the world's total shipments. The major part of the exports of copper metal went to Western Europe. Only about one fourth of the metal exports was shipped to other Circum-Pacific countries, because Japan was the only major consuming country in the region without a large domestic resource base.

Because most of the tin production in the region was by small operators, most of the tin was also traded. About 20–25% of the region's tin shipments was in the form of concentrates. These shipments, which totaled roughly 38,000 metric tons of contained tin in 1971, were primarily from Bolivia to Western Europe and from Indonesia to Malaysia. Lesser quantities were shipped from Australia, Laos, and Canada.

The major part of the export shipments of tin was in the form of refined metal. Malaysia was the largest tin producer and the largest exporter. Smaller but still appreciable quantities of metal were shipped by Thailand, Indonesia, and Bolivia. Exports of tin metal totaled 127,000 tons in 1971. About two thirds of all regional metal exports was to other countries in the Circum-Pacific region, and the United States and Japan were the largest recipients.

The two largest producers of nickel—Canada and New Caledonia—were also the largest exporters. Ores and concentrates (including matte) moved in 1971 totaled 5.4 million tons, on a gross-weight basis, and represented practically the total world trade in these materials. New Caledonia shipped most of its nickel in ores and concentrates which, including matte, totaled nearly 4 million tons in 1971. Most of this trade remained within the region, and Japan was the largest importing nation.

Canadian exports were largely in the form of unwrought nickel and nickel alloys, which in 1971 totaled about 125,000 tons; approximately 80% was shipped to the United States and 20% to the United Kingdom. New Caledonia also exported significant quantities of ferronickel—approximately 30,000 tons in 1971—primarily to France and the United States.

The Circum-Pacific region depends on imports of many important minerals in order to meet its trade requirements. In addition to oil, coking coal, which is vital for the Japanese steel industry, is imported. The United States has depended on imports for practically all of its manganese requirements, most of its aluminum raw materials, a large part of its tungsten supply, and varying quantities of other minerals.

Factors and Constraints Governing Future Mineral Trade

Although supply and demand have been the immediate factors governing mineral trade, investment climate and national priorities have played an important role in establishing trade patterns. The uncertainties facing the investors have become increasingly complex, and competition for investment capital has become more severe.

Nevertheless, all countries have the basic objective of raising the standards of living and of creating meaningful employment opportunities for their people. This has been particularly true for the developing nations. Mineral resources, which can be used to generate investment capital needed for economic advancement, represent an important basic asset for many of these nations. Unfortunately, investment capital is needed to develop the mineral resources. Thus, these nations will need foreign investment to achieve at least some of their national objectives.

As is well known, mining represents a large investment committed to a unique venture, and the investor has to look forward over many years for repayment of the original investment and an adequate return commensurate with the risk. Also, there are no absolute rules that govern the decisions of capital exporters and importers. Conditions governing transfer of capital have always been relative and subject to change with changing economic conditions in both the host country and the investor country. What were considered favorable conditions yesterday may not hold for either or both parties today, and what is favorable today may not be favorable tomorrow. Not only the availability of capital, but also the competition for capital, changes.

Governments seeking foreign investment also should appreciate that the capital invested is owned by individuals. The bulk of the capital in the United States is owned not by extremely wealthy people, but by a large population of middle-income citizens who have invested their savings earned from hard work for their future financial security and retirement. Thus, overseas capital cannot be taken for granted and must be afforded practical encouragement.

A divergence of interest is bound to exist between the developing country and the foreign investor from the developed countries. The foreign investor will be concerned primarily with business advantages, whereas the government of the developing country will seek to secure the investor's contribution in capital and expertise at the lowest possible economic and political cost. However, this divergence of interests will not be any less natural or legitimate than in other national and international economic relations—between producer and consumer or between buyer and seller.

Although relations between the foreign investor and the government of the developing country will be more complicated because of political considerations, the unavoidable differences generated by the more complicated relations will not be irreconcilable. However, they can be resolved only by determining and recognizing the reasonable and legitimate expectations and requirements of both sides; both are free agents. The government will be free to control the access of foreign investors to its territories and their operations. The foreign investor will be free to invest or not, on the basis of terms and conditions available under local laws.

The foreign investor must look for opportunities that combine economic possibilities with the greatest security. Thus, he must examine carefully the political stability of the capital-seeking country; the attitude of the government of such a country toward business, private investment, and nationalization; and its attitude toward international law.

Other important conditions that must be examined are stable currency, tendencies toward inflation or stable prices, the safeguards for foreign investment, and the ability to repatriate capital and to remit earnings without overly severe exchange-control restrictions. Also, there will be the highly important question of an equitable tax formula and stable tax rate. Other factors are the conditions under which foreign personnel may be employed, customs duties that govern the cost of imports, and taxes on the transfer of property.

In many countries, the displacement of local personnel by foreign personnel in enterprises based on foreign capital is feared. However, without the foreign investment there would be no new employment opportunities for either local or foreign personnel. Most foreign companies prefer to use local personnel as much as possible, and in many instances local personnel, as they became trained and experienced, gradually have replaced the foreign personnel initially brought into the country. Arrangements toward such an objective certainly can be worked out.

It is only natural for countries to shy away from increased foreign ownership that would come as a result of foreign investment in domestic assets, but it should be remembered that the outflow of dividends that represent the return of this investment will be but a small fraction of the income that can be generated internally by the investment.

Some years ago, the Chase Manhattan Bank, in an exhaustive study, analyzed the results of United States investment in Latin American mines. It found that in one particular year the U.S.-owned mines in this vast area received $808 million from the sale of metals. Of this sum, $601 million stayed in Latin America, in the form of wages, local materials and services, and local taxes. It would be fair to say that only a small fraction of this sum would have been generated locally if outside investment had been discouraged. Nevertheless, perhaps some arrangements for a gradually increased national participation over a number of years, at the option of the host country, can be worked out.

The problems and conditions affecting future decisions on foreign investment will be far more complex than they were a few years ago. Many of the developed countries will be experiencing conditions that at least have the tendency of restricting foreign investment. These countries will have problems of their own infrastructures, unemployment, balance of payments, social difficulties, and/or defense and security commitments. All these problems are leading to increasing restraints on capital outflows.

The nationalization of foreign investment that has taken place in several countries during the past few years certainly has had a general restraining influence on new foreign ventures. The Mideast oil embargo of 1973, followed by high prices, created serious economic disruptions in Europe and Japan and, to a lesser degree, in the United States, where it resulted in strong government action to promote self-sufficiency in energy resources and proposed legislation that would discourage investment for the development of all minerals in foreign countries.

In summary, if a developing country becomes interested in foreign investment and an increase of revenue through mineral trade, it must carefully and continuously examine the climate it will provide for foreign investment in the competition for the limited capital available. It cannot take such investment for granted; requirements for foreign investment are unlimited, whereas funds available for foreign investment are limited. Each country has its own national policies. It must evaluate the importance of foreign investment in its overall development plan and then formulate policies and conditions either to encourage or discourage foreign investment in the development of its mineral resources.

Reference Cited

Callot, F., 1968, Production et consommation mondiales des minerals en 1968: Annales Mines.

Economics of Energy and Mineral Resources—Summary[1]

ARTHUR J. ZRALY[2]

My remarks will prove more meaningful if two vitally important facts are understood. First, the satisfaction of virtually all needs for goods and services involves the use of energy. In fact, I cannot think of a single business activity that is not dependent upon energy to some degree. Because of that dependence, as much as 70% of all energy consumed in the United States is for business-related purposes. For a long time, there has been a close relation between the utilization of energy and the nation's gross national product. No significant change in that relation is now in prospect. We doubtless will be able to become more efficient in our use of energy in the future by limiting waste, but the gains are not likely to be greater than those achieved in the past. Obviously, an adverse economic impact cannot be avoided if our future supply of energy proves inadequate.

That development would create many social as well as economic problems. The level of employment in the United States is closely tied to the gross national product (GNP). If the future expansion of the GNP is slowed because of a lack of energy, the growth of employment will also be slowed, but the nation's labor force still will continue to grow at a steady rate for many years to come. That is a mathematical certainty. A lack of energy, therefore, can lead to an intolerable level of unemployment. Clearly, it is very much in the best interest of all of us that our supply of energy always be sufficient.

The second important fact that we should keep uppermost in mind is that the supply of energy is directly related to the investment made to provide that supply. That principle applies to all forms of energy—oil, natural gas, coal, water power, nuclear power, and the more exotic forms not yet in use. If the expenditures required to find, develop, or otherwise make available a supply of any form of energy are insufficient, a shortage will be the certain consequence no matter how abundant the basic energy resources may be.

The United States does not lack basic energy resources to be developed. An enormous additional supply of energy could be made available if the necessary investment were made. Although the existing energy shortage is the result of several factors, the most important, by far, is underinvestment.

It is vitally important that we understand why there has been prolonged underinvestment in respect to the supply of energy. Two related factors were involved. First, neither the petroleum industry nor the coal industry has been able to achieve adequate capital formation for the past 2 decades. Second, the climate for investment in those industries in that period compared unfavorably with other investment opportunities. That condition led to a flight of capital away from the energy industries.

If we are to resolve our energy problems, we should make the climate for investment at least as good as anywhere else—and preferably better.

Underinvestment has existed for as long as profits have been too small. Now there is a shortage of petroleum not only in the United States, but worldwide. The world's proved reserves of petroleum are much too small relative to the indicated future needs, and their geographic distribution is disproportionate. The petroleum industry is now faced with the enormous task of finding enough new oil to satisfy expanding market demand and also raise proved reserves to satisfactory levels, particularly in areas other than the Middle East.

Oil is the single most important form of energy—it will satisfy a much broader range of purposes than any of the other forms. Between 1970 and 1985, oil alone is expected to be used to satisfy nearly 60% of the total energy needs of the non-Communist world. This proportion could change, because we are faced with great uncertainty regarding not only future political and pricing trends that impact all forms of energy, but also the consumer's reaction to those trends. However, we are not going to be able to make a new, realistic measurement until conditions stabilize and we have more evidence in hand.

In terms of the volume and location of crude oil reserves throughout the non-Communist world, the current availability of oil can be called "tight"—and it is likely to become even tighter. In the period from 1970 to 1985, we indicate that the world will consume nearly three fourths of the

1Manuscript received, December 26, 1974.

2Energy Economist, Chase Manhattan Bank, N.A., Chase Manhattan Plaza, New York, New York 10015.

total proved reserves of oil now in existence. This amount is approximately two and a half times greater than the oil consumed in the preceding 15 years. Let us suppose that, as a result of higher oil prices, all future growth in oil consumption will cease. That is, we would just continue at the 1973 level. In that event, the accumulated consumption in the 1970–85 period still would be twice as large as in the preceding 15 years.

Obviously, the proved reserves will have to be much larger in 1985 than at present to accommodate a worldwide demand for oil that is expected to be more than twice as large as in 1970. Therefore, the petroleum industry will have to find at least twice as much new oil in the 1970–85 period as it did in the preceding 15 years.

In the 15 years from 1955 to 1970, the petroleum industry invested $104 billion in the search for petroleum throughout the non-Communist world. If the industry were to invest the same amount in the 1970–85 period, it could be expected to find little more than one third of what the world's markets might require in that time. In that event, the world's needs could be accommodated only by depleting the inventory of proved reserves to a dangerous level. However, it is highly improbable that the various producing nations would permit that to happen. Most likely, they would impose restraints on the rate of production instead. The world's needs for petroleum would therefore go largely unsatisfied.

There is a compelling need for a much larger investment in the search for petroleum. To find enough to satisfy the world's expanding requirements and also maintain a realistic inventory of proved reserves will require an investment of at least $450 billion in the 1970–85 period. The need to spend so much more than in the preceding 15 years reflects several factors: the world's petroleum requirements will be much larger, past underinvestment must be overcome, the search must be conducted in more costly areas, and continued inflation must be expected.

Although there is an urgent need to accelerate the search for more petroleum, the investment for that purpose would be pointless if the facilities for bringing the petroleum to market were not also expanded. To provide sufficient refineries, petrochemical plants, tankers, pipelines, and the vast distribution system required to deliver petroleum products to the world's far-flung markets, the petroleum industry will need to invest approximately $360 billion between 1970 and 1985. That is more than treble the actual investment for such purposes in the preceding 15 years.

Therefore, if the world's needs for petroleum are to be satisfied fully, the industry will have to invest more than $800 billion in the 1970–85 period. That figure is four times the amount spent in the previous 15 years.

However, large as these sums are, they do not by any means represent all the money the industry will require if it is to function as a viable business. Debt will have to be repaid, of course, along with the charges imposed on the use of the borrowed money. Shareholders will also have to be paid a realistic amount for the use of the funds they have provided. Furthermore, as the petroleum industry's worldwide activities expand, working capital must also grow. Together, these financial needs are likely to amount to $540 billion for the period 1970–85.

Altogether, the capital investment and all the other financial requirements will total nearly $1.4 trillion for the 15-year period. As a result of the inflationary effects of the actions taken by foreign governments to raise the price of their oil, the financial needs of the petroleum industry could even be substantially higher.

The task of generating such an enormous sum will be difficult, at best. Part of the money can be obtained from capital markets, but the amount will be limited by the industry's ability to repay. That ability, in turn, will be determined by the size of the industry's profits. Another portion of the money will be provided from capital recovery. The various provisions for capital recovery are subject to change by political actions, however, and all the money currently expected may not be forthcoming. Borrowed capital and capital recovery are expected to satisfy about half of the industry's financial needs between 1970 and 1985. The other half must come from net profits.

If the petroleum industry is not able to generate the required amount of profits, it obviously will be unable to make all of the necessary capital investment. In that event, a progressively worsening shortage of petroleum throughout the world will be the inevitable result. Alternate sources of energy will not be available as substitutes because the supply of any form of energy is dependent upon adequate capital investment and the profits needed to make that investment possible.

Because a sufficient supply of petroleum is so essential to all nations, there is an urgent need for a broader understanding of this capital-requirements problem. A lack of understanding has led the United States to a severe shortage of energy. It can also lead to the same predicament in other countries. Better industry and government cooperation are necessary. Clearly, a more enlightened attitude on the part of government is urgently needed. The well-being of the world might very well depend upon it.

Latin America's Outlook Toward Energy Resources—Summary[1]

M. F. BACA[2]

Latin American countries are alike in many ways, but they differ widely in respect to present and future problems in energy and mineral resources. Rapidly rising demands for energy already have created serious trade-balance problems for some countries. As development rate accelerates, Latin America will be consuming an ever-increasing share of its own mineral resources, most of which are now exported. As a region, Latin America is rich in energy and mineral resources. For a group bound by history, culture, and language, a logical approach for the future is to make a joint study of the situation, and then to propose a unified formulation of a regional development plan which will make the most effective use of each country's potential for the mutual benefit of all the Latin American countries.

[1]From Circum-Pacific Energy and Mineral Resources Conference program booklet.

[2]PetroPeru, Lima, Peru.

Marine Environment in Australia: Two Studies—Summary[1]

ROBERT J. FOSTER[2]

Introduction

In line with the rest of the world, in Australia, public apathy toward the environment has been replaced in less than a decade by public awareness. Governments are now faced with the unwelcome and recurring task of publicly choosing between competing uses of the environment—broadly, between development and preservation.

Governments are generally ill-equipped for making decisions where the only benefits are intangible or long term—particularly when opposed by determined, vocal, and widespread public opposition. This paper examines and compares two routes chosen by governments in Australia to solve this problem of decision-making in emotional areas of development/preservation confrontation. The two cases investigated are (1) the Royal Commission investigation into petroleum drilling on the Great Barrier Reef and (2) the Westernport Bay Environmental Study.

Great Barrier Reef

The Great Barrier Reef (Fig. 1) is the largest tract of coral reefs in the world. It comprises over 2,000 individual reefs and extends about 1,200 mi (1,900 km) along the northeast coast of Australia; the reef province covers 100,000 sq mi (260,000 km^2) of the Australian continental shelf. The reef fauna is typical of the Indo-Pacific and is probably no more prolific in species or individuals than that of the Indonesian–New Guinea archipelago. The Great Barrier Reef is nevertheless unique—both in its size and its accessibility to visitors; it is a resource of world importance.

Every good story has a villain. The villain on the Barrier Reef was *Acanthaster planci* (Lamarck), the crown-of-thorns starfish. This fearsome animal has 13–17 arms and grows to a diameter of about half a meter; the prominent spines on its upper surface can cause severe local pain and protracted vomiting should they penetrate the skin. *Acanthaster* eats the polyps of stony corals, particularly those of the branching and palmate forms such as *Acropora*. Because a feeding starfish can consume 20–30 sq in (130–190 cm^2) of living coral per day, the animal leaves a wake of white coral skeletons.

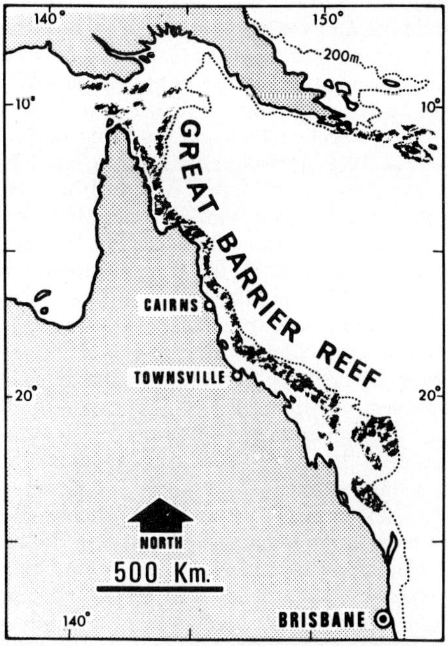

FIG. 1—Great Barrier Reef and continental shelf of northeast Queensland.

A. planci is widespread, though generally scarce on the Barrier Reef, and its range extends across the Indo-Pacific from the Red Sea to Hawaii. In the 1960s, however, it began to appear in plague proportions, both on the Barrier Reef and elsewhere. At least in Australia, it had never before been recorded in large numbers.

Much concern was expressed by a section of the Australian scientific community that *Acanthaster* posed a threat to the continued existence of the reef, and the controversy captured the imagination of the Australian public. Studies were commissioned (Endean, 1969; Walsh, 1970) which confirmed the presence of very large, though patchy, concentrations of the starfish and

[1]Manuscript received, October 21, 1974.

[2]Manager, Planning & Development, Oil and Gas Division, The Broken Hill Proprietary Co., Ltd., Melbourne, Australia 3000.

areas of intensive local destruction of the living coral.

Actually, there appears to be no conclusive evidence that *Acanthaster* causes damage to coral reefs from which they will not recover. There have, in fact, been two positively beneficial results of the plague—an inflow of money into studying the marine biology of the Australian tropical seas, and the creation of a genuine public concern for the welfare of the Great Barrier Reef.

Public awareness of the potential dangers of oil spillage had been created by reports of the Torrey Canyon shipwreck in 1966 and the Santa Barbara blowout in 1969. It was announced in late 1969 that a Japanese group, Japex, was to drill in Repulse Bay on a farmout from the Australian company Ampol Exploration. The farmout had been approved by federal and state governments, the well was to be 60 mi (100 km) from the nearest coral reef, and four previous wells had been drilled without incident in the vicinity of the reef. However, in the face of outcries from a newly alert public, plus the threat of direct trade-union action, the government embargoed all future drilling until the matter had been examined by a Royal Commission.

A Royal Commission is an inquiry held before one or more commissioners appointed, in the case of the Commonwealth of Australia, by the Governor-General in accordance with the Royal Commissions Act 1902–66. Similar statutes apply to the states of Australia. The chairman of the Commission, or the sole commissioner as the case may be, is empowered to summon witnesses to give evidence or produce documents. Evidence before a commission is given under oath, and there is provision for the taking of evidence in private. Although a Royal Commission is not a court of law, it is common practice for parties to be represented by legal counsel.

The Great Barrier Reef Petroleum Drilling Royal Commissions were created by the Australian Commonwealth and Queensland governments and run as one Commission. The three commissioners were a former judge of the New South Wales Supreme Court (Sir Gordon Wallace, Sydney) who was the chairman, and two internationally known technical experts in the fields of petroleum engineering (Vincent John Moroney, Calgary) and marine biology (James Eric Smith, Plymouth, England). The terms of reference were the following:

1. Taking into account existing world technology in relation to drilling for petroleum and safety precautions relating thereto, what risk is there of an oil or gas leak in exploratory and production drilling for petroleum in the Area of the Great Barrier Reef?

2. What would be the probable effects of such an oil or gas leak and of the subsequent remedial measures on (a) the coral reefs themselves; (b) the coastline; and (c) the ecological and biological aspects of life in the area?

3. Are there localities within the Area of the Great Barrier Reef and, if so, what are their geographical limits, wherein the effects of an oil or gas leak would cause so little detriment that drilling there for petroleum might be permitted?

4. If exploration or drilling for petroleum in any locality within the Area of the Great Barrier Reef is permitted, are existing safety precautions already prescribed or otherwise laid down for that locality regarded as adequate and, if not, what conditions should be imposed before such exploration or drilling could take place?

5. What are the probable benefits accruing to the State of Queensland and other parts of the Commonwealth from exploration or drilling for petroleum in the Area of the Great Barrier Reef, and [what is] the extent of those benefits?

The Commission was established on May 5, 1970, and the first hearing was held on May 22. Hearings continued intermittently until July 3, 1972. During this time, 88 witnesses were called, 465 exhibits were tendered, and the proceedings of the Commission were recorded in 18,256 pages of transcript. The cost of the Commission is not known, but including costs of respondents, is likely to be US$1–2 million. The report had not been released at the end of August 1974.

Full-time respondents to the Commission were the Queensland government through the Minister for Mines, a group of conservation societies, and the Australian Petroleum Exploration Association (APEA). All three were represented by counsel, and legal fees for the conservation group were paid by the Australian Commonwealth Government. Among the respondents who appeared from time to time was the Australian Labour Party, which was then in opposition in both Queensland and Federal Parliaments.

Horler (1974) described the inquiry in detail, and the views of the parties are summarized only briefly here. APEA tried to show that, although absolute security from oil spillage cannot be guaranteed (and probably never will be), the chance of large-scale accidental spillage from well blowout, or from equipment failure during production, is low. Furthermore, in at least certain parts of the area under consideration, the chance that spilled oil will actually reach and strand on a reef is also low. Evidence was called which suggested that the extent of damage to coral contacted by oil would be limited, and that substantial regeneration of damaged areas would be relatively rapid and complete ultimate recovery would ensue.

The conservation view was well summarized by McKinney (1973, p. 2). She said: "Our attitude towards the proposed development of a marine oil industry on or near the Great Barrier Reef is . . . unequivocal, and our policy of preservation would be maintained until it were shown that no alternative sources of oil existed elsewhere."

The concern of the conservationists as expressed by Brown (1972) when speaking of the crown of thorns, or by McKinney (1973) when speaking of development along the Queensland coast, is obvious. In support of her views, McKinney quoted Gilbert Voss in *Natural History*—"Florida's coral reefs are dying." So far as I know that is not the case; nor is the Great Barrier Reef dying. However, it is better that people think they are when they are not than vice versa.

WESTERNPORT BAY

Westernport Bay (Fig. 2) is a landlocked harbor with 260 sq mi (670 km^2) of water area; low-water depth is 60 ft (18 m) at the entrance bar. The catchment of the bay covers about 1,200 sq mi (3,100 km^2) with a population of about 40,000 people on farms and in small towns. About 12% of the land area is still in an uncleared natural state.

The region has a prolific avifauna because of its swamps, mangroves, and mud flats; 270 species have been recorded (Connell, 1975). One of the few remaining habitats of the potoroo, or rat kangaroo, is found in the area, as well as the only colonies of the New Holland mouse—which until recently was thought to be extinct in Victoria. At the mouth of the bay is one of Australia's largest seal colonies, and on Phillip Island the fairy penguin parade—perhaps the most remarkable sight in Australia, and a major tourist attaction—recurs nightly.

The city of Melbourne is built on the shores of Port Phillip Bay. The entrance to Port Phillip is relatively shallow, and rock makes deepening prohibitively expensive. Only 35 mi (56 km) away

FIG. 2–Westernport Bay and catchment area, Victoria.

is Westernport, where the bottom is amenable to deepening as required for modern tankers and bulk carriers. The flat hinterland of Westernport would make fine industrial land; it is near the oil and gas fields of Gippsland and the second-largest market in Australia—Melbourne, with a population of 2.6 million people.

Westernport Bay is still an unspoiled body of water, and there are only three significant industries on its shores—a BP oil refinery, an Esso/BHP crude-oil loading terminal and gas-liquids fractionation plant, and a Lysaght/BHP steel rolling mill.

The clean water and sandy beaches of Westernport make it a good recreation area for the people of Melbourne. Past land subdivision is being followed apace by the construction of unsewered holiday houses. The Melbourne people with holiday homes in the area are perhaps more vocal in their opposition to further industrial development than are the longtime residents, whose grown children now have to go to Melbourne to find jobs. Ironically, it appears to be unsewered housing development, rather than industry, that is the most pressing environmental threat to Westernport.

In the longer term, the most serious threat perhaps is the unchecked growth of the suburbs of Melbourne, whose population will be 4 or 5 million by the end of the century and will spread into much of the hinterland of the bay.

Conservation interest in Westernport was stimulated in 1969 by a proposal by Ampol and Sleigh, two Australian companies, to build an oil refinery; determined opposition stopped the project. Additionally, a proposal to build a 10-in. (25.4 cm) pipeline to carry ethane gas from the Esso/BHP fractionation plant on the bay across the Mornington Peninsula, and then across Port Phillip Bay to the Esso/Mobil petrochemical plant at Altona, attracted strong opposition from conservation groups and a faction of the trade-union movement. Ultimately the pipeline was built.

It became obvious to the Victorian government that a long-term solution was required to problems posed by the competing interests of development, recreation, residential growth, and preservation in the Westernport area. Consultants hired by the government proposed studies which evolved finally into the Westernport Bay Environmental Study. Government has embargoed further industrial development until the study is complete but, unfortunately, has not extended the ban to include subdivision of rural land and the subsequent construction of urban or holiday housing.

The study is controlled by an executive committee comprising the former Director of the Victorian Fisheries and Wildlife Department as chairman, and representatives from the Victorian Ports and Harbours Division, Westernport Regional Planning Authority, and industry. The cost of over US$2 million is funded about 30% from the three industries in the area and the balance from the Victorian State Government. The detailed plan for the study was developed by a study committee drawn from universities, government, and industry and led by J. M. Swan, Pro-Vice Chancellor of Monash University in Victoria.

The study is coordinated by a core group of specialists in several key disciplines who were hired especially to work full-time on the project. They are led by the Study Director, Maurice Shapiro, formerly Professor of Environmental Health Engineering and Professor of Urban Affairs at the University of Pittsburgh. Their task is to identify the areas in which information is required, control the groups gathering the data, and integrate all information into a coherent whole which can be presented as a written report.

The broad objectives of the study are:

1. To establish baseline data on the physical, chemical, and biological environment of Westernport Bay against which current status and future changes may be evaluated, and
2. To elucidate relationships between biota and environmental factors which can be used in planning and management of the bay and its catchment.

The study program has been divided into more than 40 separate projects which are being carried out by groups from universities or government departments, or by outside consultants. Broadly, the study areas are:

Land studies: geology, soil, climate, air quality, inputs to the bay from streams.

Marine studies: physical and chemical characteristics, biota population and community structure, mangroves and seagrass, toxicology.

Social and economic studies: land and bay uses and forecast uses.

Land-use model: to enable the potential inputs to the bay from any planned development in the catchment area to be forecast, to help formulate acceptable development plans, and to improve the understanding of the interested public.

Water-quality model: to indicate intensity, extent, and duration of contamination of the water body arising from a given input.

The study was scheduled to include a development period for staff buildup and task specification, followed by a 2-year intensive study during 1973 and 1974. Delays were experienced in recruiting some of the specialists. Nevertheless, the study report is expected on schedule.

This broad-based multidiscipline study is the first of its kind and scope to be undertaken in Australia.

Conclusions

Which of the two methods of tackling the problem is better? Both have advantages; but there is a great disadvantage common to both. Neither the Westernport Bay Environmental Study nor the Barrier Reef Royal Commission will do the government's job. Both studies will delay the evil hour for, perhaps, 4 years, but one day government will have the reports. It then will be faced with the necessity to plot a balanced, and doubtless unpopular, course between uncontrolled development and absolute preservation. I hope, though, that government will be much better informed by that time.

One advantage of a Royal Commission is that it is finite in duration; once it adjourns *sine die*, its hearings are over, and one can only await its report. A scientific study is never over in the sense that scientists never have all the information they could wish to have in order to arrive at the best possible findings. The Westernport Study, for instance, will continue into a second and more detailed phase lasting perhaps an additional 3 years, although neither funding nor objectives are finalized at this stage; and monitoring of key indicators on a routine basis presumably will continue indefinitely as a safeguard against environmental deterioration.

Another advantage of a Royal Commission is the testing of evidence by cross-examination of witnesses. A skilled barrister can expose loose logic and emotional thinking—even in the evidence of scientists. However, this strength of the forensic process can also be its weakness. A barrister is a partisan for a cause—that of the party who briefed him. The result is that polarization sets in, and a Royal Commission on a scientific topic may develop into adversary proceedings.

A scientific study brings the parties (government, industry, and research scientists) together to work for a common cause. Furthermore—and this advantage outweighs all others, in my opinion—a study such as that at Westernport spends time and money on developing new information, and not on arguing the merits of that which already exists.

However, we must not make a mistake in overestimating the role of rational thought in an emotional issue. No amount of investigation and reporting can be guaranteed to eliminate completely the genuine concern of all persons. Such continuing concern is well expressed by McKinney (1973, p. 4). Under the heading "Action Stations!", she wrote: "As soon as the Report of the Royal Commission into Oil-drilling in Great Barrier Reef waters comes out, your responsibility begins. The pressures for exploitation are ready, set and all systems go. We must be at least as ready."

References Cited

Brown, Theo, 1972, Crown of thorns: the death of the Great Barrier Reef?: Sydney, Australia, Angus and Robertson, 128 p.

Connell, D. W., 1975, A strategy for studies of the marine environment in Victoria: Royal Soc. Victoria Proc. 87, p. 95–105.

Endean, R., 1969, Report on investigations made into aspects of the current *Acanthaster planci* (crown of thorns) infestations of certain reefs of the Great Barrier Reef: Fisheries Branch, Queensland Department of Primary Industry, Brisbane, 35 p.

Horler, K. M., 1974, The oil exploration industry's role in the Royal Commission into oil drilling on the Great Barrier Reef: 2d Internat. Coral Reef Symposium Proc., Great Barrier Reef Committee, Brisbane, p. 703–704.

McKinney, J. W., 1973, The Great Barrier Reef and the future: Wildlife Preservation Society of Queensland, Brisbane, Newsletter 52, p. 1–5.

Walsh, R. J., et al, 1970, *Acanthaster planci* (crown of thorns starfish) and the Great Barrier Reef: Australian Acad. Science Rept. 11, 20 p.

Tectonic Framework of Pacific Region[1]

J. D. MOODY,[2] R. W. ESSER,[3] and D. A. HOLMGREN[4]

Abstract The tectonic framework of the Pacific is analyzed in terms of physiography, orogenic units, and linear tectonic elements. Orogenic units are classified into five categories: linear uplifts with longitudinal shears, autochthonous fold belts, vertically uplifted or tilted fault blocks, domal uplifts, and volcanic chains. Volcanic chains predominate in the open Pacific, but linear uplifts with longitudinal shears predominate around the Pacific margin.

Segmentation of the Pacific area into crustal blocks or plates bounded by geosutures is the most evident result of Pacific tectonism. The existence of the crustal blocks and of their bounding geosutures is the most significant feature of the tectonic framework of the area, regardless of the origin or importance ascribed to the geosutures in various geotectonic schemes.

INTRODUCTION

The NASA space photo (Fig. 1), taken from an unmanned craft at an altitude of 35,800 km (22,200 mi), shows with exceptional clarity the eastern margin of the Pacific Ocean basin. Certain prominent features include Baja California, the western and southern Sierra Madre of Mexico, Florida and the southeastern coastline of the United States, the Gulf of Mexico eastward from the mouth of the Mississippi River, all of Central America, and the entire South American continent from the Caribbean to the Falkland Plateau. Hawaii is located at the dot near the northwest horizon.

The enormous size of the Pacific Ocean basin can be appreciated by considering certain statistics. The Pacific Ocean basin's area of 174 million km^2 (67 million sq mi) equals about a third of the world's total surface area of 510 million km^2 (197 million sq mi). Furthermore, the basin's area equals about 48% of the world's water-covered area and more than all of the land area put together. The curvature of the earth is such that, for an area as enormous as the Pacific basin, less than half appears in Figure 1.

Most of the limited land area in the Pacific Ocean basin consists of volcanic islands and atolls. In many places these atolls have developed in association with flat-topped guyots and seamounts, most of which also are volcanic.

The photo shown in Figure 2 is a view facing southeast along the two subparallel, southeast-trending ridges of the central part of the Tuamotu island chain. This picture was taken on the NASA Apollo VII flight from an altitude of 250 km (155 mi) and covers an area greater than 150,000 km^2 (57,900 sq mi). The surficial parts of the atolls are composed entirely of reef-type material.

Figure 3, also a photo taken from the air, shows the volcanic island Moorea, just east of Tahiti. Even at this scale the curvature of the earth is apparent. Noteworthy is the surrounding reef-enclosed lagoon and the wave-cut bench onshore. Surface-outcrop information in the Pacific Ocean basin is available only from relatively limited volcanic and carbonate islands such as those shown in Figures 2 and 3. Therefore, one must rely on the adjacent continental geology, together with bathymetric, geophysical, bottom-sampling, and drilling information *within* the Pacific, to gain any real understanding of the geology of this large segment of the globe. We emphasize the real paucity and reconnaissance nature of the information available. It might well be compared with the level of knowledge of the geology of North America when James Hall first advocated the geosynclinal theory. Only a beginning has been made at detailed mapping of even the smallest areas of the Pacific Ocean bottom.

Our intent in this paper is to describe in broad terms the tectonic elements of the area in order to establish a framework that will permit a better understanding of the following papers on energy and mineral resources of the Pacific Ocean basin.

TOPOGRAPHY AND BATHYMETRY OF CIRCUM-PACIFIC REGION

The bathymetry of the Pacific Ocean basin is shown in Figure 4. The map, completed in 1964, is a compilation by the Institute of Geology and the Institute of Oceanology of the Academy of Sciences of the USSR (Anon., 1964). This map, a pseudocylindrical sinusoidal projection on an original 1:10,000,000 scale, serves as a base map for the following tectonic maps. The darkest blues

[1]Manuscript received, November 1, 1974.

[2]Consulting geologist, New York, New York 10022. Formerly Senior Vice President, Exploration and Producing, Mobil Oil Corp.

[3]Mobil Oil Corp., New York, New York 10017.

[4]Mobil Oil Corp., Princeton, New Jersey 08540.

FIG. 1—Full-disc space photograph of eastern Pacific Ocean. NASA-photo SCI-1893-ATS-1.

and browns represent the deepest and highest areas, respectively. In the ocean the trenches show up as the darkest blue areas; those in the northern and western Pacific exceed 7,625 m (25,000 ft) in depth, and those in the eastern Pacific generally are less than 6,100 m (20,000 ft) deep. In general, the northwestern part of the ocean is deepest, having an average depth of 3,000–5,000 m (9,840–16,400 ft). The continental-shelf areas are shown in the lightest colors and, except for the Gulf of Mexico and the Argentine shelf on the Atlantic side, are confined mainly to the western part of the map in areas such as the northwest shelf of Australia and the East Asian marginal seas. White-colored ocean areas, such as the Campbell, East Chatham, Lord Howe, and Ontong Java Rises and the Colville-Lau (South Fiji) and Cocos Ridges, range in depth from 300 to 1,700 m (985–5,575 ft). Light blue areas, such as the crests of the Pacific-Antarctic and East Pacific Ridges and much of the eastern margin of the Pacific, range in depth from 2,000 to 3,000 m (6,560–9,480 ft).

The greatest elevation differential in the world, 19,895 m (65,228 ft), occurs in the western Pacific area between Mount Everest in Tibet, at 8,855 m (29,028 ft), and the Marianas Trench south of Guam, at −11,040 m (−36,200 ft). However, if one considers the distance between these points, approximately 5,950 km (3,700 mi), and the effect of curvature of the earth, the earth's maximum surface relief is barely discernible. At the scale of this map, this maximum relief is less than 1/20 in. (0.13 cm). More spectacular is the elevation differential of almost 9,700 m (31,800 ft) between Mauna Kea, at 4,210 m (13,796 ft), and the base of the Hawaiian deep, at −5,490 m (−18,000 ft), across a distance of 96 km (60 mi) on the island of Hawaii. In this example, the effect of the curvature of the earth is practically eliminated and the tremendous size of the oceanic volcanoes can be appreciated.

Tectonic Map of Circum-Pacific Region

Introduction

The map shown in Figure 5 is a new tectonic map of the Pacific Ocean basin, derived by us from an exhaustive search of the literature, analysis of published bathymetry, seismicity, geology,

FIG. 2–Tuamotu atolls in Pacific Ocean. NASA-photo, Apollo VII. Exposure by W. Shirra, D. Eisele, W. Cunningham.

FIG. 3–Volcanic island of Moorea and surrounding reef-enclosed lagoon.

FIG. 4—Bathymetry of Pacific Ocean (Anon., 1964). Deeper color tones show deeper water areas.

etc, and the results of our own exploration. The tectonic map has been constructed as an overlay to the Russian bathymetric map of 1964. In certain onshore areas, only data deemed to be most significant are shown. The features shown include fractures, faults, bathymetric and other lineaments, and positive elements such as uplifts, ridges, folds, arcs, seamount lineations, and island chains. Also shown are certain negative features such as trenches, troughs, and basins. To make these features more apparent, we have shown them without the bathymetric base in Figure 6.

Physiographic Features

The most common physiographic features of the areas of oceanic crust in the Pacific Ocean basin are volcanic in origin and occur as individual seamounts or lineations of seamounts, abyssal hills, and platforms. Some platforms support a relatively thick sedimentary section. Examples are the Shatsky Rise (Deep Sea Drilling Project drillsites 305, 306), Ontong Java Rise (DSDP sites 288, 289), and Manihiki Plateau (DSDP site 317A). On these plateaus, basalt is overlain by Cretaceous and Tertiary ooze, chalk, and limestone with chert; sedimentary thickness is 985–1,310 m (3,000–4,000 ft). According to Krasny et al (1974), these broad, irregularly shaped plateaus, elevated above the surrounding abyssal deeps, yield continental-crust seismic velocities and probably were produced by faulting. Noteworthy is the relative concentration of volcanic features in the presumed older and deeper, central and western parts of the Pacific Ocean basin.

Currently there are only four areas of active volcanism in the oceanic Pacific—the island of Hawaii, the Juan de Fuca Rise near Cobb Seamount, MacDonald Seamount at the southeast end of the Tubuai (Austral) island chain, and an area near Easter Island. This very modest amount of current volcanic activity contrasts strongly with the extensive and pervasive volcanism which obviously dominated the Pacific area in pre-Holocene time. This recent marked decrease in volcanic activity must reflect a fundamental change in the earth's thermal regime.

Volcanism in seamount lineations in the Pacific commonly reveals a progression from an active or recently active volcano, to volcanic peaks with small lagoons (e.g., Moorea; Fig. 3), to eroded remnants with large lagoons, to atolls (e.g., Tuamotus; Fig. 2), and to seamounts. These lineations appear to be fault controlled; examples are the Hawaiian, Emperor, and Tubuai (Austral) island chains.

One of the main features of the Pacific Ocean basin is an ocean-girdling ridge system—the East Pacific and Pacific-Antarctic Ridges (Albatross Cordillera) and associated aseismic "45° fractures." The East Pacific Ridge differs from its presumed counterpart in the Atlantic Ocean in that it does not appear to have a median rift valley, it is less active seismically, and its crestal topography is relatively smooth compared with parts of the ridge flanks (Herron, 1972).

The Chile Ridge and the Gorda Rise resemble more closely the Mid-Atlantic Ridge, having steep, rough flanks (Pitman et al, 1968), well-developed median valleys (Herron and Hays, 1969), and pronounced seismicity. The East Pacific and Pacific-Antarctic ridge system ranges in width from 1,930 to 4,020 km (1,200–2,500 mi) and has up to 3,000 m (9,840 ft) of relief (Menard, 1964).

A series of west-southwest-striking fracture zones seems to be limited to the northeastern part of the basin (Menard, 1955, 1964). These zones are of great linear extent and have associated subparallel fracture zones. They appear as linear troughs and ridges as long as 4,500 km (2,800 mi) and have a relief of as much as 3,000 m (9,840 ft). Common topographic features include a central trough which generally is deeper than the adjacent seafloor. It is apparent that some of these zones actually are major faults with predominantly horizontal movement. The fracture zones appear to correlate with seismic activity only where they intersect the ocean rises, and earthquakes are confined to the fracture zones between offset parts of the ridge crests (Pitman et al, 1968; Sykes, 1967).

Marginal Seas, Island Arcs, and Pacific-Type Continental Margins

The marginal seas of the Pacific basin are bounded by an island-arc system on the ocean side and a Pacific-type continental margin on the landward side. All are located on the western side of the Pacific basin. Examples include the Okhotsk, Japan, Philippine, South China, Coral, and Tasman Seas.

The Pacific-type continental margin is characterized by youthful tectonic belts which coincide with a Mesozoic or Cenozoic geosynclinal area on the adjacent, parallel continental masses. Generally there are large faults along which sizable blocks have subsided, together with much associated volcanism and seismic activity.

The marginal seas are similar in some respects to the adjacent ocean basin in that they share a common island-arc system and have abyssal plains, seamounts, seamount chains, and fracture zones (Packham and Falvey, 1971). They can be of oceanic depth, and may have either extensive or narrow continental shelves. In some places, blocks of continental crust form rises or ridges which separate basins of oceanic-type crust within the sea. This is evident in the Tasman Sea, where the oceanic Tasman, New Caledonia, and South Fiji basins are separated from west to east by the continental Lord Howe Rise and the Norfolk Ridge. Oceanic-type features characterize the areas of oceanic crust (e.g., seamount lineations in the Tasman basin and the Philippine Sea). In general, most of the marginal basins appear to have a crust similar to that of normal ocean basins, but they have a much greater sedimentary thickness (Packham and Falvey, 1971). The acoustic basement of the marginal seas appears to be rougher and younger than the adjacent Pacific floor. High heat flow also is characteristic of the marginal seafloors (Karig, 1971).

Adjacent to the western Pacific margin is a broad continental shelf which comprises most of the Okhotsk, Japan, Yellow, East China, South China, and Java Seas. On many of these shelves, marginal ridges consisting of rocks of Precambrian to Mesozoic age have been formed by faulting, folding, intrusion and subsequent erosion, differential subsidence, or reef growth. Some island arcs have served as dams or sediment traps, and prograding sedimentary shelves have been ponded behind them where major rivers enter the marginal sea. During basin filling, the oceanward part of the basin received very little sediment—generally mud—until the inner basin overflowed with sediment, generally sand. Thus, those basins closer to the present shoreline are filled with progressively older sediments. Basement ridges of this type have been shown to exist off many coasts around the world. The presence of the Fukien-Reinan massif, the Taiwan-Sinzi folded zone, and the volcanic Ryukyu Ridge caused the ponding of thick (up to 5 km; 16,400 ft) sections of sediments in the Yellow Sea, the East China Sea, and the Okinawa Trough, respectively. In the South China Sea a peripheral ridge consisting of volcanic rocks forms a dam behind which a minimum of 3 km (9,800 ft) of sediments collected in the Gulf of Thailand (Parke et al, 1971) and on the Sunda shelf (Ben-Avraham and Emery, 1973), including the Java Sea. This area is one of the most extensive continental shelves in the world, in contrast to the relative narrowness of the shelves bordering the eastern Pacific.

Island-arc trenches rim the Pacific Ocean basin with one notable exception—the western North American border. Ordinarily, a trench is the first feature of an island-arc system between oceanic and predominantly continental crust. Exceptions to this configuration are the reversals seen in the Melanesian trenches—New Britain, Solomon, and New Hebrides—which are landward from the arc system. Intersections of trenches and oceanic crust generally are the sites of intense shallow seismicity, whereas less intense associated intermediate and deeper earthquakes occur landward. Where trenches do not exist, as along western North America, scattered shallow seismicity prevails. Noteworthy is the close association of 83% (113) of the world's active volcanoes with the deep-ocean trenches of the Pacific (Tanner, 1973). This high level of volcanic activity marginal to the Pacific Ocean basin is in striking contrast to the situation elsewhere in the basin.

Regmatic Shear Pattern of Circum-Pacific Region

One of the more important aspects of the tectonic map of the Pacific (Figs. 5, 6) is that the orientations of the mapped linear features are not random. These linear features compose part of the earth's regmatic shear pattern which is common to both continental and oceanic crust. The regmatic pattern of the oceanic areas appears at first glance to differ markedly from that of the continents, but this difference is very likely a function of the relative density of data available. For example, the Hawaiian area has been more intensely studied than other areas, and its regmatic patterns do appear to be more similar to those of the continents.

The Pacific oceanic crust can be divided areally into four quadrants on the basis of regmatic pattern differences. First, the east half of the Pacific basin differs from the west half, and the dividing line runs approximately from New Zealand to Hawaii to Kamchatka. Also, the north half differs from the south half, and the dividing line extends from at least Kalimantan (Borneo) to the northern coast of New Guinea, to Samoa, to the Society Islands, and thence to southern Chile along the Easter Island fault zone and the Chile Ridge.

The northeast quadrant is characterized by long west-southwest-striking fracture zones. Small horizontal displacements on these fractures were suspected (Menard, 1955), but large horizontal displacements seemingly were confirmed by magnetic patterns (Mason, 1958; Menard and Vacquier, 1958). These fracture zones appear to continue through the Emperor and Hawaii chains, apparently without large offsets. Udintsev (1973) has discovered similar fracture trends continuing

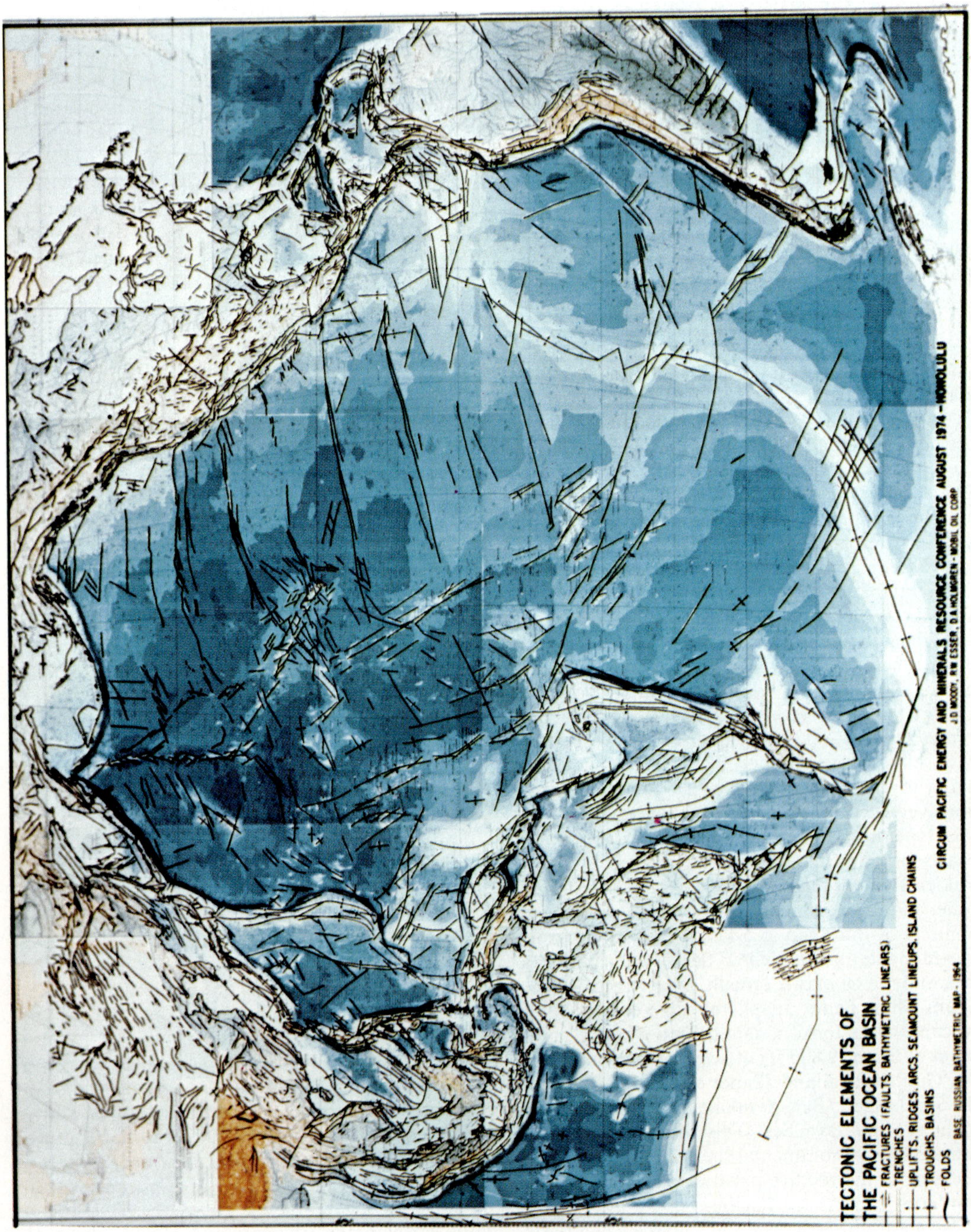

FIG. 5–Tectonic elements of Pacific Ocean basin with bathymetric base.

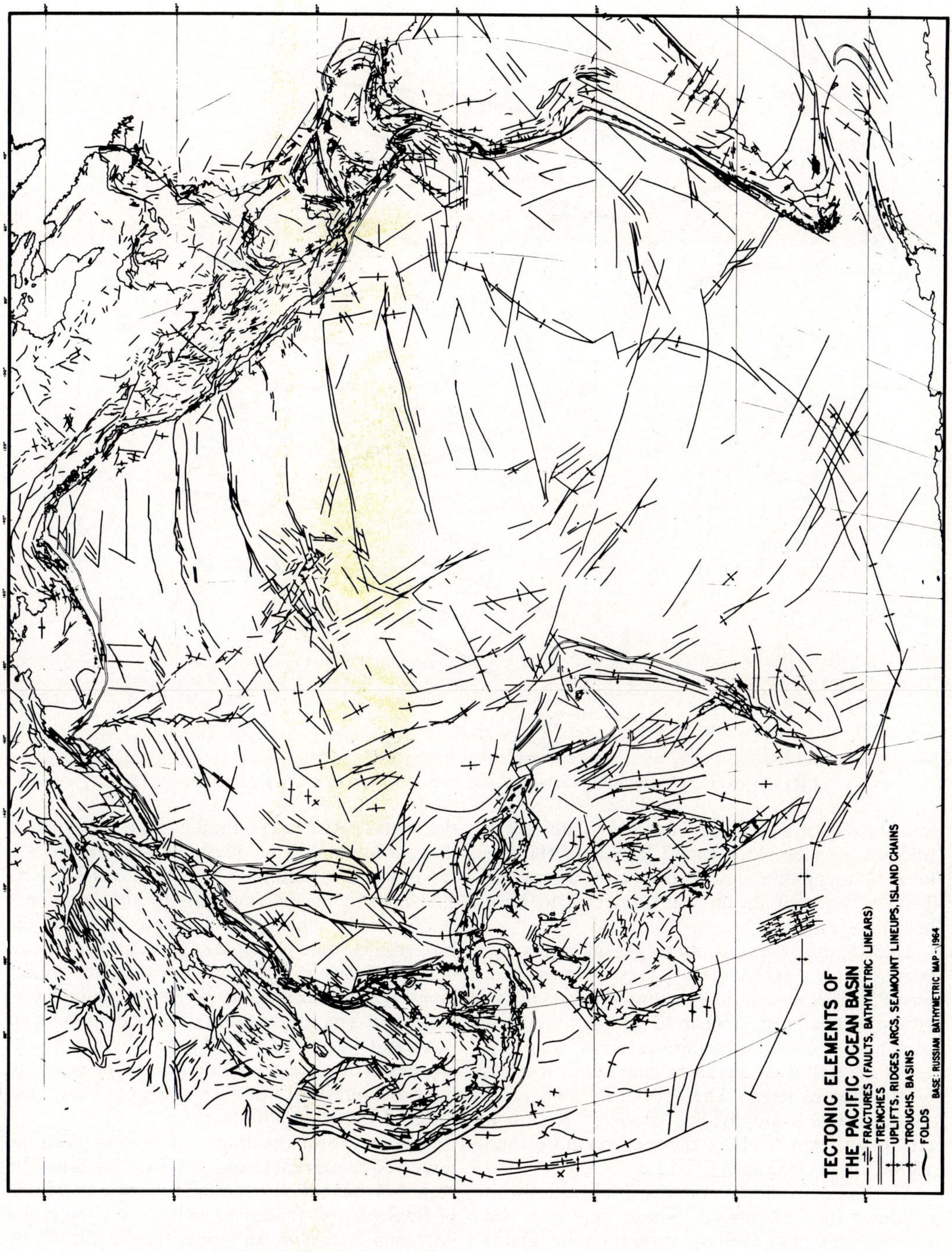

FIG. 6–Tectonic elements of Pacific Ocean basin without bathymetric base of Figure 5.

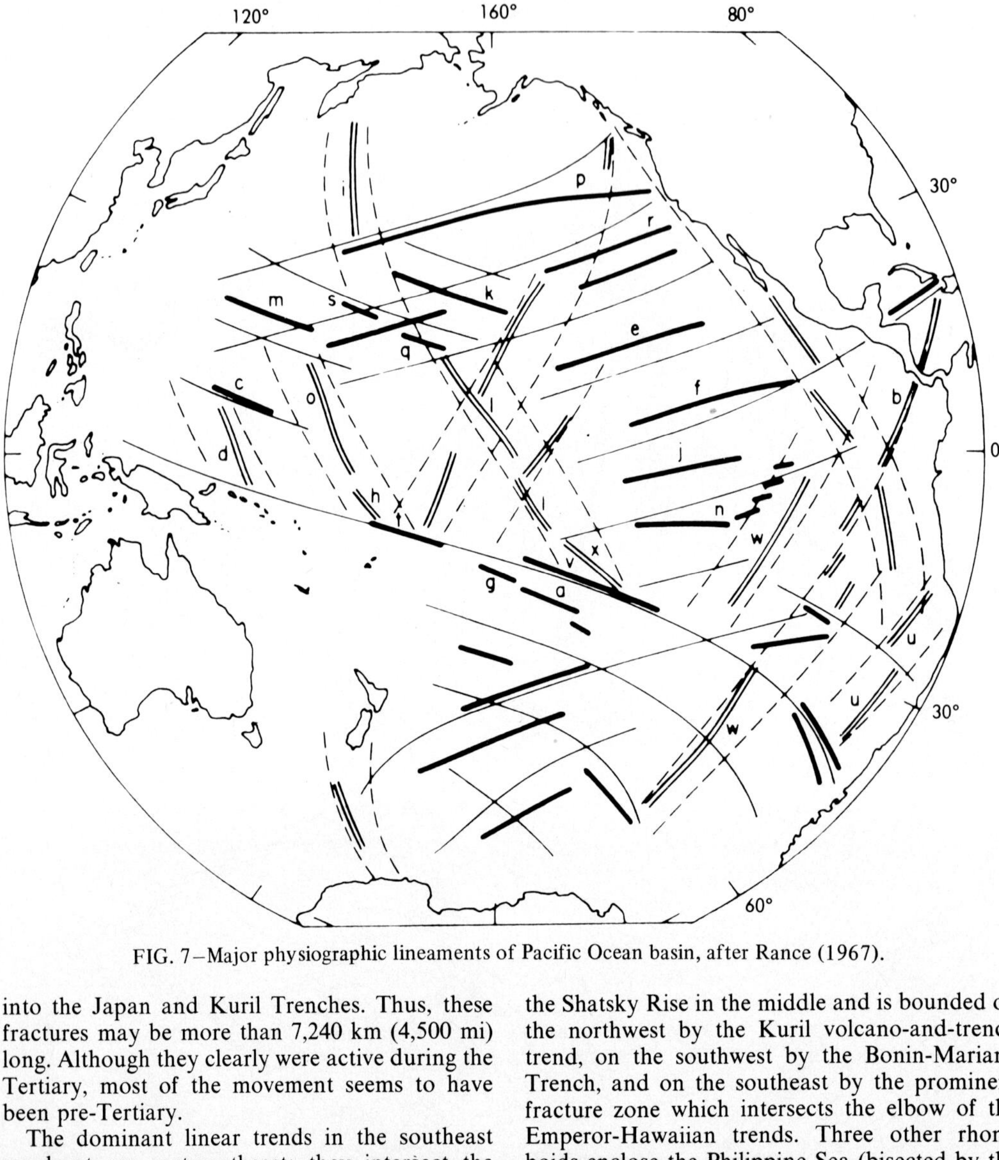

FIG. 7–Major physiographic lineaments of Pacific Ocean basin, after Rance (1967).

into the Japan and Kuril Trenches. Thus, these fractures may be more than 7,240 km (4,500 mi) long. Although they clearly were active during the Tertiary, most of the movement seems to have been pre-Tertiary.

The dominant linear trends in the southeast quadrant are east-southeast; they intersect the East Pacific Rise and, like the linear trends in the northeast quadrant, appear to be relatively uncomplicated. However, significantly less is known about the southeast quadrant than about the rest of the Pacific basin. The most profound east-southeast line is the dividing line from Kalimantan to southern Chile, a distance of more than 17,700 km (11,000 mi).

The dominant linear patterns in the northwest quadrant form rhomboids whose long axes are nearly meridional. The largest rhomboid contains the Shatsky Rise in the middle and is bounded on the northwest by the Kuril volcano-and-trench trend, on the southwest by the Bonin-Mariana Trench, and on the southeast by the prominent fracture zone which intersects the elbow of the Emperor-Hawaiian trends. Three other rhomboids enclose the Philippine Sea (bisected by the Kyushu-Palau Ridge), the Japan Sea, and the Okhotsk Sea. The region south of Hawaii to the Kalimantan-Chile trend is dominated by northwest-southeast fracture zones that seem to interrupt the west-southwest fractures of the northeast quadrant of the Pacific basin.

The southwest quadrant contains at least one large rhomboid enclosing the Tasman Sea. The region of the Solomon Islands appears to consist of fragments of rhomboids which, like the larger polygons farther north, appear to be cut off by

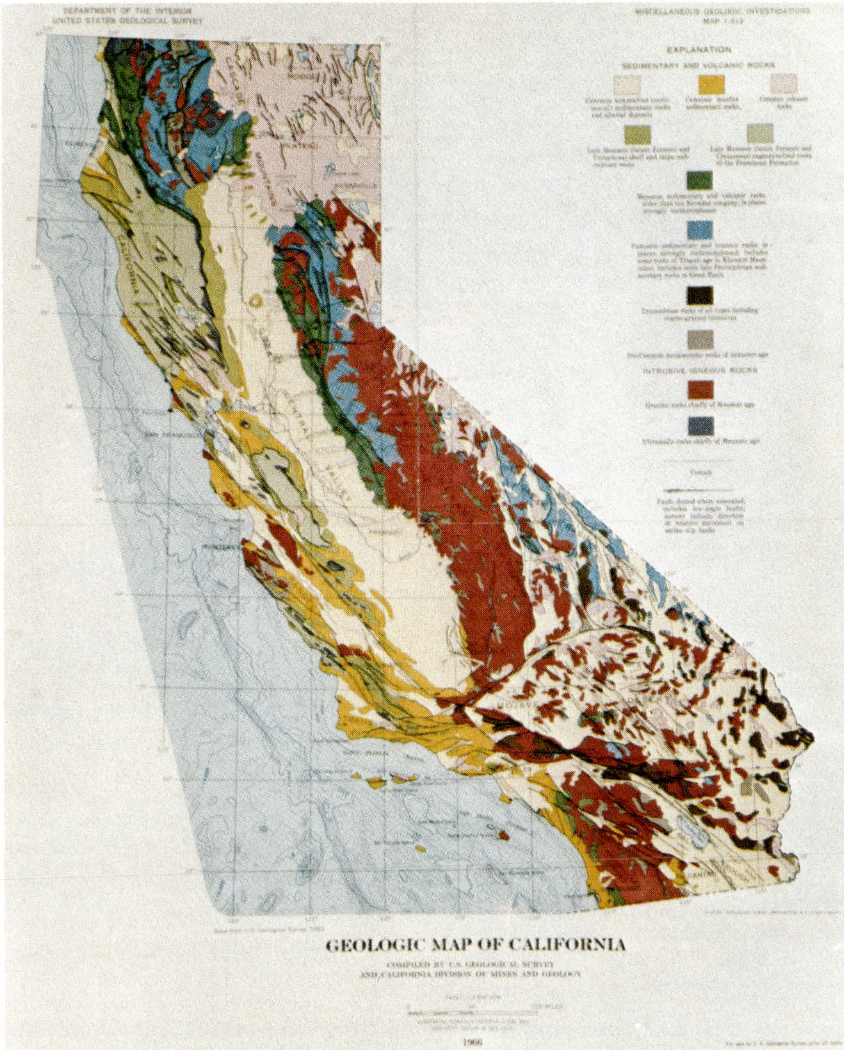

FIG. 8–Geologic map of California (after Anon., 1966). Note northwest-southeast-trending San Andreas fault system with evidence of right-lateral motion.

the Kalimantan-Chile line. The southwest quadrant differs from the others in that it contains the large continental mass of Australia and the smaller continental masses of New Zealand and adjacent submarine areas.

The rhomboid patterns of the northwest and southwest quadrants seem to have been displaced northwestward along the profound 17,700-km (11,000 mi) line from Kalimantan to southern Chile.

Figure 7 illustrates a Pacific basin tectonic pattern derived by Rance (1967) on the basis of model experiments of torsion on a global scale. The similarity of his derived patterns to our observed patterns may well be significant.

Major Fault Symmetry

In addition to the arc-trench and associated seismic-epicenter and continental-shape symmetries of both sides of the Pacific, we draw attention to a fault symmetry. Figure 8 is a geologic map of California (Anon., 1966, original scale, 1:2,500,000). Figure 9 is a geologic map of the eastern Siberian coast of Sikhote-Alin, the north end of the Japan Sea, and the islands of Hokkai-

FIG. 9—Geologic map of eastern Siberian coast of Sikhote-Alin (after Krasny, 1972). Note northeast-southwest-trending Sikhote-Alin fault with clear evidence of left-lateral motion.

do and Sakhalin (Krasny, 1972, sheet 10, original scale, 1:1,500,000). Both maps show major wrench faults. In western North America the major fault zone is the well-known San Andreas fault. This north-northwest-trending, right-lateral fault, with a total mapped length of 1,450 km (900 mi), has been active at least since Jurassic time and has had about 295 km (185 mi) of movement since the early Miocene (Turner, 1968; Huffman et al, 1973). In eastern Siberia the major fault is the relatively unknown, but just as important, Sikhote-Alin fault, which trends north-northeast for a distance of more than 700 km (435 mi). This fault, which reportedly has about 100 km (62 mi) of left-lateral offset and which may have originated in late Paleozoic time, is associated with Mesozoic and Cenozoic igneous bodies and is active today. The Sikhote-Alin fault possibly is part of a larger trend which continues south-southwest of Vladivostok under the Japan Sea into Korea, then crosses the East China Sea into eastern China, and finally ends against the major northwest-trending Red (Yüan) River fault zone (Yünnan). The length of this probably continuous wrench-fault trend is about 4,830 km (3,000 mi).

Mountain Ranges and Regmatic Pattern

Figure 10 portrays the mountain ranges of the Pacific basin and surrounding areas as they might appear if the water were drained from the ocean.

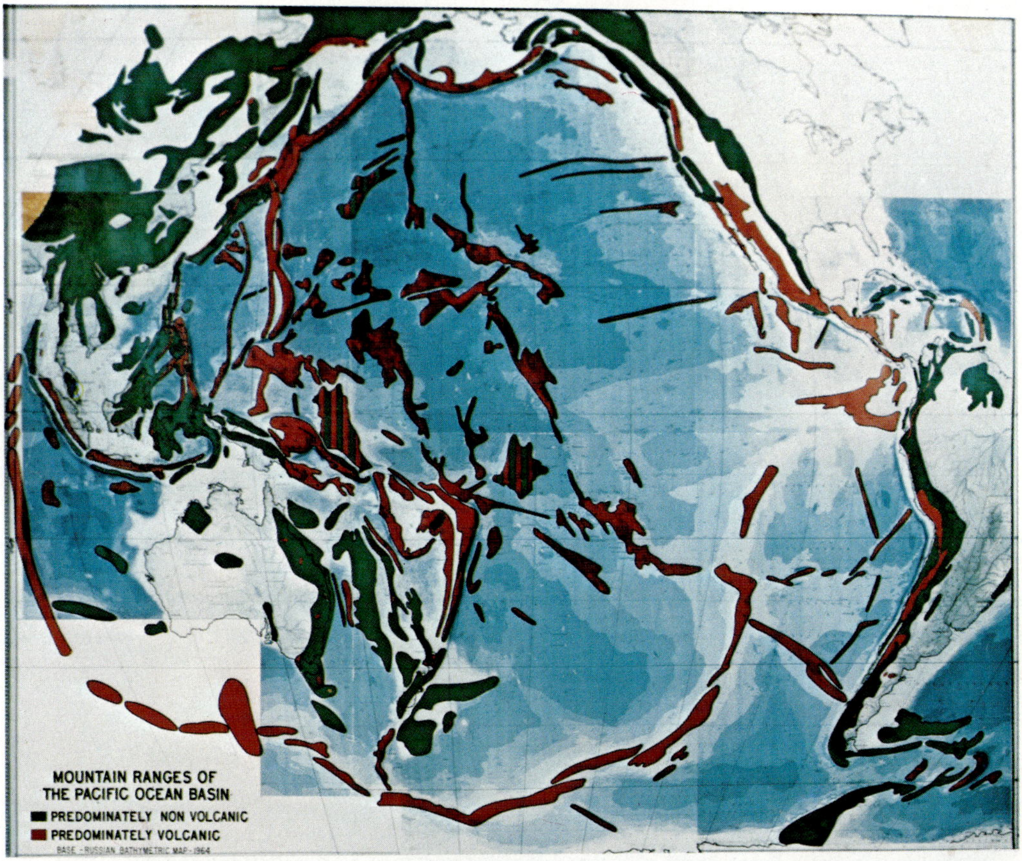

FIG. 10—Mountain ranges of Pacific Ocean basin with bathymetric base. Note close correlation among mountain ranges, zones of tectonic activity, and bathymetry. Also noteworthy is variation in tectonic style in each of the four quadrants.

The coincidence of mountain ranges and zones of tectonic activity is remarkable.

An attempt has been made to differentiate between mountains composed predominantly of volcanic material (shown in red) and all other classes of mountains (shown in green). Areas of known thicker sedimentary sections associated with possible granitic cores (Krasny et al, 1974) in the ocean basin (such as the Manihiki Plateau) are shown in red and green vertical stripes.

The differences in the tectonic style of the four quadrants of the Pacific basin are even more apparent on this map than in Figures 5 and 6. Segmentation of the Pacific Ocean area into polygonal blocks also is more apparent. The size of these major polygonal blocks is remarkable. For example, the major central ocean block between the Line Island mountains and the Marshall-Gilbert mountains is considerably larger than the United States.

Congruency Between Western Pacific and Atlantic Ocean Margins

Figure 11 illustrates an impressive congruency between the western margin of the Pacific Ocean basin and the western margin of the Atlantic Ocean basin. The fit as shown requires no further comment, but the differences in the known geology of these two ocean margins are considerable.

It thus appears that the geometry of the polygonal blocks into which the crust of the earth has been segmented transcends differences between oceans and continents, as well as other presumed fundamentally important differences in the geology of the outer parts of the earth's crust.

Conclusions

The following conclusions can be inferred concerning the tectonic framework of the Pacific Ocean basin:

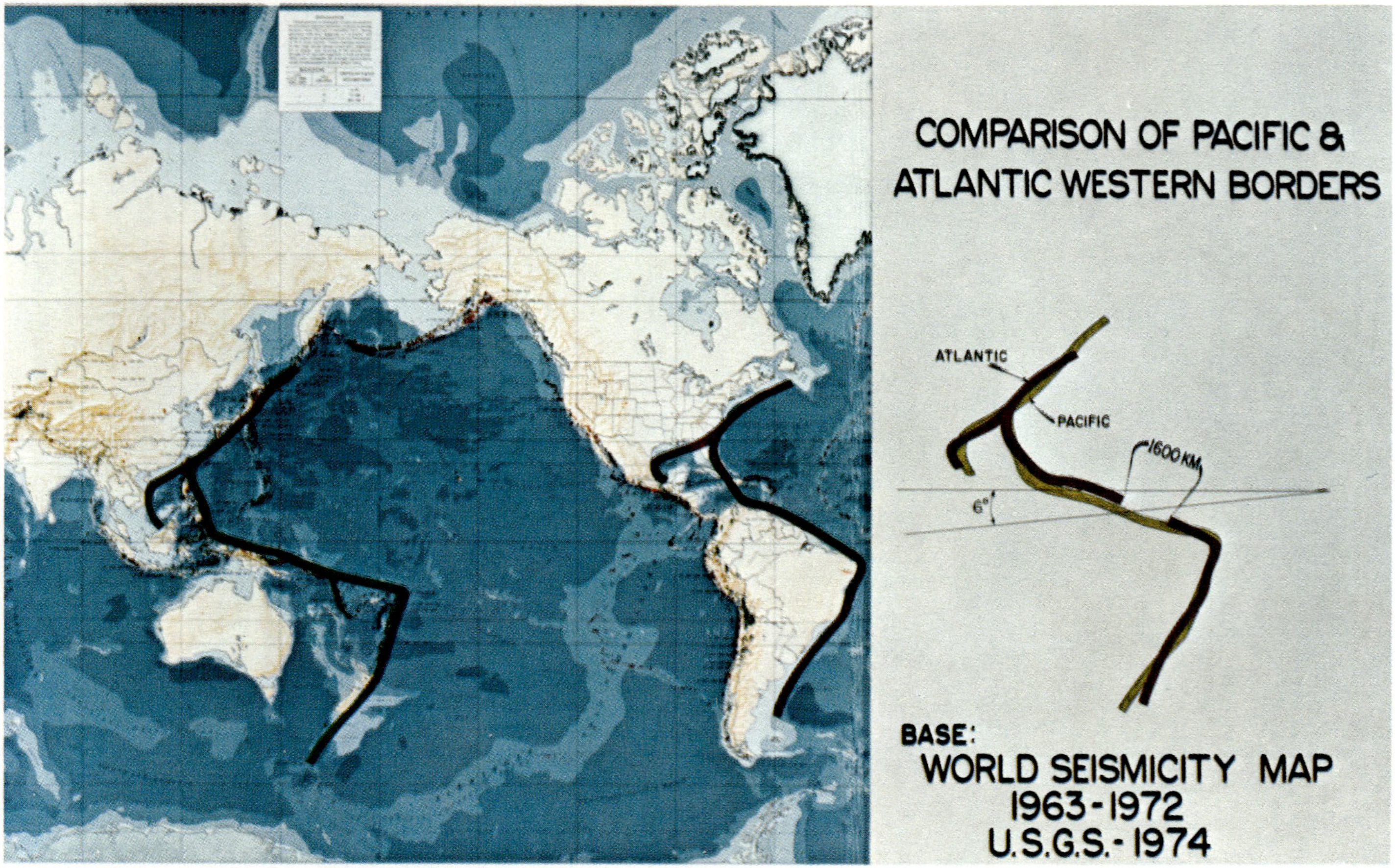

FIG. 11–Comparison of western borders of Pacific and Atlantic Oceans. Base from U.S. Geological Survey (1974). Note nearly exact similarity in size and shape of Atlantic border with addition of a 1,600-km segment and 6° rotation to Pacific border.

1. Data on which to base any but the most general conclusions are quite limited.

2. The crust of the Pacific Ocean basin has been fragmented into several polygonal blocks. The boundaries of the blocks are by definition part of the earth's regmatic shear pattern.

3. The geometry of the polygonal crustal blocks of the Pacific—size, shape, and orientation—shows more similarities than differences with the crustal blocks of better known continental areas.

4. The earth's regmatic shear pattern appears to transcend such fundamental geologic features of the earth's crust as differences between continental and oceanic crust, and differences between the western margins of the Pacific and Atlantic basins.

It is apparent that much remains to be learned about the geology of the Pacific Ocean basin and, indeed, of the earth.

References Cited

Anonymous, 1964, Bathymetry of the Pacific Ocean: Moscow, Inst. Geology and Inst. Oceanology, Acad. Sci. USSR, scale 1:10,000,000.

——— 1966, Geological map of California: U.S. Geol. Survey and Calif. Div. Mines and Geology, scale 1:2,500,000.

Ben-Avraham, Z., and K. O. Emery, 1973, Structural framework of Sunda shelf: AAPG Bull., v. 57, p. 2323–2366.

Herron, E. M., 1972, Sea-floor spreading and the Cenozoic history of the east central Pacific: Geol. Soc. America Bull., v. 83, p. 1671–1692.

——— and D. E. Hays, 1969, A geophysical study of the Chile Ridge: Earth and Planetary Sci. Letters, v. 6, p. 77–83.

Huffman, O. F., D. L. Turner, and R. N. Jack, 1973, Offset of late Oligocene–early Miocene volcanic rocks along the San Andreas fault in central California, *in* Conference on tectonic problems of the San Andreas fault system: Stanford Univ. Pubs. Geol. Sci., v. 13, p. 368–373.

Karig, D. E., 1971, Origin and development of marginal basins in the western Pacific: Jour. Geophys. Research, v. 76, p. 2542–2561.

Krasny, L. I., ed., 1972, Structure-formation map of the northwestern part of the Pacific mobile belt within the USSR, sheet 10: Leningrad, VSEGEI (All-Union Geol. Research Inst.), scale 1:1,500,000.

——— et al, 1974, The geoblock systems of Eastern Asia and the western Pacific: Acad. Sci. USSR, Doklady, v. 214, no. 2, p. 414–417 (in Russian).

Mason, R. G., 1958, A magnetic survey off the west coast of the United States between latitudes 32° and 36°N, longitudes 121° and 128°W: Royal Astron. Soc. Geophys. Jour., v. 1, p. 320–329.

Menard, H. W., 1955, Deformation of the northeastern Pacific basin and the west coast of North America: Geol. Soc. America Bull., v. 66, p. 1149–1198.

——— 1964, Marine geology of the Pacific: New York, McGraw-Hill Book Co., 271 p.

——— and V. Vacquier, 1958, Magnetic survey of part of the deep sea floor off the coast of California: U.S. Office Naval Research, Research Reviews, June, p. 1–5.

Packham, G. H., and D. A. Falvey, 1971, An hypothesis for the formation of marginal seas in the western Pacific: Tectonophysics, v. 11, p. 79–109.

Parke, M. L., Jr., et al, 1971, Structural framework of continental margin in South China Sea: AAPG Bull., v. 55, p. 723–751.

Pitman, W. C., III, E. M. Herron, and J. R. Heirtzler, 1968, Magnetic anomalies in the Pacific and sea-floor spreading: Jour. Geophys. Research, v. 73, p. 2069–2085.

Rance, H., 1967, Major lineaments and torsional deformation of the earth: Jour. Geophys. Research, v. 72, p. 2213–2217.

Sykes, L. R., 1967, Mechanism of earthquakes and nature of faulting on the mid-ocean ridges: Jour. Geophys. Research, v. 73, p. 2131–2153.

Tanner, W. F., 1973, Deep-sea trenches and the compression assumption: AAPG Bull., v. 57, p. 2195–2206.

Turner, D. L., 1968, Potassium-argon dates concerning the Tertiary foraminiferal time scale and San Andreas fault displacement: Univ. California (Berkeley), unpub. PhD thesis, 99 p.

Udintsev, G. B., 1973, Zones of fractured relief on the bottom of the northwestern basin of the Pacific: Moscow, Geomorphology, Acad. Sci. USSR, v. 2, p. 35–40 (in Russian).

U.S. Geological Survey, 1974, World seismicity map—compiled from the earthquake data of National Oceanic and Atmospheric Administration: Washington, D.C., U.S. Geol. Survey.

Geologic Map of Pacific Mobile Belt and Pacific Ocean: General Principles and Compilation Methods—Summary[1]

L. I. KRASNY[2] and P. L. BEZRUKOV[3]

The first unified geologic map of the Pacific Ocean and the continents bordering it is the result of cooperation of geologists from the USSR Ministry of Geology and the Academy of Sciences of the USSR. The map synthesizes geologic data obtained from many different countries. The principle used in map compilation is the determination of rock masses according to age and composition. The *Paleobiogeographic Atlas of the Pacific Mobile Belt and Pacific Ocean* was used as an aid in correlation of deposits of different ages. In the Precambrian shield margins, local names are given. The map emphasizes the importance of Mesozoic and Cenozoic intrusive magmatism. The distribution of endogenic mineralization is related to this magmatism.

Special attention is given to Cenozoic volcanic rocks. The following series are distinguished: alkaline and alkaline-earth, high-alumina basalts, calc-alkaline, tholeiitic, etc. Various extrusive volcanic features are indicated by separate symbols. Radiometric-age dates of oceanic basalts also are given.

The distribution of different recent complexes of sediments—clastic, argillaceous, and biogenic— is shown in the ocean and marginal seafloors. The complexes, including admixed volcanic material of different composition, have special symbols. In order to distinguish biogenic sediments, quantitative data on the most important sedimentary components ($CaCO_3$, amorphous SiO_2) were taken into consideration. Some sampled pre-Quaternary deposits (Cretaceous-Neogene) and young volcanic rocks in the ocean-bottom areas are included on the map. Deep-drilling data (*Glomar Challenger*, etc) are shown as enlarged columns in which color indicates the ages of recovered rock. The distribution of manganese nodules is contoured, and phosphorite outcrops (mostly pre-Quaternary) also are shown.

On the continents framing the Pacific Ocean, the following structures are distinguished: geosynclinal folded structures, mobile nongeosynclinal structures, and block structures (including ancient and young platforms, shelves, and massifs). The main structural elements of the transitional zone between the continents and the ocean are marginal deep-sea basins, generally double island arcs, and deep-sea trenches. Pacific Ocean structures are larger than continental and transitional structures. Their development is simpler, and they correspond clearly to ocean-floor morphology.

The Pacific mobile belt and Pacific Ocean compose a global mineralogenetic province. Thus, the Pacific ore belt, the Pacific sulfur-bearing belt, the Pacific oil-gas belt, and others are singled out. The geologic map, especially its new international edition, is intended to be the basis for predictional estimation of resources in the Pacific region.

One theoretical problem ensuing from the map is the divisibility of the Pacific segment tectonosphere into geoblocks. The latter are defined in the continental margins, the Pacific ocean, and the transitional zone between the ocean and the continents. The geoblock model for the Circum-Pacific region is more reliable than the plate-tectonic model in its original form, although it has some features of the latter. The geoblock model raises doubt regarding the significance of the East Pacific Rise as the source of ocean-floor spreading, but coincides with the idea of the underthrusting of smaller plates (geoblocks) beneath island arcs.

[1]Manuscript received, November 11, 1974.

[2]VSEGEI (All-Union Geological Research Institute), Leningrad, USSR.

[3]P. P. Shirshov Institute of Oceanology, USSR Academy of Sciences, Moscow, USSR.

Geologic History and Paleozoogeography of Pacific Ocean—Summary[1]

K. M. KHUDOLEY, M. A. RZHONSNITSKAYA, and O. I. NIKIFOROVA[2]

The *Paleobiogeographic Atlas of the Pacific Mobile Belt and Pacific Ocean* consists of a set of maps which reflect paleogeographic and paleobiogeographic conditions during the Paleozoic and Mesozoic Eras.

For the Cambrian Period (Lenian age and Middle and Late Cambrian), two zoogeographic areas were recognized: the Pacific (with Siberian, Sayhan-Altay, and Pacific provinces) and Indo-Altay areas.

For the Ordovician Period, the Siberian, Kolymian, Alaskan, Canada-Coloradan, Sino-Australian, and Andean provinces were established.

During the Silurian, the provincial boundaries became obscured and environments were favorable for a wide faunal exchange. Within North America and East Asia there was a single province, and a second province was established for South America (South American or Malvinocaffric province).

The Devonian was marked by three zoogeographic areas: (1) the Atlantic, with such subareas as the Mediterranean (tropical), including East Australian and Indochinese provinces within the Pacific mobile belt, and the Arctic, including Taymyr-Kolymian, West Canadian, and Cordilleran provinces; (2) the Pacific, with such subareas as the Dzhungaria-Okhotsk and Appalachian; and (3) the Malvinocaffric.

Three areas were established for the Carboniferous Period: (1) tropical, including Ussuriysk-Koryak, Chinese, and North American provinces; (2) boreal, including such provinces as Mongolia-Okhotsk and Verkhoyansk-Kolymian; and (3) antiboreal. These zoogeographic areas continued to exist through the Permian Period.

The zonation for the Mesozoic Era was based on ammonoid distribution. The following provinces were established for the Triassic Period: East Asian, Yakutian, Canadian, Indonesian, and Californian. Provinces established for the Jurassic and Cretaceous were the East Asian, North Pacific, North American, Mexico-Cuban, and Argentina-Chilean.

The persistence of sedimentary and partly volcanic rocks most likely implies warmer climate in the northern hemisphere.

During Paleozoic and Mesozoic time, the Pacific Ocean was larger than at present because of widespread epicontinental seas on the Asian and American continents and part of the Australian continent. The analysis of the faunal distribution suggests the presence during the Mesozoic, and perhaps the Paleozoic, of ocean currents similar to those of recent time.

[1]Manuscript received, November 11, 1974.

[2]VSEGEI (All-Union Geological Research Institute), Leningrad, USSR.

Plate Tectonics and Mineral Resources of Circum-Pacific Region[1]

PETER A. RONA[2] and LAWRENCE D. NEUMAN[3]

Abstract Distribution of selected energy (petroleum and geothermal) and mineral (precious, base, iron, and ferro-alloy metals) resources of the Pacific Ocean basin and Circum-Pacific continents appears to be related to lithospheric plate boundaries. Divergent plate boundaries (oceanic ridges) are related in time to the development of stratigraphic traps in sedimentary basins of the continents, and in space to metalliferous deep-sea sediments and the possible occurrence of massive stratabound metallic sulfide deposits in oceanic crust. Convergent plate boundaries are related in space to areas of offshore petroleum potential and to onshore deposits of precious, base, iron, and ferro-alloy metals. Models suggest genetic relations between the observed distribution of deposits and geologic processes at plate boundaries, and may lead to the discovery of new resources.

Introduction

The Pacific Ocean basin and surrounding continents provide a natural laboratory to develop and test ideas on the relation between plate tectonics and mineral resources. Our approach is as follows:

1. Outline the geologic framework of the Pacific with particular attention to boundaries of the lithospheric plates.
2. Determine the spatial distribution of various mineral resources with respect to the plate boundaries (Figs. 1–4).
3. Consider models of mineral-concentrating processes related to plate boundaries that may explain the observed distribution of mineral resources (Fig. 5).

The boundaries of lithospheric plates are delineated by narrow earthquake zones where the plates are moving with respect to each other. The theory of plate tectonics recognizes three types of plate boundaries (Isacks et al, 1968). One type, a convergent plate boundary, is where two adjacent plates move together and collide or where one plate descends under the other plate along a Benioff seismic zone and is subducted into the upper mantle. The second type, a divergent plate boundary, is where two adjacent plates move apart because new lithosphere is added to each plate by the process of seafloor spreading. The third type is the transform plate boundary, where two adjacent plates slide past one another.

A series of resource maps (Figs. 1–4), using the same base map (Van der Grinten projection) as that used by McKelvey and Wang (1969), was compiled from numerous sources (Van Royan and Bowles, 1952; Roberts and Irving, 1957; Anon., 1962, 1963, 1970, 1972a, b; Lafitte and Rouveyrol, 1965; McKelvey and Wang, 1969; Bonine et al, 1970; Dengo and Levy, 1970; Jones, 1972).

Pacific Geologic Framework

Lithospheric Plate Boundaries

Divergent plate boundaries expressed as oceanic ridges divide the Pacific into several lithospheric plates (Fig. 1). The oceanic ridge system of the Pacific is not midoceanic but is located in the eastern Pacific off Central America and South America and northwestern North America. The average half-rates of seafloor spreading about the oceanic ridges of the Pacific, determined from the geomagnetic polarity-reversal time scale, range from 1.2 cm/yr at the Gorda Rise off the northwestern United States to about 10 cm/yr off Chile and Peru (Table 1).

The Pacific is a closing ocean basin surrounded on three sides by convergent plate boundaries expressed as oceanic trenches (Fig. 1). In the global crustal budget, the Circum-Pacific convergent plate boundaries account for the major portion of crustal consumption. Linear rates of crustal consumption around the Pacific, assumed to be equal to rates of plate convergence, range between about 1 and 15 cm/yr (Table 2).

The dip of the Benioff zone at convergent plate boundaries (Table 3) is inversely proportional to the relative rate of convergence of the adjacent plates (Luyendyk, 1970). For example, the dip of the Benioff zone in the upper 100 km is about 12° beneath Peru, where the relative rate of plate convergence is 8.8 cm/yr (Table 2), and the half-rate of seafloor spreading about the corresponding section of the East Pacific Rise is 9.5 cm/yr (Table 1). The dip of the Benioff zone increases to about 55° beneath the Kermadec Islands (Table 3), where the relative rate of convergence decreases to 5.8 cm/yr, and the half-rate of seafloor

[1]Manuscript received, December 26, 1974.

[2]National Oceanic and Atmospheric Administration, Atlantic Oceanographic and Meteorological Laboratories, 15 Rickenbacker Causeway, Miami, Florida 33149.

[3]Office for Ocean Economics and Technology, United Nations, New York, New York 10017.

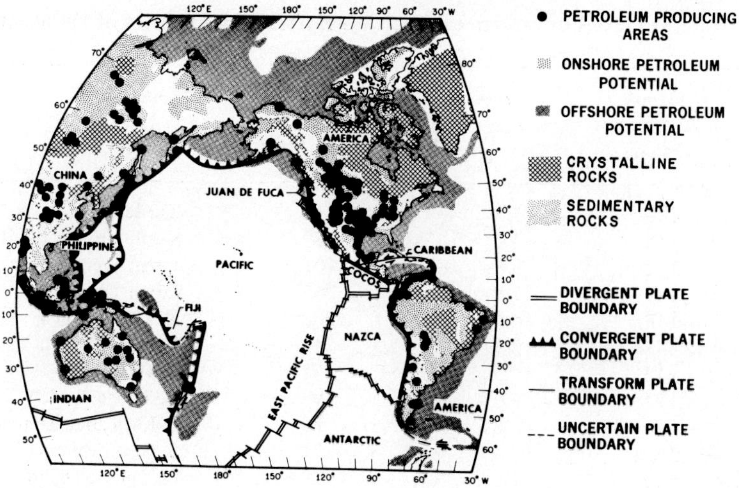

FIG. 1–Areas of petroleum production and potential of Pacific region. Lithospheric plates and plate boundaries are shown.

spreading about the East Pacific Rise decreases to about 5.5 cm/yr.

Oceanic and continental crust are juxtaposed at convergent plate boundaries in the eastern Pacific (Fig. 1). Marginal basins generally underlain by oceanic crust intervene between convergent plate boundaries and continental crust in the western Pacific. The development of marginal basins may be related to the dip of Benioff zones (Oxburgh and Turcotte, 1971; Karig, 1971; Sleep and Toksöz, 1971; Bracey and Ogden, 1972). Where the marginal basins are present in the western Pacific, the dip of the Benioff zone exceeds about 35°; where marginal basins are absent in the eastern Pacific, the dip is less than about 35° (Table 3).

Table 1. Seafloor-Spreading Half-Rates about Pacific Oceanic Ridges

Latitude (+North, –South)	*Longitude (+East, – West)*	*Half-spreading rate (cm/yr)*	*Location*
46.5	–129	1.5	Juan de Fuca Ridge[1]
41.3	–127.5	1.2	Gorda Rise[2]
12.5	–103.5	4.5	East Pacific Rise[3]
3.2	–102	6.4	East Pacific Rise[3]
3.0	– 83.5	3.2	Galapagos rift zone[3]
2.0	– 97.5	3.2	Galapagos rift zone[3]
0.75	– 87.6	3.2	Galapagos rift zone[3]
– 5.6	–106.8	6.5	East Pacific Rise[3]
– 9.2	–110.3	7.5	East Pacific Rise[3]
–19	–113	9–9.5	East Pacific Rise[3]
–28	–112	8–10	East Pacific Rise[3]
–35.6	–110.9	5.0	East Pacific Rise[3]
–43.2	– 82.5	3.0	Chile Ridge[3]

[1] Vine and Wilson, 1965.
[2] Atwater and Mudie, 1973.
[3] Herron, 1972.

Ages of Seafloor and Continents

The age of the Pacific seafloor, as determined by dating of remanent magnetic anomalies in the geomagnetic polarity-reversal time scale (Pitman et al, 1974) and dating of rock samples recovered by the Deep Sea Drilling Project (Fischer et al, 1971), ranges between Late Jurassic (about 150 m.y. ago) and recent. The distribution of ages about divergent plate boundaries is regular, and the seafloor becomes progressively older with distance from these boundaries. The distribution of ages at convergent plate boundaries is irregular, and the seafloor is of various ages at these boundaries. The distribution of ages on continents is delineated by structural provinces that reflect the youngest deformational event, as distinguished from the ages of the seafloor, which reflect the origin of the constituent rocks. Structural provinces of the eastern Pacific are predominantly Cenozoic along western South America and Mesozoic along western North America. In the western Pacific, structural provinces of eastern Asia and Australia exhibit a complex distribution and

Table 2. Rate of Plate Convergence at Convergent Plate Boundaries of the Pacific[1]

Latitude (+ North, – South)	*Longitude (+ East, – West)*	*Rate (cm/yr)*	*Azimuth*	*Location*
51	160	7.2	114	Kurile Trench
43	148	7.5	107	Kurile Trench
35	142	8.6	101	Japan Trench
27	143	7.5	265	Japan Trench
19	148	4.9	282	Mariana Trench
11	142	2.3	243	Mariana Trench
– 3	142	14.5	78	New Guinea
– 13	– 172	9.9	97	N. Tonga Trench
– 34	– 178	5.8	95	S. Kermadec Trench
– 45	169	3.5	72	S. New Zealand
– 55	159	2.6	29	Macquarie Island
– 50	– 75	3.1	240	Cape Horn
– 35	– 74	8.7	74	S. Chile Trench
– 4	– 82	8.8	77	N. Peru Trench
7	– 79	8.3	75	Panama Gulf
20	– 106	6.4	39	N. Middle America Trench
57	– 150	5.6	144	E. Aleutian Trench
50	– 178	6.9	126	W. Aleutian Trench
54	162	7.0	115	W. Aleutian Trench

[1]From Le Pichon et al, 1973, Table V.

range in age from Precambrian through Cenozoic.

Distribution of Geothermal Phenomena

The distribution of geothermal phenomena, including heat flow, active volcanoes, thermal springs, fumaroles, and geysers, is spatially related to lithospheric plate boundaries. Values of heat flow in the Pacific Ocean basin (Langseth, 1969; Sclater, 1972) are relatively high (>2 HFU) at divergent plate boundaries and decrease basinward away from these boundaries. Values of heat flow at convergent plate boundaries and in marginal basins (Karig, 1971) around the Pacific are variable. The distribution of heat-flow values on the Circum-Pacific continents is complex. Relatively high values (>2 HFU) are present in limited areas of eastern Australia, eastern Asia, and western North America. Active volcanoes, thermal springs, and fumaroles are aligned along the landward side of convergent plate boundaries on continents and island arcs around the Pacific (Kennedy and Richey, 1947; Waring, 1965; Snead, 1972). These features are not evenly spaced but are grouped (Kelleher, 1972).

It is apparent from the distribution of heat-flow values and thermal springs on islands and continents around the Pacific that numerous potential sites exist for the development of geothermal energy. Geothermal energy is being utilized at sites in New Zealand, Japan, and the western United States.

Distribution of Petroleum Resources

Areas of offshore petroleum potential conform with convergent plate boundaries around the Pacific (Fig 1; McKelvey and Wang, 1969). Areas of petroleum potential in the eastern Pacific comprise thick sedimentary accumulations underlying the continental margins of western North America and South America and partially filling oceanic trenches seaward of the margins of Central America and South America along convergent plate boundaries. In the western Pacific, island arcs directly landward of convergent plate boundaries divide the ocean basin into marginal basins partially enclosed between the islands and eastern Asia—for example, the South China Sea, the East China Sea, the Yellow Sea, the Sea of Japan, the Sea of Okhotsk, and the Bering Sea.

Table 3. Dip of Benioff Zone (upper 100 km)[1]

Dip (degrees)	*Location (counter-clockwise around Pacific)*
45	New Zealand
55	Kermadec
50	Tonga (south)
30	Tonga (north)
65	New Hebrides (south)
55	New Hebrides (north)
65	Solomons
45	New Britain
40	Sunda: Flores Sea
60	Sunda: Java
50	Sumatra
50	Burma
55	Mindanao
45	Marianas
25	Izu-Bonin
40	Ryukus
35	Kurile
30	Honshu
40	Aleutians
35	Middle America
12	Peru
15	Chile (north)

[1] Isacks and Molnar, 1971; Oliver et al, 1973.

Areas of petroleum potential in the western Pacific comprise sedimentary accumulations in the marginal basins as well as in oceanic trenches along convergent plate boundaries.

The oceanic trenches along the eastern and western sides of the Pacific Ocean basin receive deep-sea sediments that presumably are transported into the trenches on a "conveyor belt" of spreading seafloor during subduction of the oceanic lithosphere. The content of organic matter in the deep-sea sediments that are transported into the trenches varies in space and in time; for example, a zone of high organic productivity extending across the equatorial Pacific affects the composition of the sediments deposited in that region. Both the amount of organic matter delivered to the oceanic trenches and the petroleum potential are expected to vary accordingly. Temperature and pressure conditions beneath the trenches and the marginal basins may facilitate the conversion of organic matter in the sediments to petroleum (Tarling, 1973; Sorokhtin et al, 1974).

Areas of onshore petroleum production are sedimentary basins on continents with no apparent spatial relation either to divergent or convergent plate boundaries of the Pacific (Fig. 1). However, the development of the sedimentary sequences in the basins may be related to divergent plate boundaries in time, if not in space, through the influence on global sea level of reversible volume changes of oceanic ridges (Rona, 1973b; Rona and Wise, 1974). The volume of the world oceanic ridge system is not constant but appears to vary directly with rates of seafloor spreading. A volume increase in the world oceanic ridge system reduces the cubic capacity of ocean basins, resulting in transgression of the sea onto all the continents and deposition of a sedimentary sequence that may contain organic source material and reservoir rocks. Conversely, a volume decrease in the oceanic ridge system increases the cubic capacity of ocean basins, resulting in regression of the sea from all the continents and the development of widespread unconformities that may be associated with traps for the accumulation of petroleum. Reversible volume changes of the worldwide oceanic ridge system have operated on a time scale of tens of millions of years, as evidenced by the presence of six sedimentary sequences separated by surfaces of unconformity in the Phanerozoic stratigraphy of North America (Sloss, 1963). The inferred relations of the volume of oceanic ridges to sedimentary sequences and unconformities may prove useful in exploration for stratigraphic traps.

Distribution of Selected Mineral Resources

Light Metal Deposits

Deposits of light metals, including aluminum, beryllium, lithium, and titanium, exhibit no apparent relation to plate boundaries of the Pacific. These metals are associated with granitic rocks of the continents as opposed to basaltic rocks of the ocean basins. The distribution of aluminum, beryllium, lithium, and titanium on continents is related to the occurrence of particular minerals in granitic rocks and to conditions of weathering.

Metal Deposits at or Near Convergent Plate Boundaries

Precious metal deposits—Deposits of precious metals, including gold, silver, and platinum, exhibit a spatial relation to convergent plate boundaries (Fig. 2). In the eastern Pacific, precious metal deposits are found along the western margins of North America and South America. Additional deposits are present in eastern North America

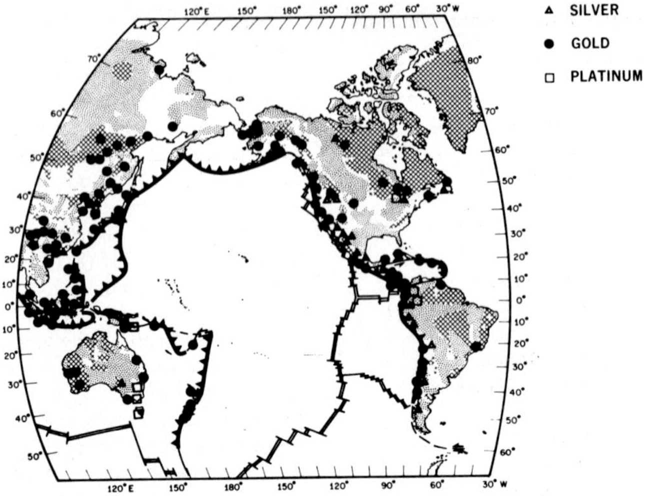

FIG. 2–Map of precious metal deposits of Pacific region. Lithospheric plate boundaries are shown.

in the area of the Canadian shield and in eastern South America in the area of the Guianian and Brazilian shields. In the western Pacific, precious metal deposits occur on island arcs (including Japan, the Philippines, and Indonesia) situated along convergent plate boundaries. Deposits also occur in eastern Asia and Australia, where they are separated by a gap from active convergent plate boundaries. Precious metal deposits have not been found along those sections of convergent plate boundaries where oceanic lithosphere is juxtaposed and island arcs are absent (e.g., the eastern side of the Philippine Sea and the section between New Zealand and Samoa).

Base metal deposits—Deposits of base metals exhibit a spatial relation to convergent plate boundaries (Fig. 3). Their distribution is grossly similar to that of precious metal deposits (Fig. 2). In the eastern Pacific, base metal deposits are present along the western margins of North America and South America. In the Peruvian Andes, porphyry copper and vein-type mineralization are associated with Neogene silicic volcanic rocks (Mitchell, 1973). Additional base metals occur in central and eastern North America including the area of the Canadian shield.

In the western Pacific, base metal deposits are found on island arcs along convergent plate boundaries. Deposits of tin, tungsten, and fluorite with minor bismuth and molybdenum occur in belts of predominantly Mesozoic granites and acidic eruptive rocks along the southern margin of Alaska and the eastern margin of Asia (Mitchell, 1973). Base metal deposits also occur in eastern Asia and Australia, which are separated by a gap from active convergent plate boundaries. As in the case of precious metals, base metal deposits have not been found along those sections of convergent plate boundaries where oceanic lithosphere is juxtaposed and island arcs are absent.

Sediment samples recovered by coring near the crest of oceanic ridges and by deep-sea drilling away from the crest reveal widespread enrichment by base metals of those strata directly overlying basalt of the Pacific Ocean basin (Boström and Petersen, 1969; von der Borch et al, 1971; Cook, 1972; Cronan et al, 1972; Dymond et al, 1973; Sayles and Bischoff, 1973; Piper, 1973). The base metals include antimony, copper, lead, mercury, and zinc, but no tin. The observation that the base metal enrichment is limited to the basal sedimentary layer overlying basalt implies that the enrichment occurred soon after the generation of the underlying basalt by seafloor spreading about an oceanic ridge at a divergent plate boundary.

Iron and ferro-alloy metal deposits—Deposits of iron and ferro-alloy metals exhibit a spatial relation to convergent plate boundaries around the Pacific (Fig. 4). Their distribution is grossly similar to that of precious and base metals. In the eastern Pacific, iron and ferro-alloy metal deposits occur along the western margins of North America and South America. Additional deposits occur in central and eastern North America, including the Canadian shield, and in the Guianian and Brazilian shields of South America. In the western Pacific, iron and ferro-alloy deposits occur on island arcs along convergent plate bound-

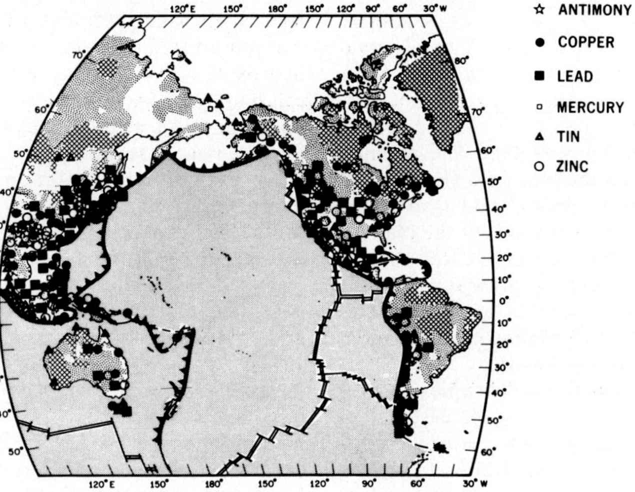

FIG. 3–Map of base metal deposits of Pacific region. Lithospheric plate boundaries are shown. Sediments enriched in base metals (shaded) overlie basalts of ocean basin.

aries and in eastern Asia and Australia, where they are separated by a gap from active convergent plate boundaries. Iron and ferro-alloy metal deposits have not been found along those sections of convergent plate boundaries where oceanic lithosphere is juxtaposed and island arcs are absent. Sedimentary strata directly overlying basalt of the Pacific Ocean basin are enriched in iron and ferro-alloy metals, in addition to base metals.

Nodules containing variable percentages of manganese, copper, nickel, and cobalt are present on about two thirds of the Pacific seafloor (Fig. 4; Strakhov et al, 1968). A zone of nodules of anomalously high copper and nickel content (1.5–2.0%) and areal density (20–50% of seafloor) extends east-west across the Pacific between latitudes 5° and 20°N. No apparent spatial relation exists between plate boundaries and either the overall distribution or the enriched zone of nodules in the Pacific Ocean basin.

Metal Deposits at Divergent Plate Boundaries

Knowledge of the distribution of metal deposits with respect to divergent plate boundaries is

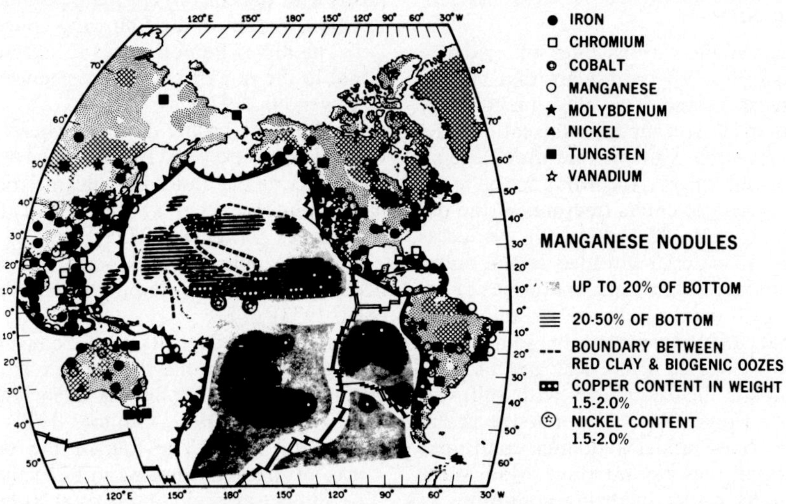

FIG. 4–Map of iron and ferro-alloy metal deposits, including manganese nodules, of Pacific. Lithospheric plate boundaries are shown.

limited because, as submerged oceanic ridges, these boundaries are less accessible to observation than are convergent plate boundaries. Sediments deposited about divergent plate boundaries in the Pacific Ocean basin are enriched in certain base, iron, and ferro-alloy metals. The recently discovered TAG hydrothermal field (R. Scott et al, 1974; Rona et al, in press), named after the Trans-Atlantic Geotraverse (TAG) of the National Oceanic and Atmospheric Administration, represents a type of metal deposit present in oceanic crust and may provide insights to processes of metal concentration at divergent plate boundaries, including those of the Pacific.

The TAG hydrothermal field occupies an area at least 15 km by 15 km including the east wall of the rift valley of the Mid-Atlantic Ridge at 26°N. Manganese oxide was recovered consistently by dredging from deposits concentrated along steps in the wall that are interpreted as faults which have acted as conduits for hydrothermal fluids. Concentration of the manganese oxide by hydrothermal processes capable of extreme chemical fractionation is evidenced by an exceptionally rapid rate of accumulation, about 200 mm per million years, and extreme purity, about 40% manganese with only trace quantities of iron and copper (M. Scott et al, 1974). Present activity is indicated by a positive temperature anomaly with an inverted temperature gradient (Rona et al, 1975) and by metal enrichment of suspended particulate matter (Betzer et al, 1974) in the seawater overlying the TAG hydrothermal field.

Model of Metal Deposits at Divergent Plate Boundaries

A model based on various lines of evidence (Spooner and Fyfe, 1973) considers that certain precious, base, iron, and ferro-alloy metals may be concentrated as deposits by sub-seafloor hydrothermal convection systems that may develop at crests of oceanic ridges. According to the model, cold dense seawater enters fractures in the basalt of an oceanic ridge. The seawater is heated as it encounters hot material intruded at the divergent plate boundary. The warm, less dense seawater rises through fractures in the basalt and leaches metals that are transported in solution as complexes with chloride in the seawater. A fraction of the metals then combines with sulfur in the seawater and precipitates as a massive stratabound sulfide body under reducing conditions beneath the basalt-seawater interface. Manganese oxide precipitates under oxidizing conditions at the basalt-seawater interface as the hydrothermal solutions discharge from the seafloor. Colloidal ferric hydroxide precipitates from the hydrothermal solutions in the overlying seawater. The ferric hydroxide scavenges the remaining metals from solution by colloidal absorption and settles to deposit a layer of metalliferous sediments on basalt of the adjacent seafloor.

An example of an economic mineral deposit interpreted as the product of a sub-seafloor hydrothermal convection system occurs in the Troodos massif of Cyprus. The Troodos massif is interpreted as an obducted slice of oceanic crust generated at a divergent plate boundary (Gass and Masson-Smith, 1963; Moores and Vine, 1971). An "umber" of metallic oxides overlies massive stratabound copper sulfide ore bodies in basaltic rocks of the Troodos massif. The manganese oxide of the TAG hydrothermal field chemically resembles the umber and may be underlain by massive copper sulfide bodies (Rona, 1973a; R. Scott et al, 1974; Rona et al, in press). Relict metallic oxide and sulfide deposits may be expected to extend in belts along flow lines of seafloor spreading transverse to the axis of an oceanic ridge; the extent of the belts would depend on the continuity of seafloor spreading and the persistence of the sub-seafloor hydrothermal convection system which concentrates the deposits (Rona, 1973a). The metal deposits may be buried by off-axis volcanism.

Model of Metal Deposits at Convergent Plate Boundaries

A model to interpret the observed association of precious, base, iron, and ferro-alloy metal deposits with the convergent plate boundaries of the Pacific is based on the following concepts:

1. The dip of Benioff zones is inversely proportional to the rate of plate convergence (Tables 1–3; Luyendyk, 1970).
2. Marginal basins develop where the dip of Benioff zones exceeds about 35°.
3. Calc-alkalic andesitic volcanic rocks and tonalitic plutons occur above steeply dipping Benioff zones (Mitchell, 1973).
4. Silicic volcanic rocks and granitic plutons occur above shallow-dipping Benioff zones (Mitchell, 1973).
5. Metals in deposits along convergent plate boundaries are primarily derived from oceanic crust descending along the associated Benioff zone (Sawkins, 1972; Sillitoe, 1972) and from continental crust. The role of the mantle as a source of material remains to be evaluated.
6. The nature and volume of volatile matter expelled from oceanic crust descending along Benioff zones influence the extraction, transport,

concentration, and deposition of metals (Rub, 1972).

The model presents two regimes to account for the past and present distribution of metals along convergent plate boundaries of the western and eastern Pacific, as follows:

1. Relatively fast seafloor spreading and plate convergence are associated with a shallow Benioff zone and the generation of silicic volcanic rocks and granitic plutons (Fig. 5a, b). High-level porphyry copper deposits occur in the silicic volcanic rocks, and deep-level tin, tungsten, bismuth, fluorite, and molybdenum occur in the granites. The copper is primarily derived from metalliferous sediments and massive stratabound metallic sulfide deposits of the oceanic crust that descends along the underlying Benioff zone. The tin and associated metals, along with a portion of the granite (Stern and Wyllie, 1973), are derived from continental crust and their segregation is aided by the rise of volatile matter expelled from the descending oceanic crust. This regime is exemplified by western South America at present and by eastern Asia in late Mesozoic time.

2. Relatively slow seafloor spreading and plate convergence are associated with a steep Benioff zone, the development of marginal basins, and the generation of calc-alkalic andesitic volcanic rocks and tonalitic plutons with associated porphyry copper deposits (Fig. 5c, d). The calc-alkalic volcanic rocks and the copper are primarily derived from the oceanic crust (Jakes and White, 1972) that descends along the underlying Benioff zone; this crust includes metalliferous strata and massive strata-bound metallic sulfide deposits. The tonalitic plutons and some copper may be derived from the base of continental or island-arc crust (Jakes and White, 1971; Brown, 1973). Granitic plutons emplaced during the first regime (Fig. 5b) are unroofed by erosion to expose deposits of tin and associated metals (Fig. 5d). The development of marginal basins (Fig. 5c, d) results in the separation of island arcs from the continent, producing a gap such as that observed between the metal deposits of eastern Asia and Australia and the active convergent plate boundaries of the western Pacific (Figs. 2–4). This regime is exemplified by eastern Asia at the end of the Mesozoic and at present.

SUMMARY

Consideration of the distribution of selected energy and material resources with respect to lithospheric plate boundaries of the Pacific (Figs. 1–4) leads to the following conclusions:

1. Divergent plate boundaries (oceanic ridges) are inferred to be related in time to the development of stratigraphic traps for petroleum in sedimentary basins on continents through the control of eustatic sea level by reversible volume changes of oceanic ridges.

2. Divergent plate boundaries are related in space to metalliferous deep-sea sediments and to the possible occurrence of massive strata-bound metallic sulfide deposits in oceanic crust.

3. Convergent plate boundaries are related in space to areas of offshore petroleum potential

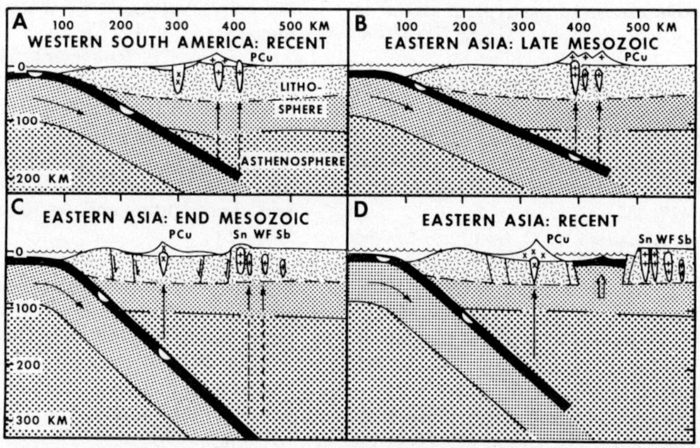

FIG. 5–Model of metallogenesis at convergent plate boundaries of Pacific (modified from Mitchell, 1973). Diagrammatic cross sections through convergent plate boundaries show Benioff zone, oceanic crust (black) incorporating metalliferous sediments and massive strata-bound metallic sulfide bodies (white semicircles in oceanic crust), rising magma (solid vertical arrows), rising volatile matter (dashed vertical arrows), silicic volcanic rocks and granitic plutons (+), andesitic volcanic rocks and tonalitic plutons (×), and known deposits of porphyry copper (PCu), tin (Sn), tungsten (W), fluorite (F), and antimony (Sb). For explanation, see text.

and to onshore deposits of precious, base, iron and ferro-alloy metals.

4. Certain precious, base, iron, and ferro-alloy metal deposits that are present in Precambrian shields and old orogenic belts distant from present plate boundaries may be related to former plate boundaries.

5. Models suggest genetic relations between the observed distribution of metal deposits and geologic processes at plate boundaries. The models are interpretative in that they attempt to explain the observed distribution of deposits. With further development these models may predict the locations of undiscovered metal deposits.

References Cited

Anonymous, 1962, Oil and natural gas map of Asia and the Far East, scale 1:5,000,000: U.N. ECAFE, no. 16351, U.N. Pub. Sales No. 62-1-16, 4 sheets.

——— 1963, Mineral distribution map of Asia and the Far East, scale 1:5,000,000: U.N. ECAFE, no. 63-157 14, U.N. Pub. Sales No. 63-1-18, 4 sheets.

——— 1970, Energy atlas of Asia and the Far East: U.N. ECAFE, U.N. Pub. Sales No. E-70-1-11, E/CN-11/900, 25 p.

——— 1972a, Principal mineral areas of Canada, scale 1:7,600,200, 22nd ed.: Canada Dept. Energy, Mines and Resources, Mineral Resources Br., and Canada Geol. Survey, Map 900A.

——— 1972b, The Times atlas of the world, 4th ed. (revised): Edinburgh, John Bartholomew and Son Ltd., Times Newspapers Ltd., 272 p.

Atwater, T., and J. D. Mudie, 1973, Detailed near-bottom geophysical study of the Gorda Rise: Jour. Geophys. Research, v. 78, p. 8665–8686.

Betzer, P. R., et al, 1974, The Mid-Atlantic Ridge and its effect on the composition of particulate matter in the deep ocean: Eos (Am. Geophys. Union Trans.), v. 55, no. 4, p. 293.

Bonine, M. E., et al, 1970, Atlas of Mexico: Univ. Texas at Austin, Bur. Business Research, 138 p.

Boström, K., and M. N. A. Petersen, 1969, The origin of aluminum-poor ferromanganoan sediments in areas of high heat flow on the East Pacific Rise: Marine Geology, v. 7, p. 427–447.

Bracey, D. R., and T. A. Ogden, 1972, Southern Mariana arc; geophysical observations and hypothesis of evolution: Geol. Soc. America Bull., v. 83, p. 1509–1522.

Brown, G. C., 1973, Evolution of granite magmas at destructive plate margins: Nature; Phys. Sci., v. 241, p. 26–28.

Cook, H. E., 1972, Stratigraphy and sedimentation, p. 933–943 *in* J. D. Hays et al, eds., Initial reports of the Deep Sea Drilling Project, v. IX: Washington, U.S. Govt. Printing Office, 1205 p.

Cronan, D. S., et al, 1972, Iron-rich basal sediments from the eastern equatorial Pacific; Leg 16, Deep Sea Drilling Project: Science, v. 175, no. 4017, p. 61–63.

Dengo, G., and E. Levy, 1970, Estudios et mapa metalogenéticos de América Central: Inst. Centroamericano Inv. y Tecnología Indus., Pub. Geol. No. 3, 57 p., scale 1:200,000.

Dymond, J., et al, 1973, Origin of metalliferous sediments from the Pacific Ocean: Geol. Soc. America Bull., v. 84, p. 3355–3372.

Fischer, A. G., et al, 1971, Initial Reports of the Deep Sea Drilling Project, v. VI: Washington, U.S. Govt. Printing Office, 1329 p.

Gass, I. G., and D. Masson-Smith, 1963, The geology and gravity anomalies of the Troodos Massif, Cyprus: Royal Soc. London Philos. Trans., Ser. A, v. 255, p. 417–467.

Herron, E. M., 1972, Sea-floor spreading and the Cenozoic history of the east-central Pacific: Geol. Soc. America Bull., v. 83, p. 1671–1692.

Isacks, B., J. Oliver, and L. R. Sykes, 1968, Seismology and the new global tectonics: Jour. Geophys. Research, v. 73, p. 5855–5899.

——— and P. Molnar, 1971, Distribution of stresses in the descending lithosphere from a global survey of focal mechanism solutions of mantle earthquakes: Rev. Geophys. and Space Physics, no. 9, 103 p.

Jakes, P., and A. J. R. White, 1971, Composition of island arcs and continental growth: Earth and Planetary Sci. Letters, v. 12, p. 224–230.

——— and ——— 1972, Major and trace element abundances in volcanic rocks of orogenic areas: Geol. Soc. America Bull., v. 83, p. 29–39.

Jones, D. B., advisory ed., 1972, Oxford economic atlas of the world, 4th ed.: London, Oxford Univ. Press, 239 p.

Karig, D. E., 1971, Origin and development of marginal basins in the western Pacific: Jour. Geophys. Research, v. 76, p. 2542–2561.

Kelleher, J. A., 1972, Rupture zones of large South American earthquakes and some predictions: Jour. Geophys. Research, v. 77, p. 2087–2103.

Kennedy, W. Q., and J. E. Richey, 1947, Catalogue of the active volcanoes of the world: Bull. Volcanol., Supp. Ser. 2, v. 7, 11 p.

Lafitte, P., and P. Rouveyrol, 1965, Carte minière du globe sur fond tectonique au 1:20,000,000 (projection équiaréale): Paris, Bur. Recherches Géol. et Minières, 74, ed. provisoire, pt. 2.

Langseth, M. G., Jr., 1969, The flow of heat from the Earth and the global distribution at the surface: Am. Inst. Aeronautics and Astronautics, Paper No. 69–589, 10 p.

Le Pichon, X., J. Francheteau, and J. Bonnin, 1973, Plate tectonics: New York, Elsevier, 300 p.

Luyendyk, B. P., 1970, Dips of downgoing lithospheric plates beneath island arcs: Geol. Soc. America Bull., v. 81, p. 3411–3416.

McKelvey, V. E., and F. F. H. Wang, 1969, World subsea mineral resources: U.S. Geol. Survey Misc. Geol. Inv. Map I-632.

Mitchell, A. H. G., 1973, Metallogenic belts and angle of dip of Benioff zones: Nature; Phys. Sci., v. 245, p. 49–52.

Moores, E. M., and F. J. Vine, 1971, The Troodos Massif, Cyprus, and other ophiolites as oceanic crust; evaluation and implications: Royal Soc. London Philos. Trans., Ser. A., v. 268, p. 443–466.

Oliver, J., et al, 1973, Dynamics of the downgoing lithosphere: Tectonophysics, v. 19, p. 133–147.

Oxburgh, E. R., and D. L. Turcotte, 1971, Origin of paired metamorphic belts and crustal dilation in island arc regions: Jour. Geophys. Research, v. 76, p. 1315–1327.

Piper, D. Z., 1973, Origin of metalliferous sediments from the East Pacific Rise: Earth and Planetary Sci. Letters, v. 19, p. 75–82.

Pitman, W. C. III, R. L. Larson, and E. M. Herron, 1974, Age of the ocean basins determined from magnetic anomaly lineations: Geol. Soc. America, map.

Roberts, R. J., and E. M. Irving, 1957, Mineral deposits of Central America: U.S. Geol. Survey Bull. 1034, 205 p.

Rona, P. A., 1973a, Plate tectonics and mineral resources: Sci. American, v. 229, no. 1, p. 86–95.

——— 1973b, Relations between rates of sediment accumulation on continental shelves, sea-floor spreading, and eustacy inferred from the central North Atlantic: Geol. Soc. America Bull., v. 84, p. 2851–2872.

——— and D. U. Wise, 1974, Symposium: global sea level and plate tectonics through time: Geology, v. 2, no. 3, p. 133–134.

——— et al, 1975, Anomalous water temperatures over Mid-Atlantic Ridge crest at 26° North latitude: Deep Sea Research, v. 22, p. 611–618.

——— et al, in press, Tectonic fabric and hydrothermal activity of Mid-Atlantic Ridge crest (lat. 26°N): Geol. Soc. America Bull.

Rub, M. G., 1972, The role of the gaseous phase during the formation of ore bearing magmatic complexes: Chem. Geology, v. 10, p. 89–98.

Sawkins, F. J., 1972, Sulfide ore deposits in relation to plate tectonics: Jour. Geology, v. 80, p. 377–397.

Sayles, F. L., and J. L. Bischoff, 1973, Ferromanganoan sediments in the equatorial East Pacific: Earth and Planetary Sci. Letters, v. 19, p. 330–336.

Sclater, J. G., 1972, Heat flow and elevation of the marginal basins of the western Pacific: Jour. Geophys. Research, v. 77, p. 5705–5709.

Scott, M. R., et al, 1974, Rapidly accumulating manganese deposit from the median valley of the Mid-Atlantic Ridge: Geophys. Research Letters, v. 1, p. 355–358.

Scott, R. B., et al, 1974, The TAG Hydrothermal Field: Nature, v. 251, p. 301–302.

Sillitoe, R. H., 1972, A plate tectonic model for the origin of porphyry copper deposits: Econ. Geology, v. 67, p. 184–197.

Sleep, N. H., and M. N. Toksöz, 1971, Evolution of marginal basins: Nature, v. 233, p. 548–550.

Sloss, L. L., 1963, Sequences in the cratonic interior of North America: Geol. Soc. America Bull., v. 74, p. 93–114.

Snead, R. E., 1972, Atlas of world physical features: New York, Wiley, 158 p.

Sorokhtin, O. G., S. A. Ushakov, and V. V. Fedynskiy, 1974, Dynamics of crustal plates and origin of petroleum deposits: Akad. Nauk SSSR Doklady, v. 214, p. 1407–1410.

Spooner, E. T. C., and W. S. Fyfe, 1973, Sub-sea floor metamorphism, heat and mass transfer: Contr. Mineralogy and Petrology, Springer-Verlag, p. 287–304.

Stern, C. R., and P. J. Wylie, 1973, Water-saturated and undersaturated melting relations of a granite to 35 kilobars: Earth and Planetary Sci. Letters, v. 18, p. 163–167.

Strakhov, N. M., et al, 1968, Geochemistry of a sedimentary manganese ore-forming process: Akad. Nauk SSSR Geol. Inst. Trans., v. 185, 495 p.

Tarling, D. H., 1973, Continental drift and reserves of oil and natural gas: Nature, v. 243, p. 277–279.

Van Royan, W., and O. Bowles, 1952, The mineral resources of the world, *in* Atlas of the world's resources, v. II: New York, Prentice-Hall, 181 p.

Vine, F. J., and J. T. Wilson, 1965, Magnetic anomalies over a young oceanic ridge off Vancouver Island: Science, v. 150, no. 3695, p. 485–489.

von der Borch, C. C., W. D. Nesteroff, and J. S. Galehouse, 1971, Iron-rich sediments cored during Leg 8 of the Deep Sea Drilling Project, *in* J. I. Tracey, Jr., et al, eds., Initial reports of the Deep Sea Drilling Project, v. VIII: Washington, U.S. Govt. Printing Office, p. 725–819.

Waring, G. A., 1965, Thermal springs of the United States and other countries of the world: a summary: U.S. Geol. Survey Prof. Paper 492, 383 p.

Geologic-Structural Synthesis of Pacific Area as Basis for Establishing Regularities of Distribution of Mineral Resources—Summary[1]

YU. M. PUSHCHAROVSKY,[2] P. L. BEZRUKOV,[3] YU. K. BURLIN,[2] L. I. KRASNY,[2] and B. A. SOKOLOV[2]

Advances in Soviet geologic studies of the Pacific Ocean and its margins include results presented on the unique published maps: (a) Tectonic Map of Eurasia, scale 1:5,000,000 (1966); (b) Geological Map of the Pacific Mobile Belt and the Pacific Ocean, scale 1:10,000,000 (1973); and (c) Tectonic Map of the Pacific Segment of the Earth, scale 1:10,000,000 (1970). Moreover, many maps and books have been published which deal with the geology and mineral resources of the northwestern segment of the Pacific including the eastern USSR.

The geologic map synthesizes the data on stratigraphy, age, and composition of sedimentary, volcanic, and intrusive rocks on the Pacific continental margins, in the island arcs, and on the Pacific Ocean floor. The arrangement of geologic complexes of different ages and character—ancient platform, perioceanic belt, and ocean—reflects effects of various tectonic styles on the inner structure. A regular rejuvenation of geologic complexes from continental nuclei toward the ocean is indicated. The map also shows that the distribution of sediment types on the Pacific floor is determined by such factors as climatic zonation, position of zones of upwelling and downwelling of waters, and depth of the ocean; whereas, in the perioceanic zones, a tectonic factor influencing intensity of supply of terrigenous and volcanic material is especially important.

On the Pacific tectonic map, the structure of the Pacific area is divided into four principal categories: fold zones of the ocean margins, recent geosynclinal zones, thalassogenes (large parts of the oceanic floor), and the East Pacific mobile belt. The structural analysis, on the whole, shows that plate tectonics fails to explain the constitution of the Pacific area. However, horizontal movements of rather large blocks and sheets of the earth's crust are apparent.

The geologic-structural synthesis creates the basis for general metallogenic constructions. Further studies on concentric zonation in the arrangement of ore deposits around the Pacific Ocean appear promising. The zone adjacent to the ocean is rich in gold and copper; the next zone continentward is characterized by rich deposits of tin and tungsten (in addition to gold) in Asia and lead and zinc in North America. The ophiolite belts are characterized by peculiar complexes of metals. Study of morphologic and geodynamic properties of ore-controlling faults poses a special problem.

The ocean floor is characterized by extensive accumulations of manganese-polymetallic ores and, locally, by metasomatic phosphorites.

The geologic-structural synthesis also is important for prediction of oil and gas prospects. About 140 known and possible gas and oil basins have been recorded in the Pacific margins. Most of them are associated with depressions and basins within the fold zones and geosynclinal zones. The basins in the perioceanic sector are composed mostly of Cenozoic deposits; the main productive strata are Miocene to Pliocene in age. Oil and gas already have been detected in about half of the basins.

[1]From Circum-Pacific Energy and Mineral Resources Conference program booklet.

[2]Geological Institute, USSR Academy of Sciences, Moscow, USSR.

[3]P. P. Shirshov Institute of Oceanology, USSR Academy of Sciences, Moscow, USSR.

Structural-Formation Analysis of Northwestern Part of Pacific Mobile Belt, USSR[1]

V. V. RUSS and D. A. KIRIKOV[2]

Abstract Analysis of the geology of the Soviet sector of the Pacific mobile belt, on the basis of a structural-formation map, permits distinction of four major complexes: Verkhoyansk-Chukotsk folded region, Amur folded region, Nippon geosynclinal region, and East Asian volcanogenic belt. Each complex is characterized by distinctive suites of rocks related to its tectonic setting and history. Study of the spatial and temporal distribution of igneous rocks shows that magmatic processes commonly develop monotypically in different tectonic settings.

The relation of mineral resources to rock types is being studied, but few predictions can be made on the basis of formation analysis at present.

The analysis of regularities of distribution of rock units and structural-stratigraphic complexes in the Soviet sector of the Pacific mobile belt, made on the basis of a structural-formation map (Krasny et al, 1967), permits more accurate definition of ideas about the tectonic zonation, history of evolution, and formation of material of the earth's crust. Four major complexes are distinguished (Fig. 1):

1. Verkhoyansk-Chukotsk folded region, comprising the Yana-Kolyma and Chukotsk folded systems and the Omolon, Okhotsk, Chukotsk, and Kolyma median masses;
2. Amur folded region, comprising the Mongolo-Okhotsk and Sikhote-Alin folded systems of the Argun', Bureya, and Khanka median masses;
3. Nippon geosynclinal region, comprising the Hokkaido-Sakhalin and Okhotsk-Koryak folded systems, the Kuril–East Kamchatka geosynclinal systems, and the central Okhotsk and Anadyr' median masses; and
4. East Asian volcanogenic belt, comprising the Chukotsk, Okhotsk, and Sikhote-Alin volcanogenes.

On the west, within the USSR, the Pacific mobile belt is limited by the Siberian platform and the Baykal folded region; on the east, its boundary is fixed by abyssal fractures expressed by deep-sea trenches which separate it from the thalassocraton of the Pacific Ocean.

The folded regions are characterized by polycyclic evolution. In the eastern USSR, Baykal, Caledonian, and Hercynian geotectonic cycles developed, and later Alpine tectonic movements (Kimmerian in the Mesozoic and middle and late Alpine in the Cenozoic) formed the present structure of the region. In geotectonic cycles, eugeosynclinal stages are subdivided into early and late substages and inversion (uplift) and post-inversion stages.

Stratigraphic analysis has shown that each folded region has its specific formation types; this differentiation is considered to be important in tectonic zonation. Thus, the Verkhoyansk-Chukotsk folded region is characterized by terrigenous-carbonate formations in the lower and middle Paleozoic and by eugeosynclinal terrigenous deposits in the upper Paleozoic–Mesozoic. The Amur folded region is characterized by spilite-diabase siliceous formations which were deposited in geosynclines, and by intensive magmatic (granitic) reworking of median masses of geanticlinal type. The Nippon folded region is distinguished by eugeosynclinal basalt and chert strata and by tuffite-diatomite and tuffite-andesite (green tuff) formations of the inversion stage. The different formation types in geosynclines probably are related to mobility of the basement. Geosynclines characterized by terrigenous-carbonate and terrigenous strata and almost devoid of magmatic products have rather simple fold structures and sedimentary thicknesses of 5,000–10,000 m. They are developed on a weakly mobile basement. Geosynclines characterized by volcanogenic-siliceous strata and by various and abundant magmatism have complex folding and great thicknesses of strata (10,000–15,000 m). They are developed on a more mobile basement. During the evolution of geosynclinal systems, terrigenous-carbonate (Verkhoyansk-Chukotsk region) and volcanogenic-siliceous (Amur region) deposition changed to terrigenous or essentially terrigenous. Folded areas of the region underwent a complex polycyclic evolution, and successions of formations changed from cycle to cycle, reflecting evolution of the earth's crust.

In terrigenous-carbonate and terrigenous geosynclines, magmatism of eugeosynclinal stages is mainly intrusive, with the deposition of gabbro-diabase and gabbro-ultrabasite formations. In the

[1]Manuscript received, August 29, 1974.

[2]VSEGEI (All-Union Geological Research Institute), Leningrad, USSR.

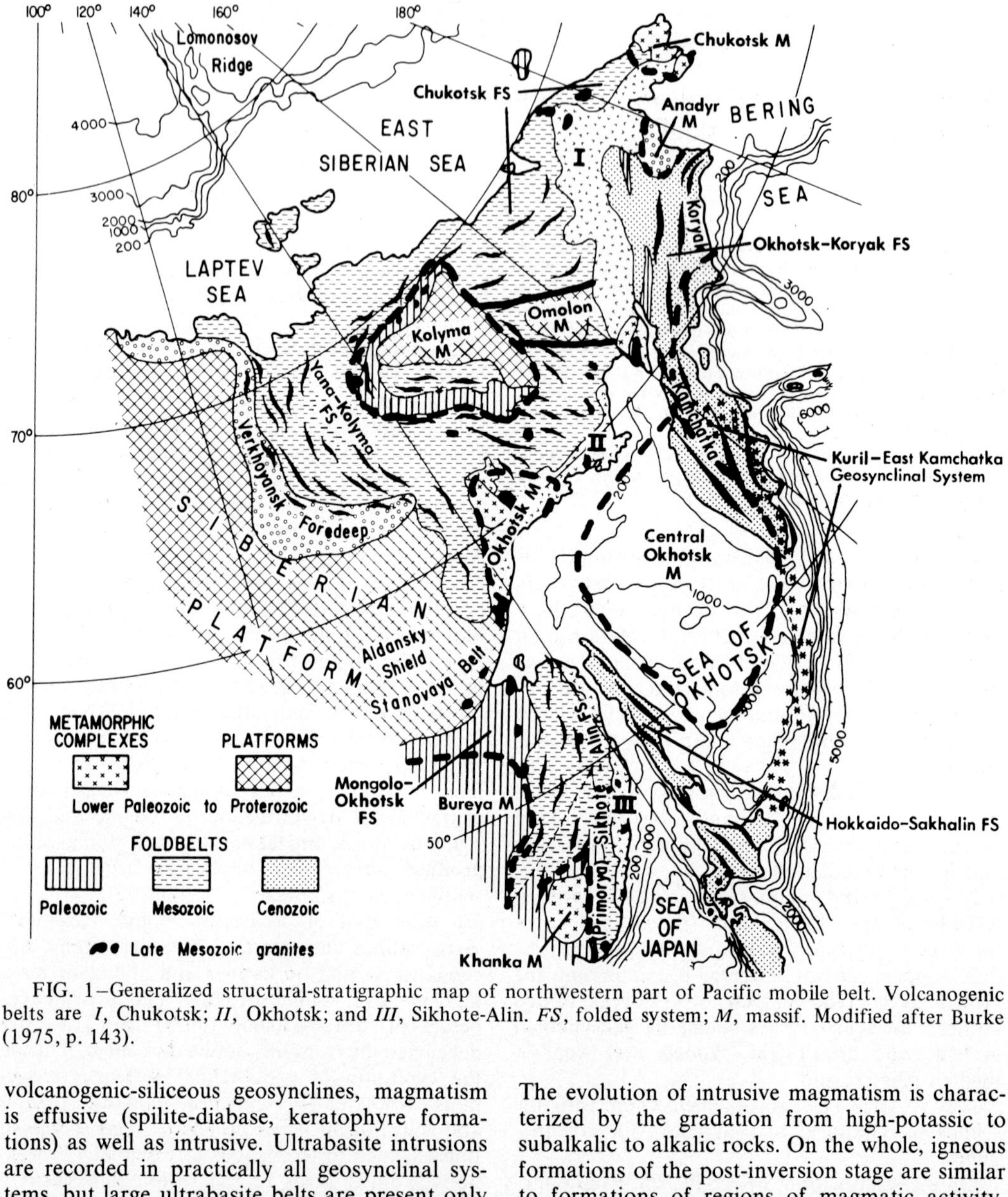

FIG. 1–Generalized structural-stratigraphic map of northwestern part of Pacific mobile belt. Volcanogenic belts are *I*, Chukotsk; *II*, Okhotsk; and *III*, Sikhote-Alin. *FS*, folded system; *M*, massif. Modified after Burke (1975, p. 143).

volcanogenic-siliceous geosynclines, magmatism is effusive (spilite-diabase, keratophyre formations) as well as intrusive. Ultrabasite intrusions are recorded in practically all geosynclinal systems, but large ultrabasite belts are present only in the Nippon region near the Pacific Ocean. Inversion magmatism for geosynclines of any type and age is practically identical and is represented by igneous formations of high sodium content including plagioliparite-basalt, plagioliparite-andesite, gabbro-plagiogranite, diorite-granite-granodiorite, and diorite formations. Also present in Mesozoic geosynclines are high-alumina granites having generally lowered alkalinity. In the post-inversion stage of Mesozoic geotectonic cycles, terrestrial effusive formations were widespread. The evolution of intrusive magmatism is characterized by the gradation from high-potassic to subalkalic to alkalic rocks. On the whole, igneous formations of the post-inversion stage are similar to formations of regions of magmatic activity. However, the volume of magmatic products carried up from depth to the surface is greater in regions of magmatic activity than in folded regions. The voluminous belts of granites of Stanovoy-Dzhugdzhur and volcanic rocks of the East Asian volcanogenic belt are examples.

Tracing the evolution of magmatism in time and the distribution of magmatic rocks in space has shown that magmatic processes commonly develop monotypically in different tectonic settings. Intrusives of alkalic-ultrabasic composition are developed in the Siberian platform and the Sikhote-Alin fold belt. Devonian and Triassic basic rocks (traps) are characteristic for the Siberian platform and the Sette-Daban pericratonic trough, whereas Triassic rocks of somewhat different composition (gabbro-diabase) characterize the Verkhoyansk-Chukotsk folded region. Monotypic granitic rocks such as gold-bearing diorite-granite-leucogranite, molybdenum diorite-granodiorite, and ores are present within folded systems, in the region of magmatic activity, and in massifs. In a relatively short interval of time (Jurassic–Late Cretaceous), granitic magmatism of similar composition took place over the vast territories of eastern Asia under different tectonic conditions. At present, the problem of relating the occurrence of useful minerals to rock types is being studied. Predictions of mineral resources on the basis of formation analysis are few, and the data are considered at present to be preliminary.

Coal is known in rocks of all ages beginning with the late Paleozoic. Great accumulations are known in the Upper Jurassic and, especially, in the Cretaceous; coal is found in great quantity in Paleogene and Neogene deposits. All the large coal deposits are related to continental molasse and lacustrine formations. Also, there is, in general, an association of coals with formations of the inversion stage of geosynclinal evolution, and of brown coals with the post-inversion stage and with magmatic activity. Molasse that was formed as a result of inversion of terrigenous geosynclines contains the most coal.

The most favorable sites for oil and gas are brachyanticlinal folds, post-inversion depressions of Mesozoic-Cenozoic age, foredeep strata, and marginal-platform folded structures (domes, anticlines) in upper Precambrian–lower Paleozoic rocks of the mantle of the Siberian platform.

Iron-manganese deposits of sedimentary origin are known in carbonate-terrigenous and volcanogenic-siliceous geosynclinal formations of late Proterozoic–Early Cambrian age. Among the deposits of this age are known concentrations of phosphorites.

In volcanogenic-siliceous sequences of the early geosynclinal stage that contain differentiated Paleozoic volcanic rocks (spilite-keratophyre and basalt-porphyry-dacite-liparite formations), pyrite mineralization may be present. According to the data at our disposal, the most prospective regions are the Sikhote-Alin and Mongolo-Okhotsk folded systems.

Gold and tin are characteristic metals of the eastern USSR. The mesothermal type of gold mineralization took place in Late Jurassic–Cretaceous time, and is related to diorite-granite-granodiorite intrusions which completed the inversion stage of evolution of Mesozoic geosynclines. Tin mineralization is present in Late Cretaceous–Paleogene rocks and is related to diorite-granodiorite-leucogranite intrusions, which took place later than the gold mineralization—that is, in the post-inversion stage. Formation analysis shows that gold-bearing and tin-bearing magmatic formations are monotypic in geosynclines of various types (terrigenous, volcanogenic-siliceous, carbonate), and do not depend on the character of preceding geosynclinal evolution.

References Cited

Burke, C. A., 1975, Sea of Okhotsk—thirteenth cruise of the *Dmitry Mendeleev*: Geology, v. 3, p. 140–144.

Krasny, L. I., editor-in-chief, D. A. Kirikov, and V. V. Russ, eds., 1967, Geological map of the northwestern part of the Pacific mobile belt: Moscow, scale 1:1,500,000.

Tectonics of Continental and Oceanic Structures of North Pacific Mobile Belt—Summary[1]

B. KH. EGIAZAROV, B. V. ERMAKOV, L. I. ANIKEYEVA, T. N. KOPYLOVA, S. I. ANDREYEV, M. I. ITSIKSEN, and V. I. BERGER[2]

The North Pacific mobile belt consists of the Koryak-Kamchatka-Kuril and Aleutian-Alaska-Cordillera geosynclinal areas. The Koryak–West Kamchatka and Olyutor-Kamchatka fold systems and recent Kuril geosynclinal systems are recognized within the former. The Koryak–West Kamchatka system formed during two tectonic cycles—one occurring in part of the Paleozoic–early Mesozoic and one encompassing the complete late Mesozoic–Cenozoic. The Olyutor-Kamchatka fold system resulted from the Late Cretaceous–Neogene tectonic cycle. The recent Kuril geosyncline represents the middle stage of the geosynclinal episode of development.

The migration of a geosynclinal process with time is from the continent toward the ocean along the mobile belt. The fold area is characterized by a polycyclic migrational style and eugeosynclinal type of development.

The Aleutian-Alaska-Cordillera geosynclinal area consists of the South Alaskan and Columbian fold systems, fold and fault systems of the Rockies, and the present Aleutian geosyncline. The Columbian fold system was formed during the early Paleozoic. The orogenic stage of the development began in the Late Jurassic. The fold and fault system of the Rockies formed as a result of the late Proterozoic–Neogene tectonic cycle. The orogenic stage covers the Middle Jurassic and Neogene. The South Alaskan fold system resulted from the two tectonic cycles—one representing the complete middle Paleozoic–early Mesozoic and one occurring in part of the late Mesozoic–Cenozoic. The accumulation of thick molasse complexes of the second cycle covered the Late Cretaceous–Neogene. The present Aleutian geosynclinal system also represents the middle of the geosynclinal episode.

The Aleutian-Alaska-Cordillera geosynclinal area is characterized by rejuvenation along the Pacific mobile belt and across the strike from the ocean to the continent. This geosynclinal area also is characterized by a polycyclic migrational style of development.

In the Pacific mobile belt, the migration of a geosynclinal process was directed from the continent toward the ocean in the Eastern Asiatic sector, whereas, in the North American sector, it was the opposite.

The North Alaska and Chukotsk fold systems of the adjacent regions of the Arctic mobile belt are separated from the Pacific system by the Mackenzie-Lena fracture zone. These systems are characterized by complete formational series, monocyclie style, and miogeosynclinal type of development. They differ from the Pacific structures by the absence of typical eugeosynclinal formations.

The differences between the fold systems of the Arctic and Pacific mobile belts result from the crustal structure. The Arctic geosyncline appears to have formed on the continental crust, whereas the Pacific systems most probably developed on crust lacking a "granitic" layer.

The available data suggest a mineralogenic autonomy of the North Pacific mobile belt (Pacifides) and adjacent areas of the Arctic mobile belt (Arctides) with a different metal content for the late Mesozoic and Cenozoic stages (Au, Hg, Cu, Sn, W, and polymetals, especially in the Arctides).

Troughs in the Arctic Canada, North Alaska, Chukotsk, Chukchi, and East Siberian seas form a single Arctic belt of oil and gas accumulation associated with the Arctides and probably with the marginal part of the submerged Hyperborean platform. The North Pacific belt of oil and gas accumulation may include troughs and basins of the Kamchatka-Koryak and Cordillera-Alaska areas.

[1]From Circum-Pacific Energy and Mineral Resources Conference program booklet.

[2]VSEGEI (All-Union Geological Research Institute), Leningrad, USSR.

Surveying Earth and Its Environment from Space[1]

W. A. FISCHER,[2] PRAYONG ANGSUWATHANA,[3] W. D. CARTER,[4] KAZUO HOSHINO,[5] E. H. LATHRAM,[6] N. R. ALBERT,[7] and ERNEST I. RICH[8]

Abstract Data from the Earth Resources Technology Satellite (ERTS) are being used for cartographic, geologic, and other earth-science purposes by a large majority of the nations of the world. At present, the Earth Resources Observation Systems (EROS) Data Center in Sioux Falls, South Dakota, is serving the data needs of 121 nations.

The cartographic qualities of the data from ERTS, especially those from the multispectral scanner (MSS), have exceeded pre-launch expectations. MSS data have been used to produce photomaps of parts of the United States that have root-mean-square positional errors of less than 150 m.

Structural geologic information has been obtained from Nimbus and later ERTS data. This information has led to revision of the metallogenic map of Alaska and is providing focus on potential locations of metallic ores in other areas of the Western Hemisphere. Tectolinear interpretations of ERTS-1 data, in conjunction with seismic epicentral data from other sources, may provide a new approach to defining hazardous areas as an aid in future development planning.

Recent work involving special processing of ERTS images of parts of the State of Nevada has led to a capability to detect and map areas of surface alteration.

INTRODUCTION

In 1966, William T. Pecora, then Director of the U.S. Geological Survey, instructed the Earth Resources Observation Systems (EROS) Program personnel to design a remote-sensing system that would yield data ". . . of utmost value to the broadest possible number of scientists and technicians throughout the world." The initial concepts resulting from this effort were first published by Fischer and Robinove in 1968, and the NASA ERTS-1 satellite is now aloft and yielding data that are beyond expectations.

CONCEPTUAL DEVELOPMENT OF ERTS*

The study team concluded that, in light of information needs and economics, the system should address problems of mapping at regional scales; the data should include the fewest possible variables, such as varying illumination and vignetting; and the system should approach real-time delivery and be repetitive in order to deal effectively with the dynamics of the earth and the rapid changes man is making to his planet. It was also concluded that the system should be multispectral and multitime to provide for the recognition of discrete objects or conditions (spectral and temporal factors are, in part, substitutional for resolution in the recognition process).

Finally, the team concluded that there was no viable aircraft alternative to an orbital system.

THE ERTS SYSTEM

The ERTS-1 satellite was launched July 23, 1972, and has performed satisfactorily for more than 2 years. There are three primary sensor systems aboard ERTS-1:

1. A three-camera Return Beam Vidicon (RBV) cluster designed to produce frame-type multispectral images;
2. The Multispectral Scanner (MSS), a 4-band multispectral system, whose data are returned to earth in a digital mode (MSS data are now available for 80–90% of the land surface of the world);
3. The data collection system (DCS) that serves as a radio relay system for information transmitted from ground sensors, such as seismographs, to a central receiving center.

The following quotations are excerpted from the statement by Leonard B. Jaffe, the U.S. representative to the United Nations task force and working group on remote sensing (February 1974):

> Data acquired in electronic form by the spacecraft are telemetered to earth by microwave links (to NASA reception facilities). . . . The first copies of the data go to a facility called the Earth Resources Observation Systems (EROS) Data Center at Sioux Falls, South Dakota, operated by the U.S. Department of the Interior. When received and receipted by the EROS Data Center, the data are considered to be in the public domain. All of the ERTS data can be purchased from the EROS Data Center.

*Now referred to as "LANDSAT." *Editor*

[1]Manuscript received, November 11, 1974.

[2]Senior Scientist, EROS Program, U.S. Geological Survey, Reston, Virginia 22090.

[3]Department of Mineral Resources, Bangkok, Thailand.

[4]Geologist, U.S. Geological Survey, Reston, Virginia 22090.

[5]Senior Geologist, Geological Survey of Japan, Tokyo, Japan.

[6]EROS Coordinator, Alaskan Geology Branch, U.S. Geological Survey, Menlo Park, California 94025.

[7]Geologist, U.S. Geological Survey, Menlo Park, California 94025.

[8]Office of the Dean, Stanford University, Stanford, California 94305.

> The extent and vigor of foreign interest is perhaps greater than we expected. Both Canada and Brazil have built and operate ground stations for receiving ERTS data directly and. . . numerous other countries have expressed interest in operating their own ground facilities.

At this time, the EROS Data Center is serving the data needs of 121 nations.

Experimental Results

Before considering examples of geologic and environmental uses of ERTS data, it is appropriate to recognize that timely maps, appropriately referenced to the configuration of the earth, are basic to these studies.

The U.S. Geological Survey has produced several image-type cartographic products using ERTS data. These include state maps of New Jersey, Arizona, and Florida; 1:250,000-scale quadrangle maps; and image-format maps on which coverage is determined by the orbit of the satellite. These maps are in black and white or "false color," in which vegetation in vigorous growth appears red, clear water black, and sediment-laden water and urban areas as shades of blue. The positional accuracy of features shown on these maps exceeds National Map Accuracy Standards, at regional scales, when referenced to the overprinted Universal Transverse Mercator (UTM) grid.

Many nations of the world are using ERTS data for cartographic purposes; South Africa, for example, is now in the process of mapping the country completely on an image-format basis.

ERTS was designed to test four geologic hypotheses:

1. That some dynamic geologic processes such as sedimentation and glacial motion could be better understood if viewed in a "time-lapse" mode;
2. That color or multispectral images properly recorded and processed would prove useful in mapping the distributions of rock types, geochemical anomalies, and alteration products;
3. That some geologic and hydrogeologic features are only intermittently visible, depending on angle of illumination or on snow, water, or vegetation distributions; and
4. That large structures are present on the surface of the earth which, by virtue of their size and subtle expression, have gone unrecognized in conventional ground and aerial surveys.

Dynamic Geologic Processes

The broad synoptic view from ERTS has enabled Post and Meier (1973) to classify glaciers with respect to their characteristics of movement, that is, uniform or surging. This classification has been tested successfully against known glacial movements in the United States and in the USSR. One series of ERTS images of the Yetna Glacier in Alaska has been converted by personnel at the Stanford Research Institute into a time-lapse motion picture which shows dramatically the shock wave that travels up the glacier at time of surge. Investigators at Stanford Research have indicated (W. Evans, oral commun., 1974) that they believe motions as small as 20 m can be observed in the time-lapse mode.

D. B. Krinsley (written commun., 1974) has used repetitive observations of water distribution in playa lakes in the Great Kavir (Dasht-e Kavīr) of Iran to arrive at an optimum road alignment through this region.

Seismic-event counters and tiltmeters have been emplaced on 15 volcanoes in Central and North America to demonstrate the technologic and economic feasibility of building a global volcano surveillance system. Not long ago, signals received via ERTS from seismic-event counters emplaced near Volcan Fuego, Guatemala, signaled its impending eruption 5 days prior to the event. The reliability, extremely long, unattended life, and low cost of the DCS ground components make it practical to install monitoring instruments in very remote areas and to study many more volcanoes than conventional land communication links would permit (Ward et al, 1974).

Recognition of Rock Types, Geochemical Anomalies, and Alteration Products

Data received from the MSS aboard ERTS are in digital form and can be manipulated by computer, without interjection of the photographic process. This procedure preserves the spectral purity and wide dynamic range of the returned data. Pioneering work by Rowan et al (1974), using computer-compatible tapes, has permitted increased color contrast and resulted in a capability to map the distributions of certain rock types in the Goldfield, Nevada, region and to identify rocks that are significantly altered. The procedure offers promise in minerals exploration. This work is based on known iron-absorption bands in the solar-infrared part of the spectrum and involves ratio processing of the various spectral bands.

Intermittently Visible Geologic Features

C. J. Robinove and Morris Deutsch (oral commun., 1972) were able to correlate the distribution of vigorous spring vegetation growth in the Texas Gulf Coast with the distribution of the Brazos River alluvium, a prolific groundwater aquifer. More recently, R. B. Morrison (written commun.,

FIG. 1–Map showing possible major lineations in United States as interpreted from a 1:5,000,000-scale mosaic of ERTS images, by W. D. Carter. Significance of many of these features is undetermined.

1974) reported that some glacial landforms are visible on ERTS-1 images only when enhanced by snow cover and low sun angle. C. F. Withington (oral commun., 1974) also noted that it is possible to extend the mapped length of major linear features in the vicinity of Washington, D.C., by using ERTS-1 images taken at a time when a light snow covered part of his area of study. Fischer and Lathram (oral commun., 1974) noted a structural feature in northern Alaska that may control part of the course of the Colville River. This feature is not visible on images of the area that are free from snow and taken when the sun illumination is at a higher angle. On the same sequence of images, several anomalous snow-free areas are visible. One set of anomalies remained snow-free on a subsequent pass and, although the evidence cannot be considered as solid at this time, one plausible interpretation of the anomalies is that they represent the surface expression of gas seeps.

At least four instances are on record showing that oil on water can be discriminated on ERTS data. The most recent published example is by Stumpf and Strong (1974). There is, of course, a high potential for these types of observations in environmental surveys, but one should not overlook the fact that the repetitive nature of ERTS may make it possible to identify natural marine oil seeps, with consequent improvement in the efficiency of offshore oil exploration.

Large, Previously Unnoticed Geologic Features

W. D. Carter (oral and written commun., 1974) has begun interpretation of lineations and curvili-

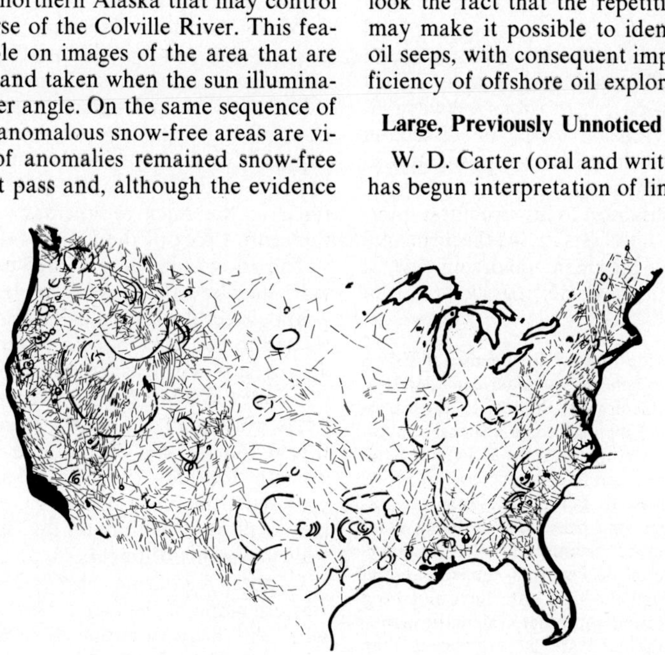

FIG. 2–Map showing possible curvilineations in United States as interpreted from a 1:5,000,000-scale mosaic of ERTS images, by W. D. Carter. Significance of many of these features is undetermined.

neations on mosaics of the United States produced from ERTS images by the Soil Conservation Service of the Department of Agriculture in cooperation with NASA. Several scientists in the U.S. Geological Survey are making interpretations independently in order to build "a level of confidence" in the result. The studies have been done at scales of 1:1,000,000 and 1:5,000,000. Figures 1 and 2 show the preliminary results of the interpretations at 1:5,000,000 scale and show only the major features that have thus far been discriminated. One of the major lineaments extends from northern Florida to the vicinity of New Madrid, Missouri, site of the most severe known earthquake on the North American continent.

Saunders et al (1973) have completed structural studies of (1) the areas of major Laramide mineral deposits, (2) the Williston–Blood Creek basin, (3) the Colorado mineral belt, (4) the uranium areas of Utah and Colorado, and (5) other mineralized and petroliferous regions of North America. They have concluded that there is a great advantage in photogeologic interpretation from the satellite viewpoint to provide a truly synoptic examination of regional geologic features—lineaments and other geomorphic features with dimensions of tens to hundreds of miles which are not generally detected on aerial photographs. They also found that the use of ERTS imagery for regional structural studies is about 1/500 as expensive as the use of conventional aerial photography. They noted that close empirical relations between lineaments and curvilinear, tonal, or dissection anomalies occur both in mineral deposits and in the structure of sedimentary basins.

W. D. Carter, in addition to his structural interpretations of the United States, is the principal investigator for an experiment involving several Latin American nations. Carter reported (written commun., 1974):

> In Chile, there is wide interest in the use of ERTS data that is fostered by the Inter American Geodetic Survey and the U.S. Geological Survey, in the cartographic applications of ERTS data and a focused investigation under the leadership of Gabriel Perez to determine the relations between lineaments seen on ERTS images and ore deposits. In Peru, Jose Pomalaza has developed a data-processing center for utilizing computer-compatible tapes to produce detailed enhancement of river drainage in the Peruvian-Amazon area at 1:25,000 scale. Although the tape data have not been corrected for earth rotation and other systematic errors, the products being produced are more accurate than existing maps of the area, which is under intense exploration for oil at the present time. Bolivian efforts under the leadership of Dr. Carlos Brockman are among the most advanced. . . in Latin America. In a cooperative experiment, Carter and Brockman have produced a mosaic of the La Paz area and mapped a large number of structural features, and compared them geographically with known minerals and oil information, to arrive at a set of exploration priorities.
>
> . . . the Mineral Resources Division of Geobol is using the interpretation of the La Paz mosaic as a guide to exploration for new mineral deposits. One geologist had just brought in a large rock sample of mineralized fault breccia from a linear feature, known to be 20 km in length, in the Corocoro Copper Mining District.

APPLICATIONS

In the process of developing this paper, the principal author has sought and received contributions from the coauthors; summaries of these contributions follow.

ERTS Prospecting in Northern California[9]

The geomorphic, lithologic, and structural interpretation of ERTS-1 imagery of northern California suggests that new data, perhaps not fully appreciated heretofore, may be available for mineral and mineral-fuel prospecting.

For example, several ERTS-detected features (Fig. 3) which have been grouped into several linear systems (Fig. 4) are important in the structural interpretation of northern California (Rich and Steele, 1974).

Within northern California, economically important minerals are distributed as follows: (1) mercury and manganese deposits are chiefly associated with the mélange-style deformation of the Coast Ranges; (2) copper, lead, zinc, and gold deposits are associated with plutonism in the Sierran arc; and (3) minerals fuels are chiefly confined to the thick sedimentary sequence within the central part of the Great Valley.

The known deposits of mercury are associated with steeply dipping dikes or large, nearly horizontal, sheetlike bodies of serpentinite commonly related to the San Andreas system of linears (Fig. 4). An east-northeast-trending system of linears, the Valley system (Fig. 4), overprints the San Andreas system in the Coast Ranges and can be traced across the Sacramento Valley into the western foothills of the Sierras. Economically productive mercury deposits are at, or near, the points of intersection of the San Andreas and Valley systems of linears.

Copper, lead, zinc, and gold deposits on the western slope of the Sierra Nevada are associated with the north-trending Sierran linear system (Fig. 4). Rich and Steele (1974) have suggested

[9]By E. I. Rich.

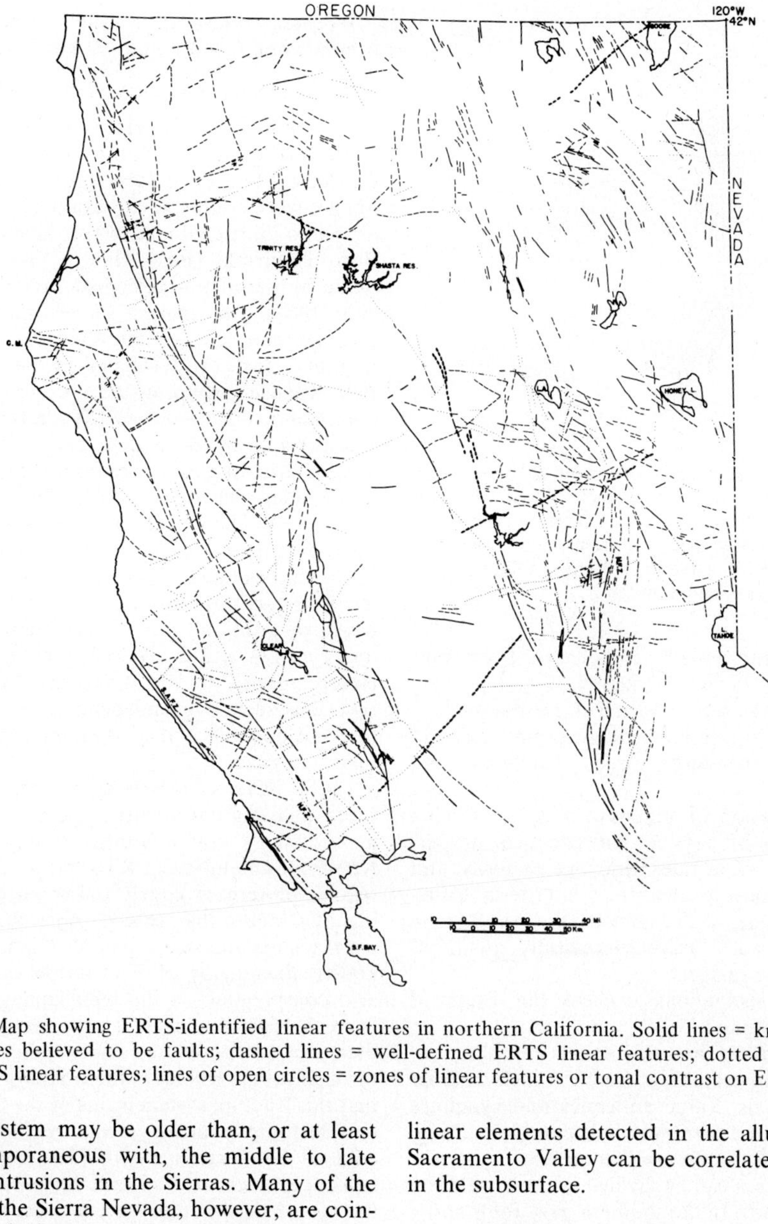

FIG. 3–Map showing ERTS-identified linear features in northern California. Solid lines = known faults or linear features believed to be faults; dashed lines = well-defined ERTS linear features; dotted lines = poorly defined ERTS linear features; lines of open circles = zones of linear features or tonal contrast on ERTS imagery.

that this system may be older than, or at least penecontemporaneous with, the middle to late Mesozoic intrusions in the Sierras. Many of the deposits in the Sierra Nevada, however, are coincident with the intersections of the Sierran system and the extension of the east-trending Valley system of linears.

Most of the oil and gas fields in northern California underlie the alluviated part of the Sacramento Valley. The locations of some of these fields are closely related to elements of the Valley system and, locally, to the northwest-trending San Andreas system. Published subsurface data from a few of the wells suggest that some of the linear elements detected in the alluvium of the Sacramento Valley can be correlated with faults in the subsurface.

Geologic Mapping in Thailand[10]

The area studied includes a part of Bilauk Taung Range and lies approximately 250 km west-northwest of Bangkok. It consists of sedimentary, metamorphic, and igneous rocks. This area was uplifted and extensively eroded in earlier geologic times and is quite mountainous. There

[10]By Prayong Angsuwathana.

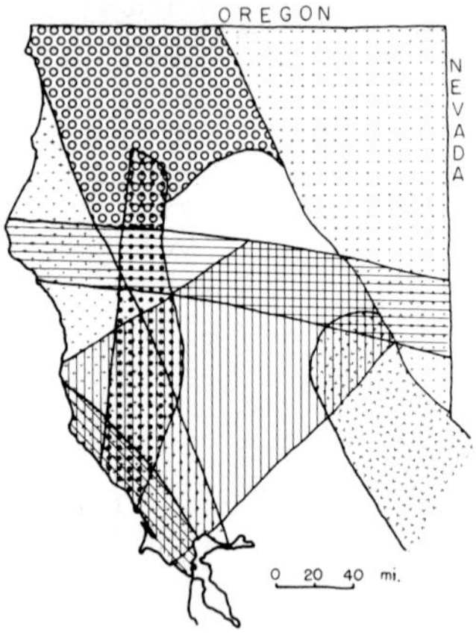

FIG. 4–Map showing location of linear systems. Plus signs = San Andreas system; NW-SE diagonal lines = Coastal system; dots = Central system; horizontal lines = Mendocino system; open circles = Klamath system; X = Modoc system; V = Sierran system; vertical lines = Valley system.

are no roads for vehicles; it has never been studied intensively by ground survey.

The work was separated into three parts:

1. Drawing of inferred geologic boundaries and mapping of structures, faults, fractures, and folds.

2. Identification of dominant rock types using considerations of physical and chemical properties, known erosion rates, drainage patterns, and surface roughness as identification criteria. Limestone, sandstone, shale, igneous rock, and some metamorphic rocks were successfully identified from the space imagery.

3. Field investigations to check the results of the interpretation.

The structure of the study area is complex, owing to frequent tectonic movement including igneous intrusions. Large anticlines and synclines were formed, and there was considerable faulting, with movements of several kilometers.

The study area can be divided into two parts by structural trends. In the southern part, folds strike parallel with the granitic intrusion along the Burma border, and dips are to the northeast. In the northern part, strikes are generally north and dips are northwest and southeast. A large anticline of old rock plunging to the south and flanked with Permian limestone is present in the central part of this area. Rocks in the northern part were folded in Ordovician and later times. Most of the folding is asymmetric, and folds are overturned to the west.

Normal, reverse, thrust, and wrench faults are present; one is more than 80 km in length.

Structural Analysis in Japan[11]

The tremendous usefulness of ERTS images for analysis of geologic structure is recognized by the Geological Survey of Japan. In this study, ERTS images were compared to results of conventional air-photo lineament analysis of the middle part of Boso Peninsula (Bōsō Hantō), east of Tokyo, made by petroleum geologists in 1967 (Shiraishi et al, 1967). This analysis showed that three fracture systems are conspicuous on aerial photographs. Type A, developed in the mostly Pliocene Kazusa Group, in the northern part of the area, trends north-northeast. Type B trends north and prevails in the middle and northern parts. Type C, trending east or east-northeast, is present in the older rocks in the south. The ERTS images clearly show these same fracture systems (Hoshino and Matsuno, 1974). In addition, the comparative studies made it possible to understand better the structural nature of the fracture systems, which was difficult to interpret from the air photos. Types A and C are clearly recognized to be closely related to the basin structure, whereas type B is seemingly independent of basin structure but possibly is related to the regional structure.

ERTS images further showed that the north-south-trending lineaments (type B) exist in other parts of the southern Kantō region, especially in Kantō mountainland. ERTS images show a very regular pattern of largely unknown north-trending lineaments that extend more than 100 km, with a constant spacing of 2–3 km. Previously known lineaments of west-northwest trend are also conspicuous on the ERTS images. Judging from the ERTS images, it now seems likely that there is a north-trending regional fracture system throughout the whole southern Kantō region, and that this fracture system is one of the most important tectonic elements in this region. It is now covered by Quaternary volcanic deposits from volcanoes such as Fuji and Hakone; therefore, the age of the fracture system is pre-Quaternary, possibly Miocene or Pliocene.

The economic significance of these north-south fractures is shown in the case of Mobara gas field, the only productive gas field in Kantō, where the reservoir body is characterized by a swarm of minor north-trending normal faults. In relation to earthquake-prediction problems, the 1930 earthquake in Kitaizu, located directly south of the

[11]By Kazuo Hoshino.

Kantō mountainland, was caused by movement along major north-trending faults about 30 km in length. Thus the north-south fracture system is still potentially active in some places.

ERTS Study of Alaska[12]

Examination of an image of Alaska taken by the Nimbus IV Image Dissecter Camera System revealed the existence of a previously unrecognized orthogonal set of northwest- and northeast-trending linears that may reflect crustal structures (Lathram, 1972). Study of ERTS imagery has been undertaken to identify these and other regional linears, determine their pattern, relate their distribution to subsurface structure, and analyze their significance in terms of various theories of the tectonic history of Alaska and of continental mobility and crustal deformation in general.

Studies to date have been focused on:

1. A regional determination of the orientation pattern of lakes in lowland areas of central and northern Alaska;
2. A study of an energy-rich area, comparing prominent linears with known surface structures and with seismic, magnetic, and gravity data; and
3. Development of a new working hypothesis relating structural features seen on space images to the distribution of mineralized areas.

A qualitative analysis of the trend of rectilinear shores of lakes in lowland areas of central and northern Alaska, and of alignment of lakes and interlake areas, reveals a relatively consistent pattern of north, northwest, northeast, and east trends (Lathram, 1973a).

No single trend is dominant in all areas, and some of the trends vary slightly in azimuth from area to area. In northern Alaska, west of the Ikpikpuk River, a dominant northerly trend is expressed by an average N9°W elongation of lakes (Black, 1969). To the east of this line, an orthogonal easterly linear trend in the lake area is dominant (Lathram, 1974). Numerous workers concur that the northerly elongation of lakes is due to erosion at right angles to the prevailing wind direction (Black, 1969). The areal restriction in dominance of the northerly elongation suggests that, although the prevailing wind direction may have been a contributing factor, it was not the primary determinant of the northerly elongation. This conclusion is substantiated by the apparent structural control of the easterly linear trend in the Umiat area and the variation in prominence of the northerly elongation of lakes in other parts of northern and central Alaska, where the wind direction is similar.

Energy resources—The Arctic coastal plain of Alaska is developed on a nearly flat-lying mantle of the Quaternary Gubik Formation, which overlies a thick Cretaceous molasse basin.

Previously unrecognized regional linears in the Arctic coastal plain north of Umiat, Alaska, trending about N78°E, are clearly shown on ERTS-1 image 1004-21395, recorded July 27, 1972. Study of geological and geophysical data in the area suggests that the east-trending linears are expressions of concealed geologic structures beneath the Gubik Formation (Fischer and Lathram, 1973; Lathram, 1973a).

If the linears are controlled by geologic structure, the total extent of the area characterized by linears is significant and linears in the lake area do parallel the alignment of magnetic anomalies in the area of ERTS image 1004-21395 and to the west. The geological and geophysical evidence of control by subsurface structure has been described in detail by Lathram and Albert (in press).

Mineral resources—Studies of Nimbus and ERTS imagery resulted in the formulation and strengthening of a hypothesis that mineralized areas in Alaska may be spatially related to a regional set of northeast- and northwest-trending linears—many previously unrecognized—that possibly reflect crustal fractures (Lathram and Gryc, 1973; Lathram et al, 1973; Lathram, 1973b). Linears representative of this set are clearly apparent on an ERTS mosaic of the Yukon-Tanana upland east of Fairbanks.

According to the conventional metallogenic hypothesis, areas deemed favorable for occurrence of various types of ore deposits are assumed to describe arcuate belts through southern and central Alaska. The new hypothesis postulates that favorable areas would form belts parallel with major northwest- and northeast-trending fractures, and deposits would be more abundant where such fractures crossed. For example, such deposits might be found in two northeast-trending belts that could extend northeast of the Denali fault.

Word has been received that the hypothesis has been substantiated (E. R. Chipp, written commun., 1973). In 1970 and 1971, independent mineral exploration using northeast and northwest fracture systems as a guide resulted in the discovery of several copper-molybdenum deposits in the area of the Tanacross quadrangle, two of which are currently being explored by drilling. These deposits lie northeast of the Denali fault.

[12]By E. H. Lathram and N. R. Albert.

Environmental Results

The potential use of ERTS data to monitor the distributions of oil on water has been previously noted. Other specific environmental uses of ERTS data that are being tested include: (1) inventory and monitoring of strip-mining activities and monitoring of the reclamation process, (2) monitoring the effects of oil development and transport in northern Alaska, (3) developing and maintaining a timely inventory of man's use of his land, and (4) using the data to assess disasters, such as floods.

Strip Mining

A contract has recently been negotiated between the U.S. Bureau of Mines and the Mathematica Corporation, working with Ford, Bacon and Davis, to undertake an inventory of all strip mines in the United States. This effort is designed to make use of ERTS data while supporting the interpretations with statistical samples of aircraft (high resolution) data and limited ground observations (Eric Hoffman, oral commun., 1974).

Several individuals and organizations have done the testing necessary to establish the validity of using ERTS data for this purpose. The results of investigations by John Rehder of the University of Tennessee are shown in Figure 5.

Oil Development and Transport in the Arctic

Examination of color composites of ERTS image 1004-21395, covering Umiat, Alaska, and areas to the west, revealed no indications of scarring of the delicate tundra as a result of the intense oil exploration in this area in the late 1940s and early 1950s. This suggested that the environmental effects of oil exploration were not spreading but rather were healing at a relatively rapid rate.

This conclusion was largely substantiated by low-level helicopter surveys undertaken by R. L. Detterman in the summer of 1973 and reported by Lathram and Albert (in press). One short, clear-dozed and repeatedly used trail near Umiat still formed a marked scar on the tundra. Subsequent microscopic examination of the band-5 black-and-white ERTS image revealed this scar and established ERTS as a viable monitoring tool for this purpose.

Man's Changing Use of Land

In an environmental sense, the most significant observation that can be made is that of man's constantly changing use of the land. ERTS was not designed to identify land use at relatively large scales, but rather to provide a national overview of land use in gross categories and to indi-

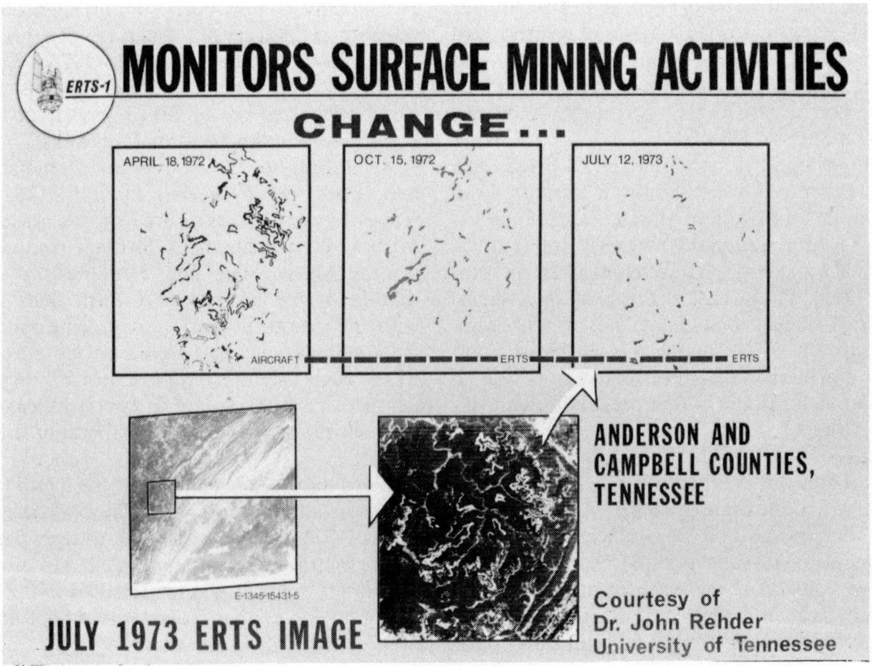

FIG. 5–Map showing changes in strip-mining activities in Anderson and Campbell Counties, Tennessee, as mapped from ERTS images. Lower, central image is an enlarged, negative print of band 5 of a part of ERTS image E-1345-15431.

cate changes in land use that would be subject to further investigation.

Analysis of information derived from computer-compatible tapes of ERTS data, however, strongly supports the contention that ERTS data may be used to map actual land cover down to an area of approximately 80 by 80 m at a level of between 2 and 3 of the National Land Use Classification System, developed by J. R. Anderson et al (1972).

Maps of current land cover have tremendous bearing on the understanding of a wide variety of environmental factors—sediment loads in streams, malaria distribution, wildlife predictions, food availability, and energy demands, to name but a few.

Observing Disasters

ERTS data were used to make preliminary studies of the effects and geologic relations that pertained to the earthquake that struck Managua, Nicaragua, in December 1972, and to monitor floods of several rivers throughout the world, including the Indus and the Mississippi. The detailed techniques used in flood analysis were described by Deutsch et al (1973). Their findings can be summarized as follows:

1. Flood maps can be made from ERTS data that have precision commensurate with map scales of 1:250,000 or smaller and are useful in (a) reaching a rapid assessment of flood damage; (b) improving flood-forecast models; (c) evaluating the effectiveness of engineering structures used for flood control; and (d) planning follow-up actions, such as redredging, or designing new flood-control measures.
2. The extent of flooding could be mapped for as long as 18 days after the crest.
3. The use of band 7 of the MSS (whose spectral sensitivity lies outside the range of photographic film) was essential to the success of the mapping effort.
4. Successive images could be overlaid without special processing to assess temporal change rapidly.

Future Outlook

ERTS-2, similar in all respects to ERTS-1, will be launched in late 1974 or early 1975 to provide greater insight into the dynamics of the earth. A major component of this experiment will be the conversion of much of the data processing from the analog to the digital mode. This is expected to improve the information content of the data reaching the user by more than 50%.

A major effort is being made to secure approval for the launch of an ERTS-C, similar to ERTS-1 and 2, but incorporating a fifth channel of information in the thermal infrared and possibly a significant improvement in the resolution of the RBV's. Such a mission would further refine our knowledge of earth dynamics and provide the continuity necessary to encourage foreign investment in interpretation and reception services and goods.

The following quotations, taken from the statement of James R. Fletcher before the Utah Section of the American Congress on Surveying and Mapping, Salt Lake City, February 21, 1974, reflect plans beyond ERTS-3.

> What we have in mind are Earth Observatory Satellites three times the size of ERTS but in the same orbit. Tentative plans, not yet approved by the President or Congress, call for us to launch our first Earth Observatory Satellite in 1978.
>
> In our new budget, which is now before Congress, we propose a new satellite to be launched in 1977 to make thermal measurements of the Earth's surface. This is called our Heat Capacity Mapping Mission satellite.

References Cited

Anderson, J. R., E. D. Hardy, and J. T. Roach, 1972, Land use classification system for use with remote-sensor data: U.S. Geol. Survey Circ. 671.

Black, R. F., 1969, Thaw depressions and thaw lakes, a review: Bull. Peryglacjalny, no. 19, p. 131–150.

Deutsch, Morris, et al, 1973, Mapping of the 1973 Mississippi River floods from the Earth Resources Technology Satellite (ERTS), *in* Remote sensing and water resources management: Am. Water Resources Assoc. Proc., no. 17, p. 39–55.

Fischer, W. A., and C. J. Robinove, 1968, A rationale for a general purpose Earth Resources Observation Satellite: Remote Sensing Symposium, Univ. Washington, Proc., 6 p.

——— and E. H. Lathram, 1973, Concealed structures in Arctic Alaska identified on ERTS-1 imagery: Oil and Gas Jour., v. 71, p. 97–102.

Hoshino, K., and K. Matsuno, 1974, Fracture pattern of southern Kanto region recognized by ERTS imageries: Chishitsu (Geological) News, no. 235, p. 13.

Lathram, E. H., 1972, Nimbus IV view of the major structural features of Alaska: Science, v. 175, p. 1423–1427.

——— 1973a, Geologic application of ERTS imagery in Alaska, *in* Third Earth Resources Technology Satellite-1 symposium: NASA Spec. Pub. 351, addendum, 12 p.

——— 1973b, Tectonic framework of northern and central Alaska, *in* M. Pitcher, ed., Arctic Geology: AAPG Mem. 19, p. 351–360.

——— 1974, Analysis of state of vehicular scars on Arctic tundra, Alaska, *in* Third Earth Resources Technology Satellite-1 symposium: NASA Spec. Pub. 351, v. 1, sec. A, p. 633–641.

——— and N. R. Albert, in press, Significance of space linears in Alaska, *in* First International Conference on New Basement Tectonics, Proc.: Utah Geological

Assoc., Salt Lake City, Utah, June 1974.
——— and George Gryc, 1973, Metallogenic significance of Alaskan geostructures seen from space: 8th Internat. Symposium on Remote Sensing of Environment Proc., p. 1209–1211.
——— et al, 1973, Preliminary geologic application of ERTS-1 imagery in Alaska, *in* Earth Resources Technology Satellite-1 symposium: NASA Spec. Pub. 327, v. 1, sec. A, p. 257–264.
Post, Austin, and M. F. Meier, 1973, Galloping glacier poses flood threat: U.S. Geol. Survey, press release.
Rich, E. I., and W. C. Steele, 1974, Speculations on geologic structure in northern California as detected from ERTS-1 satellite imagery: Geology, v. 2, p. 165–169.
Rowan, L. C., et al, 1974, Discrimination of rock types and detection of hydrothermally altered areas in south-central Nevada by the use of computer-enhanced ERTS images: U.S. Geol. Survey Prof. Paper 883, 35 p.
Saunders, D. F., et al, 1973, ERTS-1 imagery use in reconnaissance prospecting—evaluation of the commercial utility of ERTS-1 imagery in structural reconnaissance for minerals and petroleum; Type III Final Rept. to NASA: U.S. Dept. Commerce, Natl. Tech. Inf. Service, E74-10345.
Shiraishi, T., K. Hoshino, and K. Oida, 1967, The use of photographs for structural analysis of fracture system: Prof. Yasuo Sassa Memorial Vol., p. 701–718.
Stumpf, H. G., and A. E. Strong, 1974, ERTS-1 views an oil slick?: Remote Sensing of the Environment, v. 3, no. 1, p. 87–90.
Ward, P. L., 1974, Development and evaluation of a prototype global volcano surveillance system utilizing the ERTS-1 satellite data collection system; Type III Final Report to the NASA, Goddard Space Flight Center, Greenbelt, Maryland: 154 p., U.S. Dept. Commerce, Natl. Tech. Inf. Service, E74-10689.

Solar Desalination Experience of University of Arizona—Summary[1]

CARL N. HODGES and WAYNE L. COLLINS[2]

Current energy problems have stimulated interest in solar energy. Although use of solar energy has not been economically feasible in the past, it may soon be competitive with increasingly expensive conventional fuels. It is received in sufficient quantity to make a major contribution to U.S. heat and power requirements, and there are no technological barriers to its wide application.

Since the late 1800s, numerous attempts have been made to develop economical systems for solar distillation. Initially, simple single-effect, glass-covered stills were used. During the 1950s, as a result of support from the Office of Saline Water of the U.S. Department of the Interior, many efforts to develop a multiple-effect solar still were undertaken.

The University of Arizona developed a multiple-effect, humidification-cycle, solar-powered distillation unit that was installed in Puerto Peñasco, Sonora, Mexico, in 1962 (Hodges et al, 1966). The theory of operation is simple. During sunshine hours, seawater flows through solar collectors and is heated to approximately 150°F (66°C). This seawater is stored in insulated underground tanks, from which it is pumped at a continuous rate 24 hours a day through a multiple-effect distillation plant. This unit operated successfully until 1965, when it was converted from solar energy to waste thermal energy from a diesel-electric set. The most fundamental reason the plant is no longer operating is simply that the support of the research work, which was from the Office of Saline Water, was discontinued. Furthermore, a major obstacle to the development of an economical solar-distillation system was inadequate low-cost materials that would provide a reasonable service life in a solar-energy application.

The solar-collector technology gained at Puerto Peñasco has been applied to the development of controlled-environment agricultural systems using desalted seawater—but that desalted seawater is produced with fossil fuels (Fontes, 1973). Desalting of seawater appears to be a logical application for solar energy, for it is in dry, desert-type areas which typically have abundant sunlight that production of desalted seawater is most attractive.

The key to whether solar energy will be developed to meet the projections given by a 1972 NSF/NASA panel will be a function of (1) the amount of research funds allotted for a national development effort; (2) the quality of the research carried out with these funds; and (3) perhaps most critically, the time frame in which these funds are provided.

The technology of terrestrial solar energy conversion has been developed to its present limited extent through very modest government support and some private funding. With the current renaissance of interest in solar energy, it is hoped that the necessary long-term development programs will be established to improve the probability of successful solar-energy applications in the future. Funding and time for research are essential, as solar energy applications will depend on materials engineering.

References Cited

Fontes, M. R., 1973, Controlled-environment horticulture in the Arabian Desert at Abu Dhabi: HortScience, v. 8, no. 1 (February).

Hodges, C. N., et al, 1966, The University of Arizona, Solar distillation utilizing multiple-effect himidification: Report, January 31.

NSF/NASA Solar Energy Panel, 1972, An assessment of solar energy as a national energy resource: Report, December.

[1]Manuscript received, November 11, 1974.

[2]Environmental Research Lab, University of Arizona, Tucson, Arizona.

Antarctica: an Unprospected, Unexploited Continent—Summary[1]

F. ALTON WADE[2]

INTRODUCTION

The geology and geologic history of Antarctica are not unique. In all proposed reconstructions of Gondwanaland, Antarctica is positioned adjacent to and is nearly surrounded by Australia, New Zealand, India, South America, and Africa (Dutoit, 1937). The breakup of Gondwanaland began near the end of the Paleozoic Era. Separation from Australia occurred as late as 50 m.y. ago. Each of these neighboring units of continental crust has been, and is, a source of great mineral wealth, so the chance that Antarctica would be barren of mineral deposits is unlikely.

Why, then, has Antarctica not been prospected and exploited? Three reasons are obvious: (1) geographic position, (2) unfriendly environment, and (3) cost. Antarctica is isolated from the populated areas of the world (Fig. 1). Until only a few decades ago, travel to Antarctica was a most hazardous undertaking. The seas in the southern latitudes often are extremely rough, with winds up to hurricane force. A ring of ice surrounds the continent during the summer months. There are gaps in the ring and segments in which the ice is not closely packed. In the days before the advent of the powerful ice breaker, traverse of the pack ice might be made in hours, or it might require weeks.

On the continent the environment is severe. With rare exceptions, and those only in the northernmost sectors, temperatures are subfreezing. Roughly 95% of the continent is ice covered and the thickness of the ice ranges up to more than 3,500 m. The virtual absence of terrestrial flora and fauna makes living off the land impossible. The field season is only 3½–4 months long. During most of the remainder of the year, personnel who "winter-over" are confined to the vicinity of the permanent bases. Until recent years, travel by exploration parties has been slow, inefficient, and often hazardous.

Today, exploration and prospecting can proceed at an accelerated, more efficient pace. With powerful ice breakers as escorts, ships of all types can reach most of the coastal areas of the continent where cargo can be discharged efficiently and bases of operations established. Personnel and many cargo items can be flown in to certain established bases on conventional and jet aircraft in a matter of hours. Transportation to all remote areas from coastal stations is accomplished by means of ski-equipped cargo-type aircraft and track-laying over-snow vehicles. Transportation from remote or main bases to areas of exploration is by helicopter or motor toboggan. Camps from which seasonal operations are conducted can be erected and occupied in a few days. More permanent bases can be established almost anywhere in the interior of the continent. Man can now live and operate in Antarctica in reasonable comfort. Exploratory operations will always be controlled by the weather. In some areas, only a few days of work during a season will be lost because of the weather conditions, whereas in other areas the number of days of operations, particularly by aircraft, may be as few as one in four.

Exploratory operations in Antarctica have always been very expensive. As transportation, living conditions, and support operations are improved, the costs rise. Therefore, first-class privately supported expeditions are a thing of the past. The responsibility for exploration and prospecting must be, and has been, assumed by national governments. The importance of the exploration, prospecting, and possible exploitation of Antarctica, combined with scientific research in many disciplines, must be weighed against other national objectives and funds allocated accordingly.

GENERAL GEOGRAPHIC AND PHYSIOGRAPHIC FEATURES

Including its ice cover and ice shelves, Antarctica has an area somewhat in excess of 13 million km^2, or about one and three-quarters the area of Australia. If the ice were removed and there were no isostatic adjustments, Antarctica would appear to consist of a continental mass—East Ant-

[1]Manuscript received, October 21, 1974.

[2]Antarctic Research Center, The Museum, Texas Tech University, Lubbock, Texas 79409.

I am indebted to Carl A. Cathey for his drafting of all line drawings and criticism of the original manuscript, and to Charlotte A. Dillon for typing the manuscript.

The material for this paper was taken largely from published reports and from knowledge gained by the writer during seven field seasons in Antarctica.

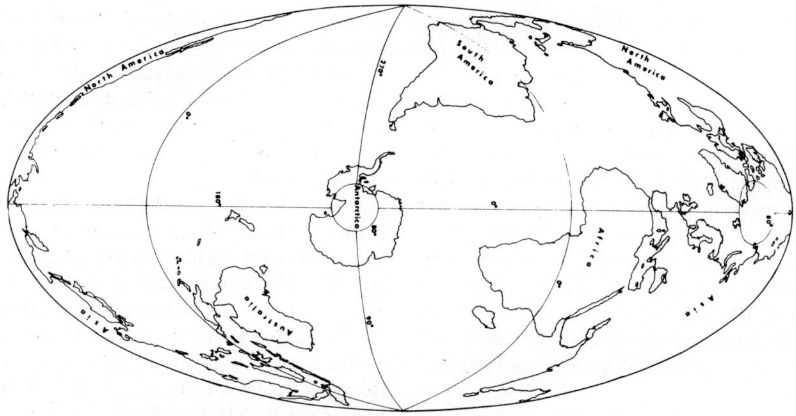

FIG. 1–Antarctica today is separated from other continents by wide expanses of southern oceans (Hammer transverse elliptical equal-area projection).

arctica—and an archipelago—West Antarctica. The former would be a continuous landmass with mountain ranges along much of the coastline, interior mountain ranges, central lowlands, and basins partially below sea level. Two or three large islands and many small ones would constitute the archipelago; the islands would be separated by deep trenches and basins. Isostatic rebound would alter this general description very little.

Geophysical observations indicate that East and West Antarctica together constitute a continuous segment of continental crust. One bit of direct evidence, a drill core from a site in the south-central Ross Sea occupied by the *Glomar Challenger* in 1973, establishes the existence of continental crust in that area (Hayes et al, 1973). Large-scale block faulting and general subsidence of a large part of West Antarctica have been suggested as the mechanisms responsible for the present land-water distribution.

Rock is exposed over only 5% of the area of Antarctica (Fig. 2). This seemingly small area is actually the equivalent of nine-tenths of the area of the State of Texas, roughly 660,000 km². The

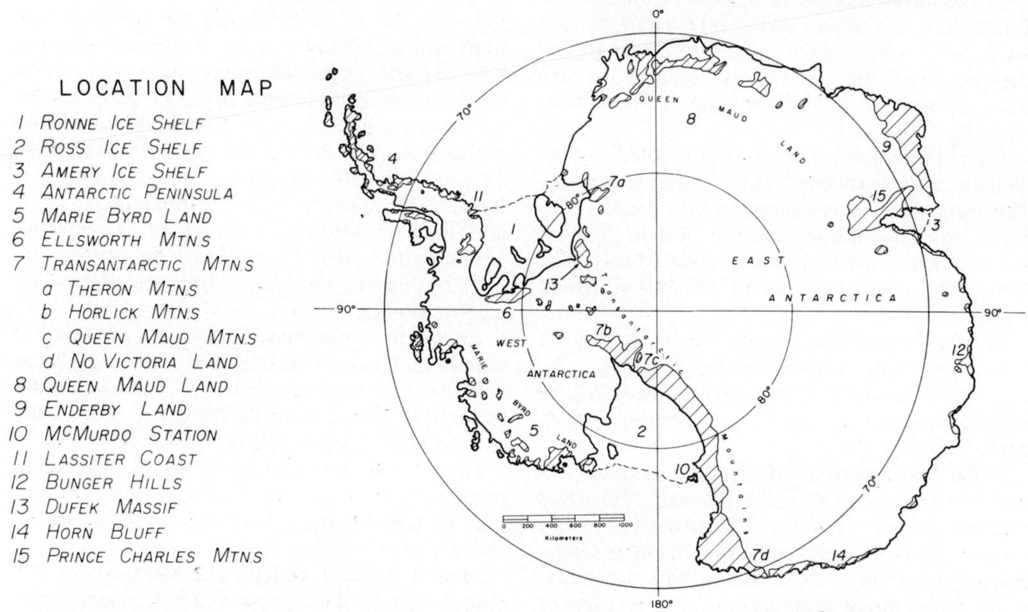

FIG. 2–Locations of principal geographic and physiographic features mentioned in text. Principal outcrop areas are cross-hatched.

rock outcrops are located mostly in the coastal areas and on mountains that penetrate the ice sheet in a coastal zone 200–300 km wide. The outstanding exceptions to this distribution pattern are those rocks which crop out along the Transantarctic Mountains, which extend across the continent for a distance of 3,000 km (Fig. 2). This fault-block mountain system acts as a dam to the spreading East Antarctic ice sheet. Rock exposures on these mountains are exceptionally good along the Ross Sea and Ross Ice Shelf side. Nearly continuous profiles of the rock exposures 2,000–3,000 m thick can be studied by ascending the outlet glaciers which transect the mountains.

Geology

Only condensed and general descriptions of Antarctic geology are presented here. For more complete descriptions the reader is referred to Craddock et al (1970) and the collection of papers in *Antarctic Geology and Geophysics* (Adie, 1972). Geographic features mentioned in the following paragraphs are located on Figure 2.

East Antarctica

An Archean basement complex has been observed only in East Antarctica and is best exposed in Enderby Land. From their composition it appears that the rocks of the complex are derived from andesitic to basaltic lavas, sandstones, shales, and carbonate rocks through regional metamorphism, migmatization, and granitization. Enderbites and charnockites are characteristic units in the complex. On the basis of a single age determination, the complex is believed to have been formed about 3,000 m.y. ago (Grikurov et al, 1972).

Early Proterozoic time is represented by regionally metamorphosed geosynclinal sequences. The metamorphic environment was amphibolite facies. Intensive deformation in mobile areas of the Archean shield produced linear fold belts. Good exposures are located in Queen Maud Land.

Middle Proterozoic time is represented by a metasedimentary sequence of the greenschist facies, which unconformably overlies lower Proterozoic formations in the Prince Charles Mountains.

In the Transantarctic Mountains the oldest exposed rocks are late Precambrian–early Paleozoic metamorphosed geosynclinal sedimentary and volcanic rocks. During the Ross orogeny in Ordovician time, intense deformation with accompanying intrusions of granodiorites and granites of batholithic proportions produced the ancestral Transantarctic Mountains. These were eroded to a peneplain during the Silurian Period. Superimposed upon this surface is the Beacon Supergroup—a Devonian-Jurassic sequence of littoral and terrigenous sedimentary rocks more than 4,000 m thick. Of particular interest are the Permian glacial beds and the Permo-Triassic coal formations. Similar formations are present in the Prince Charles Mountains, Horn Bluff, and the Ellsworth Mountains. During Jurassic time, great volumes of diabase were intruded as dikes, sills, and sheets into this sedimentary sequence. One related differentiate is of particular interest—the Dufek Massif, a gabbro-norite intrusion similar in many respects to the Bushveld, Stillwater, and Skaergard complexes (Ford and Boyd, 1968). The present Transantarctic Mountains are the result of extensive block faulting which probably occurred in early to middle Cenozoic time.

West Antarctica

No true basement rocks are exposed in West Antarctica. The oldest rocks constitute a sequence of late Precambrian–early Paleozoic metamorphosed geosynclinal sedimentary and volcanic rocks that accumulated in a marginal mobile belt. The western portion of this belt was subjected to an orogeny in middle Paleozoic time, and the eastern portion in late Paleozoic time. Deformation was intense and the intruded plutons, mostly of granodiorite, were voluminous (Wade and Wilbanks, 1972).

In the Antarctic Peninsula the oldest formations are interbedded metamorphosed volcanic and geosynclinal sedimentary rocks of late Precambrian–early Paleozoic age. On these formations lies a thick sequence of Carboniferous slates and graywackes. Both sequences were involved in an early Mesozoic orogenic event. Lying unconformably on these truncated formations are accumulations of Middle Jurassic–Early Cretaceous andesites and tuffs. These in turn are overlain locally by Jurassic terrigenous beds and Early Cretaceous marine strata.

In middle Cretaceous time, great volumes of granite and diorite intruded the older complexes, preceded and accompanied by a minimum of deformation. This Circum-Pacific event affected the entire Pacific mobile belt in West Antarctica.

Late Cretaceous–Cenozoic strata, excluding pyroclastic rocks, are known only in the northern part of the Antarctic Peninsula and nearby islands.

Intense volcanic activity characterized the Pacific margin of West Antarctica in Cenozoic time. Volcanoes are arranged in suborthogonal patterns, undoubtedly fracture controlled, from the South Shetland Islands to Victoria Land. Radiometric dating places the beginning of this activity at least 50 m.y. ago. Today there still is activity at both ends of the belt—Deception Island on the east and Mt. Erebus on the west.

The onset of glaciation occurred in West Antarctica about 40 m.y. ago (Oligocene). Evidence seems to indicate that, although there have been advances and retreats since then, the ice cover was never completely removed during this long interval of time (LeMasurier, 1972).

Mineral Resources

In the knowledge of the writer, there has never been an intensive, or even casual, search for Antarctic mineral deposits which some day may have economic value. Most geologic exploration has been of a reconnaissance nature—to map the major structural features and the distribution of rock types and formations, and to unravel the geologic history. Incidental to these efforts have been reports of mineral occurrences. Over 220 different minerals have been reported (Stewart, 1964).

Fossil Fuels

The discovery of coal in Antarctica dates back to 1908, when coal seams were noted in the Transantarctic Mountains by members of Shackleton's first expedition (David and Priestly, 1914). These discoveries were confirmed by members of Scott's ill-fated expedition, 1910–1913 (Debenham, 1921). More recently, coal seams of Permian age in the Beacon Supergroup have been located along the Transantarctic Mountains from the Theron Mountains to central Victoria Land. Seams range in thickness from 1–2 cm to more than 2 m.

Coal also has been reported in Beacon-equivalent formations in the Ellsworth Mountains (Craddock et al, 1965), Horn Bluff (Madigan, 1915), the Prince Charles Mountains (Crohn, 1959), and the Heimefront Range, Queen Maud Land (Juckes, 1972). Unfortunately, most of this coal now is a natural coke. The outstanding exception is the coal sampled in the Prince Charles Mountains. It is believed that the natural coking resulted when diabase sills and dikes were intruded into the Beacon Supergroup formations during the Jurassic Period (Schopf and Long, 1966). No such intrusions are present in the Prince Charles Mountains (Bennett and Taylor, 1972).

As far as petroleum and natural gas are concerned, the picture is not a bright one. The interior basins of East Antarctica may be underlain by hydrocarbon-bearing formations, but the world would have to be in very desperate need of oil and gas before exploration of the basins now overlain by 2,000–3,000 m of moving ice would be attempted. The continental shelf appears to present the best possibility, particularly around West Antarctica where it is wide. Gas shows were reported from three drillholes in the Ross Sea sector by scientists on Leg 28 of the *Glomar Challenger* Deep Sea Drilling Project (Hayes et al, 1973). Ice conditions in the seas, however, present a serious problem to drilling operations. Another possibility is the bedrock beneath the ice shelves. The feasibility of such an operation will be tested in the next 2 years when a multipurpose hole will be drilled through the Ross Ice Shelf and the sedimentary cover, and into the underlying bedrock.

Metallic Minerals

Occurrences of minerals from which metals can be extracted have been reported in all sectors of Antarctica. It must be remembered that a search for mineral deposits has never been a primary purpose of any exploring party in Antarctica, and that the mineral discoveries have been largely incidental.

Encouraging are the reports of iron-ore deposits in East Antarctica (Ravich et al, 1965). The ores are composed mostly of magnetite. Magnetite-bearing schists and gneisses occur in Queen Maud Land, Enderby Land, and Bunger Oasis. Magnetite is concentrated in layers ranging up to 80 m thick. Iron content ranges from 15 to 20%. Some veins, however, are nearly pure magnetite.

In Enderby Land, in brecciated zones 10-50 m thick in the metamorphic complex, the cementing material is nearly all magnetite; iron content in the brecciated zones is estimated to be between 20 and 30%. In the central mountain areas of Queen Maud Land, iron-ore occurrences are plentiful. Magnetite-bearing veins penetrate the gneiss, forming extensive stockworks which assay 20–30% iron.

Of interest are the occurrences of ferruginous quartzite boulders in the moraines along the coast between long. 78° and 93°E. The source of these boulders is not known, and the source rock is probably ice covered.

Copper apparently is ubiquitous in Antarctica, particularly in Marie Byrd Land, where copper "shows" have been noted at outcrops on nearly all mountains and nunataks. The copper generally is along or near joints, indicating a metasomatic origin. Occurrences also have been reported from many East Antarctic platform localities, the Transantarctic Mountains, and the Antarctic Peninsula. The only reported occurrence which approaches the category of a workable deposit is a recently described porphyry deposit in the mountains of the Lassiter Coast (P. D. Rowley, personal commun.).

Of the known mineralized areas, the one that appears to have the greatest economic potential is the Dufek Massif, Pensacola Mountains (Ford and Boyd, 1968). The massif, including its continuation in the Forrestal Range, is a large stratiform mafic intrusion consisting predominantly of pyroxene gabbro interlayered with pyroxenite and anorthosite and capped with granophyre. It has a minimum area of 8,000 km^2 and an estimated thickness of 7 km. The principal ore mineralization consists of copper, iron, and chromium minerals. In addition, minerals of such metals as gold, silver, the platinum group, cobalt, nickel, zinc, and vanadium occur in a disseminated state. An oxidized zone is present at the surface. Mineralization of the enclosing sedimentary rocks by contact metasomatism is to be expected.

Small deposits of galena and molybdenite have been reported at widely scattered localities, as have occurrences of uranium, thorium, niobium, tantalum, and rare earths in pegmatites.

Nonmetallic Minerals

Nonmetallic minerals which have been reported and which might occur in deposits of economic importance include micas, beryl, graphite, tourmaline, topaz, fluorite, barite, quartz, and zeolites. Most reported deposits of some economic potential occur in Queen Maud Land (Ravich et al, 1965).

Low concentrations of phosphate minerals occur in Devonian quartzose sandstones in the Neptune Range of the Pensacola Mountains (D. L. Schmidt, personal commun.).

Finally, there is fresh water. Over 90% of the world's supply of fresh water could come from ice which mantles the Antarctic continent. Determinations of the economic feasibility of towing freshwater icebergs to Australia and South America indicate that such an operation can be accomplished at a profit.

Geothermal Energy

Geothermal energy is a possibility in certain areas of the Pacific sector of Antarctica. Volcanism has occurred in relatively recent times at Ross Island, Victoria Land, Marie Byrd Land, and the South Shetland Islands. At present the only locality where the use of geothermal energy might be practical is Ross Island. On the fringe of Mt. Erebus, which is still mildly active, are located McMurdo Station and Scott Base. Electricity for these bases is provided by diesel-driven generators. If geothermal energy is available here and can be harnessed to supply electrical energy, it would mean an annual saving of 0.5–1 million gal of diesel fuel, all of which has to be imported.

Prospecting and Exploitation

It would appear that Antarctica has great potential as a storehouse of mineral wealth. Because of the accelerating rate of depletion of mineral wealth in all other continents, the resources of Antarctica will be in demand in the foreseeable future. A program of prospecting should be planned and instituted now so that, when the need arises for one or more specific minerals, exploitation can proceed in an orderly manner with a minimum of preliminary effort.

Prospecting of all other continents has been carried on over long periods of time by individuals, teams, corporations, and governments. Because of the environment and inaccessibility of the southern continent, methods used elsewhere, for the most part, are not applicable.

In the opinion of the writer, the only way that Antarctica can be systematically prospected is through a cooperative effort by various interested nations, particularly by those which already are or have been involved in the exploration of that continent.

Sectors of Antarctica have been claimed to territorial sovereignty throughout the years by various nations, and it may seem that prospecting of those sectors should be conducted by the claiming nations. However, this may not be feasible or practical.

It was demonstrated during the International Geophysical Year, 1957–58, that through cooperative effort very fruitful scientific and exploratory programs could be executed. The success of that overall program prompted the drafting and signing of the Antarctic Treaty by the 12 nations that had conducted research in Antarctica during the I.G.Y.

In essence, the treaty guarantees that the land areas south of 60°S lat. be used only for peaceful purposes, assured by a system of inspection; that Antarctica will constitute a laboratory for cooperative scientific research; and that the legal status will be maintained for the duration of the treaty. Nuclear explosions are prohibited pending general international agreement on the subject.

The treaty is open for accession by any state which is a member of the United Nations or which may be invited to accede to the treaty. Since December 1, 1959, five additional nations have acceded to the treaty. The treaty is of indefinite duration, but after 30 years any of the signatory nations may call a conference for review and amendment.

Whether or not prospecting is considered to be scientific research is open to question. The question should be resolved as soon as practicable by the Scientific Committee on Antarctic Research (SCAR), which operates under the provisions of the treaty. There are no great obstacles to approval of a well-organized program of prospecting which could proceed as the scientific research programs have proceeded. In those programs there has been a free exchange of data and reports by the scientific investigators of the various nations. Multination programs have been many. The practice of exchanging scientists for various programs has been a most successful one. The success of these programs has demonstrated that a peaceful international cooperative program is possible.

The question of exploitation of Antarctic resources is not so easily answered. Political questions and territorial sovereignty rights will have to be considered and resolved. This is not the responsibility of the scientific community but of the members of the diplomatic corps of the several nations. The treaty may be amended with the approval of a majority of the contracting parties. One or more of the treaty nations should be encouraged to initiate consideration of this problem by a committee composed of representatives of the signatory nations. By this procedure, an agreement could be reached by which the exploitation of mineral resources in Antarctica could proceed, when necessary, with little delay.

References Cited

Adie, R. J., ed., 1972, Antarctic geology and geophysics: Internat. Union Geol. Sci. Pub., Ser. B, no. 1, 875 p.

Bennett, A. J. R., and G. H. Taylor, 1972, Coals from the vicinity of the Prince Charles Mountains, *in* R. J. Adie, ed., Antarctic geology and geophysics: Internat. Union Geol. Sci. Pub., Ser. B, no. 1, p. 591–598.

Craddock, Campbell, et al, 1965, Glossopteris discovered in West Antarctica: Science, v. 148, p. 634.

——— et al, 1970, Geologic maps of Antarctica: Antarctic Map Folio Ser. (Am. Geog. Soc.), no. 12, 6 p.

Crohn, P. W., 1959, A contribution to the geology and glaciology of the western part of Australian Antarctic Territory: Australia Bur. Mineral Resources Geology and Geophysics Bull., no. 52, 103 p.

David, T. W. E., and R. E. Priestley, 1914, British Antarctic Expedition 1907–1909, Reports on the Scientific Investigations, Geology: London, William Heinemann, v. 1, p. 247–250.

Debenham, F., 1921, British Antarctic (Terra Nova) Expedition, 1910, Natural History Reports, Geology: London, Harrison and Sons, Ltd., v. 1, p. 103.

Dutoit, A. L., 1937, Our wandering continents: Edinburgh, Oliver and Boyd.

Ford, A. B., and W. W. Boyd, Jr., 1968, The Dufek intrusion, a major stratiform gabbroic body in the Pensacola Mountains, Antarctica: 23d Internat. Geol. Cong., Prague, Czechoslovakia, Rept., Proc., Sec. 2, p. 213–228.

Grikurov, B. E., M. G. Ravich, and D. S. Soloviev, 1972, Tectonics of Antarctica, *in* R. J. Adie, ed., Antarctic geology and geophysics: Internat. Union Geol. Sci. Pub., Ser. B, no. 1, p. 457–468.

Hayes, D. E., et al, 1973, Leg 28 deep-sea drilling in the southern ocean: Geotimes, v. 18, no. 6, p. 19–24.

Juckes, L. M., 1972, Geology of north-eastern Heimefrontfjella, Dronning Maud Land: British Antarctic Survey Sci. Repts., no. 65, p. 1–44.

LeMasurier, W. E., 1972, Volcanic record of Antarctic glacial history: implications with regard to Cenozoic sea levels, *in* C. Embleton, ed., Polar geomorphology: Inst. British Geographers Spec. Pub. 4, p. 59–74.

Madigan, C. T., 1915, p. 336 *in* D. Mawson, The home of the blizzard, v. 1: London, William Heinemann.

Ravich, M. G., L. V. Klimov, and D. S. Soloviev, 1965, Dokembriy Vostochvoy Antarctidy (The Precambrian of East Antarctica): Moscow, Izd. Nedra; English transl., 1968, Jerusalem, Israel Program for Scientific Translations, p. 451–455.

Schopf, J. M., and W. E. Long, 1966, Coal metamorphism and igneous associations in Antarctica, *in* R. F. Gould, ed., Coal science: Am. Chem. Soc. Advances Chemistry Ser., no. 55, p. 156–195.

Stewart, D. H., 1964, Antarctic mineralogy, *in* R. J. Adie, ed., Antarctic geology: Amsterdam, North-Holland Publishing Co., p. 395–401.

Wade, F. A., and J. R. Wilbanks, 1972, Geology of Marie Byrd and Ellsworth Lands, *in* R. J. Adie, ed., Antarctic geology and geophysics: Internat. Union Geol. Sci. Pub., Ser. B, no. 1, p. 207–214.

COAL

Status of Coal Exploration and Mine Development in Australia [1]

E. N. MILLIGAN[2]

Abstract The last decade has seen a spectacular increase in the growth of the Australian coal industry, primarily as a result of the development of new mines in the Bowen basin of central Queensland. Production from the Bowen basin increased from less than 0.5 million MT in 1963 to over 14 million MT in 1973. This growth followed the discovery of deposits of premium high-grade metallurgical coal ideally suited for Japanese coke-oven blends. These coals have recently begun to be used in European steel mills, and 2 million MT were exported to this market in 1973. Also contributing to the industry's rapid growth are the amenability of the deposits to extraction by opencut methods and the proximity of the coalfields to good harbors in relatively unpopulated areas. Exploration by state governments in the established coal basins has resulted in a dramatic upgrading of previous reserve estimates. Recent exploration has been directed to less well-known coal basins and is expected to increase the country's known reserves further. Large deposits of low-grade steam coal already have been discovered in three of these basins.

INTRODUCTION

Although coalfields are widely distributed throughout Australia and include 85 billion MT of brown coal used exclusively for power generation in the State of Victoria, 85% of Australia's black coal is produced from Permian coal deposits of the Sydney and Bowen basins in eastern Australia (Fig. 1). Until 1963, production of black coal increased steadily at the rate of 3% per annum, owing to growing demand for power generation and, to a lesser extent, to increases in iron and steel production. Production in 1963 was about 20 million MT; exports accounted for less than 3 million MT. By 1973, exports had increased to over 40% of the total production of 55 million MT. Half of these exports came from the Bowen basin, where production in 1963 had been less than 0.5 million MT. This rapid increase in exports and the increasingly large share of the exports produced from the Bowen basin are related to several factors—coal quality, markets, mining conditions, and locations of the fields with respect to the coast and to population centers.

COAL QUALITY AND MARKETS

About 20% of the recoverable reserves in the Sydney basin and 60% of those in the Bowen basin are natural metallurgical coals. In both basins these range in rank from high-volatile, poorly coking coals to low-volatile, prime-coking types. The fact that these coals could be substituted for American low- and medium-volatile coals in blends for Japanese coke ovens resulted in Japanese domination of the export market. Prime requirements for this market are low-volatile, low-fluidity coals to complement Japan's high-volatile, high-fluidity domestic coals. Sydney basin coking coals, even the high-volatile types, are mostly in the medium- to low-fluidity range and none are of the high rank of prime American coking coals. Bowen basin coals cover the complete spectrum of coking coals except for high-volatile, high-fluidity types. Medium-volatile, high-fluidity coal; medium-volatile, low-fluidity coal; and low-volatile, low-fluidity coal cover the quality range of United States medium- and low-volatile coals. Owing to the accessibility of the medium-volatile, low-fluidity coals in the Bowen basin, these coals were the first mined in the area and were comparable in market value to the medium- and low-volatile Sydney basin coals. Currently, more acceptable high-fluidity, medium-volatile coals are being exported. In the future, low-volatile, low-fluidity coals are expected to command an even higher price on the Japanese market.

Definition of premium coals on the European market varies from mill to mill, but high fluidity is evidently a desirable component in medium- and high-volatile coals. Bowen basin medium-volatile, high- to moderate-fluidity coal has found acceptance in several European countries, and over 2 million MT were exported to this area in 1972–73. Currently, 26 million MT are contracted for delivery to Italy, France, and Holland, and other sales have been made to Spain and Greece.

Australian coals do have some undesirable characteristics such as a high raw-ash content, moderately high sulfur, and, especially in some

[1]Manuscript received, October 21, 1974.

[2]Senior Project Geologist, Utah Development Company, Canberra, Australia.

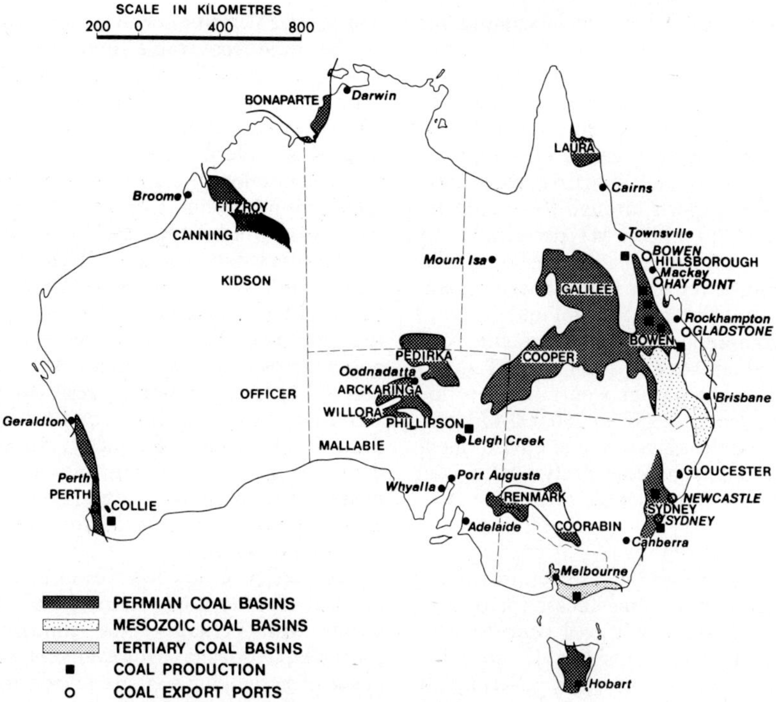

FIG. 1—Location of Australian coal basins and principal mines.

Sydney basin coals, a high aluminum content in the ash. These problems have been overcome to a considerable extent by the installation of beneficiation plants. These plants can increase the production costs significantly, as their installation can cost more than A$10 million. The return for coal mined is also affected, as the finished, specified product from these plants may be only 50–60% of the raw-coal input. Furthermore, owing to the particular banded nature of the coals, the sizing of the beneficiated coal creates wash-plant recovery difficulties and results in further increases in production costs.

Mining and Transport

Premium high-grade coals occur in the south part of the Sydney basin where there is considerable overburden. Thus, the premium-grade coal is extractable only by underground mining methods. In the Bowen basin, coal of all grades can be won by opencut mining methods. In fact, coking coal is sufficiently shallow to permit opencut mining over perhaps 80% of the margins of recognized coalfields. In June 1973, about 90% of coal produced in the Bowen basin was mined by opencut methods. By comparison, less than 20% of the Sydney basin coal mined in 1972–73 was recovered by opencut methods, and this production was predominantly noncoking coal.

In the Bowen basin, overburden is removed by walking drag lines with bucket capacities ranging from 16 to 100 m^3 and capabilities of removing over 10×10^6 m^3 of overburden a year. The coal is loaded by 10-m^3 shovels into dump trucks, some of which have capacities of over 100 MT. Because of this capability for capital-intensive opencut extraction in the Bowen basin, high annual output can be achieved with attendant economies of scale. Of the nine operating opencut mines in the basin, five have a capacity of 4–5 million MT per year.

The Bowen basin also has the advantage of being located in a relatively unpopulated area so that opencut mining can be carried out with a minimum of disturbance to agriculture. Furthermore, it has been possible to establish outlet ports in virgin areas where there is ample room for expansion. For example, Hay Point, a new port built to handle exports from two mines, initially was constructed to accommodate 100,000-MT ore carriers and to handle 10 million MT per year. It is already being expanded to handle 20 million MT per year. Similarly, planned expansion also will enable Gladstone, the second major port shipping Bowen basin coal, to accommodate 100,000-MT ore carriers. By contrast, the Sydney basin coal outlets of Newcastle, Sydney, and Port Kembla are operating at near capacity. As they are sited in areas of high population density, future development is being hindered by competing interests of residential and industrial expansion. For example, coal exports through Newcastle are projected to reach 10 million MT per year in the near future, but problems in siting onshore facilities have continued for over half a decade. Similar problems confront a A$50-million proposal for extra coal-handling and loading facilities at Port Kembla, which has a 1–2-million MT stockpile. Dredging of the harbor would increase the harbor capacity from 55,000- to 250,000-MT carriers. Recently, public agitation over environmental aspects prevented the construction of a coal loader north of the Port Kembla area.

Reserves

Theoretically,the available reserves of coking coal are a potentially limiting factor for continued expansion of Australia's coal exports. This limitation has received considerable attention over the past few years, and state governments have been taking steps to ensure the orderly allocation of established reserves and also to prove other coal areas where assessment of reserves is less reliable. The Joint Coal Board of New South Wales (1973) has categorized Sydney basin reserves as *first order* where coals are minable under current economic conditions by use of current mining technology, and *second order* where reserves are established outside these limits, or where data are less precise and are based on geologic premises rather than detailed testing. The Board has assessed reserves of the first category to be over 16 billion MT. Of this figure, mine-recoverable reserves of coking coal are 6 billion MT. Owing to loss in product beneficiation, salable tonnages may be only 60–70% of this figure.

By using a similar classification, an analysis of Queensland Coal Board (1973) figures shows that the first-order reserves of the Bowen basin can be assessed at over 8 billion MT, of which 4 billion MT is coking coal. Mine recoveries will be significantly higher than in the Sydney basin as a result of the dominance of opencut mining, which permits mine recoveries of up to 95%. Also, as there is a large opencut component in the reserves, there is a greater scope for upward reassessment. Improved economics and technology will permit opencut mining to continue beyond the limits currently accepted. However, even taking the current reserve definition with an opencut limit of 60 m, the mine-recoverable reserves of coking coal would be assessed at over 3 billion MT.

The mine-recoverable reserves of both basins will be increased to some extent by the development of underground mining technology facilitating extraction of a greater percentage of thick underground seams. Estimates of first-order reserves also will be upgraded by results from current exploration programs. This is particularly true of the Bowen basin, where geologic information is not as detailed as that in the Sydney basin. In fact, Bowen basin first-order coal reserves are now probably close to 4 billion MT. The delineation of deep second-order reserves is well advanced in the Sydney basin, and the upgrading of these is likely to come mostly from improved mine recovery. The amount of deep coking coal in the Bowen basin is less well known, but considerable reserves are inferred for downdip extensions of known seams.

Discoveries of new reserves of extractable coking coal are likely to be restricted to the east-coast basins with perhaps some contribution from southwest Australia. However, considerable deposits of noncoking coal are known to be present in other basins. Currently, exploration activity has diversified into a search for these fuel coals in anticipation of an increased demand for steam generation and production of synthetic fuels. Of the 21 other known coal basins, 11 are currently being explored both by companies and by state mines departments. As a result, large tonnages have been delineated in at least three basins—Galilee in Queensland, Oaklands in New South Wales, and Phillipson in South Australia—and considerably larger tonnages are expected to be outlined in the future. In the Galilee basin alone, *in situ* fuel-coal reserve estimates based on extrapolation of data from oil bores are of the same order as the estimated total reserves of the Sydney and Bowen basins combined. Of these vast reserves, only a small portion can be considered as extractable with present-day technology, but their magnitude is such that they form a virtually limitless source of fuel if demand should become sufficient to encourage the development of *in situ* synthetic-fuel production. Reserves of coking coal have more closely defined limits, but the developments in the field of artificially formed cokes will eventually extend the range of coals that can be used for metallurgical purposes. Constraints on exploitation of these coals will, of course, still be imposed by the limits of conventional mining techniques.

CONCLUSION

In assessing the future of coal production in Australia, it is evident that ample reserves of coal are available for steam generation and production of synthetic fuels. However, two limiting factors must be considered: (1) the economics of extracting the coal from under increasingly greater depths of overburden and at increasingly greater distances from the coast and from consumers, and (2) strong competition from alternative energy sources including other fossil fuels and coal deposits in other countries.

Until technologic advances provide a cheaper alternative to natural metallurgical coals, Australia's coking coals, because of the increasingly widespread acceptance of their characteristics in coke-oven feed, appear to have an assured and increasing market. At present, about 250 million MT of Bowen basin coal are contracted for export. For the future, in the Bowen basin alone, there are at least seven coalfields with opencut potential at the predevelopment stage. At current assessments, salable reserves of coking coal are in the order of 7 billion MT, and this figure does not make allowance for increased mine recovery with improved economics and technology. Furthermore, it does not take into account additional reserves being proved in current exploration programs and those which will be proved in the future. Even taking this figure of 7 billion MT and a depletion rate of 100 million MT a year, there are sufficient reserves to last 70 years, or well into the time when alternative metallurgical processes will be an economic reality.

REFERENCES CITED

Joint Coal Board, 1973, 26th annual report: p. 149–153.

Queensland Coal Board, 1973, 22nd annual report: p. 40–41.

Coal Resources and Potential in Indonesia—Summary[1]

MANGARA SIMANJUNTAK and SOELISTIYO U. WIJAYA[2]

INTRODUCTION

Coal, which is Indonesia's second most important source of energy (next to petroleum), has not been effectively exploited because of its relative inaccessibility to industry and its varying quality.

The coal-mining industry in this country began in the 19th century when solid fuel was needed for steamships, locomotives, local electric generating plants, and industrial heating. Indonesian coal has also been exported in considerable quantity to neighboring countries. Those demands, however, have declined gradually since 1950, when petroleum became more economical and efficient for operating modern diesel engines.

After World War II, coal mines in Indonesia never recovered their total production peak of 2 million tons in 1940, owing to various unfavorable conditions.

Recently, it has been proposed, as part of the basic fuel policy of the Indonesian government, that coal be utilized as effectively as possible to support the existing coal industries and to assume part of the role of fueling existing and future domestic industries.

HISTORY OF COAL EXPLORATION

The objectives of exploration in the pre–World War II period were to provide data for the development of coal mines and for further expansion. During 1950–65, coal exploration, carried out mainly by the Geological Survey of the Republic of Indonesia in cooperation with some donor countries, was directed toward the development of an iron and steel project which was supposed to be built in this country in that period. Since 1969, exploration activities have been conducted by the Department of Mines of the Republic of Indonesia, in cooperation with some private foreign companies, under the framework of the first 5-year national development plan.

Since 1970, exploration in central Sumatra has been conducted under a contract between the government of the Republic of Indonesia and P. T. Rio Tinto Bethlehem Indonesia. In south Sumatra (71,000 km^2), exploration since 1973 has been carried out under a contract between the Coal State Enterprise (P.N. Batubara) and the N.V. Shell Mijnbouw.

GEOLOGY AND DISTRIBUTION

Coal deposits in Indonesia are spread over all the main islands—Sumatra, Kalimantan (Borneo), Java, Sulawesi (Celebes), and Irian Jaya (West Irian). In these areas, the available data are based on deposits that have coal seams more than 2 ft (0.6 m) thick, including about 80 locations (Fig. 1).

Coal deposits in Indonesia are of Tertiary age, ranging from lignite to bituminous coal with high volatile content and of noncaking type. The Tertiary coals can be divided into Paleogene and Neogene coals. The Paleogene coals are lustrous black subbituminous and bituminous coals.

The important coal deposits are located in Ombilin (west Sumatra), Bukit Asam (south Sumatra), the Mahakam area (east Kalimantan), and Pulau Laut (south Kalimantan).

From the data available, estimation of the total bituminous and subbituminous coal resources in some principal areas is about 580 million tons (Table 1).

At the beginning of 1974, the estimated remaining coal reserves of Indonesia were 525 million tons, consisting of 80 million tons as determined by mapping and exploration and 445 million tons in relatively unexplored areas. The original estimated reserves were 580 million tons of coal, and the total quantity of coal mined to 1974 was 55 million tons (40 million tons before World War II and 15 million tons since).

The potential reserves of lignite and brown coal probably are about 1 billion tons.

Additional prospects—Evaluation studies presently are being carried out in areas where coal beds have been penetrated in oil-exploration boreholes. For example, coal layers were penetrated in the Muara Enim and Talang Akar Formations in Lematang field (south Sumatra). In the Suban-Jarigi area (about 20 km^2) in south Sumatra, coal reserves have been calculated to be probably 600 million tons to 650 m depth. This

[1]Manuscript received, November 11, 1974.

[2]Department of Mines of Djakarta, Indonesia.

Compilation of this paper was made possible through cooperation between the State Coal Enterprise and the Directorate of Mines, Department of Mines of Indonesia.

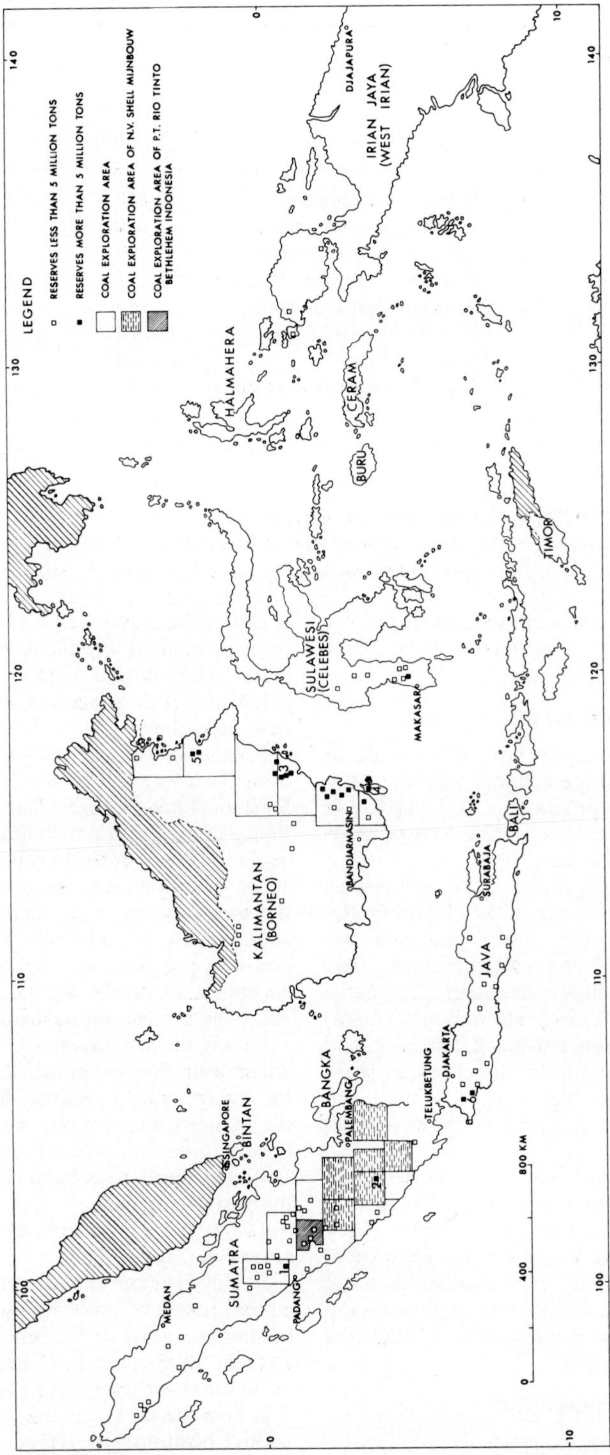

FIG. 1—Coal distribution in Indonesia.

Table 1. The Main Potential Reserves of Indonesian Coal

Number of location on Figure 1	*Location*	*Probable reserve (X 10^6 tons)*	*Proved reserve (X 10^6 tons)*
1	Ombilin–west Sumatra[1]	200	29
2	Bukit Asam–south Sumatra[2]	150	29
3	Mahakam–east Kalimantan	50	2
4	Pulau Laut–southeast Kalimantan[3]	21	11
5	Parapatan–east Kalimantan	40	?
6	South Banten–west Java	15	1.5
..	Other coal fields	24	6.5
P.T. R.B.I.	Siluar–Jambi (central Sumatra)	80	--
	Total	580	79.0

[1] According to Cekop-Poland, the reserves of coal are about 453 million tons.
[2] According to Paul Weir, the reserves of coal are about 360 million tons.
[3] According to S. Sigit (1974), the reserves of coal are about 11 million tons.

estimate was based on well records from 201 boreholes. Another locality under study is the Tanjung field in south Kalimantan.

Characteristics of Indonesian Coals

Many tests and analyses have been made of Indonesian coals, but the methods employed differ considerably. A program is still needed to classify Indonesian coals according to their characteristics and possible uses.

According to the available data, Indonesian coal is considered to belong to the clarain or clarovitrain lithotype. The useful varieties are ranked as bituminous and subbituminous. They generally possess a high percentage of volatile matter (about 30–40%, air-dried analysis). The calorific values range from 5,000 to 8,000 cal/g, and the coals may be suitable for direct use as fuels. The coals have rather high oxygen content and therefore are believed to have undergone low-degree metamorphism.

Almost all of the coals have low ash (less than 10%) and sulfur contents, except the Bayah coal (Java) and the Sigihan coal (Kalimantan). They range from noncoking to low coking. Because of their low coking capacity, it is impossible during the process of independent coking to obtain lump coke of the required strength without adding coal of good coking properties.

Exploitation and Production

The two existing coal mines are Ombilin, mainly operated by the longwall method with hydraulic sand filling, and Bukit Asam, operated by the open-pit method. Productivity of these coal mines is low. The tonnage output per man shift (OMS) is less than 0.25, compared to about 3.5 for Japanese operations.

Several surveys of the coal mines have been done. Two foreign companies—Marubeni/Kaiser Steel in 1969–71 and Charter Consolidated in 1972—investigated possibilities of developing and exploiting these mines to export coal, especially to Japan and Southeast Asian countries. However, the two projects were finally discontinued because, in the first instance, production of 3 million tons per year from Ombilin coal mine is infeasible, and, in the second, government policy made the project impracticable.

During the last decade, the amount of coal produced from the coal mines of Bukit Asam, Ombilin, and Mahakam declined. In 1970, operation of the Mahakam coal mines was discontinued. The highest point of production, 2 million tons in 1940, has never been achieved by these mines in the postwar period.

The decrease in coal production is due to several factors (Sigit, 1966):

1. Sluggishness in the work of rehabilitation, either because of excessive production costs in relation to earnings or because of the limited subsidy from the government. These factors may cause exploration for new coal to be discontinued.

2. Limitation of coal utilization and weak competitive position with respect to the cheap petroleum in the country.

3. Unestablished government policy in energy and fuel economy. A sound policy, it is believed, will help to produce more coal in the country and make it more competitive.

Coal usage in Indonesia has decreased annually, with minor upswings, since 1963. At present, Indonesian coal is used only for small-scale generation of electric power and for railways in south and west Sumatra. The important consumers are the state railway enterprise, Banka tin mines, Padang cement factory (west Sumatra), and various other minor industries.

Because of the low annual production, the percentages of production used at the mines are becoming larger and larger—for example, the Ombilin mine consumed about 25% and the Bukit Asam mine, 36%, of the 1973 production.

Production cost per salable ton of Ombilin coal was about US$10.63 per metric ton in 1970, and the average selling price was about US$7.37 per metric ton. The production cost of Bukit Asam coal was about US$14.93 per metric ton in 1970, and the selling price was about US$8.83 per metric ton.

The present plan of the government is to rehabilitate the Ombilin and Bukit Asam coal mines to produce annually 500,000 and 1,000,000 tons of coal, respectively.

FUTURE PROSPECTS

The domestic demand for coal can be expected to increase significantly, although the possibility of coal export is still uncertain.

Domestic consumption of coal is by the following (Braithwaite et al, 1972): the railway company in southern and western Sumatra; a steam-powered electric generator at the mine site, as well as regional demand; and the metallurgical industry, in the production of ferronickel, aluminum or alumina, tin, iron and steel, carbide, etc.

The latest figures on the development of industries in the area neighboring the Ombilin coal mine indicate that a good market for coal, possibly about 500,000 tons per year, may be expected in that area.

These figures include expansion of the cement industry of Padang from 330,000 tons of cement in 1974 to 1.33 million tons in 1978; establishment in Ombilin of a calcium carbide plant of 13,500 tons annual capacity, as well as a steam-powered electric plant (50 Mw) and a lime industry. Also included is consumption by railways (Mangunwidjaja et al, 1973).

It is reported that Ombilin coal is of moderate coking grade and can be used in foundries. The demand for coke in the country, estimated by the Metal Industries Development Centre, Department of Industry, is expected to be about 10,000 tons in 1980 (to produce about 40,000 tons of cast iron).

The latest development figures for industries in the area of Bukit Asam coal mine indicate that a good coal market of about 575,000 tons per year may be provided. These figures include plans to establish an electric generating plant of 2×50 Mw capacity and to establish a cement industry of 500,000 tons capacity. Also included in this estimate is use by the railway, the tin smelter of Mentok-Bangka, and the power plant of the Tin State Enterprise (P.N. Timah; Braithwaite et al, 1973; Indonesia Dept. Mines, 1973).

Because of the world energy crisis, the prospect of coal rehabilitation and development in Indonesia has become bright. The increase in price per liter of "bunker C oil" in Indonesia from 7.50 to 12.00 rupiah (ca. US$0.018–$0.029) gives a stimulus to coal-mine development. Consequently, although other factors such as additional cost in coal handling must be considered, comparing its calorific value, the price of coal per ton becomes 6,800 rupiah (ca. US$16.38) franco at the Kertapati harbor for Bukit Asam and the Teluk Bayur harbor for Ombilin.

This price increase will have a healthy influence on coal production; a break-even production of 400,000 tons per year for each mine is possible. The present annual production is approximately 70,000 tons for each mine.

N.V. Shell Mijnbouw, in cooperation with P.N. Batubara (State Coal Enterprise), has been planning a working program for production and export of about 25 million tons of coal annually.

Also, P.T. Rio Tinto Bethlehem Indonesia has discovered a coal reserve of about 80 million tons in the Siluar field, Jambi (central Sumatra), and is continuing more detailed exploration work and preparing feasibility studies for further development.

CONCLUSION

At present the problem of coal in Indonesia is still being reevaluated by the government with the aim of placing it in the proper order of priority of utilization in regard to economic activities. The goal is to utilize coal instead of oil in some industries and thus delay depletion of oil resources. Also significant is the tendency of oil to increase in price in the world market.

The success of the program in developing the utilization of coal will depend entirely on the decisions and restrictions made by the government. These must be well coordinated and well arranged in their application, beginning with the stage of rehabilitation and including all transport facilities for the coal produced.

Relative to the location of the Bukit Asam and Ombilin coal mines, the first step in the large-scale development program should deal with the local or regional development in south and west Sumatra. The rehabilitation and further development program of the Bukit Asam and Ombilin coal mines should be in line with, and supplemented by, the development programs for electricity, industry, and railways.

At the present price of fuel oil/bunker oil, which is 20 rupiah (US$0.05) per kilogram (for consumers), coal still will have a favorable situation for competition in the fuel market. Furthermore, if the price of coal does not exceed 9,600 rupiah (US$24.00) per ton (for consumers), which averages 20% cheaper per unit calorie in comparison with oil, it is quite probable that coal gradually will be able to reach its proper position in the market in the future.

For the time being, the development of coal reserves on a large scale for export is limited by the government, except for the coal reserves of central Sumatra (developed by P.T. Rio Tinto Bethlehem Indonesia) and south Sumatra (developed by N.V. Shell Mijnbouw). The possibility of discovering other potential coal reserves for export is still promising.

A systematic overall inventory and exploration work on coal resources in Indonesia are still in progress on a widespread national scale, in the framework of the national energy-utilization policy of the government. One objective is to assist small private coal mines, especially in west and central Java, to supply lime industries and local consumers with coal as a substitute for firewood.

References Cited

Braithwaite, T. A. J., D. A. M. Smith, and A. G. Murray, 1972, Coal in Indonesia: Charter Consolidated Coal Division.

Indonesia Department of Mines, 1973, Naskah Repelita II 1974/75–1978/79 Sektor Pertambangan (National Five Year Development Plan II, 1974/75–1978/79).

Mangunwidjaja, A. S., Wurjandi Hardjono, and U. W. Soelistijo, 1973, Evaluasi Tambang Batubara Ombilin (Evaluation of Ombilin coal mine): Indonesia Coal Survey Project; Department of Mines.

Sigit, S., 1966, Batubara Indonesia, Beberapa Catatan Mengenai Keadaan, Potensi dan Masalah 2 Pengusahaannya (Coal of Indonesia, Some notes on its occurrence, potential and utilization): Department of Mines, Jakarta.

——— 1974, Kemungkinan Peranan Batubara dalam Pola Pengembangan Energi Nasional (The possibility of the role of coal in the National Energy Development Policy): National Energy Conference, July.

Coal Resources of Taiwan—Summary[1]

PATZEN WU[2]

Distribution and Subdivision of Coalfields

The coals in Taiwan are mainly of Tertiary age and all the commercial deposits occur in Miocene formations. Coal lentils or indications of coal are found also in the younger and older strata, but none of them has any economic significance. The Miocene coalfields in Taiwan are grouped into a northern province and a central province. Nearly all the presently producing fields are concentrated in the northern province. Coalfields in the central province are smaller and of little economic value. In the northern coal province, the coal-bearing area stretches from the northern coast of Chinshan and Chinkuashih southwestward to the stream Taanchi, a length of 120 km; the average width of the area is 20 km. The coalfields in the northern province cover an area of approximately 2,000 km^2. In the order of importance, the principal coalfields are located in T'aipei-hsien, Hsinchu-hsien, T'aoyüan-hsien, and Mioli-hsien, in addition to the two municipalities of Chilung and T'aipei (Fig. 1).

A coalfield is defined as a group of coal belts all related to one or more geologic features that occur in a related geologic environment. The geologic features may be either structural or stratigraphic. On this basis, the coal deposits in Taiwan can be grouped into 25 leading coalfields, all located in the northern coal province.

Stratigraphy of Miocene Coalfields

The Miocene rocks in northern Taiwan include a varied sequence of shales, claystones, siltstones, and sandstones with minor amounts of limestone and tuff or tuffaceous rocks. They are composed of marine to shelf-facies sediments. The sequence has a total thickness of more than 4,000 m. At least 15 separate coal beds of economic significance are recognized in the Miocene strata, and they are distributed in three coal-bearing formations. These formations are, in ascending order: the Mushan Formation (Lower Coal Measures), the Shihti Formation (Middle Coal Measures), and the Nanchuang Formation (Upper Coal Measures). There are 3 coal beds in the Mushan Formation, 6 in the Shihti, and 4 in the Nanchuang. In north-central Taiwan, the lower part of the Nanchuang Formation contains two additional coal beds that are workable locally.

The Miocene coal-bearing rocks are characterized by an alternating sequence of white feldspathic to orthoquartzitic sandstone, dark gray to carbonaceous shale, gray sandstone, and interlaminations of shale and sandstone; they represent mainly shelf-type sediments. They alternate rhythmically with marine strata in the Tertiary geosyncline of Taiwan, and each coal-bearing formation in the Miocene section is overlain by a sequence of marine strata. The Tertiary geosynclinal basin in western Taiwan deepens from north to south. The coal-bearing rocks in the Miocene of northern Taiwan are gradually replaced by marine clastic sediments southward, beginning in different places for different coal-bearing formations. The younger the formation, the more extensive is the coal distribution toward the south. The Upper Coal Measures (youngest) are most extensively distributed toward the south; therefore, only the Nanchuang Formation is exposed in the central coal province, toward southern Taiwan. The coal beds are generally thin and lenticular, and do not have much economic value.

Structure of Miocene Coalfields

The Miocene rocks in the coalfields of Taiwan were affected by a combination of folding and thrust faulting related to the early Pleistocene orogeny. Tight, asymmetric folds and low-angle thrusts predominate in the eastern part of the coal-forming basin. The intensity of deformation decreases toward the western part, where the structural style is characterized by much gentler and more open folds. Faults also decrease in number and magnitude in a westerly direction.

Most of the important folds in the coalfields are asymmetric. Generally, the anticlines have steep or overturned northwestern flanks and the synclines have steep or overturned southeastern flanks. The folds are usually bounded by major thrust faults that are traceable for considerable distances. A series of southeast-dipping, imbricated thrust blocks is thus formed, in which each succeeding fault overrides the one northwest of it. Many of these faults are great low-angle thrusts

[1]Manuscript received, December 26, 1974.

[2]President, Chinese Institute of Mining and Metallurgical Engineers, T'aipei, Taiwan, Republic of China.

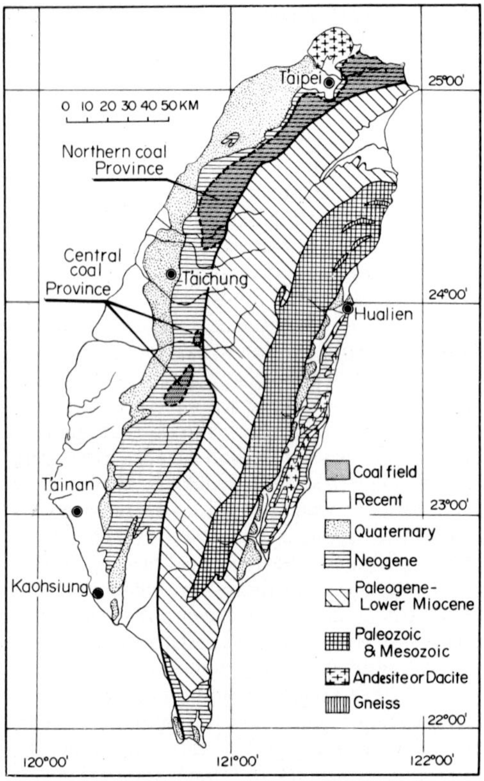

FIG. 1–Generalized geologic map of Taiwan.

and some are bedding-plane thrust faults over certain distances. Many thrusts cut upward across the bedding so that they appear on the surface to be high-angle faults. A succession of seven major overlapping thrust sheets has been mapped in the northern coalfields. All these features prove that the rocks have been deformed and pushed northwestward by a compressive force mainly from the southeast.

Because of their complicated structure, the Taiwan coalfields are not amenable to opencut mining and mechanization is infeasible in many places. Limited by such geologic conditions, most coal mining is operated on a small scale. The production from the large mines is only 150,000 to 200,000 MT per year.

Coal-Bearing Rocks Other Than Miocene

In addition to the Miocene coal-bearing formations, there are also small coal deposits found in older and younger rocks. In central Taiwan, disseminated coal lenses or discontinuous layers have been found in the Paileng Formation of possible Paleogene age. The coal is impure—mixed with large amounts of carbonaceous shale. The rocks are slightly metamorphosed and the pure coal is high in fixed-carbon content. Some may grade into subgraphite rank. However, the coal has no value for mining. Another coal-bearing formation, the Aoti Formation (formerly regarded as Oligocene, but now believed to be mostly Miocene), is present in northeastern Taiwan; discontinuous coal beds have been found in the upper part of this formation. They are generally thin and too varied in thickness to be of any mining value. Peat layers and thick beds have been found in the lacustrine deposits in several intermontane basins of central Taiwan. The most important ones are in the Puli and Toushe basins of Nant'-ou-hsien.

Rank Classification of Coals

Although the general opinion is that time is less important than other causes for advance in rank, the rank of coal in Taiwan advances with its geologic age. According to the ASTM (American Society for Testing and Materials) rank classification, the coals of the Nanchuang Formation (Upper Coal Measures) are mainly of the subbituminous *A* or *B* rank; those of the Shihti Formation (Middle Coal Measures) are largely of the high-volatile *B* bituminous rank; and those of the Mushan Formation (Lower Coal Measures) are chiefly of the high-volatile *A* bituminous rank (Table 1). Of all the coal beds in the Nanchuang Formation, those in the lower member are richer in fixed carbon and consequently higher in rank than those in the upper member. High temperature at depth, pressure of overburden, tectonism, and other depositional factors may be the important causes of coal metamorphism and increase in coal rank with geologic age in Taiwan.

Most of the Taiwan coals fall into two main groups in general rank classification: low-rank bituminous and subbituminous. True lignite was seldom found in the coal samples investigated. Peat is known to be present in some small areas in central Taiwan and on the Penghu Islands. Coals of the semianthracite group are found only in a few mines where igneous baking brought about by dacite intrusions converted the coals into high-rank varieties. Such anthracitic coals have been worked in a few restricted areas in the Chilung volcano group in northernmost Taiwan. The reserve is small and is almost depleted.

The coals in the Mushan Formation (Lower Coal Measures) are generally coking, and this formation provides most of the coking coals of Taiwan. However, the best coking coal in Taiwan is produced from some of the fields in the Shihti Formation (Middle Coal Measures) located in the southeastern part of the northern coal province,

Table 1. Typical Analysis of Coal in Different Formations

Coal-bearing Formation	*Coal Mine*	*H_2O (%)*	*V.M. (%)*	*F.C. (%)*	*Ash (%)*	*S (%)*	*Calories*	*Remarks*
Upper (Nanchuang)	Tienmei	4.81	40.25	39.29	11.65	1.20	6,400	Fuel coal
Middle (Shihti)	Juishan	4.45	42.07	45.45	8.03	1.63	6,927	Fuel coal
	Shihti	1.40	32.05	58.70	6.70	1.15	7,200	Fuel coal
	Hsinchu	1.20	19.24	67.72	8.00	0.7–0.85	8,000	Coking coal
	Nanchuang	3.05	23.25	67.80	5.90	0.75	8,079	Coking coal
Lower (Mushan)	Shuangchi	1.47	37.88	53.02	7.63	0.70	7,730	Coking coal
	Mushan	1.03	34.04	38.48	26.45	2.30	6,000	Coking coal

near the Central Range (Chungyang Shanmo) of Taiwan. The coal belongs to the medium- to low-volatile bituminous rank and has excellent coking characteristics. It is low in volatile matter—average 20–25%—and also is low in sulfur—generally below 0.8%. This is one of the best coking coals produced in East Asia, and no parallel has been found in other areas outside of continental China.

Geology of Coal Beds

Geologic Occurrence

The Miocene coals in Taiwan are generally thin, broken, and impure. The coal beds range from 0.3 to 0.6 m in average thickness, and few exceed 1 m. Only the main coal bed in the Middle Coal Measures attains a thickness of about 1 m in some northern coalfields. As a general rule, most of the coal beds vary greatly in thickness and lateral extent from one area to another, or even within the same field. Many of the coal beds are markedly lenticular or irregular and are limited in areal distribution. Some major coal beds in Taiwan are subdivided by shale or sandy shale partings into two benches. The intervening parting is composed of carbonaceous to coaly shale and is lenticular and variable in thickness. The parting of a coal bed is a characteristic feature commonly used to identify an individual coal bed over wide areas.

Nearly all the coal beds in Taiwan were formed in a paralic basin which originated near the sea. The typical sedimentary units are of sublittoral or shelf type, and marine layers may be interstratified with terrestrial clastic rocks in places. Marine beds increase in number and importance toward the south and southwest until they become the predominant lithotype and the formation loses its coal-bearing characteristics and grades completely into marine strata.

Characteristically for a coastal coal basin, most of the coal beds in Taiwan are impure. They commonly are closely mixed with clayey constituents so that the coal grade is comparatively low. The solid coal generally grades laterally into carbonaceous shale or a barren interval along the same bed. The roofs and floors of the Taiwan coal beds are commonly composed of carbonaceous to dark gray shale. Gray or white sandstones are also recognized as normal roof or floor of certain coal beds but are relatively less common. Underclay has not been found beneath the Taiwan coal beds.

Tectonic Disturbance

The island of Taiwan is located in the Tertiary orogenic belt in the Circum-Pacific region. It is complicated by numerous large-scale thrust faults and intensely folded structures. The coal beds have been compressed into irregular forms. Local pinching and swelling are common, and the beds are discontinuous. Strike and dip of many coal beds change locally, and steeply inclined coal beds are common. Dislocations have caused most coal belts to be dissected into numerous faulted blocks. Coal is generally pulverized after continued stress. The common ratio of lump coal to dust coal in most mines ranges from 2:8 to 1:9. Rarely, the lump coal exceeds one third of the total output in a coal mine.

Tectonic movements may also have favorable effects on coal resources. Coal beds originally far apart may be brought close together by faulting and folding; thus they can be mined with the same mining system and transportation facilities. Lateral pressure and temperature caused by folding and faulting may also advance the coal rank. This is best shown by the coking coal of the Shihti Formation found in the southeastern coalfields, where the intensity of tectonism is great owing to the proximity to the Central Range orogenic zone. The coal is high in fixed carbon and heating value and low in volatiles, and has very good coking properties.

Igneous Activity

Alteration of coal by igneous rocks has been observed in some parts of the Taiwan coalfields. The effect may be either beneficial or destructive. The semianthracitic coal found in the coalfields in northernmost Taiwan is definitely related to the emplacement of the dacite masses of the Chilung volcano group. More commonly, the effect of igneous activity has been disastrous, as the coal has generally been reduced to a cindered or worthless mass. Still more commonly, the coal has been disturbed, squeezed, and shattered by the intrusion of igneous bodies. Nearly all beds of the three Miocene coal-bearing formations in Taiwan have been contaminated by basaltic tuff in different places; the tuff was introduced and deposited with the coal-bearing strata.

Origin of Miocene Coal

The Miocene coal beds in Taiwan were formed in a paralic coal basin on the tidal flat or the shallow shelf of the Tertiary geosynclinal sea. The accompanying sediments are mostly of transitional or littoral type. Vegetal growth in the shelf area was not lush, and the coal beds are generally thin and irregular. During formation of the coal beds, there was persistent influx of abundant clayey or silty detritus into the peat swamp. The result was that the coal beds are generally impure and have admixed clayey components, and commonly are separated by shale partings.

The Miocene coal swamp was distributed widely over much of northern Taiwan. However, the seawater was gradually deepening toward the south and east away from the coast, inhibiting plant growth. The main coal beds thus thin southwestward as well as southeastward. No swamp condition existed in central Taiwan or farther south, and only marine sediments were deposited during the Miocene in southern Taiwan. Coal deposits are thus found mostly in the northern part of the geosynclinal basin.

Sedimentology

The clastic sediments of the Miocene coal-bearing formations in northern Taiwan were derived from an old landmass west or northwest of the island of Taiwan; this landmass served as a source area for the Tertiary geosyncline of Taiwan. Sedimentologic study suggests that the source area was composed predominantly of acidic eruptive rocks but included low-rank metamorphic or feldspar-rich sedimentary rocks. Fluvial or tidal water may have been the main transporting agent which carried the sediments to the depositional basin.

The Miocene coal-bearing formations are characterized by scarcity of marine organisms, presence of plant remains, absence of limestones, and occurrence of discontinuous thin coal beds. All these features suggest a generally transitional environment such as that of the unstable marginal shelf of a sedimentary basin subjected to constant geosynclinal deformation. The depositional conditions could be typical of a widespread coastal plain or tidal flat with a network of delta lakes, marshes, lagoons, and shifting stream channels. Coal beds were formed wherever vegetation could grow over a deltaic or marshy surface. Coal formation took place at intervals in the swampy area, which was characterized by alternating uplift or silting of the swamp, plant growth, subsidence, and sedimentation.

Coal Petrography

Petrographically, the composition of Taiwan coals is marked by the predominance of the vitrinite group and the extreme scarcity of the inertinite group. Particularly scarce is the maceral micrinite, which is a strongly decayed plant material in the coals. This may indicate that the geologic environment of the Miocene coal swamps of Taiwan was characterized by highly toxic waters that inhibited rapid decay of plants. Also, it proves that the oxidation and bacterial attack responsible for plant decay were proceeding rather slowly and less actively in the swamps.

The scarcity of fusinite and semifusinite in the Taiwan coals is also remarkable. Fusain is generally believed to be formed as a result of forest fire or special biological and chemical action during the peat and diagenetic stage of coal formation.

Petrographic studies indicate that a large proportion of the inorganic constituents in the Taiwan coals was deposited or taken into solution and redeposited during peat accumulation. Only small amounts of carbonate or clay minerals were deposited by circulating groundwater after burial of the coal beds. The high inorganic content and irregular thickness of the coal beds indicate an unstable condition of peat formation in the paralic coal basin of Taiwan.

Economic Potential of Coal Resources in Taiwan

An up-to-date appraisal of the coal reserves of Taiwan was completed in 1965. The total original coal reserves were approximately 660 million MT. As of the end of 1965, the total remaining coal reserves were 497 million MT. The estimated total recoverable coal reserves were 260 million MT. Coking coal amounts to about 98 million MT, or a little more than one third of the recoverable reserves. The rate of recovery is about 54%. Of the recoverable reserves, 87 million MT is contained in coal beds 0.6–1 m thick; 104 million MT is in coal beds 0.4–0.6 m thick; and 69 million MT is in coal beds 0.25–0.4 m thick.

Of the three Miocene coal-bearing formations in Taiwan, the Shihti Formation leads in potential coal reserves and accounts for about 75% of the total coal reserves in Taiwan. The Mushan Formation contains about 17% of the total reserves, and the Nanchuang Formation contains the remaining 8%.

Because of its highly deformed nature, the coal can be mined only by underground methods. Because of the high topographic relief, some of the mine openings are adits, but most are inclined shafts. No vertical shafts have been attempted because of the greater depth involved and the comparatively small reserve per square kilometer of the mine area, thus making vertical shafts too expensive for such thin deposits. Another feature to be noted is that some of the best coal beds extend from the coast toward the seafloor. Therefore, a considerable amount of coal has been mined from offshore areas through roadways and inclines extended undersea. Two such coal mines are located next to a major thermal electric-power plant near the coast.

Among the various kinds of minerals produced in Taiwan, coal has been the leading mineral commodity in monetary value and accounts for almost 40% of the total value of mineral production. At present, coal is still the most important basic energy resource that is produced in Taiwan. Except for hydroelectric power and some natural gas, all other types of power production have to rely on raw materials imported from abroad. Since World War II, the coal industry in Taiwan has experienced several periods of boom and depression depending on the balance between supply and demand. From right after the war until the middle 1960s, coal provided the badly needed indigenous energy source for the reconstruction and development of industry in Taiwan at a time when the country was short of foreign exchange. A small amount of coal was also exported, but exportation has been stopped for a long time. Since the middle 1960s, the rapidly expanding national industry has required an enormous energy supply which outpaces that which can be provided by the coal industry. The energy policy has thus shifted to imported oil, and the annual coal production has gradually dropped below 5 million MT, owing to depletion of favorable coal beds, increasing cost of deep mining, rising wages, and shortage of miners. The increasing cost of coal mining renders it less competitive with imported oil. Therefore, the share of coal in the energy market has decreased from a former high of 70% to the present 10%.

During the recent energy crisis, the high oil price has stimulated the redevelopment of coal as a reliable indigenous energy resource, and the coal industry has been revitalized—at least the downward trend of coal production has been checked somewhat. On the other hand, encouraging measures have been taken by the government in a program to increase coal-mine productivity and reduce operation cost so that a firm domestic fuel supply at reasonable cost can be assured. In view of the anticipated world shortage of oil in the 1980s and the limititation in domestic coal production, Taiwan is preparing to import a substantial amount of coal to complement its rapidly increasing energy requirement. This country may thus be transformed from a coal-exporting to a coal-importing region.

Paleozoic and Mesozoic Coal in Republic of Korea[1]

BONG KYUN KIM[2]

Abstract Coal produced in Korea is primarily anthracite, supplemented by small amounts of Tertiary lignite. The anthracite beds are intercalated in Carboniferous, Permian, Jurassic, and Cretaceous rocks; the most productive beds are in the Permian. Nine of the 11 anthracite fields are aligned in the Ogcheon geosynclinal zone, from the middle east coast to the southwest coast of the Korean Peninsula. Anthracite reserves in southern Korea are estimated at 1.45 billion MT, of which 90% is Permian and 10% Jurassic. Annual anthracite production in Korea has been 10–14 million MT for several years. Calorific value of the Paleozoic anthracite is 5,500–6,000 cal (9,900–10,800 Btu), and the value for Mesozoic coal is 5,000–5,500 cal (9,000–9,900 Btu), with some exceptions of 6,000–6,500 cal (10,800–11,700 Btu). Prospecting for anthracite beds is being carried out by detailed geological and geophysical surveys and by core boring or drifting.

INTRODUCTION

In southern and northern Korea, Paleozoic and Mesozoic anthracite and Tertiary lignite beds have been found; no bituminous beds are known. This paper concerns the distribution, stratigraphy, paleontology, reserves, and quality of Paleozoic and Mesozoic anthracite beds in southern Korea.

GEOLOGIC SETTING

Korea consists mostly of Precambrian granitic gneiss intruded by Jurassic (in southern Korea) and Cretaceous (in northern Korea) granites. In southeastern Korea, large Cretaceous terrestrial deposits, the Gyeongsang (or Kyongsang) Group, are present; small patches of Tertiary rock are scattered along the east coast. Present in the north-northeast-trending Ogcheon zone are Cambro-Ordovician marine deposits, marine Carboniferous and largely terrestrial Permian to Triassic strata with coal beds, and terrestrial Jurassic strata with coal beds. Two areas of coal-bearing Jurassic rock are present on the west coast and near Seoul.

COAL-BEARING STRATA

All the coalfields except the Jurassic Chungnam and Gyeonggi (or Kyonggi) fields are situated in the Ogcheon (or Okchon) geosynclinal zone (Fig. 1). The coalfields in the Ogcheon zone, from northeast to southwest in southern Korea, are the Gangneung (Kangnŭng), Jeongseon, Yeongweol, Samcheog, Danyang, Mungyeong, Boeun, Jeonbug, and Honam.

The northeastern part of the Ogcheon zone has not undergone metamorphism, but the middle and southwestern parts of the zone have been intensely metamorphosed. Lee (1973) postulated that metamorphism in the southwestern part resulted from deep burial through subsidence. The coalfields in the metamorphosed area are the Boeun, Jeonbug, and Honam.

STRATIGRAPHY AND PALEONTOLOGY

Carboniferous to Triassic Strata

The Carboniferous through Triassic sequence has been named the Pyeongan System, and is divided into the Hongjeom, Sadong, Gobangsan, and Nogam Series in northern Korea. This subdivision has been used in the anthracite coalfields in southern Korea. Cheong (1969) proposed a new division of the sequence. The correlations are shown in Figure 2.

In the Samcheog and Danyang coalfields, the Manhang Formation (190 m thick) consists of red or green shale, mottled sandstone, and light-colored limestone lenses. The Geumcheon Formation (70 m thick) consists of dark gray shale and sandstone with intercalations of fusulinid-bearing gray limestone. The presence of the fusulinids indicates that the formation is in the *Fusulina* zone and that it is correlatable with the Moscovian (middle Upper Carboniferous). The important Moscovian species described from the Manhang Formation are *Eostaffella subsolana* Sheng, *Verella proxila* (Sheng), *Pseudostaffella gorskyi* (Dutkevitch), *P. kimi* Cheong, *Aljutovella* sp., *Profusulinella pararhomboides* Rauser and Beljaev, and *Beedeina mayiensis* (Sheng). The following species are known from the Geumcheon Formation:

[1]Manuscript received, November 6, 1974.

[2]Department of Geology, Seoul National University, Seoul, Korea.

I am indebted to the Institute of Geology and Mineralogy of Korea, Korean Mining Promotion Corporation, Daehan Coal Corporation, and Dong-A Applied Geological Consultant Company, Ltd. These organizations supplied the materials for this paper. I also thank C. H. Cheong, a Permo-Carboniferous scholar of the Department of Geology of Seoul National University, for assistance in preparing this paper.

FIG. 1–Paleozoic and Mesozoic coalfields in southern Korea.

Ozawainella vozhgalica Safonova, *O. paratingi* Manukalova, *O. magna* Sheng, *Wedekendellina dutkevitchi* Rauser and Beljaev, *Neostaffella papilioformis* Cheong, *N. sphaeroidea cuboides* Rauser, *Fusulinella pseudobocki* Lee and Chen, *Beedeina siviniensis* (Rauser), *Fusulina fortissima* Rauser, *F. cylindrica domodedovi* Rauser, and *Quasifusulina paracompacta* Chang.

The Jangseong Formation, the main coal-bearing unit in southern Korea, has been divided into four cyclothems, each containing one coal bed. The minable or main coal bed is found in the third cyclothem from the bottom. The average thickness of this bed ranges from 1 to 4 m, but the local thickness varies from 0 to 50 m because of folding. The coal bed in the first cyclothem is mostly represented by coaly shale, and no minable seams are found. The beds in the second and fourth cyclothems are less than 1 m thick and are not generally minable; in places there are swells caused by folding, and these beds are mined economically.

The Jangseong Formation contains various Permian fossil plants. The important genera include *Annularia, Calamites, Callipteris, Callipteridium, Cordaites, Desmopteris, Emplectopteris, Lepidodendron, Mariopteris, Neuropteris, Odontopteris, Pecopteris, Pterophyllum, Sphenophyllum, Sphenopteris,* and *Taeniopteris.*

The rocks in the coalfields in the metamorphosed Ogcheon zone are so severely metamorphosed that the fossil records have been destroyed and accurate correlation is not possible.

Geologic Age	Northern Korea		Southern Korea (from Cheong, 1969)	
Triassic	Pyeongan System	Nogam Series	Hwangji Group	Donggo Formation
		Gobangsan Series		Gohang Formation
				Dosagog Formation
Permian			Cheolam Group	Hambaeg Formation
		Sadong Series		Jangseong Formation
Carboniferous			Gomog Group	Geumcheon Formation
		Hongjeom Series		Manhang Formation

FIG. 2–Correlation of Carboniferous-to-Triassic sequence in Korea.

However, some correlation of the sequences may be possible from the stratigraphic positions of the beds.

The Paleozoic and Triassic formations overlying the Jangseong Formation are terrestrial deposits and are insignificant with regard to coal.

Jurassic Strata

The Jurassic Bansong Group, distributed in the Ogcheon zone, contains anthracite beds. The group is divided into the Cheongna, Jogyeri, Baegunsa, and Seongjuri Formations, in ascending order. The Jogyeri Formation consists mainly of conglomerates; the other formations are mainly sandstones and shales. The main coal beds are present in the upper part of the Cheongna Formation and in the lower part of the Baegunsa Formation.

The largest Jurassic coalfield—Chungnam—is located in the middle of the west coast of southern Korea. The lower and upper coal beds in the Cheongna Formation are 0.5 and 1 m thick, respectively. In the lower part of the Baegunsa Formation, three coal beds are found. The upper bed is less than 1 m thick, and the middle and lower beds are less than 0.5 m. These coal beds are locally as thick as 20 m as a result of folding.

Along the periphery of the coalfield, the Bansong Group has been metamorphosed to form granitic gneiss, but most of the strata have not been affected by the granitization. Therefore, the Baegunsa and the other formations contain rather well-preserved Jurassic plant fossils. Some of the important genera are *Podozamites, Neocalamites, Baiera, Cladophlebis, Nilssonia,* and *Ginkgoites.*

Tertiary Strata

Nonmarine Tertiary (Miocene) deposits, the Yangbug Group, are present near Pohang. Two lignite-bearing formations are found in this group.

Table 1. Analysis of Anthracite in Coalfields

Coalfield	*Water (%)*	*Ash (%)*	*Volatile Component (%)*	*Sulfur (%)*	*Fixed Carbon (%)*	*Average Calorific Value (Kcal/Btu)*
Kangnung	0.44	33.05	3.28	0.45	63.24	5,312/ 9,600
Jeongseon	1.55	27.05	3.07	0.35	68.37	5,818/10,500
Samcheog	1.22	23.30	2.96	0.95	72.58	6,188/11,200
Danyang	1.89	24.00	3.27	0.25	72.52	5,968/10,750
Mungyeong	0.87	17.22	4.25	0.35	77.67	6,497/11,680
Boeun	2.18	13.23	5.67	0.26	78.92	6,717/12,100
Honam	2.06	23.86	5.19	0.46	68.89	5,708/10,250
Chungnam	2.34	24.81	2.53	0.36	71.32	5,960/10,700

PROSPECTING

Two organizations, the Daehan Coal Corporation and the Korea Mining Promotion Corporation, are prospecting for coal in Korea. They are conducting detailed geologic surveys, boring, and tunneling to clarify the reserves and qualities of coal. About 150 km² of the coalfields has been mapped in detail in the past 5 years, and from 10,000 m (in 1970) to 18,000 m (in 1973) has been cored. The length of adit of tunneling has been 10,000 m per year since 1970. By these methods, proved reserves have increased by about 10 million tons each year since 1900.

The detailed geologic survey of the coalfields in Korea will be completed by 1976; 50,000 m of core boring will be performed in 1975. Geophysical prospecting for the coal beds under younger deposits, for those hidden by structure, and for offshore deposits is to begin in 1975.

QUALITY OF COAL

Paleozoic and Mesozoic coals are metamorphosed and mostly pulverized to form anthracite or even graphite. The average analytical value of the coal is shown in Table 1. About 40% of the total reserves of anthracite is of low quality, having calorific values of 3,500–4,500 cal (6,300–8,100 Btu). Low-quality coal is being used as generator fuel. In the present energy crisis, intensive research is needed to determine greater use for low-quality coal.

RESERVES AND PRODUCTION

Estimated total reserves in Korea are 1.45 billion MT. Ninety percent of this coal is Permian and the rest is Jurassic. In 1973 the minable reserves were 550 million tons. The coal reserves and potential output of each field are shown in Table 2. Annual coal production in southern Korea in 1965, 1968, and 1969 was about 10 million tons; it was 13 million tons in 1971 and 14 million tons in 1973. Annual production is projected to increase to 20 million tons by 1980.

The problem of increasing coal production may be solved by (1) development of low-grade anthracite not now being mined; (2) exploitation of coal beds structurally hidden by older rocks; (3) development of offshore coal beds; (4) discovery of new beds by more detailed geologic and geophysical surveys; and (5) evaluation of the Tertiary lignite not now being mined.

Table 2. Anthracite Reserves in 1973, in South Korea (Million metric tons)

Coalfields	*Coal Reserves*	*Potential Output*
Kangnung	61	41
Jeongseon	485	73
Samcheog	435	235
Danyang	78	31
Mungyeong	76	42
Boeun	10	7
Honam	43	27
Chungnam	146	83
Others	116	6
Total	1,450	545

References Cited

Cheong, C. H., 1969, Stratigraphy and paleontology of the Samcheog coalfield, Gangweondo, Korea: Geol. Soc. Korea Jour., v. 5, no. 1, p. 13–54.

——— 1971, Stratigraphy and paleontology of the Danyang coalfield, N. Chungcheong-do, Korea: Geol. Soc. Korea Jour., v. 7, no. 2, p. 63–79.

Lee, S. M., 1973, Application of metamorphic facies and facies series to the tectonics of Korea: Geol. Soc. Korea Jour., v. 9, no. 1, p. 11–23.

Influence of Temperature on Coalification of Tertiary Coal in Japan—Summary[1]

TOSHIO SHIMOYAMA[2] and AZUMA IIJIMA[3]

INTRODUCTION

Factors which influence coalification are discussed by many researchers (e.g., Kuyl and Patijn, 1961; Elphick and Suggate, 1964; Teichmüller and Teichmüller, 1966; Kisch, 1974), most of whom agree that temperature is the most significant factor during metamorphism.

In Japan the main productive coal deposits are Tertiary, particularly Paleogene. The ranks of Paleogene coal range from lignite to anthracite. The vegetal substances are quite similar throughout every coalfield; Tanai (1967) and Tanai and Huzioka (1967) discussed the Tertiary floras of Japan and concluded that the Paleogene floristic components are subtropical or warm-temperate types which maintained themselves during Paleogene time. The extensive distribution, the wide range of coal ranks, the uniformity of vegetal substances, and the young geologic age make the Paleogene coals ideal for studying comparative ranks of coalification.

Authigenic zeolites are extensively distributed in the post-Jurassic strata in Japan (Iijima and Utada, 1972). Alkali zeolites and albite were formed progressively by burial diagenesis of felsic vitric tuffs, resulting in regional zeolite zoning. Because these zeolitic reactions are temperature-dependent, zeolite zoning can be used as a geothermometer (Iijima and Utada, 1971).

JAPANESE COAL DEPOSITS

The Japanese coal deposits, which are widely distributed, range from Permian to Tertiary in age (Fig. 1). Productive coal deposits, however, are largely Tertiary, and those in the Paleogene formations are especially significant. The upper Paleozoic and Mesozoic deposits have little economic value except the Triassic Omine coalfield in western Honshu. The coal reserves of Japan are shown in Table 1.

Japanese coal ranges from lignite to anthracite. Various parameters have been used for evaluating coal ranks. The Japanese Industrial Standard (JIS) classification of coal is based on calorific values for low ranks and on the fuel ratio for high ranks. On the basis of the JIS classification, about 90% of the Paleogene coal ranges from bituminous coal (B1) to high-rank lignite (F1).

DESCRIPTION OF COAL RANKS

For purposes of genetic classification of coal ranks, the reflectance of vitrinite is more acceptable than the JIS parameters. It was measured on polished surfaces of vitrinite in oil immersion on the Leitz microscope photometer using zincblend as the calibrating standard. The Krevelens coal band is used where the reflectance data were not available. The coal is subdivided into two groups from the standpoint of use; lignite and subbituminous coal are noncoking, whereas bituminous coal is coking. Tempoku and Chikubetsu yield the Miocene lignite and subbituminous coal, respectively. Mitsui-Bibai, Kushiro, and Jōban yield the Paleogene subbituminous coal. All other localities provide Paleogene bituminous coking coal.

The critical reflectance to separate coking from noncoking coal is about 0.6. The critical atomic ratio of oxygen to carbon to distinguish the coking from the noncoking coal is about 0.120 within a range of the atomic ratio of hydrogen to carbon from 0.650 to 0.900.

The reflectance of the Tertiary coal increases in inverse proportion to the H/C ratio below a H/C ratio of 0.8, and it decreases gradually with the increase of the H/C ratio above 0.8 (Sugimura et al, 1966). Fujii et al (1966) noted that the H/C ratio of the Japanese Tertiary coal is higher than that of Euramerican Paleozoic coal of the same O/C range.

PALEOGENE COAL DEPOSITS

Description of Representative Coalfields

Coal seams are widespread in Paleogene strata in the Japanese Islands except in the Setogawa-Nakamura geosyncline on the Pacific side of southwest Japan. Paleogene coal measures consist of thick sections of marine and freshwater sedi-

[1]Manuscript received, November 6, 1974.

[2]Mitsui Mining Co., Ltd., Muromachi 2-1, Nihonbashi, Tokyo 103, Japan.

[3]Geological Institute, University of Tokyo, Hongo, Tokyo 113, Japan.

T. Tanai of the Hokkaido University gave us valuable advice and read the typescript. R. L. Hay of the University of California at Berkeley read the typescript. Akiko Wakabayashi typed the manuscript.

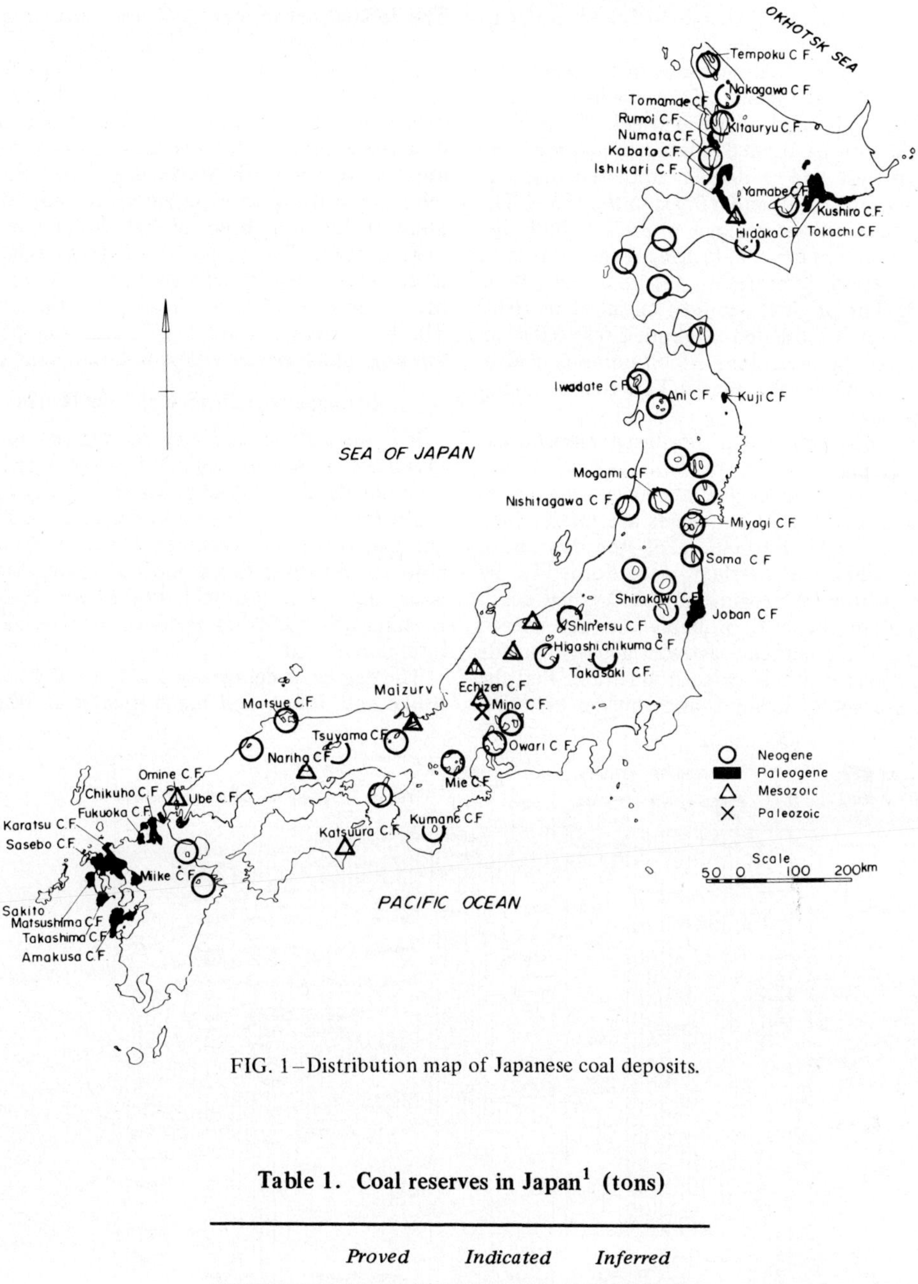

FIG. 1–Distribution map of Japanese coal deposits.

Table 1. Coal reserves in Japan[1] (tons)

	Proved	*Indicated*	*Inferred*
Hokkaido	2,592,899	1,484,719	5,989,606
Honshu	808,550	444,819	934,206
Kyushu	2,458,215	1,007,623	4,525,149
Total	5,859,664	2,937,161	11,448,961

[1] After *Mining Year Book of Japan 1971.*

ments. The geologic correlation of the main coalfields is shown in Figure 2.

The Paleogene coal measures in Hokkaido are represented by the Ishikari Group, which is thickest (4,000 m) in the Ishikari coalfield. The Ishikari coalfield extends about 85 km north-south and 30 km east-west and is divided into two districts, Sorachi in the north and Yūbari in the south. The schematic east-west geologic profiles of both districts reveal that the coal-bearing formations have been intensely deformed by tectonic movements (Fig. 3). The original depth of burial of the Ishikari Group is estimated as about 5,000–6,000 m. Except for the noncoking subbituminous coal in the Mitsui-Bibai area, the Ishikari coal is coking bituminous.

The Jōban coalfield in northeast Honshū extends about 70 km north-south and 15–25 km east-west. The Oligocene strata have a maximum thickness of 700 m. Coal seams are interbedded with the Iwaki Formation of the Shiramizu Group, which is overlain unconformably by about 1,000 m of Neogene strata. Original depth of burial of the coal probably did not exceed 2,000 m. The schematic east-west geologic profile of the Jōban coalfield (Fig. 3) illustrates that the strata are gently inclined and thicken offshore. The Jōban coal ranges from noncoking subbituminous to lignite.

The Paleogene strata in northern Kyushu are distributed in several separate coal basins and have a maximum thickness of 2,000 m. The Miike coalfield extends about 20 km north-south and 15 km east-west beneath Ariake Bay. This field contains the Miike offshore colliery—the largest coal mine in Japan—whose annual production is 6 million tons. The Miike coalfield is relatively undeformed (Fig. 3). The Miike coal was probably buried by overburden less than 2,000 m thick. The coal measures are the Ōmuta and Manda Groups, which produce coking bituminous coal.

Geologic Significance of Coal Ranks

It is difficult to find any definite relation between coal rank and original depth of burial. For example, depth of burial at Miike and Takashima coalfields is estimated as approximately 2,000 m, and that at Ishikari is estimated as 5,000–6,000 m; however, all three fields produce coking bituminous coal. Jōban coalfield, where depth of burial is estimated as 2,000 m, produces noncoking subbituminous coal.

The degree of deformation is about the same at Miike and Jōban and much greater at Ishikari.

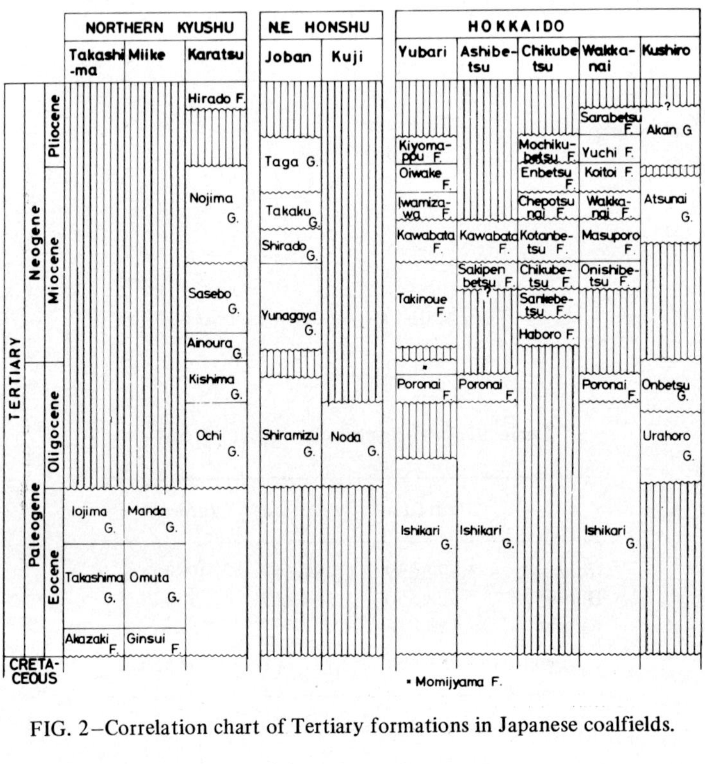

FIG. 2–Correlation chart of Tertiary formations in Japanese coalfields.

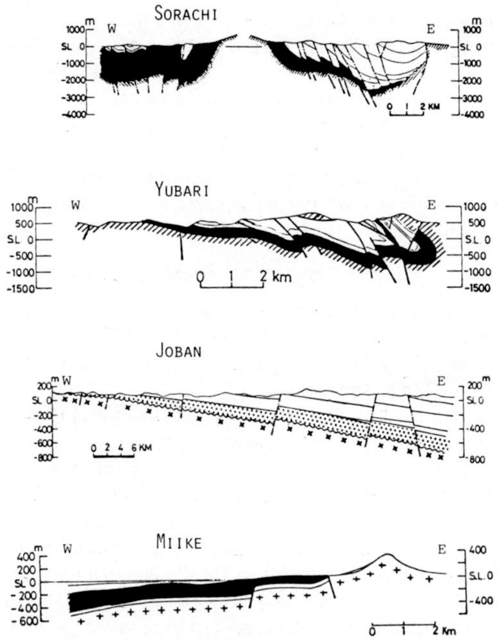

FIG. 3–Geologic profiles of representative Paleogene coalfields of Japan. Host formations of coking coal are marked in black, and those of noncoking lignite and subbituminous coal in dots.

However, Miike and Ishikari both yield coking bituminous coal, whereas Jōban produces noncoking subbituminous coal.

These facts strongly suggest that coalification is not significantly affected either by the depth of burial (confining pressure) or by the degree of tectonic deformation.

Authigenic Zeolites in Tertiary Coalfields

Five species of authigenic zeolites have been found in the Paleogene coal measures and their underlying and overlying strata since clinoptilolite was reported by Iijima (1961) from the Harutori mine in the Kushiro coalfield. They are clinoptilolite, mordenite, analcime, heulandite, and laumontite. These zeolites, except for laumontite, occur exclusively as an alteration product of felsic glass in vitric tuffs which are sparsely interbedded with coal seams, shales, and sandstones as persistent layers usually 10–100 cm thick. No distinction in the mode of occurrence of zeolites was recognized among freshwater, brackish, and marine deposits.

Zonal Distribution of Zeolites

Authigenic zeolites are present in most of the Paleogene coalfields in Japan. They are distributed systematically in a vertical zonal arrangement within a limited local stratigraphic sequence (Fig. 4).

Clinoptilolite and mordenite always appear in the shallowest zone. This clinoptilolite-mordenite zone corresponds to the following deposits: the Miocene-Oligocene deposits in Kushiro (Iijima, 1961), the Miocene of the MITI-Wakkanai drill cores in Tempoku, the Miocene and Upper Cretaceous deposits in the Chikubetsu district of Tomamae (Iijima, 1971), the lower Pliocene and upper Miocene of the MITI-Nanporo cores in the Yūbari district of Ishikari, the Oligocene and Upper Cretaceous in Kado (Iijima, 1972) and Kuji, the Miocene-Oligocene in Jōban, the upper and middle Miocene in the Karatsu coalfield, and the upper part of the Yotsuyama Formation (Eocene) in Miike.

Analcime characterizes the zone underlying the clinoptilolite-mordenite zone. Heulandite and laumontite occur locally in the upper and lower parts, respectively, of the analcime zone. The following deposits belong to this zone: the Utsunai Formation (Oligocene) of the MITI-Wakkanai cores in Tempoku, the middle and lower Ishikari Group (Eocene) and the Upper Cretaceous deposits in the Sorachi district of Ishikari (Iijima,

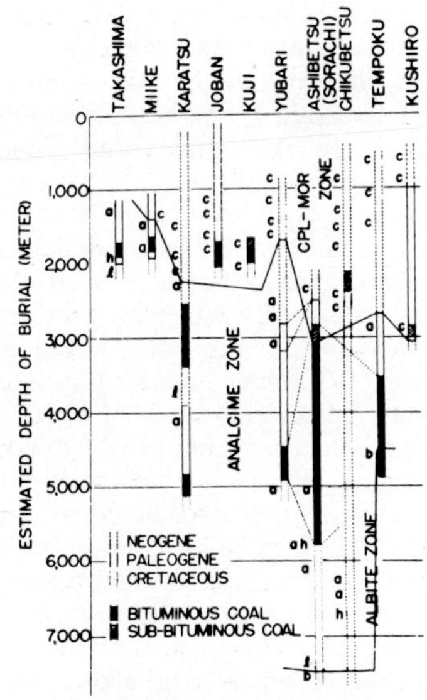

FIG. 4–Relation among coal ranks, zeolite zones, and burial depth in Tertiary coalfields in Japan. Black columns designated as bituminous coal show host formations of coking coal, and hatched ones designated as subbituminous coal are those of noncoking lignite and subbituminous coal. *c* = clinoptilolite and/or mordenite, *a* = analcime, *h* = heulandite, *l* =laumontite, and *b* = albite.

1973), the middle and lower Miocene and Oligocene of the MITI-Nanporo cores and the Hakobuchi Group (Upper Cretaceous) in the Yūbari district of Ishikari, the middle and lower Miocene and Paleogene in Karatsu, the Paleogene (except for the upper part of the Yotsuyama Formation) in Miike, and the Paleogene and Upper Cretaceous in Takashima.

The deeper parts of the Tempoku coalfield and the Sorachi district probably represent the albite zone, because plagioclase grains in the lower Paleogene and the Albian sandstones in the MITI-Wakkanai and the MITI-Sorachi cores (Iijima, 1973), respectively, are extensively albitized.

Burial Zeolitic Diagenesis

There are many localities of sedimentary zeolites in Japan (Iijima and Utada, 1966, 1972). The ages of host rocks containing the zeolites range from Pleistocene to Jurassic. Glassy tuffs, particularly felsic tuffs and tuffaceous sandstones, are important as host rocks, for they are abundant and widespread.

Zeolite Zoning Resulting From Burial Diagenesis

In thick sedimentary piles of marine and freshwater deposits, various zeolites are successively formed either from felsic glass or precursor zeolites or, less commonly, from clay minerals and clastic plagioclase as the sediments are buried; thus, this phenomenon is called "burial metamorphism" (Coombs, 1954) or, preferably, "burial zeolitic diagenesis" (Iijima and Utada, 1966; Hay, 1966). As a result, zeolites are progressively formed in a vertical zonation. These diagenetic zones differ somewhat in the assemblage of authigenic minerals and in thickness in different areas owing to lithology of host rocks, geothermal gradient, etc. However, the same zonation can be recognized in thick sedimentary sequences throughout Japan. The depth of burial and the temperatures increase from zone I to zone IV in the same area. Zone I is characterized by relict felsic glass which is partly altered to montmorillonite and low cristobalite. Within zone II, the glass has reacted with interstitial water to form alkali clinoptilolite and/or alkali mordenite. Both zeolites are replaced by analcime in zone III, in the upper half of which (IIIa) alkali clinoptilolite is replaced by heulandite and, locally, by calcium clinoptilolite. In the lower half (IIIb), laumontite substitutes for heulandite, calcium clinoptilolite and, in places, mordenite. Laumontite commonly replaces clastic grains of plagioclase and, less commonly, montmorillonitic clay. In zone IV, analcime reacts to form albite. Albitization of plagioclase also occurs in this zone, though it may progress in the lowermost part of zone IIIb, where laumontization of plagioclase is extensive.

Nonzeolitic minerals such as quartz and various clay minerals are progressively formed in the same manner as the zeolites.

It is significant that the zeolite zoning in the coalfields is essentially the same as the zoning described here.

Factors Controlling Zeolite Zoning

The most typical burial zeolitic diagenesis has been documented in the Niigata oil field on the coast of the Sea of Japan, central Honshū (Iijima and Utada, 1971).

Three 5-km-deep wells, the MITI-Ōbuchi, -Shimoigarashi, and -Masugata, penetrated the nearly flat-lying Quaternary and Neogene strata beneath Niigata Plain, which has been a site of deposition since Miocene time. There, diagenesis is progressing within a pile of marine geosynclinal deposits more than 5,000 m thick. Subsurface data prove that the zeolite zoning of burial diagenesis is essentially temperature-dependent—it is controlled by the geothermal gradient. The temperature is estimated from measured bottomhole temperature. It is 85–90°C at the top of the analcime zone (zone III) where alkali clinoptilolite and mordenite start to change to analcime. It is 120–125°C at the top of the albite zone (zone IV) where the analcime begins to change to albite. Experimental studies seem to support the field evidence.

Influence of Temperature on Coalification

Relation of Coal Ranks to Zeolite Zoning

Figure 4 illustrates the relation between coal ranks and zeolite zonation in representative coalfields. The specific correlation of coal ranks with zeolite zones is recognized throughout the Japanese Tertiary coalfields. All of the lignite and subbituminous coal studied belongs to the clinoptilolite-mordenite zone; examples are the Kushiro, Kuji, and Jōban fields. The coking bituminous coal belongs exclusively to the analcime zone, as in the Ishikari, Karatsu, Miike, and Takashima fields.

Main Factor Controlling Coalification

Burial diagenesis of felsic vitric tuffs in thick sedimentary sequences of marine and freshwater deposits results in vertical zonation of authigenic zeolites and feldspar as follows (from the surface downward): alkali clinoptilolite-mordenite zone, analcime zone, and albite zone. This diagenetic zoning is essentially controlled by the geothermal gradient at lower water pressures. The coal ranks of the Japanese Tertiary coal correlate well with the zeolite zoning. Consequently, the coal ranks should be essentially influenced by the geothermal gradient.

The stability fields of alkali clinoptilolite-mordenite, analcime, and albite in altered felsic vitric tuffs in thick sedimentary piles have been determined on the basis of data from the MITI deep cores drilled in the Niigata, Akita, and Tempoku Neogene oil fields. It is reasonable to conclude that the degree of coalification of the Japanese Tertiary coal is dependent on temperature, as is the zeolite zoning, and that the formation temperatures of noncoking lignite and subbituminous coal are as high as about 85°C, whereas those of coking bituminous coal range from approximately 85° to 125°C at water pressure below 2 kb.

References Cited

Coombs, D. S., 1954, The nature and alteration of some Triassic sediments from Southland, New Zealand: Royal Soc. New Zealand Trans., v. 82, p. 65–109.

Elphick, J. O., and R. P. Suggate, 1964, Depth/rank relations of high-volatile bituminous coals: New Zealand Jour. Geology and Geophysics, v. 7, p. 594–601.

Hay, R. L., 1966, Zeolites and zeolitic reactions in sedimentary rocks: Geol. Soc. America Spec. Paper 85, 125 p.

Iijima, A., 1961, Diagenetic alteration of some acidic tuffs in the Kushiro coal basin: Japanese Jour. Geology and Geography, v. 32, p. 507–522.

——— 1971, Composition and origin of clinoptilolite in the Nakanosawa Tuff of Rumoi, Hokkaido: Am. Chem. Soc. Advances Chemistry Ser., no. 101, p. 334–341.

——— 1972, Latest Cretaceous–Early Tertiary lateritic profile in northern Kitakami massif, northeast Honshu, Japan: Tokyo Univ. Fac. Sci. Jour., sec. II, v. 18, p. 325–370.

——— 1973, Reversed diagenetic zoning in the MITI-Sorachi cores (abs.): 80th Ann. Mtg. Geol. Soc. Japan Abstract Papers, p. 117 (in Japanese).

——— and M. Utada, 1966, Zeolites in sedimentary rocks, with reference to the depositional environments and zonal distribution: Sedimentology, v. 7, p. 327–357.

——— and ——— 1971, Present-day zeolitic diagenesis of the Neogene geosynclinal deposits in the Niigata oil field, Japan: Am. Chem. Soc. Advances Chemistry Ser., no. 101, p. 311–316.

——— and ——— 1972, A critical review on the occurrence of zeolites in sedimentary rocks in Japan: Japanese Jour. Geology and Geography, v. 42, p. 61–84.

Kisch, H. J., 1974, Anthracite and meta-anthracite coal ranks associated with "anchimetamorphism" and "very-low-stage" metamorphism; I, II, III: Koninkl. Nederlandse Akad. Wetensch. Proc., ser. B, v. 77, p. 81–118.

Kuyl, O. S., and R. J. H. Patijn, 1961, Coalification in relation to depth of burial and geothermic gradient: 4th Congr. Avanc. Études Stratigraphie et Géologie Carbonifère, Heerlen, 1958, Compte Rendu, v. 2, p. 357–365.

Sugimura, H., et al, 1966, Chemical structure and properties of heat treated coal in the early state of carbonization: Jour. Fuel Soc. Japan, v. 45, p. 198–208.

Tanai, T., 1967, Tertiary floral changes of Japan: Prof. Yasuo Sassa Memorial Vol., p. 317–334.

——— and K. Huzioka, 1967, Climatic implications of Tertiary floras in Japan, *in* Tertiary correlations and climatic changes in the Pacific: 12th Pacific Sci. Cong., Tokyo, 1966, Symp., no. 25, p. 89–94.

Teichmüller, M., and R. Teichmüller, 1966, Geological causes of coalification: Am. Chem. Soc. Advances Chemistry Ser., no. 55, p. 133–155.

Coal Resources of Alaska—Summary[1]

GEORGE GRYC and E. H. COBB[2]

Estimated coal resources of Alaska are 130 billion short tons, of which about 19.5 billion tons is bituminous and 110.5 billion tons is subbituminous and lignite. These estimates are very conservative and include only coal near known exposures of drillholes. Anthracite is known in two coalfields, but in such structurally complex settings that no valid resource estimates can be made.

More than 90% of the known resources are in Cretaceous rocks in an area of about 58,000 sq mi (150,220 km^2) in northern Alaska. Most of the rest are in Tertiary coal-bearing rocks south of the Yukon River in the Nenana and Susitna coalfields. All Alaskan coal is low in sulfur, but much that is easily accessible is difficult to ship because of spontaneous ignition.

Although coal was mined as early as the 1850s, practically all commercial production has been since 1916; the total is about 25 million tons, most of which was used by the Alaska Railroad and military bases, and in the Fairbanks area. Only the Nenana coalfield is now being mined on a large scale, and recent annual production has been less than 1 million tons.

[1]From Circum-Pacific Energy and Mineral Resources Conference program booklet.

[2]U.S. Geological Survey, Menlo Park, California 94025.

Coal Resources of Canadian Cordillera—Summary[1]

B. A. LATOUR[2]

INTRODUCTION

Coal deposits of various ages and ranks are widely distributed throughout the Canadian Cordillera. Although separated from the Pacific Ocean by about 350 mi (563 km) of rugged mountainous terrain, the coal deposits of the eastern part of the region are of considerable significance to the Circum-Pacific area.

The current production from all other coal deposits in this region is sufficient only to meet local domestic needs, although some deposits have undergone extensive exploitation in the past. In view of the current, rapidly changing energy situation, some of these deposits may become more significant. In particular, the Hat Creek and Groundhog coalfields of British Columbia appear to have potential.

DISTRIBUTION OF COAL DEPOSITS

Jurassic, Early Cretaceous, Late Cretaceous, and Tertiary coal occurs in the Canadian Cordillera (Fig. 1). There is no general pattern of age distribution but, for descriptive convenience, the region is grouped into three north-south geographic belts. The eastern belt extends from Fernie northward to Peace River, the central belt includes the area from Princeton coalfield northward through British Columbia to Dawson in the Yukon, and the western belt comprises Vancouver Island and Queen Charlotte Islands.

All ranks of coal are present; rank generally increases with the age of the coal. Sulfur in most of the coal is consistently less than 1%, except in Comox coalfield on Vancouver Island, where the sulfur content ranges up to 2%.

The only known deposits of high-quality metallurgical coal in the Canadian Cordillera are located in the eastern belt. Because the deposits are extensive and the demand for this type of coal has been high over the past few years, they are the most actively exploited deposits in the Cordillera; in 1973 production rose to about 11 million short tons (10 million MT). This coal was all shipped to Japan and represented about 15% of that country's metallurgical coal requirements. The total coal tonnage produced in 1973 from the central belt was only about 25,000 short tons (23,000 MT), and there was no production from the western belt.

COAL RESOURCE ESTIMATES

General

The resource estimates presented in this paper are based on a coal-resource study carried out by the National Department of Energy, Mines and Resources in 1969–70 (Latour and Chrismas, 1970). The estimates are categorized by the terms *measured, indicated,* and *inferred,* and are grouped according to rank. The resources are *in situ* coal contained in seams not less than 5 ft (1.5 m) thick and not more than 2,500 ft (762 m) below the surface.

Definitions

Measured resources—Measured resources are those for which tonnage is computed from dimensions revealed in outcrops, trenches, mine workings, and drillholes. The points of observation are generally about 0.5 mi (800 m) apart, and the computed tonnage is judged to be accurate within 20%.

Indicated resources—Tonnage of indicated resources is computed partly from specific measurements and partly from projection of visible data for a reasonable distance on the basis of geologic evidence. The points of observation are about 1 mi (1.6 km) apart, but they may be as far as 1.5 mi (2.4 km) apart for beds of known continuity.

Inferred resources—Inferred resources are those for which quantitative estimates are based largely on broad geologic knowledge of the bed or region and for which few measurements of bed thickness are available. The estimates are based primarily on assumed continuity, in areas remote from outcrops, of beds that were used to calculate measured or indicated tonnage in areas near outcrops. In general, inferred coal lies more than 2 mi (3.2 km) from the outcrop or from points for which mining or drilling information is available.

Resource Estimates

The Canadian Cordillera contains an estimated 87 billion short tons (79 billion metric tons) of coal. Almost all of this is located in the eastern belt, but some is located in the central belt (Table

[1]Manuscript received, November 6, 1974.

[2]Institute of Sedimentary and Petroleum Geology, Geological Survey of Canada, Calgary, Alberta.

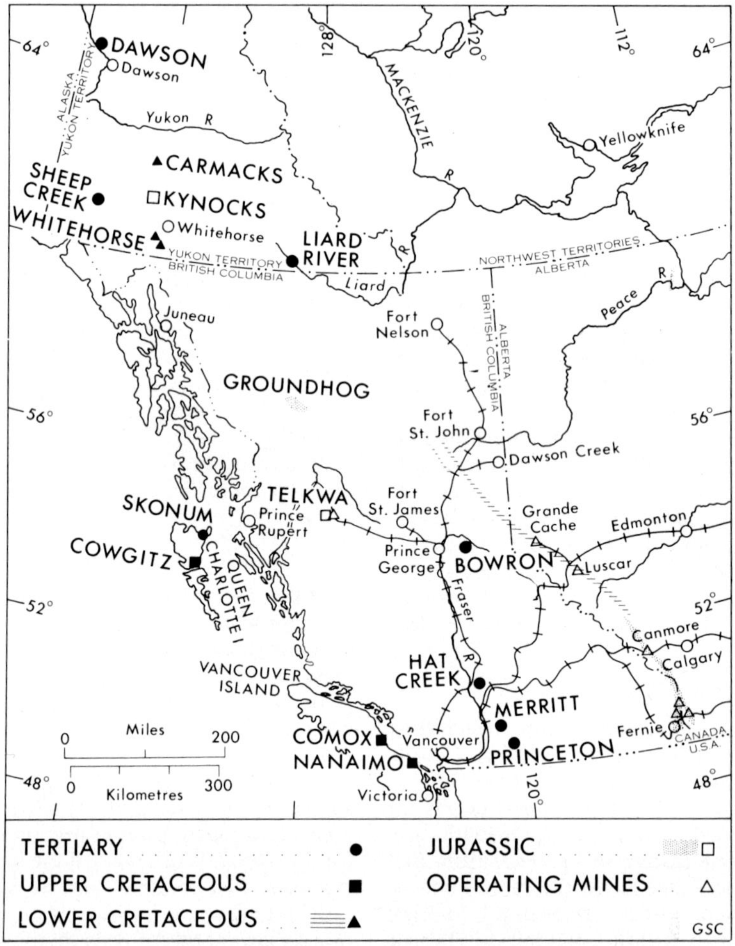

FIG. 1—Distribution by age of coal deposits of Canadian Cordillera.

1). The large amount of coal attributable to the eastern belt is a reflection of the large areal extent of the deposits and the greater amount of information available. The central and western belts probably contain significantly greater amounts of coal than those recorded here, but data are insufficient to permit estimates to be made within the defined categories.

Description of Deposits

Eastern Belt

The eastern belt, along the eastern front of the Rocky Mountains, extends northwestward about 600 mi (966 km) from lat. 49°N to the vicinity of lat. 56°N. The width of this belt is normally about 20 mi (32 km) and does not exceed 35 mi (56 km). Coal-bearing formations occur almost continuously throughout the belt; they have their greatest development along the western border and thin abruptly to the east.

In the southern part of the belt, between lat. 49°N and 52°N, the coal seams are present in the Jurassic Kootenay Formation. In the area of Fernie, British Columbia, the Kootenay Formation is at least 3,500 ft (1,067 m) thick and contains 13 coal seams ranging in thickness from 5 to 40 ft (1.5–12 m). Near Blairmore, Alberta, 25 mi (40 km) east of Fernie, the Kootenay Formation is only 400 ft (122 m) thick and contains two seams, each having a thickness of 5 ft (1.5 m) or more. These two sections of Kootenay Formation are situated on opposite sides of the Lewis thrust fault, which has a lateral displacement of about 40 mi (64 km). The 3,100 ft (942 m) of thinning

Table 1. Coal Resources of the Canadian Cordillera by Rank[1] and Geographic Belt[2]

Rank	*Measured*	*Indicated*	*Inferred*	*Total*
Eastern Belt				
Low-and medium-volatile bituminous	7,925,100	30,395,100	47,846,600	86,166,800
Central Belt				
Lignitic	340,000	300,000	300,000	940,000
High-volatile bituminous	45,600	100,400	172,900	318,900
Central belt total	385,600	400,400	472,900	1,258,900
Grand total	8,310,700	30,795,500	48,274,500	87,425,700

[1] American Society for Testing and Materials (1966).
[2] Thousands of short tons.

between the two sections takes place, therefore, over a palinspastic distance of about 65 mi (104 km). Northward along this part of the belt the seams are fewer and thinner until, at about lat. 52°N, the Kootenay Formation contains no seams of economic interest.

North of lat. 52°N, coal seams are present both in the Lower Cretaceous Luscar Formation and its equivalents in northeastern British Columbia—the Gething and Commotion Formations. The best development of coal seams in these formations appears to be in the area of Luscar and Grande Cache, Alberta.

Throughout the eastern belt the coal is mainly medium-volatile bituminous rank though it does range upward into the low-volatile bituminous rank; coal of semianthracite rank is present at Canmore, Alberta. The coal is invariably low in sulfur and generally has good coking qualities.

The entire eastern belt is characterized by numerous major folds and faults associated with the tectonism that produced the Rocky Mountains. The coal seams normally are steeply inclined and have a general northwesterly trend parallel with the mountain front. In addition, the seams may be extremely contorted, thickened or thinned, truncated by faults, and severely crushed so as to make the coal exceedingly friable. In some places, however, the seams may be so near the surface or so thickened as to permit strip mining. Figure 2 shows typical geologic conditions in the eastern belt. At the locality indicated by the mine symbol, a seam having a normal thickness of 20 ft (6 m) was thickened to 100 ft (30 m) in the syncline and to 60 ft (18 m) in the adjacent anticline.

Central Belt

This belt extends through the central part of British Columbia and Yukon Territory and contains widely scattered coal deposits of various ages and ranks (Fig. 1). Because of their diversity, the deposits are described individually or in small groups.

Princeton-Merritt coalfields—These Tertiary coal deposits range from lignite to high-volatile *C* bituminous in rank. The deposits are relatively small, irregular erosional remnants. The strata were deposited on much older sedimentary or crystalline rocks and are intruded or overlain by volcanic rocks. The coal seams have been mined extensively, especially in the Princeton coalfield. Additional information is required before resource estimates of the remaining coal can be made.

Hat Creek coalfield—This field warrants separate mention because of its unique assemblage of coal seams. At least 3,000 ft (914 m) of Tertiary sedimentary rocks contains five seams of lignitic coal that range in thickness from 100 to 700 ft (30–213 m); the total known thickness of all five seams is about 2,000 ft (610 m). Owing to the nature of the terrain and the fact that the deposits are partly overlain by volcanic rocks, neither the top nor the bottom of the Tertiary section has been determined. Near the surface the beds dip about 80° to the west, but drilling data indicate that, at a depth of around 800 ft (244 m), the dip becomes 30–45°, and it may become nearly horizontal at greater depth. Reserves are estimated conservatively at 1 billion short tons (900 million

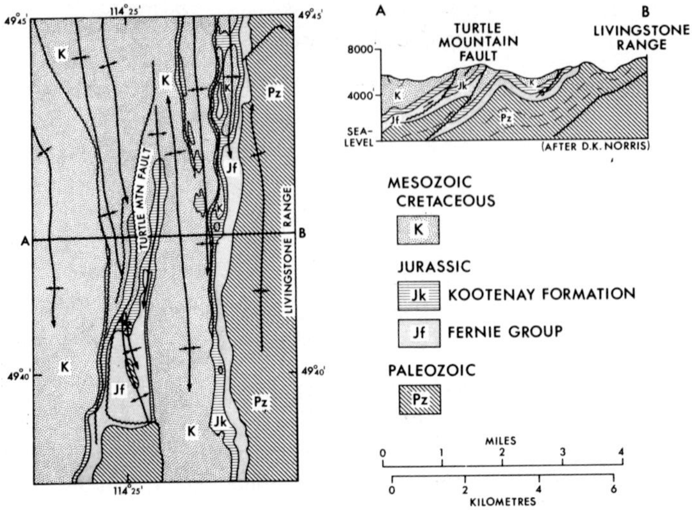

FIG. 2–Typical geologic map and cross section, inner foothills coal belt, Crowsnest area, Alberta.

MT) for all three categories, and much of this is recoverable by surface mining methods.

Bowron River coalfield—This is a small basin of Tertiary rocks, at least 2,000 ft (610 m) thick, which in the basal 200 ft (61 m) contains three seams with thicknesses of 8–12 ft (2.4–3.6 m). The coal beds are poorly exposed, and only very limited drilling has been done. Where exposed, the beds dip at about 45°, but drilling shows them to flatten in a short distance to dips on the order of 10–20°. The coal rank is high-volatile *C* bituminous.

Telkwa coalfield—This is a basin of about 7 sq mi (19 km^2) underlain by Jurassic–Lower Cretaceous coal-bearing rocks as thick as 500 ft (152 m). Folding and faulting and subsequent erosion have divided the Telkwa deposit into several small, separated coal areas in which the seams commonly are folded and truncated by faults and cut by volcanic rocks. Exploratory work is hampered by the thick cover of alluvium and glacial drift, and mining operations have encountered serious problems. The coal is good-quality, high-volatile *A* bituminous, but it has poor coking properties.

Groundhog coalfield—Extensive deposits of coal-bearing rocks of Jurassic age occur in this remote area. The beds are commonly faulted, overturned, and folded in flat recumbent folds. In places the coal is excessively sheared and crushed to a powder. The coal is mainly anthracitic, and generally is high in ash. Much exploration is required before this field can be assessed properly.

Liard River, Sheep Creek, and Dawson coalfields—These three areas in Yukon Territory are all little-known basins of Tertiary coal-bearing sedimentary rocks. Seams of lignitic coal of minable thicknesses are known to be present, but their number and extent have not been determined. The Dawson field appears to offer the most potential because it has the greatest area (5 by 80 mi [8 × 129 km]) and the seams there are reported to be thicker (up to 11 ft [3.3 m]) than those in the other two areas.

Whitehorse coal areas—Two small areas are underlain by the Lower Cretaceous Tantalus Formation, which contains seams of low-volatile bituminous coal. The seams are cut by intrusive rocks and have a higher rank of coal than is found elsewhere in the Tantalus Formation. Seams 6, 9, and 10 ft (1.8, 2.7, and 3.0 m) thick are reported from outcrop observations.

Kynocks coal areas—Limited drilling and trenching have revealed seams of high- and medium-volatile bituminous coal in the Jurassic Laberge Formation. One seam, 43 ft (13 m) thick, contained 37 ft (11 m) of coal, and two others were 8 ft (2.4 m) thick. The seams have been traced along strike for about 4 mi (6.4 km); they dip 70° to the southwest.

Carmacks coal area—Near Carmacks, a small area on each side of Lewes River is also underlain by the Tantalus Formation. This is the only area in Yukon Territory that has a history of coal production; a small mine has operated there since 1905. At least 3 seams with thicknesses of 7–9 ft

(2.1–2.7 m) have been reported, but the one seam presently being worked shows abrupt variations in thicknesses. The seam dips to the west at 45° and is cut by minor faults. The coal is high-volatile bituminous.

Western Belt

Vancouver Island—High-volatile *A* bituminous coal of Late Cretaceous age occurs in both the Nanaimo and Comox coalfields. The deposits at Nanaimo were depleted by operations that extended from 1852 to 1953; 54 million short tons (49 million MT) were produced. The Comox coalfield also has been extensively mined. The field is not depleted, but the remaining coal is not easily accessible and estimates require additional drilling information. Three seams were worked in the Comox field, and all were extremely variable in thickness and quality.

Queen Charlotte Islands—Tertiary deposits occur on the north end of Graham Island, the most northerly of the Queen Charlotte Islands. The coal is lignite, is of poor quality owing to high ash content, and is generally present in thin seams.

At the south end of Graham Island there are coal deposits of Late Cretaceous age. The coal generally ranks as low-volatile bituminous, but some anthracitic coal occurs where the measures have been intruded by volcanic rocks. The deposits are in five small basins, all of which are intensely folded and faulted. Few of the coal seams are more than 3 ft (0.9 m) thick, and the coal generally is high in ash.

References Cited

American Society for Testing and Materials, 1966, Book of standards, Pt. 19, Gaseous fuels, coal and coke.

Latour, B. A., 1972, Coal deposits of western and northern Canada, *in* 1st geological conference on western Canadian coal, proceedings: Alberta Research Council Inf. Ser., no. 60, p. 1–8.

——— and L. P. Chrismas, 1970, Preliminary estimate of measured coal resources including reassessment of indicated and inferred resources in Western Canada: Canada Geol. Survey Paper 70-58, 14 p.

MacKay, B. R., 1947, Coal reserves [Canada]: Chap. 1, Royal Comm. Coal Rept. 1946, p. 1–56, App. A, p. 601–649; reprinted 1947, 113 p.

Norris, D. K., 1955, Blairmore, Alberta (geologic map with marginal notes): Canada Geol. Survey Paper 55-18, scale 1:63,360.

GEOTHERMAL ENERGY

Geothermal Prospects Around the Pacific[1]

JAMES HEALY[2]

Abstract The fact that major geothermal fields are commonly located above zones of major crustal structures suggests that thermal resources are large, but heat extraction depends on hydrologic conditions which constitute the limiting factor.

The heat is believed to originate by conduction in areas of anomalously high heat flow, and is transmitted and concentrated by deep circulating groundwater. The existence of horizontal flow of hot groundwater has been demonstrated in several fields, and large hot-water fields tend to be located in discharge basins with large intake areas.

In zones of horizontal flow, stored heat resources are probably minimal and, even in large systems such as Wairakei, the dominant resource is groundwater recharge. Consequently, preproduction estimates of resources based on stored heat and natural heat flow may be unreliable. Production from Wairakei has been characterized by decreases in pressure, temperature, and output, though recharge has increased threefold.

Vapor-dominated systems tend to be located in elevated areas of groundwater recharge, where an important factor appears to be a potent heat source beneath the reservoir. The most important resource is probably recharge, and extraction of stored heat from rock depends on reservoir permeability.

Conservation of heat and disposal of saline waters by reinjection may be impracticable where recharge is an important contribution to water resources.

INTRODUCTION

Most areas explored so far for geothermal energy lie within regions of Quaternary volcanism, many of which are within and around the Pacific Ocean. Major geothermal fields commonly lie at plate margins, above subduction zones or spreading ridges (Muffler and White, 1972), and thus are linked with major geologic structures of global significance.

Discussion of Pacific prospects for geothermal steam would be incomplete without a review of basic ideas on resources and exploitation of geothermal systems.

GEOTHERMAL ENERGY

Geothermal energy is simply the heat of the earth. The world's mean heat flow (Lee and Uyeda, 1965) is 1.5 ± 10% cal/cm^2/sec (62.802 × 10^{-3} watts/m^2), which amounts to 32 × 10^6 Mw over the entire surface of the earth. This estimate does not include heat discharged from volcanoes and hot springs. The average thermal gradient is only 30°C/km, yet in many geothermal systems temperatures above 200°C exist within 1 km of the surface, showing them to be areas of thermal anomaly.

Temperature alone is not sufficient to indicate geothermal prospects. It is a measure of the grade of heat present, but this must be extracted in water or steam from which the energy can be obtained by conventional methods. A geothermal system suitable for energy production should contain rocks with adequate permeability, saturated with hot water or steam, within economic drilling depth. For the generation of electric power, temperatures above 200°C are preferred, though lower temperatures are satisfactory for various other purposes. The economics of development of hot-water fields, particularly, is improved if the heat normally discharged in the effluent water can be used for secondary purposes. Most of the heat stored in a geothermal system is in the rock, and how much of this can be extracted depends on the nature of the system and on field management during exploitation.

GEOTHERMAL SYSTEMS

Geothermal systems so far developed fall into two main types. A hot-water system is one in which the rocks are saturated with hot water, commonly at temperatures considerably above 200°C. Steam is flashed from the water by reduction of pressure and then separated from it. A vapor-dominated system is one in which the steam phase is continuous, though hot water may be present in pore spaces. Wells drilled into a vapor-dominated system produce undersaturated steam. Characteristics of the two types of systems have been compared by James (1968) and White et al (1971).

[1]Manuscript received, October 21, 1974. Published with permission of the Director, New Zealand Geological Survey.

[2]Department of Scientific and Industrial Research, New Zealand Geological Survey, Rotorua, New Zealand.

Origin of Geothermal Systems

Many geothermal systems occur in volcanic districts. Following the classic work by Allen and Day (1935), it was generally accepted that in fields like Yellowstone National Park the heat was derived in the form of steam from cooling magma. The acid condensate in groundwater then reacted with the rocks to produce the dissolved mineral content in the thermal waters discharged at the surface. This concept had to be modified when isotope chemistry (Craig et al, 1956) demonstrated absence of magmatic water except for a possible undetected amount of 5%.

Evidence for the presence of magmatic content in the thermal waters of the New Zealand Taupo volcanic zone was finally dispelled by Mahon (1967), who showed experimentally that their chemical composition could be attained by the solvent action of hot water on the local volcanic rocks. The heat was still considered to be of magmatic origin, and the favored model was a body of cooling magma heating deeply circulating groundwater by conduction. To explain the long lives of some of the known large geothermal systems, convection within the magma was postulated (e.g., Grindley, 1960; Banwell, 1963).

Healy (1970, Fig. 1) proposed a model to concentrate abnormally high heat flow by lateral flow of deep groundwater as part of a regional system. Drilling at El Tatio in northern Chile has confirmed the existence, for at least 5 km, of horizontal flow of the hot water discharged by the geysers and hot springs (Healy and Hochstein, 1974). It has been suggested (Healy and Hochstein, 1974) that in New Zealand hot water flows at a depth of 1,000 m from the Waiotapu geothermal system for 6 km to Reporoa.

It is believed that regional and intermediate-depth groundwater movement (Töth, 1972) is an important factor in the origin and mode of occurrence of geothermal systems (Fig. 1), though it has been largely neglected in favor of models in which convection cells of heated groundwater are located vertically above isolated heat sources. White et al (1973) have described thermal waters of nonmeteoric origin in the California Coast Ranges, but the hydrology of these is still little known.

Hot-Water Systems

Hot-water geothermal systems are by far the most common. They range in magnitude from isolated, single warm springs with their feeding channels to large systems like the Wairakei field in New Zealand which, including the contiguous Tauhara field, has a cross-sectional area of 35 km^2, or to large systems like that of the Salton Sea which has little or no surface expression. Most contain and discharge saline waters, though compositions are affected by the types of rock through which the fluids pass and by the superficial processes of oxidation and boiling.

Hot-water systems are of many types, some of which have little or no discharge. Boldizsár (1964) has reported heat flow nearly twice normal in the Hungarian basin, where he estimated that there is 4,000 km^3 of groundwater heated by the geothermal gradient to temperatures of 60–200°C between depths of 1,000 and 3,000 m from the surface. The Salton Sea system might appear similar, yet Craig (1969) considered that some source fluid is meteoric water circulating downward. Possibly the Hungarian geothermal system, too, is not static, but heat flow higher than normal is common to both systems.

Geothermal systems which discharge hot water are typically located in discharge areas in large basins or in valleys. In New Zealand (Fig. 2) the Wairakei and Broadlands geothermal systems are located in a large depression 85 km long. The Waikato River, the largest on the North Island, flows northeast from Lake Taupo for 30 km along the depression before leaving it, crossing both geothermal systems which discharge into the river. Electrical resistivity surveys show the margins of both systems to be vertical, defining columns of hot water rising from depths of about 2,000 m. Farther northeast the Waiotapu geothermal system lies within the same depression, though other areas of hydrothermal activity within 10 km to the north and northwest appear to be linked to it at depth. As mentioned, there is believed to be a deep flow of hot water from Waiotapu south to Reporoa along the axis of the depression.

The geothermal systems of the Taupo volcanic zone (Fig. 2) occur within a region of Quaternary acid volcanism associated with relatively shallow crustal melting (Healy, 1962). Studt and Thompson (1969) reported zero thermal gradient over much of the zone, and suggested that the combined natural geothermal discharges, which are equivalent to 20 times the normal heat flow for the whole area, represent meteoric intake over large areas. The regional geologic background is similar to that of geothermal areas in Central America, northern Chile, Yellowstone National Park, and parts of Japan and Indonesia (Sumatra).

In New Zealand the main geothermal areas are not related to individual volcanoes; however, in northern Chile, Central America, Kamchatka,

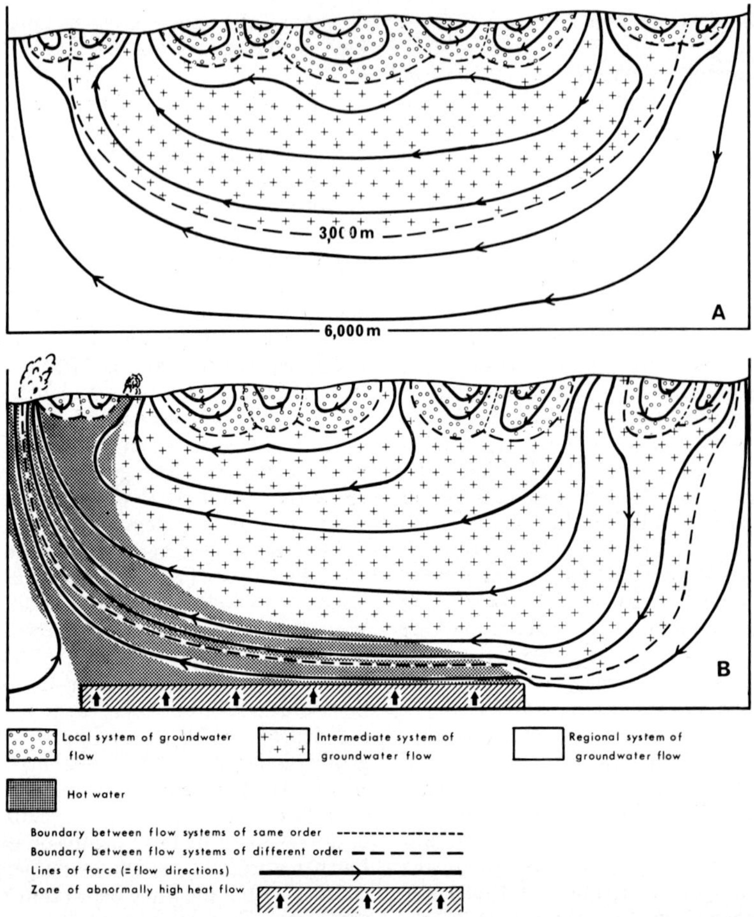

FIG. 1–Two-dimensional theoretical flow pattern of groundwater into a drainage basin underlain by homogeneous rock; **A,** under normal conditions (after Töth, 1972); and **B,** with abnormally high heat flow beneath part of basin.

Japan, the Philippines, and Indonesia, there is a much more direct relation. In northern Chile, for example, the waters of the geothermal systems become heated beneath large volcanoes of the Andes and flow out from these. Some water flows at relatively shallow depth, and certain characteristics are common to many fields. At the foot of a volcano where the slope changes, the hot water approaches the surface more closely, and may boil and give off steam, giving rise to fumaroles and steam-heated thermal springs. Farther out the hot water may flow directly to the surface and produce boiling springs. Water which has lost steam and gas may emerge still farther away as warm springs. Systems like this have been tapped in El Salvador, Japan, and the Philippines, and are being investigated in Indonesia, Guatemala, and Nicaragua.

Vapor-Dominated Systems

Vapor-dominated systems tend to be located in elevated areas which commonly are capped by rocks of limited permeability that are water saturated down to the top of the steam zone (Fig. 3). A conductive thermal gradient may exist through the cap if it is impermeable, or, if the cap is permeable, boiling-point temperatures may occur. The maximum pressure in a vapor-dominated system is about 30 atm, and the maximum temperature about 235°C, both controlled by the thermodynamic properties of steam and water (James, 1968). An underlying water table has

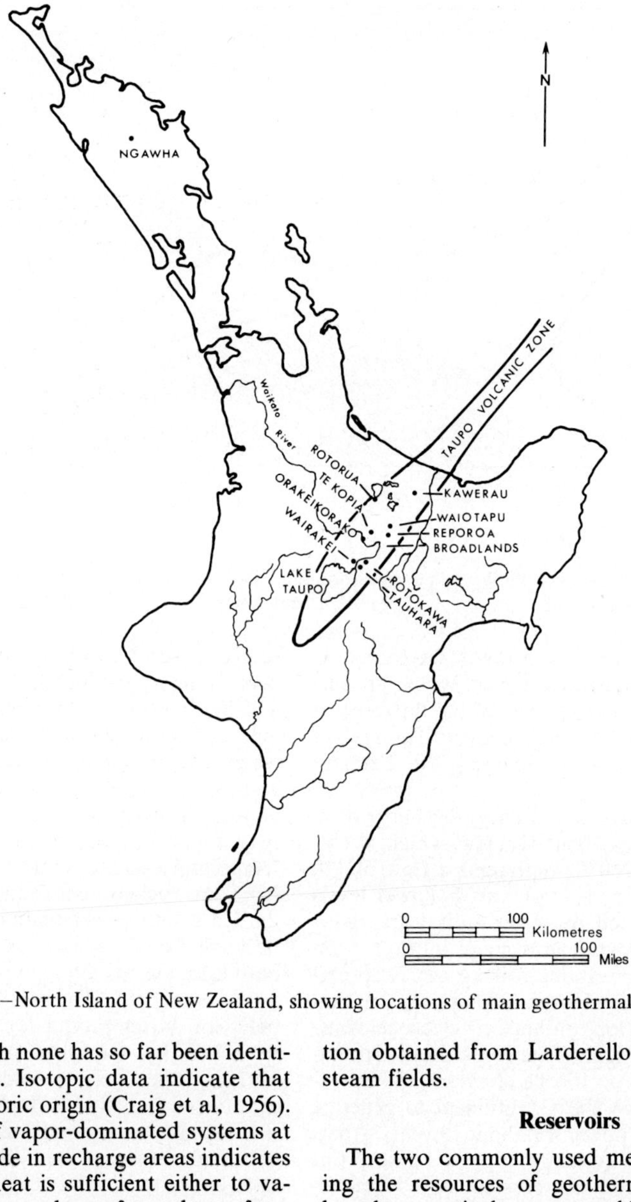

FIG. 2–North Island of New Zealand, showing locations of main geothermal fields.

been inferred, though none has so far been identified in drilled wells. Isotopic data indicate that the water is of meteoric origin (Craig et al, 1956).

The occurrence of vapor-dominated systems at relatively high altitude in recharge areas indicates that the supply of heat is sufficient either to vaporize or exclude any recharge from the surface. White et al (1971) suggested that vapor domination commences when discharge starts to exceed recharge. Whereas the source of heat in a hot-water geothermal system may lie outside the area of the system, it seems more likely that the heat source for a vapor-dominated system is located beneath it (Fig. 3), and that the heat flow is large relative to the rate of groundwater recharge. This concept is consistent with the large steam production obtained from Larderello and The Geysers steam fields.

Reservoirs

The two commonly used methods for estimating the resources of geothermal reservoirs are based, respectively, on natural heat discharge and stored-heat content. Comparison of the methods is perhaps best considered with regard to two fields of different character. The El Tatio field in northern Chile is fed by hot water at temperatures less than 265°C moving horizontally through ignimbrite sandwiched between impermeable tuff-breccias. The Wairakei field in New Zealand is fed by hot water at about 260°C rising vertically through stratified volcanic rocks and a thick ig-

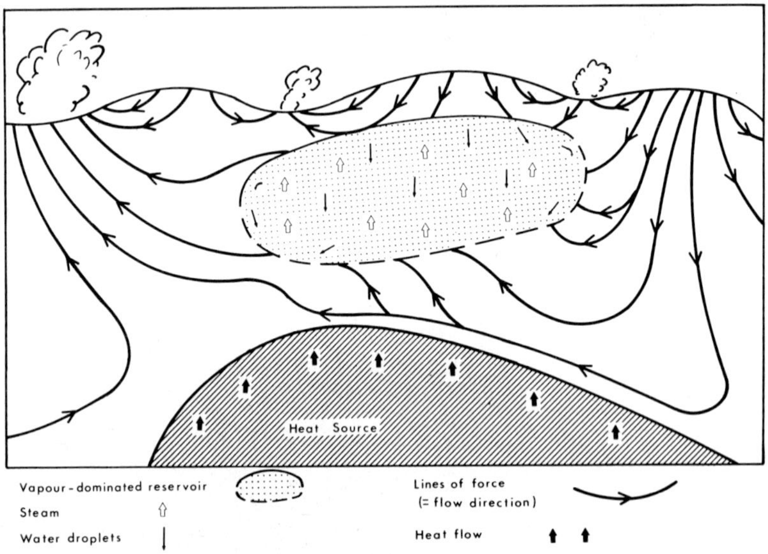

FIG. 3—Two-dimensional profile through a vapor-dominated system fed by heated groundwater underlain by a potent heat source.

nimbrite into tuff-breccia in which the fluids expand beneath a mudstone cover. Wells tap permeable zones in the lower part of the tuff-breccia and the upper part of the ignimbrite. The area of the hot zone is 20 km^2 excluding the Tauhara field.

El Tatio field—The natural heat discharge from the El Tatio field is about 125 Mw, which, if carried by water at 260°C, represents a flow of 110 kg/sec. Cold water also flows at shallower levels and becomes heated by water and steam rising from the main flow. The average outflow from the whole basin is about 350 kg/sec; approximately half is derived from cold springs. Within the hot area of 10 km^2 outlined by electrical resistivity surveys, the ignimbrite contains an estimated quantity of 2.5×10^{11} kg of water at temperatures mostly above 200°C, sufficient to generate 20 Mw of electric power for about 19 years. Apart from the natural recharge of 110 kg/sec, this might be considered as a possible stored-heat resource with the ignimbrite as a reservoir.

In fact, the water in the ignimbrite occupies pore space, whereas permeability is related to joints and cracks confined to a narrow zone of maximum temperature. Welded ignimbrites normally possess high joint permeability, but at El Tatio the joints generally are sealed by deposition of silica and calcite except along the flow channels with maximum temperature (Healy and Hochstein, 1974). Consequently, in this field the concept of a reservoir is fictitious unless a large reservoir which can be directly tapped by wells is present nearer the source. The scale of production will be determined by recharge—that is, by the amounts by which the recharge of 110 kg/sec of water and its temperature of 260°C can be increased by lowering the pressure in the main aquifer during production. The lowered pressure may cause flashing of steam and reduction of temperature, so that some heat could be extracted from the rock by conduction, but owing to the limited extent of permeability the amount would be small. So far no data are available on which to base estimates of ultimate production.

Wairakei field—The geology of the field and its behavior during production until 1969 have been described by Grindley (1965) and by Bolton (1970), respectively. Stored-heat estimation was introduced initially by Healy (1952). In 1958 further estimates of stored heat down to a depth of about 600 m were based on accumulated drilling data but, after several years of increased production with no observed change in marginal temperatures, it was generally considered that recharge was occurring.

Pressures in wells measured at a level 275 m below sea level fell by 2,068.4 kilopascals (kPa; 300 psi) in the western area and by 2,413.2 kPa (350 psi) in the east; the falling pressures were accompanied by falling water levels (Bolton, 1970). Bolton considered the uniform fall in the west to indicate good lateral permeability through the permeable production zone at a depth of about 610 m; in the east the ignimbrite surface falls and the greater drop in pressure is related to thickening of the overlying breccia, which has lower permeability and hence greater drawdown. Pressure and hydraulic gradients at the margin of the field are steep, indicating almost impermeable boundary conditions. Bolton suggested mineral deposition to account for this.

Total discharge from all wells reached a peak in 1962, fell relatively rapidly until 1965, and then fell gradually. The enthalpy of the discharge remained constant at 1,090 joules/g until 1962, increased to a peak of 1,140 joules/g in 1965, and then fell gradually to less than 1,115 joules/g by 1969. Over the same period, temperatures did not change in the outer western area, but fell by 12°C in the western production area. In the eastern production area temperatures remained unchanged below 305 m below sea level, but decreased progressively at shallower levels.

During a partial shutdown for 3 months in 1968, measurements confirmed the existence of an inflow equivalent to two thirds of the average total field discharge of 1,830 kg/sec at that time and having a temperature equal to or higher than the maximum measured in the field (Bolton, 1970). Bolton presented evidence for inflow from the southwest.

Calculations by Healy (1956), based on field measurements by Gregg (1958) but later amended following direct measurements of seepage of hot water to the Waikato River, gave a value for natural heat discharge of 475 Mw. This would be carried by 420 kg/sec of water at 260°C, so that during production in 1968 the well discharges totaled approximately 4.5 times the natural recharge rate, and the induced recharge rate was 3.5 times the natural recharge.

This performance is consistent with the concept of the Wairakei geothermal system as a discharge channel for regional groundwater, and emphasizes recharge as the most important factor in its power life. The total mass of water already discharged has come from recharge and loss of storage water indicated by the fall in water level. Of the total heat discharge, some represents heat lost from the rock, because fall of water and steam temperatures as a result of flashing induced by lowered pressure causes some heat to be transferred from the rock. However, only the western part of the main production area has generally decreased temperature at depth (Bolton, 1970).

General Observations

In systems like those of the Salton Sea region the main resource may lie in the large stored volumes of hot water. No data are available on the magnitude or mode of recharge which might occur there during production.

Steam reservoirs like those of Larderello and The Geysers have considerable volume. Permeability is fairly high, and is indeed required to maintain the circulation of steam necessary for the uniform temperature throughout each system. Consequently, if temperature should fall during production, transfer of heat from rock to steam may add appreciably to the power life. Pressure falls in the reservoirs during production, but White et al (1971) suggested that steep pressure gradients develop at the margins, and a combination of chemical and physical factors tends to reduce inflow. If, as they supposed, a liquid phase underlies each system, and if the inferred potent heat source does exist, it seems possible that recharge is from hot water beneath the steam zone (Fig. 3). Continued existence of a steam zone during a high level of production may depend on a combination of permeability in the recharge zone and the potency of the heat source.

Reservoir Life

The permanence of geothermal reservoirs depends on heat and supplies of groundwater. Location of fields in zones of major earth structures and large-scale volcanism suggests that the heat sources are immense, so that the ultimate limit is that of groundwater supply, which is determined initially by precipitation and secondarily by permeability. The latter can be affected by processes operating in the fields. Production from a hot-water field can cause water level to fall until only steam is discharged, and mineral deposition in the zone of steam separation may gradually decrease permeability.

Bolton (1970) pointed out that drilling can reach a stage when discharge from new wells lowers pressure and decreases output from existing wells until no overall increase in discharge is obtained. The deeper production is obtained in a field, the greater is the possibility to reduce pressures and increase recharge, though drilling depth may be limited by cost. Because of high permeability in the Wairakei field, he considers (personal commun.) that, at the present level of production, conditions will stabilize without change to a steam field.

Estimates of field resources and life based on stored-heat calculations are unreliable, as no effective reservoir may exist. The ultimate resource may be recharge. The extraction of heat stored in a large volume of rock depends on permeability, which may limit the flow of fluid or steam through the mass.

General Discussion

The role of regional hydrology in relation to geothermal fields is important. Preliminary identification of the type of field should be an aim of initial geologic reconnaissance, so that geophysical surveys to outline the extent of the field can be most efficiently applied.

A particular problem is the location of zones of maximum permeability in which good production can be obtained, particularly within a field characterized by horizontal flow. Development of a geophysical technique to do this is regarded as an urgent requirement.

Particular problems in many producing fields are effluent disposal, prevention of subsidence, and conservation of resources. Most effluents consist of hot saline water, and many fields now being developed previously discharged this water into existing drainage channels. In some places increased discharge resulting from geothermal production might be acceptable in existing stream waters, but it would render others unfit for use downstream.

In the example of Wairakei, the amount of dewatering of the country within the field is considerably less than the amount of water drawn in by increased recharge. It is obvious that the effluent could not be reinjected in the immediate vicinity without either reappearing in the drainage system or increasing pressure in the field and reducing the recharge rate. Nor can it be recycled to conserve heat, as the recharge area is likely to be a considerable distance away at higher elevation and not accurately known. Some effluent might be reinjected at shallow depth to alleviate the subsidence that accompanies the dewatering process, but it should not be injected where it could reenter the hot aquifer feeding the wells.

The best long-term solution where reinjection is not feasible appears to be to extract as much of the mineral content and heat as possible from geothermal fluids before disposing of them into normal channels. Only a relatively small amount of energy is extracted from geothermal fluids for generation of electricity, and utmost use should be made of the available heat, even at the expense of generating less electric energy. In the Tasman Pulp and Paper Company's mill at Kawerau, New Zealand, geothermal steam is used primarily for process purposes, and only the surplus is used for power generation.

Few hydrologic data are available from areas surrounding geothermal fields, and data within them are usually obtained from wells with long lengths of uncased hole connecting aquifers at different levels, so that measured pressures do not represent true aquifer pressures. There are practically no reliable data on marginal conditions, which have an important and continuing influence on the development and life of a field. Field management is still a new technique. Geochemical and hydrological studies are important during the lives of all fields and, although general principles may apply in geothermal fields, each still has its individual characteristics determined by local geologic factors.

References Cited

Allen, E. T., and A. L. Day, 1935, Hot springs of the Yellowstone National Park: Carnegie Inst. Washington Pub. 466.

Banwell, C. J., 1963, Thermal energy from the earth's crust; introduction and part 1: New Zealand Jour. Geology and Geophysics, v. 6, p. 52–69.

Boldizsár, T., 1964, Heat flow in the Hungarian basin: Nature, v. 202, no. 4939, p. 1278–1280.

Bolton, R. S., 1970, The behaviour of the Wairakei geothermal field during exploitation, *in* United Nations symposium on the development and utilization of geothermal resources: Proc., v. 2, pt. 2, Geothermics, Spec. Issue, no. 2, p. 1426–1439.

Craig, H., 1969, Discussion: source fluids for the Salton Sea geothermal system: Am. Jour. Sci., v. 267, p. 249–255.

——— G. Boato, and D. E. White, 1956, Isotopic geochemistry of thermal waters, Chap. 5 *of* Nuclear processes in geologic settings: Natl. Research Council, Comm. Nuclear Sci., Nuclear Sci. Ser. Rept., no. 19, p. 29–38.

Gregg, D. R., 1958, Natural heat flow from the thermal areas of Taupo Sheet district (n94): New Zealand Jour. Geology and Geophysics, v. 1, p. 65–75.

Grindley, G. W., 1960, Geologic map of New Zealand, 1:250,000: New Zealand Dept. Sci. and Indus. Research, 1st ed.

——— 1965, The geology, structure, and exploitation of the Wairakei geothermal field, Taupo, New Zealand: New Zealand Geol. Survey Bull., no. 75, 131 p.

Healy, J., 1956, Preliminary account of hydrothermal conditions at Wairakei, New Zealand: 8th Pacific Sci. Cong. Proc., Philippines, v. 2, p. 214–227.

——— 1962, Structure and volcanism in the Taupo volcanic zone, New Zealand, *in* Crust of the Pacific basin: Am. Geophys. Union Geophys. Mon. Ser., no. 6, p. 151–157.

——— 1970, Pre-investigation geological appraisal of geothermal fields, *in* United Nations symposium on the development and utilization of geothermal resources: Proc., v. 2, pt. 1, Geothermics, Spec. Issue, no. 2, p. 571–577.

——— and M. P. Hochstein, 1974, Horizontal flow in hydrothermal systems: New Zealand Jour. Hydrology, v. 12, p. 71–82.

James, R., 1968, Wairakei and Larderello: geothermal power systems compared: New Zealand Jour. Sci., v. 11, p. 706–719.

Lee, W. H. K., and S. Uyeda, 1965, Review of heat flow data, Chap. 8 *in* Terrestrial heat flow: Am Geophys. Union Geophys. Mon. Ser., no. 8, p. 87–190.

Mahon, W. A. J., 1967, Natural hydrothermal systems and the reaction of hot water with sedimentary rocks: New Zealand Jour. Sci., v. 10, p. 206–221.

Muffler, L. P. J., and D. E. White, 1972, Geothermal energy: Sci. Teacher, v. 39, no. 3.

Studt, F. E., and G. E. K. Thompson, 1969, Geothermal heat flow in the North Island of New Zealand: New Zealand Jour. Geology and Geophysics, v. 12, no. 4, p. 673–683.

Töth, J., 1972, Properties and manifestations of regional groundwater movement: 24th Internat. Geol. Cong. Proc., Hydrogeology, sec. 11, no. 24, p. 153–163.

White, D. E., L. P. J. Muffler, and A. H. Truesdell, 1971, Vapor-dominated hydrothermal systems compared with hot-water systems: Econ. Geology, v. 66, p. 75–97.

——— I. Barnes, and J. R. O'Neil, 1973, Thermal and mineral waters of nonmeteoric origin, California Coast Ranges: Geol. Soc. America Bull., v. 84, p. 547–560.

Environmental Effects of Geothermal Energy Development—Summary[1]

RICHARD G. BOWEN[2]

Introduction

An appraisal of the environmental effects of geothermal development should do more than list potential insults to the environment; it should also weigh the benefits to mankind, and scale these insults and benefits against those from other energy sources.

Possibly the best way to compare these effects is to look at the total cycle of power production. The geothermal plant is unique in that all of the steps in the fuel cycle are localized at the site of the power-production facilities. At the other end of the power spectrum is the nuclear-reactor plant, in which the actual power-production facilities are a small fraction of a cycle requiring a complex industrial-support system for each reactor. Intermediate in complexity, and varying somewhat with the type of fuel used, are the fossil-fuel plants.

Like other thermal power plants, the geothermal plant involves the production and use of steam, expanding it through a turbine and condensing it at the turbine exhaust. The geothermal plant differs from the conventional-fuel or nuclear plant in its method of steam production and the quality of its steam. At the geothermal plant the steam is produced by the natural circulation of water in contact with hot rocks at depth. Depending on conditions within the reservoir, the geothermal fluids may be in the form of either slightly superheated dry steam or pressurized hot water. If the fluid is in the form of steam, it can be passed through the turbine directly; if in the form of hot water, it can be used directly by "flashing," or the thermal energy in the fluid can be transferred by means of heat exchangers to another liquid for expansion through a turbine. In all cases the energy-transfer method is far simpler than that needed for the combustion of fuel or for containing a nuclear reaction.

The simplicity of the whole geothermal cycle contributes greatly to its reliability. With the geothermal system there are no complex mining, processing, and transportation steps. To the extent that geothermal resources can be substituted for the more conventional fuels, both fossil and nuclear, the overall detrimental effects of energy production can be reduced.

Direct Use of Geothermal Energy

The direct use of geothermal energy for space heating is practiced in several parts of the world. The best-known example is at Reykjavík, Iceland, where a large district heating system provides nearly all the space heating for this city of 85,000. Extensive use is made of geothermal waters in Hungary; in Budapest alone, 5,600 flats are supplied with natural hot water. Not only does this save fossil fuels but, as Boldizsár (1975) reports, the heating costs are less than half those of conventional fuels. In Klamath Falls, Oregon, several hundred homes and numerous schools use geothermal waters.

Utilization of geothermal fluids for space and process heating may involve little more than the drilling of a well and circulating the fluids through a radiator in the home. In some cases heat exchangers are used, and in others it is necessary to make minor changes to the chemistry of the waters to keep minerals from plugging the pipes. The disposal of these fluids depends on the region and quality of the fluids; some fluids are put directly into surface streams, and some are disposed of underground. Commonly the spent fluids are of sufficient quality to be used directly for irrigation or stock watering. If aquifer conditions are appropriate, there is no need to bring the geothermal fluids to the surface; down-hole heat exchangers can be used and clean secondary fluids circulated to the surface installations. Nowhere has it been necessary to resort to complex techniques of effluent capture and disposal.

The direct use of geothermal energy has been largely unheralded because it lacks the glamour and large revenue aspects of the production of electric power. However, the total amount of geothermal energy produced for direct use currently exceeds the geothermal energy produced for electric power production. The direct use of geothermal fluids for space heating is particularly attractive in areas of arctic and subarctic climates, where winter heating is a major economic burden and where the winter high-pressure weather sys-

[1]Manuscript received, November 6, 1974.

[2]Economic geologist, State of Oregon Department of Geology and Mineral Industries, Portland, Oregon 97201.

tems often create a pall of lingering smoke and fog from the burning of fossil fuels.

GEOTHERMAL ENERGY FOR PRODUCTION OF ELECTRICITY

The potential environmental insults from geothermal power developments are similar to those from any other industrial operation. Construction of roads, drilling of wells, and installation of pipelines and power plants all contribute to the changes in land-use patterns for a particular site.

The two basic types of geothermal reservoirs—vapor-intensive or "dry-steam" and liquid-intensive or "hot-water" reservoirs—have been used sufficiently for electric power production to develop a history of environmental effects. The three productive dry-steam fields—Larderello, Italy, The Geysers, California, and Matsukawa, Japan—all produce steam of similar thermodynamic properties (White et al, 1971) but slightly differing chemical properties. The steam is made up chiefly of water vapor (90–99.9%) and varying amounts of noncondensable gases (carbon dioxide, methane, hydrogen sulfide, hydrogen, nitrogen, and ammonia). Other dry-steam fields that may be found will probably have similar characteristics and can be handled in a manner similar to that presently developed; furthermore, environmental effects are predictable.

The characteristics of hot-water fields are much more variable, both in thermodynamic and chemical character. Temperatures range from slightly above ambient to 388°C (Mercado, 1969), and chemical content ranges from very low to the saline brines of the Salton Sea (California) sump which have minerals in solution totaling up to 33% of the mass. Production of electricity from the liquid-intensive field, therefore, utilizes more different techniques, and it is in this area that the greatest efforts are being expended on new mechanical methods for harnessing the energy.

The amount of land needed for productive fields, pipeline rights-of-way, and power-production facilities also varies. At Wairakei, New Zealand, the steam field, supplying the plant of 192-Mw capacity, and the plant and pipelines located about a mile from the field take up only about 1 sq mi (2.6 km^2) of land. At Cerro Prieto, Mexico, approximately 1 km^2 is used for plant and production facilities for 75 Mw. At Larderello, the field covers many square miles, but the density of development is so low that numerous other productive activities are carried on within the geothermal field, making the total amount of land dedicated to electric power production very small. Larderello is an outstanding example of multiple land use, for within the bounds of the producing field is a diversified agricultural economy with many farms, orchards, and vineyards interspersed among the wells, pipelines, and power plants. The Geysers field, which may be typical of geothermal fields yet to be discovered and developed, is presently being developed with plants of 110-Mw capacity spaced about a mile apart, each supplied by a steam field about a mile in diameter. The field that produces 1,000 Mw approximates 10 sq mi (25.9 km^2) in total area.

The amount of land dedicated to a nuclear or fossil-fuel power plant of similar size ranges from a minimum of about 100 acres to several thousand acres if plant-site fuel-storage facilities and cooling ponds or lakes are needed. Fossil-fuel and nuclear plants require extensive offsite facilities to mine, process, transport, and, in some cases, guard waste disposal. By contrast, all of the production facilities of the geothermal plant are located at the site; thus, the total amount of land dedicated to the production of electricity from a geothermal power plant is less than for other types of thermal plants.

ENVIRONMENTAL INSULTS

Water-Resource Degradation

Two questions commonly raised about geothermal power development are:

1. What is the potential of contamination to surface water and groundwater?

2. To what extent would waters drawn from geothermal reservoirs deplete presently developed aquifers?

In the early days of geothermal exploration and development, several improperly cased wells blew out during drilling or after abandonment, permitting geothermal fluids to enter nearby streams and shallow aquifers. In the Salton Sea region, extremely saline geothermal brines, which contain up to 33% dissolved solids after flashing, constitute a hazard if allowed to mix with irrigation water.

With the use of modern drilling, casing, and completion practices, there is no reason to believe that geothermal production presents any more of a hazard than does petroleum production. It is a routine practice to drill through freshwater aquifers to produce petroleum from underlying zones. By casing through these zones and then firmly cementing the casing into place, one can effectively isolate the aquifers from the deeper production zones. Such isolation of thermal waters is necessary for the protection of both the aquifers and the geothermal system.

Discharge of geothermal fluids into surface streams possibly would degrade stream quality. However, this practice is prevented by law in the United States wherever surface waters contain fewer contaminants than the geothermal fluids. Even accidental discharge of brines from a geothermal well into surface waters is not a serious hazard. In most parts of the world, the geothermal fluids are potable or do not exceed the salt content of ocean waters.

In some instances, geothermal fluids can be added to surface waters because they are equal or better in quality than the existing water resources. Geothermal developments may be a significant source of useful waters in some areas because the cost of producing them will be borne mostly by the power-production facilities. Electricity, minerals, and usable water are produced at the U.S. Bureau of Reclamation project in the Imperial Valley (Laird *in* Kruger and Otte, 1973).

Arguments have been made that development of geothermal reservoirs may deplete existing developed aquifers. In a given area or basin there is only a certain quantity of water in storage. Aside from surface waters, the most useful waters are in shallow aquifers. Geothermal water provides a new source of water from depths as great as 9,000 ft (2,750 m).

All thermal power plants require water for the cooling process. Geothermal plants use the geothermal waters, whereas other plants require the use of conventional sources such as rivers, lakes, or groundwater reservoirs.

Geothermal dry-steam reservoirs are isolated from the groundwater aquifers, as is shown by their very low reservoir pressures—about 500 psi even at depths over 9,000 ft (2,750 m), where hydrostatic pressures normally would be about 4,500 psi (White et al, 1971). Such extremely low pressures can be explained only by isolation; therefore, there is no reason to believe that production of steam from a vapor-dominated system would affect overlying aquifers. At the vapor-dominated systems of The Geysers, Larderello, and Matsukawa, there has been little or no change in shallow-water zones.

In hot-water fields such as Wairakei, withdrawal of geothermal fluids can accelerate the depletion of shallow aquifers; for example, since production began at Wairakei in the 1950s, flow of the surface springs has decreased. To check the possibility of intercommunication between aquifers, monitor wells could be established in regions of geothermal production.

Each geothermal project must be considered within its geologic environment. Where existing water uses will be affected, the higher value of the water when used as the medium for transporting geothermal energy may permit it to pay for the displacement it causes. This potential for displacement of existing uses also has to be weighed against the potential to develop new sources of useful water from geothermal projects.

Noise from Geothermal Operations

Drilling operations in a geothermal field are comparable to construction activities in noise impact and are equally episodic. The noise problem is associated mainly with drilling operations and steam escape during testing. Once the field is in production, the noise level declines to that of other power plants. The drilling of dry-steam wells requires special techniques; at present, air is used for the circulating medium that removes the drill cuttings from the hole. This results in a temporary "controlled blowout" of the well that can be quenched at any time by pumping water down the drill string. When the well is completed, it must again be allowed to blow until the accumulated dust and rocks are removed from the borehole. During this clean-out period, and until the well is completed, it can be only partially muffled. The silencers now in use at The Geysers field during drilling and testing operations have significantly lowered the noise level in the field, and new developments promise further decline of noise levels.

Both solid fossil fuel and nuclear fuel require continual mining operations with blasting, operation of heavy machinery, and railroad or truck transportation of the fuels and waste products—far more individual sources of noise per unit of power than from the geothermal cycle.

Air Pollution from Hydrogen Sulfide

Gases are released to the air from each type of thermal power plant. Because the geothermal plant operates without combustion, the noxious gases produced are of much smaller quantity, and of a different nature, than those from a fossil-fuel plant (Bowen, 1973). The natural steam is predominantly water vapor—that at The Geysers yields 99.5% water vapor. The noncondensable gases present in the steam are about 80% carbon dioxide and lesser amounts of methane, hydrogen, nitrogen, ammonia, and hydrogen sulfide (Bruce, 1971). Of these, hydrogen sulfide presents the most serious environmental problem.

At The Geysers, hydrogen sulfide averages 1.5% of the noncondensable gases from the producing wells (Goldsmith, 1971, p. 31), or about 225 ppm of the steam. Because of the remoteness and the relatively small size of the power plants at The Geysers, and because the release per unit of power is lower than from fossil-fuel plants, the hydrogen sulfide emission has not caused the producers much concern in the past. However, the power company and the steam producers are making efforts to lower the hydrogen sulfide emission. Field studies and testing have shown that the hydrogen sulfide can be removed either by the addition of a metal catalyst to the circulating cooling waters or by burning the hydrogen sulfide to sulfur dioxide and then scrubbing it out. Long-term tests have shown that these methods will remove about 75% of the hydrogen sulfide. Hydrogen sulfide–abatement equipment is being added to Unit 11, due for operation in October 1974 (Allen and McCluer, 1973).

Subsidence in Geothermal Fields

Subsidence can occur whenever support is removed from beneath the ground. It has been noted in oil fields, in mines, and in areas being pumped for subsurface waters.

Because of the geologic circumstances under which dry-steam fields develop, subsidence should not be expected. The production reservoir of a dry-steam geothermal field consists of fracture zones, solution channels, or other permeable cavities filled with vapor at pressures much lower than normal hydrostatic pressure. Because this vapor does not add support to the overlying rocks, subsidence has not been a problem at any dry-steam field.

Hot-water fields could behave more like an unconsolidated petroleum reservoir, and subsidence might occur unless pressures are maintained by fluid return. Indeed, this has occurred in Wairakei, New Zealand (Hatton, 1970), where the water is not returned to the reservoir. Much has been learned about subsidence from the exploitation of petroleum reservoirs and, with the proper understanding and practices, any geothermal area where this could be a problem can be stabilized.

The hazard of subsidence from geothermal developments is minor compared with the total subsidence produced by other thermal plants (e.g., those in which coal is used).

Induction of Seismic Activity from Fluid Reinjection

The unstable conditions in the earth's crust leading to the presence of geothermal phenomena are also those conditions producing faults and earthquakes. Thus, geothermal and seismic phenomena are geographically inseparable. In fact, high seismic incidence is one of the exploration clues used in the search for geothermal reservoirs (Clacy, 1968). However, the intensity of individual shocks within the thermal areas is usually of a relatively low order, much lower than that associated with major crustal movements along faults (Ward, 1972). There is much to be learned about the relations of thermal and seismic phenomena, and the drilling and exploitation of geothermal fields should add new information to this field of knowledge.

There is no evidence that geothermal production has increased the seismicity of an area. Concern over seismic hazards arises in part from the process of reinjection of the spent geothermal fluids. Incidents of seismic activity resulting from the injection of fluids in waste-disposal operations (Evans, 1966) and from waterflooding operations to repressurize declining oil fields (Raleigh et al, 1970, 1971) have involved injection at pressures exceeding hydrostatic. Geothermal reservoirs are at subnormal pressures, and the return of fluids merely maintains preexisting pressures. The low pressure existing in geothermal reservoirs facilitates the reinjection of fluids into the field in two ways: (1) because reservoir pressure is less than hydrostatic, the water's weight produces sufficient head to ensure its entry into the formation without pumping; and (2) the returning water seeks the area of lowest pressure, thus minimizing the chances of migration of geothermal fluid into other aquifers.

Climate Modification from Release of Vapors

Geothermal power plants, because of their lower thermodynamic efficiencies, use more water per unit of power than do either fossil-fuel or nuclear plants; this water vapor could cause increased fog or other climatic alterations. However, the possibilities of climate modification by the discharge of steam from cooling towers should be considered from another perspective—the beneficial effects gained by adding moisture to the air. The activities of man through agriculture, construction, logging, urbanization, etc., all tend to increase runoff and lower the natural evaporation. It is possible that such activities in the high-rainfall belt of the Pacific Coast have significantly reduced the natural evapotranspiration, thereby reducing the water available for rainfall in the Rocky Mountains and midwestern United States. Thermal power plants in regions where water is relatively abundant may provide needed moisture to those downwind, but more investigation is needed on this subject.

Net Efficiency of Power Plants

In relation to power production, the term "efficiency" has come to mean the energy input:output ratio of fuel to electricity. The natural thermodynamic constraints placed on any electric generating machine require the rejection of 60–80% of the total energy input, leaving only 40–20% of the energy to do useful work. In the past, when cost of fuel was the main variable in the price of electricity, every effort was made to increase the thermodynamic efficiencies of power plants, because over the life of the plant a small increment of efficiency could amount to savings of millions of dollars. Because thermodynamic efficiencies have been the main variable, plant efficiency has become nearly synonymous with thermodynamic efficiency.

A major factor is the cost difference between types of power plants, which now varies more widely than in the past. Also, because different steps of the electric-power cycle are performed at localities other than the power plant, it is necessary to consider the off-plant part of the power cycle as well as the thermodynamic efficiencies of the generating plant. The net efficiency gives a true measure of the power cycle (Odum, 1971) and, in all cases, the more complex the conversion system, the lower is the net efficiency.

Recently, in Oregon, the staff of the Governor's Office of Energy Research and Planning made a detailed study of the net efficiencies of several modes of electric power production including geothermal steam, mine-mouth coal, natural gas, oil, and nuclear fission. This study measured energy inputs at all steps of the cycle: exploration, development and exploitation of fuels, transportation systems, construction of power plants, and transmission of electricity. Although this system is recognized as imperfect because of the great number of variables involved, it puts the different cycles on a basis that can be compared. From these data, two significant energy-input factors have been quantified: (1) the total energy input to achieve an output of 1,000 Btu's of electricity; and (2) the amount of fossil fuel as secondary inputs at various stages in the cycle. Table 1 shows these relations for coal, nuclear, and geothermal power systems. The relatively high net efficiency of the coal plant shows why its use has been so widespread and why it will continue to be the major source of electricity for many years.

A factor little considered when all forms of fuel were thought to be abundant is illustrated in the amount of external fuel subsidy for each of the systems. This represents the direct energy input in the construction, mining, processing, and transportation to build and operate the power plants. The simplicity of the geothermal system is apparent, as it needs less than a tenth of the fossil-fuel subsidy required by the more complex systems.

An understanding of the net efficiencies of the various methods of power production and conversion should cause a reevaluation of the priorities for energy development, which are presently dominated by the proponents of high technology. An example is Japan, where the 1974 research budget for development of geothermal and solar energy is 2.3 billion yen (US $7 million) versus a nuclear research budget of 67 billion yen (US $230 million; Mukaibo, 1974). Similar ratios exist in other "advanced" countries. If the energy crisis is to be solved, this ratio should be reversed.

Hazards to the Public

The simplicity of the geothermal cycle limits the number of potential hazards—there is no mining, no transportation of inflammable or dangerous fluids, and no wastes to be guarded. The maximum hazard from geothermal developments comes from well blowouts or spillage of brines. A geothermal well ejects a mixture of steam and water; there is no fire hazard and the water-pollution hazard is minimal. The unexpected blowout of a geothermal well is hazardous to workmen on the rig but not to the general public.

Conclusions

The environmental impact of any power-production system is reflected in the number and complexity of the steps in the fuel and production cycle. Because geothermal power plants utilize the energy occurring naturally in geothermal fluids, they do not need the complex generating equipment or extensive mining, processing, storage, or transportation facilities that other thermal power plants require.

The energy captured from geothermal fluids, whether it is used directly for space or process heating or is converted to electric power, will in effect cause an improvement in the environment to the extent that it displaces energy from more polluting sources.

The use of geothermal energy has a long history of successful operation. Seventy years of operating experience at Larderello and 20 years at Wairakei have shown geothermal systems to be economically and environmentally successful. It is known that geothermal energy exists in vast quantities; its resource base is second only to so-

Table 1. Energy Consumption and Net Efficiency[1]

	Btu Input for 1000 Btu's Electric-Power Output (293 watt hours)				*Net Efficiency*	*External Fuel Subsidy*
	A Primary Resource	*B External Resource*	*C Total Energy*	*Loss in Processing and Conversion*	*1,000 / C (%)*	*B / 1,000 (%)*
Mine-mouth coal	3,478	464	3,942	2,942	25.4	46.4
Nuclear fission	8,333	560	8,893	7,893	11.2	56.0
Geothermal steam	6,547	44	6,591	5,591	15.2	4.4

[1] From preliminary report by the Office of Energy Research & Planning, Governor's Office, State of Oregon.

lar energy. However, how much geothermal energy is sufficiently concentrated to be utilized economically is not known, and can only be answered by serious exploration. Delays and harassment are causing environmental degradation and rapid depletion of the world's store of valuable fossil fuels that should be used for purposes other than boiler feed. Geothermal energy could provide an important supplementary source of clean energy.

References Cited

Boldizsár, Tibron, 1975, Geothermal energy use in Hungary, *in* P. J. Lineau and J. W. Lund, eds., Multipurpose use of geothermal energy; proceedings of the international conference on geothermal energy for industrial, agricultural and commercial-residential uses: Geo-Heat Utilization Ctr., Oregon Inst. Technology.

Bowen, R. G., 1973, Environmental impact of geothermal development, *in* Paul Kruger and Carel Otte, eds., Geothermal energy; resources, production, stimulation: Stanford Univ. Press, p. 197-216.

Bruce, A. W., 1971, Geothermal power: on line, *in* Papers presented at the First Northwest Conference on Geothermal Power, May 21, 1971, Olympia, Washington: Washington State Dept. Natural Resources.

Clacy, G. R. T., 1968, Geothermal ground noise amplitude and frequency spectra in the New Zealand volcanic region: Jour. Geophys. Research, v. 73, p. 5377.

Evans, D. M., 1966, Man-made earthquakes in Denver: Geotimes, v. 10, no. 9, p. 11–18.

Hatton, J. W., 1970, Ground subsidence of a geothermal field during exploitation, *in* United Nations symposium on utilization of geothermal resources, Pisa, Italy: Elmsford, N.Y., Maxwell Scientific International, Inc.

Laird, A. D. K., 1973, Water from geothermal resources, *in* Paul Kruger and Carel Otte, eds., Geothermal energy; resources, production, stimulation: Stanford Univ. Press, p. 177–196.

Odum, H. T., 1971, Environment, power and society: New York, Wiley-Interscience.

Raleigh, C. B., et al, 1970, Earthquakes and water flooding in the Rangely Oil Field (abs.): Geol. Soc. America Abs. with Programs, v. 2, p. 660–661.

——— et al, 1971, Earthquake control at Rangely, Colorado (abs.): EOS, v. 52, p. 344.

Ward, P. L., 1972, Microearthquakes: prospecting tool and possible hazard in the development of geothermal resources: Geothermics, v. 1, no. 1, p. 3–12.

White, D. E., L. P. J. Muffler, and A. H. Truesdell, 1971, Vapor-dominated hydrothermal systems compared with hot-water systems: Econ. Geology, v. 66, no. 1, p. 75–97.

Magma Tap: the Ultimate Geothermal Energy Program—Summary[1]

JOHN L. COLP[2]

INTRODUCTION

The objective of the Direct Magma Tap Research Program now under way at Sandia Laboratories, Albuquerque, New Mexico, is to investigate the feasibility of extracting energy directly from deeply buried, circulating magma sources. These buried sources, with temperatures of the order of 1,000°C, represent great amounts of high-quality energy. Plan outlines call for a program whereby a fully closed heat-exchanger system is inserted directly into such a magma source to bring the heat energy to the surface with minimal environmental impact. The conversion of this heat energy to a usable form, such as electricity, at the surface could utilize fairly conventional techniques (Fig. 1). To determine the possibility or feasibility of these plans and whether they are economical is the objective of the Sandia program.

To accomplish such a program would be difficult. Nevertheless, if the energy of molten-rock sources could be tapped, the prize would be well worth the effort.

A program of this magnitude must be approached in an orderly, methodical manner if it is to achieve any degree of success. The problems must be well defined and must be approached in a manner that will permit their evaluation as the program proceeds. The program must be organized so that, should it become apparent that a real physical law precludes the possibility of successful completion, the regret costs of the entire program will be minimal. Any program devoted to tapping the energy of a buried magma source is bound to be a long-term one. Therefore, the program at Sandia is not directed toward any nearby deadline for power production. Steps are being taken cautiously, until the initial obstacles have been shown to be surmountable.

It is unlikely that the expertise for a program comprising such diverse fields will be available at any single installation. Therefore, Sandia Laboratories has formed a Magma Tap Advisory Panel of persons selected from installations throughout the United States for expertise in relevant fields. Their purpose is to provide advice on the directions in which the Sandia program should proceed, and each panel member is to provide advice in his specialty.

Three areas of this program currently require comprehensive analysis: (1) the composition of *in situ* magma, (2) techniques for locating and configuring subsurface magma sources, and (3) processes of drilling into molten-rock sources. Sandia proposes a series of workshops to assess the current state of knowledge, to decide what additional knowledge is needed, and to suggest directions for future research. These workshops will consist of a very small number of selected participants believed able to contribute to a meaningful analysis in each area of interest.

The Sandia Direct Magma Tap Program has been divided into five discrete program facets for this period of initial study: source, magma properties, material compatibilities, energy extraction, and source tapping.

SOURCE

Of critical importance to the program is the ability to locate and identify a source of molten rock in the subsurface. It is imperative that the existence of a magma source, its depth, its areal extent, and its general form (whether a finite pool or a honeycomb of crevices filled with molten material) be well known before proceeding with plans for drilling. Little is known about the precise locations and physical configurations of magma sources at this time. Existing knowledge is based almost entirely on intuitive evidence from areas near active volcanoes and eroded structures of ancient volcanoes. In the past, such information was of interest only to volcanologists.

Although many methods have been developed for the remote sensing of various subsurface geologic structures, until very recently none of these has been directed intentionally toward locating molten-magma sources.

Current Programs

In programs proposed for locating and identifying a subsurface magma source, most of the existing remote-sensing methods—seismic, microseismic, resistivity, gravity, magnetic,

[1]Manuscript received, October 21, 1974. This work is being carried out under the auspices of the U.S. Atomic Energy Commission.

[2]Technical Staff, Sandia Laboratories, Albuquerque, New Mexico.

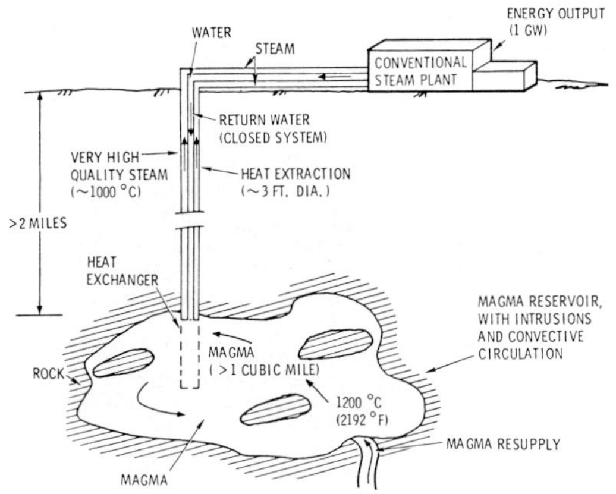

FIG. 1—Generalized drawing of magma-tap concept.

magnetotelluric, infrared, thermal—are being considered.

The most extensive program proposed is that of the U.S. Geological Survey, Menlo Park, California. The Survey plans to search for the molten magma sources on the Island of Hawaii, in the Yellowstone National Park area, in The Geysers area in northern California, and in the San Francisco Peaks area near Flagstaff, Arizona. This agency currently has work under way at some of these locations and plans to begin work at the others in the near future.

The Hawaii Institute of Geophysics at the University of Hawaii is proposing subsurface sensing surveys of magma sources in the Kilauea Volcano rift zone, and the Colorado School of Mines is proposing a similar program in the Kilauea summit area. The University of Alaska proposes magma surveys on Augustine Volcano in Cook Inlet.

A preliminary study was initiated by Sandia to evaluate possible methods for locating and identifying magma sources. A literature survey revealed that, although there had been much previous work in the field of subsurface instrumental sensing, very little work had been directed toward the search for molten rock. Only one instance of recognition of a molten region was found in the Free World literature, and very few references to Soviet work were noted.

A preliminary survey of possible locations of magma sources in the western United States followed these guidelines for selecting locations: (1) knowledge of a volcanic eruption in historic times, (2) knowledge of a volcanic flow less than 10,000 years ago, and (3) surface evidence of a very great thermal anomaly in the subsurface. About 35 possible locations in the western United States, Alaska, and Hawaii were selected.

Proposed Programs

The major thrust of the Sandia effort in source location and identification for the next year will be to cooperate closely with other agencies which are addressing this problem with subsurface sensing programs. The U.S. Geological Survey program is foremost in this area, and Sandia will work closely with its geothermal group.

In areas where Sandia has unique capabilities, a more active contribution will be made. Sandia believes that the earth-penetrator technology developed over the past several years may be applied directly to the thermal survey program proposed by the University of Alaska for implementation on Augustine Volcano. A study is under way to design a remotely implanted earth penetrator that will have thermal sensors at various locations along the penetrator body. At various intervals after implantation of the penetrator, the thermal gradient will be measured and telemetered to a receiving station for recording. The penetrator will be flown into virtually inaccessible areas. By use of a grid of such penetrators, a three-dimensional plot of thermal gradients near a suspected geothermal source can be determined.

One of the proposed workshops will be devoted to location and configuration of magma sources. The objectives will be (1) to determine the current methods and systems suitable for locating magma sources and for determining their subsurface configuration, (2) to discuss the validity of previous results, (3) to assess the requirements for additional knowledge, and (4) to suggest directions for future studies.

Magma Properties

Magma is known to be a high-temperature, highly corrosive environment. Much definitive information must be obtained about the character of this environment and its effects on engineering materials before heat-exchanger equipment is inserted into it for long periods.

Previous studies of magma have been performed by petrologists and geochemists to determine age, source, content, methods of formation, etc. Most of this work has been done on samples from past (ancient to recent) lava flows.

To our knowledge, there are no current or planned programs by others to study the *in situ* properties of magma as it relates to the compatibility of engineering materials with the magma environment. Sandia plans to direct much of its initial research efforts to this problem. The Material Science group has had extensive experience in material-compatibility studies and in the development of materials to survive in extreme environments.

Current Programs

The little information available about the *in situ* properties of magma has been obtained largely from submarine pillow-lava samples and from bombs that have been propelled from great depths and quickly cooled, thus sealing in the contained gases. Sandia has designed and built an instrumented penetrator capable of measuring properties of *in situ* magma during eruption, before the lava cools or is exposed to air (Fig. 2). Two penetrators were built to measure deceleration and temperature. The temperature-measuring instrumentation consists of two radiometers mounted along the sides of the penetrator and two thermocouples mounted on the nose. Each penetrator weighs 200 lb (90 kg), is designed to be air-dropped into a volcano, and is parachute-retarded to control impact velocity and trajectory angle. The intended use of these devices is to drop them into sources of erupting magma, permit them to penetrate the molten material, and telemeter out data on the thermal gradient of the material. Although the instrumentation and data-transmission equipment are expected to last only about 30 seconds in this environment, this is adequate time to get several feet of thermal-gradient measurements. These penetrators are ready for application whenever the opportunity arises to insert them into an erupting magma source.

A laboratory study of the gas evolution and vaporization of magma and lava at high temperatures and pressures is under way. As a result of Sandia's experience with the high-temperature vaporization of materials, a multiphased program is being conducted to characterize the volatilization and vaporization of solid and molten magmas and lavas. The program consists of the following:

1. Characterization of the noncondensable volatile gases as a function of origin of the magma and as a function of the hydrostatic pressure at the time or place of original solidification of the sample.

2. High-temperature vaporization properties of molten magma/lava with quantitative measurements of vapor pressures and heats of vaporization of the vaporizing species.

3. Pulsed-laser-induced and high-temperature, high-pressure vaporization of solid lava for determination of the relative abundances of vapor species.

4. Design of a technique for the measurement of gases in high-pressure equilibrium with magma or under conditions of high hydrostatic pressure at moderately high temperatures (1,000° or 2,000°K).

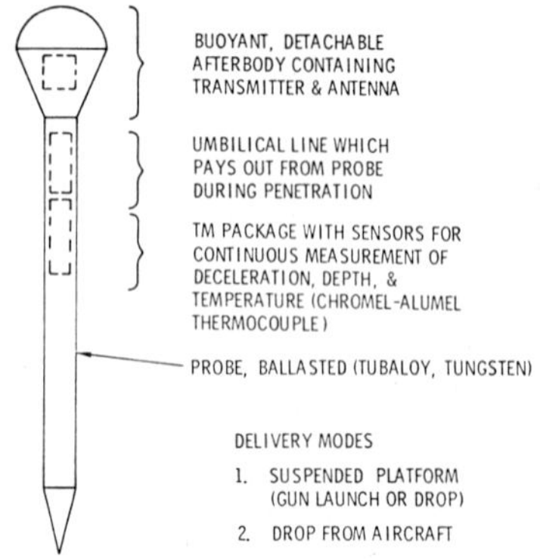

FIG. 2—Schematic drawing of Sandia Magma Penetrometer for measuring *in situ* properties of molten rock in lava lakes.

Also under way is laboratory study of electrochemical methods to measure oxygen activities, to increase the efficiency of energy extraction, and to protect materials used in a magma-tap application. The three proposed areas of research are:

1. Development and use of a solid electrolyte oxygen probe capable of making *in situ* oxygen-activity measurements in magma, using the Sandia penetrator technology. The resulting measurements will be the first *in situ* oxygen determinations of magma.

2. Development of a fuel cell which uses the reduced iron present in magma as the fuel and/or direct oxidation of magma to increase the amount and efficiency of energy withdrawal.

3. Investigation of electrochemical methods for materials protection in magma. This research must be conducted in conjunction with detailed material-compatibility tests. When all corrosive reactions are determined, various cathodic and anodic protection schemes can be proposed and tested.

Proposed Programs

Another workshop will address the current state of knowledge on magma material characteristics. The objectives of this workshop are (1) to determine the state of knowledge concerning the material characteristics (physical, thermal, electrical, chemical, mineralogical) of representative types of magmas, particularly in the *in situ* state; (2) to discuss the validity of the results constituting this state of knowledge; (3) to assess the requirements for additional knowledge; and (4) to suggest directions in which future studies should proceed.

Material Compatibilities

The ability to build a heat-transfer device that can be inserted directly into a magma source is critical to the success of the Direct Magma Tap Program. The materials developed for building a heat exchanger must retain their physical and structural properties sufficiently to allow the heat exchanger to perform for many years if the installation is to have economic value.

Little is known about the compatibilities of engineering materials in this environment. What is known has resulted chiefly from empirical applications. Samples of molten magma in metal pots have been obtained from lava eruptions. Observations have been made of the performance of some types of materials plunged into lava lakes through holes drilled in their surfaces. To my knowledge, no laboratory analyses of the mechanisms involved in the degradation of engineering materials in the magma environment have been performed.

Current Programs

The problem of developing materials compatible with a magma environment does not appear insurmountable, as there are analogies between the composition of magma and that of slags encountered in metallurgical processing. Existing materials which may meet the requirements are the heat-resistant metal alloys and the refractory compounds. Combinations of these two classes of materials in the form of cermets, if commercially available, may also be worthwhile for this application.

Preliminary data on the interaction of various representative alloys with degasified molten basalt indicate that a good correlation may be made between reactivity and the free energies of formation of the various metal oxides. These data further indicate that the nickel- and cobalt-based alloys may have special qualities in regard to the magma environment.

The refractory compounds, primarily Al_2O_3, MgO, and ZrO_2, have had extensive application in metal production processes at temperatures up to 1,700°C and exhibit good resistance to the corrosive action of slag. Their application in the magma environment may be limited by their relatively poor tensile strengths.

Magma is largely composed of silicates of Ca and Mg, but contains lesser amounts of compounds of Fe, Al, Ti, Na, and K. Within the earth, magma also contains dissolved substances, primarily H_2O, CO_2, and SO_2, which volatilize from the magma as it approaches the surface. These latter substances are anticipated to have significant effects on the corrosiveness of the magma. A case in point is the accelerated hot-gas corrosion of gas turbines, termed "hot corrosion," induced when superficial melts of Na_2SO_4 attack the protective metal-oxide film.

The current program is a survey of the high-temperature reactivity of metals with degassed molten basalt and will include refractory compounds and commercially available cermets. Individual specimens will be immersed in degassed molten basalt at temperatures between 1,100 and 1,300°C in a low-pressure atmosphere of argon. The extent and composition of any reaction products will be determined by using metallographic procedures in conjunction with analyses by the electron microprobe. These results will permit the formulation of criteria for more sophisticated materials selection and testing.

A second series of tests is being performed on those materials exhibiting the most resistance to degassed molten basalt. Specimens in contact with the basalt will be placed in high-pressure, high-temperature bombs together with varying amounts of H_2O, CO_2, and SO_2 (both individually and in combination) in order to quantify the effect of these compounds on both the rate and mechanism of corrosion and on product composition.

Proposed Programs

So little is known about material compatibilities with the magma environment that the future directions of the program are almost totally dependent on the results of the laboratory studies now under way, and on the results of the magma-characterization studies. It is fairly certain, however, that a part of the future program will involve insertion of material samples into natural lava lakes for extended periods and, hopefully, into areas of erupting lavas.

Energy Extraction

The convection and heat-transfer processes operational in molten rock at extremely high geodynamic pressures must be understood before suitable energy-extraction methods can be devised and meaningful estimates of their efficiencies can be made. No previous work along this line is known.

Sandia possesses excellent calculational and experimental facilities for approaching such a problem. A study of radioactive-waste disposal in deep rock cavities and the design of a heat-sink pit beneath a nuclear reactor to contain the molten core in the event of a major reactor meltdown accident are currently under way. Both of these problems are concerned with the convection processes in molten rock. Both theoretical studies and laboratory experiments, as well as a small-scale field experiment, of this convection process are in progress.

Current Programs

Preliminary calculations of long-term heat-extraction rates for assumed magma fields show that rates could run from 1 to 26 kw/sq ft (11–280 kw/m^2) of tube area, depending on the assumptions used. These calculation studies are proceeding, and the assumptions are being refined.

A long-tube boiler–heat-exchanger device ha been designed and built. Facilities are available at Sandia to melt 100 lb (45 kg) of basalt and maintain it in a molten state. The long-tube boiler–heat-exchanger unit will be assembled for insertion into this pot of molten basalt (Fig. 3). The primary purpose of this experiment is to study such phenomena as the temperature gradient across the magma and into the heat exchanger. This experiment should show that the temperature of the metal tube and the magma adjacent to it can be maintained low enough that corrosion of the tube is not a critical problem. The thermal transient problem in the tube associated with start-up will also be examined. Design concepts for a more efficient boiling heat exchanger, compatible with the long tube shape required to place heat-extraction pipes in a magma source, are being considered. Other heat-exchanger concepts include a Brayton cycle in which a closed cycle with gas is used to extract heat from the magma. The extraction tube in the magma would use a recirculating gas as the cooling fluid in the heat exchanger. This hot gas would run conventional gas turbines (Brayton cycle) to generate electricity.

If studies determine that the configuration of magma sources is likely to resemble large liquid pools, it will be necessary to calculate convection rates in the pool and adjacent to the heat-extraction pipes. Analytical and numerical techniques must be developed for convection problems. This development work has begun at Sandia in support of the reactor-safety and waste-management proposals.

Proposed Programs

Assuming success in the present laboratory heat-exchanger program, the next step will be to begin design of a demonstration heat exchanger using materials determined to be compatible with

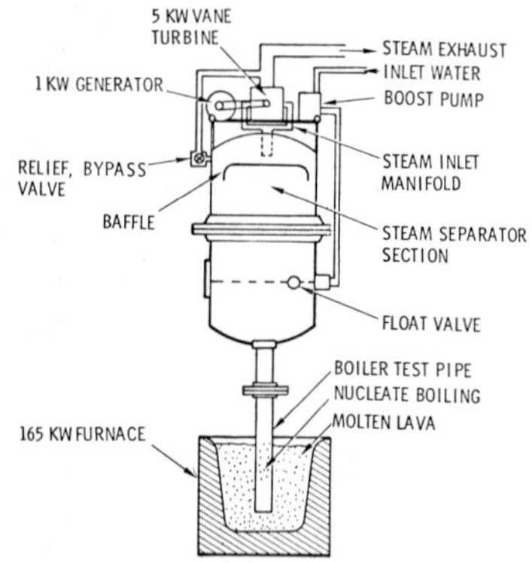

FIG. 3–Schematic drawing of laboratory test apparatus for demonstrating magma boiler tube concept.

the magma environment. The eventual objective will be a heat exchanger suitable for use in a pilot plant installed on an existing lava lake to demonstrate the utility of the magma-tap method of extracting energy.

SOURCE TAPPING

To be able to reach the source and insert the heat exchanger is a problem whose solution is vital to a magma tap. In many locations, magma sources will be at depths greater than 10 km. Conventional drilling technology at this time is limited to depths of about 30,000 ft (10 km). Difficulty has been encountered in past attempts at deep drilling in areas thought to be near magma sources because of the increasing temperature with depth. The Sandia magma-tap program is not directly involved in developing deep-drilling methods. The drilling subcommittee of the U.S. Geodynamics Committee currently recognizes the desirability of developing deep-drilling techniques that may reach the magma environment. The Los Alamos Scientific Laboratory has an ongoing program to develop a rock-melting drill that, if successful, would be an ideal means for tapping a magma source.

Current Programs

Although not a part of the magma-tap effort, Sandia does have an ongoing program in drilling research and development. The objective of this program is to devise methods for improving the speed and depth of drilling. Methods being investigated under this program may be applicable to the magma-tap problem.

Near the magma source, the rocks will have increasingly higher temperatures and will approach a plastic state. Sandia Laboratories is involved in a research program with the Center for

Tectonophysics at Texas A&M University to investigate the physical properties of rocks under dynamic stress conditions at pressures up to 4 kb and at temperatures of approximately 1,000°C.

Proposed Programs

Active participation in deep-drilling programs conducted by others may become feasible in the near future. It may be several years before the magma-tap program has achieved enough success to warrant any deep drilling into magma sources. Much valuable information can be gained from conventionally drilled holes in existing lava lakes for pilot-plant operations.

CONCLUSIONS

Tapping the energy stored in the molten rocks below the surface of the earth has been the dream of many for a long time. There is no doubt that these subsurface reserves of heat energy are enormous. However, no serious program has yet addressed the problems necessary to release this energy for man's use. That is the stated objective of the Direct Magma Tap Program.

The problems associated with locating, identifying, and configuring the sources of this subsurface energy are recognized and are being faced. Programs are under way to evaluate this problem and, eventually, to overcome it.

The severity of the magma environment and its highly detrimental effects on materials placed in it are also well recognized. Although solutions are not now known, persons knowledgeable in modern materials technologies believe that the problems are not insurmountable.

Tapping a magma source with a drill has been achieved only in surface lava lakes. Although problems lie ahead before deep sources can be drilled, these are basically engineering problems that may be solved.

Extraction of heat energy from a magma source also faces problems that require engineering solutions. The current programs indicate that the desired objectives can be achieved.

Geothermal Energy in New Zealand—Summary[1]

JAMES HEALY[2] and RUSSELL JAMES[3]

Introduction

Drilling for thermal water for domestic purposes commenced in Rotorua in the 1930s and rapidly developed on a commercial scale in the 1940s, when it was discovered that steam could be obtained by drilling deeper. Exploration for steam to generate electric power commenced at Wairakei in 1950 during a period of power shortage. Since then, the use of geothermal energy has been developed at Wairakei and Kawerau, and investigations have been extended into other fields.

The initial aim at Wairakei was to obtain steam at high pressure, but drilling soon showed that the young volcanic strata beneath the field are saturated with water in which the temperature increases downward in accord with the boiling point for the increasing pressure, to a limiting value of about 260°C. Steam flashes from the hot water rising in the wells as the pressure falls, and is separated at selected pressures at the surface and put through turbines.

At Wairakei the sole product is electric power; however, at Kawerau the steam is used for process purposes in the Tasman Pulp and Paper Company mill, and only surplus steam is used for the generation of electric power up to a maximum capacity of 10 Mw. Other fields prospected by drilling are Waiotapu, Reporoa, Te Kopia, Orakeikorako, Tauhara, Rotokawa, and Ngawha. With the exception of Ngawha, which is located in the Northland area, all the fields lie within the Taupo volcanic zone in the central North Island (see Healy, this volume, Fig. 2).

General Background

Steam was initially sought at Wairakei in the belief that the heat came from steam emitted from magma at depth. It is now believed that the water is of meteoric origin and has circulated deeply and widely in the volcanic zone, becoming heated by passage through hot zones in which the heat flow is many times normal (Healy and Hochstein, 1974; Healy, this volume). Horizontal flow from Waiotapu to Reporoa has been inferred by analogy with Chilean examples on the basis of evidence from down-hole measurements in both areas and chemical similarity of the waters (Healy and Hochstein, 1974).

Indirect evidence of the part played by regional groundwater movement comes from heat-flow measurements by Studt and Thompson (1969) in widely separated areas of the North Island. Whereas zero geothermal gradients were found to exist in and adjacent to the Taupo volcanic zone, the sum of the estimated heat outputs from the known geothermal areas in the zone is equivalent to 20 times the normal heat flow over the entire zone. They concluded that much of the volcanic zone constitutes the recharge area for the concentrated upflows of hot water.

Geophysical methods applied to geothermal prospecting have been described by Hatherton et al (1966). Most important has been electrical resistivity survey. The entire strip from Waiotapu to Lake Taupo has been surveyed using a Wenner constant-electrode spacing of 550 m. The geothermal fields of Wairakei-Tauhara, Rotokawa, Broadlands, Wharepaina (Reporoa), and Waiotapu lie within areas of resistivity of less than 10 ohm-m along the large valley extending throughout the belt. The floor of the valley generally is within the 100–ohm-m zone, outside of which resistivity rises above 200 ohm-m.

This would suggest that the valley is underlain by warm water, which may represent discharge to the main drainage channels from a deeper level as part of the regional groundwater system. Deeper penetration by greater electrode separation and by dipole surveys at Broadlands indicates that the boundary of the hot zone there is almost vertical. However, as lateral flows of hot water can lead to the accumulation of high temperatures beneath them owing to normal heat flow, it does not follow that low resistivities to great depth imply upward flow of hot water. Risk et al (1970) found low resistivity to extend down to at least 3 km at Broadlands, though it increases gradually near the bottom. The western boundary is sharp, and the eastern is less so. At Kawerau, Macdonald and Muffler (1972) found similar relations. The

[1]Manuscript received, October 21, 1974. Published with permission of the Directors of New Zealand Geological Survey and Chemistry Division.

[2]Department of Scientific and Industrial Research, New Zealand Geological Survey, Rotorua, New Zealand.

[3]Chemistry Division, Department of Scientific and Industrial Research, Wairakei, New Zealand.

western boundary, which is fault-defined, is sharp, but the eastern is diffuse. No systematic measurements have been made outside the geothermal fields to examine possible indications of inflow at depths of 1,000–2,000 m, though areal measurements are presently under way in the Waiotapu area, where several superficial hydrothermal areas are considered likely to be connected at depth.

Minor temperature reversals are found in the lowest 150 m of a hole drilled to 2,250 m at Wairakei. Stabilized temperatures down to 2,160 m in a hole at Broadlands have values of 307°C at 1,340 m, 297°C at 1,770 m, and 308°C at 2,160 m. Temperatures in the Kawerau field also indicate reversals. Temperature data of this type are believed to indicate that the flow of hot water has a horizontal component. Bolton (1970) presented evidence for a deep inflow of hot water to the Wairakei field at its southwest margin where, at the level of production within the field, the country rock is cold.

WAIRAKEI GEOTHERMAL FIELD

The Wairakei geothermal field has been exploited for a longer period than any other large hot-water field. The geology of the field was described by Grindley (1965) and Healy (1965). Changes during production were summarized by Bolton (1970), and include falling pressure, falling temperature in the production area, decreasing mass discharge, and subsidence of ground level. There was no change in natural heat discharge, though thermal springs ceased to flow and fumarole discharges increased. During a partial shutdown in 1968, evidence indicated the existence of an inflow to the field of water at least as hot as the highest temperature measured in the wells and equal to two thirds the mass discharge at the time, which was at least 4.5 times the original natural recharge of hot water at 260°C.

All geothermal fields under exploitation undergo a decline in reservoir pressure and, to a lesser degree, a fall in deep temperatures. At Wairakei, the operating wellhead pressure has declined by about 4–5 psi or 27.6–34.5 kPa (kilopascals) per year, and the pressure of the steam entering the high-pressure turbines has dropped from its design figure of 180 psig (psi, gauge pressure) or 1,241 kPag (kPa, gauge pressure) to the present value of 118 psig or 813.6 kPag during a period of 15 years. Under a given rate of draw-off, the decline in the reservoir pressure gradually is reduced and may finally stabilize at some fixed value; a greater draw-off rate results in a lower final reservoir equilibrium pressure. Extending the power plant at Wairakei, and thus increasing the discharge rate, could result only in a further decline in reservoir pressure, necessitating more wells and leading to diminishing economic returns. These factors have arrested further plant extensions beyond the installed capacity of 192 Mw of generated electricity in the field. The present rate of power generation at Wairakei is 148 Mw.

Investigative studies, however (James, 1967; 1968a, b), show that the plant could have been extended to about 300 Mw of generated electricity if the turbine inlet pressure had been designed at the much lower value of 50 psig (344.75 kPag) instead of 180 psig (1,241 kPag), and if the whole mass of well fluid had been transmitted simultaneously as a steam-water mixture instead of using only steam. The same draw-off rate would give greatly increased power, and the operational life of the field would be much increased; the reason is that the wellhead pressures could then be permitted to decline to about 75 psig (517 kPag) before affecting the turbines (a pressure drop of 25 psi [172.4 kPa] is required between the wellheads and turbine owing to the long overland transmission pipeline). On reaching the powerhouse, the steam-water mixture would be separated in two pressure stages so that the steam would enter the turbines at an inlet pressure of 50 psig (341.75 kPag) and a pass-in pressure of 2 psig (13.8 kPag).

The evidence for recharge of hot water to the field emphasizes the major role of recharge in the life of a field. Early estimates of the geothermal resources at Wairakei were based on calculations of stored heat, held mostly in hot rock. The amount of this heat extracted has been relatively small.

Bolton (1970) emphasized the apparent lack of permeability in the marginal walls of the field, and suggested mineral deposition as the probable explanation. Development of a two-phase mixture also can greatly reduce permeability. However, by analogy with known fields with horizontal flow of hot water (Healy and Hochstein, 1974), mineral deposition in the marginal zone of cooling appears likely.

KAWERAU GEOTHERMAL FIELD

The Tasman Pulp and Paper Company mill was located at Kawerau to make use of geothermal steam. In 1956–57, the company drilled 8 wells, 4 of which proved productive. They tapped an aquifer in rhyolite lava and breccia with interbedded strata in which the temperature and pressure had declined appreciably by 1960. Deeper drilling obtained increased and sustained production from a fractured andesite aquifer which is presently being developed further. On the basis of down-hole temperatures, it appears that recharge is horizontal through the andesite from an unknown source direction.

The natural heat discharge from the Kawerau field was 105 Mw (25,000 kcal/sec). The present mass output of 180 MT/hr carries a heat discharge of half the natural amount. By comparison with Wairakei, the present mass output of the wells might be increased by a factor of 8. However, no field data are available on which future behavior could be predicted with any degree of accuracy.

Maximum temperature measured in the field was 285°C at 829 m in well 8. However, the maximum well temperatures range generally between 260° and 270°C. Although the highest temperature was measured in the eastern side of the production area, there appears to be some encroachment of water, at temperatures below 250°C, from that direction during production. The area of the field outlined at a depth of 0.5 km by electrical resistivity survey is 6–10 km^2 (Macdonald and Muffler, 1972). The minimum stored heat above 100°C down to 2 km is 3.9×10^{18} to 6.5×10^{18} joules, and between 200°C and 250°C it is 1.1×10^{18} to 1.9×10^{18} joules (Macdonald and Muffler, 1972). However, owing to poor permeability in the rock above and below the andesite aquifer, there is no indication that much of this can be extracted.

Broadlands Geothermal Field

Broadlands has undergone the most complete predevelopment investigation of the geothermal fields in New Zealand. The area of the field is 10 km^2, and the maximum temperature is 308°C at a depth of 2,421 m in well 15. Production has been obtained from two levels—one at 500–550 m in tuff and breccia in the western part, and the other at 750–950 m in volcanic breccia, generally farther east.

Conditions in the field were summarized by Studt (1968). The field is visualized as a vertical cylinder in which hot water moves slowly upward. There is no certainty as to whether inferred fault zones provide channels for water moving upward, or merely provide local permeability. A positive gravity anomaly covering the field is believed to be due to densification of the rock by deposition of secondary minerals. On the basis of chemical evidence, a major upflow was inferred in the western area where most wells produce from the upper aquifer. Increased boron content in the waters of wells 7 and 10 in the southeast part of the field was ascribed to passage of the water through basement graywacke, found in well 7. However, graywacke conglomerates overlying the basement could equally well be the source. Temperatures above 300°C exist in the basement, but thermal gradients appear to be conductive and not indicative of permeability.

Permeability in the field is relatively low, although large discharges have been obtained from a few wells. Relatively large drawdowns occurred during discharge, and discharge enthalpies higher than those of the measured temperatures indicated that flashing was occurring. Healy (1968) suggested that this indicates that the fissures are of limited extent, and Grindley and Browne (1968) considered that the field could be discharging dry steam after a few years of production.

The natural heat discharge was 83.7 Mw (Dawson and Dickinson, 1970). The combined measured heat discharge of the wells to date at Broadlands is 820 Mw (G. W. Hitchcock, personal commun.), which is 10 times the natural heat discharge, and about twice the overdrawal rate in operation at Wairakei. Production at this level would seem likely to lower pressure rapidly in the reservoir and cause the wells to discharge steam. Although steam has a much higher enthalpy than water, and permits a greater extraction of energy for a lower rate of recharge, its formation will result in mineral deposition at the water table, which will tend to reduce permeability. However, it will occur at lower levels as the water level falls, and the steam will permeate the rock at higher levels and extract stored heat without further reduction of permeability in that zone, provided the reduction in pressure does not permit colder water to enter.

In the southeast part of the field, well 7 is presently producing a wetter discharge, and the temperature at the top of the lower production aquifer (670 m) has fallen, indicating inflow of colder water, presumably from the east.

Other New Zealand Fields

The Tauhara field, separated from the Wairakei field by the Waikato River, was investigated by 4 deep wells and 4 shallow monitor wells. It is now regarded as part of the Wairakei field, because falling pressures and water levels indicate communication between the two. Superficial temperatures have increased in a few localities.

Four wells were drilled in the Orakeikorako field and two in the adjoining Te Kopia field, both of which are located on the major Paeroa fault and its ramifications. It is now considered, on the basis of hydrologic and chemical evidence, that the two are connected, and that there are two hot aquifers in which horizontal flow of geothermal fluids occurs. Drilling to intersect fault planes at Orakeikorako proved unsuccessful on this account.

General Comments

This paper is concerned chiefly with the physical aspects of geothermal energy in New Zealand. However, geochemical studies are important in geothermal exploration, and even more important during the development and productive life of a field, when chemical data are used increasingly to explain the changes occurring in the aquifers. Geochemical work in New Zealand fields was reviewed by Mahon (1970).

Apart from exploring for and developing geothermal fields, study of their behavior as a means of improving field management during their productive life is a relatively new art.

Maximizing the power life of a geothermal hot-water reservoir by designing the turbine inlet pressure to be about 50 psig (344.75 kPag) would be equally applicable to dry-steam fields such as The Geysers, which operates with a turbine inlet pressure of 100 psig (689.5 kPag). This figure of 100 psig was derived from measured well-output curves which correlate mass discharge with well-head pressure; such an approach would be satisfactory if these curves never changed—but the latest information (Budd, 1973) shows a decline in output to one half the original discharge from some typical wells in a period of 5 years. Thus, the best approach is to design for low pressures to sustain output. However, because the specific volume of steam increases markedly with fall in pressure, turbine and other plant sizes must be large, thereby increasing costs at low pressures. For this reason, it is uneconomical to reduce the turbine inlet pressure much below 50 psig.

These arguments apply equally to concepts utilizing a low-vapor-pressure fluid in a closed circuit. Such binary systems also would have to be capable of operation at temperatures not much higher than 150°C in order to have a life comparable to that of the more conventional flashing system. However, even when the power lives are the same, the binary system would be much more costly to build, necessitating a very large and expensive heat exchanger. This situation may change if the binary system is designed in conjunction with a deep-well pump capable of sustained operation at temperatures up to about 300°C (such a pump is not presently available), as this may permit the power plant to be pressurized to a high value, independent of declining reservoir pressure.

For nuclear and fossil-fuel power stations, the cost per kilowatt installed decreases with increasing size of the power plant. For geothermal power, it has been found that small stations at the sites of wells can be erected at a cost per kilowatt which is no greater (and in some cases is smaller) than that of large stations drawing fluid from many wells (James, 1970b); the reason is that overland transmission pipelines are not required.

It is not yet possible to estimate the power life of geothermal reservoirs by indirect means, and so the method employed commonly is to exploit them in steps. A first step might be a power station of about 50 Mw of generated electricity, which is later increased by 50-Mw increments until decline in the underground pressure is considered serious enough to halt further expansion. Continuous monitoring of the underground pressures and temperatures indicates the changing power potential of the field. Discharge from a field may, as at Wairakei, stimulate a considerable inflow which helps to sustain production and which is virtually unpredictable prior to fairly large discharges over several years.

The very latest nuclear and fossil-fuel power stations have a thermal efficiency of about 40%. The thermal efficiency of hot-water reservoirs is about 10%, and that of dry-steam reservoirs, about 20%. These efficiencies are not strictly relevant to power economics, as the important figures are the two cost criteria—capital cost/Kw installed, and generating cost in cents/Kw-hr.

The value of thermal efficiency gives an estimate of how much heat has to be rejected at the power plant, but this must be taken into account in the design of a station anyway. As the price of fossil fuels rises, there undoubtedly will be technical advances in the design of stations that burn these fuels, in order to raise the thermal efficiency above the present maximum of 40%. However, these improvements will only hold the costs to reasonable (or competitive) levels.

The effect of the so-called energy crisis has been to raise fossil-fuel prices, which undoubtedly will lead to price increases in nuclear fuels also. These price escalations will improve the competitive position of geothermal power and any other forms of energy conversion not dependent on a bought commodity.

For geothermal power based on hot-water reservoirs, considerable cost improvement can also be made if the hot water rejected from cyclone separators is sold for domestic heating or for industrial use. As the price of heating oils increases, hot water should soon become more attractive. Superheating of geothermal steam, using fuel oil or pulverized coal, results in lower generating costs. However, because of the low pressure of the steam employed (James, 1970a), this procedure does not increase the thermal efficiency of the fuel portion of the cycle, and will become less attractive with increased fuel costs.

References Cited

Bolton, R. S., 1970, The behaviour of the Wairakei geothermal field during exploitation, *in* United Nations symposium on the development and utilization of geothermal resources: Proc., v. 2, pt. 2, Geothermics, Spec. Issue, no. 2, p. 1426–1439.

Budd, C. F., Jr., 1973, Steam production at The Geysers power plant, *in* P. Kruger and C. Otte, eds., Geothermal energy; resources, production, stimulation: Stanford Univ. Press, p. 129–144.

Dawson, G. B., and D. J. Dickinson, 1970, Heat flow studies in thermal areas of the North Island of New Zealand, *in* United Nations symposium on the development and utilization of geothermal resources: Proc., v. 2, pt. 2, Geothermics, Spec. Issue, no. 2, p. 1310–1322.

Grindley, G. W., 1965, The geology, structure, and exploitation of the Wairakei geothermal field, Taupo, New Zealand: New Zealand Geol. Survey Bull., no. 75, 131 p.

——— and P. R. L. Browne, 1968, Geology of the Broadlands field, *in* Report on geothermal survey at Broadlands 1968: New Zealand Dept. Sci. and Indus. Research Geothermal Rept., no. 5, p. 5–43.

Hatherton, T., W. J. P. Macdonald, and G. E. K. Thompson, 1966, Geophysical methods in geothermal prospecting in New Zealand: Bull. Volcanol., v. 15, p. 485–498.

Healy, J., 1965, Geology of the Wairakei geothermal field: 8th Commonwealth Mining and Metallurgical Cong. Proc., v. 7, paper 218, 13 p.

——— 1968, Geological report on Broadlands geothermal field: New Zealand Geol. Survey, open file rept., 51 p.

——— 1976, Geothermal prospects around the Pacific: this volume.

——— and M. P. Hochstein, 1974, Horizontal flow in hydrothermal systems: New Zealand Jour. Hydrology, v. 12, p. 71–82.

James, R., 1967, Optimum wellhead pressure for geothermal power: New Zealand Eng., v. 22, p. 221–228.

——— 1968a, Pipeline transmission of steam-water mixtures for geothermal power: New Zealand Eng., v. 23, p. 55–61.

——— 1968b, Second generation geothermal power: New Zealand Eng., v. 23, p. 230–236.

——— 1970a, Superheating of geothermal steam for power: New Zealand Eng., v. 25, p. 325–330.

——— 1970b, The economics of the small geothermal power station, *in* United Nations symposium on the development and utilization of geothermal resources: Proc., v. 2, pt. 2, Geothermics, Spec. Issue, no. 2, p. 1697–1704.

Macdonald, W. J. P., and L. J. P. Muffler, 1972, Recent geophysical exploration of the Kawerau geothermal field, North Island, New Zealand: New Zealand Jour. Geology and Geophysics, v. 15, p. 303–317.

Mahon, W. A. J., 1970, Chemistry in the exploration and exploitation of hydrothermal systems, *in* United Nations symposium on the development and utilization of geothermal resources: Proc., v. 2, pt. 2, Geothermics, Spec. Issue, no. 2, p. 1310–1322.

Risk, G. F., W. J. P. Macdonald, and G. B. Dawson, 1970, D.C. resistivity surveys of the Broadlands geothermal region, New Zealand, *in* United Nations symposium on the development and utilization of geothermal resources: Proc., v. 2, pt. 1, Geothermics, Spec. Issue, no. 2, p. 287–294.

Studt, F. E., 1968, Summary and conclusions, *in* Report on geothermal survey of Broadlands 1968: New Zealand Dept. Sci. and Indus. Research Geothermal Rept., no. 5, p. 1–4.

——— and G. E. K. Thompson, 1969, Geothermal heat flow in the North Island of New Zealand: New Zealand Jour. Geology and Geophysics, v. 12, p. 673–683.

Possible Energy Sources in Indonesian Volcanic Areas—Summary[1]

DJAYADI HADIKUSUMO, L. PARDYANTO, and M. ALZWAR[2]

INTRODUCTION

We recognize two types of geothermal areas—volcanic and nonvolcanic. This paper concerns only the volcanic geothermal areas, which show surface manifestations in the main volcanic belts of the Indonesian archipelago (Fig. 1): (1) The Sunda-Banda mountain system, extending from the northern tip of Sumatra to the Banda Sea; (2) the Minahasa-Sangihe-Ragay zone, extending from the northern part of Sulawesi (Celebes) into the Philippines; and (3) the Halmahera-Ternate zone on the southwest border of the Pacific.

Within these areas the temperature gradient (3°C/100 m) is above normal. Generally the heat transfer takes place convectionally, and conduction occurs only under special conditions. The convectional heat flow might be caused by intrusion of magma at shallow depth.

In these thermal areas, hot springs indicate subsurface temperatures ranging from 100 to 300°C (Healy, 1970). Such phenomena are generally observed in areas of late Tertiary and Quaternary volcanoes, such as New Zealand, Iceland, Japan, Taiwan, the Philippines, and Indonesia.

Distinct hydrothermal activity also occurs at older, eroded volcanoes. Such semivolcanic activity is closely related to shallow igneous activity, which causes a temperature increase in the circulating groundwater. These manifestations cover a vast geothermal area, including parts of Sumatra, Java, Sulawesi, and Flores.

Judging from the distribution of the geothermal surface manifestations in the volcanic belts, it might be assumed that about 90% of the geothermal fields in Indonesia are of the volcanic type.

HISTORY OF GEOTHERMAL EXPLORATION IN INDONESIA

Although the need for electric power in Indonesia before World War II was relatively low, geologists and volcanologists working in this country at that time were stimulated by the development of geothermal areas in Italy.

Van Dijk (1919) was the first to propose the utilization of geothermal energy for electric power in Indonesia, although he had made only a brief survey. However, Escher (1920) doubted that the solfatara and fumarole fields could be utilized for power because their distribution in Indonesia was not extensive and they were located at high altitudes far from the cities. He also believed that volcanic gases are highly corrosive.

Taverne (1925) and van Bemmelen (1928) supported the idea of developing geothermal potential in volcanic areas. Detailed investigations were followed by shallow exploration drilling at five localities in Kamojang geothermal field (west Java) in 1926–28. Continuous investigations on the pressure and temperature and other characteristics of these wells were pursued until 1942.

Stehn (1929) calculated that, with a steam discharge of 8,000 kg/hr, a temperature of 123°C, and a pressure of 2.5 atm abs, a generator could be driven to produce electricity to a capacity of 900 kw. The actual experimental use of steam to generate electric power seems not to have been carried out in Kamojang. One of the five shallow holes (well 3) has produced steam continuously for about 48 years. Study of the utilization of geothermal power stopped during World War II and was not pursued again for about 22 years.

A report by Kusumadinata (1964) renewed interest. A UNESCO Volcanology Mission to Indonesia in 1964 made preliminary studies and recommended two geothermal volcanic areas—Dieng (central Java) and Kamojang—for further exploration.

In 1968 a team from the Power Research Institute (PRI), together with the Bandung Institute of Technology (ITB), made reconnaissance surveys of obvious geothermal areas in Java. A mission from France, "Eurafrep," joined the PRI and the Geological Survey of Indonesia (GSI) teams in Java and Bali the same year. Their recommendations were similar to those of the former team but, unfortunately, they did not lead to further activities. Nevertheless, the report provides background information for further work. Geothermal studies were carried out by GSI during the first Five Year Development Program, 1969–74 (PELITA I). A joint feasibility study of geothermal energy was carried out by New Zealand experts and GSI in 1971, under a Technical Assistance Program of the government of New Zealand to

[1]Manuscript received, February 10, 1975.

[2]Geological Survey of Indonesia, Bandung, Indonesia.

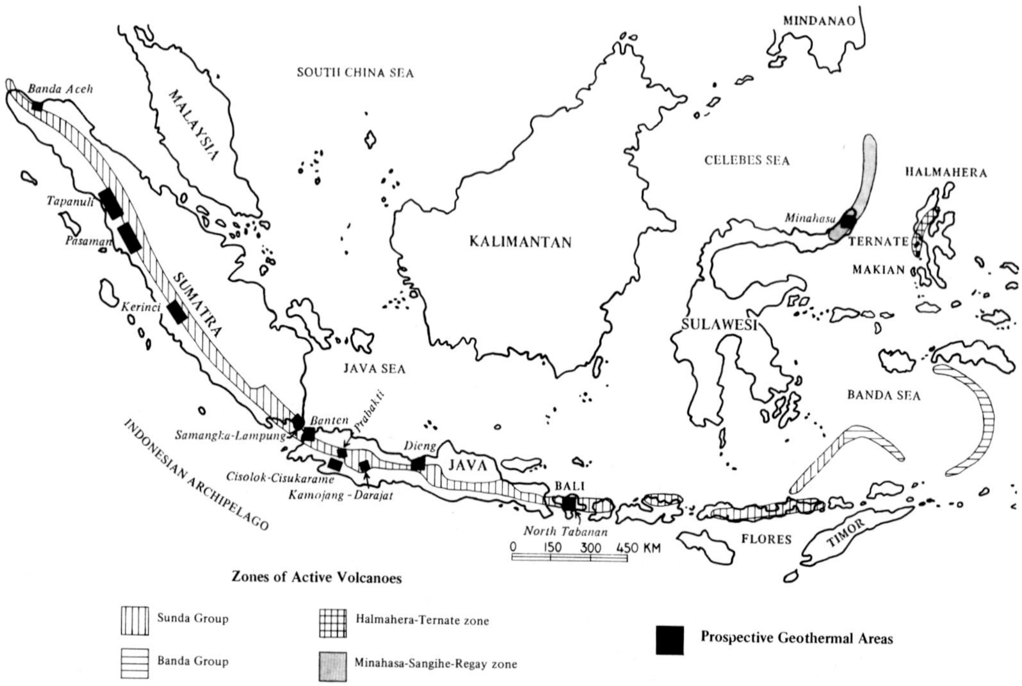

FIG. 1—Map of Indonesian archipelago showing zones of active volcanoes and prospective geothermal areas (see Table 1).

the government of Indonesia under the Colombo Plan. Exploratory drilling of geothermal localities (Fig. 1) will be carried out at Kamojang-Manuk-Darajat, Cisolok-Cisukarame, and North Tabanan toward the end of 1974.

In view of the present world energy crisis, the Indonesian government is making an effort to explore all potential natural energy resources. First priority has been given to the geothermal project under the Colombo Plan. It will be continued not only in areas already selected, but also in other geothermal areas remote from the centers of conventional power supply. By Presidential decree, participation by Pertamina commenced early in 1974.

Volcanic Geology

The Indonesian archipelago is situated geographically between the Asian and Australian continents; geologically, it is composed of parts of three crustal plates (van Bemmelen, 1949): (1) the Asiatic plate on the north which subsided to the south as far as the Sunda Shelf; (2) the ancient Gondwanaland on the west and south which divided the Indian mainland, the Indian Ocean floor, and Australia, and which includes the subsided Sahul Shelf on the north; and (3) the Pacific Ocean floor on the northeast.

The main physiographic units are the continental platform, the oceanic basins, and the orogenic belts. Of these units, three mountain systems can be distinguished in the Indonesian archipelago (van Bemmelen, 1949): (1) the Alpine-Sunda (Tethys) mountain system, extending from the Himalayas through Arakan Yoma (Burma), Andaman Islands, Sumatra, Java, and Lesser Sunda Islands to the Banda Sea; (2) the East Asiatic system, forming part of the Circum-Pacific system and extending from the Philippines to Kalimantan (Borneo) and Sulawesi; and (3) the Circum-Australian system, extending from New Zealand through New Caledonia and Irian (New Guinea) to Halmahera. The existence of young active volcanoes and other phenomena, such as earthquakes and anomalous gravity highs, shows that the earth's crust in Indonesia, approximately between the Sahul and Sunda Shelves, is still "active."

Volcanism in Indonesia is closely related to orogeny, as can be seen from the differences in magmatic suites. Geosynclines formed late in Paleozoic time. Ultramafic bodies (ophiolites) were intruded in the geosynclinal basins. Uplift was followed locally by intrusion of granitic batholiths accompanied by intense acidic volcanism of Mesozoic time.

The Barisan Mountain Range in western Sumatra is an example of the close relation between volcanism and geologic structure. Two stages of uplift occurred in Plio-Pleistocene time. Differential uplift formed the Samangko rift zone. Folding is seen in the Barisan Mountain Range and in the longitudinal fault parallel with the axis of the island of Sumatra. Above the anticline, active volcanoes were formed. In Java these events took place later than in Sumatra.

After the middle Miocene orogeny, intense volcanism produced rocks of intermediate (diorite) to acidic composition (granodiorite, granite) mainly along the southwest part of Sumatra, the southern mountain range of Java, Flores and Wetar in the Lesser Sunda Islands, and northern parts of west Sulawesi. Volcanic islands extending eastward from the island of Bali, forming the outer part of the volcanic inner arc and connecting the inner arc of Java and Banda, are a manifestation of volcanism in folded zones. Part of that area is covered by Quaternary volcanic material. Brouwer (1939) believed that the folding extended to Solor, Adonara, Lomblen, and Pantar, because there are active volcanoes on those islands.

The Indonesian archipelago contains one of the world's most extensive volcanic areas, which covers about 333,450 km^2, or 17.48% of the total area of the archipelago. The volcanic area includes about 500 young volcanoes, which are especially numerous (about 300 volcanoes) in the Sunda island complex. Among them, 127 are still considered active, representing 14% of the active volcanoes of the world. Of 127 centers of volcanic activity, 77 volcanoes (60.0%) have had magmatic or semimagmatic eruptions since 1600 A.D. (class A), 29 volcanoes (22.8%) are in the solfataric-fumarolic stage (class B), and 21 volcanic centers can be classified as solfatara and fumarole fields (class C). Volcanism in Indonesia is highly explosive; the explosive index is about 99%. In the northern part of Indonesia (e.g., Minahasa and Halmahera), the explosive index is relatively low (80–90%).

The centers of volcanic activity can be grouped into three systems: (1) Sunda-Banda, (2) Minahasa, and (3) Halmahera. About 80.3% of the volcanic centers are found in the Sunda group, extending from Sumatra to the Banda Sea—60 volcanoes (47.2%) are grouped in class A, 26 volcanoes (20.5%) in class B, and 16 (12%) in class C. Many submarine volcanoes rise from the floor of the Banda Sea. The Minahasa group extends farther north toward the island of Mindanao in the Philippines. The Halmahera group extends between Makian and Tobelo. Extinct volcanoes are found in the Bonthain region of south Sulawesi.

VOLCANIC-GEOTHERMAL POTENTIAL

Types of Volcanic Thermal Phenomena

Most of the geothermal manifestations in Indonesia are closely related to volcanism (volcanic thermal areas).

Nonvolcanic geothermal areas occur along the Samangko rift zone (Sumatra); in southern Java (Cisolok, south Tasikmalaja); and in the Masepe, Tana Toraja, and Meranda-Poso coastal area in south and central Sulawesi. These areas are closely related to local intrusions of granitic or granodioritic bodies. Both kinds of geothermal areas may contain hot springs, hot-mud pools, geysers, and steam vents. They also emit H_2S and SO_2 gases, which come mainly from the magma.

In volcanic areas, such geothermal phenomena as fumaroles, which mainly contain water vapor, generally are found on extinct volcanoes or eroded volcanic remnants. Solfataras contain H_2S, and the percentage is high, especially in gases from active volcanoes.

Two theories deal with the formation of geothermal areas. According to one theory, the steam comes from groundwater which has been heated by magmatic gases or hot rocks. Convection flow occurs if the quantity of steam is balanced by the amount of water entering the rocks. In tropical regions, this theory is likely to be valid. The second theory states that the hot water and the steam, and other gases, come from the cooling magma. The two types of steam differ in chemical composition. Juvenile waters from magma contain more Cl, HCl, H_2S, H_2CO_3, CH_4, H_2, NH_4Cl, and S than heated groundwater.

Volcanic Thermal Areas Prospective for Electrical Power

The dominant structure in the Indonesian archipelago is young and is closely related to intense volcanic activity, both past and recent. The volcanoes in Indonesia are very explosive, indicating the liquid nature of the magma and its presence at shallow depths; these factors stimulate the accumulation of natural steam and volcanic gases. It is possible that the rock units, because of the thick deposits of volcanic material, may form geothermal reservoirs and caprocks. Plentiful rainwater is available for recharge.

Several factors that favorably influence the feasibility of developing specific geothermal sites in Indonesia for electric power are that (1) no eruptive activity has occurred in historic times; (2) the location is not too far from cities, towns, and industrial areas; and (3) surrounding areas are favorable for development. The volcanic thermal areas in Indonesia which satisfy these conditions

Table 1. Tentative Evaluation of Prospective Thermal Areas

Thermal Area	*Area (km^2)*	*Surface Temp. (°C)*	*Type of Thermal Phenomena*[1]	*Est. Power and Energy (mw)*
		SUMATRA		
North Sumatra				
Banda Aceh	12	50-96	f, hs, hm	30
Tapanuli	300	39-123	f, hs, g	200
Toba				
Tarutung				
Pahei				
Central Sumatra				
Pasaman	1,360	38-101	f, hs, g	20
Sibangor Jae				
Kayulaut				
Kerinci	630	60-101	f, hs, g	20
Sumerup				
Sei Pagu				
South Sumatra				
Samangko-Lampung	120	70-103	f, hs, hm, g	100
Suoh-Antatai				
		JAVA		
West Java				
Banten	200	35-63	hs	?
Danau Caldera				
Cisolok-Cisukarame	20	47-104	hs	20
Prabakti	25	36-93	f, hs, hm	50
Cibeureum				
Salak				
Kamojang	0.5	67-106	f, hs, hm	60
Darajat				
Manuk				
Central Java				
Dieng	44	42-94	f, hs, hm	30
		BALI		
North Tabanan	30	39-48	hs	?
		SULAWESI		
Minahasa	300	50-100	f, hs, hm, g	30
Tempang				
Langoan				
Tamposo				
Sarongsong				

[1] f = fumarole; s = solfatara; hs = hot spring; hm = hot mud; g = geyser.

are class-C solfatara-fumarole areas of volcanic ruins.

At present, the infrastructure for development is most advanced in Java and north Sulawesi. The promising geothermal areas are Kamojang-Manuk-Darajat, Patuha, Telagabodas, Karaha, Lamongan (Java), and Sarongsong-Tamposo (north Sulawesi). In Sumatra, the infrastructure still needs to be developed, except in a few areas such as Toba-Tarutung. In general, the geothermal areas in Sumatra are outstanding with respect to the extent of the geothermal phenomena, the high surface temperature, and the ideal geologic conditions.

The use of geothermal energy is considered most important for volcanic islands where conventional power sources are lacking (e.g., Bali and the islands east of it).

Since the beginning of the first Five Year Development Program, feasibility studies for the development of geothermal energy have involved inventory and exploration work. At the end of PELITA I (1974), 65% of the work had been completed, mainly for Sumatra, Java, Bali, north and central Sulawesi (nonvolcanic thermal area), and Halmahera.

The same type of work will be continued in the second Five Year Development Program (1974–79) in the eastern part of Indonesia—Lesser Sunda Islands and Moluccas, and the nonvolcanic geothermal areas of Irian, south and southeast Sulawesi, and Kalimantan.

Detailed exploration has been carried out in five selected areas in Java and Bali—Kamojang-Manuk-Darajat, Cisolok-Cisukarame, Prabakti-Cibeureum-Salak, Dieng, and North Tabanan. The work at Dieng was carried out from 1970 to 1972 through a contract between USAID/USGS and PRI-GSI. Owing to technical problems, the exploration work which had proceeded to the stage of test drilling (to about 150-m depth) was terminated in 1972. Other work in these areas has been conducted to date (1974) through technical assistance from the government of New Zealand. Test drilling to 500–700-m depths will commence soon at five locations in the Kamojang area.

A tentative evaluation has been made of the inventory data to determine the selected areas which are promising for future development (Table 1; Fig. 1).

References Cited

Brouwer, H. A., 1939, Tectonic evolution of the Lesser Sunda Islands: Geol. Soc. London Quart. Jour., no. 377, v. 95, pt. 1, p. iii–vi.

Escher, B. G., 1920, Over de mogelijkheid van dienstbaarmaking van vulkaangassen: Bandung, De Mijning, v. 1.

Healy, J., 1970, Pre-investigation geological appraisal of geothermal fields, *in* United Nations symposium on the development and utilization of geothermal resources: Proc., v. 2, pt. 1, Geothermics, Spec. Issue, no. 2, p. 571–577.

Kusumadinata, K., 1964, Kemungkinan menggunakan gas-gas volkanik untuk tenaga listrik di Indonesia: Direktorat Geologi, unpub.

Stehn, Ch. E., 1929, Kawah Kamojang: 4th Pacific Sci. Cong.

Taverne, N. J. M., 1925, Omzetten van vulkanische in electrische energie: Bandung, De Mijning, v. 6.

van Bemmelen, R. W., 1928, Over de toekomst van een met vulkanischen stoom gedreven centrale in Nederlansch Indie: Bandung, De Mijning, v. 9.

——— 1949, The geology of Indonesia; v. IA, General geology of Indonesia and adjacent archipelagoes: The Hague, Govt. Printing Office, 732 p.

Van Dijk, J. Z., 1918, Krachtbronnen in Italie: Koloniale Studien, v. 2, no. 3.

Geothermal Exploration and Development in the Philippines—Summary[1]

ARTURO ALCARAZ[2]

Introduction

The power needs of the Philippines are increasing at the unprecedented rate of about 8.4% per year. Indications are that this rate will increase even further owing to the following growth factors: (1) improved conditions of peace and order; (2) improvement in the national economy brought about by changes in the tax and tariff structure, increased internal-revenue and custom-duty collections, and introduction of banking and financial reforms; and (3) increased foreign investments resulting from liberalized regulations, particularly on the granting of incentives and guarantees on repatriation of investments and remittances of profits and dividends of foreign investors.

As a consequence of the global energy crisis, the economic planning body of the country was forced to make a drastic change in its power-development program. Planning previously was centered on thermal plants providing base-load requirements, especially in the provincial service areas. Now the main thrust is toward hydroelectric and geothermal resources, combined with nuclear-power development.

Essentially, the present power-development program of the Philippines, as recently enunciated by the President upon recommendation of the Power Development Council, is aimed at attaining self-sufficiency by minimizing dependence on oil. One step in the proposed program (Ravanzo, 1974) is to intensify investigations of geothermal fields all over the country and speed up the construction of geothermal power plants. Fortunately, the Philippines appears to have abundant geothermal potential.

Geothermal Conditions

Study of the geologic setting of the Philippines, and especially post-Tertiary volcanism, permits a general evaluation of geothermal implications (Alcaraz, 1974a, b):

1. The potential geothermal areas of the Philippines will be found in the inner plutonic and volcanic areas bounded by three primary-arc chains of peridotite (Santos-Yñigo, 1966).

2. The most promising areas for geothermal exploration and development will be in the vicinity of inactive Pliocene and Quaternary volcanoes that dot the archipelago from north to south.

3. The predominant crystalline rocks and their derivatives in the geothermal areas will most likely be acid to intermediate in composition, and will in general be intercalated with relatively thin clastic sedimentary rocks and/or reef-limestone lenses.

4. The reservoir rocks will be hydrothermally altered pyroclastic and clastic sedimentary beds, highly fractured formations, and andesitic or dacitic pyroclastic flows.

5. Lava flows will invariably provide capping over the reservoir rocks. Geophysically, therefore, many geothermal reservoir areas will be characterized by resistivity lows below high resistivity values that are correlatable with relative magnetic highs.

6. Arcuate faulting, brought about by the amassed weight of the volcanic deposits and/or by recession of magmatic material or by pressure, will provide some degree of structural control that may delineate the reservoir areas. In some cases, as in geothermal areas located close to the Philippine fault zone, structural control may be provided by the zone itself or by secondary faults formed as a consequence of major fault displacements.

7. Except where a geothermal area is the result of convective heat transfer from the magma chamber itself (e.g., volcanic craters), the geothermal mechanism will most likely be groundwater circulating in a porous formation and heated upon contact with hot rocks overlying the magma chamber.

8. Most, if not all, of the potentially major geothermal areas of the Philippines will be hot-water systems in accordance with the concept of White (1973).

9. The thicknesses of Tertiary and recent volcanic rocks in studied areas of the archipelago,

[1]Manuscript received, November 6, 1974.

[2]Chief, Geothermal Division, National Power Corporation, Manila, Philippines.

The writer expresses his profound gratitude to the organizers of the 1974 Circum-Pacific Energy and Mineral Resources Conference for their invitation to participate in its technical sessions. Thanks are also tendered to the National Power Board of the Philippines for its support.

plus thermodynamic considerations, indicate that most of the productive zones will be at depths between 3,500 and 5,000 ft (1,066–1,524 m). Minor steam zones may exist at much shallower depths, however.

10. The rate of groundwater recharge will be high, considering the high annual rainfall. This condition will insure long productive life of the fields.

Distribution and Potential of Geothermal Areas

The presence of numerous hot springs throughout the Philippines indicates that the country is well endowed with geothermal resources. Scattered throughout the archipelago are about 39 thermal fields (Fig. 1) of probable importance, as well as 36 presently little-known thermal spots (hot springs) that could have geothermal significance.

Geothermal fields under varying stages of exploration and development are Tiwi of Albay province, the Mt. Maquiling–Mt. Banahao area of south-central Luzon, Tongonan of Leyte, the Palimpinon-Dauin and Mambucal areas of Negros, and some geothermal prospects of Mindanao Island.

Tiwi Geothermal Field

The geothermal possibilities of Tiwi (15, Fig. 1) in Albay province, Luzon Island, were the first to be investigated (Alcaraz, 1966) by the Commission on Volcanology (COMVOL) with financial assistance from the National Science Development Board. However, because exploration work at Tiwi had reached the stage when emphasis should be shifted from research to commercial development, the Government gave the National Power Corporation (NPC) the responsibility for implementing further work there. Under Presidential Proclamation No. 739, dated August 14, 1970, about 17,661 ha (43,623 acres) in the Tiwi geothermal area were reserved for exploration, development, and exploitation by NPC. In September 1971, NPC entered into a service contract with Philippine Geothermal Incorporated (PGI), a subsidiary of Union Oil Company of California, for the development and exploitation of the field for electric power generation.

The Tiwi field is a hot-water geothermal system in Quaternary andesites and subsidiary dacites. Faults are the main structural features of the area. From morphologic studies and geophysical-survey results (AFMAG survey), four main structural trends can be recognized clearly: (1) northwest-southeast trend corresponding to the regional tectonic axis; (2) northeast-southwest trend transverse to the regional system; (3) radial volcano-tectonic system; and (4) concentric volcano-tectonic system. It is believed that the microfractures related to these trends, together with the joint systems and horizontal discontinuities between rock units, have provided the permeability of the reservoir (Electroconsult, 1974).

A resistivity survey of the area using a dipole-dipole electrode configuration was undertaken by McPhar Geoservices for PGI in May and June of 1972. The survey outlined approximately 2,300 ha (5,680 acres) with low resistivity. On the basis of this finding, PGI then drilled deep exploration/production wells with remarkable results. Of five holes drilled so far, the first four (Naglagbong 1, 2, 3, and 4), sited within the low-resistivity area, reached the reservoir and yielded geothermal fluids at high flow rates. The fifth hole, which was purposely sited between low- and high-resistivity areas, was unproductive. These results confirmed the resistivity findings.

An AFMAG electromagnetic survey was also carried out. Results of the survey agree with the data obtained from the wells and indicate several potential geothermal sections within the area.

Intensive testing of the productive wells has been carried out by PGI to establish the reservoir capacity, initial pressure, well productivity, formation change, flow capacity, and heat availability and transfer (Philippine Geothermal, Inc., 1973).

No depletion of reservoir pressure has yet been detected as a result of production from the existing wells. However, productivity of some of the wells has declined owing to carbonate scaling. Enthalpy has also varied with time and production rate.

The first geothermal power plant, with 50-Mw capacity, is to be operative by early 1977.

Mt. Maquiling–Mt. Banahao (Los Baños) Geothermal Prospect

This geothermal prospect is in south-central Luzon, about 70 km southeast of Manila (35, Fig. 1). Thermal (Abear and Española, 1973), geochemical (Fajardo and Tansinsin, 1973), and magnetic (Villalva and Lagasca, 1973) surveys were performed by COMVOL, with financial support by NPC, in the second half of 1973. At the same time, reconnaissance resistivity surveys were carried out by McPhar Geoservices for PGI to outline the major resistivity trends of the area.

Preliminary findings disclosed the presence of favorable geothermal parameters in the prospect area. It is suspected that the best sections of the large area may be "blind,"—that is, there are no adequate surface leakages to indicate their pres-

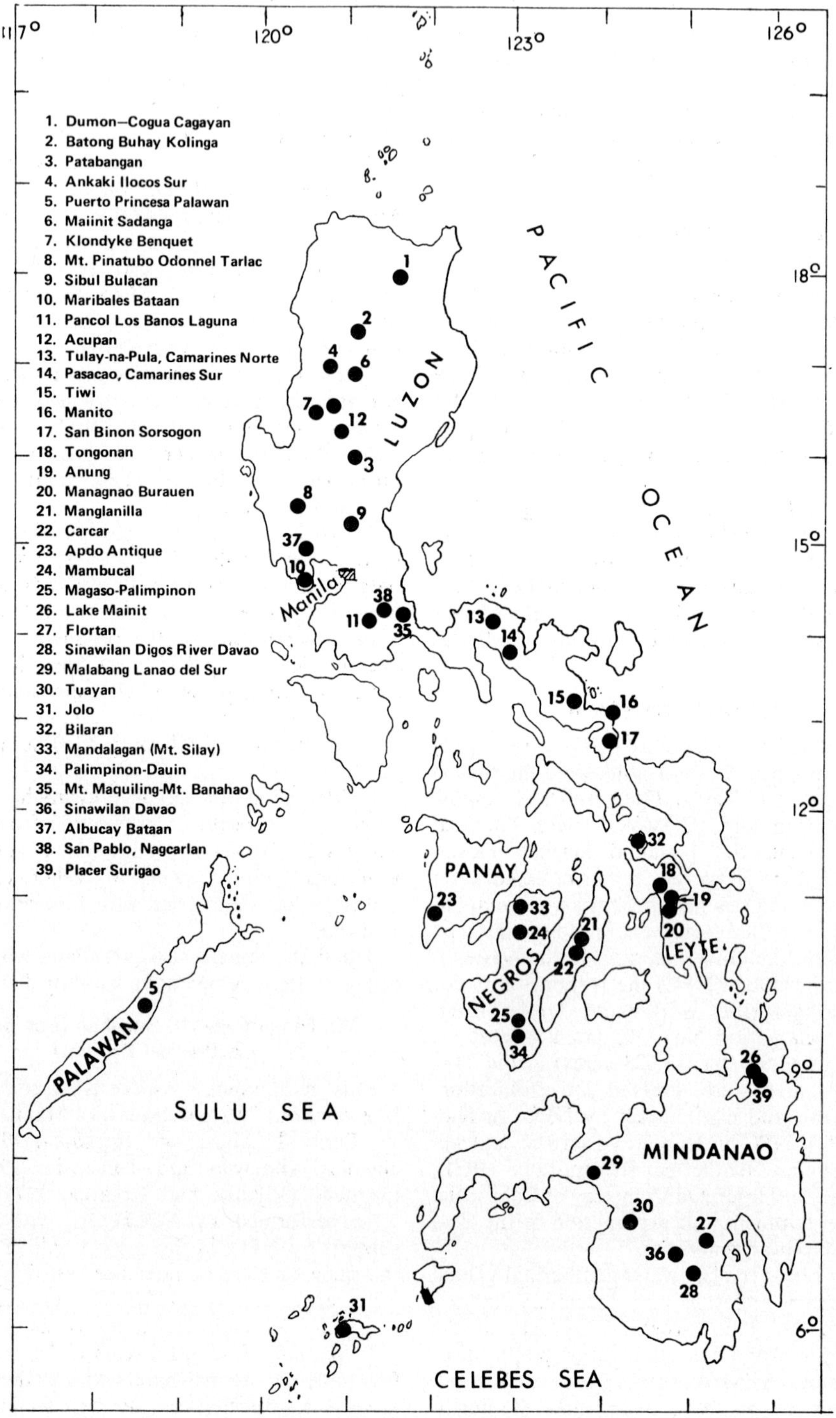

FIG. 1–Distribution of Philippine thermal fields.

ence. However, detailed geophysical-survey techniques should help detect such "blind" geothermal reservoirs.

The area lies between two dormant Quaternary volcanoes, Mt. Maquiling and Mt. Banahao, and includes several diatremes or maars. The presence of these single-explosion craters was taken as geothermally significant and was one of the considerations taken into account in determining the size of the geothermal reservation (1,620 km^2). It was learned subsequently that Marinelli (1969) earlier had related the Larderello geothermal area to past phreatic explosions.

Lithologic units, as can be expected in young volcanic provinces of the Philippines, consist of basaltic-andesitic lava flows, tuffs, laharic deposits, and agglomerates. Cinder cones and tuff rings abound in the area. A few, but apparently younger, dacitic domes and associated pyroclastic flows have also been identified.

The structural control has been provided by faults whose north-south trend corresponds to that of the main regional trend in the area. However, volcano-tectonic secondary faults, both radial and peripheral, seem to be the dominant control of the geothermal plumbing system.

Analyses of thermal spring waters in the Mt. Maquiling sector of the prospect area show neutral to slightly alkaline saline waters of high silica content. Practically all springs show more than 300 ppm of SiO_2, indicating a temperature at the source of possibly more than 200°C.

The encouraging results of these studies of the prospect area make it a most attractive geothermal area for exploitation; it is not only close to Manila, but its generated power could easily be hooked into the Luzon grid of the NPC for ready consumption. Therefore, fast development of the prospect is planned.

Tongonan Geothermal Prospect

The Leyte geothermal area is undergoing active exploration and development for exploratory drilling operations by the NPC with the assistance of the Government of New Zealand.

The geothermal prospects of Leyte have been included in a government geothermal reservation which is a northwest-southeast-trending rectangle about 8 km wide and 40 km long. In the northwest is the Tongonan (18, Fig. 1) geothermal prospect, and the Managnao Burauen (20, Fig. 1) prospect is in the center. These prospects lie along the Philippine rift zone (Allen, 1962), which extends from Mindanao in the south to Luzon in the north.

Of the two prospects, Tongonan appears, at least from surface thermal manifestations, to be more promising. The largest natural discharge of heat is in this area, where numerous boiling springs occur at the intersections of major faults with subsidiary fractures and joints in andesite and sedimentary rocks (Cardoso, 1973). These springs (Fernandez, 1968; Abiog, 1969) are distributed in an area of approximately 2.3 km^2 within the Bao River valley. The area includes the relatively flat river valley, a rolling upland, and mountainous terrain; elevations range from 150 to more than 1,000 m above sea level. Chemical analyses of water samples taken from the Tongonan hot springs are typical of geothermal waters which have been exploited in other countries. The chloride content is high and the concentration of silica indicates source-water temperatures of 190–200°C.

Four shallow exploration drillholes have been sunk in the Tongonan prospect with assistance from the New Zealand government. However, the holes, which range in depth from 1,200 to 1,900 ft (365–579 m), did not reach the reservoir. It is possible that these holes were drilled at the edge of the reservoir area. More exploratory holes are planned, together with deeper exploration/production holes.

Geothermal Prospects of Negros Island

In consonance with the electrification program of the Visayas region, NPC and COMVOL initiated in 1973 a program of investigations of the geothermal prospects of Negros Island. These prospects are the Palimpinon-Dauin, Mambucal, and Mandalagan (Mt. Silay) areas. Investigations consisted of reconnaissance geologic surveys, thermal surveys and samplings, ground geothermal prospecting in restricted areas, and hydrochemical investigations.

Palimpinon-Dauin prospect—This area (34, Fig. 1) is located at the southeastern extremity of Negros Island (Negros Oriental province) and is considered the most promising on the island. The thermal manifestations are essentially controlled by geologic structures such as faults, fractures, and probably buried permeable channels. At Palimpinon, the hot-water seepage, mud pools, and warm grounds are located inside a graben and along obvious faults and fault intersections. At Dauin, a large area showing a thermal anomaly was revealed by 1-m probes.

The heat sources of the prospect area are considered to be two contemporaneous, dormant Quaternary volcanoes—Cuernos de Negros and Balinsasayao. Dominant rock types are generally andesitic, although younger extrusive and pyroclastic flows are dacitic.

Samples from 12 hot springs show favorable SiO_2 content. It ranged from 100 to 200 ppm in 8 springs, 200 to 300 ppm in 3 springs, and exceeded 300 ppm in 1 spring, indicating high temperature at depth.

Mambucal prospect—This prospect area (24, Fig. 1) is situated in Negros Occidental province in northern Negros Island. No definite control has yet been established concerning the surface thermal manifestations at Mambucal. However, the area apparently lies on a nearly circular depression with collapse-structure characteristics.

The geology of the area is characterized by andesite lava flows and pyroclastic deposits, generally intensely fractured and markedly altered in the vicinity of the hot springs, most of which are concentrated in the restricted area of the Mambucal depression. Chemical analyses of two of these hot springs show a moderately high SiO_2 content of approximately 150 ppm. The area, therefore, may be considered as only moderately promising though worthy of further detailed investigation.

Mandalagan (Mt. Silay) thermal area—The location of this thermal area (33, Fig. 1) was previously placed as in Silay, but field investigation revealed that it really represents the crater area of Mt. Mandalagan, a dormant volcano about 25 km north-northeast of the active Canlaon Volcano of northern Negros Island.

The thermal activity is largely solfataric. Superheated steam at 106°C issues under strong pressure from a vent inside the crater area and from other smaller vents and blowholes. Sulfur deposits form around these vents. However, the utility of these manifestations is negated by the high acidity of the issuing thermal fluids. It is anticipated that this condition will pose tremendous corrosion problems.

Geothermal Prospects of Mindanao

In Mindanao, the second largest island of the Philippine archipelago, several geothermal prospects should be studied further, because the hydroelectric potential of the Agusan River, according to NPC estimates, will be fully tapped by 1980. Of these geothermal prospects, those of the Lake Mainit region (26, Fig. 1) and the Flortan area (27, Fig. 1) of Cotabato province in the central part of Mindanao Island seem to be best.

Investigations of these areas have been started by COMVOL, with financial support from NPC. However, data so far gathered are meager and do not permit any preliminary assessment of the prospects beyond an indication that an enlarged investigation program is worthy of consideration.

Other Geothermal Prospects

Figure 1 shows other probable geothermal areas in the Philippines. In consonance with the power-development program of the Government, the NPC is now drawing up plans to accelerate the exploration and exploitation of the more promising of these areas.

Summary

In a recent appraisal of geothermal resources in the Philippines by Electroconsult (1974) of Milan for the National Power Corporation, the assessment was made that the geothermal potential of the Philippines is indeed enormous, and the probabilities are high that study of the known prospects will result in the discovery of numerous geothermal fields. In addition, many other prospects may be discovered through a systematic survey of the Philippine territory, including the sedimentary areas.

The appraisal also echoed the thinking of Government power-program planners that, as a result of the energy crisis, there is a pressing need for the quick exploration and utilization of the Philippines' geothermal potential.

There are, however, certain constraints that work against a speedy geothermal development program—unavailability of suitable drilling rigs and casings, and other funding requirements. To the extent that these constraints could be eased or solved, the Government has decided to subvert the academic approach to a bolder, though unorthodox method of development. Such a plan will of necessity involve more "risk capital," but is perhaps justified in the interest of attaining self-sufficiency.

References Cited

Abear, M., and O. Española, 1973, Thermal and geochemical survey of Makiling-Banahao geothermal field: Commission on Volcanology Rept., August.

Abiog, D. B., 1969, Geothermal study of Tongonan springs, Ormoc City, Leyte: Geol. Soc. Philippines Jour., v. 23, no. 2. p. 78–97.

Alcaraz, A., 1966, Study of Tiwi geothermal area for power utilization: Commission on Volcanology Ann. Rept., 1966–67.

——— 1974a, Potential utilization of heat energy from Philippine volcanic areas: Paper read before U.S.-Japan Cooperative Science Seminar on "Utilization of Volcano Energy," Hilo, Hawaii.

——— 1974b, Status and potentials of geothermal power in the Philippines: Sci. Rev., v. 15, no. 1, p. 12–15.

Allen, C. R., 1962, Circum-Pacific faulting in the Philippine-Taiwan region: Jour. Geophys. Research, v. 67, no. 12, p. 4795–4812.

Cardoso, M. Tan., 1973, Heat flow measurements of the

Tongonan and Burauen geothermal fields in Leyte: Commission on Volcanology Ann. Rept., 1972–73.
Electroconsult, 1974, Geothermal resources of the Philippines; appraisal and development: Rept. to Natl. Power Corp.
Fajardo, A. C., and L. G. Tansinsin, 1973, A memorandum report on the chemistry of thermal waters from Los Baños, Laguna; Negros Oriental and Negros Occidental; Asin, Benguet; and Balungao, Pangasinam: Commission on Volcanology Ann. Rept., 1972–73.
Fernandez, J. C., 1968, Geophysical reconnaissance at Tongonan geothermal field, Ormoc City, Leyte del Norte: Mineral Engineering Mag., v. 20, no. 1.
Marinelli, G., 1969, Some geological data on the geothermal areas of Tuscany: Bull. Volcanol., v. 33, no. 1, p. 319–334.
Philippine Geothermal, Inc., 1973, Tiwi project, interim report on test results: Rept. to Natl. Power Corp.
Ravanzo, R. R., 1974, The role of the National Power Corporation in meeting the energy crisis: unpub.
Santos-Yñigo, L., 1966, Island arc features of the Philippine archipelago: Philippine Geologist, v. 20, no. 3, p. 79–92.
Villalva, E., and R. Lagasca, 1973, Preliminary report on magnetic and gravity investigations in Makiling-Banahao geothermal field: Commission on Volcanology Rept., 1972–73.
White, D. E., 1973, Characteristics of geothermal resources, *in* Geothermal energy: Stanford Univ. Press, p. 69–94.

Origin and Potential of Some Geothermal Areas in Taiwan—Summary[1]

WENG-TSE CHENG and YUNG-JUH WU[2]

Taiwan is a spindle-shaped island located at the junction of the Ryukyu and Taiwan-Luzon arcs. Nearly 100 thermal springs are scattered throughout the island. Genetically, they can be classified into volcanic and nonvolcanic types. The hottest and most impressive thermal springs and fumaroles of volcanic origin are distributed in the Tatun volcanic region north of T'aipei, which was given priority in geothermal exploration. The nonvolcanic thermal springs are widely distributed over the island. They occur mostly in the submetamorphic Paleogene to lower Miocene rocks and less abundantly in the Neogene terrane. The heat sources of these nonvolcanic thermal springs can be attributed mainly to the geothermal gradient; however, a hidden intrusive body or magma chamber at depth may account for the heat supply of some thermal springs. It is unknown if, and to what extent, magmatic heat has played a role in the nonvolcanic thermal springs of Taiwan.

Geothermal exploration in the Tatun volcanic region revealed that the central part is most prospective. Subsidence of the Miocene sedimentary formations may have occurred as a consequence of the volcanic activities, and a graben probably exists in the prospective area. The Miocene sandstones below the thick volcanic cover are considered to be the main reservoirs for geothermal fluids. The highest subsurface temperature recorded was 293°C. The power potential of the Tatun region has been estimated to be 5,000–14,000 Mw/year. Unfortunately, the high acidity of the hot water presents a serious corrosion problem, and development of this region is suspended at the present time.

A province-wide geothermal exploration program was started in July 1972. After a preliminary investigation, 10 prospective thermal areas are thought to justify further exploration. Among them, the Tuchang thermal area, about 40 km southwest of Ilan in the northeastern part of Taiwan, seems most favorable and is presently being studied.

[1]Manuscript received, October 21, 1974.

[2]T'aipei, Taiwan, Republic of China.

We are grateful to C. S. Ho of the College of Chinese Culture for pertinent comments and critical reading of the manuscript. We also express our gratitude to K. K. Huang, Chief, and C. H. Chen, Deputy Chief, of the Energy Resources Division for their valuable discussions, and to F. P. Cheng and C. F. Lee for the preparation of geochemical and geophysical data.

Geothermal Resources and Energy in Japan Estimated from Geophysical Data—Summary[1]

MASAMI HAYAKAWA[2] and KENZO BABA[3]

At present, heat in the form of volcanic steam from dormant volcanoes is used for electric power generation in Japan. However, it may eventually be possible to make direct use of heat (the thermal energy is most significant) from active volcanoes; but the necessary techniques must be developed.

In early Miocene time, Japan apparently entered into a new stage of geologic evolution. Subsidence, violent volcanism, and some igneous intrusion took place rather suddenly on land, and the Miocene sea began to transgress and finally covered almost all of Japan. The deep depression of Fossa Magna formed and separated northeast Japan from southwest Japan. In the Pleistocene, the sea largely regressed, but volcanism has continued to the present, and many volcanic cones and lava plateaus have been constructed.

The writers have estimated the heat generated and discharged since the earth was formed. The total energy gain during the earth's evolution is 4.9×10^{38} ergs (accretion, 2.5×10^{38} ergs; formation of core, 1.5×10^{38} ergs; radioactive energy emitted by the long-lived radioactive elements, 0.93×10^{38} ergs). The energy lost from the earth's interior during the 4.5-billion-year period is estimated as 2.9×10^{38} ergs, of which mantle convection accounts for 2.1×10^{38} ergs. Most of the heat brought from the upper mantle into the earth's crust is expended in orogenesis, continental drift, and heat flow from the surface. Volcanoes, hot springs, etc., represent a relatively small part of the total heat discharged. However, the amount of energy discharge by volcanic activity in Japan still is large enough to make it worthwhile to try to utilize the steam energy from volcanic sources. Heat-flow values in wells in geothermal areas in Japan are 30–40 HFU at shallow depths and 4–5 HFU at depth. The average is 12–13 HFU.

Surface-temperature patterns and geothermal-gradient distributions were used in making assumptions about thermal conductivities. The differences between so-called "running mean values" were used, and spherical sources were assumed for the heat-source models. Some significant correlation of heat-flow anomalies with past igneous centers can be demonstrated.

[1]Manuscript received, November 6, 1974.

[2]Tokai University, Shizuoka-ken, Japan.

[3]Geological Survey of Japan, Tokyo, Japan.

Geological and Geochemical Characteristics of Geothermal Resources in Japan—Summary[1]

HISAYOSHI NAKAMURA,[2] KIYOSHI SUMI,[3] and TAKEJIRO OZAWA[4]

The distribution of thermal areas in Japan is geologically divided into the following units: (1) Neogene green tuff areas, (2) Neogene sedimentary basins, (3) Neogene plutonic areas, (4) Pliocene to Quaternary sedimentary basins, and (5) Quaternary volcanic areas.

Among these, the Pliocene to Quaternary sedimentary basins are areas of normal terrestrial heat flow. The Tertiary (Neogene) plutonic areas have locally abnormal heat flow, and heat flow from the Neogene sedimentary basins and green tuff areas is generally higher than normal.

There are about 100 geothermal areas with water temperatures higher than 80°C in the areas mentioned, and most of these are distributed in the Quaternary volcanic areas, near active volcanos. Some hot springs issuing from the active volcanic areas are characterized by a low pH owing to contained hydrochloric acid. These are typical volcanic hot springs. The volcanic rocks have good reservoir characteristics as well as high heat flow.

Most of the geothermal areas are concentrated in southwestern Hokkaido, the western half of northern Honshū, the Izu-Hakone region of central Honshū, and central and southern Kyūshū.

To promote geothermal development, the Geological Survey of Japan is conducting geothermal investigations in 30 promising areas located mainly in the Neogene and Quaternary volcanic areas. These investigations commenced in 1973 and will be continued for a period of 3 years. Exploratory wells will be drilled in the areas selected by detailed surveys. These wells will be financed by special government funds commencing this year (1974). The present investigations are being made to find new geothermal areas consisting of hot-water and/or dry-steam systems. However, if a study on the extraction of heat from hot dry rocks is commenced in the near future, the Quaternary dacite areas as well as the active volcanic areas will be selected as promising areas.

[1]Manuscript received, October 21, 1974.

[2]Japan Metals & Chemicals Co., Ltd., Tokyo, Japan.

[3]Geological Survey of Japan, Tokyo, Japan.

[4]Tokyo Institute of Technology, Tokyo, Japan.

Potential for Geothermal-Energy Development in Alaska—Summary[1]

THOMAS P. MILLER[2] and IVAN BARNES[3]

INTRODUCTION

The association of the major known geothermal systems of the world with such features as plate boundaries, spreading ridges, and rift valleys is well known (Muffler and White, 1972; White, 1973). These areas of high heat flow are characterized by recent volcanism and the occurrence of thermal springs; it is therefore essential, in evaluating the geothermal energy potential of any Circum-Pacific area, to know the distribution and nature of late Cenozoic volcanic rocks and of thermal springs of the region.

Two areas in Alaska—Pilgrim hot spring (Pilgrim Springs) on the Seward Peninsula and Geyser Spring Basin and Okmok caldera on Umnak Island in the Aleutian Islands (Fig. 1)—have been classified as "Known Geothermal Resource Areas" (KGRA's) by the U.S. Geological Survey (Godwin et al, 1971). In addition, about 75 known thermal springs and volcanic areas have been classified as having prospective value for geothermal energy. In 1972, the Geological Survey began a 3-year reconnaissance program in Alaska designed to evaluate these and other areas containing potential geothermal resources.

RECENT VOLCANIC ROCKS

Aleutian Volcanic Arc

One of the most obvious areas of high geothermal potential in Alaska is the Aleutian volcanic arc (Fig. 1), which is part of a seismically active volcanic arc-trench system that extends for more than 2,500 km across the northern Pacific. Of 76 major volcanoes of late Cenozoic age that occur along the northern edge of the arc, at least 40 have been active within the past 200 years (Coats, 1950). The Quaternary volcanic rocks are predominantly pyroxene andesite but vary in composition from olivine basalt to rhyolite (Coats, 1962).

All the known late Cenozoic volcanoes of the Aleutian volcanic arc could be considered as potential geothermal target areas because they indicate anomalously high heat flow in geologically recent time. The maximum depth at which geothermal systems might be exploited is thought to be less than 10 km (Muffler and White, 1972; White, 1973). Geothermal target areas can therefore be reduced, as a first approximation, to those volcanoes that have, or had, large magma chambers relatively close to the surface. A possible criterion for selection of such sites is the presence of a collapsed caldera, which in itself implies a magma chamber at relatively shallow depth. Furthermore, Smith and Shaw (1973) have postulated that most large (10^2 to 10^5 km^3) magma chambers in the upper 10 km of the crust either are silicic or have silicic differentiates. Silicic rocks in andesitic arc systems are commonly associated with calderas, and silicic ejecta have been found at several calderas in the Aleutian volcanic arc; the Valley of Ten Thousand Smokes, with Katmai caldera and Novarupta dome, is a well-documented example (Fenner, 1920; Griggs, 1922; Curtis, 1968). Recent reconnaissance mapping by the writers and R. L. Smith has also shown the occurrence of dacitic and rhyodacitic ash flows associated with calderas or silicic domes at Mt. Peulik volcano, Aniakchak caldera, Black Peak volcano, Mt. Veniaminof volcano, Emmons caldera, and Dana Volcano, all on the Alaska Peninsula, and Fisher caldera on Unimak Island in the eastern Aleutians.

Powers (1958) listed 20 known calderas in the Aleutian volcanic arc. One of these, on Buldir Island, has since been referred to as a volcano-tectonic depression rather than a true caldera (Marlow et al, 1970). Recent work has shown that at least three more calderas on the Alaska Peninsula—Black Peak, Peulik, and Emmons—can now be added to the list (Fig. 1).

Because of the remoteness of the Aleutian Islands, the calderas on the Alaska Peninsula and those on the easternmost islands, perhaps as far west as Okmok caldera on Umnak Island, may be considered as having more immediate potential for commercial development.

Wrangell Mountains Volcanic Pile

The thick calc-alkaline volcanic pile, occupying much of the Wrangell Mountains in eastern Alaska (Fig. 1), is one of the largest andesitic piles in

[1]Manuscript received, February 10, 1975.

[2]U.S. Geological Survey, Anchorage, Alaska.

[3]U.S. Geological Survey, Menlo Park, California 94025.

Calderas of Aleutian Volcanic Arc

	Name	*Location*	*Approximate diameter (km)*
1	Kaguyak	Alaska Pen.; 58°37′N, 154°05′W	2.8
2	Katmai	Alaska Pen.; 58°16′N, 154°59′W	2.1 × 2
3	Peulik	Alaska Pen.; 57°45′N, 156°21′W	5
4	Aniakchak	Alaska Pen.; 56°53′N, 158°10′W	9.7 × 8.4
5	Black Peak	Alaska Pen.; 56°32′N, 158°37′W	2.7 × 2.6
6	Veniaminof	Alaska Pen.; 56°10′N, 159°23′W	8.4
7	Dana	Alaska Pen.; 55°37′N, 161°12′W	2
8	Emmons	Alaska Pen.; 55°20′N, 162°01′W	18 × 10
9	Frosty	Alaska Pen.; 55°04′N, 162°51′W	5.6
10	Fisher	Unimak Is., 54°38′N, 164°25′W	10 × 11
11	Akutan	Akutan Is.; 54°08′N, 166°00′W	2
12	Makushin	Unalaska Is.; 53°52′N, 168°56′W	2.4 × 3.2
13	Okmok	Umnak Is.; 53°25′N, 168°03′W	12
14	Yunaska	Yunaska Is.; 52°39′N, 170°39′W	2.9 × 2.4
15	Seguam 1, 2	Seguam Is.; 52°19′N, 172°29′W	8; 5.5
16	Kliuchef	Atka Is.; 53°19′N, 174°09′W	4.5±
17	Great Sitkin	Great Sitkin Is.; 52°04′N, 176°07′W	1.6
18	Kanaton	Kanaga Is.; 51°55′N, 177°10′W	4
19	Takawangha	Tanaga Is.; 51°52′N, 178°00′W	3.2
20	Tanaga	Tanaga Is.; 51°53′N, 178°07′W	11±
21	Semisopochnoi	Semisopochnoi Is.; 51°56′N, 179°35′E	6.1 × 7.5
22	Little Sitkin	Little Sitkin Is.; 51°57′N, 178°32′E	4.5
23	Davidof	Davidof-Kvostof Is.; 51°58′N, 178°20′E	2.4

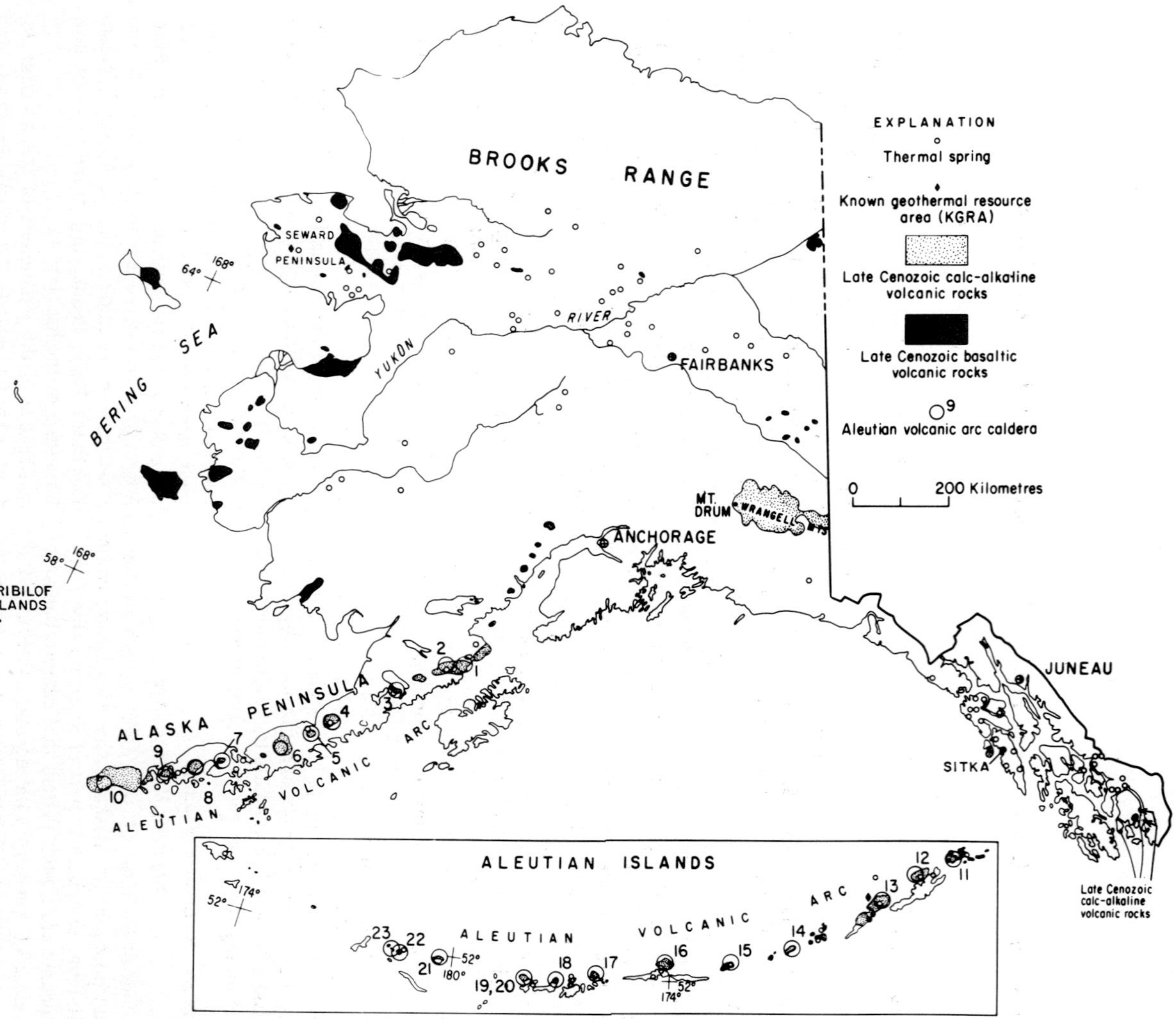

FIG. 1–Distribution of late Cenozoic volcanic rocks and thermal springs in Alaska. Calderas of Aleutian volcanic arc are indicated by numbers.

North America (Forbes et al, 1969) and contains some of the highest stratovolcanoes in the world. They include Mt. Sanford at 4,950 m, Mt. Drum at 3,662 m, Mt. Wrangell at 4,320 m, Mt. Blackburn at 5,037 m, and Regal Mountain at 4,210 m. The basaltic to rhyolitic volcanic rocks of the pile underlie an area of about 10,000 km^2.

The Wrangell Lava, a name applied by Mendenhall (1905) to all the volcanic rocks in the pile, unconformably overlies Paleozoic and Mesozoic rocks of the eastern Alaska Range. The base of the thick volcanic sequence contains rhyodacite flows, endogenic domes, and pyroclastic cones (Richter and Jones, 1973). These silicic volcanic rocks are overlain by basalt, andesite, and dacite flows and pyroclastic rocks erupted from large shield volcanoes and stratovolcanoes. Deininger (1972) gave a minimum whole-rock K/Ar age of 13.5 ± 0.8 m.y. for some old flows on the north side of the volcanic pile, and Richter and Jones (1973) suggested that the volcanic activity began in early Miocene time. K/Ar ages as young as 1.97 m.y. (Richter and Schmoll, 1973) and 1.63 m.y. (Richter, oral commun., 1974) have been obtained from volcanic rocks east of Mt. Sanford. Activity continuing into the Holocene is indicated by the fumaroles in the largely ice-filled summit caldera on Mt. Wrangell and by the White River Ash Bed (Péwé, 1975), which covers over 320,000 km^2 of eastern Alaska and southern Yukon. This rhyodacite ash-fall deposit has been traced by Lerbekmo and Campbell (1968) to a probable source area in the eastern Wrangell Mountains, about 8 km from the Canadian border. Capps (1915) dated this eruption at 1,400 years B.P.; radiocarbon dates on peat and wood samples above and below the ash fall are in the general range of $1,200 \pm 140$ years (Lerbekmo and Campbell, 1968).

Of specific interest from a geothermal standpoint are the stratovolcanoes at the west end of the Wrangell Mountains, in particular Mt. Drum and Mt. Sanford. Recent geologic mapping on Mt. Drum has shown that the central andesitic cone of this stratovolcano is at least partly ringed by a group of young silicic dacite domes. As pointed out by Smith and Shaw (1973), the age, composition, and distribution of volcanic rocks provide clues concerning the volume, depth, age, and state of evolution of magma chambers. These, in turn, permit estimation of the size and remaining "life-span" of the thermal anomaly. The silicic domes ringing the central core of the Mt. Drum stratovolcano indicate that a large magma chamber once existed at relatively shallow depth, possibly within 10 km. We have instituted a program of detailed geologic mapping and K/Ar age studies on Mt. Drum and nearby Mt. Sanford to determine if the volcanism was recent enough for significant magmatic heat to have been retained. Preliminary K/Ar ages indicate that some of the silicic domes are younger than 200,000 years.

Interior and Southeastern Alaska

Late Cenozoic volcanic rocks underlie a total area of about 25,000 km^2 in interior Alaska, mostly in the western part of the region (Fig. 1). These rocks are chiefly olivine tholeiite and alkalic basalt; no intermediate or silicic rocks of Pliocene or younger age have been reported in this part of Alaska. K/Ar dates obtained on basalt from the western part of this province range from 6.12 ± 0.18 m.y. to $30,000 \pm 20,000$ years B.P. on Nunivak Island (Hoare et al, 1968) and from 2.19 ± 0.1 m.y. to $65,000 \pm 100,000$ years B.P. on the Pribilof islands (Cox et al, 1966). Stratigraphic and morphologic evidence suggests that volcanic activity has continued almost to the present; Hopkins (1963) stated that the Lost Jim Lava Flow on the Seward Peninsula may be as young as a few hundred years.

This volcanic province may still be volcanically active. However, no hot springs, fumaroles, or other manifestations of current volcanic activity occur within areas underlain by these basalts, and the known thermal springs in interior Alaska show no spatial association with them.

Late Cenozoic volcanic rocks occur in a few scattered localities in southeastern Alaska. The most important occurrence is at Mount Edgecumbe, an inactive composite volcano about 980 m in elevation located about 26 km west of Sitka (Fig. 1). The volcano is part of a Quaternary volcanic field that covers about 260 km^2 on the southern end of Kruzof Island and ranges in composition from olivine-augite basalt to rhyolite. Radiocarbon dates indicate that eruptions which deposited ash and lapilli as far away as 225 km occurred as recently as 9,000–11,000 years B.P. (Brew et al, 1969; McKenzie, 1970). Volcanic rocks of probable late Cenozoic age also crop out in several other small areas in southeast Alaska. Little or no spatial relation is shown between the thermal springs of southeastern Alaska and the young volcanic rocks.

The geothermal energy potential of areas underlain by late Cenozoic volcanic rocks in interior and southeastern Alaska appears to be low. Most geothermal areas now exploited in other parts of the world are associated with calc-alkaline volcanic deposits, whereas the Quaternary volcanic rocks of interior Alaska are primarily basalt and basaltic andesite which probably originated from considerable depth. These volcanic areas display no signs of current activity such as fumaroles, geysers, or thermal springs. However, the Mount Edgecumbe area, having had relatively recent calc-alkaline volcanism, has higher geothermal potential.

Thermal Springs

The most recent compilation of thermal springs in Alaska lists 94 known and reported thermal springs (Miller, 1973) located in three main areas (Fig. 1): the Aleutian volcanic arc; southeastern Alaska; and interior Alaska, including the Seward Peninsula. The thermal springs in the Aleutians and on the Alaska Peninsula are closely related spatially, and probably genetically, to the geologically young volcanoes. Most of the thermal springs, however, are in southeastern and interior Alaska and are not spatially related to recent volcanism. Rather, as first pointed out by Waring (1917), the springs are along the fractured and faulted margins of granitic plutons.

In interior Alaska, approximately 30 thermal springs occur in a variety of geologic provinces including volcanogenic and regionally metamorphosed crystalline terranes. The geologic feature common to all provinces is the presence of granitic plutons of late Mesozoic and early Cenozoic age. Every thermal spring whose bedrock geology is known occurs within 5 km of either side of the contact of a granitic pluton. The distribution of thermal springs shows no correlation with the age, composition, or size of the plutons; thus, these thermal springs probably are not related genetically to the magmatic processes that formed the plutons. Rather, the association of thermal springs with pluton contacts, and in some cases with known faults and lineaments, suggests that the pluton contacts are zones of well-developed "open-fracture systems" that permit hot water to rise to the surface. Chemical and isotopic analyses of these hot-spring waters indicate that they consist entirely of deeply circulating meteoric water (Miller et al, 1975).

Thermal springs in southeastern Alaska also appear to be related to fractured granitic masses, according to Waring (1917). Of 19 hot springs where the bedrock geology is definitely known, 11 occur within or very close to granitic plutons. Twenhofel and Sainsbury (1958) found that 14 of the 19 hot springs are located on prominent linears which are visible on aerial photographs and which are probably faults.

In interior and southeastern Alaska, the highest reservoir temperature estimated by using either the SiO_2 conductive geothermometer or the Na-Ca-K geothermometer is below the 180°C minimum temperature generally thought necessary for the generation of electricity using existing technology (Muffler and White, 1972).

For the Aleutian Islands and Alaska Peninsula, however, reservoir temperatures above 180°C are indicated for two of the six springs from which chemical analyses have been obtained. A thermal spring at Geyser Spring Basin on Umnak Island has estimated reservoir temperatures of 234°C (Na-Ca-K geothermometer) and 210°C (SiO_2 conductive geothermometer), and Chiginigak hot spring on the Alaska Peninsula yields estimated reservoir temperatures of 205°C (Na-Ca-K geothermometer) and 180°C (SiO_2 geothermometer; Miller, 1973).

Conclusions

The only use now being made of geothermal resources in Alaska is for local recreational purposes, space heating, and agriculture. The widespread occurrence of young volcanic rocks and thermal springs, however, points to numerous areas of anomalous heat flux and indicates that Alaska has a relatively large potential for geothermal energy resources. Meaningful quantitative estimates of the potential geothermal energy and the selection of individual sites will require additional studies.

The Aleutian volcanic arc and the western end of the Wrangell Mountains appear to have the highest potential for development of geothermal energy. The western end of the Wrangell Mountains lies adjacent to major transportation routes and could provide energy for use within Alaska. The potentially large geothermal areas in the remote Aleutians and the Alaska Peninsula are expected to be of most economic use to industries with high energy requirements for processing raw material (e.g., aluminum), rather than to the limited indigenous population or to power transmission outside the area.

The thermal springs in interior and southeastern Alaska appear to represent deeply circulating meteoric water. Despite the relatively low subsurface temperatures, the demand for cheaper energy in the remote regions of interior Alaska may cause these small thermal springs to be the first geothermal areas exploited in Alaska. Hot water could be used for space heating as in Iceland, or, if the use of heat exchangers becomes technologically practicable for the generation of electrical power, these springs could provide limited electrical power for local use by remote villages.

REFERENCES CITED

Brew, D. A., et al, 1969, Reconnaissance geology of the Mount Edgecumbe volcanic field, Kruzof Island, southeastern Alaska, *in* Geological Survey Research 1969: U.S. Geol. Survey Prof. Paper 650-D, p. 1–18.

Capps, S. R., 1915, An ancient volcanic eruption in the upper Yukon basin: U.S. Geol. Survey Prof. Paper 95-D, p. 59–64.

Coats, R. R., 1950, Volcanic activity in the Aleutian arc: U.S. Geol. Survey Bull. 947-B, p. 35–49.

——— 1962, Magma type and crustal structure in the Aleutian arc, *in* The crust of the Pacific basin: Am. Geophys. Union Geophys. Mon. Ser. 6, p. 92–109.

Cox, Allan, D. M. Hopkins, and G. B. Dalrymple, 1966, Geomagnetic polarity epochs—Pribilof Islands, Alaska: Geol. Soc. America Bull., v. 77, no. 9, p. 883–909.

Curtis, G. H., 1968, The stratigraphy of the ejecta from the 1912 eruption of Mount Katmai and Novarupta, Alaska, *in* R. R. Coats, R. L. Hay, and C. A. Anderson, eds., Studies in volcanology: Geol. Soc. America Mem. 116, 678 p.

Deininger, James, 1972, Petrology of the Wrangell volcanics near Nabesna, Alaska: Univ. Alaska, unpub. M.S. thesis, 66 p.

Fenner, C. N., 1920, The Katmai region, Alaska, and the great eruption of 1912: Jour. Geology, v. 28, no. 7, p. 596–606.

Forbes, R. B., et al, 1969, The comparative chemical composition of continental vs. island arc andesites in Alaska, *in* Andesite conference, Eugene and Bend, Oregon, 1968, Proc. (Internat. Upper Mantle Proj. Sci. Rept. 16): Oregon Dept. Geology and Mineral Industries Bull. 65, p. 111–120.

Godwin, L. H., et al, 1971, Classification of public lands valuable for geothermal steam and associated geothermal resources: U.S. Geol. Survey Circ. 647, 18 p.

Griggs, R. F., 1922, The Valley of Ten Thousand Smokes [Alaska]: Washington, D.C., Natl. Geog. Soc., 340 p.

Hoare, J. M., et al, 1968, Geology, paleomagnetism and potassium-argon ages of basalts from Nunivak Island, Alaska, *in* R. R. Coats, R. L. Hay, and C. A. Anderson, eds., Studies in volcanology: Geol. Soc. America Mem. 116, p. 377–414.

Hopkins, D. M., 1963, Geology of the Imuruk Lake area, Seward Peninsula, Alaska: U.S. Geol. Survey Bull. 1141-C, 101 p.

Lerbekmo, J. F., and F. A. Campbell, 1968, Distribution, composition, and source of the White River Ash, Yukon Territory: Canadian Jour. Earth Sci., v. 6, p. 109–116.

McKenzie, G. D., 1970, Some properties and age of volcanic ash in Glacier Bay National Monument: Arctic, v. 23, no. 1, p. 46–49.

Marlow, M. S., et al, 1970, Buldir depression, a late Tertiary graben on the Aleutian Ridge, Alaska: Marine Geology, v. 8, no. 1, p. 85–108.

Mendenhall, W. C., 1905, Geology of the central Copper River region, Alaska: U.S. Geol. Survey Prof. Paper 41, 133 p.

Miller, T. P., compiler, 1973, Distribution and chemical analyses of thermal springs in Alaska: U.S. Geol. Survey Open-File Rept., 5 p.

——— Ivan Barnes, and W. W. Patton, Jr., 1975, Geologic setting and chemical characteristics of hot springs in central and western Alaska: U.S. Geol. Survey Jour. Research, v. 3, no. 2, p. 149–162.

Muffler, L. J. P., and D. E. White, 1972, Geothermal energy: Sci. Teacher, v. 39, no. 3.

Péwé, T. L., 1975, Quaternary stratigraphic nomenclature in unglaciated central Alaska: U.S. Geol. Survey Prof. Paper 862, 32 p.

Powers, H. A., 1958, Alaska Peninsula-Aleutian Islands, *in* Williams Howel, ed., Landscapes of Alaska: Berkeley and Los Angeles, Univ. California Press, p. 61–75.

Richter, D. H., and D. L. Jones, 1973, Structure and stratigraphy of eastern Alaska Range, Alaska, *in* M. G. Pitcher, ed., Arctic Geology: AAPG Mem. 19, p. 408–420.

——— and H. R. Schmoll, 1973, Geologic map of the Nabesna C-5 quadrangle, Alaska: U.S. Geol. Survey Geol. Quad. Map GQ-1062.

Smith, R. L., and H. R. Shaw, 1973, Volcanic rocks as geologic guides to geothermal exploration and evaluation (abs.): Eos, v. 54, no. 11, p. 1213.

Twenhofel, W. S., and C. L. Sainsbury, 1958, Fault patterns in southeastern Alaska: Geol. Soc. America Bull., v. 69, no. 11, p. 1431–1442.

Waring, G. A., 1917, Mineral springs of Alaska: U.S. Geol. Survey Water-Supply Paper 418, 114 p.

White, D. E., 1973, Geothermal resources, *in* Paul Kruger and Carel Otte, eds., Geothermal energy: resources, production, stimulation: Stanford, California, Stanford Univ. Press.

Deep Research Drillhole at Kilauea Volcano, Hawaii—Summary[1]

GEORGE V. KELLER,[2] JOHN C. MURRAY,[3] CHARLES J. ZABLOCKI,[4] ROBERT I. TILLING,[5]
DONALD W. PETERSON,[5] and ROBERT L. CHRISTIANSEN[6]

Commercial geothermal power plants throughout the world are located in areas of recent silicic volcanism, but the obvious supply of heat associated with Hawaii's active basaltic volcanoes is attractive as a possible future source of power.

Between April 6 and July 9, 1973, the first deep borehole at the summit of an active volcano was drilled at Kilauea Volcano, Hawaii. The 1,262-m-deep hole was drilled to test predictions based on surface geophysical surveys and to obtain information on the hydrothermal regime above a postulated magma reservoir. Surface deformation during inflations by magma and deflations (via eruptions) had been interpreted to indicate a complex magma-reservoir system 2–4 km below Kilauea's summit. An electromagnetic sounding survey had defined a dome-shaped zone of low resistivity 900–2,000 m below the summit. This zone was believed likely to contain saline groundwater domed into a hydrothermal convection cell.

During drilling, fluid loss into permeable zones prevented return circulation of mud to the surface, and no cuttings were retrieved. Intermittent coring recovered about 47 m of core or 3.7% of the total hole depth. Initial megascopic observations of the core show that the entire hole is in basalt, but three broad lithologic zones appear to be represented in the depth intervals 0–300 m, 300–600 m, and 600–1,262 m. The uppermost 300 m appears to be composed mainly of thin pahoehoe flows characterized rather uniformly by about 20–45 volume-percent vesicles and a few phenocrysts. Certain ash zones in the interval may correlate with named units in the vicinity. The 300–600-m interval is generally coarser grained and more crystalline than the upper section, and is more competent than the rock above or below. The bottom zone seems lithologically distinct from those above; the core is darker, locally has a greenish cast, and lacks phenocrysts. It appears to contain alternating thick and thin flows, including some aa, as well as the densest rocks recovered, which may represent chilled zones of sills or thick flows. This interval contains a greater variety of secondary minerals than higher cores.

Zeolites and calcite partially filling vesicles first appear at 488 m. Geophysical data indicate that the standing fluid level in the hole is also about 488 m, and this level probably is the local water table.

Temperature measurements indicate (1) an essentially isothermal interval between the surface and 488 m, (2) a rapid rise in temperature between 488 m and 732 m, (3) a decline in temperature between 732 m and 976 m, and (4) an increasingly steep rise in temperature between 976 m and 1,262 m. The gradient toward the bottom of the hole is about 0.37°C/m and the maximum temperature is 137°C, at the hole bottom. Conductive heat flow at the bottom of the hole is estimated—admittedly on meager data—as about 25 cal/cm^2/sec.

No formation fluids were obtained, but geophysical logging gives an indication of salinity. For a single-salt solution of sodium chloride, the resistivities are equivalent to a concentration of about 5,000 ppm for fluid in the upper interval and as much as 30,000 ppm at the bottom. A decrease in resistivity of the contained water is evident near the 1,189-m level, and it becomes more pronounced toward the bottom of the hole.

Geophysical data indicate that permeability generally decreases gradually with depth below the water table.

Analysis of results indicates that the temperatures encountered can be explained either by transient heat flow from fairly recent intrusions or by a hydrothermal convection cell. Critical evidence is the existence of a water table at 488 m in

[1]Manuscript received, December 26, 1974.

[2]Colorado School of Mines, Golden, Colorado 80401.

[3]Colorado School of Mines. Present address: Tenneco Oil Co., Denver, Colorado 80203.

[4]U.S. Geological Survey, Hawaiian Volcano Observatory. Present address: Branch of Theoretical and Applied Geophysics, Denver, Colorado 80225.

[5]U.S. Geological Survey, Hawaiian Volcano Observatory, Hawaii National Park, Hawaii 96718.

[6]U.S. Geological Survey, Hawaiian Volcano Observatory. Present address: U.S. Geological Survey, Menlo Park, California 94025.

This study was supported by Grant No. GI-34993, Program for Research Applied to National Needs (RANN), National Science Foundation.

the hole, an elevation of 614 m above sea level at a distance of 13 km from the shore. A water table in hydrostatic equilibrium would be elevated 1–2 m per 1 km distance from the shoreline. In one possible model, the Kilauea water table is postulated to be uplifted locally by thermal drive above a magma reservoir in a zone of reduced permeability near the magma body. Another possible model postulates that geologic factors, not necessarily related to a shallow magma reservoir, trap and contain high-level water reservoirs. These two models are not necessarily mutually exclusive, and other models may be postulated.

The general reduction in permeability toward the hole bottom is important in evaluating geothermal potential. If all the basalt layers were as permeable as those above the water table, heated groundwater would move quickly through the rock, removing the heat from a magma reservoir too quickly for an economically viable system to form. Alteration in the rocks below the water table may represent the self-sealing believed to take place in geothermal reservoirs; migrating thermal water causes alteration, reducing permeability and trapping the waters in a reservoir in which the temperature can reach economic levels.

Considering the steep temperature gradient at the bottom of the hole, it is possible that temperatures suitable for production of high-energy steam might be found if the hole were deepened.

Geothermal Potential of Western United States—Summary[1]

JOHN E. KILKENNY[2]

INTRODUCTION

Most of the commercial geothermal potential of the United States is in Washington, Idaho, Montana, Oregon, Wyoming, California, Utah, Colorado, Arizona, and New Mexico. Figure 1 shows the geologic and geomorphic provinces of the western United States, whose boundaries are influenced by geothermal characteristics. This area is characterized by above-average heat flow, numerous hot springs, recently active volcanoes and calderas, large areas of young volcanic rocks, and numerous active faults.

The distribution of volcanic rocks in the area is shown in Figure 2. The Quaternary volcanic rocks are predominantly basalts but include some andesites and other less basic rocks. Two areas in particular stand out—the Cascade Range and the Snake River downwarp. Smaller areas of Quaternary volcanic rocks are found east of the southern Sierra Nevadas—the Mono Lake and Mammoth areas, California; southwestern Utah; the San Francisco Peaks, Arizona; an area around the southern Arizona–New Mexico border; and the Valles (Baca) Caldera area, north-central New Mexico. Two other important areas of Holocene volcanism are too small to be shown—the Clear Lake area just north of The Geysers steam field in northern California and the southeastern edge of the Salton Sea in the Imperial Valley, southern California. Older volcanics rocks, mostly late and middle Tertiary, are also mainly basic but include some andesites, dacites, and pyroclastic rocks. Some of the early Tertiary volcanic rocks are coarser textured and more silicic. Recently active volcanoes and calderas are considered favorable for geothermal prospecting. The recently discovered geothermal field on the Baca Ranch, Valles Caldera, New Mexico, is a notable example. The late Mesozoic batholiths are drawn to include associated older metamorphic rocks. The Franciscan Complex (Berkland et al, 1972) has a higher than normal heat flow and is the producing formation in The Geysers field.

Figure 3 shows some of the more important hot springs in the western United States. Those springs shown by solid black dots have had at least one exploratory well drilled. None of these wells have been commercially productive because of lack of sufficient temperature, lack of reservoir, or insufficient permeability. Some of the areas, such as Klamath Falls in southern Oregon, are currently producing hot water for space heating. It is possible that some of these areas can be commercially productive with the perfection of heat-exchanger or binary-fluid techniques. Deeper drilling in some areas offers promise of higher temperatures and possibly greater permeability.

The large circles (Fig. 3) represent geothermal fields that are producing or capable of producing geothermal power commercially. In addition to The Geysers and the Valles Caldera, the Cerro Prieto field (in the Salton trough), just south of the California-Mexico border, is shown.

Most of the hot springs in the region are associated with faulting (Fig. 3); normal and lateral faults are most common.

GEOTHERMAL PROVINCES

The presence of significant faults, young volcanic rocks, and hot springs, plus temperature information from wells and gradient holes, permits us to conclude that the Cascades, Snake River downwarp, Basin and Range province, Rio Grande rift, and Salton trough (Fig. 1) all appear to have geothermal potential. The California Coast Ranges, especially the areas of Franciscan outcrops, are warm and interesting. The Coast Ranges of Oregon and Washington are cool. The Transverse Ranges and the White-Datil volcanic areas are warm.

Unfavorable areas include the young sediment-filled basins of the Rockies and the Great Valley of California. The batholithic areas have different heat values. The Sierras are cool, whereas the Peninsular Ranges offer some potential, especially along the Elsinore and San Jacinto faults where there are many hot springs. The Idaho batholith has numerous hot springs, most of which are along Miocene igneous dikes. All of these granitic areas have one unknown factor—permeability.

All of the important, presently known geothermal fields produce from fracture or secondary permeability, and this type of permeability is dif-

[1]Manuscript received, September 19, 1974.

[2]Union Oil Company of California, Los Angeles, California 90051.

FIG. 1–Geologic and geomorphic provinces of western U.S. High lava plains of eastern Oregon are included with Snake River downwarp.

ficult to predict. Even in sedimentary basins, such as the Salton trough, the porosity of young strata is altered by the high-temperature fluids.

Plate Tectonics

Some of the recently identified oceanic crustal features (Fig. 4) may have a bearing on the source of heat underlying the western United States. The East Pacific Rise extends toward the continent and the Gulf of California. It continues northwesterly up the gulf, offset at intervals by northwest-trending transform fracture zones. Exploratory work on the floor of the gulf indicates that the rise is hot. Elders et al (1972) projected the rise onshore through the Cerro Prieto field and through some boiling mud pots across the border into California, where it is offset by continental "transform" faults such as the San Jacinto. It continues to the Salton Sea and is the source of heat for the Niland geothermal brine field. Just north of this field the rise intersects the San Andreas fault. Menard (1964), Wilson (1970), and Atwater (1970) have made further postulations regarding the northerly continuation of the East Pacific Rise.

A trench is believed to exist along the edge of the continental shelf north of the Mendocino fracture zone parallel with the Oregon and Washington coastline (Silver, 1971). This trench is proposed as an active current subduction zone. The Cascade Range, with its numerous recently active volcanoes, lies about 330 km to the east parallel with the trench. A similar feature—the Middle America Trench—is present off the west coast of Mexico, and an andesitic volcanic chain lies onshore at about the same distance as that between the Cascades and Silver's proposed trench. This volcanic chain is marked by active volcanoes and several geothermal areas including Los Negritos, which is thought to have prospects of developing into a large geothermal field like Cerro Prieto.

Lowell (1974) proposed a subduction zone along the common boundary of the Basin and Range province and the Colorado Plateau. This boundary is also marked by considerable volcanism and the presence of hot springs. Lowell's belief that the Basin and Range crust is underthrusting the Colorado Plateau crust, possibly by as much as 100 km, is substantiated by magnetic data. He proposed another belt of subduction along the Rocky Mountain front where the Rockies meet the Great Plains.

Another interesting theory recently advanced is that of a "fixed-mantle hot spot" (Morgan,

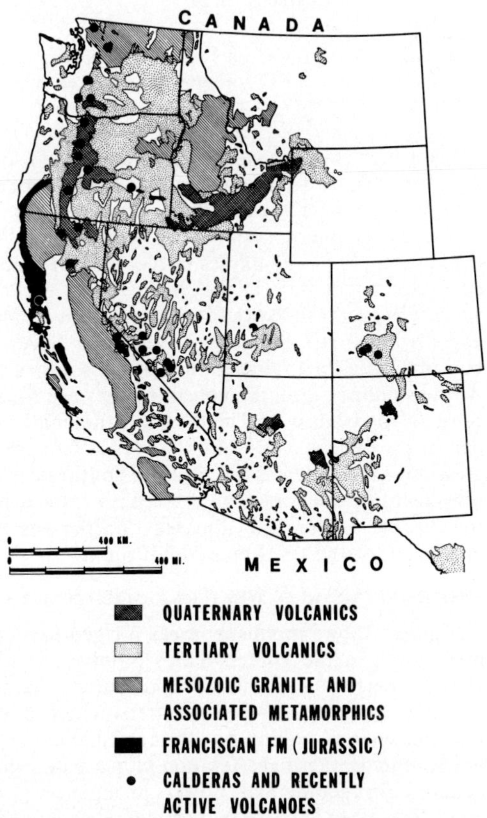

FIG. 2–Distribution of volcanic rocks.

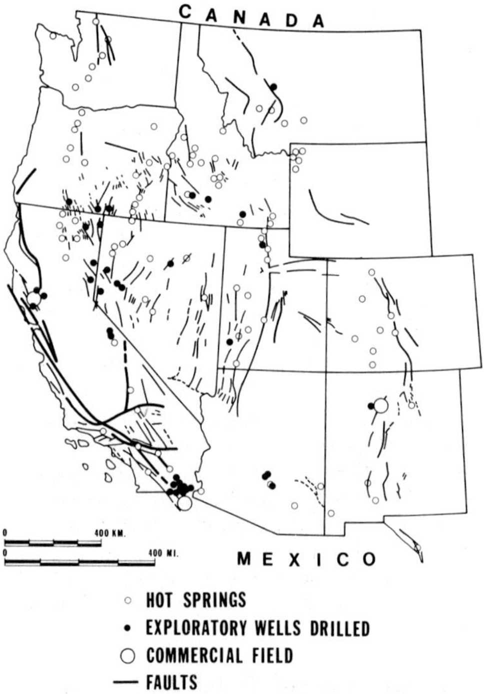

FIG. 3—Hot springs and faults. Most springs with temperatures over 50°C are shown. In places several springs are combined because of map scale.

1971), which explains volcanic trends such as the Hawaiian Island Chain, where volcanism is progressively younger toward the southeast. This is explained as plate motion over a fixed-mantle convection plume. Examples in the western United States might be: (1) the Snake River downwarp, (2) southern Nevada and southern Utah, and (3) the Coast Ranges of California. Several volcanic areas are aligned northwest-southeast. At the southeast end, the volcanic rocks are Miocene. In the vicinity of The Geysers, the volcanic deposits are Pleistocene and Holocene. This progression is explained as the result of southeasterly movement of the American plate with respect to the Pacific plate, and apparent northwesterly movement of the fixed-mantle hot plume.

Promising Areas in Western United States

Figure 5 shows promising areas for geothermal exploration in the United States (Godwin et al, 1971). Known geothermal resource areas (KGRA, category *1*) represent areas where phenomena such as boiling hot springs, shallow test holes, or hydrothermal alteration indicate that the chances for geothermal power production are good. The other areas shown (category *2*) contain lesser hot springs, young volcanic rocks, fumaroles, or hydrothermal alteration of the surface rocks.

In January 1974, the U.S. government opened up lands for geothermal prospecting to individuals and private companies. Lands in category *1* were put up for competitive sealed bids. Lands in category *2* have been made available for simultaneous filing. If two or more companies file on the the same acreage, it is automatically put into category *1* and then put up for competitive bidding.

Explored and/or Producing Areas

The Geysers field—The first recorded drilling for geothermal power at The Geysers, in northern California, was in the 1921–25 period. A group of local businessmen drilled eight shallow wells, the deepest of which went to 200 m. Although steam was encountered, local demand for electricity was insufficient to justify further drilling. Thirty years later, Magma Power Co. began development of the field as a dry-steam reservoir. Deep drilling began in 1966 with Union Oil Co. as operator. The Geysers now is the world's largest geothermal field. Present capacity is over 500 Mw and average well depth is 2,000 m. The reservoir is in fractured graywacke of the Franciscan.

Valles (Baca) Caldera field—Exploration drilling in the Valles Caldera, in Sandoval County, New Mexico, began in 1960 by Westates Petroleum Co. The objective was oil and gas believed to be trapped in the Rio Grande graben in Paleo-

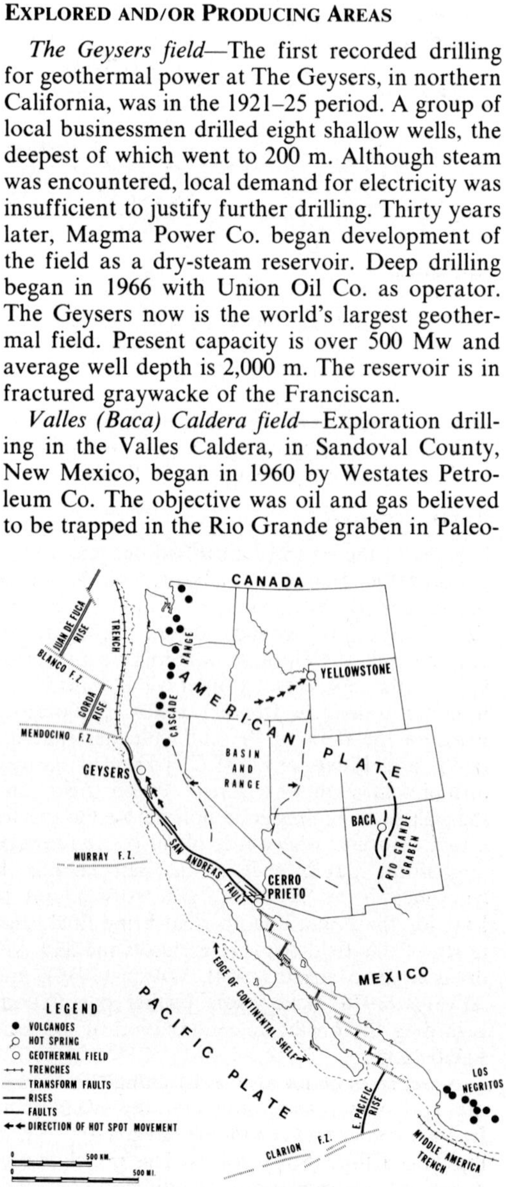

FIG. 4—Plate-tectonic map.

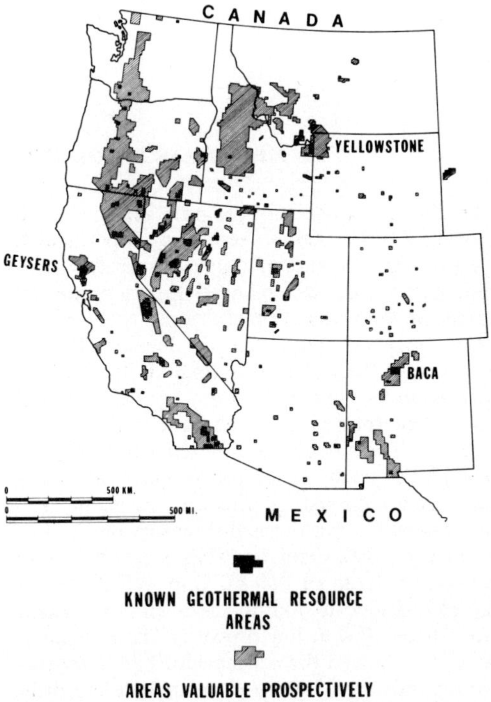

FIG. 5–Promising geothermal areas of western U.S. (after Godwin et al, 1971).

zoic beds truncated updip by the core of the Valles rhyolite plug, but drilling encountered hot water and steam. Union Oil Company leased the acreage in 1970 and has drilled several wells in this hot-water geothermal field.

Cerro Prieto and environs—Cerro Prieto field, located 29 mi (47 km) south of the U.S. border, is an excellent example of utilization of geothermal energy. The Salton trough (Imperial Valley) area has been the site of much exploratory drilling in southern California. More than 20 wells have been drilled in the Niland wet-steam field. Temperatures range as high as 350°C. The high brine content causes a severe corrosion problem and has prevented commercial production to date.

Exploration

Exploratory wells drilled for geothermal prospects, including wells drilled for oil and gas on geothermal anomalies, number 174 in the western United States, in 62 different areas. Of this number, 61 have been drilled below 1,500 m and only 4 have been drilled below 3,000 m. Several deep wells have been drilled for oil and gas in areas of interbedded sedimentary and volcanic rocks with a high heat flow. In addition, hundreds of shallow (less than 100 m deep) water wells, mainly in Idaho, Oregon, and California, have encountered warm or hot water, and some have been used for space heating. Government agencies are conducting geophysical surveys and drilling temperature holes and a few deep tests in selected areas such as Marysville, Montana; Battle Mountain, Nevada; and Los Alamos, New Mexico.

Conclusion

Geothermal exploration has an advantage over early-day oil and gas exploration because of the availability of geophysical methods such as electrical resistivity, gravity, magnetics, seismic noise, magnetotellurics, micro-earthquake recordings, and infrared surveys. Although these tools have been of value in selecting locations, the problems of complexity of the geology, the irregular permeability of igneous and metamorphic rocks, and lack of knowledge of the character of geothermal traps make exploration more risky than searching for oil or gas. So far, the best place to look for geothermal accumulations has been around hot springs.

Any conclusion as to the success ratio and the role that geothermal energy will play in the future must await further drilling. Considerable exploratory drilling should take place in the next few years, because a large amount of exploratory acreage has been leased. Another factor in future exploration will be the perfection of equipment and methods of utilization of lower temperature (150–225°C) fluids with heat-exchanger techniques.

References Cited

Atwater, Tanya, 1970, Implications of plate tectonics for the Cenozoic tectonic evolution of western North America: Geol. Soc. America Bull., v. 81, p. 3413–3536.

Berkland, J. O., et al, 1972, What is Franciscan?: AAPG Bull., v. 56, no. 12, p. 2295–2302.

Elders, W. A., et al, 1972, Crustal spreading in southern California: Science, v. 178, p. 15–24.

Godwin, L. H., et al, 1971, Classification of public lands valuable for geothermal steam and associated geothermal resources: U.S. Geol. Survey Circ. 647, 18 p.

Lowell, J. D., 1974, Plate tectonics and foreland basement deformation: Geology, v. 2, no. 6, p. 275–278.

Menard, H. W., 1964, Marine geology of the Pacific: New York, McGraw-Hill, 271 p.

Morgan, W. J., 1971, Convection plumes in the lower mantle: Nature, v. 230, p. 42–43.

Silver, E. A., 1971, Tectonics of the Mendocino triple junction: Geol. Soc. America Bull., v. 82, no. 11, p. 2965–2977.

Wilson, J. T., 1970, Some possible effects if North America has overridden part of the East Pacific Rise (abs.): Geol. Soc. America Abs. with Programs, v. 2, no. 7, p. 722–723.

Geothermal Potential of Mexico[1]

ALFREDO MANON MERCADO[2]

Abstract Geothermal surveys have been conducted for 10 years at Cerro Prieto, Baja California, and for 15 years in other areas of Mexico. In April 1973, commercial operation of the first power plant of 75,000-kw capacity, using natural geothermal steam, was initiated at Cerro Prieto.

The future of geothermal energy in Mexico is very attractive, owing to the presence of approximately 130 known areas that have potential for geothermal development. Cerro Prieto is the only area developed at present; five more areas are being investigated.

INTRODUCTION

The geothermal potential of Mexico is related mainly to intense volcanism in the central part of the country. Most geothermal activity is in the belt called "Eje Volcanico de Mexico," which crosses the country from the Pacific Ocean to the Gulf of Mexico. In this volcanic belt many exploratory surveys have been made, mainly in nine geothermal areas: Ixtlán de los Hervores, Los Negritos, and Los Azufres in Michoacán State; San Marcos, Los Hervores de la Vega, La Primavera, and La Soledad in Jalisco State; Pathe in Hidalgo State; and Los Humeros in Puebla State. The only significant field outside of this belt is Cerro Prieto in Baja California.

GEOTHERMAL AREAS IN MICHOACÁN

Ixtlán de los Hervores—Located east of Lago de Chapala at 1,500 m above sea level, this field covers a large hydrothermal area. Springs in the area have a high flow of water at 90°C and the following principal chemical content: 450 ppm chlorides, 200 ppm sulfates, 400 ppm sodium, 35 ppm potassium, and 150 ppm silica.

Geological, geochemical, and geophysical surveys have been made in an area of 30 km^2; they include gravimetric, seismic, resistivity, and thermal-flow studies, and infrared photography.

Many small-diameter shallow wells and three deep, large-diameter exploratory wells were drilled. One well, drilled to 1,000 m, produced a steam-water mixture with an approximate enthalpy of 160 cal/g, a gauged wellhead pressure of 6 kg/cm^2, and a steam flow of 5 tons/hr.

The heat-flow survey showed 310 kcal/sec in an area of 450 km^2. The silica content and the sodium/potassium ratio in the water of the producing well show that temperatures over 230°C exist at greater depth.

The future exploratory program includes drilling wells 2,000–3,000 m deep to reach permeable layers of high pressure and high temperature. Isotopic surveys are also planned to determine the sources of the water supply.

Los Negritos—This field is located 20 km southwest of the Ixtlán field. Its hydrothermal activity includes mud volcanoes, hot-water pools, and a lake formed by a phreatic explosion 70 years ago. The size of the explored area is 30 km^2. Geological, geochemical, and geophysical exploration has included gravimetric, magnetic-seismic, resistivity, and thermal-flow surveys.

Thirteen small-diameter wells were drilled to an average depth of 300 m. One well 27 cm in diameter, drilled to 700 m, yields an intermittent water-steam flow at low pressure. The geochemical survey showed the existence of high-temperature groundwater. Future plans include the drilling of wells up to 1,500 m deep.

Los Azufres—This field is located west of Hidalgo City in Michoacán State. Preliminary geological and geochemical surveys have been made and resistivity surveys are being initiated. Hydrothermal manifestations include hot-water springs with high flow, boiling-water pools, and dry-steam vents. A large exploration program is under way in an area of 180 km^2 in the vicinity of the hydrothermal activity.

GEOTHERMAL AREAS IN JALISCO

San Marcos—After preliminary geological, geochemical, and geophysical surveys, 7 small-diameter wells with maximum depths of 448 m and 19 shallow wells of 7-cm diameter were drilled in this area.

Hervores de la Vega—In this area, geological and geochemical surveys were made and seven 7-cm wells were drilled. Of these, six are shallow and one is 382 m deep.

La Primavera—Only preliminary geological, geochemical, and geophysical surveys have been made in this area.

[1]Manuscript received, November 11, 1974.

[2]Suptte. del Laboratorio, Residencia de Estudios Geotérmicos, Comision Federal de Electricidad, Mexicali, Baja California, Mexico.

La Soledad—Geological and geochemical surveys have been made and a 7-cm exploratory well was drilled to 424 m.

The future program for all these areas is to continue the geophysical surveys and the drilling of deep exploratory wells.

Pathe Geothermal Area (Hidalgo)

This field is 150 km from Mexico City in Hidalgo State. Its hydrothermal activity is represented by small hot springs and fumaroles. Drilling included 18 wells, only 5 of which were producers; the deepest was 1,300 m. Two of these wells produced dry steam and the other three produced a mixture of water and steam.

A 3,500-kw experimental plant installed in 1959 was operated for 13 years at 500-kw capacity; in 1972, it was shut down owing to its low commercial production. Possibly in the future new exploratory techniques may permit operation of this plant with steam sources at greater depths.

Cerro Prieto Geothermal Area (Baja California)

History

The first geologic and photogeologic surface surveys for geothermal resources in this area were made in 1959, and the same year three exploratory wells were drilled to 523, 755, and 403 m. Exploration has proceeded systematically and, in April 1973, Cerro Prieto became the first commercial geothermal power plant in Latin America.

Gravimetric and refraction seismic surveys were made from 1961 to 1963.

Four deep exploratory wells drilled in 1964 reached the granitic basement at 2,500 m; two of these wells produced a steam-water mixture with an enthalpy over 300 cal/g. In 1965, an isothermal-surface survey was made by drilling 50 shallow wells. A geochemical survey of the hot-water springs and steam-gas fumaroles was made from 1965 to 1966.

From 1966 to 1968, 14 wells were drilled to provide steam for the first geothermal power plant.

In 1968, a geohydrologic survey was made to determine the source of the geothermal water that is supplying the Cerro Prieto wells. In 1969, a material corrosion survey was made, soil testing was done, and the site for the power plant was selected.

Equipment for the construction of the power plant was manufactured and delivered from 1970 to 1973. This included the separation and conduction system for steam and water, the evaporation and cooling towers, and the power transmission system from Cerro Prieto to Mexicali.

An aeromagnetic survey was made in 1972, and a resistivity survey was initiated in the Mexicali Valley, which covers an area of approximately 400 km^2.

From 1972 to the present, 12 more wells have been drilled, 4 of them to supply the first addition to the power plant. There now are 34 drilled wells, of which only 5 are not producers.

A precision leveling survey was initiated in 1973 to record subsidence. After a year of operation the first turbine inspection was made, and the turbines were found to be in good condition.

Geologic Environment

Regional geology—The Cerro Prieto geothermal field is located along the projection of the San Jacinto fault into the Salton trough. Regional gravity surveys show that sediments up to 6,400 m thick have accumulated and that crustal thinning has occurred in the region.

Surface geology—Topographically, the local area is dominated by the Cerro Prieto Volcano at the northwest and the Sierra de los Cucapas to the west. Hydrothermal activity includes hot-water springs, mud volcanoes, and fumaroles. To the southwest, a phreatic-explosion crater and the volcano lake are predominant.

Subsurface geology—Relatively unconsolidated Tertiary and Quaternary sediments lie on a granitic-metamorphic basement that, in turn, lies on dense crystalline mantle material. The impermeable clays and shales overlying water-saturated sediments, combined with a thin and highly fractured granitic basement, provide an excellent environment for the geothermal reservoir of the Salton-Mexicali geothermal province.

Geohydrology—Isotopic surveys show that meteoric water from the Colorado River travels across the Salton trough as underflow to supply the Cerro Prieto geothermal area.

Permeability is relatively high in the sandstone of the producing zones penetrated by the wells; the porosity averages more than 15%.

Fluid chemistry—The water of the Cerro Prieto geothermal wells is saline and has high concentrations of carbon dioxide, potassium, silica, and calcium. A typical analysis of the formation water is: 4,800 ppm Na, 8,800 ppm Cl, 1,200 ppm K, 300 ppm Ca, 550 ppm SiO_2, 3,800 ppm HCO_3, and 800 ppm H_2S. Concentrations of the elements in solution are controlled mainly by the temperature at which the water-rock reactions occur.

Characteristics of Production Wells

The wells in operation at Cerro Prieto have the following average characteristics: 245 tons/hr steam-water production, 63 tons/hr steam production, 300 cal/g enthalpy; production casing is 20 cm in diameter.

Separation and Transmission of Steam and Hot Water

With some modifications, the separators used at Cerro Prieto are similar to those used at Wairakei, New Zealand—750 tons/hr of steam-water mixture was the maximum capacity tested in a 137-cm-diameter separator.

The separated hot water is carried to an evaporation lake by noninsulated pipes as long as 800 m. The separated steam is sent from the wells to the power plant, using a system of four main insulated pipes with a maximum diameter of 86 cm. The longest pipe is 1,000 m. Before the steam is admitted to the turbines, the moisture is eliminated in secondary separators.

Power Plant

The power plant has two 37.5-Mw turbine generators. The inlet pressure is 5.27 kg/cm^2 (gauge), and the vacuum discharge is 671 mm of mercury. The condensers are the barometric type; the vacuum and gas extraction is made with steam ejectors which use approximately 10% of the total steam.

The materials of the main parts of the turbines are: rotor—alloy steel, Cr, Mo, V; rotor blades—low-carbon steel, 12% Cr; nozzle blades—alloy steel, 13% Cr, Al; outer and inner casing—carbon steel, labyrinth packing, alloy steel, Cr, Mo.

Environmental Problems

The principal environmental problem from exploitation of the Cerro Prieto geothermal reservoir was disposal of the geothermal water. This problem has been partly solved by the construction of an evaporation lake of approximately 16 km^2. A canal approximately 50 km long is planned to carry the brine to another, larger evaporation lake near the Gulf of California.

Other environmental problems, such as noncondensable gases and the high noise level during the free discharge of the wells to the atmosphere, have already been solved.

Reservoir Capacity

Estimates of reservoir capacity have been based on (1) the volume of hot water in the reservoir, considering its average porosity values, and (2) temperatures within the area actually explored by deep wells. The estimated reservoir capacity for an area of 12 km^2 was approximately 34 million Mw-hr. This estimate does not take into account the heat contained in the rock; it was calculated with the specific water and steam rate of the power plant. The last information obtained in the exploratory well M-53, 1,300 m east of the power plant, confirms that the capacity is larger than 34 million Mw-hr.

Plans for the Mexicali geothermal area call for expansion to a total capacity of 500 Mw in 1982.

Exploration for Geothermal Energy in Nicaragua—Summary[1]

JOHN B. THIGPEN[2]

INTRODUCTION

A geothermal exploration project was conducted in 1969–70 over an area of about 38,000 km^2 in the Cenozoic volcanic zone of western Nicaragua (Fig. 1). The project included a regional geological and geochemical evaluation of the geothermal potential of the area and more detailed geological, geophysical, and borehole studies at two geothermal prospects—San Jacinto–Tisate and Momotombo.

Western Nicaragua is a region of widespread late Cenozoic volcanism, with activity locally continuing to the present (McBirney and Williams, 1965; McBirney, 1958). Tertiary acidic ash-flow tuffs predominate east of the Nicaraguan depression (Fig. 1), whereas Quaternary andesites and basalts are present to the west. An impressive chain of Quaternary volcanic centers is localized along the southwestern margin of the Nicaraguan depression, which is a regional structure of probable volcano-tectonic origin (McBirney and Williams, 1964). This geologic setting is very favorable for development of hyperthermal conditions and, thus, geothermal potential.

Relatively little was known about Nicaraguan thermal activity until recently, when brief descriptions were given by McBirney (1958) and Del Giudice (1959). An inventory, with temperature and some chemical data, of most of the water wells and many of the springs in the project area, by the Catastro i Inventario de Recursos Naturales de Nicaragua project was helpful as a preliminary indication of several thermal manifestations.

REGIONAL EVALUATION OF THERMAL MANIFESTATIONS

The initial phase of the project was an inventory and evaluation of all thermal manifestations, including thermal springs, wells, and fumarolic areas. Expectably, all the fumaroles and most of the higher temperature thermal waters are associated with the area of Quaternary volcanism. Additionally, higher silica concentrations and lower Na/K ratios, both suggesting high subsurface temperatures, are present in thermal waters near the Quaternary volcanic zone. A detailed inventory of all thermal manifestations of western Nicaragua has been prepared by the writer (Thigpen, 1970a; 1971).

Several of the more significant thermal manifestations have been grouped together as thermal areas (Fig. 2). A brief synopsis of each area is presented in Table 1. Thermal activity at the designated areas ranges from moderately intense fumarolic activity and boiling thermal water at Momotombo to relatively low-temperature water at Agua Caliente.

Based on geological and geochemical data, the designated thermal areas are ranked from best to poorest in probable geothermal potential as follows: Momotombo, San Jacinto–Tisate, Jiloa-Apoyeque, Mombacho, northeast Telica–Najo, Casita, Tipitapa, Agua Caliente, Cerro Colorado, and San Luis.

SAN JACINTO–TISATE GEOTHERMAL PROSPECT

Geology—Two small fumarolic areas and several thermal springs (32–66°C) are localized along a north-south-trending fault. Quaternary andesitic-basaltic lava flows and pyroclastic deposits have piled up against a probable fault-line escarpment of older volcanic rocks. Reservoir permeability and recharge potential appear favorable.

Moderately intense thermal activity, including fumaroles (100°C), mudpots, and thermal ground, is present at San Jacinto in a local zone of intense hydrothermal alteration. The current thermal activity at Tisate, 2.5 km north, is restricted to a few feeble fumaroles, but considerable hydrothermal alteration exists. Geophysical data indicate that San Jacinto and Tisate are probably connected at depth.

The only previous geothermal investigations in Nicaragua were made at Tisate in 1953, when three holes were drilled to a maximum vertical depth of 94.2 m and a maximum temperature of

[1]Manuscript received, October 21, 1974; revised, August 1, 1975. Published with permission of the Servicio Geologico Nacional de Nicaragua and Texas Instruments Inc.

[2]Consulting geologist, Golden, Colorado 80401. The described work was done while the writer was with Texas Instruments Inc.

Appreciation is expressed to other project participants, including N. Harthill, F. L. Doyle, R. Bennett, M. Mazzolini, and D. E. Heilner. Some financial assistance for preparation of this paper was received from Texas Instruments Inc.

FIG. 1–Index map of western Nicaragua. Numbered volcanoes are: *1*, Casita; *2*, Telica; *3*, San Jacinto; *4*, Santa Clara; *5*, Rota; *6*, Momotombo; *7*, Apoyeque; and *8*, Mombacho.

142°C was recorded (A. R. McBirney, written commun., 1953). The investigations were terminated when the mining-type drilling equipment proved unsuitable (D. E. Spencer, oral commun., 1969).

Geochemistry—The small (0.3 l/sec) 66°C spring southeast of Tisate is the only significantly mineralized spring in the area. That spring is displaced from the probable locus of thermal activity suggested by the dipole resistivity-mapping survey.

The 140 ppm of SiO_2 and the Na/K ratio of 13.3 from that spring are both favorable, indicating possible reservoir temperatures of about 155°C and 215°C, respectively (after curves of Fournier and Truesdell, 1970; Ellis, 1970). The chemical data are inconclusive, but a hot-water geothermal system is suggested.

Chemical analyses of fumarolic gases were not particularly diagnostic.

Geophysics—Several types of geophysical surveys were conducted over the San Jacinto–Tisate area. Much of the following discussion is after Keller and Harthill (1970).

Resistivity profiling and depth soundings with a Schlumberger array were not very effective because of difficulties in penetrating the highly resistive surficial lavas. A dipole resistivity-mapping survey was much more effective in defining the areal extent of apparently lower resistivity rocks. The survey suggests that a geothermal cell, about 3.0 km long by 0.5–1.5 km wide at the 6-ohm-m contour, may be localized in the San Jacinto area. The contour pattern indicates a north-northeast elongation of the lower resistivity zone, with the 9-ohm-m contour extending to just northeast of Tisate. The southern extent was incompletely defined. Extended depth-sounding calculations suggest that the near-surface zone of relatively high resistivity is about 100 m thick; beneath this zone, resistivities decrease to a few tens of ohm-meters outside the area of thermal activity and to 2–3 ohm-m within. A possible electrical basement is present at a depth of 1.0–1.5 km.

Temperature-gradient boreholes—Six boreholes drilled near the San Jacinto thermal area to obtain temperature-gradient (Table 2) and geologic data penetrated alternating lavas and pyroclastic deposits with minor hydrothermal alteration except for moderate alteration in hole 3.

Momotombo Geothermal Prospect

Geology—Five areas of fumarolic activity and thermal ground are located on the lower southern slopes of Momotombo Volcano (Fig. 3A). This part of Momotombo is an older volcanic structure, separated from the presently active Momotombo cone by a somma ridge, which together with an eastward breach in the summit crater has served to deflect recent lava flows. The last significant eruption of Momotombo was the 1905 lava flow down the northern flank (McBirney, 1958), but high-temperature fumarolic activity continues

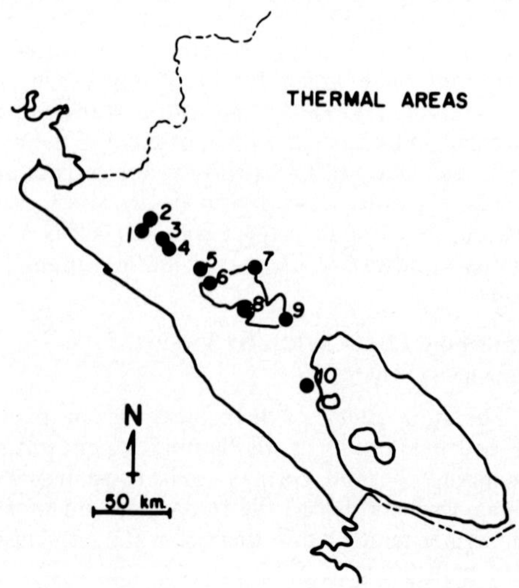

FIG. 2–Index map of designated thermal areas: *1*, Casita; *2*, Agua Caliente; *3*, NE Telica-Najo; *4*, San Jacinto-Tisate; *5*, Cerro Colorado; *6*, Momotombo; *7*, San Luis; *8*, Jiloa-Apoyeque; *9*, Tipitapa; and *10*, Mombacho.

Table 1. Summary of Thermal Areas

Area	*Thermal Activity and Geology*	*Size*	*Elev.*	*Remarks*
Casita	Fumarolic areas localized along faults in pyroclastic deposits on upper slopes of dormant Casita Volcano. Much alteration, locally severe.	0.4 km^2 within overall 3 × 4 km	650-1200 m	Poor potential, unfavorable chemistry, gases mostly H_2O; probable synvolcanic origin, altitude and access unfavorable.
Agua Caliente	Thermal water (max. 55°C) in numerous wells and a few springs. Quaternary alluvium and volcaniclastic sediments in Nicaraguan depression.	3 × 10 km overall	40	Poor to fair potential; perhaps involves deep circulation of meteoric water from volcanic slopes.
NE Telica-Najo	Small, scattered fumarolic areas, often localized along faults in pyroclastic deposits on middle slopes of San Jacinto Volcano.	0.04 km^2 active in overall 1.5 × 3 km	290-500	Fair potential, more study needed.
San Jacinto-Tisate	Vigorous fumaroles, mudpots, and thermal ground with intense alteration in 0.02 km^2 area at San Jacinto. Few feeble fumaroles in moderately altered 0.06 km^2 at Tisate. Localized along N-S fault, with thermal springs, on lower eastern slopes on San Jacinto and Santa Clara Volcanoes.	0.08 km^2 two active areas 2.5 km apart	160-190	Good potential, favorable surface display, chemistry from one spring, and geophysics. Environmental problems due to adjacent village.
Cerro Colorado	Weak fumarolic activity with widespread alteration in eroded cinder cone tephra. Thermal spring (48°) 3.5 km SW.	0.2 km^2	160	Poor potential at synvolcanic fumaroles. Favorable spring chemistry.
Momotombo	Five areas of strong fumarolic activity; prevalent alteration, locally severe on lower southern slopes of "older" Momotombo Volcano in lava-tephra overlying pyroclastics. Strong fault control. Boiling springs and seeps along Lake Managua shoreline.	0.8 km^2 active in overall 2 × 2.3 km	40-360	Best potential of sites investigated. Favorable geology, chemistry, geophysics, with few environmental problems.
San Luis	Numerous thermal springs (max. 89°C) in fracture or fault controlled alignments. Several associated travertine mounds. In Quat. alluvium near Tert. volcanics.	1 × 3 km overall	40	Poor potential; travertine is a qualitative indicator of low temperature at depth (White, 1970), but favorable SiO_2.
Jiloa-Apoyeque	Very small fumarolic area and several thermal springs near summit crater and lateral phreatic crater of eroded and inactive Apoyeque Volcano.	2 × 5 km overall	40	Fair potential; favorable chemistry. Close to major city of Managua.
Tipitapa	Several localized thermal springs (max. 96°C) in Quat. alluvium and pyroclastics perhaps on concealed fault.	Small, other nearby springs	40	Poor to fair potential; silica and boron favorable, but Na/K ratio unfavorable (?).
Mombacho	Very small fumarolic area and small thermal springs near top of Quat. Mombacho Volcano. Other thermal springs (max. 55°C) around base.	Small, but with basal springs 8 × 10 km	30-1000	Fair potential. Synvolcanic(?) features on the volcano are less favorable than those around base.

to the present in the summit crater. The thermal areas are in andesitic-basaltic pyroclastic deposits and lavas, whereas the younger lavas from Momotombo are basalts. Suitably permeable reservoir rocks are present in the sequence and recharge potential is good.

The surface thermal manifestations—extensive fumarolic-solfataric and thermal-ground areas—have the most activity of any site in Nicaragua. Hydrothermal alteration ranges from moderate to severe, locally resulting in a siliceous residue within the fumarolic areas. Minor boiling springs (101°C) discharge about 5 l/sec from permeable alluvium-pyroclastic layers at the southeastern site. The presence of minor siliceous sinter at some of these springs is a qualitative indicator of

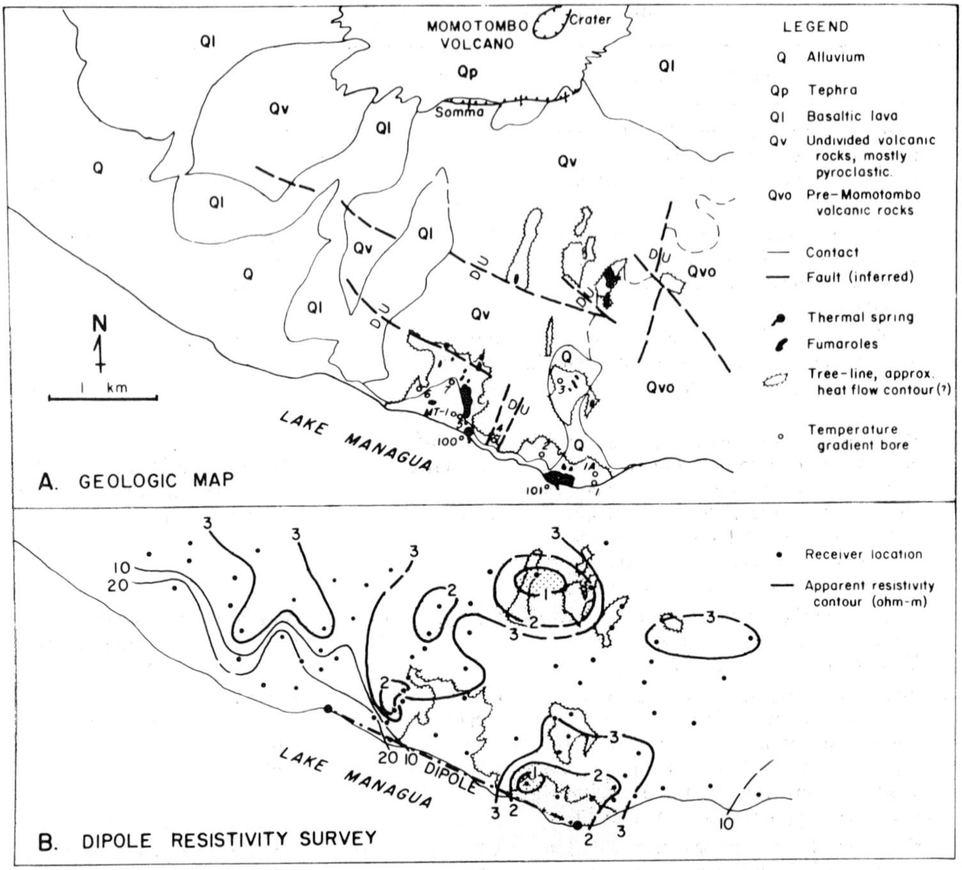

FIG. 3–Momotombo thermal area: **A.** Geologic map (Thigpen, 1970b); **B.** Dipole resistivity survey (after Keller and Harthill, 1970).

temperatures above 180°C (White, 1970). Boiling thermal seeps occur at the southwestern site and, especially in the wet season, numerous thermal springs and seeps occur nearby in the waters of Lake Managua.

Several faults, partly inferred, are interpreted by the writer to exert strong control on the localization of thermal activity by serving as the conduits from depth. The two northern thermal areas are adjacent to an east-west fault system (Fig. 3A), and the three southern thermal areas are, for the most part, either adjacent to or downslope from another east-west fault system.

The thermal manifestations are thought to represent late-stage solfataric volcanic activity, perhaps originating at considerable depth from a cooling magma intrusion.

Geochemistry—Thermal waters from the boiling springs are relatively highly mineralized. Although chemical data are inconclusive, a hot-water geothermal system is suggested.

The thermal water contains an average of 145 ppm SiO_2, with a maximum of 180 ppm, and an average Na/K ratio of 16.7. These values suggest subsurface temperatures of 170°C and 185°C, respectively (Fournier and Truesdell, 1970; Ellis, 1970). A higher temperature of 209°C was measured in a deep borehole. The discrepancy could be due to chemical inequilibrium and/or geographic separation (Fig. 3A).

Geophysics—Dipole resistivity-mapping surveys revealed an extensive area of apparently low-resistivity rocks with much lower values than at San Jacinto (Fig. 3B). Part of the following discussion is after Keller and Harthill (1970). Closure of the low-resistivity contours was not attained, either northward toward Momotombo or, in the southeastern part of the area, southward underneath

Table 2. Temperatures Measured Near San Jacinto Thermal Area

Hole	*Total Depth (m)*	*Max. Temp. (°C)*	*Depth (m)*	*Temp. Gradient (°C/10 m)*
1	13.1	No data	—	—
1A	12.2	No data	—	—
2	61.9	50	60.9	5.9
3	61.9	105	60.9	9.5
4	61.9	30	15.2	Isothermal
5	61.9	30	Uniform	Isothermal

Lake Managua. The resistivity contour pattern suggests two separate zones trending generally east-west. The northern zone practically coincides with the previously mapped fault system to the west, but not to the southeast. The incomplete 3-ohm-m contour encloses a minimum area of about 4 km^2.

Extended depth-sounding calculations and electromagnetic-sounding data show high resistivity in the near-surface rocks and, in the underlying rocks, low resistivities apparently extending to considerable depth (at least 2 km). There was no strong indication of electrical basement.

Temperature-gradient boreholes—Eight holes penetrated an alternating sequence of predominantly lavas and lesser pyroclastic deposits. Hydrothermal alteration ranges from slight to severe, but is more intense than at San Jacinto.

Temperature data, obtained with a thermister array, revealed favorable temperature gradients. These data, with maximum credible temperatures and highest gradients from several observations, are summarized in Table 3.

Deep borehole—One nominal 14.3-cm hole (MT-1) was drilled with a slim-hole drill rig to a depth of 608.2 m (Fig. 3A). A sequence of basalt flows with minor tuff intercalations was penetrated to a depth of 65 m. Below that depth the apparently pre-Momotombo lithologic sequence is mostly tuffs and other pyroclastic deposits with minor andesitic and basaltic lava flows. Hydrothermal alteration ranges from moderate to locally severe; severely altered zones occur at 150–195 m, 225–260 m, and in most of the section from 460 to 608.2 m. Much of this pyroclastic sequence appears to have good reservoir potential.

A maximum temperature of 209°C was measured at a depth of 213 m where a cave-in prevented deeper wire-line temperature measurements. That temperature confirmed predictions from thermal gradients as in nearby hole 5 and showed a favorable correlation between low resistivity and high temperature at depth. No production tests were attempted.

Summary and Conclusions

Western Nicaragua has very good geothermal potential. Ten prospects were defined during this evaluation. Most geothermal prospects and other thermal manifestations are associated with the Quaternary volcanic zone, where hyperthermal conditions appear to be widespread. Other hidden or "blind" geothermal reservoirs are likely to be present in the Quaternary zone and possibly scattered through the Tertiary volcanic zone.

The San Jacinto–Tisate and Momotombo geothermal prospects both show very favorable potential after detailed preliminary evaluation. The Momotombo prospect shows more positive indications of geothermal potential, particularly in larger size and higher temperature. Geochemical

Table 3. Temperature Data for Momotombo Thermal Area

Hole	*Depth (m)*	*Max. Temp. (°C)*	*Depth (m)*	*Temp. Gradient (°C/10 m)*	*Remarks*
1	61.9	152	57.9	13.8 av., 27.5 max.	
1A	61.9	—	—	—	No data, hole erupted
2	61.9	96	60.2	2.6	Near isothermal
3	61.9	140(?)	53.4	7.7 av., 23.0 max.	
4	47.2	84	23.5	7.5	Inversion below 23 m
5	55.8	165	49.2	9.8	Hole later erupted
6	31.4	92	19.2	17.0	Shallow data only
7	61.9	150	59.5	11.8	Inversion below 44 m, later erupted

and geophysical data are moderately favorable at San Jacinto–Tisate and very favorable at Momotombo.

References Cited

Del Giudice, Daniele, 1959, Informe sobre las investigaciones de las "fuerzas endogenas" en Nicaragua: Servicio Geologico Nacional de Nicaragua Bol., no. 3, p. 49–84.

Ellis, A. J., 1970, Quantitative interpretation of chemical characteristics of hydrothermal systems: Geothermics, Spec. Issue 2, p. 516–528.

Fournier, R. O., and A. H. Truesdell, 1970, Chemical indicators of subsurface temperature applied to hot spring waters of Yellowstone National Park, Wyoming, U.S.A.: Geothermics, Spec. Issue 2, p. 529–535.

Keller, G. V., and Norman Harthill, 1970, Electrical geophysical surveys in relation to the location of geothermal power resources in northwestern Nicaragua: Unpub. rept., Texas Instruments Inc., Dallas.

McBirney, A. R., 1958, Active volcanoes of Nicaragua and Costa Rica, *in* Catalogue of the active volcanoes of the world: Napoli, pt. VI, p. 107–146.

——— and Howell Williams, 1964, The origin of the Nicaraguan Depression (abs.): Bull. Volcanol., v. 28, p. 63.

——— and ——— 1965, Volcanic history of Nicaragua: California Univ. Pubs. Geol. Sci., v. 55, 65 p.

Thigpen, J. B., 1970a, Regional geology; thermal manifestations of western Nicaragua: Unpub. rept., Texas Instruments Inc., Dallas.

——— 1970b, Surface geology; South Volcan Momotombo fumaroles (Nicaragua): Unpub. rept., Texas Instruments Inc., Dallas.

——— 1971, Geology of thermal manifestations and potential geothermal resources of western Nicaragua: Unpub. M.S. thesis, Stanford Univ.

White, D. E., 1970, Geochemistry applied to the discovery, evaluation, and exploitation of geothermal energy resources; Rapporteur's report, Sec. V, U.N. Symposium on Development and Utilization of Geothermal Resources: Pisa.

Geothermal Exploration in Northern Chile—Summary[1]

ALFREDO LAHSEN[2]

Introduction

Geothermal exploration in Chile started in 1968 as a result of a project subscribed by the Chilean Government and the United Nations Development Program. This project is being administrated by the Corporación de Fomento de la Producción together with the United Nations (CORFO/UNDP). The purpose of this project is to determine the geothermal possibilities of an area of 108,000 km² in the high Andes of Tarapacá and Antofagasta provinces approximately between 18° and 25°S lat. and east of 69°30′W long.

The first stage of geothermal investigations has been restricted to the northernmost part of the country mainly because this region possesses no other energy resources. Limited hydroelectric resources have already been totally developed, and oil exploration has not been successful.

Although there was some previous geothermal exploration in the area, this paper is based on the writer's experience from 1968 to 1973 while in charge of the project's research in Tarapacá and Antofagasta provinces. Many of the results presented were obtained by a team of geologists working in this project, in which J. Healy, of the New Zealand Geological Survey, had a very important part.

Geothermal Areas of Northern Chile

Recent geothermal exploration in Chile began with systematic studies of the geology, geophysics, and geochemistry of the El Tatio thermal area. At the same time, geologic surveys were carried out in most of the other hydrothermal areas in Tarapacá and Antofagasta to select the best areas for further investigations (Healy, 1969). The distribution of the main hydrothermal areas investigated is shown in Figure 1. Their physical characteristics and geologic settings appear in Table 1.

All of these hot-spring areas are located in the high Andes, along the Pliocene-Quaternary volcanic zone. This fact strongly suggests that the heat source derives from volcanic activity. However, the tectonics of this zone also control the genesis of the geothermal manifestations. Pliocene-Quaternary east-west extension initiated many small grabens along the Andean range in which the hot water may be contained.

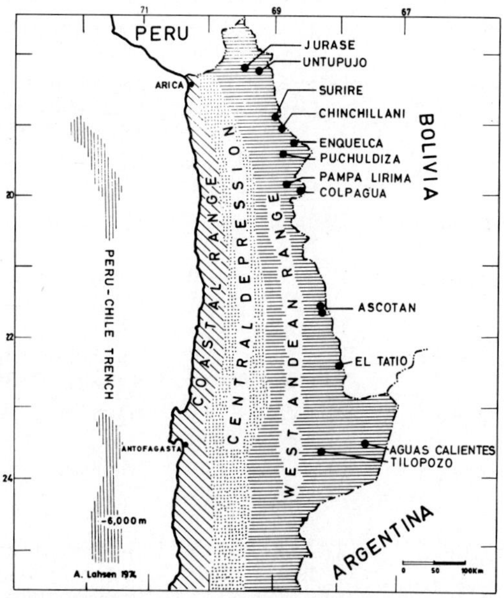

FIG. 1—Distribution of hot-spring areas in northern Chile.

Tectonics and Related Late Cenozoic Volcanism

Along the Circum-Pacific orogenic belt is a well-established earthquake focal zone down to 600 km. This focal zone separates the supposedly upthrusting continental crust from the downthrusting oceanic crust. Orogenic (taphrogenic) movements accompanied by very active Cenozoic volcanism characterize this belt.

In the Andes of northern Chile, late Cenozoic volcanism (Fig. 2) includes two principal episodes: (1) pre-Pliocene or Miocene volcanism, represented by several sheets of rhyolitic to daci-

[1]Manuscript received, October 21, 1974.

[2]Departamento de Geología, Universidad de Chile, Santiago, Chile.

The writer is greatly indebted to J. Healy for his great experience in the subject and for his many valuable suggestions. Reynaldo Charrier, of the Department of Geology of the University of Chile, Coordinator for Chile for the Circum-Pacific Energy and Mineral Resources Conference, discussed with the writer several aspects of this paper and improved the form and style of the manuscript.

Table 1. Characteristics of Some Hydrothermal Areas of Tarapaca and Antofagasta Provinces

Hydrothermal Area	*Location*	*Surface Activity, Temperature*	*Geologic Environment*
Jurase	18°12′S, 69°32′W Elev. 4,000 m	Hot springs, alteration zones, silica deposits; 68°C	Tuffs and sediments near Quaternary volcanoes
Untupujo	18°13′S, 69°17′W Elev. 4,350 m	Warm springs; 12-15°C	Pleistocene ignimbrites and andesitic volcanoes
Surire	18°55′S, 68°59′W Elev. 4,300 m	Fumaroles, hot springs, boiling pools, alteration zones, sinter deposits; 20-80°C	Plio-Pleistocene dacitic volcanoes
Chinchillani	19°08′S, 68°55′W Elev. 4,200 m	Warm springs, altered zones; 10-30°C	At foot of an eroded Pliocene volcano
Enquelca	19°15′S, 68°47′W Elev. 3,750 m	Warm springs; 30°C	Andesitic lava flows
Puchuldiza	19°23′S, 69°01′W Elev. 4,250 m	Geysers, fumaroles, boiling pools, hot springs, altered zones, deposits; >86°C	Volcano-tectonic depression surrounded by Plio-Pleistocene volcanoes
Pampa Lirima	19°51′S, 68°56′W Elev. 3,900 m	Hot springs, bubbling pools, salt deposits; 40-73°C	Upper Tertiary sedimentary and volcanic rocks
Colpagua	19°53′S, 68°38′W Elev. 3,800 m	Warm springs; 44°C	Upper Tertiary volcanic rocks
Ascotán	21°36′S, 68°15′W 21°42′S, 68°14′W Elev. 3,780 m	Warm springs; 25-33°C	Downfaulted block bordered by Pleistocene andesitic volcanoes
El Tatio	22°20′S, 68°01′W Elev. 4,300 m	Geysers, fumaroles, boiling water and mud pools, hot springs, sinter deposits, altered zones; 86°C	N-S graben filled by Miocene to Pleistocene ignimbrites and andesitic volcanoes
Aguas Calientes	23°32′S, 67°35′W Elev. 4,400 m	Warm springs and pools; 15-27°C	Flat depression of ignimbrites, bounded by dacitic and andesitic volcanoes
Tilopozo	23°48′S, 68°15′W Elev. 2,300 m	Warm springs, sinter deposits; 25-27°C	Folded and faulted ignimbrites

tic ignimbrites and andesitic stratovolcanoes, and (2) Pliocene-Quaternary volcanic activity that includes several units of ignimbrites and the andesitic stratovolcanoes that crown the highest part of the Andes.

Volcanic activity corresponding to the oldest unit started at least 18.7 m.y. ago (Ruiz, 1965). The ignimbrites and lava flows of this episode were folded during the late Miocene compressive phase, which was the last tectonic compression of the Andean orogenic cycle. East-west extension began at the end of this phase and is still occurring. Volcanic activity decreased and probably stopped during the compressive phase and resumed immediately afterward.

The Pliocene-Quaternary extension, which reached its maximum intensity in the late Pliocene, produced differential uplift of blocks along a nearly north-south fault system. The major morphostructural units generated by the extension process are, from west to east: (1) the Peru-Chile Trench, (2) the coastal range, (3) the central depression, and (4) the west Andean range, where the late Cenozoic volcanism took place.

In the high Andes of northern Chile are about 470 volcanoes, of which 17 are presently active (Ruiz, 1965; González, 1972). Late Cenozoic volcanic rocks cover approximately 80,000 km^2 (Fig. 2). Many of the volcanoes are higher than 5,500 m; between their summits and the Peru-Chile Trench is an elevation difference of about 12,000–14,000 m over a distance of 350–400 km. The continental crust beneath the highest part of the Andes reaches a thickness of 60–70 km (Dragicevic, 1962, 1974).

Parallel with the main morphostructural features, a well-defined Benioff zone is deduced by the depth distribution of earthquake foci, which are deepest below the highest part of the Andes (Kausel and Lomnitz, 1968).

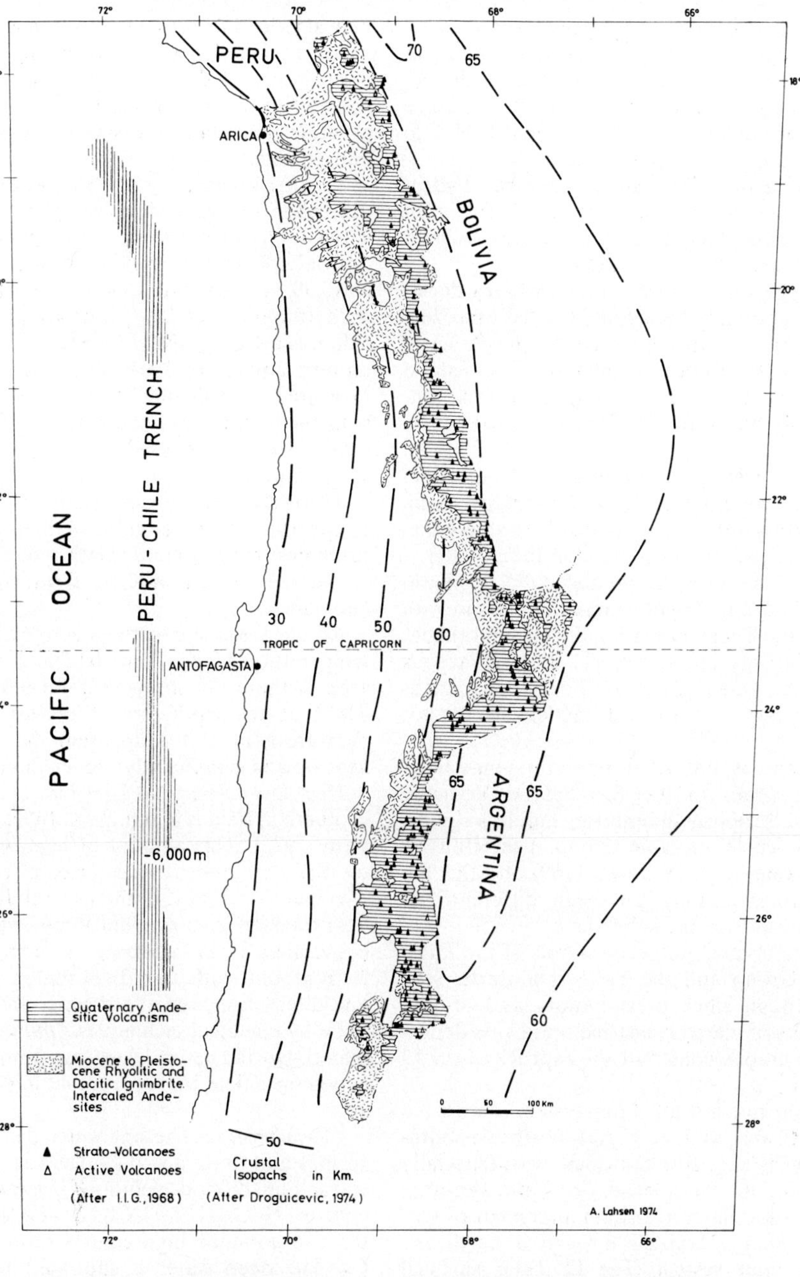

FIG. 2–Distribution of late Cenozoic volcanism in northern Chile.

A synchronism between the Pliocene-Quaternary volcanism, the extension period in this orogenic region, and the present spreading activity in the oceans is evident. Moreover, the parallelism of the volcanic belt of the high Andes, the Peru-Chile Trench, and the Benioff zone suggests a close causal relationship.

El Tatio Geothermal Field

El Tatio geothermal field is located in the northeasternmost part of Antofagasta province, at an altitude of 4,300 m (Fig. 1). It lies in a north-south downfaulted block (El Tatio graben) between the horst of Serranía de Tucle to the west and the volcanoes of the high range to the east.

Geothermal exploration at El Tatio started in 1968. Six exploration drillholes were made between 1969 and 1971 to depths up to 750 m. Since 1973, six production wells have been drilled to depths ranging from 870 to 1,820 m.

Geologic setting—El Tatio, like other hydrothermal fields of northern Chile, is located in an intense volcanic zone. Its origin is mainly related to the Pliocene-Quaternary extension.

The late Cenozoic volcanism in this area developed over a strongly folded and faulted Mesozoic basement. In late Miocene time the pre-Pliocene unit was gently folded, so that the Rio Salado Volcanic Group and the Sifon Ignimbrite lie unconformably below the Pliocene-Quaternary volcanic unit.

The older unit was deposited contemporaneously with the intense block movement along nearly north-south normal faults. Volcanic emission centers also are located along these faults.

The block movements produced the El Tatio graben, which has dropped more than 600 m with respect to the Tucle–Loma Lucero structural belt. In this relatively uplifted block, the argillaceous rocks of the Quebrada Justo Formation constitute a barrier to westward movement of hot fluids.

The steam or hot-water reservoir consists of permeable zones in the Rio Salado Volcanic Group, the Puripicar Ignimbrite, and, less significantly, the Tucle Volcanic Group. The Sifon Ignimbrite (amend. from Guest, 1969) and the Toconce Formation have not been differentiated from other units in the subsurface.

Some impervious tuffaceous levels of the Tucle Volcanic Group and the Tatio Ignimbrite, self-sealed through silica precipitation, and argillaceous layers of glacio-lacustrine sediments deposited in the graben constitute the caprock of the El Tatio field.

The main north-south fault system is affected by a northwest-southeast and northeast-southwest wrench-fault system which was especially active during the Pleistocene. The Tatio Volcano, located outside the north-south alignment of volcanoes, probably erupted through a northeast-southwest fault system. The El Tatio surficial thermal activity is mainly controlled by northwest-southeast and northeast-southwest fractures. The deep hot fluids may reach the surface through these same systems (Healy and Lahsen, 1970).

Hydrothermal activity—The surficial thermal activity at El Tatio is scattered over an area of about 10 km^2. Most of the hot springs emerge at the lowest levels of the El Tatio graben. The thermal activity includes geysers, fumaroles, boiling water and mud pools, and flowing springs. The temperature of these springs ranges up to 86°C, the boiling point at this altitude, and higher temperatures have been measured in some fumaroles and geysers.

Resistivity surveys—Geophysical exploration to determine the subsurface conditions of the hot-water reservoir has been based on electric methods only. Resistivity surveys show an extent of at least 30 km^2 for the apparent resistivity of less than 10 ohm-m at a depth of about 1,000 m. Inside this area, a resistivity anomaly of less than 5 ohm-m covers 15 km^2. The thermal gradients measured in drillholes down to 30 m, together with the temperature, indicate that the low resistivity values are caused by the presence of hot saline fluids.

Thermal measurements—Shallow (0.15 and 2 m) temperature surveys and thermal gradients in drillholes reveal a close relation of the isotherms to surface activity, altered areas, and electrical anomalies.

All the areas with surface activity have ground temperatures higher than 10°C at 2 m depth. The areas without hot springs exhibit temperatures of 4–8°C at this depth. Since the mean annual temperature is 0°C, it is apparent that the regional heat flow is significantly higher than normal.

Heat loss—The total heat loss of the El Tatio geothermal field is about 26×10^6 cal/sec (Hochstein, 1972). This amount of heat was estimated by direct methods from an area of about 10 km^2 which encloses most of the natural discharge (i.e., heat transfer from evaporation of water surfaces, convection from steaming ground, conduction from ground with a gradient higher than 1°C/m, and direct discharge of hot water and fumaroles).

Using chemical techniques, Mahon (1970) estimated that the natural heat loss from the El Tatio geothermal field ranges from 40 to 60×10^6 cal/sec.

Geochemistry—The hot water discharged from springs and drillholes at El Tatio is a nearly neutral sodium chloride solution. The major constituents are NaCl, KCl, $CaCl_2$, B, and SiO_2. The water also contains high concentrations of Li and Cs. The deep water is saturated with silica in equilibrium with quartz. The major dissolved gases are CO_2 and H_2S. Constituent concentrations vary as deep hot water is diluted with shallow cold water.

Prior to drilling, a minimum underground temperature of about 210°C was estimated from Na/K ratios and SiO_2 contents. The isotopic content of deuterium and oxygen-18 of different waters of the area supports a meteoric origin.

Drilled wells—Six exploratory holes (4-in. or 10-cm diameter) and six production holes (8-in. or 20-cm diameter) have been drilled at El Tatio. Several permeable zones were detected by losses of circulation during drilling and by injection of cold water and its increase in temperature after drilling. These permeable zones are located in the Tucle Volcanic Group, the Puripicar Ignimbrite, and the Rio Salado Volcanic Group. The thickest permeable zones are present in the last two units. The permeability of these formations is due to fractures produced by tectonic processes and to columnar joints formed during cooling.

The discharge of the wells is a steam-water mixture with a dryness ranging between 8 and 17%.

Measurements in wells in the west and northwest part of the field show that temperatures reach maximums of 212° to 230°C and then decrease downward. In other wells in the southeast part of the field, the temperatures rise to maximums ranging from 178° to 256°C at the bottom without reversal.

The temperature inversions may be explained by lateral and nearly horizontal movements of hot water from the southeast part of the field, where the fluids rise and flow through the permeable formations toward the west and north (Lahsen, 1971).

Puchuldiza Geothermal Field

The Puchuldiza geothermal field is located in the high Andes in east-central Tarapacá province (Fig. 1, Table 1). It lies in a volcano-tectonic depression surrounded by Plio-Pleistocene andesitic volcanoes.

Geothermal exploration began at the end of 1969 and was interrupted about the middle of 1972. Surveys included geological, geophysical, and geochemical studies, as well as an account of hot springs and a thermal survey by means of gradient holes to 30 m depth.

Geologic setting—The oldest rocks that crop out in the geothermal area of Puchuldiza correspond to the Churicollo Formation, a continental sequence of argillaceous and tuffaceous sandstones with conglomerate intercalations. These deposits, of fluvial and lacustrine character, have tentatively been assigned to the Lower to middle Cretaceous. The generally argillaceous matrix of the sediments and the high degree of compaction make this formation practically impermeable.

The Churicollo Formation was affected by compressive movements referred to the Paleocene Laramide orogenesis. Periods of extension followed the Laramide phase and also the late Eocene–early Oligocene phase (Charrier and Vicente, 1970). This extension produced north-south normal faults which defined tectonic depressions. These depressions were filled from Miocene to Pliocene time by thick volcanic deposits and fluvial and lacustrine sediments.

The oldest known ash flows of the Puchuldiza area form the Utayane Ignimbrite, which unconformably overlies the Mesozoic units. This formation consists of 160 m of densely welded and partly vitreous rhyolitic tuffs.

The Utayane Ignimbrite appears to be strongly fractured and leached and might constitute a good aquifer at depth. It is overlain by the Chojña-Chaya Formation, 60–80 m of well-stratified sandstones with a few thin intercalations of fine conglomerates.

The Chojña-Chaya Formation is overlain by the Condoriri Formation, which consists of dacitic ignimbrites and pumice tuffs with some augite and olivine basaltic andesites at the top. This formation seems to interfinger toward the west with the Puchuldiza Formation, which in its lower part consists of augite-bearing andesites and, in its upper part, of andesitic ignimbrites with fiamme.

An erosional disconformity separates the Puchuldiza Formation from the overlying Lupe Formation. The Lupe consists, in its lower part, of sandstones and conglomerates of fluvial character. The upper part consists of pumice tuff breccias and pumice lapilli pyroclastic sediments.

The late Cenozoic formations are gently folded; structures are oriented N30–50°W. The folding phase has been assigned to the middle to late Miocene compressive phase. After this compression more tectonic extension resulted in a nearly north-south fault system associated with northwest-southeast and almost east-west fault systems.

The major vertical displacements of blocks were produced along the north-south fault system. The Puchuldiza geothermal area is located in a block which has sunk 500 m with respect to the horst of the Cerro Condoriri. These north-south faults have been cut by the wrench faults of the northwest-southeast and east-west systems, to which the Quaternary volcanic centers of the area are related.

The hydrothermal area of Puchuldiza is limited both to the north and south by several Plio-Pleistocene stratovolcanoes reaching altitudes higher than 5,000 m. In most of these volcanoes, pronounced hydrothermal alteration zones were found. The lavas are biotite-hornblende andesites with thin pyroclastic intercalations.

A Pleistocene-Holocene fracture system characterized by northeast-southwest faults has affected the Plio-Pleistocene volcanoes and produced, in part, the surficial hydrothermal activity of Puchuldiza.

Hydrothermal activity—The Puchuldiza thermal springs emerge in two main areas: an eastern area including Puchuldiza (s.s.) and Tuja, and a smaller area about 5 km to the west. Both areas are covered by white sinter and evaporites precipitated from the hot springs. Most of these springs are flowing springs or clear pools of hot water, but some geysers and fumaroles are present. The temperatures of the hot springs commonly exceed 86°C, the boiling point at the altitude of 4,200 m.

Most of the springs emerge at relatively low levels along the Rio Puchuldiza and its small tributaries. The hot springs reach the surface through fissures mainly of north-south and northeast-southwest orientation. White, hydrothermally altered rocks extend along the traces of these faults.

No superficial hydrothermal activity is observed east of the Condoriri fault, the eastern border of the downfaulted block of Puchuldiza. The reservoir must be contained in the Utayane Ignimbrite and the Puchuldiza Formation. The permeability of these volcanic formations is due to cooling joints and tectonic fissures.

Geochemistry—Detailed chemical analyses were made by Mahon (1970). As at El Tatio, the hot waters supplied by the springs of Puchuldiza are a nearly neutral sodium chloride solution. The extent of mineralization of the Puchuldiza springs, however, is less than half that of the El Tatio springs. The waters also contain high concentrations of Li, Cs, and Rb; the major dissolved gases are CO_2 and H_2S.

The Na/K ratios suggest an underground water temperature of 200–212°C. Silica content indicates an original aquifer temperature between 180° and 200°C.

Surire Thermal Area

The Surire thermal area is located in the southeast part of the Salar de Surire (18°54′S, 69° 00′W) in Tarapacá province near the Bolivian border. The mean altitude of the salar is 4,280 m. It lies in the central part of the western Andean range.

Geologic setting—The Salar de Surire is a depression filled with fluvial and lacustrine sediments, as well as glacial outwash and moraine deposits. This salar extends over an area of 570 km^2 and contains important evaporitic minerals. Borax deposits have been exploited in the past.

Many volcanoes rise above 5,500 m around the salar; some lavas have flowed into it and are now partially buried by sediments. The volcanoes are dacitic and andesitic; the dacites generally occur as domes and the andesites form stratovolcanoes. The thermal area is located at the southeastern extreme of the salar.

Hydrothermal activity—In an area of 45 km^2, about 230 hot springs are present; most of them are clear hot pools and flowing springs. Water temperatures range from 20° to 80°C, considerably above the 0°C mean annual temperature of the area.

Almost all the hot springs, especially those with the highest temperatures, are located along northeast-trending shear fractures. Some are present along north- and northwest-trending fractures. All these fractures are very recent. In some places, open fissures about 1 m wide cut across the sintered and indurated deposits of the salar.

Heat loss—The natural heat loss of the Surire hydrothermal area was determined by direct measurements of conductive heat transfer at the surface, heat loss due to evaporation of hot pools, and heat loss due to direct discharge of flowing springs and fumaroles. It is as high as 28×10^6 cal/sec (Trujillo, 1972). The heat loss in the Surire area thus is greater than that in any of the thermal areas of Japan ($5–20 \times 10^6$ cal/sec; Fukuda et al, 1970) and less than that in many of the New Zealand thermal areas ($20–135 \times 10^6$ cal/sec; Dawson and Dickinson, 1970).

By comparison of the natural heat loss in Surire with that in other thermal areas of the world, a minimum power potential of 50 Mw has been estimated for the area (Trujillo, 1972).

Power Potential

Estimates of possible power potential are commonly made by comparison with other producing fields. In New Zealand the natural heat discharge measured at the Wairakei field was 100×10^6 cal/sec (Dawson and Dickinson, 1970), whereas the present output from the geothermal wells is about 6 to 7 times that amount (Healy, 1970). In the Japanese geothermal field (Fukuda et al, 1970) and in The Geysers, California (McNitt, 1964), the output of the steam wells is many times the total amount of the natural heat loss.

Thus, a minimum power power potential of 50 Mw may be estimated for the El Tatio field as well as for the Puchuldiza and Surire geothermal areas. At El Tatio, a power potential of 17 Mw has already been proved in the drilled wells.

Utilization of Geothermal Fluids

In addition to utilization of geothermal fluids for electric power, the possibilities of desalinization of the thermal water and extraction of minerals of economic value have been examined. A pilot plant for desalinization of geothermal steam is being installed at El Tatio.

Another process for which the use of geothermal steam seems ideally suited is that of refining sulfur at El Tatio, because several volcanoes near this field contain sulfur deposits.

References Cited

Charrier, R., and J. C. Vicente, 1970, Liminary and geosynclinal Andes: major orogenic phases and synchronical evolutions of the Central and Magellan sectors of the Argentine-Chilean Andes: Buenos Aires, Solid Earth Problems Conf., p. 451–470.

Dawson, G. B., and D. J. Dickinson, 1971, Heat flow studies in thermal areas of the North Island of New Zealand, *in* United Nations symposium on the development and utilization of geothermal resources: Proc., v. 2, pt. 1; Geothermics, Spec. Issue, no. 2, p. 446–473.

Dragicevic, M., 1962, Interpretaciones de las anomalías gravimétricas de las fosas marinas de Valparaiso, Antofagasta y Arica: Dept. Geof., Univ. Chile Pub. 28.

——— 1974, Carta gravimétrica de América del Sur.: Dept. Geof., Univ. Chile Pub. 167.

Fukuda, M., et al, 1970, Some geothermal measurements at the Otake geothermal area, *in* United Nations symposium on the development and utilization of geothermal resources: Proc., v. 2; Geothermics, Spec. Issue.

González-Ferran, O., 1972, Distribution, migration and tectonic control of upper Cenozoic volcanism in West Antarctica and South America, *in* Antarctic geology and geophysics: Internat. Union Geol. Sci. Ser. B, no. 1, p. 173–179.

Guest, J. E., 1969, Upper Tertiary ignimbrites, northern Chile: Geol. Soc. America Bull., v. 80, p. 337–362.

Healy, J., 1969, Geological reconnaissance of hot springs localities in Tarapacá and Antofagasta provinces: Santiago, Chile, unpub. rept., Com. Geot. CORFO.

——— 1971, Pre-investigation geological appraisal of geothermal fields, *in* United Nations symposium on the development and utilization of geothermal resources: Proc., v. 2, pt. 1; Geothermics, Spec. Issue, no. 2, p. 571–577.

——— and A. Lahsen, 1970, Geology of El Tatio geothermal field, Antofagasta province: Santiago, Chile, unpub. rept., Com. Geot. CORFO.

Hochstein, M., 1972, Geophysical survey of El Tatio Geothermal area; results from measurements from Jan. to Dec. 1971: Santiago, Chile, unpub. rept., Com. Geot. CORFO.

Kausel, E. and C. Lomnitz, 1968, Tectonics of Chile, *in* Simposio Panamericano del manto superior; Grupo II, manto superior, petrología y tectónica: Inst. Geofís., U.N.A.M., Mexico, v. 2, p. 47–67.

Lahsen, A., 1971, Algunas consideraciones sobre el sistema geotérmico de El Tatio y recomendaciones para futuros posos de producción: Santiago, Chile, unpub. rept., Com. Geot. CORFO.

Mahon, W. A., 1970, Preliminary reports on the geochemistry of the hot springs in the Puchuldiza-Lupe area and the fluids discharged from well No. 1 El Tatio: Santiago, Chile, unpub. rept., Com. Geot. CORFO.

McNitt, J. R., 1964, Geology of The Geysers thermal area, California, *in* United Nations conference on new sources of energy, Rome, 1961: Proc., v. 2, Geothermal Energy, 1, p. 292–303.

Ruiz, C., 1965, Geología y yacimientos metalíferos de Chile: Santiago, Chile, Inst. Invest. Geol., 305 p.

Trujillo, P., 1972, Estudio de las manifestaciones termales de Suriri: Santiago, Chile, unpub. rept., Com. Geot. CORFO.

HYDROCARBONS

Oil- and Gas-Bearing Basins of Circum-Pacific Belt—Summary[1]

YU. K. BURLIN,[2] YU. M. PUSHCHAROVSKY,[3] and B. A. SOKOLOV[2]

The Circum-Pacific belt, more than 50,000 km long and from 600 km wide in the east to 5,000 km wide in the west, is a complex of tectonic structures developed at different times since the late Precambrian. The Circum-Pacific belt is a ring system of interrelated tectonic zones—the tectonic belt. Volcanic, seismic, ore, and granitoid ring belts also can be distinguished around the Pacific Ocean. About 40% of all the known basins of the world are within the Circum-Pacific belt. Some of the sedimentary basins in the belt contain oil and gas and account for almost a third of world oil production. These basins contain thick sedimentary sequences which form lenslike bodies. A sedimentary sequence more than 2 km thick is thought to be a general prerequisite for petroleum content.

Basins in the Circum-Pacific belt can be divided into five types on the basis of their tectonic position. They are as follows: (1) basins situated at the juncture between folded areas of the Circum-Pacific belt and ancient platforms; (2) basins within intermontane folded areas; (3) basins at continental margins, or pericontinental basins; (4) basins within recent geosynclinal zones, or perioceanic basins (including island-arc structures); and (5) basins present on young platforms with Mesozoic-Cenozoic cover. Most significant for oil and gas prospects are (1) geosynclinal depressions filled with thick Mesozoic-Cenozoic beds and extending along the strike of the perioceanic zones (e.g., depressions of north Sumatra, Taiwan, and the Great Valley in California); (2) intermontane depressions of various types, especially those of the late-geosynclinal stage of development (Maracaibo, Cook Inlet); and (3) foredeeps and submontane depressions (Sub-Verkhoyansk, North Alaska, and sub-Andean depressions).

Basins situated at the juncture between folded areas of the Circum-Pacific and ancient platforms account for 21 of the basins in the Circum-Pacific belt. Most are in the eastern part of the belt, extending along the Cordilleras and Andes; all are composed of thick Mesozoic-Cenozoic deposits and are characterized by high oil and gas content.

More than 40 basins in the Circum-Pacific belt are classified as intermontane. All of them overlie crust of continental type and are composed of folded Mesozoic and Cenozoic sequences as thick as 3–5 km. Oil and gas are produced from many of these basins, and future petroleum potential is promising.

Most of the 10 pericontinental basins occur along the American coast of the Pacific Ocean, and many lie beneath the ocean. The basins are composed mostly of Cenozoic rocks, and thicknesses range from hundreds of meters to 10 km, or even more. Oil production is significant, and future prospects are good.

The perioceanic basins number more than 40. Although of no commercial importance to date, their potential capabilities should be studied further. Gas reserves have been developed from Neogene strata in the East Lugovsk uplift on the shore of Aniva Bay (Okhotsk Sea), and oil and gas have been proved within the Neogene of the marginal coastal parts of the deep-water basin of the Indonesian-Philippine area. The island-arc basins also hold some potential for petroleum prospecting.

There are few platform basins in the Circum-Pacific belt because, tectonically, the region is still mobile. Some oil and gas are present in eastern Australia basins in upper Paleozoic and Mesozoic rocks. Also, oil is produced from Mesozoic and Paleozoic rocks in the Rocky Mountain basins of the North American platform.

The formation of oil and gas basins around the Pacific Ocean is closely related to development of geosynclinal areas and their transformation into orogenic folded areas. The processes of thick sedimentation are favorable for generation and accumulation of oil and gas. Oil and gas content is especially characteristic of structures of the late stages of geosynclinal development proper and of orogenic-type (late geosynclinal) structures.

[1]Manuscript received, November 11, 1974.

[2]Geological faculty, Moscow University, Moscow, USSR.

[3]Geological Institute, USSR Academy of Sciences, Moscow, USSR.

Plate Tectonics in Petroleum Exploration of Convergent Continental Margins[1]

THOMAS L. THOMPSON[2]

Abstract Plate tectonics, the hypothesis of multiple crustal plates which float on a viscous layer called the "asthenosphere," provides reason for viewing the earth's outer shell as a system of shifting continents and changing ocean basins. The idea of diverging plates (or seafloor spreading) implies that plates converge elsewhere at compatible rates. Estimated convergence rates range up to 10 cm/yr, or 224 km in the 2 million years since the beginning of Pleistocene time.

Convergence between oceanic and continental crust may result in thermal generation of oil and gas in sediments as young as Pleistocene because of rapid deep burial associated with subduction. Mountainous source areas for sediment plus steep continental slopes favor rapid burial of organic material with turbidites. Rapid subduction of oceanic crust under continental margins may carry sediments to depths which provide requisite thermal environments for generation of oil and gas from organic matter disseminated in the sediment. Continued subduction of oceanic crust under continental slopes may cause reverse faulting such that oil and gas accumulations are uplifted toward the ocean bottom.

Core samples obtained adjacent to the Aleutian Trench in the western Gulf of Alaska apparently show effects of subduction and reverse-fault uplift on a Pleistocene sedimentary sequence. Although this Pleistocene sequence lies only a few hundred feet below the ocean bottom, organic-matter carbonization suggests previous burial of at least 8,500 ft (2,600 m) and a late pregeneration state for petroleum. In contrast, noncommercial oil production from uplifted deep-water strata of early Tertiary age at Katalla, Alaska, suggests that formerly significant accumulations may have been dissipated by faulting, uplift, and erosion. Late Tertiary rocks beneath outer continental shelves and/or upper continental slopes at convergent margins may be in the optimum stage of current petroleum expulsion but still buried deeply enough for entrapment of giant oil accumulations.

On the continent side of volcanic ridges and mountain trends, parallel with subduction zones, geographic coincidence of intermediate and deep-focus earthquakes with regions of high earth heat and giant petroleum accumulations suggests an empirical means of predicting earth heat trends that would favor petroleum accumulation in block-fault basins.

INTRODUCTION

Wegener (1924) assembled much geologic evidence in support of the concept of continental drift. However, he did not have an acceptable answer for how the continents moved through the ocean basins. Therefore, even though Du Toit (1937) strongly supported the concept with more geologic evidence, continental drift remained unacceptable in the minds of most North American geologists until the last decade. During the intervening 3 decades, ship surveys have provided much new information on the bathymetry, gravity, magnetics, and seismicity of the ocean regions. These new data led to concepts of seafloor spreading (Hess, 1962), transform faulting (Wilson, 1965), reassessment of continental drift (Blackett et al, 1965; Takeuchi et al, 1967), and subduction (a result of crustal convergence)—all within the context of plate tectonics (Maxwell, 1970; Wilson et al, 1972; Bird and Isacks, 1972; Shagam et al, 1972; Dewey, 1972). Since 1968, these concepts have been strongly supported through deep-sea coring by the JOIDES expeditions and have been widely accepted by the scientific community.

Guidelines for oil and gas exploration evolve as natural extensions of the plate-tectonics hypothesis, because plate tectonics provides an explanation for the origin of continental margins and suggests the basis by which marine deposits are formed into sedimentary rocks and eventually elevated as parts of the continents. The marine sedimentary origin and accumulation of most oil and gas are well recognized, and most marine sediments have accumulated on past and present continental margins. Most of the known reserves of oil and gas (roughly 95%) have accumulated within the past 250 million years, essentially concurrent with formation of the present ocean basins, marginal island ridges, and linear mountain chains of Andean and Alpine types.

Continental margins comprise huge frontiers for oil and gas exploration extending from shorelines to water depths of 15,000 ft (4,900 m). The total area of offshore oil and gas potential of 28,000,000 sq mi (72,500,000 km^2) exceeds the total area of onshore oil and gas potential of 20,000,000 sq mi (51,800,000 km^2.)

Problems and Objectives

The problem, as with most exploration, is to find the greatest concentration of economically recoverable petroleum with the least expenditure of time, manpower, and capital. Reasons for wildcat drilling seldom include scientific proof; instead, producible reserves usually are discovered with a working hypothesis based on limited data.

[1]Manuscript received, December 26, 1974; revised, January 20, 1975.

[2]Amoco Production Co., Tulsa, Oklahoma 74102; present address: University of Oklahoma, Norman, Oklahoma.

The objectives of this report are to describe the nature and suspected origins of convergent (Pacific-type) continental margins on the basis of plate tectonics, to indicate the potential for rapid generation of oil and gas near subduction zones at continental margins in regions of lithospheric plate convergence, and to discuss the association of giant petroleum accumulations with convergent plate margins.

Procedures

The data for this study consist of scattered geophysical and core information derived from published sources, ideas from oceanographic research workers, and some data from the oil industry. Plate tectonics furnishes the broad perspectives which help explain origins of continental margins, and which are useful in estimating their potential for oil and gas.

Hypothesis of Plate Tectonics and Its Impact on Continental Margins

Plate-Tectonics Concept

Comprehension of plate tectonics requires a global viewpoint. The atmosphere swirls about the earth at rates of several hundred miles per hour. Giant currents in the hydrosphere, such as the Gulf Stream, move at rates of several miles per hour in response to the earth's gravity, axial spin, and heat distribution. In this context, movement of several inches per year in the so-called solid earth seems reasonable, particularly on a global scale wherein the earth apparently behaves as a viscous fluid (Hubbert, 1939).

The earth's outer shell, or lithosphere (about 60 mi or 100 km thick), apparently floats on a viscous material called "asthenosphere," thus allowing divergences and convergences between lithospheric plates (Fig. 1). The lithosphere, in essence, pulls apart to form continental rifts, such as in east Africa, and oceanic ridges, such as the Mid-Atlantic and Mid-Indian ridges and the East Pacific Rise. The lithosphere converges to form subduction zones such as those along Japan and the west coast of South America. The Himalaya Mountains apparently resulted from the collision of two continental plates, India and Asia.

Distribution of earthquakes suggests breaking of the lithosphere into a mosaic of semirigid plates that move relative to each other (Fig. 2). The plates diverge, converge, and otherwise slip parallel with or oblique to their boundaries.

Classification of Continental Margins

These relative movements of lithospheric plates provide the basis for classifying plate boundaries and continental margins as divergent, convergent, or transform—in essence, representing extension faults, compression faults, and strike-slip faults on a global scale.

Transform faulting implies dominance of strike-slip displacements as manifest in the San Andreas and Fairweather fault systems of Pacific North America; it further implies change from one sense of displacement to another, for example, transformation from divergence at the East Pacific Rise to convergence in the western Gulf of Alaska. Thus, transform margins include combinations of strike slip and divergence (as with the separation of Baja California from mainland North America), as well as combinations of strike slip and convergence (as in the transition from the Fairweather tectonic system of the eastern Gulf of Alaska to the Aleutian tectonic system of the western Gulf of Alaska).

The different types of continental margins based on plate tectonics display contrasting physiography. Prominent continental rises, gentle continental slopes, broad continental shelves, and coastal plains characterize divergent margins. In contrast, deep oceanic trenches, steep continental slopes, narrow shelves, and island ridges or mountainous uplifts characterize convergent margins. Peninsula and island separations and continental borderlands of subsided continental fragments characterize transform margins such as offshore southern California. An estimated world distribution of margin types shows that divergent margins dominate the Atlantic Ocean whereas convergent margins dominate the Pacific Ocean (Fig. 2).

Divergent Plates and Seafloor Spreading

Interpretation of magnetic lineations and dating of the oldest rocks overlying basaltic crust have substantiated concepts that new oceanic crust forms at oceanic ridges in a process called "seafloor spreading," which suggests that the present ocean basins are quite young (Dietz, 1961; Hess, 1962; Vine and Mathews, 1963). Plate divergence along spreading trends apparently results in upwelling of magma from the asthenosphere to form new oceanic crust. This void-filling process forms a differentiated sequence of rocks (ophiolite sequence) about 3 mi (5 km) thick, including gabbro at the base, dike swarms above gabbro, and pillow basalts above dike swarms (Vine and Moores, 1972). With further plate divergence, the newly formed oceanic crust splits and more magma wells up.

The symmetrical magnetic lineations characteristic of the crustal material on either side of the seafloor-spreading trends reflect reversals in the earth's magnetic field. As the upper parts of the

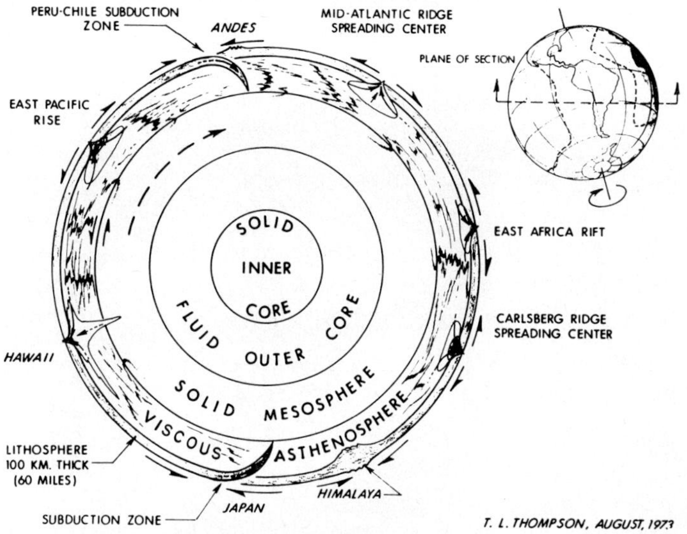

FIG. 1—Cross section of earth.

newly formed crust cooled through the curie point, they preserved the magnetic orientation of the earth's magnetic field which existed at the time of formation (Vine, 1969). Oceanic ridges may relate to the volume increase attendant on the processes of magmatic differentiation. The high heat flow characteristic of these trends proceeds from the magmas at shallow depths under the oceanic ridges. Increase in density of newly formed oceanic crust as it gradually cools perhaps explains why the former ridges have since become deep oceanic crust (Sclater and Francheteau, 1970).

Continuation of seafloor spreading at an average rate of 2.5 cm/yr for 200 million years, since the beginning of Jurassic time, would form an ocean basin 3,000 mi (4,800 km) wide, the present width of the North Atlantic. Thus we find successively older oceanic lithosphere with increasing distance from spreading trends, but the oldest known is Jurassic. Within the context of an earth history known to span at least 4 billion years, seafloor spreading seems quite rapid because 60% of the earth's present lithosphere apparently formed during the past 200 million years.

Convergent Plates and Lithospheric Subduction

The formation of new oceanic lithosphere along the oceanic ridge system implies destruction of lithosphere elsewhere unless one postulates rapid expansion of the earth. Apparently, destruction of lithosphere occurs in subduction zones at converging plate boundaries. Known lithospheric divergence at the oceanic ridges implies rapid convergence and destruction of lithosphere at subduction zones. Le Pichon (1968) estimated convergence rates ranging up to 11 cm/yr, or 140 mi (224 km) in the past 2 million years, since the beginning of Pleistocene time.

Types of convergence include oceanic-to-oceanic (e.g., the Tonga trench-island-ridge system), continental-to-continental (the Himalaya system), and oceanic-to-continental (the western Gulf of Alaska). The oceanic-to-continental type may evolve in some cases by consolidation of island ridges associated with oceanic-to-oceanic plate convergence. Essentially, this report concerns convergence of cooled and subsided oceanic lithosphere with less dense continental lithosphere.

Convergence of oceanic lithosphere with continental lithosphere forms two contrasting tectonic trends having apparent potential for petroleum on opposite sides of volcanic island ridges of western Pacific type (Fig. 3). Shallow-focus earthquakes, low earth temperatures, and thrust-fault basins under continental slopes adjacent to oceanic trenches contrast with deep-focus earthquakes, high earth temperatures, and block-fault basins on the continental side of volcanic island ridges.

Hypothesis for Petroleum Generation in Thrust-Fault Trends of Subduction Zones

Characteristics of Lithospheric Subduction

Apparent effects of subduction include formation of a deep oceanic trench, steepening of the continental slope, uplift of deep-water sediment as an island or mountain trend adjacent to the continental slope, and formation of a linear trend

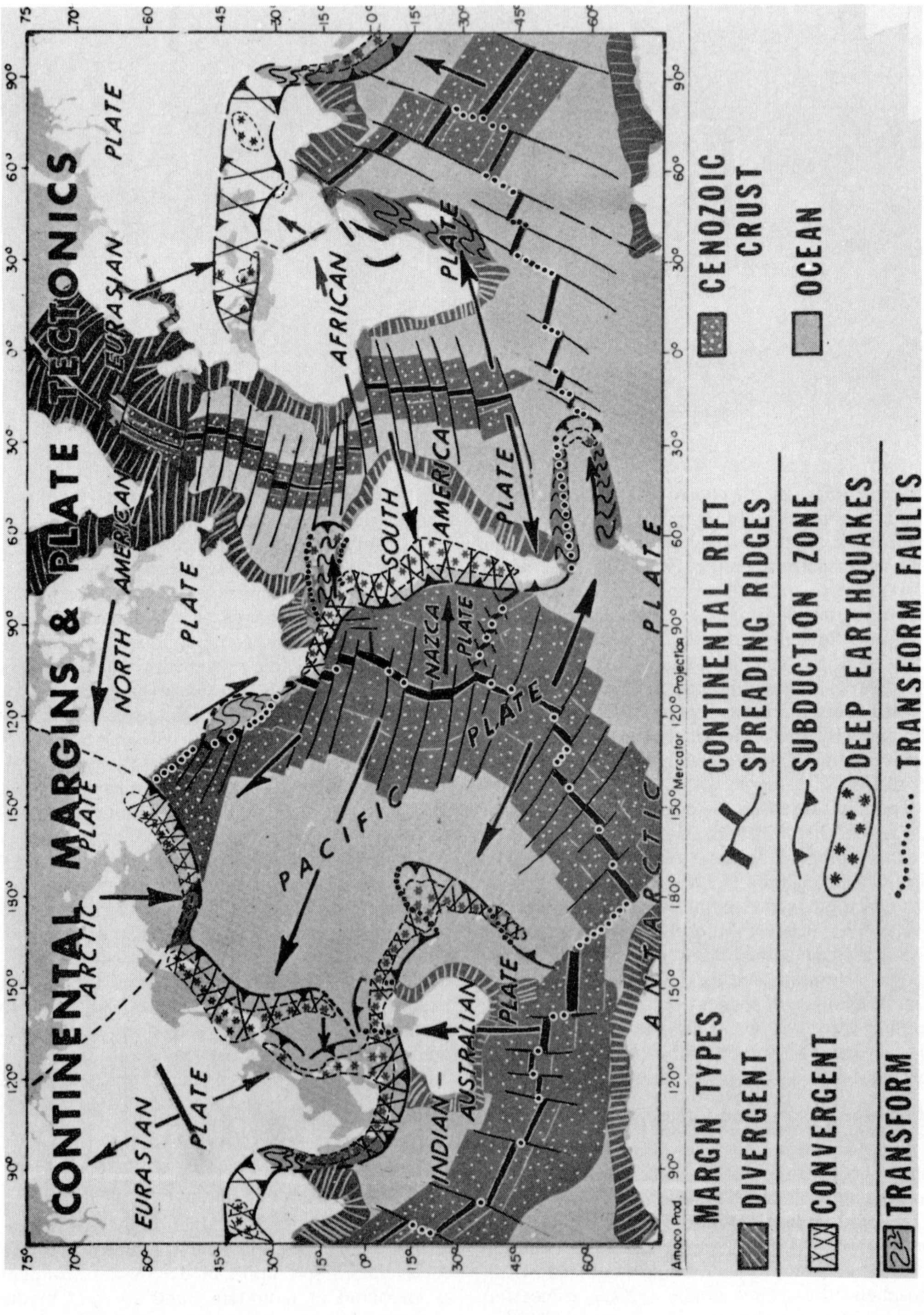

FIG. 2–Classification of continental margins based on plate tectonics.

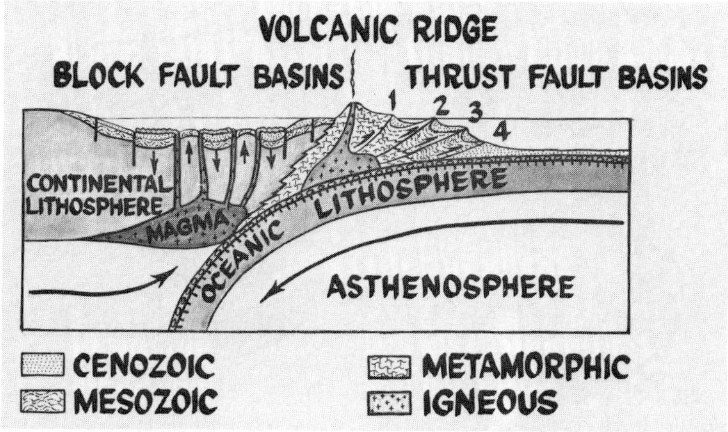

FIG. 3—Elements of western Pacific convergent margins.

of volcanoes on the continent side of the uplifted sediment trend. Isostatic gravity lows along the continental slopes suggest dynamic downwarp of the oceanic crust (Vening Meinesz et al, 1943; Bowin, 1972; Grow, 1973). Distribution of shallow, intermediate, and deep-focus earthquakes along an inclined plane which proceeds landward from the trench suggests reverse faulting (Benioff, 1955; Isacks, 1968; Stauder, 1968).

The Sediment Pile as a Potential Source of Oil and Gas

Subduction zones of converging plate boundaries seem to fulfill the basic requirements for generating oil and gas. Formation of a deep trench provides a natural repository for sediment, particularly where it is near a source of continental sediment. This would not apply in the case of oceanic trenches far removed from continents until the uplift of island ridges had progressed to subcontinental proportions, as in the west and southwest Pacific.

Continental margins generally favor prolific marine life, the organic remains of which are the recognized source of most oil and gas. Upwelling of deep, cold water from trenches provides the mineral ingredients necessary for biotic proliferation that may make convergent margins particularly favorable for petroleum generation. JOIDES cores near trenches commonly show concentrations of organic carbon in excess of 0.5%, which is adequate for thermal generation of petroleum in commercial quantities.

Burial of organic matter prior to oxidation is a third factor essential to oil and gas generation. Converging plate boundaries apparently provide optimum conditions for rapid burial to protect the organic matter, because mountainous regions provide a ready source of sediment and steep continental slopes accelerate the transport of sediment to the lower slope and trench by slumping and turbidity flow.

Compaction, Diagenesis, and Metamorphism of the Sediment Pile

In the western Gulf of Alaska, trench and lower-continental-slope deposits of Cretaceous and early Tertiary age have been tightly folded, metamorphosed, and uplifted to form land areas adjacent to the Aleutian Trench on the Kenai Peninsula, Kodiak Island, and Shumagin Islands (Moore, 1973a, b). One wonders by what process these rocks were transformed from their unconsolidated state as trench deposits to their present state of metamorphosed and uplifted rocks; and of more significance, at what stage in this process oil and gas could have been generated.

This question seemingly has been answered in part with the JOIDES discovery of tightly folded and compacted mudstone of Pleistocene age at shallow depths on the lower continental slope in the western Gulf of Alaska (Fig. 4). This compacted mudstone came from JOIDES corehole 181, just a few hundred feet (170–396 m) below the ocean bottom, at a water depth of 3,086 m on the north wall of the Aleutian Trench east of Kodiak Island (lat. 58°26.30′N, long. 148° 27.88′W). Specific properties of one core segment (core 29, depth 340 m) included dry bulk density of 2.1, porosity of 23%, and water saturation of 76% (Lee et al, 1973). Samples of this core segment were analyzed for organic matter (kerogen) by the Geochemistry and Palynology Groups at the Amoco Production Company Research Center. Principal interest was in the thermal history of these sediments, which would be reflected in

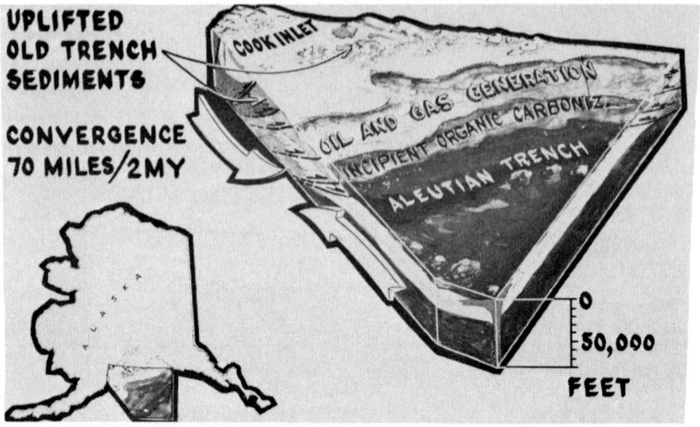

FIG. 4—A possible trend of oil and gas generation at a convergent margin in western Gulf of Alaska.

the carbonization level of the kerogen. By comparison of the carbonization level of the kerogen with that from Gulf Coast Tertiary sections, something about past temperatures could be inferred.

Grayson (1972) noted low-carbonization-level organic material and darkened, more carbonized organic material. Translucency measurements of psilate, trilete spores (20 grains) showed a range of 35–44% translucency with an average of 40.2%, indicating a carbonization level of approximately 70–73% carbon. This carbonization level is equivalent to that of Pleistocene sediments in the U.S. Gulf Coast province at approximately 8,500 ft (2,600 m) subsea and a log temperature of 155°F (68°C). The abundant darker (more carbonized) organic material noted by Grayson was in an advanced stage of diagenesis and probably was carbonized in older sediments, then transported to the site and redeposited. This conclusion seems reasonable in the setting of mountain uplift and rapid erosion, sediment transport, and burial. Elemental analysis of the kerogen by LaPlante (1972) indicated 82.9% carbon, 3.8% hydrogen, 11.8% oxygen, and 1.5% nitrogen. This percent carbon (carbonization level) is equivalent to that of Oligocene sediments in the U.S. Gulf Coast province at a burial depth of approximately 17,000 ft (5,200 m) and a log temperature of 280°F (138°C). However, it probably reflects an averaging of the two stages of diagenesis indicated above.

How did the compacted Pleistocene mudstone reach this stage of diagenesis when it lies at a depth of only a few hundred feet below the ocean bottom? One might postulate deposition on hot oceanic crust at a spreading trend, but spore-translucency determinations on other JOIDES samples from the northeast Pacific averaged 47%, suggesting a carbonization level of only 65–70% (Grayson, 1972). Alternately, one might postulate a present high geothermal gradient, but trenches generally are cold thermal regions. Furthermore, no recent volcanism has been noted in the vicinity of JOIDES hole 181. I suggest that rapid lithospheric convergence at the Aleutian Trench subduction zone resulted in rapid, deep burial and subsequent uplift of the Pleistocene sediment.

Thrust-Fault Hypothesis for Generation of Oil and Gas

Convergence between oceanic lithosphere and continental lithosphere may result in thermal generation of oil and gas in sediments as young as Pleistocene because of rapid, deep burial associated with subduction. Mountainous source areas and the presence of steep continental slopes favor rapid burial of organic material with turbidites. Rapid subduction of oceanic crust under continental margins may carry sediments to depths which provide requisite thermal environments for generation of oil and gas from organic matter disseminated in the sediment. Continued subduction of oceanic crust under continental slopes may cause thrust faulting such that oil and gas accumulations are uplifted toward the ocean bottom.

Cooled and subsided oceanic lithosphere in the western Gulf of Alaska apparently has wedged 70 mi (110 km) under the continent since the beginning of Pleistocene time, and may have contributed to conditions suitable for petroleum accumulation along the outer continental margin (Fig. 4). Carbonization measurements on organic matter from a JOIDES core of compacted Pleistocene sediment from a few hundred feet below the ocean bottom on the north flank of the Aleutian Trench indicate a late pregeneration stage of petroleum and a former burial depth of at least 8,500 ft (2,600 m; Grayson, 1972).

Thrust faulting wherein mechanical burial plays a dominant role may explain the rapid burial and subsequent uplift implied here (Fig. 5). Trench deposits may have been buried to depths of several miles, heated to account for observed carbonization, and subsequently uplifted to the ocean bottom on the continental slope. Sediment may slump and flow down an oversteepened continental slope, accumulate in the Aleutian Trench area, and then be carried downward along a subduction zone by the oceanic plate and be buried deeply enough to be compacted into rock. The increased temperatures at this depth (estimated as 15,000 ft or 4,600 m) would cause diagenesis of organic matter to form oil and gas. Burial by sedimentation may be augmented with mechanical burial by thrust faulting. Sequential formation of new thrust faults at the base of the continental slope would cause thickening of the sediment pile, steepening of the continental slope, and seaward advancement of the slope and trench. Continued thrust faulting could transport compacted sediment (and perhaps oil and gas accumulations) to shallower depths.

This process could satisfy the general requirements of crustal material balance from a plate-tectonic viewpoint, could explain the gravity minima associated with continental slopes, and could explain how sediment deposited beneath the sea is transformed to sedimentary rock and subsequently uplifted in islands and mountains above the sea (Fig. 6).

Supporting evidence for the application of a thrust-fault interpretation in the western Gulf of Alaska includes (1) an inclined zone of earthquake epicenters (Benioff zone) dipping from the vicinity of the slope landward and (2) landward thickening of refraction seismic-velocity units mapped by Shor and Von Huene (1972). The suggested sequence of velocity units includes mantle (8 km/sec), gabbros and dikes (6.8 km/sec), pillow basalts (5 km/sec), compacted early Tertiary and Miocene sediments (4.4–5.4 km/sec), and uncompacted Pliocene and Quaternary sediments (2–2.9 km/sec).

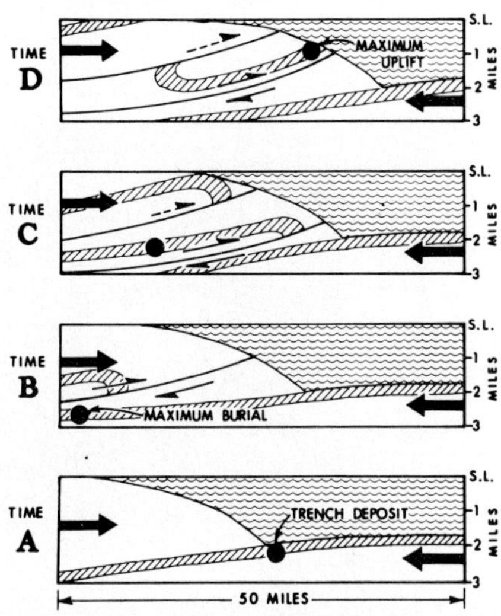

FIG. 5–A thrust-fault mechanism for the mechanical burial and uplift of deep-water sediment.

Potential Applications in Petroleum Exploration

Regional seismic profiles and geologic sections crossing the continental margin in the western Gulf of Alaska seem explainable by thrust faulting (Fig. 7). Important corollary effects may include shale diapirs and near-surface normal faults. However, net results indicate thickening of the sediment pile and uplift of successively older rocks inland from the continental slope. Thus we find Miocene units adjacent to Pliocene, early Tertiary adjacent to Miocene, and Cretaceous adjacent to early Tertiary. Perhaps the noncommercial oil discovery in early Tertiary rocks at Katalla reflects a formerly more valuable accumulation that has been dissipated by faulting, uplift, and erosion. Such dissipation may explain the apparent absence of pre-Cenozoic oil and gas accumulations along Pacific margins in general.

The subduction process with thrust faulting as described in the preceding paragraphs amounts to continental accretion but may include mechanisms for generating petroleum. This suggestion could lead to a method of predicting trends of maximum oil and gas generation based on organic-matter diagenesis at convergent continental margins in general. JOIDES drilling adjacent to the Aleutian Trench recovered Pleistocene sediment in a late pre–oil-generation stage. This contrasts with early Tertiary rocks which accumulated in the same general depositional environment (i.e., continental slope and oceanic trench), but which have since been metamorphosed beyond the stage of petroleum generation to the stage of rock metamorphism and even to the evolution of igneous rocks. Somewhere between the stage of pre–oil generation and the stage of rock metamorphism should be the ideal place for peak generation and maximum expulsion of oil and gas.

Visualize a trend of oil and gas generation somewhere between the lower-continental-slope trends of incipient organic carbonization and shore trends of metamorphosed and uplifted older rocks with dissipated oil accumulations like Katalla, perhaps along outer shelves and upper slopes. Late Tertiary rocks beneath outer continental shelves and/or upper continental slopes at

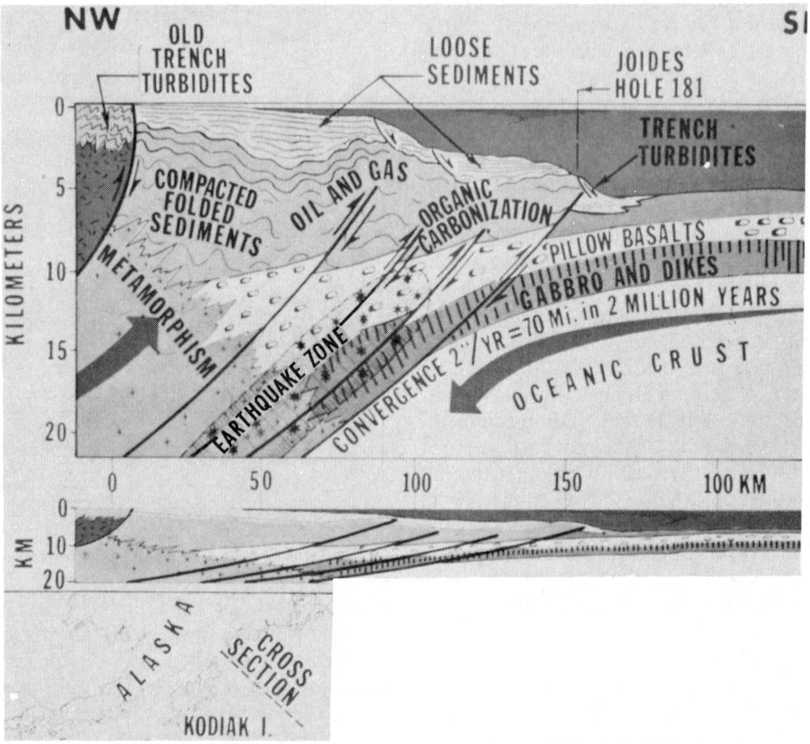

FIG. 6–Continental-margin subduction and petroleum generation.

convergent margins may be in optimum stages of petroleum expulsion but still buried deeply enough for entrapment of giant oil accumulations.

Petroleum present offshore east of Trinidad may come from sediments uplifted from hotter levels by thrust faulting along a northeast trend that represents the same pattern of structure evident onshore in Trinidad (Fig. 8). This offshore trend can be projected northeastward into deeper water between the Puerto Rico Trench and the Antilles island arc. A reflection profile across the north end of the Puerto Rico Trench indicates oceanic crust wedging under the Antilles island arc. A reflection profile near the south end of the island arc suggests thrust faulting in thick sequences under the continental slope. These thick sequences probably were derived from the Orinoco River to the southwest.

Other petroleum-productive margins perhaps reflecting rapid thrust-fault burial, heating, and uplift above subduction zones—analogous to the western Gulf of Alaska—include the eastern offshore of central Japan and offshore Peru (Fig. 9).

Other interesting regions closely analogous to the western Gulf of Alaska include: the east margin of the Philippines; the south margins of Indonesia, New Britain, and the Solomon Islands; and the west margins of Central and South America. In each of these localities, as with the western Gulf of Alaska, substantial land areas for sediment source lie close to the associated trench. The potential of these sites will depend, in part, on the lithologic nature of the sediment source and on whether there is a substantial component of sand for the formation of turbidite reservoirs.

Oceanic crust converging with oceanic crust should result in mechanical characteristics similar to those of the above regions in the sense of a subduction zone—similar favorability from the standpoint of organic production, but less favorable sources of land-derived sediments. Such regions include the island margins adjacent to other deep oceanic trenches. Examples are the outer Aleutian Trench; the Kuril, Ryukyu, Mariana, New Hebrides, Kermadec, Tonga, and South Sandwich Trenches; and the northern part of the Puerto Rico Trench, east of the Lesser Antilles. Although these areas generally receive less land-derived sediment, they may include reef carbonate rocks that would provide petroleum-reservoir potential in proximity to subduction zones.

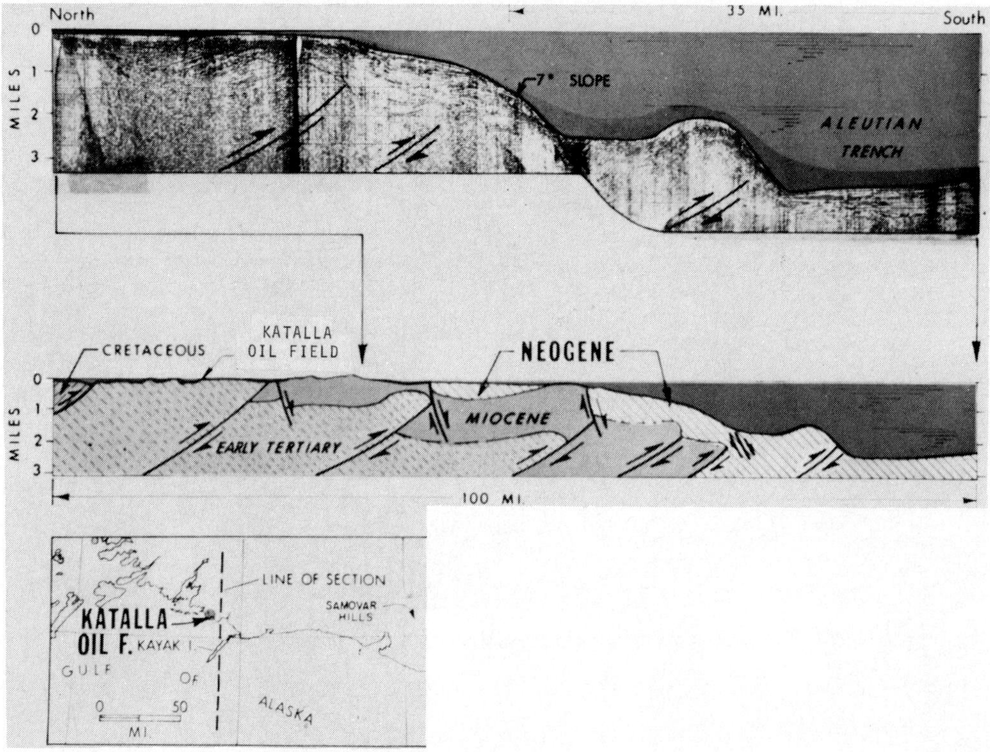

FIG. 7–Convergent continental margin in western Gulf of Alaska.

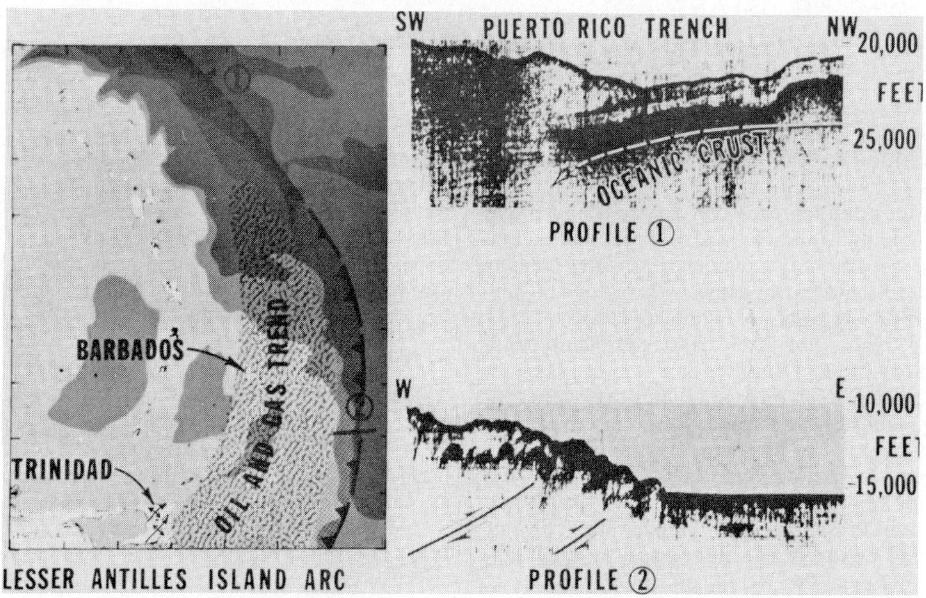

FIG. 8–Subduction zone at convergent margin of Trinidad (after Chase and Bunce, 1969).

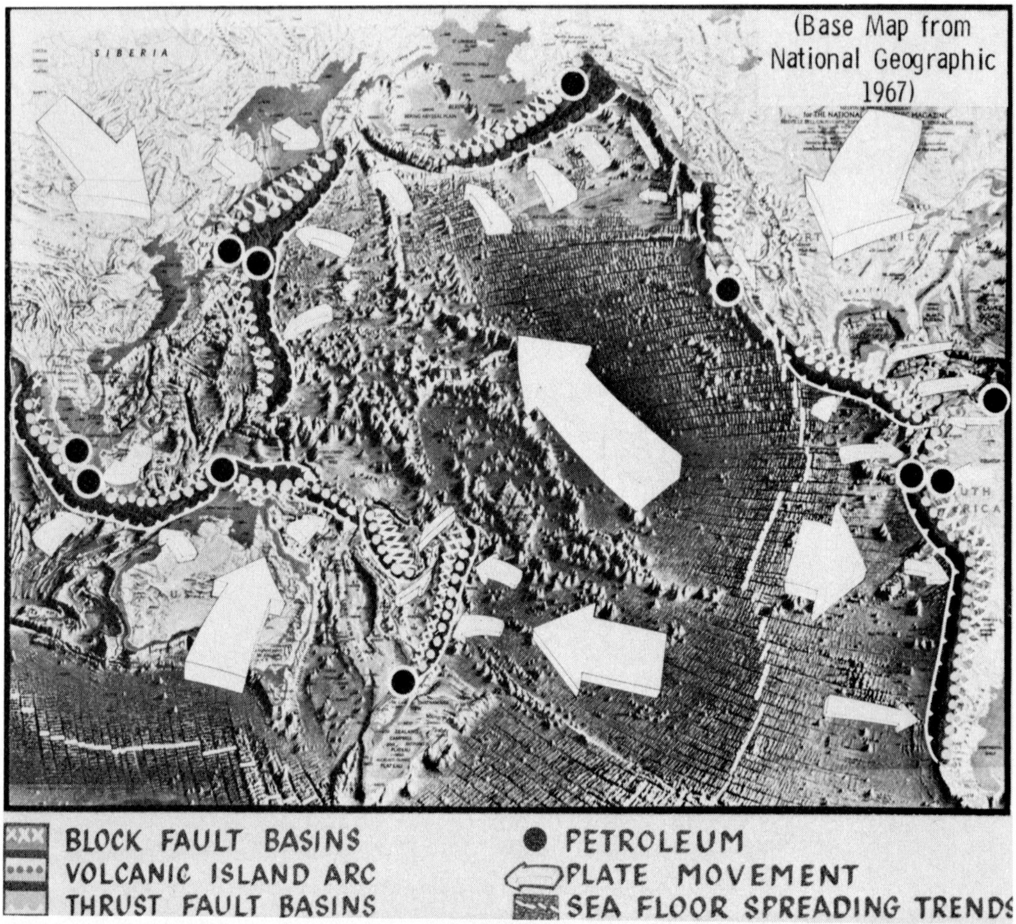

FIG. 9—Oil and gas at convergent margins of Pacific and Indian Oceans.

Petroleum in Geothermal Trends of Back-Ridge Basins

On the continent side of volcanic island ridges and mountain trends, above deep subduction zones, geographic coincidence of intermediate and deep-focus earthquakes with regions of high earth heat suggests an empirical means of predicting trends that would favor petroleum accumulations in block-fault basins.

Petroleum reservoirs with major discoveries include: sandstones of Miocene age discovered by Standard Oil of California and Atlantic Richfield in central Sumatra and the Java Sea, respectively; reefs of Miocene age discovered by Trend and Phillips in West Irian, New Guinea; and volcanic rocks of Cenozoic age discovered by Shell offshore between the North and South Islands of New Zealand.

The giant Minas field in central Sumatra and recent discoveries in the Java Sea demonstrate oil potential associated with block faulting and heat trends on the continent side of volcanic island ridges (Fig. 10). Temperature gradients several times higher than normal reasonably explain shallow production (3,000–4,000 ft or 910–1,200 m) from rocks of Miocene age.

Summary and Conclusions

Convergent continental margins represent promising frontiers for satisfying energy requirements of the Circum-Pacific region. Low earth temperatures and thrust faulting under continental slopes adjacent to oceanic trenches contrast with deep-focus earthquakes, high earth temperatures, and block faulting of back-ridge basins.

1. Convergence between oceanic crust and continental crust may result in thermal generation of

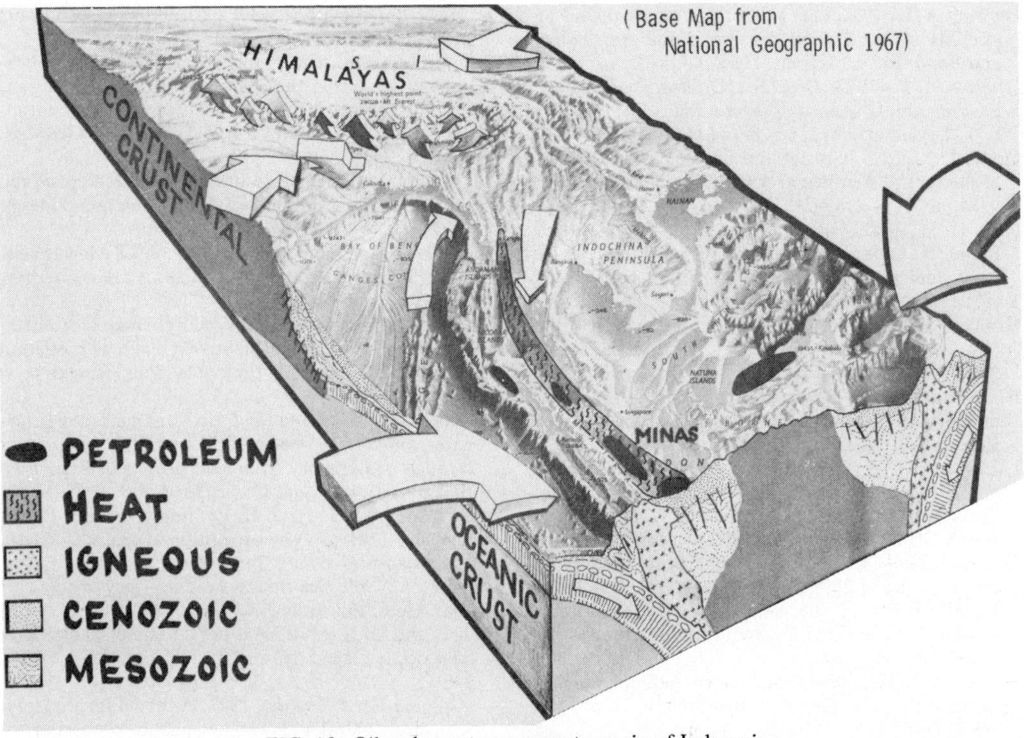

FIG. 10–Oil and gas at convergent margin of Indonesia.

oil and gas in sediments as young as Pleistocene because of rapid deep burial associated with subduction.

2. Rapid deep burial of sediments involves both rapid deposition of turbidites and mechanical burial by reverse faulting.

3. The most probable trend of oil and gas accumulations would be somewhere between the lower continental slope and the shoreline, perhaps in the region of the outer shelf and upper slope.

4. Seeking trends of probable current expulsion from source rocks associated with active uplift may be important because the same dynamic conditions that favor early generation of oil and gas also tend to rupture and dissipate oil and gas accumulations.

5. In the western Gulf of Alaska, the JOIDES core of Pleistocene sediment did not quite reach expulsion stage; the oil accumulation in early Tertiary rocks at Katalla apparently has been uplifted and dissipated; upper Tertiary rocks beneath the outer shelf and/or upper slope may be in the optimum stage of current expulsion but still buried deeply enough for entrapment of giant oil accumulations.

6. Distribution of intermediate and deep-focus earthquakes provides an empirical means of predicting heat trends that would favor petroleum accumulations on the continent side of volcanic island ridges and mountain trends.

References Cited

Benioff, H., 1955, Seismic evidence for crustal structures and tectonic activity, *in* A. Poldervaart, ed., Crust of the earth—a symposium: Geol. Soc. America Spec. Paper 62, p. 61–75.

Bird, J. M., and Bryan Isacks, eds., 1972, Plate tectonics: selected papers from the Journal of Geophysical Research: Washington, D.C., Am. Geophys. Union, 563 p.

Blackett, P. M. S., Edward Bullard, and S. K. Runcorn, 1965, A symposium on continental drift: Royal Soc. London Philos. Trans., ser. A, v. 258, no. 1088, 323 p.

Bowin, C., 1972, The Puerto Rico Trench negative gravity belt: Geol. Soc. America Mem. 132, p. 339–350.

Chase, R. L., and E. T. Bunce, 1969, Underthrusting of the eastern margin of the Antilles by the floor of the western North Atlantic Ocean, and origin of the Barbados Ridge: Jour. Geophys. Research, v. 74, p. 1413–1420.

Dewey, J. F., 1972, Plate tectonics: Sci. American, v. 226, p. 56–67.

Dietz, R. S., 1961, Continent and ocean basin evolution by spreading of the sea floor: Nature, v. 190, no. 4779, p. 854–857.

Du Toit, A. L., 1937, Our wandering continents; an hypothesis of continental drifting: Edinburgh, Oliver and Boyd, 366 p.; reprint. 1957.

Grayson, J. F., 1972, Deep Sea Drilling Project N. E. Pacific: Amoco Research memoranda, J. F. Grayson to T. L. Thompson, June 20 and October 23.

Grow, J. A., 1973, Crustal and upper mantle structure of the central Aleutian arc: Geol. Soc. America Bull., v. 84, no. 7, p. 2169–2191.

Hess, H. H., 1962, History of ocean basins, *in* A. F. J. Engel, H. L. James, and B. L. Leonard, eds., Petrologic studies—a volume in honor of A. F. Buddington: New York, Geol. Soc. America, p. 599–620.

Hubbert, M. K., 1939, The problem of earth deformation (abs.): Earthquake Notes, v. 11, nos. 1–2, p. 19–20.

Isacks, Bryan, Jack Oliver, and L. R. Sykes, 1968, Seismology and the new global tectonics: Jour. Geophys. Research, v. 73, no. 18, p. 5855–5899.

LaPlante, R. E., 1972, Source rock evaluation—JOIDES deep sea core sample from the Alaska continental margin: Amoco Production Company Research Center Tech. Serv. 8421C (Addendum).

Lee, H. J., H. W. Olsen, and R. von Huene, 1973, Physical properties of deformed sediments from Site 181, *in* Initial reports of the Deep Sea Drilling Project: Washington. D. C., U.S. Govt. Printing Office, v. 18, p. 897–901.

Le Pichon, Xavier, 1968, Sea-floor spreading and continental drift: Jour. Geophys. Research, v. 73, p. 3661–3697.

Maxwell, A. E., ed., 1970, The sea—ideas and observations on progress in the study of the seas, v. 4, New concepts of sea floor evaluation: New York and London, Wiley-Interscience, pt. 1, 791 p.; pts. 2–3, 664 p.

Moore, J. C., 1973a, Cretaceous continental margin sedimentation, southwestern Alaska: Geol. Soc. America Bull., v. 84, p. 585–614.

——— 1973b, Complex deformation of Cretaceous trench deposits, southwestern Alaska: Geol. Soc. America Bull., v. 84, p. 2005–2020.

National Geographic Society, 1967, Indian Ocean floor: Map, scale 1:30,000,000 (approx.).

Sclater, J. G., and Jean Francheteau, 1970, The implications of terrestrial heat flow observations on current tectonic and geochemical models of the crust and upper mantle of the earth: Royal Astron. Soc. Geophys. Jour., v. 20, p. 509–542.

Shagam, R., et al, 1972, Studies in earth and space sciences (a memoir in honor of Harry Hammond Hess): Geol. Soc. America Mem. 132, 683 p.

Shor, G. G., and Roland von Huene, 1972, Marine seismic refraction studies near Kodiak, Alaska: Geophysics, v. 37, p. 697–700.

Stauder, William, 1968, Tensional character of earthquake foci beneath the Aleutian Trench with relation to sea-floor spreading: Jour. Geophys. Research, v. 73, p. 7693–7701.

Takeuchi, H., S. Uyeda, and H. Kanamori, 1967, Debate about the earth—approach to geophysics through analysis of continental drift: San Francisco, Freeman, Cooper and Co., 253 p.

Vening Meinesz, F. A., J. H. F. Umbgrove, and Ph. H. Kuenen, 1943, Gravity expedition at sea, v. 2: Delft, Netherlands Geodetic Commission.

Vine, F. J., 1969, Sea-floor spreading—new evidence!: Jour. Geol. Education, v. 17, p. 6–16.

——— and D. H. Matthews, 1963, Magnetic anomalies over ocean ridges: Nature, v. 199, no. 4897, p. 947–949.

——— and E. M. Moores, 1972, A model for the gross structure, petrology, and magnetic properties of oceanic crust: Geol. Soc. America Mem. 132, p. 195–205.

Wegener, Alfred, 1924, The origin of continents and oceans; 1966, translated by John Biram from fourth (1929) revised German edition: New York, Dover, 246 p.

Wilson, J. T., 1965, A new class of faults and their bearing upon continental drift: Nature, v. 207, p. 343–347.

——— et al, 1972, Continents adrift: San Francisco, W. H. Freeman, 172 p.

Some Shallow Tectonic Consequences of Subduction and Their Meaning to the Hydrocarbon Explorationist[1]

PETER A. MONTECCHI[2]

Abstract Good-quality multichannel seismic data have revealed structural details across some of the major tectonic alignments in the western Pacific and eastern Indian Oceans along which lithospheric subduction has been proposed.

The data were gathered by R/V *Gulfrex* for Gulf Oil Corporation as part of the reconnaissance survey of the petroleum potential and tectonic framework of the region.

The evidence indicates that the alleged lithospheric subduction at oceanic trenches is largely restricted to the crystalline substratum, whereas the sedimentary layers undergo a "reorganization" as they are scraped off the substratum and piled on a surface much smaller than that on which they were originally deposited. This tectonic disturbance of apparently compressive nature affects the sedimentary sequence through the oceanward migration of a "tectonic front" which abruptly separates the undisturbed strata from the severely disrupted sequence. The disrupted rock mass is largely homogeneous seismically and could be construed as mélange.

It can now be seen why interpreters of earlier and low-resolution seismic data concluded that trench deposits are "undisturbed." They were obtaining reflection events only from those markers not yet reached by the migrating tectonic front. They had no way of postulating a lithologic continuity across the front itself.

Besides trench-related compression, the *Gulfrex* data also indicate that a similar phenomenon probably occurs along pseudocontinental margins in the northeastern Sulu Sea and offshore southeast of Honshū, Japan. No bathymetric trench is present in either area, and the sedimentary rocks scraped off the substratum contribute to the buildup of the lower continental slopes.

The significance of this phenomenon to the explorationist is that, although hydrocarbons may be synthesized from dispersed organic material as the result of "tectonic milling," it is highly unlikely that reservoir or seal conditions could exist in these highly disturbed areas unless the mélange were covered by, or juxtaposed against, undisturbed sedimentary units in which the fluids could be trapped.

INTRODUCTION

The main purpose of this article is to make public some of Gulf Oil Corporation's proprietary information on shallow structural details in areas critical to understanding lithospheric evolution. The secondary purpose is to present my personal, somewhat speculative interpretation of the observed phenomenon. This interpretation is not necessarily shared by any individual in Gulf Oil Corporation, but it is offered here with the hope that it will be kept in the minds of some interpreters as one of the possible alternatives, and be submitted to a fair test of its validity. The implications of my speculations may affect the way in which future regional hydrocarbon exploration is conducted.

Most data presented here are from Circum-Pacific areas where the boundaries between contiguous plates are supposed to be associated with subduction.

SELECTED DATA

Figure 1 shows the location of those deep-water regional seismic lines containing the segments that were selected for this paper. The selection was made primarily on the basis of the quality of the data and an evaluation of their contribution toward effective illustration of the phenomena described in the text.

Although seismic information is the main basis for my observations, gravitational, bathymetric, and magnetic data have also been systematically considered, and have yielded valuable contributions.

The processing and presentation of the data were neither standardized nor, in many cases, optimized. The displays are multichannel seismic data, with the exception of Figure 15, and were produced from a 24-fold common-depth-point (CDP) coverage.

Figures 2–5 display samples of the kind of phenomenon on which much of the following reasoning is based. A thick and little-disturbed sedimentary sequence dips toward the left side of each display—toward the axis of the bathymetric trench. Before reaching completely across the display, however, the bathymetric slope reverses, losing its apparent parallelism with the deeper seismic reflectors, and the shallow seismic markers become unresolvable through a maze of diffraction patterns. Hardly any coherent seismic event is resolvable toward the left edge of the display, where the rock mass appears as an acoustically homogeneous medium. The basement is resolved

[1]Manuscript received, November 11, 1974. Published by permission of the Exploration and Production Department of Gulf Research and Development Co.

[2]Senior Research Geophysicist and Onboard Chief Scientist with the Marine Exploration Program of Gulf Research and Development Co.. Pittsburgh, Pennsylvania 15230.

Sincere appreciation is expressed to Direktorat Jenderal Minyak Dan Gas Bumi, Departemen Pertambangan, Jakarta, Indonesia, for permission to display the seismic sections which are Pertambangan's property. Appreciation is also expressed to O. F. Huffman and J. M. Wageman, who reviewed the manuscript.

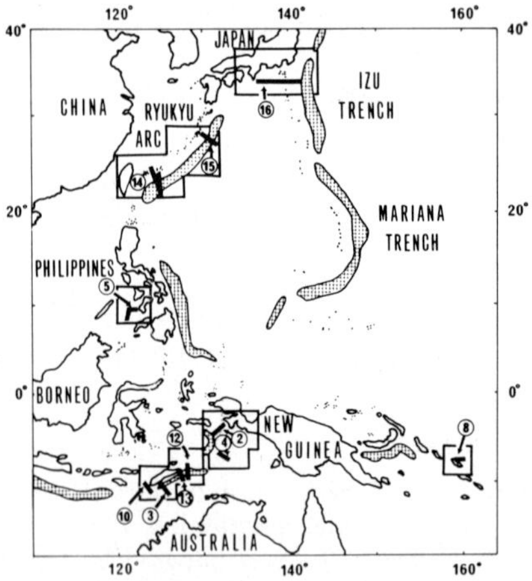

FIG. 1—*Gulfrex* seismic lines selected for study. Circled numbers indicate locations of following figures.

only in Figure 5, but it is expected that the deeper sedimentary markers of the other three figures reflect the attitude of the buried crystalline rocks at least for the right half of the displays.

The exact locations of these four lines are shown in Figures 2B, 3B, 4B, and 5B. A plot of the pertinent potential field data is included only with Figure 5, but these data are qualitatively representative of the potential fields as measured across all the regions described herein.

The interval velocities derived from fair-quality velocity analyses are shown in Figures 2C, 3C, and 4C. None of these line segments was saturated with velocity analyses, and most of the velocity determinations in the shallow left portions are unreliable. However, the presence of fair-quality deeper markers permits us to conclude that the seismically homogeneous masses are composed of low-velocity rocks, and that the main difference between the left and right sides of the displays is a pronounced discrepancy in the volume of the low-velocity material.

Gravity data also show an overall, increasing mass deficiency for the left half of the displays. This is probably caused by one or more of the following: (1) leftward deepening of the basement; (2) lithologic change in the basement; or (3) a volumetric preponderance of low-density sedimentary rocks.

Figures 3–5, besides describing the common structural theme, offer an opportunity for illustrating additional details. Figure 3D, for instance, reproduces that portion of Figure 3A enclosed by a rectangle. The data, presented after two-dimensional migration, successfully resolve a structure of obvious compressive origin which was entirely hidden by the diffraction patterns in the original data; thus they implicitly validate our velocity information. Figure 4D reproduces that portion of Figure 4A enclosed by a rectangle. The depth scale gives an idea of the thickness of the units involved. The lower part of Figure 5A also helps to indicate the type of structures hidden behind the scrambled seismic events. It also seems to indicate, for the first time, that a zone of relative lithospheric shortening may be present on the southwestern side of the Philippines in the northeastern Sulu Sea.

INTERPRETATION

Fundamental to an understanding of the mechanism which produced the tectonic style illustrated in Figures 2–5 is the concept that *rock layers transmit stress in proportion to their rigidity and thickness.* Also fundamental is the observation that, when a sedimentary sequence is being compressed laterally, the tectonic or diastrophic disturbance begins at the point where the compressive forces are applied. The disturbance then migrates progressively outward as a tectonic or diastrophic front. Away from this front, the strata show virtually no evidence of the compressive forces acting at the front (Fig. 6C). The common but erroneous idea of the response of an undisturbed rock layer (Fig. 6A) to the action of a compressive stress field is represented in Figure 6B. The wavelengths of individual convolutions appear to be primarily a function of the layer thickness (Fig. 6D), and the amplitude of the folds is expected to be primarily a function of the layer's competence. The amplitude of a fold continues to increase as long as the hydrostatic pressure of the rock column (applied at the base of the convoluted layer, and transmitted laterally) remains smaller than the resistance to diastrophism of the contiguous and yet-undisturbed portion of the layer. The development of the next convolution will start as soon as the growth of the previous fold is inhibited.

Note also that decollement surfaces (Fig. 6D) can develop both at the base of the sedimentary sequence and anywhere within the sequence where significant changes of the bulk physical properties occur. Thus very complex structural and stratigraphic situations are produced.

In a real situation, the very poor elasticity of layered rocks may impede the development of the convolutions of layers, and faulting may occur.

Figure 6E shows a slightly more realistic model of the deformation in trenches, and summarizes in one display three of the possible initial styles of sedimentary reorganization that I have recog-

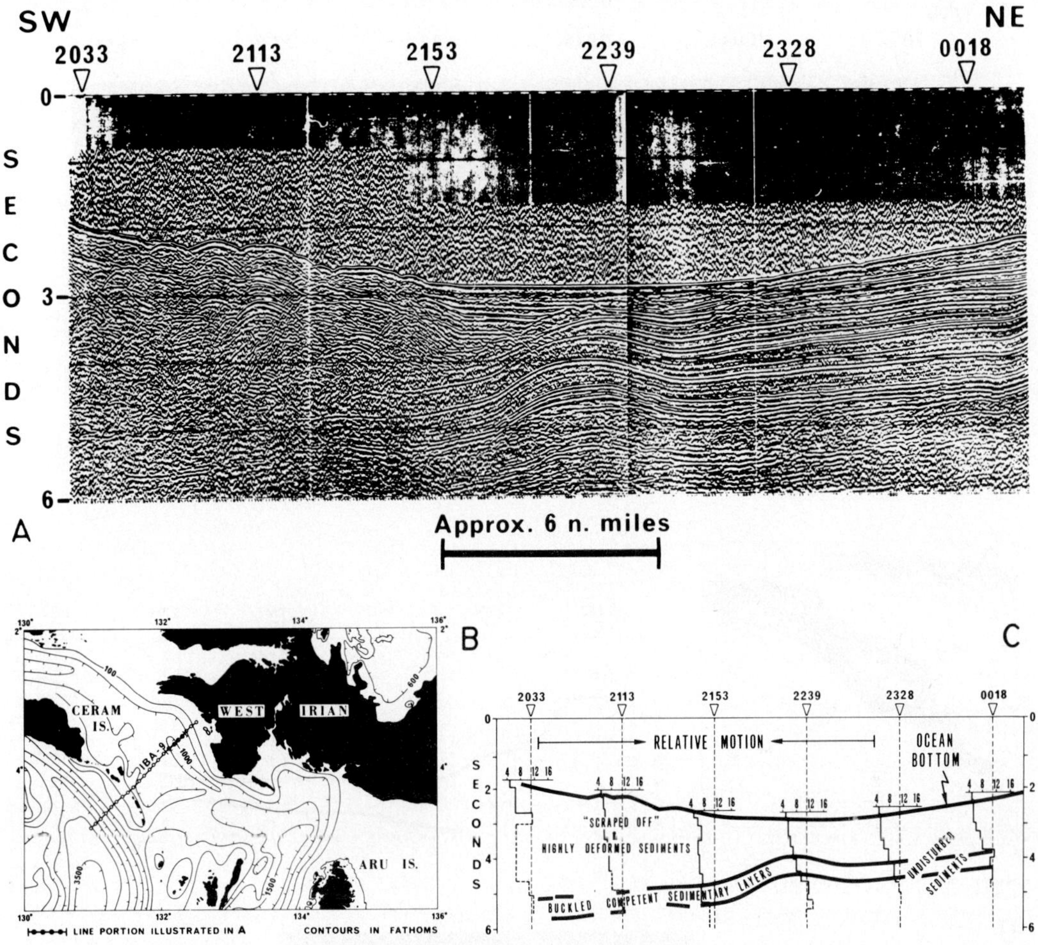

FIG. 2—**A.** A portion of line IBA-9, stacked data. **B.** Location of line IBA-9. **C.** Line IBA-9 interval-velocity distribution (ft x 10^3/sec).

nized from the *Gulfrex* data. In this illustration, sedimentary rocks tend to pile against a stationary obstacle, and the layers are carried "piggyback" toward it by the moving substratum, which is not flat or horizontal, but convex upward and simultaneously deepening toward the left in a stepped-down fashion (see an example of some indirect evidence in Fig. 7). Thus, some of the basal and more competent layers may not undergo folding and possibly are caught by the faulted irregularities of the moving substratum which, in turn, "stuffs" them toward the base of the lithosphere. The superficial layers may become an imbricated sequence (Fig. 8), but the dipping of the layers (or scales) is opposite to that of the basal layers, and the diastrophism happens more readily because of the absence of confining pressure by overlying strata. Figures 3D and 9 are migrated portions of seismic sections which have successfully resolved the structure of some "intermediate layers." They show that the complete convolution of layers is possible for that part of the sedimentary sequence between the basal and superficial complexes, where the confining pressure is approximately uniform spatially.

Other Evidence Supporting the Compressional Regime of Trenches

Figures 10–13 illustrate further, through specific details, how the sedimentary sequence may record the effects of tectonic activity along plate boundaries.

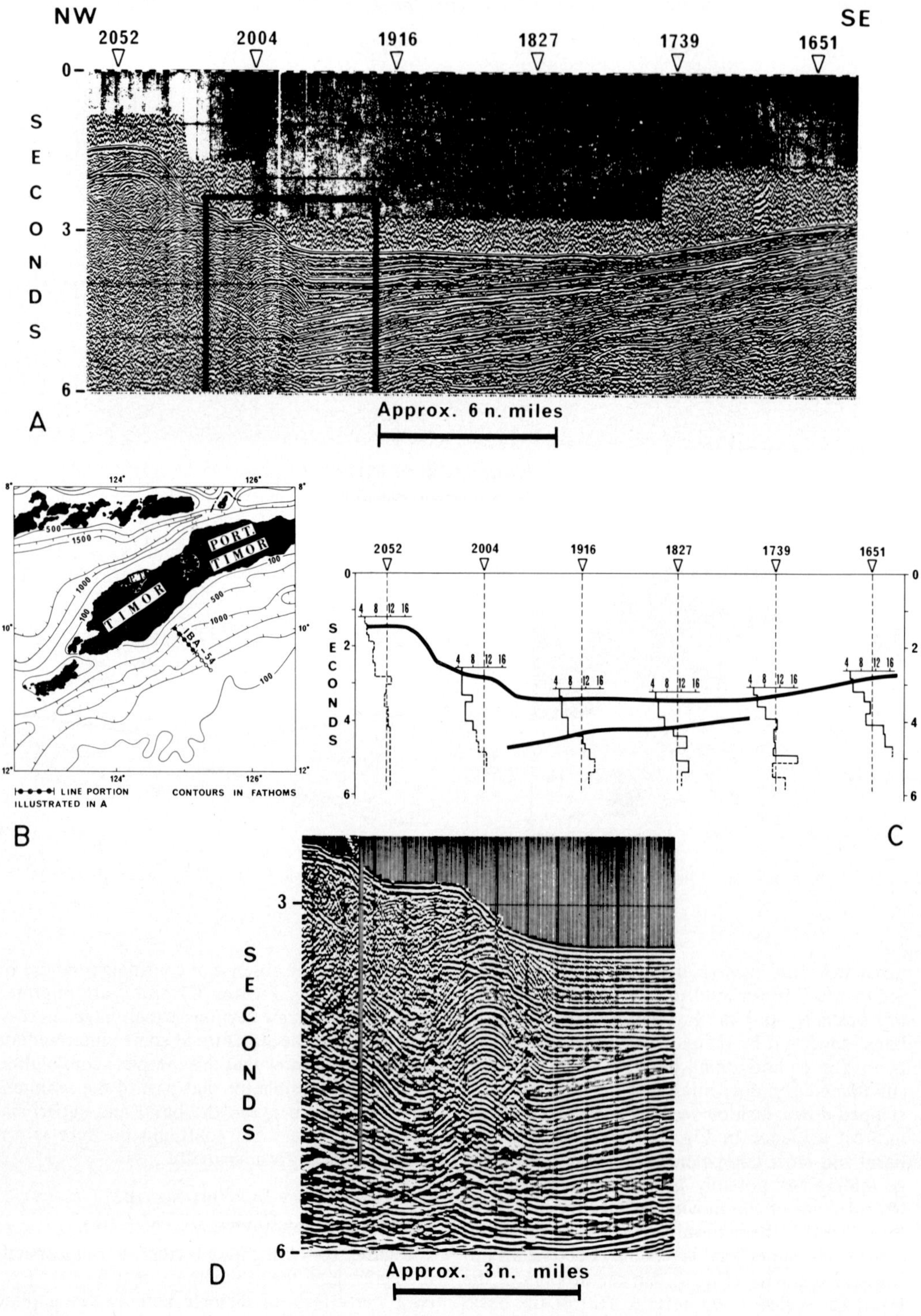

FIG. 3–**A.** A portion of line IBA-54, stacked data. **B.** Location of line IBA-54. **C.** Line IBA-54 interval-velocity distribution (ft x 10^3/sec). **D.** Migrated stacked data for section of line IBA-54 indicated by box in **A.**

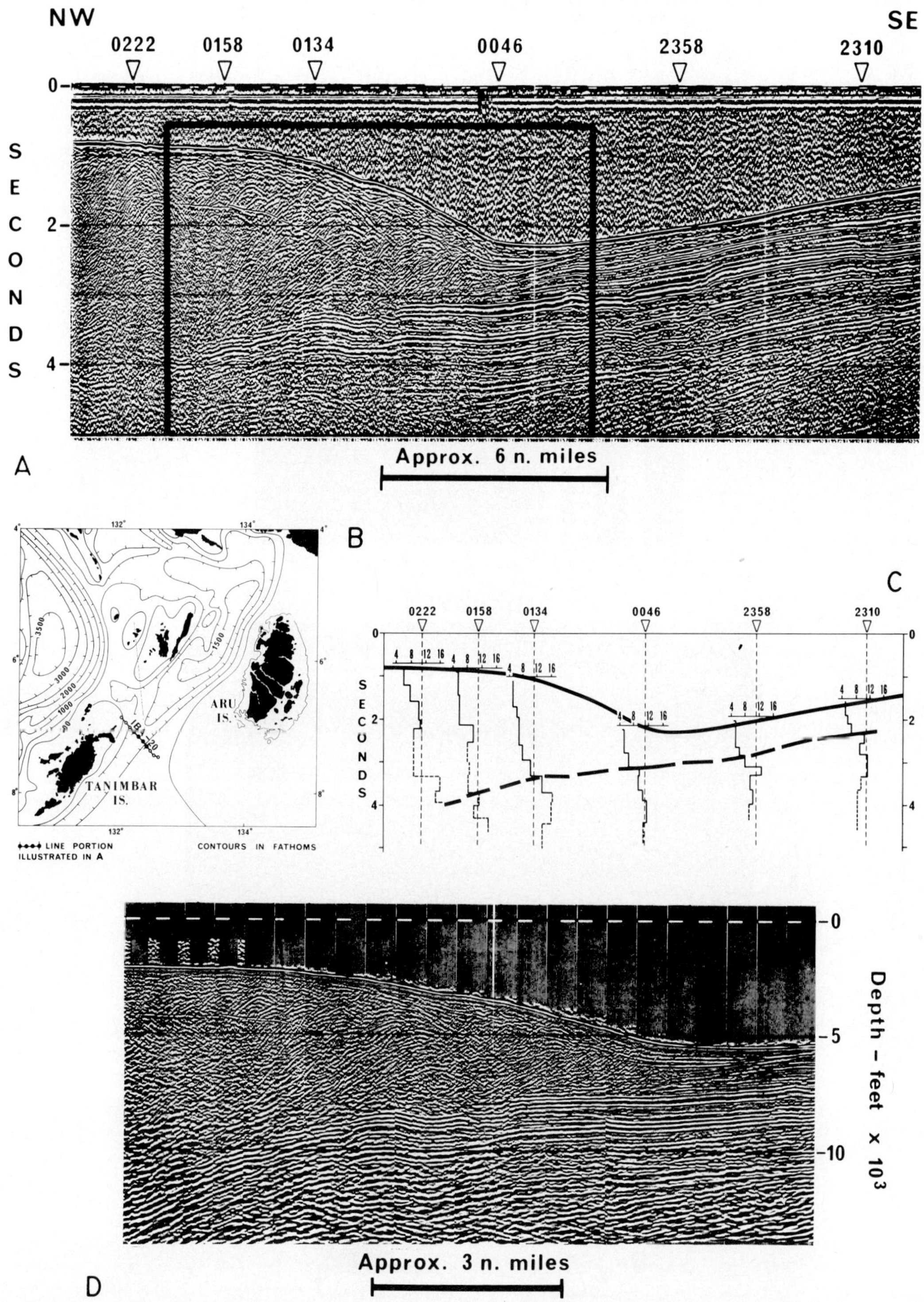

FIG. 4—**A.** A portion of line IBA-20, stacked data. **B.** Location of line IBA-20. **C.** Line IBA-20 interval-velocity distribution (ft x 10^3/sec). **D.** Depth-converted section of line IBA-20; section indicated by box in **A.**

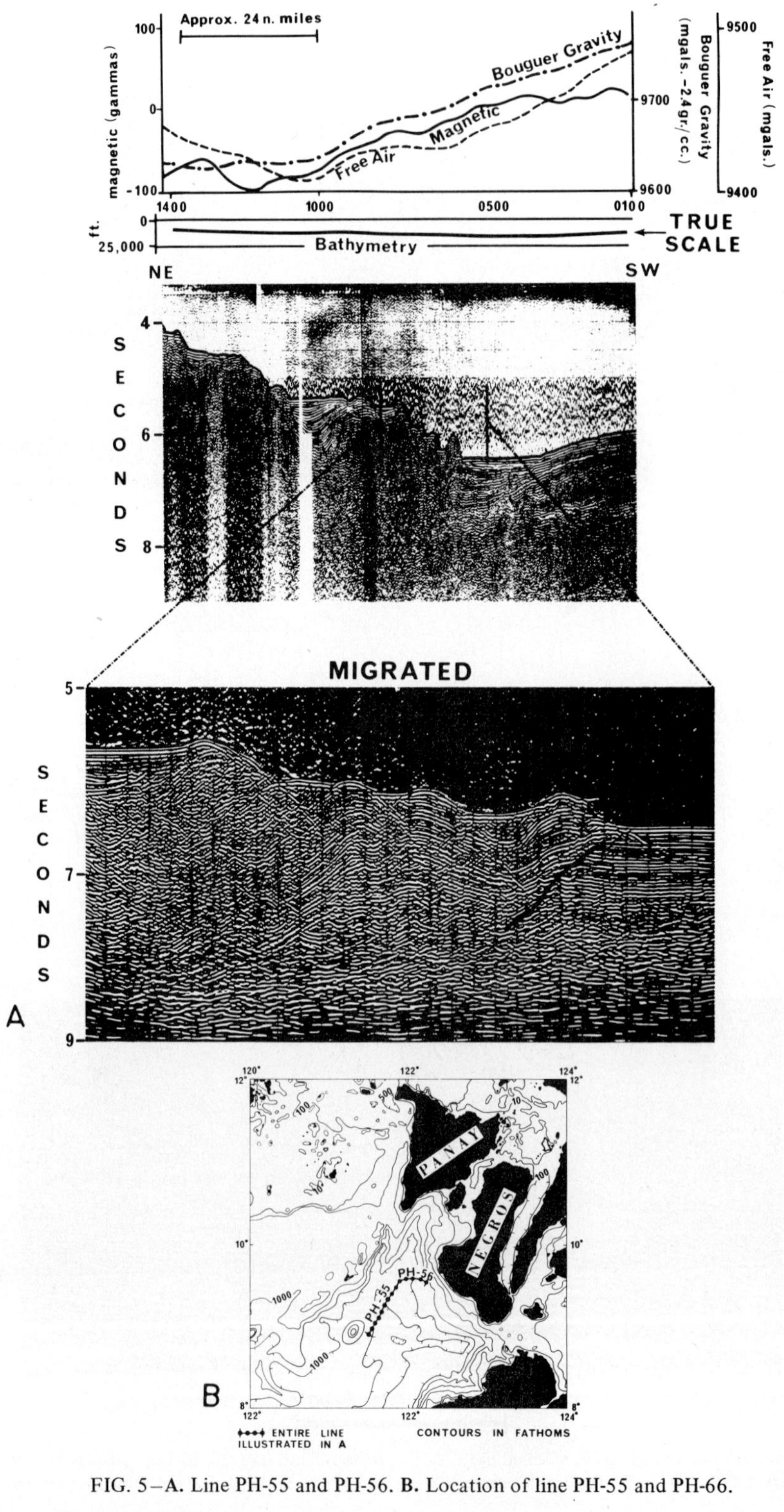

FIG. 5–**A.** Line PH-55 and PH-56. **B.** Location of line PH-55 and PH-66.

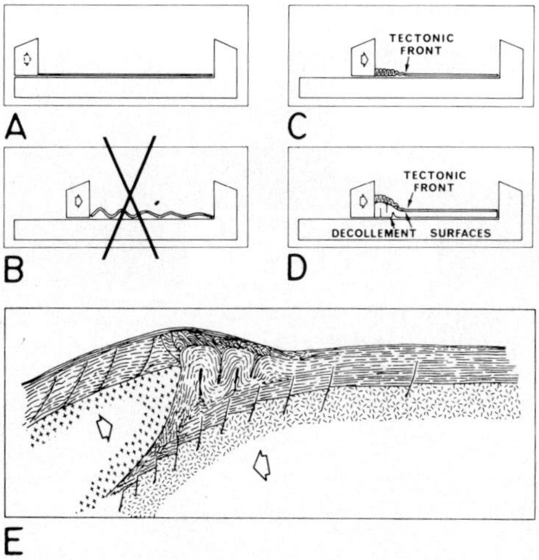

FIG. 6–**A-D.** Migration of a tectonic front. **E.** Summary of three possible styles of sedimentary reorganization.

Figure 10 provides unquestionable evidence of a recent compression which affects the total sedimentary sequence, here obviously divisible into two major units. The deeper one is the sequence of "normal" sediments deposited on the continental slope of northwestern Australia; the shallower unit is a wedge of turbidite deposits.

Figure 11 is an alternate version of line IBA-54 (Fig. 3; IBA indicates Indo-Banda arc). It shows the "dramatized" version of the entire line and a migrated version of the trench fill. As in Figure 10, two sedimentary units can be seen here, but additional structural detail can also be tentatively resolved. Three features are significant: (1) the expected migration toward the right of the "diastrophic front" (i.e., of the first obvious disturbance of the layered sequence shown by heavy lines on the lower profile), moving from the deeper layers toward the ocean floor, is somewhat discontinuous at the base of the turbidite fill; (2) the contact between the wedge of turbidites and the underlying slope sediments is not as smooth as expected on the basis of the appearance of the exposed slope surface farther up to the right; and (3) the wedge of turbidites exposed on the ocean

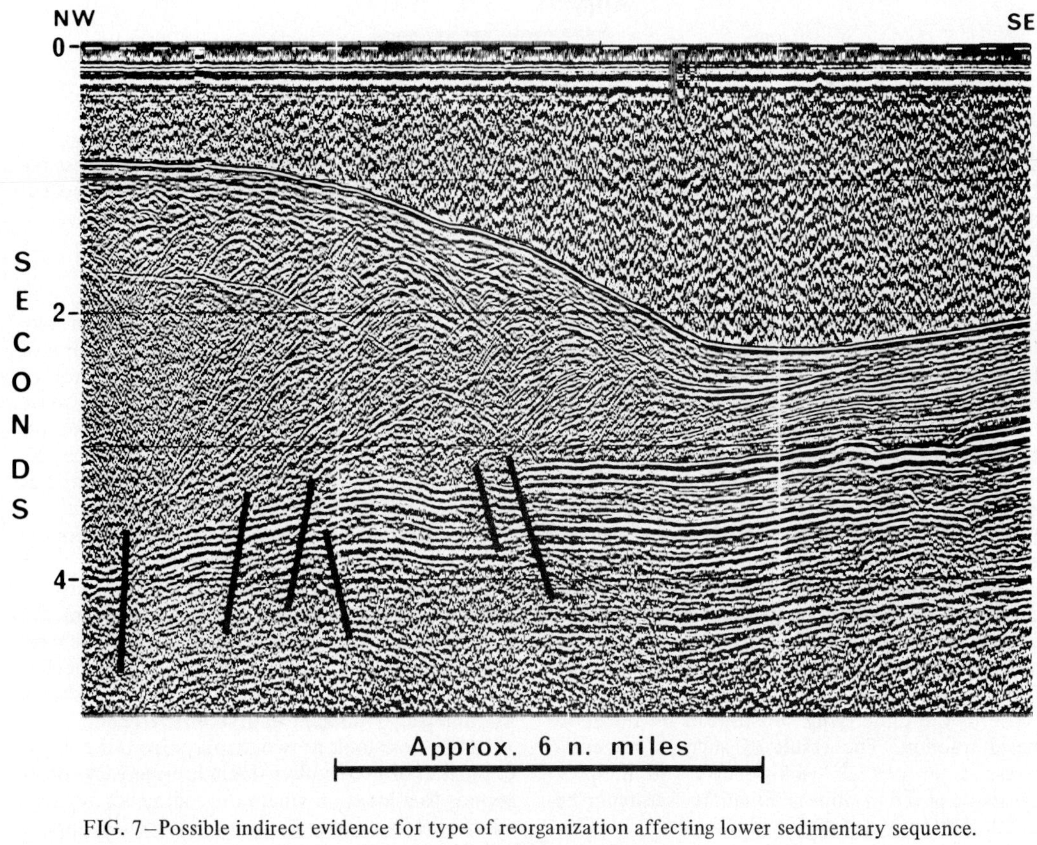

FIG. 7–Possible indirect evidence for type of reorganization affecting lower sedimentary sequence.

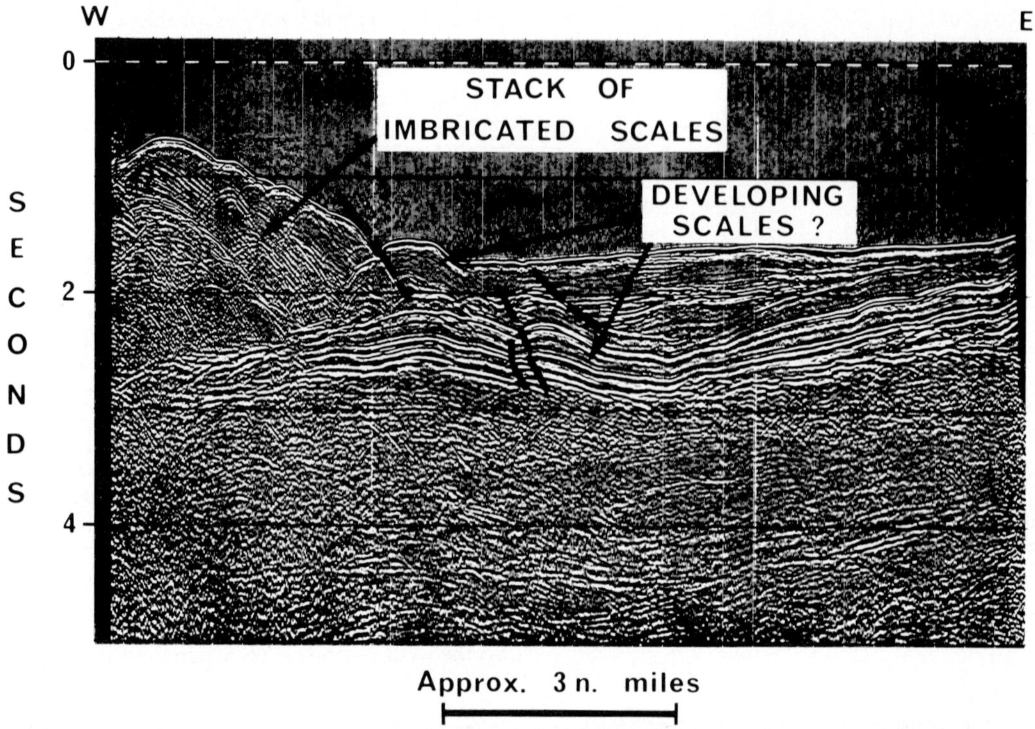

FIG. 8–Line SI-30, stacked data. Possible evidence for type of reorganization affecting upper sedimentary sequence.

floor exhibits morphologic details which cannot be entirely explained by differential compaction and/or subsidence. These three observations suggest that the wedge of turbidites was affected by crumpling along with the rest of the sedimentary sequence for some time. However, it recently has decoupled partially from the lower unit and presently tends to lessen the effect of the antithetic forces by permitting some differential motion along its contact with the lower unit.

Figure 12, also from the Indo-Banda arc, illustrates one type of behavior of a whole "packet" of shallow sedimentary units under compressive stress. The block confined between the trench axis and S.P. 306-0230 seems to have developed enough crumpling resistance to become partly decoupled through the establishment of a decollement surface possibly coincident with one of its deeper layering discontinuities. In contrast, the sedimentary sequence to the right of S.P. 306-0230 is still well coupled and receives the full impulse of the underlying substratum by means of basal traction. The result of such a tangential stress is the partial "jacking up" and incipient thrusting of the upslope sedimentary sequence beneath the lower decoupled block.

Figure 13 is a further portrayal of a possible remnant of a wedge of turbidites which supposedly was affected for some time by severe diastrophism as it was caught between two converging rock masses. The curved envelope described by the apparent left-hand terminations of the individual turbidite layers could mean one of two things: either it marks the staggered beginning of the individual turbidite layers and the location of the diastrophic front at that time ("steady-state model" proposed by Helwig and Hall, 1974), or, more simply, it shows the manner in which the diastrophic disturbance is transmitted upward at any one time through the turbidite sequence.

The observations reported effectively support the thesis of a localized shortening of the outer lithospheric layer. The key to the availability of evidence supporting this thesis is largely in the presence of a thick sedimentary cover which records the superficial effects of some deep mechanism and prevents the development of a sharp, asymmetric, V-shaped bathymetric trench.

Let us now look at two displays from the deeper portion of the Ryukyu Trench. Figure 14 corresponds to a location where the sediments deposited on the oceanward side of the trench appear

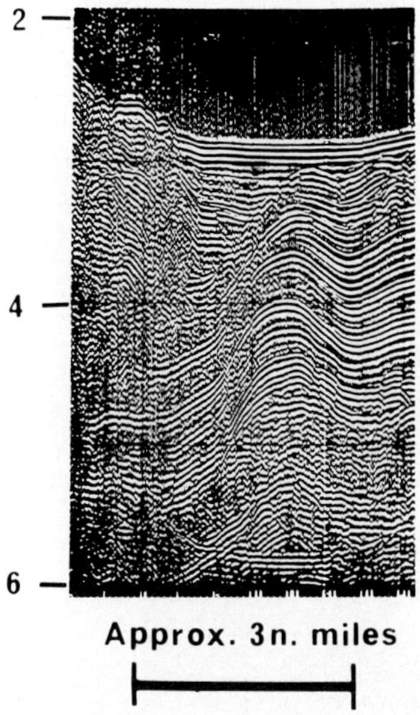

FIG. 9—Possible evidence for type of reorganization affecting intermediate sedimentary sequence.

only as a very thin, discontinuous veneer, barely blanketing an irregular crystalline substratum. The strong return of energy at about 10 sec in the middle of the display permits some fairly accurate seismic-velocity determinations (not displayed here) which again indicate that the low-velocity sediments originally deposited on the oceanic basement have been "bulldozed" off the substratum and chaotically accreted to the foot of the landward scarp of the trench.

Figure 15, also from the Ryukyu Trench, again portrays a sedimentary accumulation which is apparently insignificant. The seismic data are presented as a migrated single-fold version of data recorded 12-fold. A band of reflected energy slopes to the left, beginning just at the bathymetric minimum, and is thought to represent the shallow trace of the main thrust zone within the crystalline basement. This particular detail probably is not more apparent in general for one or more of the following reasons: (1) most seismic energy is reflected at the interface between water and crystalline rocks with very little effective penetration; (2) the acoustic contrast at the thrust zone is probably negligible; (3) the slope angle of the thrust zone abruptly becomes too steep for

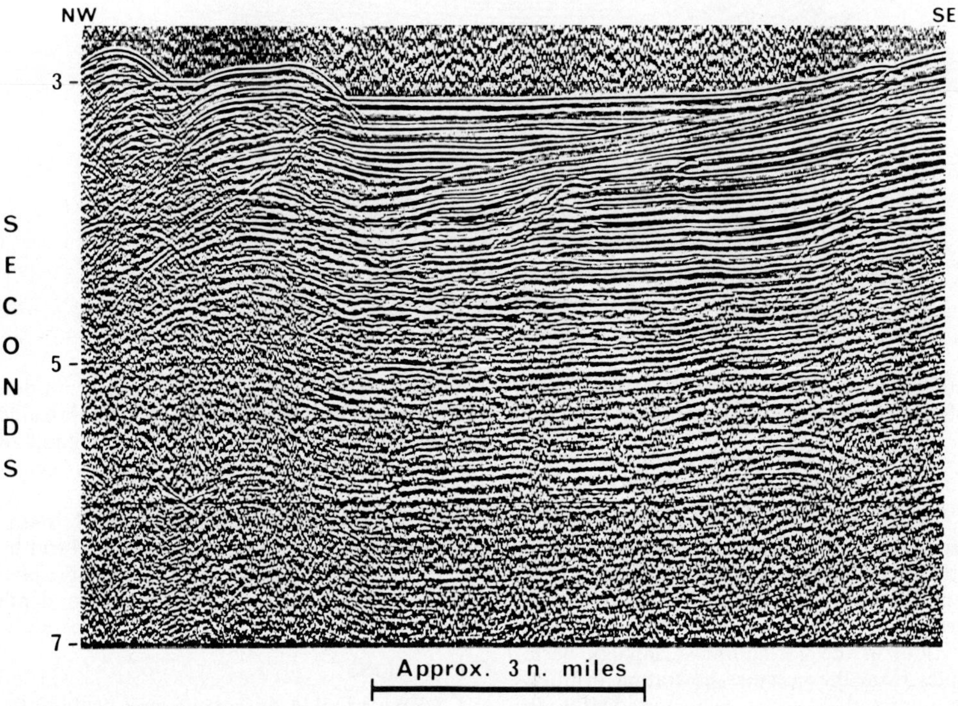

FIG. 10—A portion of line AU-38F, stacked data. An example of a recent turbidite fill involved in crumpling of total sedimentary sequence.

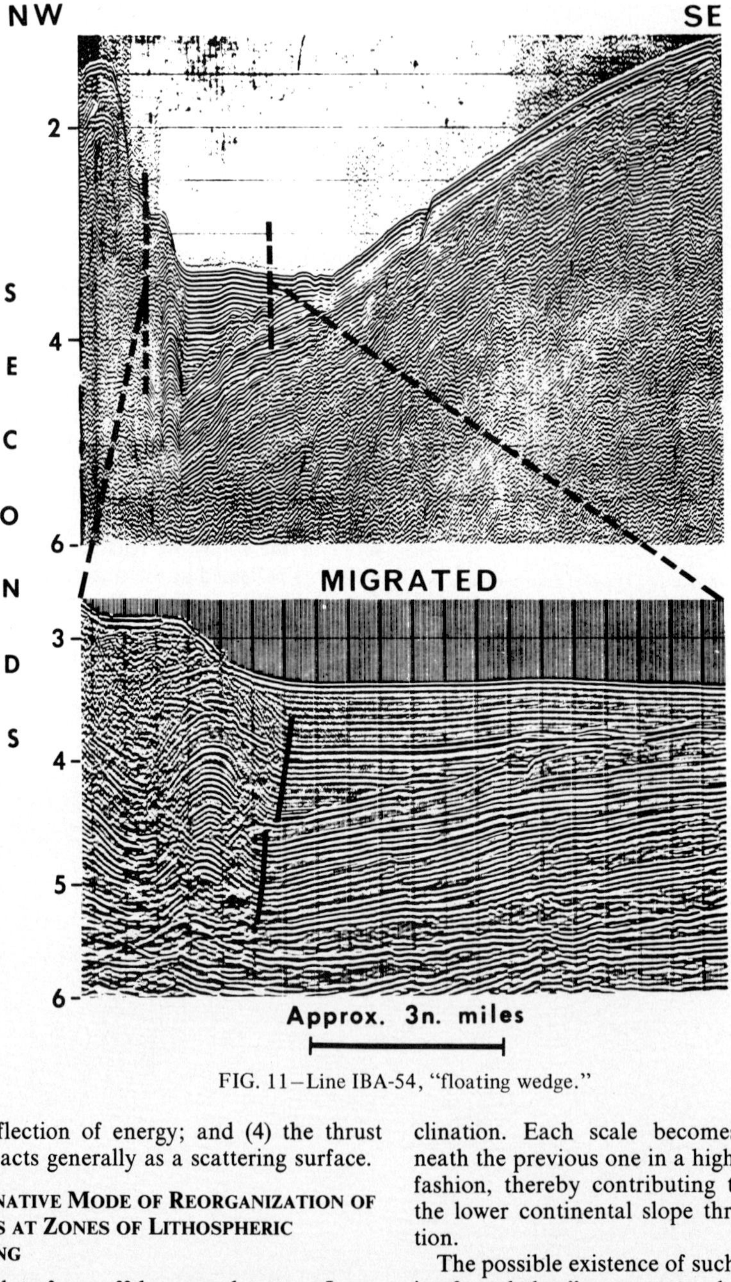

FIG. 11—Line IBA-54, "floating wedge."

upward reflection of energy; and (4) the thrust zone itself acts generally as a scattering surface.

An Alternative Mode of Reorganization of Sediments at Zones of Lithospheric Shortening

Gulfrex data from offshore southeastern Japan (Fig. 16) also suggest that an alternative style of reorganization of the scraped materials may exist. The sedimentary sequence originally deposited on the floor of the northernmost Shikoku basin decouples from the oceanic substratum in huge, largely undisturbed scales which gradually tilt landward until they attain a certain maximum inclination. Each scale becomes stacked underneath the previous one in a high-angle imbricated fashion, thereby contributing to the buildup of the lower continental slope through mass accretion.

The possible existence of such a style of "stacking from below" was proposed in sketch form by Oxburgh and Turcotte in 1971. However, the phenomenon probably extends to far shallower depth than they indicated.

Conclusions

1. I have adopted, with some modification, the concept of the existence of elongated "subcrustal

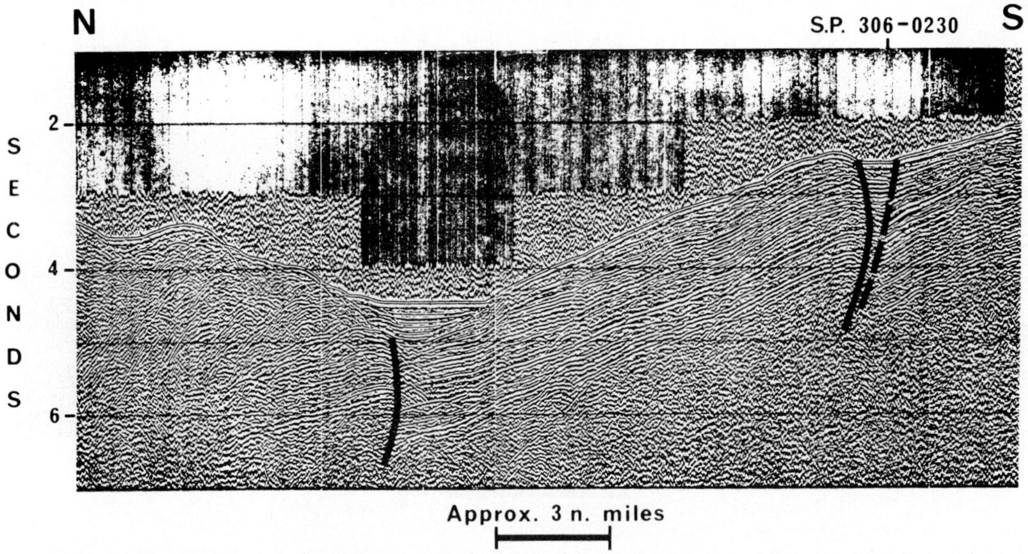

FIG. 12–A portion of line IBA-25, stacked data.

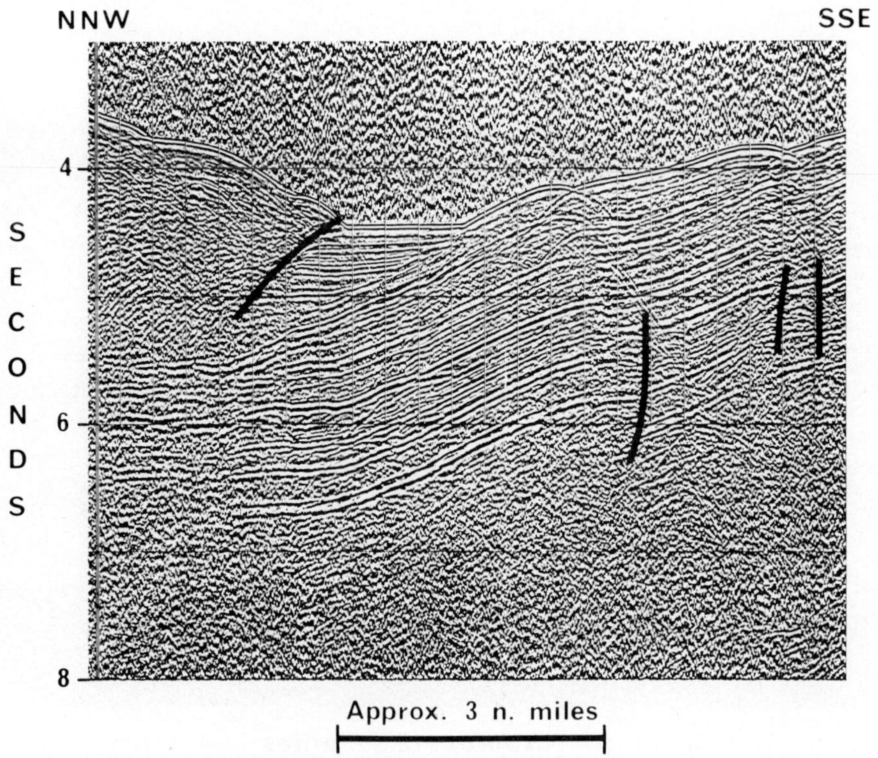

FIG. 13–A portion of line TIS-2, stacked data.

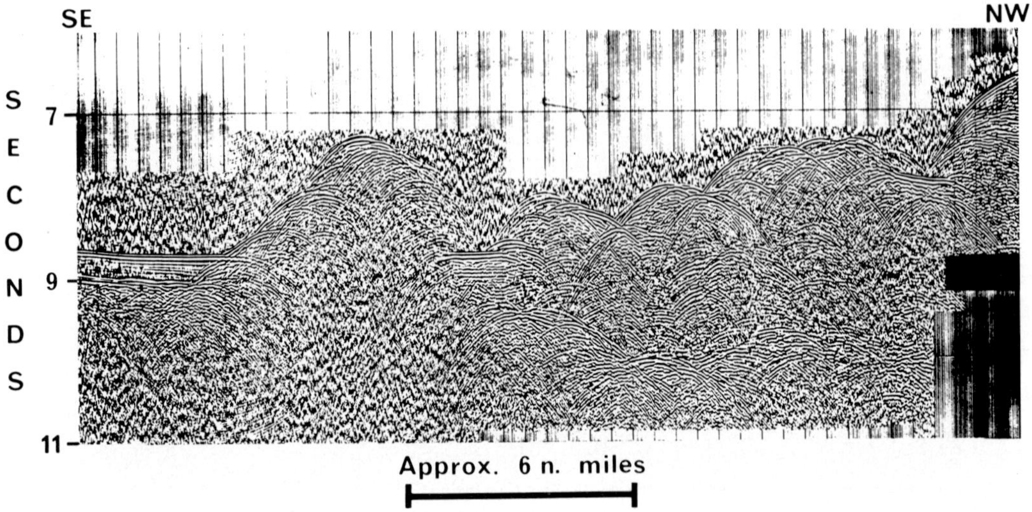

FIG. 14–A portion of line RKU-5S, stacked data. A smaller volume of sediments than that in previous figures is involved in crumpling, but it changes into an equally chaotic mass and builds up base of trench.

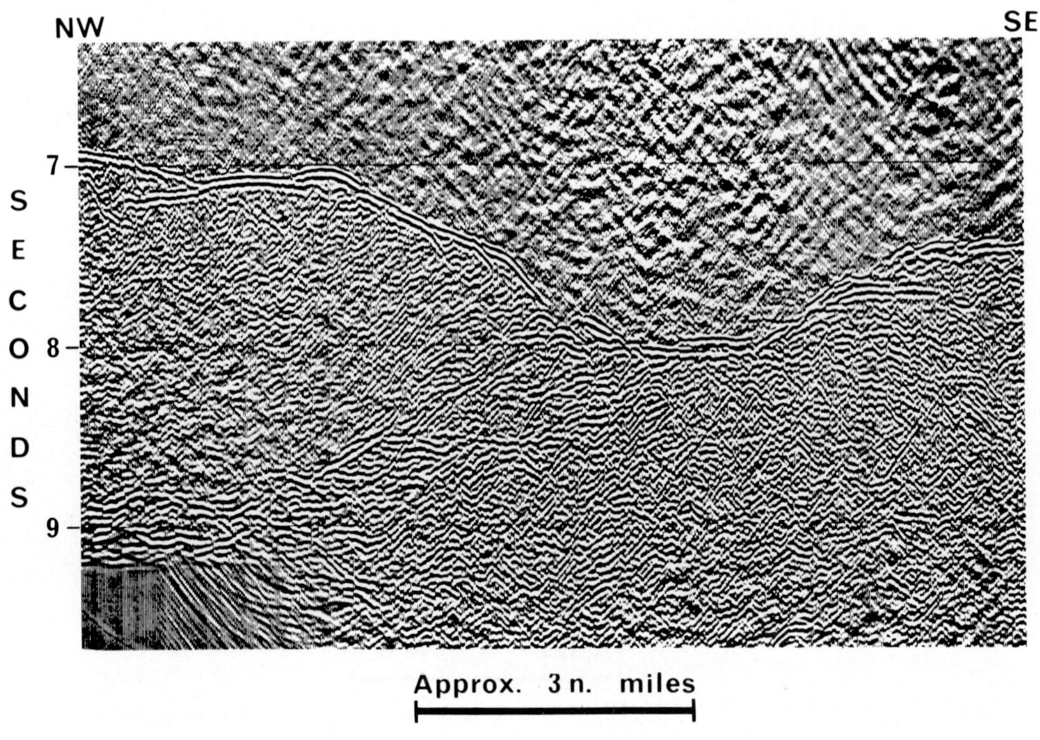

FIG. 15–Migrated single coverage. Possible trace of a thrust surface within crystalline rock at bottom of Ryukyu Trench.

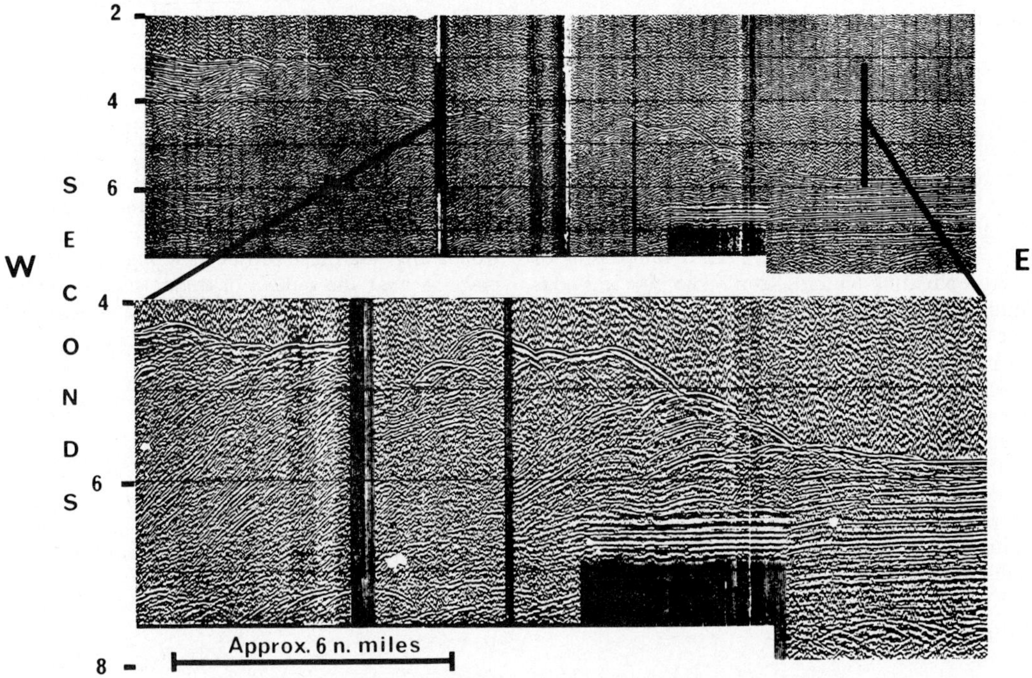

FIG. 16–Line JPN-23; an alternative mode of sedimentary reorganization.

sink zones" along large sections of the boundaries between the major plates. The lithosphere participates in a poorly understood global mechanism which involves its horizontal translation at a geographically nonuniform rate toward the zones of subcrustal sinking. Materials which are not typically oceanic are passively carried atop moving lithosphere.

2. The outer lithospheric complex undergoes a reorganization near these sink zones by a partial "skimming" and piling of the sedimentary layers, and by the subduction of portions of the crystalline substratum and the remaining sedimentary section. The tectonic style by which the reorganization develops may vary from place to place; it seems to be controlled by the total thickness of the sedimentary sequence, its predominant lithologic types, and the rate of deepening of the oceanic substratum toward the subcrustal sink zones.

In all cases, a tectonic front is thus established on the oceanward side of the "sink zone." It sharply separates the undisturbed sedimentary rocks from the piled layers, and it migrates outward with the progressive accretion of scraped material. A major and complex thrust zone connects the tectonic front with the subcrustal sink zone. The thrust zone coincides with one or more of the basal decollement surfaces, and it accommodates the differential motion between the crystalline substratum and the essentially stationary scraped layers.

3. A bathymetric trench probably is always closely associated with an active subcrustal sink zone. However, when the volume of the sediments is great, the trench may become obliterated and the sink zone well hidden below and behind a chaotic mass of scraped material.

4. Some relatively buoyant continental masses may have attained geographic stability, after extensive passive drift, by being trapped on top of active subcrustal sink zones. Such continental blocks are now essentially decoupled from the oceanic crust, which continues to creep toward the sink zones and underride the edges of the arrested continents. Thus, many of the world's continental slopes could be composed essentially of reorganized pelagic materials which eventually undergo metamorphosis and become the foundation for the later construction of continental shelves by the accumulation of largely land-derived sediments.

Overall Significance for the Hydrocarbon Explorationist

From the preceding speculative interpretation of the data, it appears that at many locations around the world there exists a major phenomenon by which huge amounts of sedimentary rocks undergo a process of "tectonic milling," whose intermediate and final products can possibly be construed, respectively, as mélanges and gneisses.

It is here proposed that the explorationist can consider these masses of uprooted strata as one of the major sources of hydrocarbons. It is expected that the diagenetic processes, which normally and progressively affect the sediments in an irreversible manner, become accelerated by the passing of the propagating tectonic front. The physical continuity of the individual layers is disrupted and the diagenetic fluids are mobilized. Sharp increases in the internal pressure and temperature may eventually cause the synthesis of hydrocarbons from the organic matter dispersed in the sediments. These hydrocarbons would eventually join with the other diagenetic fluids and follow the complex pressure gradients through the essentially plastic and partially homogenized rock mass until they reached the ocean floor, where they subsequently would be dispersed.

The explorationist should therefore concentrate his efforts in locating those mélange masses (both ancient and modern) which were "reorganized" during rapid deposition of large quantities of waterborne particulate matter. Under these circumstances, at least part of the fluids escaping from the mélanges may have been trapped in structures located upslope within the blanketing complex.

The adoption of such a selective criterion would upgrade those prospects along continental margins near the outlets of major river systems, and it would downgrade those associated with the oceanward side of island arcs.

References Cited

Helwig, J., and G. A. Hall, 1974, Steady-state trenches?: Geology, v. 2, no. 7, p. 309–316.

Oxburgh, E. R., and D. L. Turcotte, 1971, Origin of paired metamorphic belts and crustal dilation in island arc regions: Jour. Geophys. Research, v. 76, p. 1315–1327.

Overview of Exploration Geophysics: Recent Breakthroughs and Challenging New Problems[1]

B. S. FLOWERS[2]

Abstract Changes in geophysical technology have been rapid and dramatic in recent years so that work done only a year or two ago is referred to as the "old" geophysics. As might be expected, each new method gives rise to additional problems, often unforeseen, which limit its use and thus require a new round of improvements. Three topics are chosen to demonstrate this: (1) problems involving velocity-structure, (2) elimination of noises after stacking, and (3) direct detection of hydrocarbons. The examples are not all from the Circum-Pacific area, but the problems are general and will surely be encountered in the exploration of the Pacific province.

VELOCITY-STRUCTURE

An example of the velocity-structure problem from the Michigan basin (Fig. 1) typifies the special case of near-surface "sonic lenses." Highly irregular deposits of glacial till, with abrupt thickness variations of hundreds of feet and velocities ranging from 2,000 to 6,000 ft/sec (610–1,829 m/sec), cause differential time delays and ray-path distortions which result in misalignments of the common-depth-point (CDP) stack. Until this problem was solved a few years ago, the Michigan basin was essentially a "no data" area. The disruptions of seismic events caused by the small Silurian pinnacle reefs, a few hundred feet high and roughly 160 acres (64 ha.) in areal extent, commonly were indistinguishable from disruptions caused by stack misalignments under glacial trenches. New correction methods which eliminate misalignments in the stack have led to the current reef-detection success ratio of about 8:10. This is a special case of the velocity-structure problem, because (1) the sonic lenses are known to be at the surface; (2) the wavelengths of the variations are reasonably short relative to shooting distances; (3) there are strong reflections from a shallow Devonian reflector which is structurally simple and can be used as a correction plane; and (4) the interpretation objective is stratigraphic resolution rather than structural definition, so precise depth determinations at the Silurian level are not required. These methods do not work well in such general situations as (1) areas where the sonic lenses are not at the surface (e.g., buried karst topography and submarine gorges), (2) instances where the wavelengths of the variations are long compared to shooting distances (e.g., permafrost changes near large bodies of water), and (3) provinces where the interpretation objective is accurate structural definition.

Another velocity-structure problem, also a special case, has been encountered in the Gulf Coast province. The problem is one of mapping highly contorted beds associated with a mobile salt layer. It is impossible to map the complex structures without migration and, because the steep dips are misaligned in migration-after-stack procedures, it is necessary to migrate the data before they are stacked. This procedure works quite well if the seismic data are acquired along true dip directions so that there are no cross-line components of dip. Here the problem is straightforward because the velocity overburden is simple: there are no sonic lenses to distort the ray paths and confuse the velocity and the dip determinations. Migration-before-stack is currently the standard migration technique and is valid where these conditions obtain. The method can be extended to situations involving cross-line components if sufficient cross-line information is acquired to determine uniquely all dip components.

The third velocity-structure example (Fig. 2), from west Africa, represents the general case. The objective is to map beneath a highly deformed Cretaceous salt layer. In addition to the ray-path distortions caused by the salt layer, distortion is also caused by two other layers above it. A Cretaceous carbonate layer directly above the salt has variable thickness and velocity, and the overlying low-speed Tertiary clastic section has large lateral velocity changes caused by salt solution and collapse. With continuous RMS velocity measurements, the pertinent reflections are stacked well enough to be visible, but accurate velocity measurements that will give proper depth conversions are not possible. The ordinary time-migration section produced with migration-before-stack techniques does not provide a satisfactory solu-

[1]Manuscript received, October 21, 1974.

[2]General Manager–Exploration, Shell Oil Company. New Orleans, Louisiana 70160.

I would like to express my appreciation to the management of Shell Oil Company for permission to publish this paper and for their support of geophysical exploration and research. I also express my appreciation to my Shell colleagues whose work I have included.

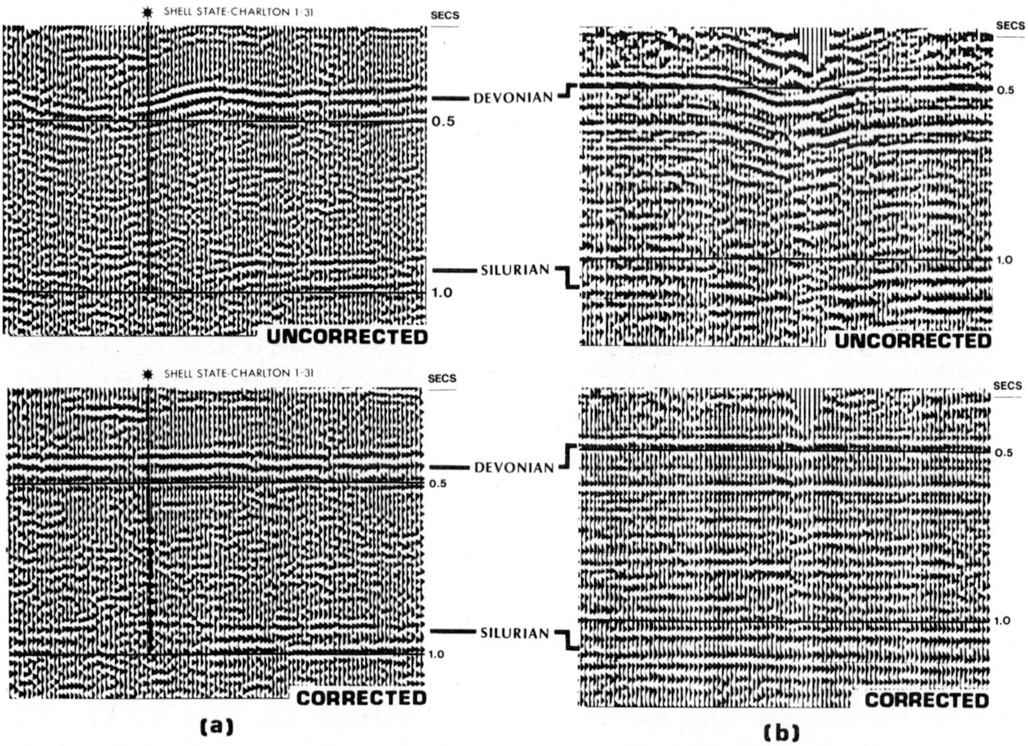

FIG. 1—Michigan basin seismic sections showing valid Silurian reef anomaly **(a)** and fictitious Silurian anomaly **(b).** Top panels show effects of misalignments due to variations in near-surface glacial till, and bottom panels show data after proper alignment.

tion, for the normal assumption of simple velocity overburden is not valid. Rather, the velocity overburden contains three layers of sonic lenses: the salt, carbonate rocks, and clastic rocks. The raypath distortion caused by these layers can be corrected for a good stack, migration, and depth conversion provided an accurate model for each layer can be constructed. This model construction requires the correct measurement of velocity gradients, in the presence of structural dip, for several layers—hence the nomenclature "velocity-structure." All the pertinent dips are visible on sections processed in this way and the depth conversion looks reasonable. Obviously, improved reflection quality and reasonable-looking depth sections do not guarantee a proper migration and depth conversion; there must be better methods to verify the accuracy of the model.

In actual practice, the third dimension must be taken into account. The assumption has been that the data were acquired in the dip direction, with no cross-line components. This assumption must be valid for all three layers of the model. Obviously, it is not possible to satisfy this requirement with one line if the various layers dip in different directions. Three-dimensional analysis and processing are difficult, and the acquisition of cross-line information must be sufficient not only to measure cross-line dips and to prevent aliasing, but also to measure velocity gradients in the presence of these dips. Cross-line dips generally are present, and simple wide-line recording methods are not likely to provide sufficient density of control to solve the problem uniquely.

Elimination of Noises after Stacking

Common-depth-point (CDP) stacking is a velocity filter that causes attenuation of multiple reflections and other noises. In an example chosen from the eastern Pacific offshore (Fig. 3), the multiples are so strong that the primary reflections are barely visible on single records and the stacking velocity for primary reflections is not easy to determine. With the correct velocity, a 24-fold stack reveals an anticline, but details are obscured by residual multiples not eliminated by the stacking process. Cancellation "beyond the stack" is needed. A method for cancelling multiples "beyond the stack" was applied to the data in this example. Details of the anticline thus be-

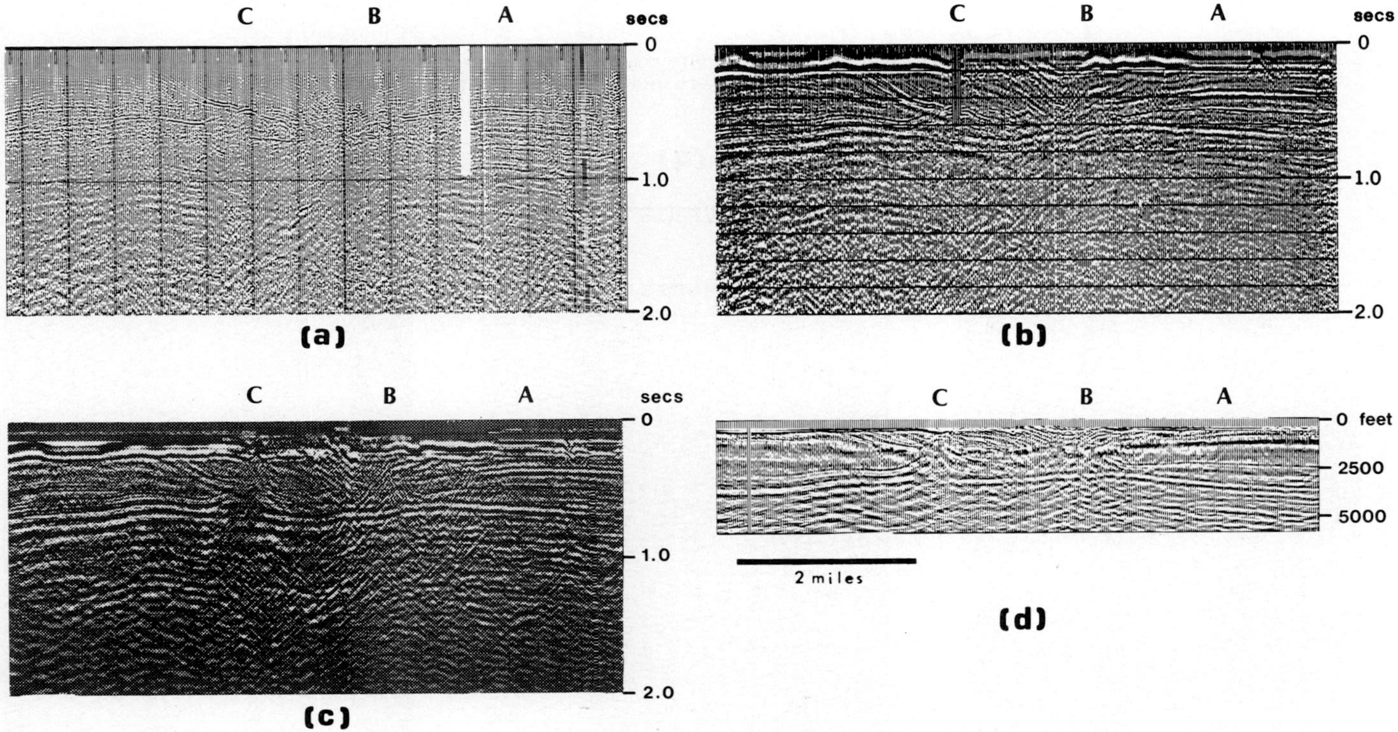

FIG. 2–Example from west Africa illustrating general velocity-structure problem. **(a)** Preliminary stack which shows high at *A* and indications of salt dome at *C*. **(b)** Refined stack which shows continuous base-of-salt event between 0.6 and 0.8 sec. **(c)** Migration-before-stack time section. **(d)** Migrated depth section showing (1) no deep high at *A*, (2) indications of salt dome with overlying collapse at *B*, (3) salt dome at *C*, and (4) continuous, smooth base-of-salt reflection between depths of 2,500 and 4,000 ft (762-1,219 m).

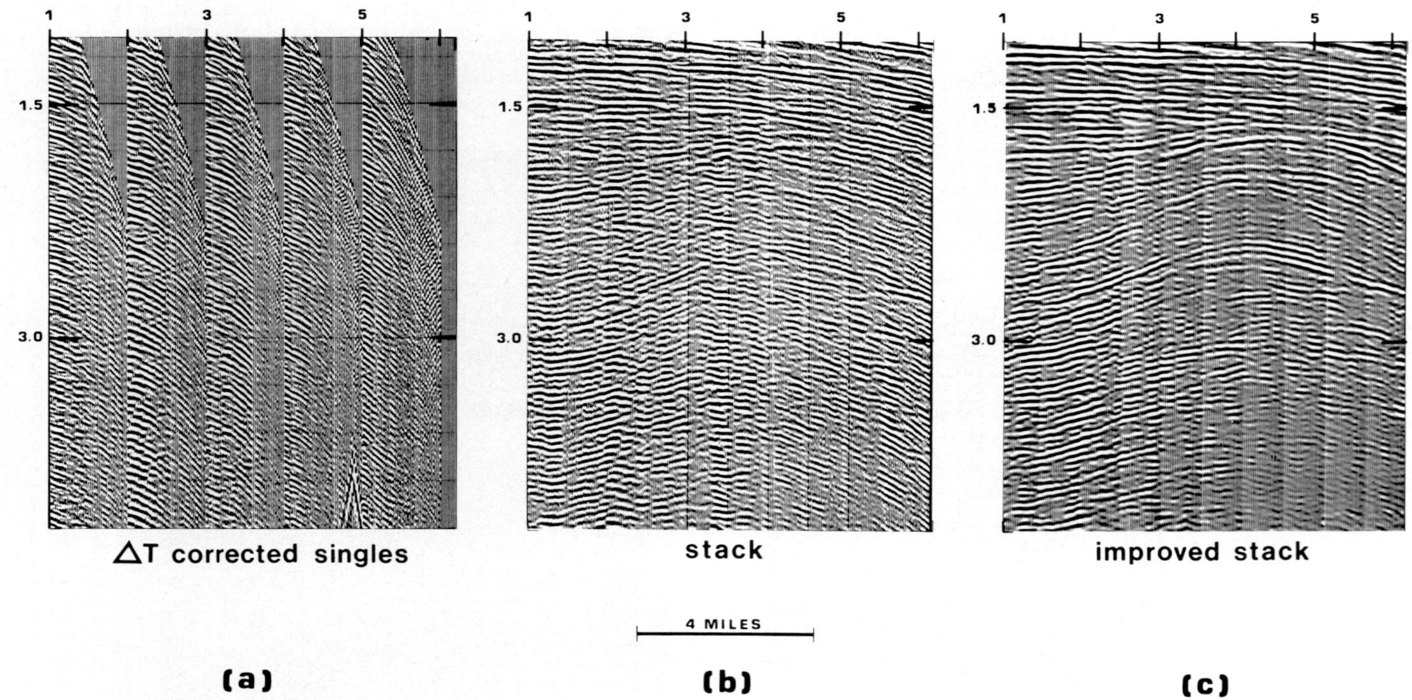

FIG. 3—Example from eastern Pacific offshore: **(a)** multiple-reflection problem area; **(b)** some multiple attenuation obtained with stacking; **(c)** additional rejection obtained with processing "beyond the stack."

come clear—for example, evidence of structural growth, unconformities, and a shift in the crest of the anticline at depth.

There are two basic ways to attack multiples "beyond the stack": (1) use of a velocity filter or (2) use of a predictive technique. In the first method, use of a velocity filter stronger than the stacking filter may eliminate the multiples, but only if there is sufficient velocity separation in the earth between multiples and primaries. Even with a sufficient velocity separation, however, dips of the multiple generators may in some cases cause the multiples to coincide with the primaries; hence any velocity filter would be insufficient. The second method involves predicting the positions of multiples and subtracting them. This requires either (1) a detailed model of the multiple generators which must be obtained from the reflection record by accounting for all of the reverberations, simple multiples, and peg-leg multiples generated by the reflecting interfaces; or (2) a predictive model derived from the periodicity of multiply reflected events through autocorrelation. In either method of modeling, the periodicity and amplitude of the multiples change with time and distance, respectively. Moreover, the water-bottom reverberations of multiples and of primaries have distincly different phase characteristics. Because the required models are complex and are dynamic in space and time, these techniques must be used with great discretion.

Direct Detection of Hydrocarbons

For the past few years, geophysicists have been routinely predicting hydrocarbons from "bright-spot" amplitude analyses. The amplitudes of seismic reflections are governed by differences in the acoustic impedances (velocity × density) of the rock layers at reflecting interfaces. Because the acoustic impedance of a loosely cemented rock filled with hydrocarbons is different from that for a similar water-filled rock, the presence of hydrocarbons may be detected by amplitude changes under certain conditions.

In an example from the Gulf Coast province, there are several bright-spot anomalies (Fig. 4a) near the 5,000-ft (1,524 m) depth on the right side of the profile at well location 1. Borehole logs in the well confirmed five hydrocarbon "pays" as indicated by the seismic data, and a substantial field was discovered. On the left side of Figure 4a is an equally good bright spot at the same depth; it was tested by wells 3 and 4. The only anomaly on the borehole logs in these wells is cycle skipping on the sonic logs at the level of the bright spot. There is a sandstone at this level which correlates with one of the "pay" sandstones in well 1, just across the syncline. It was suspected that the bright spot and cycle skipping are caused by low-saturation gas in this unit. A theoretical paper presented by Domenico (1974) at the 1973 annual meeting of the Society of Exploration Geophysicists demonstrated that a small percentage of free gas in a reservoir (too small to affect the resistivity measurements on borehole logs) should give a strong velocity change and, hence, a bright-spot amplitude anomaly. A second example of the same problem in the Gulf Coast is shown in Figure 4b. The deep bright spots below 2.0 seconds are confirmed in wells 1 and 3 as good gas "pays." The two shallow bright spots near 1.0 second are confirmed in both wells by resistivity logs as wet sandstones, but pronounced cycle skipping is apparent on the sonic logs. Fluid-entry tests were made in these sandstones with a recovery of 9.7 cu ft/bbl (0.3 m^3/bbl) of methane. Inasmuch as this quantity is close to the maximum amount of gas that can be dissolved in the formation water, it cannot be used as positive evidence of free gas in the reservoir. Conventional (rubber sleeve) cores from these sandstones were analyzed in the laboratory; the results are presented in Figure 4c. A small fraction of porosity occupied by a gas (air in this case) has a large effect on the velocity in the core. It is thought that the mild disagreement between the theoretical curve and the measured curve is caused by inhomogeneity of gas distribution in the sample, which is more noticeable at 75 kHz than it would be at seismic frequencies. All the evidence indicates that low-saturation gas is present in reservoirs at depths of interest and, although the quantity is too small to be seen on the resistivity log, unfortunately it can be seen all too easily on the seismic data.

In another instance, at a Gulf Coast salt-dome prospect, two tests were drilled into a fair bright-spot target. It was found that the amplitude anomalies were caused by two thin, broken gas sandstones, with good saturation, separated by a thin, hard streak of silty shale. The gas in these poor-quality sandstones was noncommercial. High-resolution seismic data and a good geologic prognosis of the reservoir quality are necessary to predict the commerciality of "pays" associated with such amplitude anomalies. In another prospect in the Gulf Coast onshore, a good bright spot was seen in the Oligocene section. Well data show that this amplitude anomaly, which has a good fit with structural contours, is related to a thin carbonate buildup (nonhydrocarbon bearing) that apparently developed on a low-relief to-

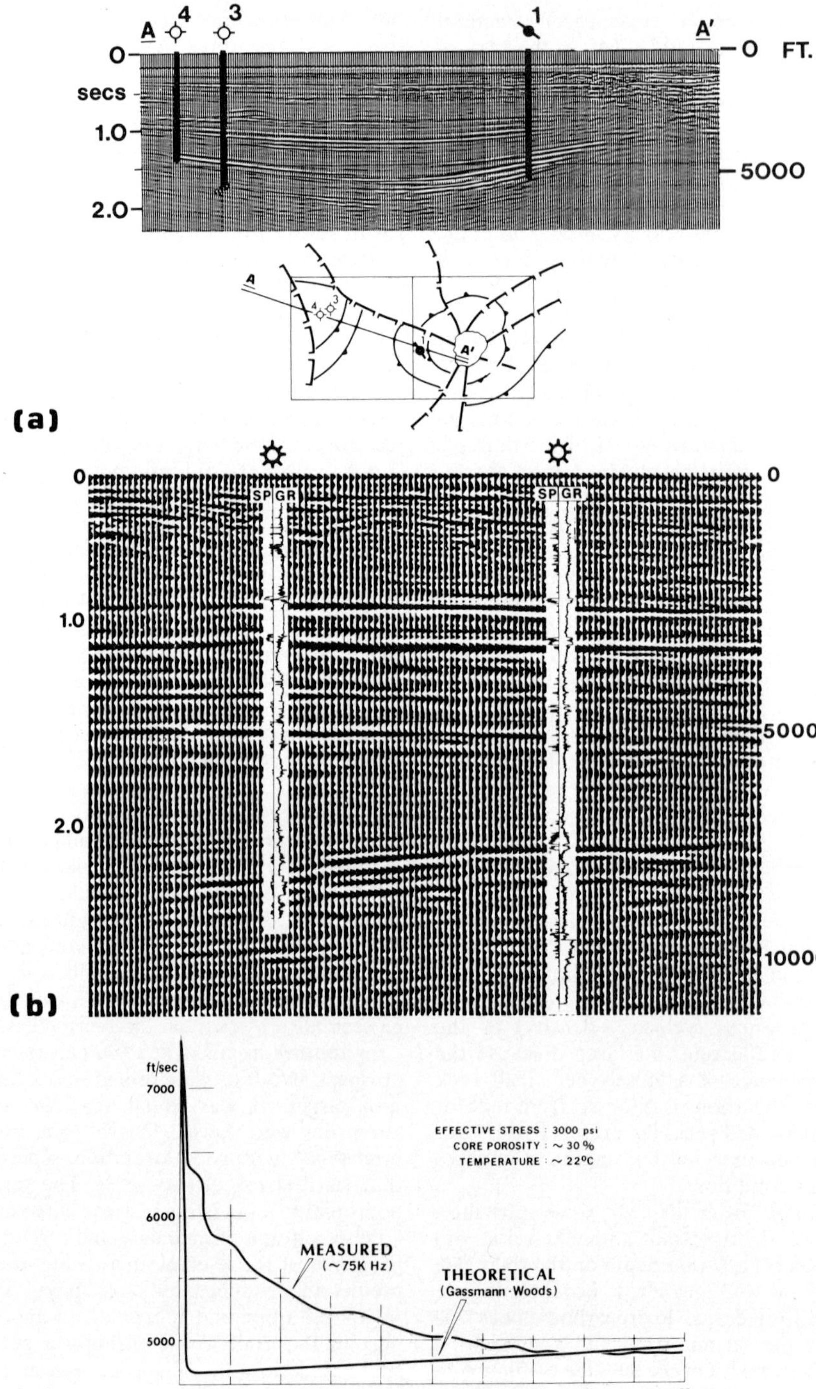

FIG. 4–**(a, b)** Examples from Gulf Coast province showing seismic-amplitude anomalies associated with both commercial and noncommercial hydrocarbon accumulations. **(c)** Plot of theoretical and measured velocities as function of gas saturation.

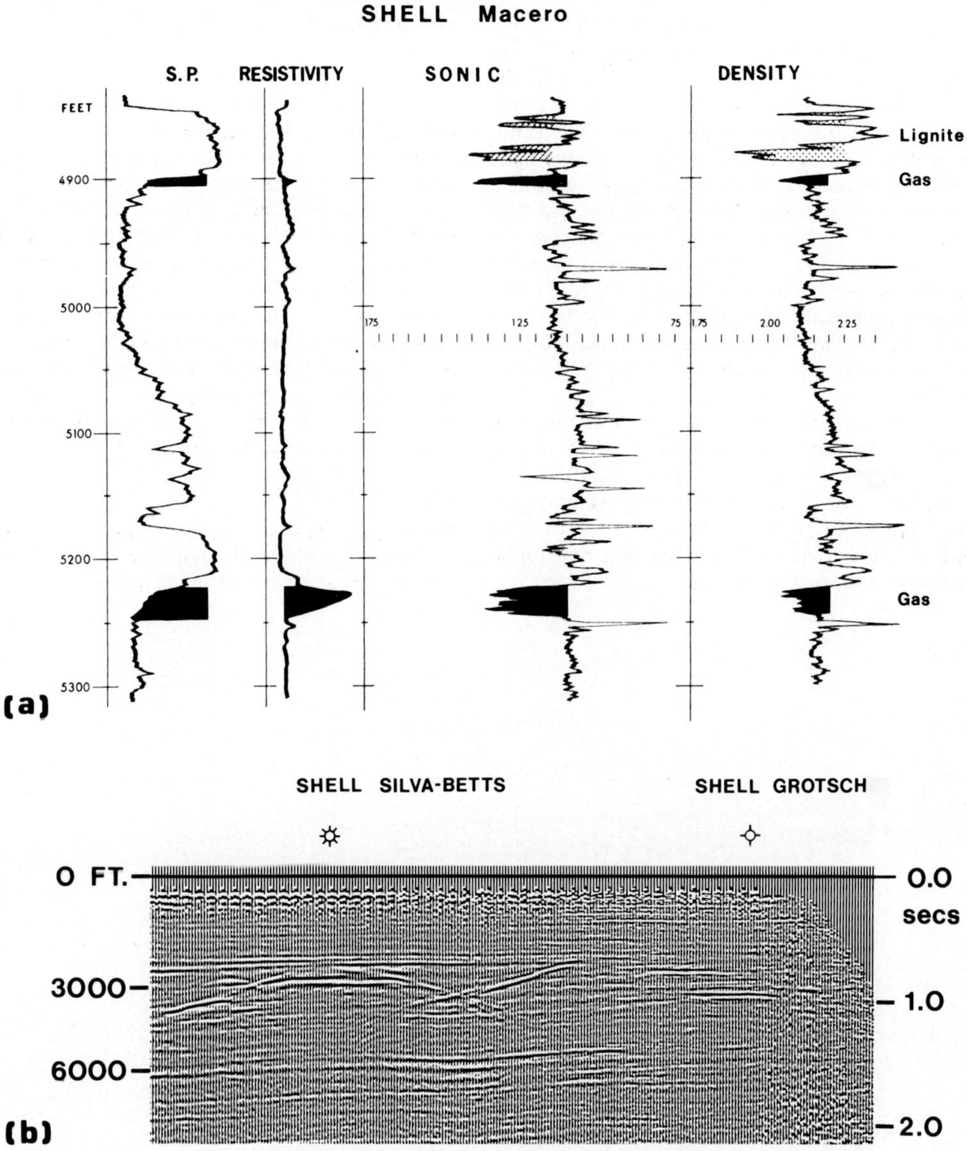

FIG. 5–California examples. (**a**) Velocity and density data for lignite and gas-bearing layers. (**b**) Seismic section showing "bright spots" associated with both commercial and noncommercial gas accumulations.

pographic high. As this paleohigh coincides with present-day structure, the structural configuration of the bright spot is the same as would be expected with a hydrocarbon accumulation. However, if the reflection polarity can be determined, this kind of amplitude anomaly can be ruled out as a hydrocarbon indicator.

In California, Upper Cretaceous sandstone targets indicated by two bright spots were drilled. Logs from this well are shown in Figure 5a. The upper bright spot at 4,850 ft (1,478 m) was slightly better on the seismic data than the lower bright spot at 5,220 ft (1,591 m). The upper one was caused by lignite; the gas-bearing interval at 4,900 ft (1,494 m) was too thin to produce an amplitude anomaly. The lower bright spot was caused by a good gas "pay." It is clear that the sonic and density responses of lignite are very close to those of gas-bearing sandstones. As a consequence, in this area of stratigraphic traps the significance of bright spots is commonly difficult to determine. Another example from California (Fig. 5b) shows two wells drilled into bright-spot targets in the Upper Cretaceous. The well on the right, at a seismic time of 1.0 second, encountered a wet sandstone with low resistivity but a good low-velocity sonic anomaly. Estimated free-gas saturation in this reservoir is about 10%. Moreover, the logs seem to indicate a water level below the low gas saturation. A good gas "pay" was encountered in the well on the left at a depth of 3,000 ft (914 m) where a pronounced bright spot occurs. In many respects these two bright spots look alike, but one led to a commercial gas discovery and the other, a dry hole.

These examples were chosen to illustrate that interpretations of amplitude anomalies are not always straightforward. There are ambiguities, and great care must be exercised in interpretation because not all bright spots indicate hydrocarbons and, conversely, not all hydrocarbon accumulations give rise to bright spots.

Summary

All of the examples discussed in this paper deal with the interpretation of bona fide seismic signals and not with other classes of problems such as spurious seismic events and processing artifacts. There are still ambiguities in the interpretation of the best data. Even if we describe the earth in precise layers of velocity and density contrasts, with known attenuation properties and correct structural attitudes, the geologic interpretation may be ambiguous in some instances. However, these ambiguities will be fewer as we recognize and solve each new complexity and use good geologic reasoning in our interpretations, thus making the "new" geophysics of today the "old" geophysics of tomorrow.

Reference Cited

Domenico, S. N., 1974, Effect of water saturation on seismic reflectivity of sand reservoirs encased in shale: Geophysics, v. 39, no. 6, p. 759–769.

Kapuni and Maui Gas-Condensate Fields of New Zealand—Summary[1]

D. M. McBEATH[2]

Introduction

The onshore Kapuni and offshore Maui gas-condensate fields, which to date contain New Zealand's only commercially exploitable hydrocarbon reserves, occur within the Taranaki basin in the west of the North Island. Kapuni, discovered in 1959, first produced in May 1970. Maui, 25 mi (40 km) offshore in 360 ft (110 m) of water, was discovered in 1969 but is not due to produce until October 1978. Production from both fields is from Eocene sandstone reservoirs within the coal-bearing Kapuni Formation, which was deposited in a widespread coastal-plain and fluviomarine environment.

The fields were discovered by Shell BP and Todd Oil Services Limited; the two discovery wells were the first drilled onshore and offshore by the company. Shell BP and Todd Oil Services Limited is a nonprofit services company owned 37½% by Shell Petroleum Mining Company Limited, 37½% by BP (Oil Exploration) Company of New Zealand Limited, and 25% by Todd Petroleum Mining Company Limited. It carries out exploration and production operations in the west of the country. Ownership of production from Kapuni is in the above proportion, but that from Maui is shared equally with the government through its mining company, Offshore Mining Company Limited.

Gas from the Kapuni field is dedicated exclusively to the Natural Gas Corporation of New Zealand (a government-owned corporation) for domestic and industrial use. That to be produced from Maui is to be sold primarily under a supply contract to the New Zealand government for electricity generation.

Regional Setting

The Taranaki basin is one of the largest of the Upper Cretaceous–Cenozoic sedimentary basins in the New Zealand area (Fig. 1). It is the central basin of a series of adjoining and interconnected basins which extend onshore and offshore along the west coasts of both the main islands (Katz, 1974). The seaward limit of the west-coast basins extends beyond the edge of the continental shelf. They are limited landward by the highly indurated geanticlinal mass of Permian to Jurassic graywackes and argillites forming the axial ranges of New Zealand.

The Taranaki basin is separated from the younger Cenozoic Wanganui basin to the east by the horst of the Patea-Tongaporutu high, which is bounded on its western side by the Taranaki fault (Fig. 2). This fault has 23,000 ft (7,010 m) of vertical throw.

Within the northern part of the Taranaki basin, the Taranaki graben to the west of the Taranaki fault is separated from the Maui block, a relatively stable crustal segment, by the Cape Egmont fault zone (Fig. 2). Cumulative maximum throw is about 7,000 ft (2,134 m) on several subparallel faults. The basin extends south to the coastal regions of the northwest South Island.

All faults in the Taranaki basin are normal, although reversal of movement has occurred along some of them. The time of reversal appears to correspond to late Oligocene fault reversals in the west-coast region of the South Island (Wellman, 1956, p. 9). Little, if any, horizontal movement can be recognized where Tertiary strata have been displaced, despite the occurrence of the dextral transcurrent Alpine fault, with a lateral displacement of about 300 mi (483 m), just beyond the southern margin of the onshore portion of the basin.

Economic basement within the Taranaki basin and to the east consists of metamorphosed rocks ranging in age from late Precambrian to Jurassic

[1]Manuscript received, October 21, 1974. Published with the kind permission of Shell Petroleum Mining Company Limited, BP (Oil Exploration) Company of New Zealand Limited, Todd Petroleum Mining Company Limited, and Offshore Mining Company Limited (Maui field only).

[2]Manager, Exploration, BP (Oil Exploration) Company of New Zealand Limited, Wellington, New Zealand.

Information for the paper was obtained mainly from unpublished reports by present and former staff of Shell BP and Todd Oil Services Limited, New Plymouth, New Zealand, and Shell Internationale Petroleum Maatschappij B. V., The Hague, The Netherlands; notably D. C. K. Allen, H. C. Arnold, G. Bakker, R. N. Cope, R. A. Couper, G. J. W. de Boer, M. E. Geiger, K. W. Glennie, P. F. Hicks, J. Hoorn, R. Jaeckli, E. W. Jonkers, J. F. Kaan, M. Lutz, G. H. A. Magnée, E. G. O'Fee, L. Ritsema, A. G. Schröder, K. C. Short, R. Smith, P. W. Stainton, D. van den Abeele, J. W. C. M. van der Sijp, H. A. van Wijlen, H. D. Waterland, J. S. Watson, and A. B. Wood. Sincere thanks are expressed to the above and also to P. Lawrence, who drafted the illustrations at considerable personal inconvenience.

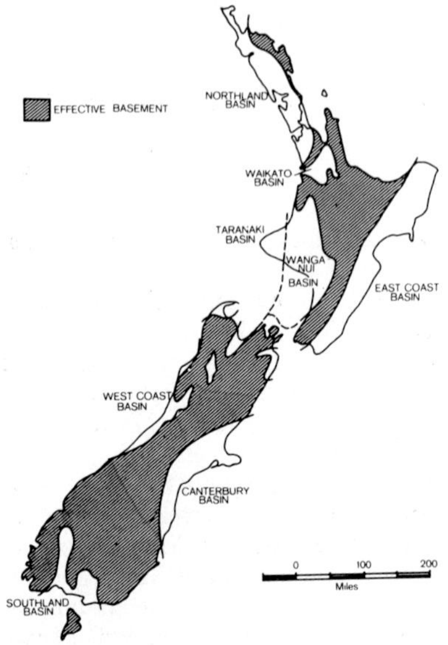

FIG. 1—New Zealand Upper Cretaceous - Cenozoic sedimentary basins. Taranaki basin is located near west coast of North Island.

(Fig. 2). The later Paleozoic and Mesozoic sediments were deposited as turbidites in a rapidly deepening trench (the New Zealand geosyncline) that was everted during the Late Jurassic–Early Cretaceous Rangitata orogeny, which included igneous phenomena such as emplacement of granite batholiths. By the Late Cretaceous, the geanticlinal landmass resulting from the Rangitata movements had been peneplaned.

Upper Cretaceous freshwater beds, including coal measures laid down before and during the period of peneplanation, are preserved in the northwest of the South Island, and a 3,000-ft (914 m) thickness of Upper Cretaceous sandstones and shales with numerous coal seams is present in the Maui-4 well. In the field wells farther north, thin Upper Cretaceous or Paleocene strata overlie basement. To the east, in the onshore portion of the Taranaki basin, Eocene freshwater beds may have been the first deposited, although none of the wells drilled in the Taranaki graben has penetrated to the base of the Cenozoic–Upper Cretaceous succession. On the Patea-Tongaporutu high, Miocene strata overlie basement graywacke; older units may have been removed following mid-Tertiary uplift (Cope and Reed, 1967). Within the onshore portion of the graben an almost complete Cenozoic sequence is preserved up to Quaternary Egmont volcanic rocks (Fig. 3).

Although these volcanic rocks completely cover the surface of the Taranaki Peninsula except in the east, the underlying Tertiary formations where investigated by drilling have been found to be extensive and thicker than 23,000 ft (7,010 m). Formation boundaries are diachronous to some degree. In general, formations which are predominantly arenaceous alternate with those which are chiefly argillaceous. The Kapuni Formation (Short, 1962; Watt, 1965), which consists mainly of deltaic sandstones and coal measures, is nowhere exposed at the surface. It lies at the base of the succession but has not been completely penetrated by any of the exploratory wells drilled onshore. Where drilled, it is of late Eocene to earliest Oligocene age and, near the Taranaki fault, consists mainly of freshwater deposits. Deposition began earlier, and the depositional environment became more marine, in a general westerly direction.

These onshore formations appear to have equivalents offshore to the west of the Cape Egmont fault zone.

Basement structure controlled Late Cretaceous–early Tertiary deposition, and growth structures connected with the differential movement of basement faults are significant for hydrocarbon exploration.

History of Exploration

More interest has been shown in the Taranaki area than in other parts of New Zealand owing to oil seepages near the seashore at New Plymouth which were first recorded in 1839. Other seepage areas on both islands were discovered later.

Attempts at exploitation of the New Plymouth seepages by hand-dug shafts, and then by drilling which commenced in 1866, were unsuccessful, although a small production was established on a continuous basis from about 1934. The production at Moturoa oil field was from sandstone lenses in the Matemateonga Formation (lower Pliocene) between about 2,000 and 2,200 ft (610 and 671 m). When the field was abandoned in July 1972, it had produced only about 216,000 bbl.

Although interest in the Taranaki area waxed and waned over the years, the favorable petroleum legislation of 1937, whereby all oil- and gas-bearing subsurface lands became Crown property, and the stimulation due to World War II led to a revival of interest in exploration. During the war years, seven exploration wells were drilled in the Taranaki and adjoining Wanganui basins by overseas interests (Superior and Standard-Vacuum), all without success.

In 1954, Todd Brothers, a New Zealand company, took out prospecting licenses over several onshore sedimentary basins, including the Taranaki and Wanganui basins. Shell and British Pe-

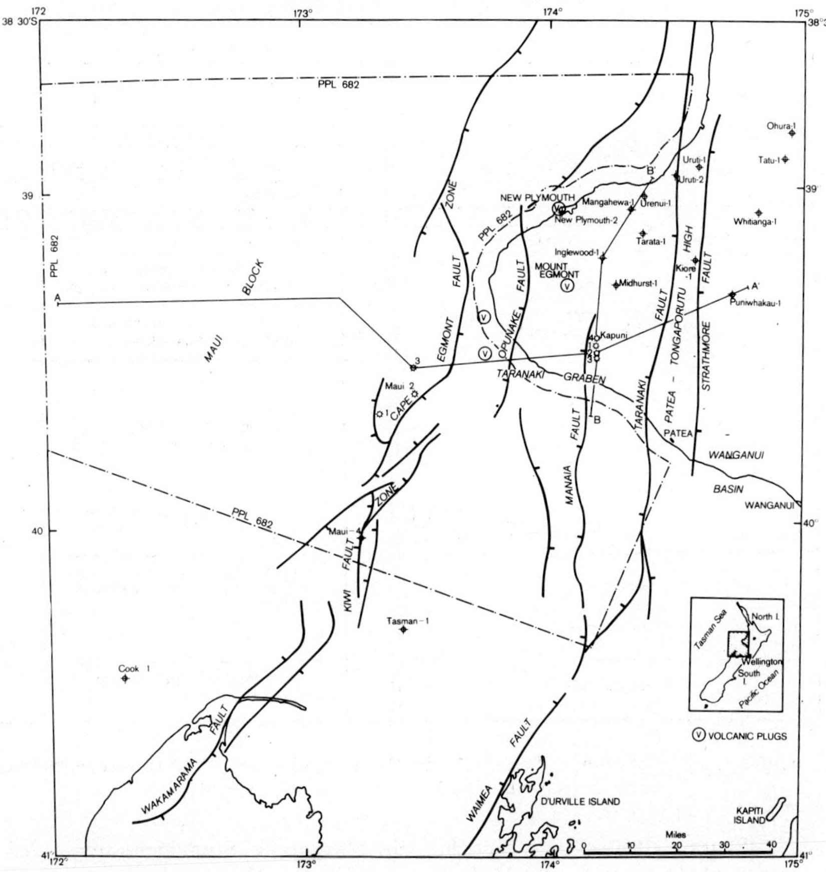

FIG. 2—Structural elements and locations of significant wells in Taranaki basin.

troleum (as D'Arcy Exploration) joined them in 1955 in exploration of the licenses. Shell BP and Todd Oil Services (initially Shell D'Arcy and Todd) was organized to carry out exploration over the Taranaki Peninsula and areas to the east and south. Eventually, 32 petroleum prospecting licenses totaling 5,733 sq mi (14,848 km^2) were obtained over this region.

Geologic field work began in late 1955 and was followed by gravity and magnetic surveys. A reconnaissance seismic program was started in February 1956 in the Taranaki basin and continued until August 1957. Additional seismic surveys were carried out mainly in the Wanganui basin until October 1958.

In March 1957, when much of the initial seismic reconnaissance had been completed, the first definite information of potential interest was observed near Manaia in the southern part of the peninsula. A detailed survey over the Manaia structure led to the discovery of the Kapuni field.

KAPUNI FIELD

The Midhurst-1, 15 mi (24 km) north-northeast of the Kapuni structure, drilled by New Zealand Petroleum Company (Stanvac) to 10,925 ft (3,330 m) in 1940–42, indicated a continuous sedimentary succession from Oligocene (Mahoenui Formation) at total depth to lower Pliocene (Matemateonga Formation), which in turn is overlain unconformably by Quaternary volcanic rocks. However, the well did not penetrate the Oligocene Te Kuiti limestone, which was thought to represent the most prominent seismic reflector in the Manaia area, and which is exposed north of the Taranaki Peninsula below the Mahoenui Formation. In the prognosis for the Kapuni well, the Mangaotaki Formation, which underlies the Te Kuiti Formation at outcrop and is arenaceous and coal bearing (but which was not penetrated in the Midhurst well), was correlated with the Eocene–lower Oligocene quartzose coal measures of

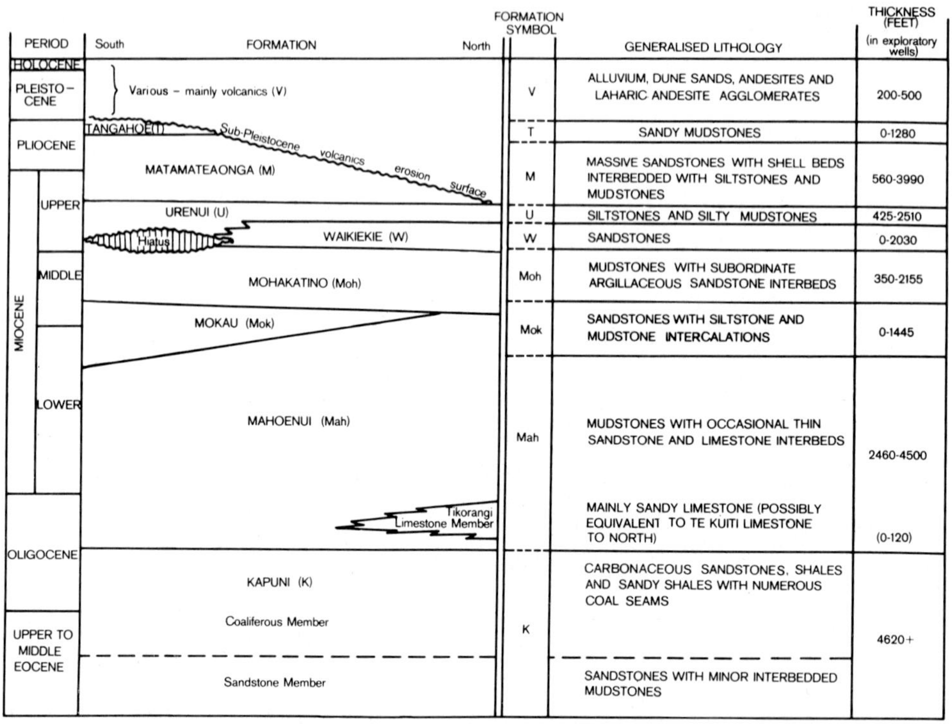

FIG. 3—Time/rock relations of onshore Taranaki basin. Drilled thicknesses of formations penetrated by all exploratory wells drilled to date are shown at right.

the South Island (Suggate, 1950). It was thought that similar strata would be present in the Manaia area and that they should be considered potential source and reservoir rocks for oil.

The Manaia (Kapuni) anticline is a north-south-trending, slightly asymmetric feature with a faulted western flank. A rather shallow saddle determines the closure in beds of pre-Miocene age on its northern plunge. A fault is thought to be present across this saddle between wells 1 and 4. Closure on the seismic horizon considered to represent the Oligocene Te Kuiti limestone at 10,000 ft (3,050 m) is 2,460 ft (750 m) and covers an area of 22,000 acres (90 km^2). Closure on a prospective upper horizon within the lower Miocene Mokau Formation at 6,890 ft (2,100 m) is only about 195 ft (59 m).

The Kapuni-1 well was drilled in 1959 and was suspended as a potential gas-condensate producer in October after being drilled to 13,040 ft (3,975 m). The first production test over the interval 12,366–12,395 ft (3,769–3,778 m) at the end of August produced 1.9 MMCFD of gas with a high carbon dioxide content (ca. 40%) and waxy condensate on a 13⁄16-in. choke from the lower part of the Mangaotaki Formation (later called the "Kapuni Formation").

The well essentially proved the geologic forecast, although erosion had removed the upper part of the Mohakatino Formation, and the overlying Waikiekie Formation apparently had wedged out on the flanks of the structure. Also, calcareous mudstones about 400 ft (122 m) thick were present in place of the Te Kuiti limestone; the mudstones were subsequently assigned to the basal part of the Mahoenui Formation.

The hydrocarbon reservoir was found about 1,400 ft (427 m) below the top of the Kapuni Formation, and two production tests of individual sandstones within the interval 12,268–12,395 ft (3,739–3,778 m) proved the accumulation.

The Kapuni Formation was found to consist of a series of carbonaceous sandstones, siltstones, and very carbonaceous shales intercalated with beds of bituminous coal (Elphick and Suggate, 1964) ranging in thickness from a few inches to 23 ft (7 m); a glauconitic medium-grained sandstone about 30 ft (9 m) thick underlies the Mahoenui calcareous mudstones. Major coal beds were concentrated in the upper part of the formation between 10,800 and 12,000 ft (3,292–3,658 m). A marked increase in sandstone was apparent below 12,050 ft (3,673 m), and the sandstone became less carbonaceous below 12,600 ft (3,840 m). Sandstone beds within the upper part of the formation above 12,049 ft (3,673 m) were generally "tight." Provisionally, a gas-water contact was assumed to exist at about 12,500 ft (3,810 m).

In all wells a hydrocarbon-bearing zone similar to that in Kapuni-1 was penetrated, but in Kapuni-2 and -3 wells, south of Kapuni-1, a second productive interval, the "Upper Sands" as distinct from the "Main Sands," was encountered about 300 ft (91 m) higher. About 75 ft (23 m) of Microlog porosity was indicated to be hydrocarbon bearing in the "Upper Sands" in each of these two wells.

The original recoverable reserves of the 4-well Kapuni field producing from below 11,300 ft (3,444 m) are estimated at 460 trillion BTU of dry gas (630 Bcf at 730 BTU/cu ft) and 34 million bbl of condensate. Production in 1973 averaged 46.3 MMCFGD (CO_2 content 43%) and 3,500 bbl/day of condensate. Cumulative production to December 31, 1973, was 49,264 MMcf of gas and 3,699,346 bbl of condensate. Condensate has averaged 75 bbl/MMcf. Production of propane from the Natural Gas Corporation's plant, with a maximum capacity of 3,000 long tons per annum, commenced June 1, 1973, and is sold to the mining companies. To the end of 1973, 9,260 bbl (734 long tons) had been recovered. Recovery of butane and natural gasoline is planned.

Additional drilling in 1974 should increase maximum deliverability to 260 MMCFGD and 18,000 bbl/day of condensate, so that 108.7 trillion BTU (about 24% of the original reserve and two thirds of the production) can be supplied as raw gas with its original carbon dioxide content to the New Plymouth power station during 1975–78 pending supply from Maui.

MAUI FIELD

After reflection and refraction seismic work, including the 3-mi (4.8 km) offshore limit, 3 exploratory wells were drilled in the onshore Taranaki basin and 6 wells, including 2 stratigraphic holes, in the Wanganui basin. All were dry.

Preparations for exploration in the offshore Taranaki area included 1,367 mi (2,200 km) of 4-fold reflection seismic coverage. The results showed that the license area covered two major fault blocks divided by the north-south-trending Cape Egmont fault zone. West of the fault zone, the structurally higher Aotea block (now known as the Maui block) appeared to be more tectonically stable than the Taranaki graben, which formed the eastern block and where basement lay at about 17,000 ft (5,182 m).

Three structural leads (apart from small structures probably associated with igneous intrusions near the tip of the Taranaki Peninsula) were outlined, of which one was structure "C," now known as the Maui structure. Seismic interpretation indicated the structure to be a large feature about 30 by 9 mi (48 × 14 km) at a reflector level considered to be equivalent to the Cobden and Takaka limestones in the northwest of the South Island or the Te Kuiti limestone of north Taranaki, all of Oligocene age. At this level there is 1,400 ft (427 m) of vertical closure. Below the postulated limestone, it was considered that there might be a more marine equivalent of the Eocene Kapuni Formation, thus upgrading the prospect of an oil accumulation.

The Maui-1, about 32 mi (51 km) from the coast, was spudded on January 27, 1969, in 363 ft (111 m) of water. It reached total depth of 11,515 ft (3,510 m) on March 6 and, after testing, was suspended on March 26.

The geologic interpretation of the Maui structure made before drilling proved to be accurate; as forecast, the Kapuni Formation equivalent was penetrated from 8,625 to 11,400 ft (2,629–3,475 m) and proved to be gas-condensate and oil bearing.

Two hydrocarbon-bearing intervals are present: 8,828–9,136 ft (2,691–2,785 m), with 242 ft (74 m) of net gas sandstone and 5 ft (2 m) of net oil sandstone; and 9,815–9,897 ft (2,992–3,017 m), with 36 ft (11 m) of net gas sandstone and 26 ft (8 m) of net oil sandstone. A production test through a 7-in. liner in the interval 9,885–9,895 ft (3,013–3,016 m) flowed 44.5° API oil at the rate of 1,645 bbl/day through a 12/32-in. surface choke; gas/oil ratio was 1,450 cu ft/bbl. There were also shows in sandstones above the Kapuni Formation in the interval 6,150–7,630 ft (1,875–2,326 m), but logs showed the sandstones to be water bearing.

The stratigraphy is described in terms of the equivalents of the onshore formations. Notably, the Takaka–Te Kuiti limestone equivalent of coccolithic-foraminiferal "lime" packstones belonging to a deep-water outer-shelf facies in the Oligocene and lower Miocene contrasts with the onshore Oligocene Takaka and Te Kuiti limestones which belong to a shallow-shelf facies of argillaceous limestones and minor calcareous sandstones. A 200-ft (61 m) bed of hard calcareous shale underlies the limestone and represents the upper bed of the Kapuni Formation. A series of major sandstone units interspersed with minor shale beds persists to 11,400 ft (3,475 m). Above 9,998 ft (3,047 m), the sandstones are medium to fine grained and friable with angular to subangular quartz grains. Below 9,998 ft (3,047 m), they are generally finer grained, argillaceous, and carbonaceous, and there are increasing amounts of kaolinite and fine quartz pebbles below about 10,800 ft (3,292 m). Coal seams 5–7 ft (1.5–2.1 m) thick are present in the intervals 10,000–10,300 ft (3,048–3,139 m) and 11,000–11,220 ft (3,353–3,420 m).

Below the base of the Kapuni Formation at 11,400 ft (3,475 m) is a 20-ft (6 m) weathered zone containing pebbles of quartzite, schist, and granite in a coarse sandstone matrix which grades downward into fresh conglomeratic beds of the same petrologic composition to 11,515 ft (3,510 m) total depth. Tentatively, this unit is correlated with the Pakawau Group of the South Island; the Pakawau is of Late Cretaceous to possibly Eocene age (Grindley, 1961).

Drilling of two more wells on the Maui structure satisfactorily defined it for further development.

Maui field reservoirs are below 8,700 ft (2,652 m). Reserves are estimated at 5,730 trillion BTU of dry gas (5,590 Bcf at 1,025 BTU/cu ft), 161 million bbl of condensate, and 51 million bbl of recoverable LPG. Two platforms, each with 12 wells, will be constructed consecutively 23 and 33 mi (37 and 53 km) offshore. Processing facilities will handle a throughput of 750 MMCFGD and 20,000 bbl/day of condensate. The gas will provide fuel for over 3,000 Mw of generating capacity in North Island power stations.

Kapuni raw gas, without condensate, is sold by Shell, BP, and Todd for NZ$0.175/million BTU at the field to the Natural Gas Corp. for removal of carbon dioxide and distribution in the North Island. Maui gas will be sold at NZ$0.37/million BTU delivered at the power stations. Prices are subject to escalation. Development of the Maui field and associated processing and distribution facilities will be undertaken jointly by the New Zealand Government (50%) and by Shell, BP, and Todd.

References Cited

Cope, R. N., and I. J. Reed, 1967, The Cretaceous paleogeology of the Taranaki-Cook Strait area: Australasian Inst. Mining and Metallurgy Proc., v. 222, p. 63–72.

Elphick, J. O., and R. P. Suggate, 1964, Depth/rank relations of high volatile bituminous coals: New Zealand Jour. Geology and Geophysics, v. 7, no. 3, p. 594–601.

Grindley, G. W., 1961, Golden Bay, Sheet 13 *of* Geological map of New Zealand: Wellington, New Zealand Dept. Sci. and Indus. Research, scale 1:250,000.

Katz, H. R., 1974, Offshore petroleum potential in New Zealand: APEA Jour., v. 14, pt. 1, p. 3–13.

Short, K. C., 1962, The stratigraphy of the Taranaki Peninsula structure: Shell BP and Todd Oil Services Limited, unpub. rept.

Suggate, R. P., 1950, Quartzose coal measures of west Nelson and north Westland: New Zealand Jour. Sci. and Technology, v. B31, p. 1–14.

Watt, D. S., 1965, Natural gas and oil in western New Zealand: Commonwealth Mining and Metall. Cong., 8th, Proc., v. 5, p. 77–88.

Wellman, H. W., 1956, Structural outline of New Zealand: New Zealand Dept. Sci. and Indus. Research Bull., no. 121, 36 p.

Sedimentary Basins and Petroleum Prospects, Onshore and Offshore New Zealand[1]

H. R. KATZ[2]

Abstract Rocks older than the Late Jurassic Rangitata orogeny are highly indurated, deformed, and/or metamorphosed, and constitute effective basement. Sedimentary basins prospective for petroleum began to form in middle to Late Cretaceous time, but their main development occurred during the Cenozoic. In the west and southeast, epicontinental basins or marginal geosynclines resulting from basement collapse and extension are filled with sedimentary rocks several kilometers thick—locally up to 5–10 km—which in their basal portion are terrestrial and include coal measures. In the northeast, the marginal basin corresponds to a complex orogenic system characterized by repeated diastrophic cycles and several marine transgressions and regressions, locally accompanied by intermittent erosion.

Throughout New Zealand, sediments are commonly of a sandstone-shale facies with only minor carbonate rocks. Potential reservoirs generally are in sandstones near the base of the sedimentary sequence (Late Cretaceous to early Tertiary) that is transitional between shallow-marine and nearshore deltaic to estuarine-brackish and nonmarine (sandstones in coal measures). Locally, reservoirs may also occur in limestones. Structural traps, including various growth structures, occur widely in most areas; good stratigraphic-trap conditions are also widespread, resulting from local unconformities, onlap, pinchout, or lateral facies changes. There seems to be a good potential for petroleum generation in all basins (oil and gas seeps are widespread throughout New Zealand), and prospects for accumulations are particularly favorable where structural and stratigraphic conditions combine. The maximum prospective area is about 50,000 sq mi (129,500 km^2) on land, and roughly 100,000 sq mi (259,000 km^2) offshore, to an arbitrary depth limit of 1,000 m.

Commercial production has been obtained in the Taranaki basin, where the proved recoverable reserves of the Kapuni and Maui fields amount to over 6 Tcf of gas and about 200 million bbl of condensate. Offshore, only 10 wells have been drilled to date, of which 3 established the large Maui gas field. Another well tested oil at a rate of 600 bbl/day, but was abandoned as noncommercial.

INTRODUCTION

The geologic history of New Zealand is dominated by events associated with the Late Jurassic Rangitata orogeny. Rock formations older than Cretaceous, in particular the very thick sequence of the New Zealand geosyncline (Fleming, 1970), are in most places so indurated, deformed, and partly metamorphosed that, for the petroleum geologist, they constitute effective basement. By Early Cretaceous time a landmass much larger than present-day New Zealand was established; considerable erosion resulted in widespread peneplanation. No marine Neocomian is known anywhere in New Zealand; the oldest Cretaceous faunas are Aptian-Albian (Speden, 1973), and rocks of this age are restricted to the East Coast fold belt and Northland; in the first area their relation with older formations, where this is known, is markedly unconformable, whereas in Northland they may belong to large-scale olistostrome deposits of unknown origin (Katz, 1968). More extensive marine transgressions occurred in the Campanian-Maestrichtian, both in the east and the west (Lord Howe Rise), but, more commonly, the sedimentation during Senonian-Maestrichtian and earliest Tertiary times was terrestrial and included widespread coal deposits.

Thus the sedimentary basins here considered are all post–Rangitata orogeny. Although they began forming in some areas in the middle to Late Cretaceous, they are mainly of Cenozoic age. The general stratigraphy and facies of the sedimentary fill, as exposed on land, were described by Katz (1968). Additional studies, based to a great extent on recently available data[3] from extensive offshore work by oil companies, permit discussion of these basins in the wider context of their regional tectonic framework.

TECTONIC SETTING AND STRUCTURAL HISTORY

The continental landmass established after the Rangitata orogeny was undoubtedly much more extensive than present-day New Zealand; the Challenger Plateau, Lord Howe Rise, Campbell Plateau, and Chatham Rise were all part of it, as possibly was the Norfolk Ridge, northwest of Northland peninsula (Fig. 1). Little is known of the early history of the deeply depressed basins and troughs between these blocks (New Caledonia Basin, Bounty Trough, and Hikurangi Trough), but they, too, may have formed part of this landmass.

The subsequent tectonic events mainly resulted in a pronounced differentiation and fragmenta-

[1]Manuscript received, November 6, 1974.

[2]New Zealand Geological Survey, Lower Hutt, New Zealand.
I thank Penny Williams and other staff of the Cartographic Section, D.S.I.R., for their cooperation in preparing the illustrations, and Ray Farmer for assistance in compiling data. M. Ruthven typed the manuscript.

[3]All oil-company reports are held by the New Zealand Geological Survey where they are, if released and on Open File, available to the public (see references).

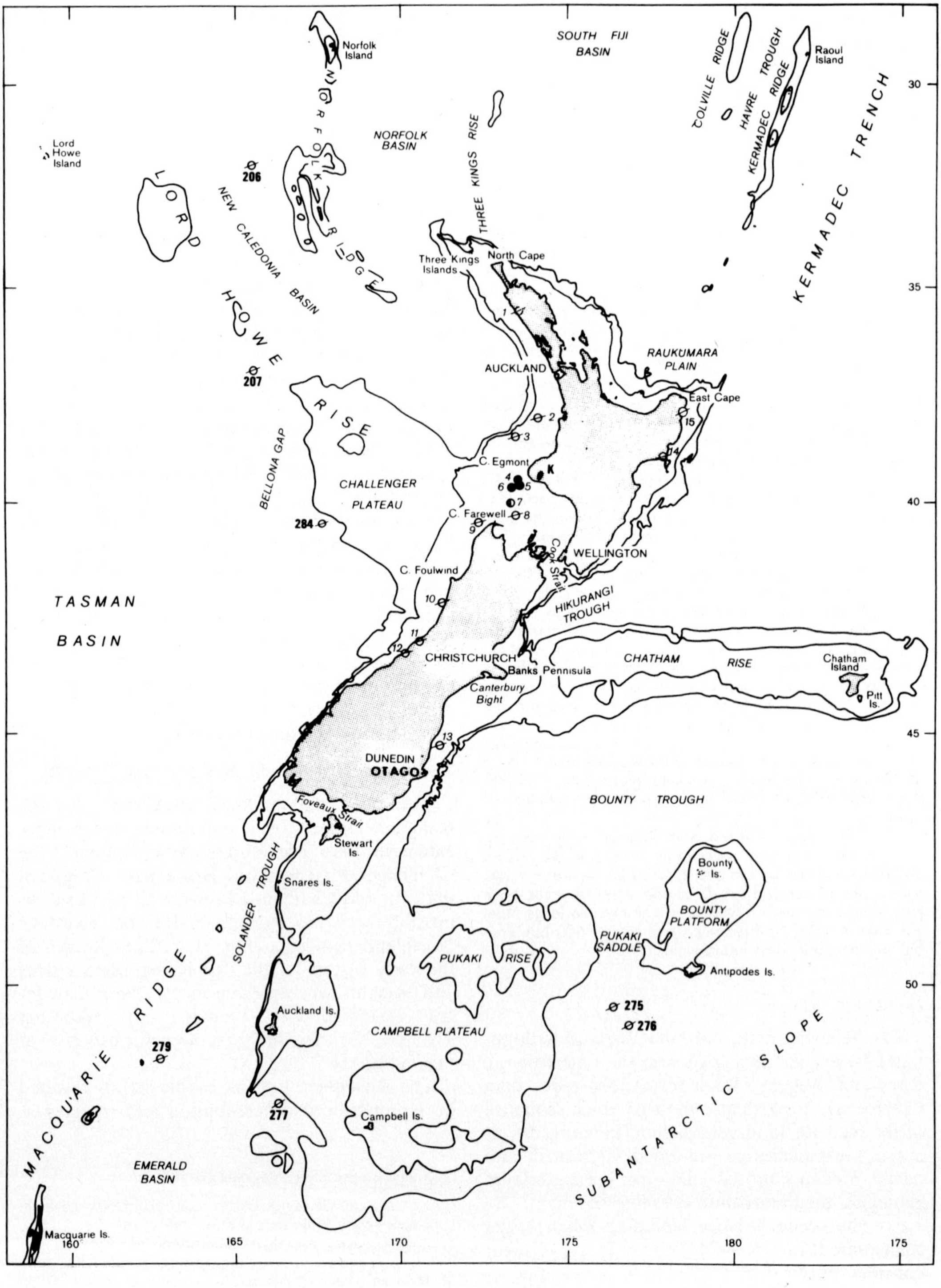

FIG. 1—New Zealand region. Bathymetric contours are at 500 and 1,000 m. *Glomar Challenger* drilling sites of DSDP Leg 21 (*206, 207*) and Leg 29 (*275-284*). *K* = Kapuni gas-condensate field. Wells drilled for oil exploration are: *1*, Waimamaku-2; *2*, Mangaa-1; *3*, Moa-1B; *4*, Maui-3; *5*, Maui-2; *6*, Maui-1; *7*, Maui-4; *8*, Tasman-1; *9*, Cook-1; *10*, Haku-1; *11*, Harihari-1; *12*, Waiho-1; *13*, Endeavour-1; *14*, Opoutama-1; *15*, Te Horo-1.

tion of the continental area; large portions of relatively stable cratonic masses became detached and temporarily or semipermanently submerged. Along the west coast of both islands, basement collapse and extension tectonics have dominated since middle Cretaceous time, giving rise to the formation of a complex marginal geosyncline. On the east side of New Zealand, a similar zone of extension tectonics, associated with widespread Cenozoic volcanism producing mainly alkali basalt, appears to have controlled sedimentation along the extensive downwarp parallel with the present coast from the Canterbury Bight south toward the central depression of the Campbell Plateau (Katz, 1974a; Figs. 1, 2). A large sedimentary wedge which fills a marginal basin 700–800 km long and more than 100 km wide (the "Great South Basin")[4] occupies a position between the uplifted landmass of the present-day South Island and the foundered cratonic block of the Campbell Plateau (Katz, 1974b); its structure and depositional history are little known, but deformation in general has been slight.

Separated from this province by the east-west-trending structural high of Banks Peninsula–Chatham Rise, the East Coast fold belt is marked by a very different tectonic history and much greater mobility. Its complex array of fragmented basins and highs formed by repeated phases of diastrophic events indicates an active orogenic system.

Western Marginal Geosyncline

The region west of the Alpine fault and its northward continuation along the Waimea and Taranaki faults (Figs. 3, 4) is distinguished as a separate tectonic province; in its northern part an additional area east of the Taranaki fault (i.e., the Wanganui and Waikato basins) is also included. Basement collapse began in mid-Cretaceous time when locally thick Hawks Crag Breccia was deposited in graben-type features (M. G. Laird, personal commun.). A rich microflora indicates an Albian age (Norris, 1968; Norris and Waterhouse, 1970). The Senonian and Maestrichtian Paparoa and Pakawau coal measures indicate areas of local subsidence in the northwestern South Island; isopach maps of the Paparoa Group (Gage, 1952) suggest repeated fault-block movements during this time. More widespread terrestrial (sandstone, shale, coal) deposition occurred during the Eocene in the Greymouth to Cape Farewell region, across the western entrance of Cook Strait to Taranaki, in Northland, and in the Waikato basin south of Auckland. During the Oligocene, terrestrial deposition spread south from here into northern Taranaki. All these basins were subsiding between elongate high horsts and collected sediments many thousands of feet thick. As marine transgression gradually advanced from the northwest and southwest, large delta fans developed in some of these basins (e.g., the Kapuni Formation in the northern Taranaki graben).

The Taranaki graben, which on a regional scale represents a complex marginal rift, is confined in the north between the Taranaki and Cape Egmont faults (Fig. 4); southward it continues between the Waimea and Alpine faults, on the east, and the Cape Foulwind fault zone on the west (Fig. 3). The latter marks the site of a Late Cretaceous to early Tertiary high from which sediments were shed to the east. At Cape Foulwind, crystalline basement is exposed, and coal measures are very thin or absent southward along this fault zone. A general, marked increase in sediment thickness in an easterly direction is noted from west of Taranaki down to Greymouth (Fig. 3).

Within this regional graben structure or marginal rift, differential block movements resulted in pronounced thickness and facies changes. Local unconformities are abundant, and several subbasins are present. In many places, reversal of tectonic movement occurred during the Cenozoic (Wellman, 1956; Laird, 1968); in Taranaki, some faults reversed their sense of movement several times (Cope, 1965). The greatest area of New Zealand covered by the sea probably was during the Oligocene (Fleming, 1962), but the regional tectonic fragmentation and collapse along the western margin continued during the Miocene and Pliocene and spread farther east and north in the North Island. In northern Taranaki and the Auckland region, comparably large areas developed marine turbidite sequences (Glennie, 1959; Ballance, 1964), and at least part of the Northland olistostromes were emplaced during this time (Kear and Waterhouse, 1967). The formation of olistostromes in Northland possibly began as early as late Eocene (Katz, 1974c). Widespread Miocene andesitic volcanism occurred in the northwestern part of the North Island, and is also suggested from seismic evidence on the Challenger Plateau (Andrews and Eade, 1973; Bentz, 1974). In the late Miocene the Patea high, east of the Taranaki fault (Fig. 4), finally became submerged. By this time, the fault block, which had been growing at least since the Eocene, ceased to be active. However, regional southward tilting in the Pliocene further modified the Taranaki graben west of the fault, and the block on the east

[4]Name suggested by Robert M. Sanford.

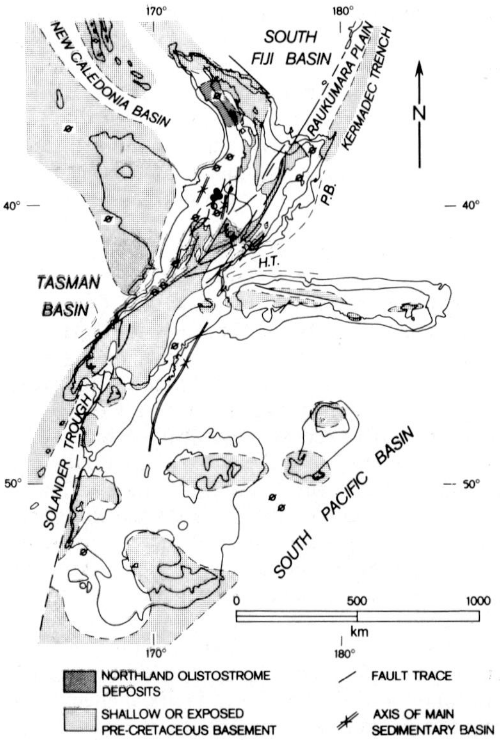

FIG. 2–Geologic-tectonic map of New Zealand region (for well locations see Fig. 1). *H.T.* = Hikurangi Trough; *P.B.* = Poverty Basin.

became markedly downthrown to form the Wanganui basin. This basin contains up to 4,000 m of Plio-Pleistocene marine strata (Katz, 1974a, c). Extreme subsidence also occurred in the small Murchison basin (Fig. 3), where more than 6,000 m of sediment was deposited during the Miocene (Mangles and Longford Formations; Fyfe, 1968). Here and farther south (i.e., in the narrowing wedge of the original marginal rift), considerable folding and thrust faulting took place locally, possibly in association with compression along the Alpine fault. The greatest deformation, and the most extreme induration, of Tertiary rocks in New Zealand is found in the extreme southwest along the coast from Paringa to Milford Sound, in a very narrow strip between the Alpine fault and the oceanic basin of the Tasman Sea. This very intense tectonism apparently occurred in one pulse in the latest Miocene, immediately before deposition of Pliocene beds. Near Cascade Point, the Pliocene rocks are virtually undeformed and hardly indurated at all (S. Nathan, personal commun.). The final uplift of basement horsts and stripping of the Tertiary cover in the northwestern South Island (Fig. 3) must have occurred, at least in part, as late as Pleistocene time.

Late-stage developments also caused the foundering of the regional block west of the Cape Foulwind fault zone (Fig. 3). In strong contrast to the repeated block faulting and deeply subsiding basins on the east, which had developed throughout the Tertiary, the Late Cretaceous–Cenozoic interval in the western block is represented by a comparatively thin and virtually flat-lying, tectonically little-disturbed sedimentary sequence which gradually thins westward to less than 1 second two-way reflection time (Hematite Petroleum Ltd., 1969). This thinning is particularly pronounced in the younger series where giant foreset bedding—which can be followed north over hundreds of kilometers (Andrews and Eade, 1973; El Paso Natural Gas Co., 1973)—contributes to the buildup of a thick sedimentary wedge along the eastern part of this tectonically more stable block.

Basins South and East of South Island

Solander Trough—As suggested by Katz (1974b), this basin (Figs. 1, 2) is part of the initial marginal rift which separated New Zealand (Campbell Plateau) from Australia. Downwarping began in the Eocene. Northwest of Auckland Island the basin is filled with turbidite-type deposits 2.5–3 km thick. These rocks are deformed and, locally, are slightly eroded down the center of the trough. Toward its head they are pierced by younger volcanic complexes, such as the hornblende andesites of the Solander Islands (Harrington and Wood, 1958) which are thought to be late Pleistocene. The small onshore Waiau basin (Katz, 1968) is slightly offset but structurally related to the Solander Trough.

Canterbury–East Otago–Campbell Plateau basin or "Great South Basin"—This basin (Figs. 1, 2) is mainly offshore, and its axis of greatest deposition is probably about 50–100 km east of the coastline. A thickness of Late Cretaceous and Tertiary rocks in excess of 3 km is anticipated.

Available outcrop and well data are restricted to areas near the northwestern margin of the basin (Fig. 5), where the Late Cretaceous to Miocene sequence generally is little more than 1,000 m thick. There, the basal portion commonly is developed in a terrestrial sandstone-shale-coal facies, but marine transgression appears to have reached most places between latest Cretaceous and early Tertiary times (Katz, 1968). The thickest coal measures are in the south; at Kaitangata they are over 500 m, and possibly 600–900 m, thick (Harrington, 1958). The marine sequence comprises glauconitic siltstone-sandstone, shale,

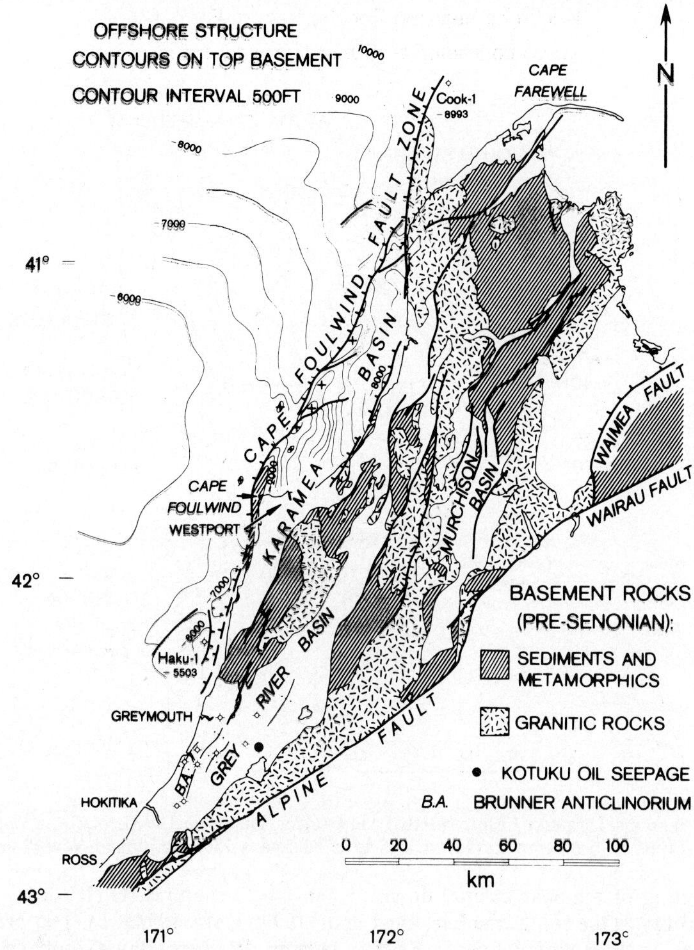

FIG. 3—Structural map, northwest South Island. Offshore contours from Esso Exploration and Production N.Z. (1969).

and marlstone. Locally, limestone is intercalated in the Upper Cretaceous, Eocene, Oligocene, and Miocene sections. Many of these limestones are biogenic and very pure. Some are massive and dense, but there are also highly porous varieties like the upper Eocene polyzoan Totara Limestone near Oamaru, which has up to 30% porosity (Gage, 1957; column B, Fig. 5). At Oamaru, a complex interfingering of biogenic and pyroclastic units with pillow lavas and intrusives occurs in upper Eocene to lower Oligocene strata. Bryozoan limestones grade laterally and vertically into tuffaceous limestones and agglomerates and, locally, into diatomaceous mudstone with intercalated tuffs (A. R. Edwards, personal commun.).

Stratigraphic gaps, disconformities, and/or erosional unconformities are widespread in the Oligocene, occurring locally at its base, within it, or at the top (Fig. 4). Since regional truncation occurred during this period on both the Campbell Plateau and Chatham Rise, where the Oligocene is absent (Katz, 1974a, b), a comparatively large gap in the stratigraphic column can be expected also in the area offshore from Canterbury and Otago.

The beginning of the Miocene was characterized by increased sediment supply from a rising hinterland (Katz, 1968; Fig. 5). The sections exposed on land include shallow-marine to nearshore silty mudstone, sand-gravel, and shelly limestone. Locally, the facies may become terrestrial and include coal measures, but toward the center of the basin offshore a facies change and a considerable increase in thickness are anticipated.

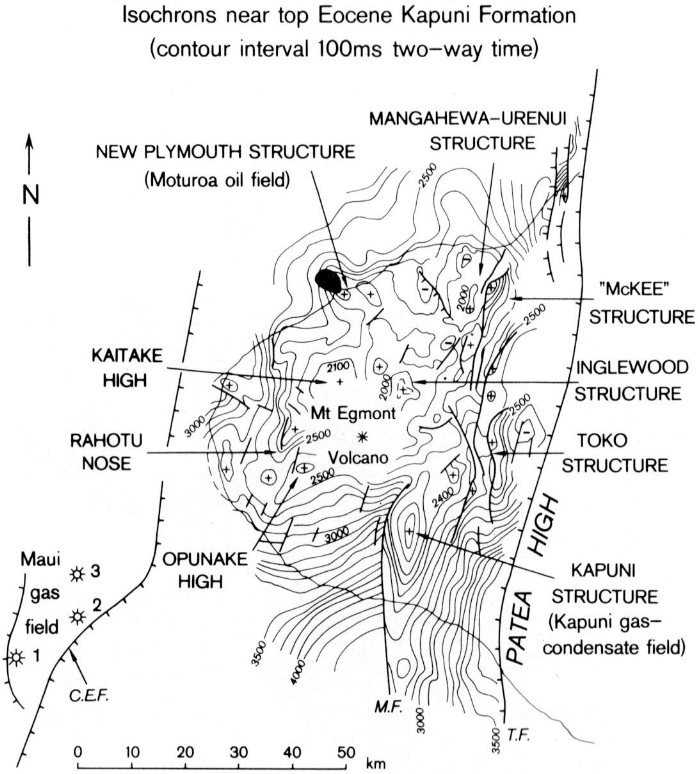

FIG. 4–Structural map of Taranaki basin, North Island (after Pinchon, 1973). *C.E.F.* = Cape Egmont fault; *M.F.* = Manaia fault; *T.F.* = Taranaki fault. Cross-hatched area is andesite plug at New Plymouth.

In a general setting of regional basinal downwarp from Canterbury to the south, the Leeston-1 and J. D. George-1 well sections (Fig. 5; Katz, 1971a) appear totally anomalous. However, based on the extremely reduced and incomplete stratigraphic section and the accompanying volcanic units, it is concluded that this area of Canterbury is part of the same tectonic province as the Chatham Rise and had a history closely comparable to that of Banks Peninsula (F, Fig, 5) and the Chatham Rise. The area is a comparably stable, arched region which, since Late Cretaceous time, has remained structurally high with respect to the surrounding regions. The downwarped, thick sedimentary wedge has developed only to the south from here. During Late Cretaceous and Paleocene-Eocene times, it appears to have plunged farther southward and projected into the central depression of the Campbell Plateau (Fig. 1).

Chatham Rise—Although there is no regional sedimentary basin developed over this feature, the mainly longitudinal tension faulting which occurred in Late Cretaceous, post-Eocene, and post-Miocene times produced local basins in graben-type structures (O'Halloran, 1971; Austin et al, 1973; Katz, 1974a, b). The pre-Senonian Cretaceous Waihere Bay Group (Hay et al, 1970) may be preserved in basement fault blocks, whereas the deposits above the main regional unconformity appear to be mainly Eocene and early Miocene.

East Coast Fold Belt

Principally developed along the east coast of the North Island, onshore and offshore, this complex major basin extends south into the northeastern South Island (Fig. 1). Thick middle to Late Cretaceous marine deposits (Aptian to Maestrichtian) continue upward into still thicker Cenozoic sequences (Katz, 1968). In Opoutama-1, nearly 2,000 m of Late Cretaceous strata were drilled but basement was not reached (Katz, 1974c); the lower Tertiary there measures 1,000 m. In the central Hawke Bay and Gisborne regions the Miocene is reportedly up to 9,000 m thick and covered by a maximum of about 3,000 m of Plio-Pleistocene in the center of the basin.

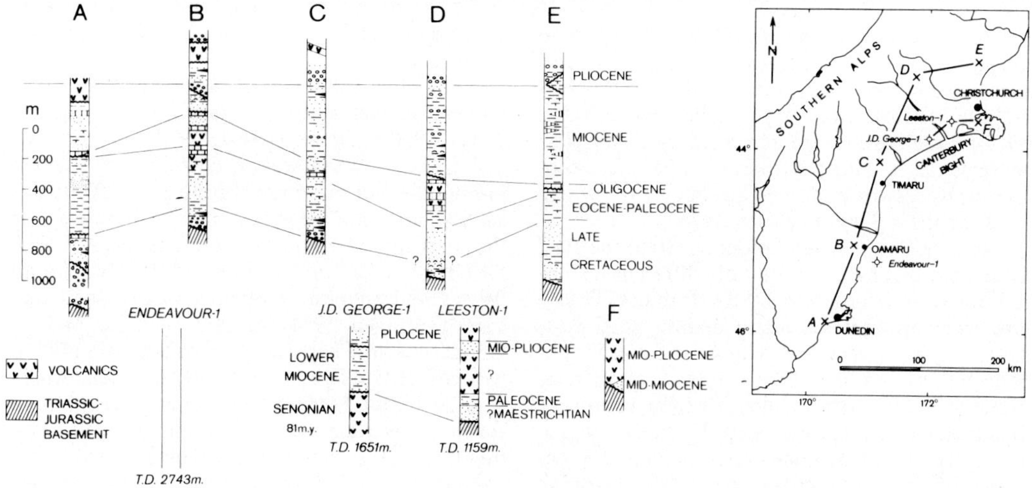

FIG. 5—Correlation of columnar sections, Canterbury - East Otago (after Benson, 1968; Gage, 1957, 1970; Hulston and McCabe, 1972; Katz, 1971a; Wellman, 1953; Wilson, 1963).

The East Coast fold belt has been affected repeatedly by diastrophic movements, causing complicated patterns of facies changes, thickness variations, and large-scale unconformities which reflect regional transgression-regression cycles. The whole region is characterized by great crustal mobility as suggested by abrupt changes, both laterally and vertically, between deep-water and shallow-neritic facies, and by repeated local occurrences of thick turbidite sequences. The rapid burial of thick, fine lower Tertiary siltstone to claystone sequences resulted in widespread undercompacted shales which impede seismic penetration below the Miocene (Katz, 1975). Associated with these phenomena are structural complications such as decollement, gravity gliding (Ridd, 1968; Stoneley, 1968), and diapirism (Ridd, 1970).

Although the Cretaceous history is still little known, recent studies by Speden (1973; 1975) and others have made it clear that important movements had already occurred within the Aptian-Senonian interval, as reflected by facies distribution and pronounced unconformities. The Campanian-Maestrichtian Mata Series is transgressive on a regional scale, cutting across the older Cretaceous formations as well as the Triassic-Jurassic basement. In the north, this transgression does not appear to have reached as far west as the Aptian-Senonian one, but the boundaries apparently overlap in the south. Widespread siltstone-sandstone and conglomerate deposits mark neritic-shelf environments during the Mata transgression. The massive, siliceous Whangai Shale, characteristic of deeper water environments, transgresses the Cretaceous-Paleocene boundary, and in most places no lithologic boundary is discernible between the Cretaceous and Tertiary.

The Paleocene-Eocene sequence is characterized by mudstone with some glauconitic siltstone sandstone, thin but persistent black and chocolate shale, and, in the south, siliceous limestone. During the early Oligocene the sea probably reached its maximum transgression, as is suggested by a remnant of glauconitic siltstone near Paraparaumu on the west side of the Central Ranges. Increased tectonic activity began during the Oligocene. It is particularly well manifested in the south of the east coast where a "coastal high" is apparent and the upper Oligocene lies in some places directly on Upper Cretaceous Whangai Shale. Regionally, intense erosion occurred between the Oligocene and Miocene, and the base of the Miocene is generally time-transgressive and marked by an angular unconformity.

During the Miocene, the sea spread westward again. Thick, shelf-type siltstones-sandstones are widespread in the Miocene, and local flysch-type turbidite sequences are present. Tectonism at the end of the Miocene caused regionally profound deformation and erosion (Katz, 1973; 1974c, Fig. 24-8). An entirely new basinal pattern evolved in the Pliocene. Along the present coast south of Hawke Bay, a wide belt with many anticlinal structures, some tightly compressed, was uplifted. Farther inland, a long narrow basin filled with thick bathyal siltstones-mudstones (locally including turbidites), and with extensive shelf limestones along its flanks, was rapidly subsiding over a distance of at least 400 km. This basin reaches its greatest depth around Hawke Bay. Offshore to the east, several similar elongate basins developed between narrowly compressed and steeply uplifted Miocene high ridges (e.g., Lachlan ridge; Katz, 1975); these basins were deformed by Pleistocene movements. A wide fold belt of such late Cenozoic structures extends 100 km out to sea where water depth is 3,000 m (Katz, 1974c).

North of Poverty Bay (about 38°30′S) the East Coast fold belt had a profoundly different tectonic history, as is reflected by its structure. Extensive block movements along mainly northwest-southeast faults occurred, with several episodes of concomitant gravity sliding, from Late Cretaceous to late Tertiary time. Drilling results of Opoutama-1, Te Horo-1, and Te Puia-1 (Katz, 1974c) suggest that such mass movements occurred in Eocene-Oligocene time; field mapping has proved large-scale imbrication by gravity sliding of early Miocene (Stoneley, 1968) and late Miocene to early Pliocene age (Ridd, 1968). This movement and the associated widespread late Cenozoic diapiric intrusions and mud-volcano activity (Ridd, 1970) apparently were facilitated by abnormally high pore-fluid pressures in the generally gas-charged montmorillonitic claystones which are known to occur within the Upper Cretaceous to lower Tertiary sequence in this region.

North of East Cape the sedimentary basin drops down to occupy the vast Raukumara plain below more than 2,000 m of water, where strata several kilometers thick are shown by seismic data to be only very gently deformed (Karig, 1970; Katz, 1974a, b).

Undeformed Late Cenozoic Basins

Except for the Wanganui basin, undeformed late Cenozoic basins are found only offshore at water depths of 2,000–4,000 m.

1. Adjacent to the East Coast fold belt, the Hikurangi Trough and Poverty Basin (Fig. 2) are filled with deposits 4.5 km and 3 km thick, respectively—probably mainly turbidites (Katz, 1974b). The Hikurangi Trough strata unconformably overlie Miocene of the Chatham Rise, and thus are Plio-Pleistocene. Tectonically, these basins are considered a late-stage addition or foredeep of the mobile belt; they are comparable to other elongate Plio-Pleistocene basins within it, but so far have remained mainly undeformed.

2. A thick sequence of flat-lying strata (over 2.5 seconds two-way reflection time) in the southernmost part of the South Fiji Basin unconformably overlies but generally sharply abuts a possible fault scarp of the extensive basement platform which stretches far northeast from the Northland peninsula.

3.The New Caledonia Basin is filled with flat-lying and well-stratified units which may be 3,200–3,900 m thick (Bentz, 1974). DSDP hole 206 drilled a 734-m section near 32°S lat., reaching middle Eocene rocks (Burns et al, 1973); this section may not be representative, however, as the location is somewhat off the main basin. DSDP hole 207 on the Lord Howe Rise found 94-m.y.-old rhyolitic basement (McDougall and van der Lingen, 1974) overlain by 357 m of Late Cretaceous to Pleistocene deposits; a significant gap covers late Eocene to middle Miocene time.

4. At about 4,000-m water depth, the Tasman Basin is overlain by horizontally bedded sediments about 2 seconds thick which sharply abut against the tectonically complex, steep and narrow continental slope. The latter appears as acoustic basement on seismic sections, but with indications of intense faulting and folding. It probably is the continuation of a belt of pronounced late Miocene deformation exposed along the coast. Conceivably, the horizontal layers could represent Plio-Pleistocene sediments derived by turbidity currents from the east.

This sedimentary basin coincides with a large negative gravity anomaly which has been interpreted as being due to a thick sequence (>30 km) of crustal rocks downthrown offshore. These would probably in part be composed of sediments derived from the uplifted Fjordland block (Woodward, 1972).

Petroleum Prospects

Source and Reservoir Rocks

Petroleum generation must have been widespread in New Zealand. Potential source rocks are found in virtually all the basins. They are located mainly near the base of the sedimentary sequence (Katz, 1968); active oil seeps occur onshore along the west coast in Taranaki basin at New Plymouth, at Kotuku near Greymouth (Fig. 3), and in the southernmost Tertiary outcrop at Madagascar Beach just north of Milford Sound. The offshore well Maui-4 (Katz, 1971b; Fig. 1), which was abandoned as noncommercial, tested a flow rate of 600 bbl/day. Taranaki, so far the only producing area, presently has proved recoverable reserves of over 6 Tcf of gas and 200 million bbl of condensate. On the east coast of the North Island, oil shows and oil-impregnated sandstones are known in the Dannevirke area south of Hawke Bay (Laing, 1961; Brunstrom, 1962), and good oil seeps are present at Totangi, Waitangi, and Ruatoria in the north. Gas seeps are numerous in many parts of the country. In the east-coast region alone there are over 100, many of which contain higher hydrocarbons as well as methane (de Caen, 1970). Also, the widespread existence of mud volcanoes associated with gas-charged undercompacted shale indicates an environment which was favorable to the generation of gas and perhaps oil (Hedberg, 1974). In the south of the South Island, small Tertiary basins contain minor amounts of good oil shale (Katz, 1968; Williams, 1974).

Reservoir rocks are mainly found in siltstone-sandstone sequences. Sandstones are widespread and thick in most basins but, owing to the tectonic framework, they generally are clay-rich and have poorly sorted and poorly rounded grains; therefore, porosity is low. Most are first-generation sandstones derived directly from a graywacke-schist hinterland. Prolonged winnowing in lowland coastal plains, paralic and nearshore channel and bar environments, and delta build-ups are the most common settings for good sandstone development, such as that in the producing late Eocene Kapuni Formation of Taranaki and correlative formations elsewhere in the western marginal geosyncline. Apparently as a result of pronounced weathering and leaching of previous land surfaces, basal coal measures commonly include good quartzose sandstone, as in the Greymouth area of the South Island or the Canterbury area, where locally thick and very pure, friable quartz sandstones, probably derived from a rhyolitic source, appear particularly attractive (Katz, 1968). The "Little Totara Sand" in the Cape Foulwind area, deposited in a tidal sandbar environment during late Eocene to early Oligocene time (S. Nathan, personal commun.), is another example of extremely clean quartz sandstone; it is 220 m thick near Charleston and interfingers laterally with the Kaiata Mudstone.

Regionally, lenticular sandstones of probably deltaic origin also occur in the Miocene of Taranaki (Mokau Sandstone), but their potential has been little explored. However, they do have some potential as reservoirs. Some oil and gas have been produced from Pliocene sandstones in the Moturoa field at New Plymouth (Fig. 4; Katz, 1974c). Shallow-marine to nearshore sandstones in the Pliocene Wanganui basin—particularly shoal and beach sandstones, shell beds, and other strandline deposits of a high-energy environment—constitute potential reservoirs of comparatively good porosity.

Sandstones on the east coast, North Island, mostly are either of turbidite or shelf type. The latter, which seem to have a greater potential as reservoir rocks, are restricted to the basal sandstones of the Cretaceous (Aptian-Albian) and to the Campanian-Maestrichtian and Miocene (mainly early Miocene) transgressions. Both the transgressive formations which are directly below and above extensive and thick, gas-charged undercompacted shales could attain significance as potential reservoirs in appropriate locations.

Most sandstones are derived from sources relatively poor in quartz, but in a few areas quartz-rich granitic derivation is indicated. One such area is the northwest South Island (Fig. 3), where basement consists of extensive granitic-plutonic and high-grade metamorphic rocks which have been the source for much of the Tertiary sediments. This same complex appears to underlie the offshore area around Cape Farewell and farther north (Wodzicki, 1974), and may have given rise to better sandstone qualities. In addition, possible sediment transport from a rhyolitic terrane with tuffs, agglomerates, ignimbrites, etc, in the west (Lord Howe Rise–Challenger Plateau; Fig. 1) may have contributed to a facies distribution of greater promise.

The shallow basement ridge south of the South Island which, for over 500 km, extends along the western margin of the Campbell Plateau (from Stewart Island to Auckland Island; Fig. 1) consists mainly of mid-Cretaceous granitic rocks (Katz, 1974b). This basement high may have been a sediment source in Late Cretaceous and early Tertiary times for much of the "Great South Basin" (Katz, 1974a), and in the early Tertiary also for the Solander Trough.

Other potential reservoir rocks are the biogenic limestones which occur mainly in the Oligocene. Similar limestones also occur locally in the latest Cretaceous, Paleocene-Eocene, and Miocene, particularly from Canterbury to Southland (Fig. 5); the thick biohermal limestones of early Miocene age in Southland (Invercargill and Waiau basins) appear to constitute a particularly interesting type of reservoir (Katz, 1968).

Perhaps the best limestone reservoirs are present in the Plio-Pleistocene on the east coast of the North Island. These very extensive and thick, mainly coquina-type limestones have good primary intergranular porosity and some vugular porosity with values typically of the order of 20–30%, and permeabilities up to 2 darcys (de Caen and Darley, 1968). They are associated with calcareous siltstones-sandstones and conglomerates, and may pass into mudstone laterally over short distances. Five dry holes have been drilled in this prospect (Leslie and Hollingworth, 1972; Katz, 1974c), although one (Mason Ridge-1) appears to have produced some substantial gas shows from thin sandstone intercalations in the lower Pleistocene (W. C. Leslie, personal commun.).

Trap Conditions

Although both anticlinal and fault-block structures are common in the western marginal rift and the East Coast fold belt, the complex sedimentation pattern resulting in abrupt facies changes as well as stratigraphic gaps is a factor of far-reaching consequences. In the Kapuni field (Fig. 4), the structural trap apparently is enhanced by the occurrence of comparatively favorable deltaic or channel-type sandstone bodies; but the same Kapuni Formation was impermeable and had an excessively high clay content in two wells drilled on other structural closures. The Miocene Mokau Sandstone, another potential reservoir near Kapuni, is entirely replaced by mudstone in the Mangahewa structure.

In the northern updip, land portion of the Wanganui basin, wells drilled on structural highs (Katz, 1968; 1974a) found very reduced, incomplete sequences. Since these structures are caused by drape over buried hills and/or growth of basement highs, sequences are more complete and thicker in the troughs between the highs, and older members actually pinch out against their flanks.

Growth structures, particularly long-lived in the western marginal geosyncline, may have been favorable for early entrapment during primary migration of hydrocarbons; however, they may also have limited the available drainage area, as is possibly the case in the Kapuni field, where the reservoir which is wedged between the Manaia and Taranaki faults (Fig. 4) is only partially filled.

The unfavorable timing of structural-trap formation as related to migration is probably the reason that several good structures in the Greymouth area of the west coast (Brunner anticlinorium, Fig. 3) are empty. All indications point to the fact that oil was generated in the basin (Katz, 1968; 1974c), but early compaction may have caused it to move updip and out of the basin long before the anticlinal updoming along its central axis began. On the east coast of the North Island, some hydrocarbons may have escaped along faults because of repeated basement faulting and diapirism. Alternatively, lack of good reservoir formations and the early sealing of overpressured shale formations may have impeded large amounts of generated hydrocarbons from migrating and concentrating in the first place.

Good stratigraphic-trap conditions do exist. Probably foremost are deltaic and sandbar deposits such as those in the western marginal geosyncline, where they are associated with growth faults. The rapid burial of large quantities of organic material, mainly land-derived, is a particularly attractive situation. Potential sand accumulation in front of, and downdip from, major deltas where broad structures may have formed as a result of gravity tectonics may provide additional targets of interest in a deeper environment; source material most likely would be derived from organic-rich deep-sea sediments. Other attractive prospects may have resulted from the shifting shorelines in the northern updip area of the Wanganui basin and from differential compaction over buried basement hills with concomitant onlap and wedging out of older sequences against the flanks of such hills. Shallow-shelf carbonate platforms, which mainly are composed of bioclastic limestones as in the Pliocene of the Hawke Bay area, and which laterally pass into deep-water mudstone, should provide good stratigraphic-trap conditions, too. Also, the shelf and nearshore siltstone-sandstone deposits formed by repeated marine transgressions on the east coast may be prospective.

Owing to the relative scarcity, on a regional scale, of widespread and uniformly developed reservoir formations, it is obvious that the best chance for petroleum accumulations is in areas where good timing of structural-trap formation coincides with the existence of favorable conditions for stratigraphic traps. However, in basins where little tectonic disturbance has occurred—as in the western offshore areas and in the major downwarp southeast of the South Island—regional unconformities, pinchout, and onlap may have produced favorable stratigraphic-trap conditions. The major clastic wedges, including carbonate rocks, of the marginal geosynclines which developed between the west coast of both islands and the Challenger Plateau, and between the southeast of the South Island and the Campbell Plateau, are therefore considered to have particularly high petroleum potential (Katz, 1974a).

The total area of sedimentary basins with petroleum potential in New Zealand is probably less than 50,000 sq mi (129,500 km^2) on land. Offshore, to an arbitrary depth of 1,000 m, about 100,000 sq mi (259,000 km^2) is prospective.

References Cited

Andrews, J. E., and J. V. Eade, 1973, Structure of the western continental margin, New Zealand, and Challenger Plateau, eastern Tasman Sea: Geol. Soc.

America Bull., v. 84, no. 9, p. 3093–3100.
Austin, P. M., R. C. Sprigg, and J. C. Braithwaite, 1973, Structure and petroleum potential of eastern Chatham Rise, New Zealand: AAPG Bull., v. 57, no. 3, p. 477–497.
Ballance, P. F., 1964, The sedimentology of the Waitemata Group in the Takapuna section, Auckland: New Zealand Jour. Geology and Geophysics, v. 7, no. 3, p. 466–499.
Benson, W. N., 1968, Dunedin district, 1:50,000: New Zealand Geol. Survey Misc. Series Map 1.
Bentz, F. P., 1974, Marine geology of the southern Lord Howe Rise, Southwest Pacific, *in* C. A. Burk and C. L. Drake, eds., The geology of continental margins: New York–Heidelberg–Berlin, Springer-Verlag, p. 537–547.
Brunstrom, R. G. W., 1962, Dannevirke revision survey: BP Shell and Todd Petroleum Development Ltd., Tech. Note 29, New Zealand Geol. Survey Open-File Petroleum Rept., no. 347B.
Burns, R. E., et al. 1973, Initial reports of the Deep Sea Drilling Project: Washington, D.C., U.S. Govt. Printing Office, v. 21.
Cope, R. N., 1965, The Taranaki Basin—a subsurface review: Shell BP and Todd Oil Services Ltd., New Zealand Geol. Survey Open-File Petroleum Rept., no. 456.
de Caen, R. F. B., 1970, Register of oil and gas manifestations, Hawkes Bay, east coast, North Island, New Zealand: BP Shell Aquitaine and Todd Petroleum Development Ltd., New Zealand Geol. Survey Open-File Petroleum Rept., no. 379.
——— and J. H. Darley, 1968, An examination of Wanganui carbonates, Hawkes Bay, North Island, New Zealand: BP Shell Aquitaine and Todd Petroleum Development Ltd., New Zealand Geol. Survey Open-File Petroleum Rept., no. 382.
El Paso Natural Gas Co., 1973, Marine seismic survey, P.P.L. 800: New Zealand Geol. Survey Open-File Petroleum Rept., no. 623.
Esso Exploration and Production N.Z. Inc., 1969, Final Report of 1968 and 1969 marine seismic surveys, Licence 711: New Zealand Geol. Survey, Open-File Petroleum Rept., no. 401.
Fleming, C. A., 1962, New Zealand biogeography: Tuatara, v. 10, no. 2, p. 53–108.
——— 1970, The Mesozoic of New Zealand—chapters in the history of the Circum-Pacific mobile belt: Geol. Soc. London Quart. Jour., v. 125, p. 723–751.
Fyfe, H. E., 1968, The geology of Murchison subdivision: New Zealand Geol. Survey Bull. 36, 52 p.
Gage, M., 1952, The Greymouth coalfield: New Zealand Geol. Survey Bull. 45, 240 p.
——— 1957, The geology of Waitaki subdivision: New Zealand Geol. Survey Bull. 55, 135 p.
——— 1970, Late Cretaceous and Tertiary rocks of Broken River, Canterbury: New Zealand Jour. Geology and Geophysics, v. 13, no. 2, p. 507–559.
Glennie, K. W., 1959, The graded sediments of the Mahoenui Formation (King country, North Island): New Zealand Jour. Geology and Geophysics, v. 2, no. 3, p. 613–621.
Harrington, H. J., 1958, Geology of Kaitangata coalfield: New Zealand Geol. Survey Bull. 59, 131 p.
——— and B. L. Wood, 1958, Andesitic volcanism at the Solander Islands: New Zealand Jour. Geology and Geophysics, v. 1, no. 3, p. 419–431.
Hay, R. F., A. R. Mutch, and W. A. Watters, 1970, Geology of the Chatham Islands: New Zealand Geol. Survey Bull. 83, 86 p.
Hedberg, H. D., 1974, Relation of methane generation to undercompacted shales, shale diapirs, and mud volcanoes: AAPG Bull., v. 58, no. 4, p. 661–673.
Hematite Petroleum (N.Z.) Ltd., 1969, Seismic survey Karamea (Licence 711): New Zealand Geol. Survey, Open-File Petroleum Rept., no. 555.
Hulston, J. R., and W. J. McCabe, 1972, New Zealand potassium-argon age list-1: New Zealand Jour. Geology and Geophysics, v. 15, no. 3, p. 406–432.
Karig, D. E., 1970, Kermadec arc–New Zealand tectonic confluence: New Zealand Jour. Geology and Geophysics, v. 13, no. 1, p. 21–29.
Katz, H. R., 1968, Potential oil formations in New Zealand, and their stratigraphic position as related to basin evolution: New Zealand Jour. Geology and Geophysics, v. 11, no. 5, p. 1077–1133.
——— 1971a, Oil exploration in New Zealand—past and future trends: APEA Jour., v. 11, pt. 1, p. 35–42.
——— 1971b, Petroleum development in New Zealand during 1970: AAPG Bull., v. 55, no. 9, p. 1657–1661.
——— 1973, Pliocene unconformity at Opau Stream, Hawkes Bay, New Zealand: New Zealand Jour. Geology and Geophysics, v. 16, no. 4, p. 917–925.
——— 1974a, Offshore petroleum potential in New Zealand: APEA Jour., v. 14, pt. 1, p. 3–13.
——— 1974b, Margins of the Southwest Pacific, *in* C. A. Burk and C. L. Drake, eds., The geology of continental margins: New York–Heidelberg–Berlin, Springer-Verlag, p. 549–565.
——— 1974c, Recent exploration for oil and gas, *in* G. Williams, ed., Economic geology of New Zealand: Australasian Inst. Mining and Metallurgy Mon. Series 4, chap. 24, p. 463–480.
——— 1975, Ariel bank off Gisborne—an offshore late Cenozoic structure and the problem of acoustic basement on the east coast, North Island, New Zealand: New Zealand Jour. Geology and Geophysics, v. 18, no. 1, p. 93–107.
Kear, D., and B. C. Waterhouse, 1967, Onerahi Chaos—breccia of Northland: New Zealand Jour. Geology and Geophysics, v. 10, no. 3, p. 629–646.
Laing, A. C. M., 1961, The geology and petroleum prospects of the Dannevirke area: BP Shell and Todd Petroleum Development Ltd., New Zealand Geol. Survey Open-File Petroleum Rept., no. 318.
Laird, M. G., 1968, The Paparoa tectonic zone: New Zealand Jour. Geology and Geophysics, v. 11, no. 2, p. 435–454.
Leslie, W. C., and R. J. S. Hollingsworth, 1972, Exploration in the East Coast basin, New Zealand: APEA Jour., v. 12, p. 1, p. 39–44.
McDougall, I., and G. J. van der Lingen, 1974, Age of the rhyolites of the Lord Howe Rise and the evolution of the Southwest Pacific Ocean: Earth and Plan-

etary Sci. Letters, v. 21, p. 117–126.
Norris, G., 1968, Plant microfossils from the Hawks Crag Breccia, south-west Nelson, New Zealand: New Zealand Jour. Geology and Geophysics, v. 11, no. 2, p. 312–344.
——— and J. B. Waterhouse, 1970, Age of the Hawks Crag Breccia: Royal Soc. New Zealand Trans., v. 7, no. 13, p. 241–250.
O'Halloran, N. G., 1971, Interpretation report, marine seismic reflection survey within Licence Area 878, Chatham Rise, New Zealand: BP Petroleum Development Ltd., Tech. Note 354; New Zealand Geol. Survey Open-File Petroleum Rept., no. 301A.
Pinchon, D., 1973, Te Popo seismic survey interpretation report: New Zealand Aquitaine Petroleum Ltd., New Zealand Geol. Survey, Open-File Petroleum Rept., no. 618c.
Ridd, M. F., 1968, Gravity gliding on the Raukumara Peninsula: New Zealand Jour. Geology and Geophysics, v. 11, no. 2, p. 547–548.
——— 1970, Mud volcanoes in New Zealand: AAPG Bull., v. 54, no. 4, p. 601–616.
Speden, I. G., 1973, Distribution, stratigraphy and stratigraphic relationships of Cretaceous sediments, western Raukumara Peninsula, New Zealand: New Zealand Jour. Geology and Geophysics, v. 16, no. 2, p. 243–268.
——— 1975, Cretaceous stratigraphy of Raukumara Peninsula: New Zealand Geol. Survey Bull. 91, 70 p.
Stoneley, R., 1968, A lower Tertiary décollement on the east coast, North Island, New Zealand: New Zealand Jour. Geology and Geophysics, v. 11, no. 1, p. 128–156.
Wellman, H. W., 1953, The geology of Geraldine subdivision: New Zealand Geol. Survey Bull. 50, 72 p.
——— 1956, Structural outline of New Zealand: New Zealand Dept. Sci. and Indus. Research Bull. 121, 36 p.
Williams, G. J., 1974, Cenozoic geology of the Lower Nevis basin with special reference to oil shale deposits: New Zealand Dept. Sci. and Indus. Research Bull. 212, 36 p.
Wilson, D. D., 1963, Geology of Waipara subdivision: New Zealand Geol. Survey Bull. 64, 122 p.
Wodzicki, A., 1974, Geology of the pre-Cenozoic basement of the Taranaki–Cook Strait–Westland area, New Zealand, based on recent drillhole data: New Zealand Jour. Geology and Geophysics, v. 17, no. 4, p. 747–757.
Woodward, D. J., 1972, Gravity anomalies in Fjordland, south-west New Zealand: New Zealand Jour. Geology and Geophysics, v. 15, no. 1, p. 22–32.

Phanerozoic History of Australia and Surrounding Oceanic Region—Summary[1]

GORDON PACKHAM[2]

The Tasman orogenic zone has been accreted onto the eastern side of a Precambrian shield. Superimposed on these entities is a series of transitional to postorogenic basins.

The Tasman orogenic zone, which was mobile from Cambrian to Early Cretaceous time, underwent progressive but stepwise cratonization with culminations of orogenesis in the Early Ordovician, Middle Devonian, Early Permian, and Early Cretaceous. Dispersed relicts of this orogenic zone crop out in New Zealand and New Caledonia, and are inferred in the foundered continental blocks of the Queensland plateau, Lord Howe Rise, Norfolk Ridge, and the Campbell Plateau. The Tasman orogenic zone differs considerably from the classic model. The tectonism can be explained generally by the presence of an active plate boundary between a "Paleo-Pacific" oceanic plate and a continental plate. Mountain building has been on a modest scale, and no great crustal thicknesses have been developed.

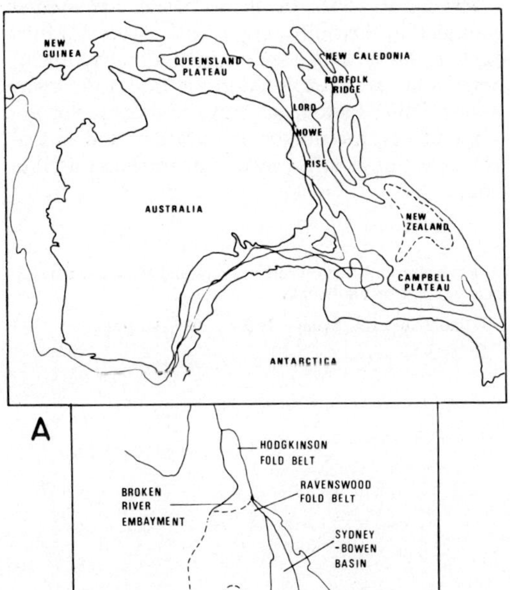

FIG. 1—**A.** Reconstruction of Australia, Lord Howe Rise, and Antarctica prior to Late Cretaceous rifting. **B.** Main fold belts of Tasman orogenic zone in eastern Australia.

The sediment filling the various basins is mainly clastic and, in the early and middle Paleozoic at least, had a predominantly southerly source. Antarctica must have been a significant contributor, and probably remained so even in the Permian and early Mesozoic when elevation of parts of the eastern highlands took place. Lower Paleozoic carbonate rocks are more abundant in the north of the continent.

Fragmentation of this part of Gondwanaland commenced in the west in the Late Jurassic to Early Cretaceous with the formation of the Wharton basin. In the east, the Tasman Sea started to form in Late Cretaceous time. Although an incipient rift formed across southern Australia in the middle Mesozoic, new seafloor formation and northward motion of Australia away from Antarctica did not commence until the Paleocene. Eastward accretion of oceanic crust took place in the Eocene and the Oligocene (the Coral Sea and the South Fiji Basin). A phase of plate convergence in the Oligocene resulted in overthrusts of oceanic crust. Relocation of the Pacific plate boundary inside the Melanesian arcs—possibly as a result of changes in the relative motion of the Pacific and Indian plates and the influence of the Ontong Java plateau—caused some of the seafloor formed during extension to be consumed from the Miocene onward. Folding and uplift in the island of New Guinea also took place in the later Tertiary and Quaternary.

The continental shelf and marginal sedimentary basins containing thick sediment accumulations locally have formed as a response to the process of rifting.

[1]Manuscript received, December 6, 1974.

[2]Department of Geology and Geophysics, University of Sydney, N.S.W. 2006, Australia.

Structural Style and Hydrocarbons of Bass, Gippsland, and Otway Basins—Summary[1]

E. A. HODGSON and W. F. THRELFALL[2]

The Bass, Gippsland, and Otway basins lie mainly offshore southeast of Australia. Offshore fields in the Gippsland basin contain recoverable reserves of approximately 8.0 Tcf of gas and 1.7 billion bbl of oil, and currently supply two thirds of Australia's oil requirements. No commercial hydrocarbons have been discovered in the adjacent Bass and Otway basins.

The basins were created during the Jurassic to mid-Cretaceous continental breakup of Antarctica, Australia, and New Zealand. Their initial tectonic framework was primarily tensional, and basins were produced by normal faulting.

The Otway basin has a classic "pull-apart" structural style with a series of normal faults parallel with the spreading axis. The Gippsland and Bass basins are similar, but their complete separation was prevented by the influence of the Tasmanian continental block (subplate). The separation of New Zealand from Australia had only a minor effect on the structural framework of the Gippsland basin in the shallow slope areas.

Following the continental breakup and margin collapse, the Otway basin remained structurally quiescent, but the Bass and Gippsland basins continued to subside and accumulate thick Tertiary sections. However, during the late Eocene and Miocene, all basins were subjected to an east-west right-lateral shear deformation. In the Gippsland basin, the shearing generated large en-echelon anticlines and rejuvenated some of the old tensional faults. Hydrocarbons accumulated in these anticlines or their eroded remnants. In the Bass and Otway basins the shear was less severe and was confined to the northern margins. The deformation in these basins was correspondingly less intense.

Structural styles of these basins are similar enough that their differing intensity does not fully explain the lack of success in finding hydrocarbons in the areas explored in the Bass and Otway basins. Other factors, such as the presence or absence of organic matter, maturation, and migration, must have influenced hydrocarbon distribution.

[1]Revised from Circum-Pacific Energy and Mineral Resources Conference program booklet.

[2]Exxon Australia, Sydney, N.S.W. 2000, Australia.

Australian Northwest Continental Shelf—Results of Ten Years of Exploration[1]

G. A. ROBERTSON, D. E. POWELL, and G. M. EDMOND[2]

Abstract The Woodside-Burmah Group had drilled 45 wells, totaling 469,692 ft (143,155 m), in this remote offshore region to the end of 1973. Ten of the 37 wildcat wells discovered new fields, and all 8 appraisal wells were successful. Permit areas total 140,000 sq mi (360,000 km²) and water depths are from 150 to 6,000 ft (46–1,830 m).

Major tectonic elements include three essentially onshore Proterozoic cratons separated by the Canning and Bonaparte Gulf sedimentary basins. The sedimentary basins cover both onshore and offshore areas. Other major depocenters—the Dampier and Beagle subbasins and the Browse basin—lie entirely offshore from the cratonic blocks. All five offshore basins are filled with thick Mesozoic to recent sediments. The basins probably formed as a result of continental rifting initiated during Permian time. Many of the basins are subdivided as a result of major block-faulting movements which occurred from Late Triassic to Middle Jurassic time.

Permian and pre-Permian strata have been penetrated in only a few places. Thick marine Early Triassic shales are succeeded by later Triassic to Middle Jurassic fluviodeltaic units. Late Jurassic to Early Cretaceous transgressive marine shales and sandstones cover an irregular block-faulted paleotopography in the Dampier and Beagle subbasins, forming important structural-stratigraphic traps. Early Late Cretaceous fine clastic units grade through later Cretaceous to early Tertiary marlstones and calcilutites to coarse-grained carbonate rocks in the post-Oligocene section. Several Cretaceous and Tertiary unconformities are present.

Recoverable gas reserves found in the Dampier subbasin to date total approximately 18 Tcf. About 13.5 Tcf of the reserves are in the proved and probable categories; the giant North Rankin field alone contains 7.9 Tcf. Another potentially large gas-condensate discovery has been made in the Browse basin.

INTRODUCTION

Discoveries of major gas and condensate fields by the Woodside-Burmah Group have focused increasing attention on the Australian northwest continental shelf. This paper summarizes the results of 10 years of exploration of offshore concessions comprising over 140,000 sq mi (360,000 km²) in total area (Fig. 1). Water depths in the region range from 150 ft (46 m) in nearshore areas to over 6,000 ft (1,830 m) on the continental rise, although a few shoal-water reefs occur near the shelf edge. To the north, the Australian shelf is separated from the island of Timor by the Timor Trough, which is 10,000 ft (3,050 m) deep.

An ambitious exploration program has been carried out on the shelf despite some severe operational problems. The Australian northwest shelf is geographically remote. Western Australia is an area of approximately 1 million sq mi (2.6 million km²) with a population of just over 1 million people. About 700,000 of these are concentrated in the Perth region, about 800 mi (1,290 km) south of the nearest part of the Group's concessions. The sparsely populated coastal areas and hinterland, coupled with the enormous lateral spread of the isolated concession areas, pose considerable supply and communication problems.

Further obstacles are the lack of harbors along the coast and the large tidal range of up to 35 ft (10.7 m). Broome was selected as the principal supply base owing to its good harbor and jetty, and its central location (Fig. 1) with respect to Group permits. Tiny remote islands must be employed as helicopter refueling stations and navigation control bases, and currently the larger island of Timor is used as an intermediate transit location for drilling operations in the Bonaparte Gulf basin, in the north of the concession areas.

Geophysical exploration commenced in 1963 with a limited reconnaissance aeromagnetic survey which indicated the presence of a sedimentary thickness sufficient for petroleum exploration. Seismic survey work was begun in 1964 and was intensified after 1967, when data quality was dramatically improved by the use of digital recording and other new techniques. Approximately 67,350 line-mi (108,390 km) of seismic coverage is in the Group's possession.

The first well—Ashmore Reef No. 1—was spudded in late 1967, and 45 wells had been drilled by the end of 1973 (Fig. 2). Of these, 37 were exploratory and 8 were appraisal wells. By the end of 1973, a total of 469,692 ft (143,155 m) had been drilled.

Exploration group partners are Woodside-Burmah Oil N.L., 50%; and Shell, B.P., and Cal-Asiatic, 16⅔% each. North of 13°30′S lat. Cal-Asiatic has no interest, and Shell's interest is 33⅓%.

STRUCTURAL FRAMEWORK

The onshore geology of northwestern Australia is dominated by three large Proterozoic cratons or blocks (Fig. 3). From south to north they are

[1]Manuscript received, November 6, 1974.

[2]B.O.C. of Australia Limited, Perth, Western Australia.

The writers thank the management of B.O.C. of Australia Limited and associated Group companies for permission to present this paper. They also acknowledge assistance received from many colleagues in preparation of the manuscript.

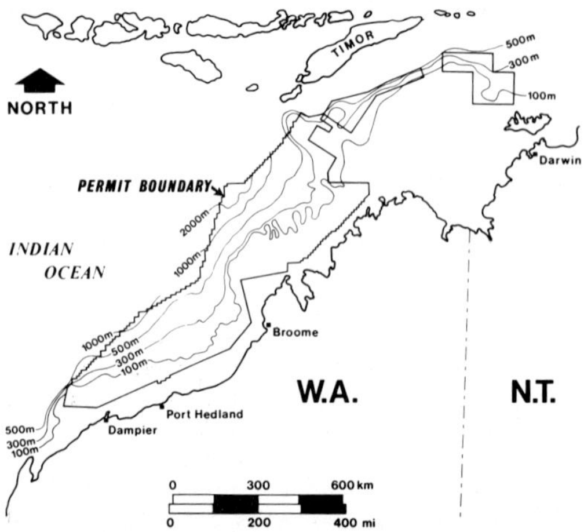

FIG. 1–Concession area and water depths in meters, northwest continental shelf, Australia.

the Pilbara, Kimberley, and Sturt blocks. These blocks have exerted strong control on the tectonic history of the region, and have provided much of the younger sediment deposited in the offshore areas.

Onshore, between the Proterozoic blocks, lie two large, essentially Paleozoic basins—the Canning and Bonaparte Gulf basins (Fig. 3). Offshore, the Paleozoic rocks plunge steeply and are overlain by thick Mesozoic to Holocene strata deposited in several major structural downwarps. The downwarps within the Group's concessions are, from southwest to northeast, the Dampier subbasin, the Beagle subbasin, the offshore Canning basin, the Browse basin, and the offshore Bonaparte Gulf basin (Fig. 3). All probably result from continental rifting initiated during Permian time.

Dampier Subbasin

This major sedimentary downwarp, containing at least 20,000 ft (6,100 m) of mostly Mesozoic to Holocene deposits, is a northeastern offshore extension of the Carnarvon basin. It grades southwestward into the Barrow subbasin with no obvious boundary.

The tectonically complex subbasin includes several prominent positive trends aligned parallel with the present coast. The outermost, known as the Rankin platform, is a large gravity-positive feature composed of a series of upfaulted blocks of Triassic to Middle Jurassic rocks unconformably overlain by a blanket of thin Cretaceous and thick Cenozoic deposits. Cretaceous shales are draped directly over the blocks and provide both top and lateral seals for major hydrocarbon accumulations.

Other prominent positive trends inshore from the Rankin platform are the Madeleine-Dampier

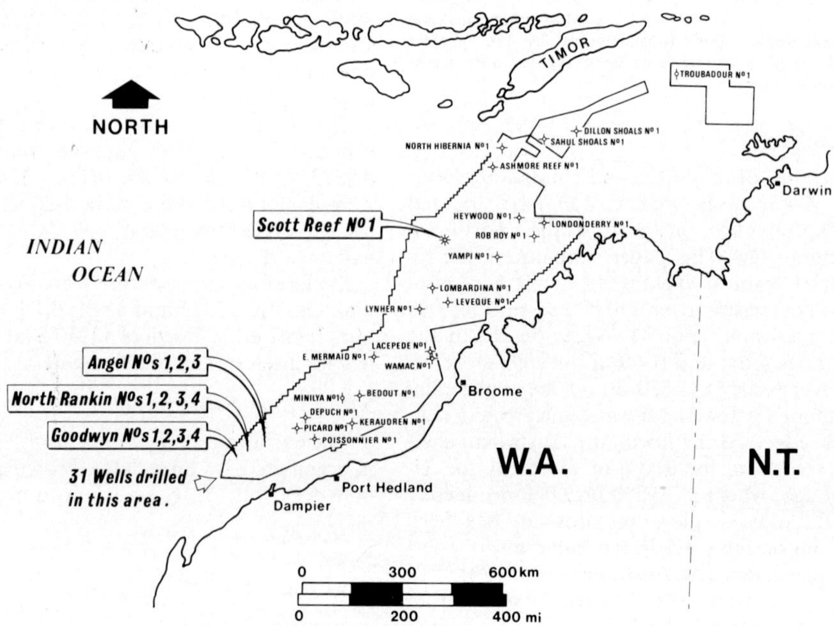

FIG. 2–Well locations.

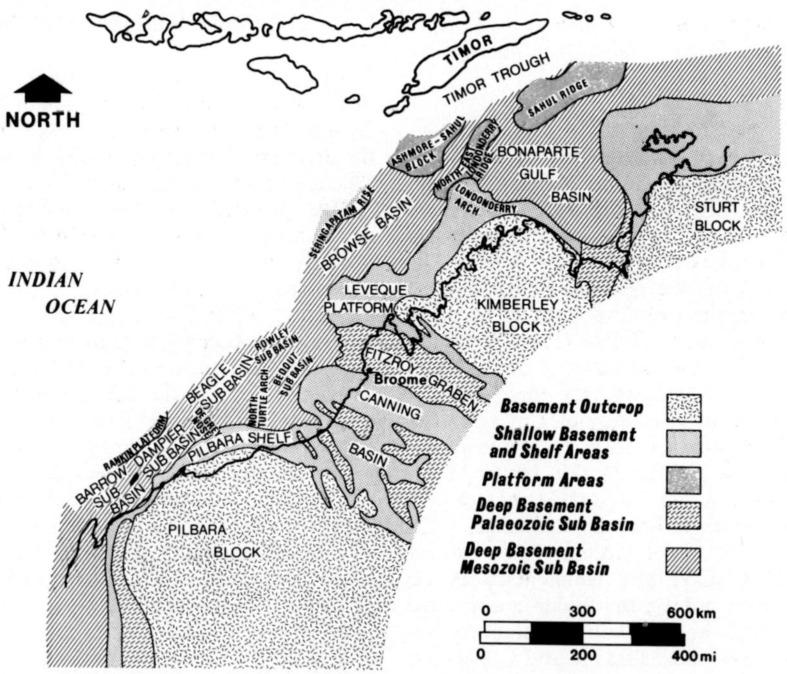

FIG. 3–Main tectonic elements, northwest continental shelf, Australia.

trend, which includes the important Angel field; the Rosemary-Legendre trend, and the complexly faulted Enderby trend. Inshore from the Rankin and Madeleine-Dampier trends, respectively, lie the faulted Kendrew trough and the simpler, deep but narrow Lewis trough.

The Dampier subbasin is separated from the Beagle subbasin by the De Grey nose (Fig. 3), an offshore extension of the Pilbara shelf.

Beagle Subbasin

This subbasin is geologically similar to the Dampier subbasin, and is therefore regarded as the northernmost extension of the Carnarvon basin. Like the Dampier subbasin, it is a typical linear "pull-apart" feature of Mesozoic tensional origin, lying adjacent to a cratonic margin. A thick, faulted Mesozoic sequence is present within the basin, and is overlain by a seaward-thickening Tertiary sedimentary wedge. Paleozoic rocks are almost certainly present at depth. An outer margin of the subbasin has not yet been defined; seismic evidence suggests that the sedimentary wedge continues to thicken northwest toward the abyssal depths of the Indian Ocean.

From northwest to southeast, important tectonic elements of the Beagle subbasin include an outer platform area which is similar in many ways to the Rankin platform, a major line of positive structures known as the Picard trend, and the Beagle trough. Southeast of the latter, a wide area referred to broadly as the southern Beagle subbasin extends shoreward to the Pilbara shelf, and is intensely faulted at depth. A complex faulted hinge line, the North Turtle arch, separates the Beagle subbasin from the Canning basin (Fig. 3).

Offshore Canning Basin

This large basin includes three major interrelated depocenters: the Bedout and Rowley subbasins and the offshore Fitzroy graben (Fig. 3).

Structurally simpler than the adjacent Beagle subbasin, the Bedout subbasin is the offshore extension of an onshore Paleozoic intracratonic basin and contains Mesozoic and Tertiary rocks up to 16,500 ft (5,030 m) thick. The Rowley subbasin constitutes a major depositional wedge of Mesozoic and Tertiary rocks up to 23,000 ft (7,010 m) thick in the area of the Rowley Shoals. The northwest-southeast-trending Fitzroy graben is structurally complex, since Late Triassic folding was accompanied by extensive faulting. Onshore, the sedimentary section comprises mainly Paleozoic and Triassic rocks, but in the offshore extension of the graben a thin Mesozoic sequence is covered by seaward-thickening Tertiary units. The broad Leveque platform and shelf (Fig. 3) separate the offshore Fitzroy graben from the Browse basin.

Browse Basin

Both structure and stratigraphy of this large (approximately 20,000 sq mi or 51,800 km^2) sedimentary basin (Fig. 3), lying entirely offshore from the Kimberley block, were entirely unknown until the present phase of oil exploration. Limited seaward by the Seringapatam rise (Fig. 3), the basin contains over 35,000 ft (10,700 m) of Permo-Carboniferous to Holocene rocks.

Structurally, the Browse basin is divided into four main tectonic elements: the inner margin, basin shelf, basin proper, and outer margin. The inner margin is characterized by shallow pre-Permian basement overlain by thin Upper Mesozoic and Tertiary strata. The shelf is an area with a somewhat thicker sedimentary sequence, limited basinward by major normal faults downthrown to the basin. A thick sedimentary section typifies the basin proper; the folded and faulted pre-Jurassic rocks are overlain by relatively undisturbed younger units. Jurassic and Cretaceous strata are believed to thin gradually toward the outer margin of the basin, whereas Tertiary strata continue to thicken.

Offshore Bonaparte Gulf Basin

This elongate northwest-southeast-trending basin, bounded seaward by the Timor Trough (Fig. 3), was formed by early Paleozoic rifting between the Sturt and Kimberley blocks. It contains thick Paleozoic and younger rocks.

The present dominant northeast-southwest structural grain was imposed in the Late Triassic to Early Jurassic by block faulting caused by incipient continental breakup. The Ashmore-Sahul block and North-East Londonderry ridge, which separate the Bonaparte Gulf and Browse basins, were emergent during the Late Jurassic and Early Cretaceous, as was the Sahul ridge (Fig. 3); adjacent negative areas received up to 4,000 ft (1,220 m) of fine marine clastic fill. Coarser sediments were also deposited farther south in the basin proper. The final phase of continental breakup, which occurred in the Neocomian, was followed by a broad regional subsidence. Late Cretaceous shales and Tertiary carbonate rocks then prograded across the shelf.

STRATIGRAPHY

Figure 4 shows generalized composite columns for the five depositional areas. Within a region the size of the Australian northwest shelf there

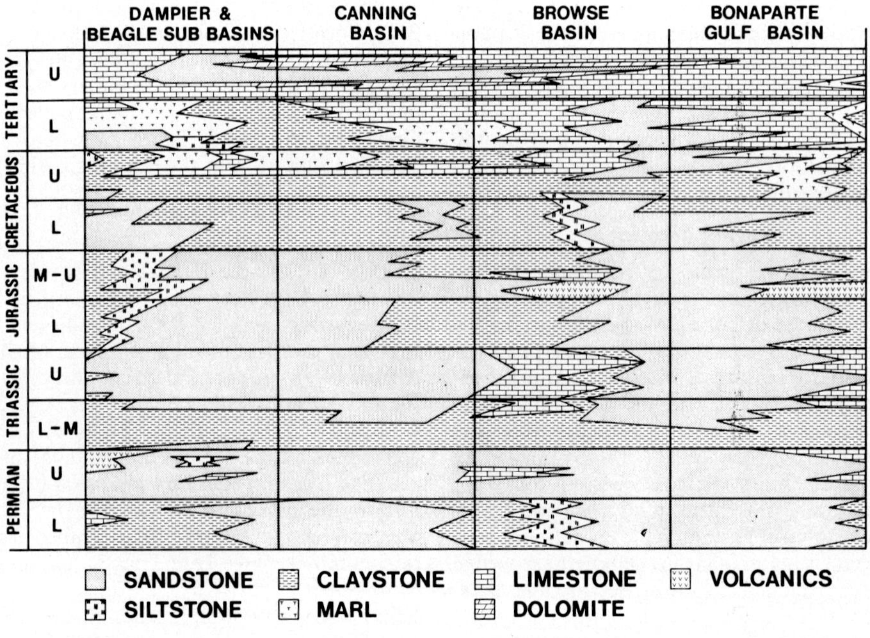

FIG. 4–Generalized stratigraphic columns for Australian northwest shelf basins.

are numerous facies variations. However, certain broad patterns of post-Permian sedimentary history which are common to all the basins are apparent.

Permian and Pre-Permian

Permian and pre-Permian rocks are deeply buried and thus have been penetrated in only a few places. Predominantly clastic units—mainly nonmarine sandstones and shales—have been penetrated to date.

Triassic

Over much of the northwest shelf, sedimentation appears to have continued from Permian through Triassic times without a major tectonic break. Where encountered, the Lower Triassic is represented by a thick marine sequence comprising mainly shales, the result of a major transgression. This transgression was replaced later in the Early Triassic by gradual regression, which caused a change to predominantly nonmarine sedimentation.

Upper Middle to Upper Triassic sediments of the Dampier subbasin are over 5,000 ft (1,500 m) thick, comprising fluviodeltaic interbedded sandstones and claystones. Many of the sandstones have high porosities and permeabilities, and assume great importance as the principal reservoirs for the large hydrocarbon accumulations of the Rankin platform.

Thick Upper Triassic fluviodeltaic clastic rocks are also present in the other basins. In addition, some marginal-marine sedimentary units are present in outer parts of the Browse and Bonaparte Gulf basins.

Jurassic

There is no major stratigraphic break between Triassic and Jurassic rocks of the Dampier and Beagle subbasins. On the Rankin platform the Lower to Middle Jurassic is represented by a claystone-siltsone sequence totaling more than 3,300 ft (1,000 m) in thickness. Elsewhere in these subbasins, thick fluviodeltaic sandstones are present.

In the offshore Canning, the Browse, and the offshore Bonaparte Gulf basins, epeirogenic movements toward the end of the Triassic were followed by deposition of a thin Lower Jurassic redbed sequence, apparently of wide areal extent. Thick Lower to Middle Jurassic strata above the redbeds consist of fluviodeltaic channel and bar sandstones with minor claystones, siltstones, and some coal. These are over 3,500 ft (1,070 m) thick in central parts of the Browse and Bonaparte Gulf basins, but they are absent near cratonic margins and thin or absent near the seaward margins of these basins.

Late Middle Jurassic tensional movements, the precursors of a major continental breakup, were particularly intense in the Dampier subbasin and, to a lesser extent, the Beagle subbasin. In these areas the ensuing major Late Jurassic to Early Cretaceous transgression took place over a highly irregular, block-faulted paleosurface, and the post-tectonic strata thus commonly exhibit marked drape over the more positive blocks.

In all basins the initial Upper Jurassic transgressive units are mainly marine shales, although important sandy intervals occur, particularly near basin margins and toward the top of these sequences. The thick shales are rich in organic material and probably constitute a major hydrocarbon source. Many Upper Jurassic sandstones have excellent reservoir properties, as in the Angel gas-condensate field in the Dampier subbasin.

Cretaceous

Cretaceous rocks have maximum thicknesses of about 3,300 ft (1,000 m) in the Dampier, Beagle, and Canning basins, and up to 9,000 ft (2,740 m) in central parts of the Browse and Bonaparte Gulf basins.

Contemporaneous deposition of marine muds in the basin centers and shallow marine sands toward their margins continued from Late Jurassic to Early Cretaceous time. In the Dampier and Beagle subbasins, many of the high areas resulting from the late Middle Jurassic block faulting remained emergent throughout the Late Jurassic and were finally transgressed during the Early Cretaceous. In some areas Lower Cretaceous shales lie directly on the Upper Triassic. Similarly, the Ashmore-Sahul block, at the northwest margin of the Bonaparte Gulf basin (Fig. 3), remained emergent during Late Jurassic to earliest Cretaceous time and was finally transgressed during the late Early Cretaceous. Widespread unconformities during the Neocomian and Albian-Aptian are important at the inner margins of all the basins but have little effect in deeper basinal areas.

During the early Late Cretaceous, thick marine clay sedimentation continued over almost the entire shelf; the only exception was minor marine to fluviodeltaic sands deposited locally toward basin margins. The first evidence of the gradual transition from predominantly clastic to predominantly carbonate sedimentation is the development of a calcilutite facies (Toolonga Calcilutite) during the early Senonian, above a slight Cenomanian-Turonian unconformity. Uppermost Cretaceous rocks are entirely marine, grading from deep-water calcilutites and marlstones to shallower water claystones and siltstones.

Tertiary and Quaternary

Cenozoic sediments form a seaward-thickening wedge, 8,000 ft (2,440 m) thick at the outer limits of the main depositional areas and over 10,000 ft (3,050 m) thick in the Browse basin. This massive blanket of sediment was deposited everywhere but the northwestern Browse basin. Except for deep-water areas within that basin, it is largely unaffected by the earlier structural configuration.

Most of the Tertiary and Quaternary rocks of the northwest shelf are carbonate rocks. Calcilutites and marlstones characterize the lower Tertiary; some Paleocene and lower Eocene claystones are present in the Dampier and Beagle subbasins. Calcilutites and calcisiltites accumulated in the Dampier and Beagle subbasins during the Oligocene, but in the three northern basins there was a reversion to clastic sedimentation, as is characterized by marlstones and minor claystones. A widespread unconformity occurs at the base of the Miocene; above it, calcarenite is predominant in all basins.

Hydrocarbon Discoveries

Although the early discovery of oil at the Legendre No. 1 in the Dampier subbasin was encouraging, the find was noncommercial. A flow of over 1,000 bbl/day was recorded, but the thin Lower Cretaceous sandstone reservoir is of limited lateral extent.

Scott Reef

The first potentially large hydrocarbon discovery was made in mid-1971 at the Scott Reef No. 1. This extremely remote well, located within the Scott Reef Atoll about 260 mi (418 km) north of Broome (Fig. 2), bottomed in the Triassic at 15,520 ft (4,730 m). A test of Lower to Middle Jurassic sandstones at 14,100 ft (4,298 m) produced a gas flow of 18,200 Mcf/day through a ½-in. choke; the gas was accompanied by condensate at the rate of 20 bbl/MMcf. A test of the underlying Triassic sequence at 14,370 ft (4,380 m) yielded gas at 9,800 Mcf/day through a ⅜-in. subsurface choke; condensate again flowed at the same rate. In both cases the condensate gravity was 54° API. The well proved an effective net "pay" of 91 ft (28 m) but did not penetrate a hydrocarbon-water contact. In the Scott Reef No. 1, a marked angular unconformity is present between the Triassic and Jurassic.

The economic potential of the Scott Reef structure could be very high. A large area of closure is possible, and the productive Lower to Middle Jurassic section is expected to thicken downflank. However, this economic potential depends on the feasibility of establishing an LNG plant and associated storage facilities within the atoll area, and on government permission to export the gas. Additional wells are required to evaluate both the areal and vertical extent of the hydrocarbon column at Scott Reef, so that no reasonable estimate of reserves can yet be given.

North Rankin

The first major discovery in the Dampier subbasin—the North Rankin No. 1 (Fig. 5)—was also made in mid-1971. This wildcat was located in 400 ft (122 m) of water over a large Triassic fault block, one of many similar features which together form the Rankin platform.

The well penetrated a thick Tertiary section and a thin Cretaceous section and reached a total depth of 11,593 ft (3,534 m) in a Triassic sandstone-shale sequence. The gross Triassic hydrocarbon column totaled 1,852 ft (564 m), and sandstones within this section exhibit excellent reservoir characteristics. Porosities average more than 20%, and permeabilities are as high as 2,200 md. Gas flow rates through ½-in. subsurface chokes ranged between 7,000 and 11,420 Mcf/day, and all gas flows were accompanied by 52° API gravity condensate at the average rate of 28 bbl/MMcf.

The North Rankin Nos. 2 and 3 were drilled to evaluate the main fault block. Beneath the caprock of unconformable Lower Cretaceous shales, both wells penetrated a Lower Jurassic sandstone-shale sequence lying conformably on Triassic rocks. Net "pay" of Triassic and Lower Jurassic sandstones totals 1,255 ft (383 m) at well No. 2 and 321 ft (98 m) at well No. 3. The presence of a near-common water line was confirmed, and the excellent reservoir characteristics in the discovery well were found to be laterally extensive. A drill-stem test in the interval 10,145–10,296 ft (3,092–3,138 m) produced a gas flow of 28,000 Mcf/day through a ¾-in choke, accompanied by condensate at the rate of 28 bbl/MMcf. This is the largest gas flow recorded on the Australian northwest shelf.

The North Rankin No. 4, located 1¼ mi (2.0 km) west of the discovery well, tested a downthrown block adjacent to the main North Rankin compartment. The well bottomed in Lower Juras-

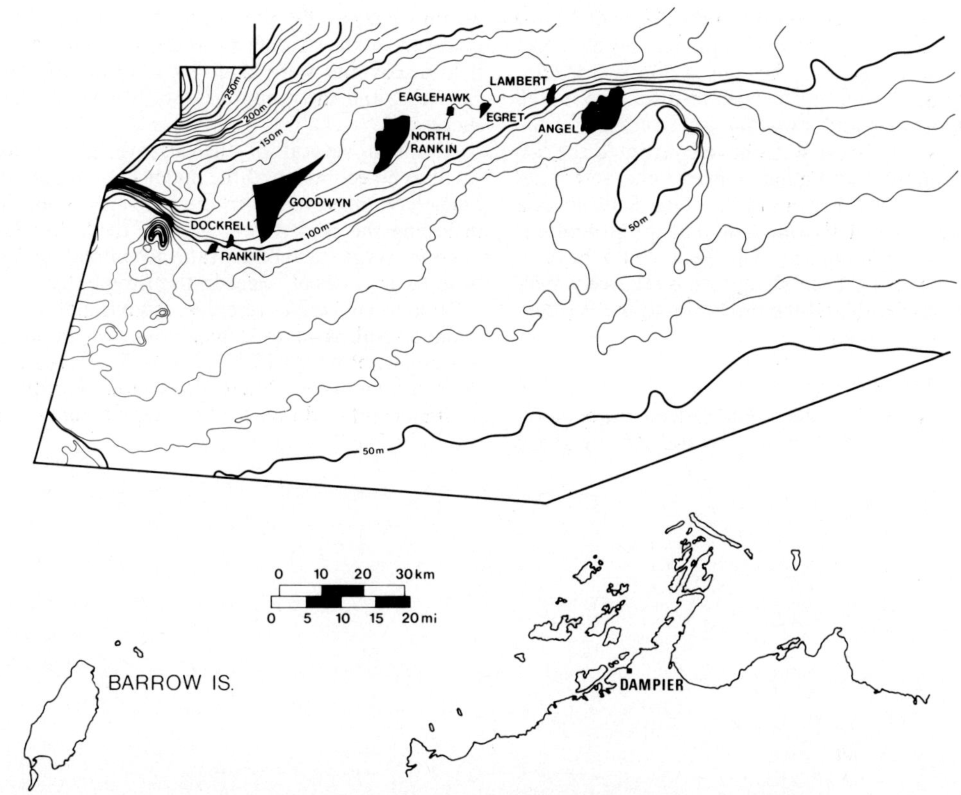

FIG. 5—Dampier subbasin water depth and hydrocarbon discoveries.

sic rocks at 13,328 ft (4,062 m), having penetrated a thick Middle Jurassic section not encountered in the main fault block. This section consists predominantly of shales, but sandstones in the interval 9,766–9,836 ft (2,977–2,998 m) produced gas at 7,200 Mcf/day during a drill-stem test.

Reservoir engineering studies indicate that North Rankin can be classed as a giant gas field by world standards; it contains proved and probable recoverable reserves of at least 7.9 Tcf of gas.

Goodwyn

The Goodwyn fault block (Fig. 5) is a Rankin platform feature which is geologically similar to the North Rankin field. Four Goodwyn wells have have established further significant gas-condensate reserves. In addition, a southern fault compartment of the main Goodwyn block contains a thin oil column which flowed 2,730 bbl/day on test. Drilling has revealed irregular vertical and lateral variations in the fluid content of "pay" sandstones, which is in direct contrast to the North Rankin field.

In addition to the two major discoveries, lesser hydrocarbon reserves have been found at five other locations on the Rankin trend. The Rankin No. 1 and Dockrell No. 1 (Fig. 5) are gas-condensate discoveries, and the Eaglehawk No. 1, Egret No. 1, and Lambert No. 1 are oil discoveries. Although individual flow rates tested as high as 2,700 bbl/day, the hydrocarbon columns are too thin to be commercial in the deep-water setting.

Angel

A fourth significant discovery was made in late 1971 on the Angel feature (Fig. 5). The well is located in approximately 250 ft (76 m) of water, inshore from the major fields of the Rankin trend and on a parallel structural feature—the Madeleine-Dampier trend. The structure at Angel field is a fault-controlled anticline. Gas-condensate reserves are contained in Upper Jurassic massive marine sandstones.

The Angel No. 1 was drilled to 11,190 ft (3,411 m) in Upper Jurassic strata. A net sandstone "pay" of 276 ft (84 m) was encountered in the interval 8,712–9,006 ft (2,655–2,745 m). Core analysis indicates that the sandstone has excellent reservoir properties; porosities average 20% and permeabilities range up to 700 md. Two drill-stem tests using 5/8-in. chokes produced flows of 12,800 and 13,200 Mcf/day, accompanied by 57° API gravity condensate at the rate of 54 bbl/MMcf.

Appraisal wells have been drilled in 5 mi (8 km) northwest and 3 mi (4.8 km) southeast of the Angel No. 1. These wells have confirmed the lateral continuity and good reservoir characteristics of "pay" sandstones in the structure. Seismic data and well control also indicate that the structure is filled to the spillpoint. The field could be economically attractive, owing to the relatively shallow water depth and the proximity to the Western Australia mainland.

Conclusions

During the first decade of petroleum exploration on the Australian northwest shelf, the group venture headed by Woodside-Burmah has been involved in the drilling of about 45 wells. Newfield discoveries were made by 10 of the 37 wildcat wells, and the 8 appraisal wells have all been successful.

Although several of the discoveries are too small to be economic, three major gas-condensate fields have been found in the Dampier subbasin, including the giant North Rankin field. Another possibly large gas-condensate accumulation has been found at Scott Reef in the Browse basin.

Total recoverable reserves established in the Dampier subbasin to date amount to about 18 Tcf of gas, of which 13.5 Tcf is in the proved and probable category. These discoveries constitute an important addition to the energy inventory of the Indo-Pacific region.

Recent Knowledge of Hydrocarbon Potentials in Sedimentary Basins of Indonesia[1]

A. PULUNGGONO[2]

Abstract Although the search for hydrocarbons in Indonesia was initiated about 8 decades ago, exploration work is still at a high level. Recent studies have resulted in a new understanding of the Tertiary sedimentary basins, and knowledge concerning offshore sedimentary basins has been updated significantly. More important, however, is the current knowledge of the mechanisms of basin formation which seems to enhance the validity and applicability of the new global tectonics to the geology of Indonesia.

Recent exploration surveys and subsequent drilling have shown that the southern part of the Sunda Shelf consists of many sedimentary basins and intervening uplifts. Major faults are common throughout the area and clearly control the distribution and shapes of the basins. Block faulting apparently broke up the periphery of the Sunda Shelf at the beginning of Tertiary time. The chief crude oil production in western Indonesia is from the regressive and deeper transgressive sandstone series of Oligocene-Miocene age, except in East Kalimantan, where producing zones range from Eocene to Pliocene.

Prospective areas have changed considerably since oil and gas in economic amount have been proved from an interbedded limestone formation of Tertiary age. Additional reserves are anticipated in stratigraphic traps.

Oil and gas discoveries within deltaic sandstones, notably in East Kalimantan, have upgraded significantly the onshore and offshore potentials of the area. Carbonate rocks are becoming a prime objective in the search for oil, especially in the East Java–Madura basinal area.

Although eastern Indonesia was chiefly the site of late Paleozoic and Mesozoic sedimentation, oil has been proved only within the strata of Tertiary age, notably in the Salawati basin. Of particular importance was the recognition of the tremendous potential that reefs, particularly Tertiary reefs, possess as hydrocarbon reservoirs. A similar basin and environmental model is anticipated for the Bintuni basin.

Recent scientific cruises have indicated the presence of several potential basinal areas between the Sunda and Sahul Shelves.

The sedimentary basins in Indonesia can be classified as arc-front, foreland, and deltaic basins.

Introduction

The search for hydrocarbons in Indonesia was initiated in the 1890s. After almost 80 years of exploration on land, offshore activity was begun in 1967. Since then, increased oil and gas exploration has resulted in the accumulation of numerous data, particularly for the prolific Tertiary sedimentary basins. The first offshore oil discovery was made in 1969.

At present, 41 foreign oil companies are actively engaged in on-land as well as offshore oil exploration in cooperation with Pertamina, which also has its own on-land exploration activities. Seven oil companies, including Pertamina, have had discoveries on land since activities were resumed in 1966, whereas six oil companies working in offshore regions were successful in finding oil and gas in economic quantities. This efficient exploration program is attributed mainly to the improved operating conditions in Indonesia since 1966. The 1973 statistics showed 29 successful exploration and appraisal wells, including 5 Pertamina wells. This means a success ratio of 1:3. Within the first 6 months of 1974, six new oil discoveries were made on five separate structures in five different basins.

Indonesia oil production reached 1 million BOPD in early 1972 and 1.5 million BOPD in July 1974. World demand for Indonesian oil is increasing at a similar pace, mainly because of the low sulfur content, which averages less than 0.1%. Indonesian oil-export revenues have accounted for about 50% of the state's foreign income to July 1974 (24% in 1969, 36.1% in 1971), reflecting the important role of the oil industry in the country's development program.

Regional Tectonics

There is a fundamental difference in the geologic settings of western Indonesia, which includes Sumatra, Java, and Kalimantan (Borneo); and of eastern Indonesia, which includes Sulawesi (Celebes), Halmahera, Ceram, Timor, the islands of the Banda arcs, and Irian Jaya (West Irian).

Western Indonesia

The major tectonic features of western Indonesia in relation to oil-basin areas are (Fig. 1):

1. An outer island arc represented by the offshore islands of west Sumatra and a submarine ridge flanking the Java Trench south of Java.

2. A complex of several sedimentary basins situated in front of the magmatic arc, between the outer arc and the islands of Sumatra and Java.

[1]Manuscript received, October 21, 1974.

[2]Exploration Coordinator, Perusahaan Pertambangan Minyak dan Gas Bumi Negara (Pertamina).

Acknowledgment is made to R. P. Kusumadinata, lecturer at the Bandung Institute of Technology, who contributed ideas and discussion on various aspects of this paper.

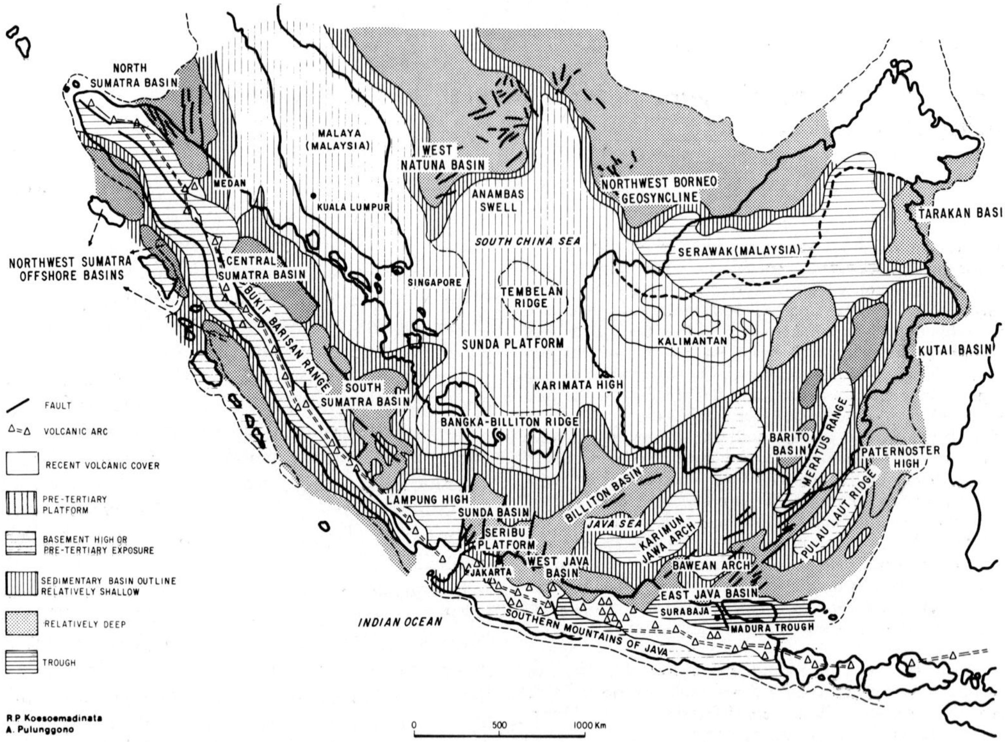

FIG. 1—Major tectonic framework of Tertiary sedimentary basins in western Indonesia.

This basinal complex is also known as the "arc-front basins."

3. An inner volcanic arc—the Sunda mountain system—which forms the backbone of Sumatra, Java, and the Lesser Sunda Islands. The Sumatra section is the Barisan mountain range, which started to develop during the Miocene and is made up mostly of folded Paleozoic and Mesozoic sedimentary rocks, metamorphic rocks, and granitic and mafic intrusive and volcanic rocks. Active volcanism characterizes this range. A 1,650-km transcurrent fault zone extends the entire length of Sumatra along and through the Barisan mountain range. The Sunda mountain system on Java also stretches along the length of the island.

4. A basin complex behind the volcanic arc represented by a series of oil-productive Tertiary basins in eastern Sumatra and northern Java which extend into offshore regions. These "foreland basins" are the North Sumatra, Central Sumatra, South Sumatra, West Java, East Java–Madura, and Java Sea basins. In the Java Sea, the existence of the western and eastern Java Sea basinal areas has been proved by recent surveys. The basins in southeastern and eastern Kalimantan are, from south to north, the Barito basin, the Balikpapan-Kutai basin, and the Tarakan-Sajau basin.

The East Java–Madura basin is a conspicuous east-west-trending basin system which is regarded as a young Tertiary feature superimposed on an older configuration. Several highs cross this system, breaking it into separate basins. In south-central Java an arc-front sedimentary basin extends into the offshore region. The Tertiary sedimentary strata within the basins lie unconformably on basement, which is generally of pre-Tertiary age.

5. The Sunda Shelf proper or Sunda platform (also called "Singapore platform") is regarded as the continental core of western Indonesia and consists of intensely deformed sedimentary and crystalline rocks of primarily pre-Tertiary age. It has been stable tectonically since Tertiary time. The Sunda Shelf, which includes the Malay Peninsula, flanks the Tertiary foreland basins of Sumatra to the west and the western Java Sea basins to the south. From south to north it extends from the islands of Bangka and Billiton (Belitung)—Bangka-Billiton ridge—to the Anambas and Natuna ridges. Its eastern extension covers part of

the area of western Kalimantan. Recent seismic surveys and subsequent drilling have further detailed the outline of the Sunda Shelf in this region. Ben-Avraham and Emery (1973) divided the Sunda Shelf into three major units: the northern Sunda Shelf basinal area, the Singapore platform, and the Java Sea basinal area. Two main basins can be recognized within the northern Sunda Shelf basinal area: the Northwest Borneo geosyncline in the northeast and the West Natuna basin (Gulf of Thailand basin) in the northwest. The north-south-trending Natuna ridge is a prominent structural feature which separates the two basins. The basement within the Northwest Borneo geosyncline, consisting of low-grade metamorphic rocks of Cretaceous-Eocene age, deepens abruptly toward the northeast, whereas southwestward onlap of late Tertiary units toward the platform has been recognized. In the West Natuna basin, a similar system has been recognized.

Eastern Indonesia

The tectonic setting in eastern Indonesia, notably in Irian Jaya, is in part associated with the stable Sahul Shelf or Arafura platform. Whereas the Sunda Shelf in western Indonesia is mainly the result of various Mesozoic tectonic phases, the Sahul Shelf in eastern Indonesia was formed mainly during the Variscan orogeny. Western Indonesia is chiefly an area of Tertiary sedimentary deposition, whereas eastern Indonesia was mainly the site of late Paleozoic and Mesozoic sedimentation.

Most oil exploration in eastern Indonesia is being carried out in Irian Jaya (Fig. 2), where two important basins are situated in the Vogelkop area. These basins, the Salawati and Bintuni basins, initially took shape during the Tertiary; they form the western extension of the Central Irian miogeosyncline. Both basins are situated northwest of the stable Arafura platform and presumably are separated by the Ayamaru platform. Another main basinal area, the North Irian eugeosyncline, is situated east of the Vogelkop area.

One prominent geologic feature is a major left-lateral transverse fault—the Sorong fault zone. This fault zone trends east-west, gradually curving southwestward, and bounds the northern end of the Salawati basin. Considerable strike slip, as well as vertical movement, has occurred along this fault.

The Sahul Shelf area is distinctly separated from the Sunda Shelf area. Much of the intervening vast area is occupied by deep ocean basins and numerous islands, commonly characterized

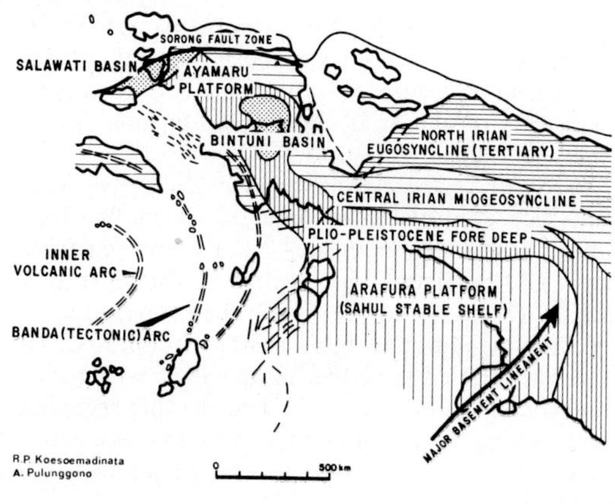

FIG. 2–Major tectonic framework of Tertiary basins of Irian Jaya.

by complex geologic relations that have only been studied in a reconnaissance manner.

It appears that the explanation of these geologic complexities may lie principally within the new global tectonics, involving collision of moving crustal plates and subduction zones between opposing plates.

Although further study is needed, the following major tectonic features are considered to be of importance for oil exploration.

Arafura platform or Sahul Shelf—This stable shelf area, underlying the Arafura Sea, Aru islands, and southwestern New Guinea, is part of the subcontinent of Australia. The shelf area is characterized by a thin section of probably Cambrian, Cretaceous, and Tertiary sedimentary strata with little folding and faulting.

Central Irian unstable shelf area (Central Irian miogeosyncline)—Several small basins in this province contain considerable thicknesses of late Tertiary and Pleistocene marine to paralic clastic deposits. The present central mountain range of New Guinea lies to the north of the Central Irian unstable shelf area. The central range is composed of uplifted metamorphic rocks and intensely to moderately folded Paleozoic to Tertiary sedimentary strata intruded by granite and ultramafic rocks which form the spine of Irian Jaya; its westerly extension is the North Vogelkop uplift.

The Salawati and Bintuni basins are in trend with the Tertiary basins within the unstable shelf area. These two basins are bounded on the north by the North Vogelkop uplift along the Sorong fault zone. The Lengguru fold belt and the Onin-Kumawa uplift border this basinal area on the east and south, respectively.

Sorong fault zone—This zone is a major left-lateral transverse fault, obviously a result of opposing movements of crustal plates. It offsets remnants of the Irian Jaya shelf terrane for at least 700 km.

Geologic province of the greater Moluccas—This province comprises the area occupied by deep ocean basins and numerous islands between Irian Jaya and Sulawesi. The Australian shield and its stable (Sahul) and unstable shelf components provide a relatively stable eastern boundary. To the west, the edge of the Sunda Shelf within Sulawesi constitutes a western limit to this complex mobile area. The greater Moluccas geologic province may be divided into several major tectonic subdivisions.

1. The Banda arc system is considered to be a continuation of the outer island arc and inner volcanic arc of western Indonesia; it is characterized by a very strong negative gravity field and associated ocean deeps and trenches both on the outer and inner sides and between the component elements. The outer orogenic arc passes through Sumba(?), Timor, the Tanimbar islands, and the Kai islands northwestward to Ceram and probably Buru.[3] It is characterized by intense folding and thrust faulting. Orogenic sedimentary deposits include flysch and turbidites. The sedimentary section ranges in age from Permian to Tertiary. Localized Plio-Pleistocene basins have developed on some of the islands. The inner volcanic arc passes through the islands of Bali, Lombok, Sumbawa, and Flores to end at the Banda islands.

2. The Sula spur is composed of metamorphic and granitic rocks and extends from the Sula islands westward to eastern Sulawesi. It is regarded to be remnants of the Irian Jaya shelf terrane which were dragged along the Sorong fault zone in a westerly direction.

3. The Halmahera arc consists mostly of ultramafic and mafic intrusive rocks associated with volcanic rocks and volcanic-derived sedimentary strata. It may be the site of fossil subduction zones associated with underthrusting of the Asian continental plate by the Pacific plate.

4. Sedimentary basins within the greater Moluccas have been recognized, notably on the islands of Timor, Ceram, and Sula. Pre-Pliocene to Permian rocks in Timor and the Sula islands have intensely deformed and, in places, mildly metamorphosed. Thus they form effective economic basement over most of the area. The section of interest for hydrocarbon prospects is the Plio-Pleistocene, which unconformably overlies the deformed rocks. Basins containing these Plio-Pleistocene strata have been recognized in Timor, Ceram, and probably Buton. Measured sedimentary-unit thicknesses are 500–3,000 m, but commonly average between 500 and 1,000 m.

[3]This view is presently being contested by Audley-Charles (1974), who proposes that a major tectonic discontinuity separates eastern Indonesia, consisting of Sumba and the Banda Arc (including eastern Sulawesi), from western Indonesia, composed of western Sulawesi and the islands west of Sumba. Furthermore, other conclusions reached are that the outer Banda arc is underlain by Australian continental crust on which the Permian and Mesozoic authochthonous facies accumulated.

Regional Stratigraphy

The Tertiary sedimentary strata in the main oil-basin areas in Indonesia unconformably overlie the basement, which in most places is pre-Tertiary. Thicknesses of these Tertiary sedimentary sequences in the various basins range from a maximum of about 6,000 m in northern Sumatra to about 3,500 m in central Sumatra and about 4,000 m in southern Sumatra. In west Java, wells drilled to about 2,500 m generally have not penetrated the pre-Tertiary basement. In east Kalimantan, wells about 3,500 m deep are still in Miocene or possibly younger sections. Exceptions are one anticlinal structure in the Barito basin (Tanjung), where a basement high was penetrated at about 1,200 m, and one other inland structure in the West Java basin with an intrusive body at about 1,950 m. The onshore portion of the West Java basin, notably the basin "lows," contains an estimated 5,000 m of Tertiary sedimentary and volcanic rocks. Basement seems to rise sharply farther north in the offshore portion of the basin where several wells have penetrated it at about 1,500 m.

Basement block faulting appears to have broken up the Sunda Shelf area, which was consolidated toward the end of Mesozoic time.

Three major sets of faults appear to dissect the pre-Tertiary landmass: northwest-southeast faults in the western part (particularly in Sumatra) and the northeast edge of the Sunda Shelf, as well as in northwest Borneo; northeast-southwest-trending faults in the eastern Java Sea area; and a north-south set, especially well developed in the western Java Sea (Sunda basin) and in the offshore and onshore portions of the West Java basin.

The first two sets may have originated as strike-slip faults, but considerable vertical movements must have taken place, resulting in several areas of basinal subsidence and intervening highs. The north-south-trending faults of the Sunda basin area also show differential vertical movements.

An east-west-trending fault set is also particularly well developed in east Java and is responsible for the troughlike subsidence of the Randublatung-Madura strait.

These fault systems apparently did not affect the Sunda Shelf during the Tertiary Period. The generation of clastic sediment from this stable platform continued throughout the period.

Fault movements appear to have been continuous from the beginning of basin subsidence, and to have controlled sedimentation (growth faults). Less differential movement took place in late Tertiary time, as the sedimentary strata of this age are generally not dissected, but folding apparently was localized by the fault trends. Uplifted fault blocks are characterized by clastic sedimentation, whereas carbonate deposition took place away from basement highs. Less is known about the northern part of the Sunda Shelf, but faulting apparently also influenced its sedimentation history. Although additional data are still needed concerning the east Indonesian oil basins, faulting is a convincing mechanism of basin formation.

According to Ranneft (1972), the traditional concept of western Indonesia as a sinuous island arc is probably incorrect. Instead, western Indonesia seems to consist of a series of generally straight segments that are offset in places as a result of transverse faulting. Thus, the foreland (Sumatra) basins consist of a series of subbasins separated by structural highs which in turn are caused by the "Bantam trend" structural pattern, a series of north-northeast-trending transcurrent or normal faults.

In the foreland basins of western Indonesia, rapid sedimentation occurred during the Tertiary Period in partially closed marine environments. Moderate to intense folding followed at the end of the Tertiary.

Development of individual basins may have differed in the timing of tectonic events, but a similar basin evolution and sedimentation cycle may be inferred for each basin (Koesoemadinata, 1969). Most of the oil fields of Indonesia are in these basins.

Sediments were deposited initially on an eroded surface cut into Late Cretaceous strata. Early Tertiary faulting and faulting contemporaneous with erosion and deposition controlled sedimentation throughout Tertiary time. This is obvious in the South Sumatra and Java Sea basins. Similarity in the stratigraphic succession, especially for the western Indonesian foreland basins, is apparent. A transgressive facies lies at or near the base; deposition of this facies was preceded in some basins by volcanic activity or terrestrial sedimentation. Littoral to neritic deposition was followed by bathyal deposition (shale).

Basinward, sedimentation appears to have been continuous. At the end of the Tertiary this series of deposits was followed by sedimentation in regressive environments, resulting in basin filling. An exception is the East Java basin, which is characterized by a high carbonate content. As noted, pre-Pliocene to Permian rocks generally constitute economic basement in the eastern Indonesian sedimentary basins. The oil-prospective basins in eastern Indonesia are relatively young, having a history beginning in late Miocene time and extending into the Pleistocene. The Plio-Pleistocene section is of special interest in Timor and Ceram, where it unconformably overlies the deformed rocks. The sedimentary basins are usually structural in character.

North Sumatra Basin

The North Sumatra basin (Fig. 3) is situated between the Sunda Shelf (Malacca platform) in the northeast offshore area and the Barisan range on the southwest. The Asahan arch, which extends offshore northwestward from the Asahan area, is believed to be the boundary separating the North Sumatra basin from the Central Sumatra basin. To the northwest the basin presumably is terminated by the so-called Samalanga uplift in the northern Aceh area.

The basin was formed in early Tertiary time. The lower Tertiary sequence is composed mostly of micaceous quartz sandstones with some transgressive carbonate units overlying the pre-Tertiary basement. Several subbasins were developed: the Paseh depression in the northern area, the Tamiang and Medan depressions in the central area, and the Pematangsiantar deep in the southeast. During middle Miocene time, most of the area was covered by the open sea, resulting in the deposition of up to 1,500 m of Baong marine shales.

Starting approximately at the end of middle Miocene time, the Barisan mountains were uplifted, yielding clastic materials to the North Sumatra "backdeep" and depositing the regressive, predominantly sandstone and shale sequence of the Keutapang and Seurula Formations. The filling of the basin ended in late Pliocene time with deposition of terrestrial and lagoonal facies of the Julu Rayeu Formation. The whole area was later affected by Plio-Pleistocene folding, which produced the present structural configuration. A deep-seated, northwest-trending fault is postulated whereby these subbasins are shown to be downthrown relative to the eastern coastal-plain

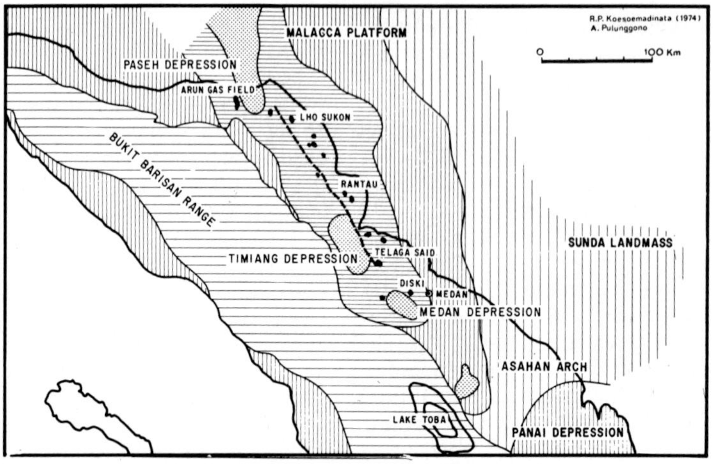

FIG. 3–North Sumatra basin configuration.

area. The western limit of the structural basin is formed by the Barisan foothills, and is characterized by block-faulted areas.

Central and South Sumatra Basins

The Central and South Sumatra basins (Figs. 1, 4) show very similar and related histories and could be considered one large basin with many troughs and grabens. The boundary between the basins is presumably a northeast-southwest-trending area through the northern part of the Tigapuluh mountains.

These asymmetric basins are bounded on the southwest by faults and uplifted exposures of pre-Tertiary rocks along the front of the Barisan mountains, on the northeast by the sedimentary or depositional boundaries of the Sunda Shelf, on the south and east by the Lampung high and by an arch that parallels the east coast of Sumatra, and on the north and northwest by the Asahan arch and outcrops of pre-Tertiary rocks northwest of Pekanbaru. Both the Asahan and Lampung arches were positive elements through much of Tertiary time, and separated the Central and South Sumatra basins from the adjoining North Sumatra and Sunda basins, respectively. The two arches were covered by shallow-marine seas during early and early-middle Miocene time to form temporary connections to the neighboring basins.

Latest exploration (seismic and aeromagnetic) results indicate the existence of subbasins in the area of the Lampung high.

The Tertiary depositional basins extended farther west and, at times, were connected by seaways to the ancestral Indian Ocean.

Structurally, the South Sumatra basin is subdivided into the Jambi and Central Palembang basins (Fig. 4). The basins are, for the greater part, bounded by major faults of which the Lematang and Muaratembesi fault zones are the most conspicuous (Fig. 4). The structural features present in the two basins are the result of at least three separate episodes of orogenic activity.

The mid-Mesozoic orogeny was the main cause of metamorphism of Paleozoic and Mesozoic strata. These strata were faulted and folded into large structural blocks and subsequently intruded by granite batholiths—with postulated extensions in the basin subsurface—forming the basic "structural grain" of Sumatra. The second significant tectonic event occurred probably in Late Cretaceous–early Tertiary time, when major tensional structures that include grabens and fault blocks were formed in the basins of Sumatra and in the adjoining Sunda basin. The general trend of these faults and grabens is north-south and NNW-SSE. These tensional features and the remnant structures from the mid-Mesozoic orogeny, coupled with rugged paleotopography developed by differential weathering of the pre-Tertiary rocks, compose the older structural elements of the basin and represent the pre-Tertiary unconformity surface, which controlled the deposition of the lower Tertiary formations. Other "old" features include stable platforms which remained essentially unaffected by structural movements during the late Tertiary orogeny (de Coster, 1974).

The most prominent structural features in the basins are the northwest-trending folds and faults formed during the Plio-Pleistocene orogeny. The convergence of the Indian Ocean plate against the Sumatra portion of the Southeast Asia plate is postulated to have been the cause of the final uplift of the Barisan mountains, the development

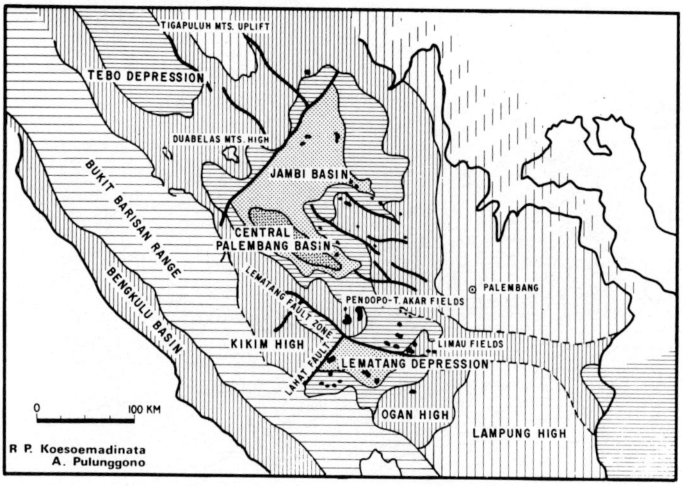

FIG. 4–South Sumatra basin configuration.

of the major right-lateral wrenching through the length of these mountains, and the formation of the associated fold-fault structures in the basin (de Coster, 1974). These Plio-Pleistocene features are the most dominant features of Sumatran geology.

Northwest Java Basin Area

The Northwest Java basin area—including a series of Tertiary basins bordering the stable Sundaland in the south—consists of the West Java, Sunda, and Billiton (Belitung) basins (Fig. 5). The Java volcanic range forms the southern rim of the basin area. The Seribu platform or horst is the structural divide between the Sunda and West Java basins. Within the West Java basin, three subbasins can be distinguished—the Jatibarang, Ciputat, and Pasir Putih (Soejitno and Yahya, 1974).

Volcaniclastic and fluvial deposits constitute the sedimentary rocks. The quantity of sediment influx was partly responsible for differential sinking of fault blocks during early basement development, and resulted in thick deltaic deposits during late Oligocene and early Miocene times, notably in the Sunda basin. Three sedimentary environments have been postulated in the West Java basin—shelf, shallow marine, and possibly deep marine.

Downwarp of the "backdeep" continued as the sea invaded the geanticline on the south through passages between high tectonic blocks which separated segments of the basin area during early basin development. As denudation and downwarp continued, Miocene transgression expanded over the Sunda basin.

As basin segments became filled, a nearly flat depositional surface developed. Slight rise in the borderland and lowering of sea level resulted in regional regression followed by major transgression in late mid-Miocene time. During this second transgressive period, a connection developed between the Sunda basin area and the South Sumatra basin. Final regional regression occurred during the Pleistocene when worldwide melting of ice caused a resubmergence to the present marine conditions.

Basement in the basin area consists primarily of low-grade pre-Tertiary metamorphic and Late Cretaceous igneous rocks. A major north-south block-fault system on this rugged surface governed erosion, deposition, and tectonic growth throughout Tertiary time in the Sunda and West Java basins.

As a result of continued Tertiary movement along these fault lines in the Sunda and West Java basins, structural growth and density are unusually high. Structures are commonly large in areal extent. The greatest Tertiary fault movement occurred in Oligocene and earliest Miocene time, when at least 120 m of displacement took place along the east flank of the Sunda basin.

Differential movement along many fault zones continued through the middle Miocene. Many features of the Sunda and West Java basins exhibit structural growth during Tertiary deposition as a result of fault movement. The north-south alignment of the Sumatra coastline has been proved to be a fault zone on the west flank of the Sunda basin.

Basin downwarp in the Billiton (Belitung) basin appears to have been influenced by a north-

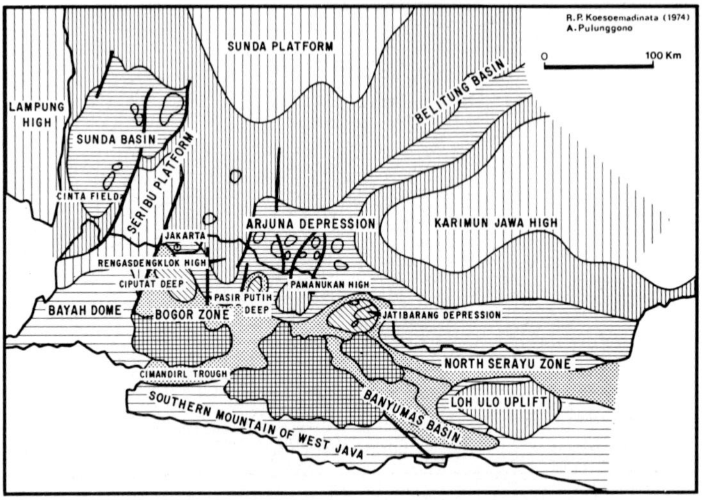

FIG. 5—Northwest Java basin configuration.

east-southwest fault zone parallel with the Karimunjawa arch.

East Java Basin

The exact limits of the Tertiary East Java basin are still ill-defined, but the northern shelf of the basin merges with the southern shelf of the East Kalimantan basin (Fig. 6). The shelf runs south from Kalimantan to an irregular edge north of Madura, where the basin plunges sharply into a very deep trough beneath the island. Actual thickness of the Tertiary sequence apparently exceeds 7,600 m. The East Java basin merges on the west into the West Java basin in the vicinity of the Karimunjawa and Bawean arch. The Tertiary stratigraphic succession is similar to that of the West Java basin. Strata thicken toward the east, and on the north a great carbonate bank-reef complex extends into the Barito basin.

East Kalimantan Basins

Kutai basin—This is one of several Tertiary basins which are present along the eastern margin of Kalimantan (Fig. 1). The offshore portion of the basin lies along the east coast of Kalimantan in Makassar Strait. The northern boundary of the Kutai basin is the Mangkalihat ridge, and the eastern boundary is the trough of the Makassar Strait. On the south the basin is bounded by the slope of the Paternoster shelf. The Kutai basin was formed in early Tertiary time by predominantly clastic sediments. The axis of maximum sediment thickness shifted progressively eastward. The maximum thickness of late Miocene to Holocene strata is found offshore, where more than 3,000 m of variable lithologic and faunal facies are present. These strata include a prominent delta system, extensive shallow-marine shelf deposits, and deeper outer-shelf and bathyal deposits. The delta prograded eastward beginning in middle Miocene time, and reached its greatest extent during late Miocene to early Pliocene time.

Barito basin—This basin (Fig. 1) is bounded on the west by the Sunda Shelf and on the east by the Meratus Range. The southern limit extends offshore and merges with the eastern Java Sea basins. Tertiary sedimentary strata range in age from Eocene to Pleistocene, and are generally thin ($<$1,500 m) except locally along the front of the Meratus Range near Tanjung (about 6,000 m).

East Indonesia Basins (Irian Jaya)

Vogelkop Peninsula and the associated southern shelf form a distinct geologic province bounded by the Lengguru fold belt on the east, by the Sorong fault block on the north, and by the Ceram Sea on the south. The western margin of the province is characterized by a westward-thickening wedge of Plio-Pleistocene clastic strata.

Tertiary deposition was characterized by a Late Cretaceous–Eocene transgression, an Oligocene regression, an early-middle Miocene transgression, and a late Miocene regression. During middle Miocene time, much of the province was inundated and a pronounced downwarp of the western and eastern margins formed the Salawati and Bintuni basins, respectively (Fig. 2).

Over 6,000 m of marine strata is present in the Salawati basin. As the basin filled and water

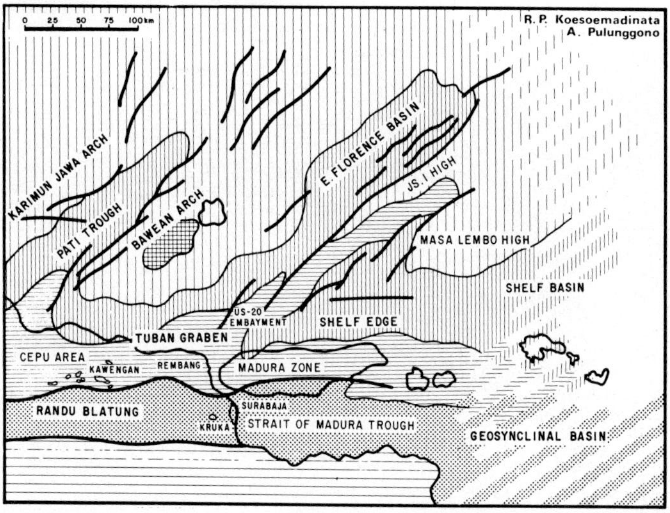

FIG. 6—East Java basin configuration.

depths shallowed, the shelf facies of the Kais Formation built a platform out into the basin over the Klamogun basinal limestones (Fig. 7). On this platform, pinnacle reefs grew to heights 300–750 m above the platform. The intervening areas were filled with a basinal shale and interbedded argillaceous limestone facies called the "Klasafet Formation." The calcareous shales of the Klasafet Formation eventually buried the pinnacle reefs, providing excellent side and top seals.

Shelfward, the pinnacle reefs and intervening shale-filled areas are replaced by a continuous shelf-carbonate facies. The shelf margin, which defines the boundary between the continuous carbonate deposition on the shelf and the area of the basinward-occurring pinnacle reefs, is extremely irregular in outline.

At the end of Miocene time, marine seas inundated the province and fine clastic sediment flooded the Salawati basin from the northern positive source area, thereby terminating the Tertiary carbonate deposition.

Plio-Pleistocene deposition was predominantly clastic facies, and several orogenic movements occurred during this time. In the east, the Lengguru belt was tightly folded and uplifted. Large-scale left-lateral movements took place along the northern margin (Sorong fault zone). This fault zone bounds the Salawati basin in the north and represents the boundary between the northern part of the Australian plate and the southern part of the Pacific plate.

Late Pliocene–Pleistocene tectonic activity within the rest of the Salawati basin was restricted to vertical movements along normal faults and structural draping of sediments over deep-seated basement faults.

Basin Classification

On the basis of regional tectonics and observed structural phenomena within each basin, the sedimentary basins in Indonesia can be tentatively classified on the basis of (1) their position relative to former subduction zones, magmatic arcs, and other tectonic features (Ben Avraham and Emery 1973), and (2) the major structural elements which were controlling features in the formation of the basins.

Where an oceanic plate underthrusts other plates, two sedimentary basins usually are present—the *arc-front basin* between the trench and the volcanic arc, and the *foreland basin* behind the volcanic arc. This classification is particularly useful in the western Indonesia basins where geologic control is good. Arc-front basins in Indonesia are the west Sumatra offshore basins and the South Java basin.

Foreland basins are generally characterized by troughs and grabens, and may be subdivided into (1) the Sumatra-type and (2) the West Java–type basins. The Sumatra-type basins are clearly influenced by strike-slip movements, whereas West Java–type basins have resulted from tensional forces. The basins in the eastern part of the Java Sea might be associated with strike-slip faulting, but the volcanic arc is notably absent; instead, a continental-shelf separates the basin area from the deep oceanic environment. Foreland basins in Indonesia are the North Sumatra basin, the Central and South Sumatra basins, the Northwest

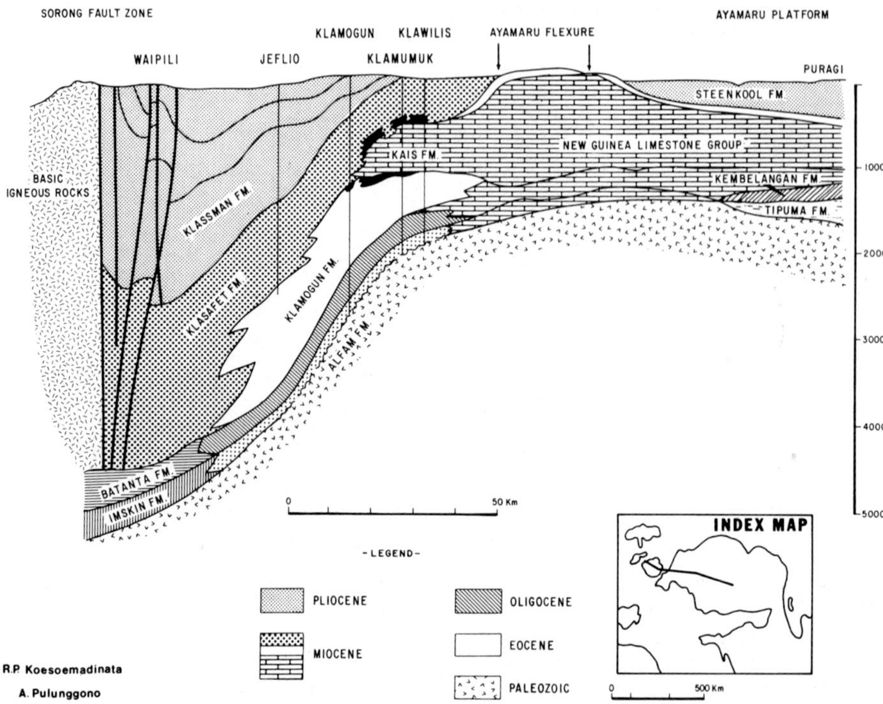

FIG. 7–Schematic geologic cross section through Salawati basin, Irian Jaya.

Java basin area, and the East Java and Madura basins.

The Barito basin and the East Java Sea basin area may be regarded as transitional forms between the foreland basins and shelf-edge basins.

A third category—*deltaic basins*—includes the East Kalimantan basins, notably the Kutai and Tarakan basins. These basins are situated at the continental margin.

The Salawati basin in East Indonesia is a combination of a half-graben basin and a depositional basin. Our present knowledge concerning tectonically complex East Indonesia does not lend itself to a proper classification of the basins.

Hydrocarbon Occurrence and Potential

New insights concerning the basinal structural framework, mechanisms of basin formation, and environmental and depositional conditions have considerably changed the outlook for petroleum prospects in each basinal area.

The faulted character of the foreland (Sumatra and Java) basins makes them excellent prospective areas. Where the subbasins are each separate entities, delineated by major faults (Central Sumatra, South Sumatra, West Java, and possibly North Sumatra basins), different environmental and depositional conditions within the subbasins have opened up new facets in prospecting for hydrocarbons. For example, from north to south in the North Sumatra basin, the Oligocene-Miocene facies grade from restricted-marine to carbonate to deltaic-coastal to supralittoral facies. This relatively abrupt lateral change in facies presents possibilities for stratigraphic traps.

The rugged paleotopography—developed by differential weathering of the pre-Tertiary rocks which constitute basement (Central Sumatra, South Sumatra, Sunda, and West Java basins)—which prevailed preceding the time of the major transgression provides obvious sites for nonmarine sedimentation, including fluvial and deltaic deposits.

Deep portions of troughs, accentuated by fault control, were likely sites of first deposition of continental clastic material. Subsequent deposition occurred in deltaic environments in the differentially subsiding subbasins and troughs (Central Sumatra, South Sumatra, Sunda, and possibly North Sumatra basins), with influx of sediments by way of paleostreams from the neighboring Sundaland.

Basal transgressive series deposited when the seas encroached farther into the basin areas may be prolific in the deeper portion of the basin (e.g., the Banuwati Member in the Sunda basin). Subsequent transgressive series locally may constitute pinchouts and wedgeouts against the rough, topographically high areas which remained relatively stable, positive blocks during Oligo-Miocene deposition.

In early to middle Miocene time, the widest extent of the basins was receiving sediment in shallow to deep-marine environments. By the early middle Miocene, a reduction in the amount of clastic material, coupled with possible episodic stillstands, allowed carbonate material to be deposited on the platform and shelf area (South Sumatra basin and Sunda basin—Baturaja Limestone Formation). Initially, these were extensive platform or bank deposits, but they later became detrital, reefal, and bank buildups—many supported by a framework of colonial corals.

The Baturaja Limestone Formation has been a prolific and potential reservoir for oil and gas in the South Sumatra and western Java Sea basins. In the East Java Sea basin, the shelf environment northward encouraged widespread limestone formation, and the middle Miocene to Pliocene limestones, especially the reefs along the irregular hinge line south of the shelf edge, have created an ideal situation for hydrocarbon reservoirs.

In East Indonesia, the middle Miocene pinnacle reefs in the Salawati basin are becoming a prime target for hydrocarbon accumulations. Prospects within this carbonate facies also may be envisaged in the Bintuni basin and the Tertiary basins within the unstable shelf area farther southeast.

Late Miocene to early Pliocene extensive delta systems, such as the Mahakam delta, are postulated to be present farther north in the Tarakan–Sajau area. Especially in the Kutai basin, where sedimentation was from west to east, the existence of older delta systems toward the west is strongly anticipated.

Part of the stable shield of the Australian continent which lies southeast of Irian Jaya supplied clastic material to the basin areas during the Mesozoic, especially during Jurassic and Cretaceous times, creating likely prospects within the Kembelangan Formation, notably in the Bintuni basin and within the unstable-shelf-area basins farther southeast.

Hydrocarbon prospects within the Plio-Pleistocene are possible in the sedimentary basins on Timor and Ceram, although prospects within the Eocene and Miocene are not wholly excluded on some other islands within the greater Moluccas.

Sedimentary basins which have been outlined as a result of reconnaissance scientific cruises in the offshore area between the Sahul and Sunda Shelves are likely future prospect areas.

Although more detailed exploration work is necessary to delineate the basins within this tectonically complex region, identical phenomena as seen in most of the other (Tertiary) basins of Indonesia seem to be present within the "greater Molucca offshore basins"—namely, a block-faulted character which formed a series of horsts and grabens on which sedimentary formations have been deposited. Transverse faults also seem to dominate the structural framework in the eastern part of the Banda Sea area.

Selected References

Audley-Charles, M. G., 1974, Banda arcs, *in* Mesozoic-Cenozoic orogenic belts; data for orogenic studies; Alpine-Himalayan orogens: Geol. Soc. London Spec. Pub. No. 4, p. 235–253.

Ben-Avraham, Zvi, and K. O. Emery, 1973, Structural framework of Sunda Shelf: AAPG Bull., v. 57, no. 12, p. 2323–2366.

De Coster, G. L., 1974, The geology of the Central and South Sumatra basins: 3d Annual Conv. Indonesian Petroleum Assoc., June 1974.

Gribi, E. A., 1973. Tectonics and oil prospects of the Moluccas, eastern Indonesia: Geol. Soc. Malaysia Bull. 6, July.

Hollis, D. L., 1973, East Java comes into its own: Petroleum News Southeast Asia, April.

Koesoemadinata, R. P., and A. Pulunggono, 1971, Offshore Tertiary sedimentary basins in Indonesia (abs.): Pacific Science Cong., no. 12, v. 1, p. 442.

Ranneft, T. S. M., 1972, The effects of continental drift on the petroleum geology of western Indonesia: APEA Jour., v. 12, no. 2, p. 55–63.

Soejitno, P., and Ibrahim Yahya, 1974, The basement configuration of the Northwest Java area: 3d Annual Conv. Indonesian Petroleum Assoc., June 1974.

Structural Evolution of Tertiary Basins of Southeast Asia—Summary[1]

R. W. Murphy[2]

Tertiary to recent basins of four main types underlie 75% of Southeast Asia's 10 million km^2. Structural style is related to (1) basement composition; (2) petrotectonic assemblages; (3) relative balance between extensional, compressional, and transcurrent forces; (4) availability of preexisting fabric; (5) susceptibility to impact of changes in plate motions; and (6) sedimentary diapirism developed in overpressured clastic wedges.

Shelf basins have continental crust on all flanks. Examples are north, central, and south Burma; central and south Sumatra; Gulf of Thailand; Sunda; west and east Java; Billiton (Belitung); and Barito. Plate-margin activity is not obviously reflected in shelf basins but may exert fundamental influence. Shelf basins develop basement-controlled tectonic patterns except where sedimentary diapirism overprints flowage structures at shallow levels in deep-basin areas.

Continental-margin basins have continental crust on one flank and oceanic crust on the other. Examples include the Gulf of Martaban; Sunda outer-arc basins; and the Mekong, Brunei, and East Kalimantan basins. These basins mirror principal movement changes in adjacent plates; thus, fairly satisfactory correlation of internal unconformities to changes in spreading direction and rate is possible. The 13-m.y. and 26-m.y. events are particularly widespread. Outer-arc basins (continental-margin basins associated with island arcs) were filled mainly by volcanic material from andesitic inner volcanic arcs, but also by subduction accretion from buoyant inner-trench-wall wedges with fan-shaped internal structure. Structural style results from interplay of compressional and transcurrent movements.

Archipelago island-arc basins are intraoceanic arc-basin complexes. Philippine examples include the Cagayan valley, the Central Luzon trough, and the Agusan-Davao and Cotabato basins. They form between intricate subparallel arc-trench systems with strong vertical mobility and relatively stable map-view configuration. Their structural style is exceedingly complex.

Small ocean basins form by interarc spreading or by trapping of older oceanic crust behind newly risen intraoceanic island arcs. Examples include the Andaman Sea, South China Sea, southeast Sulu Sea, Celebes Sea, and Banda Sea.

Transitional types of basins are known, and some basins change category during their evolution.

[1]From Circum-Pacific Energy and Mineral Resources Conference program booklet.

[2]Esso Exploration Inc., Singapore.

Petroleum Geology of Thailand (Gulf of Thailand and Andaman Sea)—Summary[1]

CHARAN ACHALABHUTI[2]

Introduction

Petroleum exploration in the Gulf of Thailand has been active since the Government, in early 1968, granted to six oil companies the right to explore for, and produce, petroleum in 17 exploration blocks. Two additional blocks were granted in 1972 to one other company. Drilling activities were not started until the promulgation of the Petroleum Act and Petroleum Income Tax Act on March 26, 1971. Up to mid-1973, seven exploratory wells had been drilled, two of which had substantial shows and discoveries of oil and gas in Tertiary strata at depths less than 10,000 ft (3,000 m).

Many major and minor closed structures suitable for hydrocarbon accumulation within Tertiary sections beneath the Gulf were located by geophysical means. Because of the presence of these structures, 15 additional wells were scheduled for late 1973 and 1974. As of May 1974, six wells were completed, and gas and condensate were found in two of them.

In addition to the concessions in the Gulf, the Government granted petroleum concessions in the Andaman Sea off the west coast of Thailand; four blocks covering the shallow water were awarded to three oil companies in 1971 and 1972, and three deeper water blocks were granted in May 1974 to three companies and one consortium. No wells had been drilled on these blocks to mid-1974.

Regional Framework

Land Areas

Thailand is flanked on the east by the Indochina massif and on the west by the Himalayan belt. Three major physiographic and geologic provinces are recognized (Fig. 1): (1) Paleozoic-Mesozoic folded belt of the Thai-Malay Peninsula in the west, (2) Cenozoic Chao Phraya depression in the central region, and (3) Mesozoic Khorat plateau in the northeast.

The main structural trend of the Thai-Malay Peninsula is north to northwest. In peninsular Thailand the strike changes along the west side of the Gulf (about 9°N lat.) to a northeast trend and then back to a northwest trend onshore near the head of the Gulf. The folded belts, resulting chiefly from the late Paleozoic Hercynian orogeny but modified by the Tertiary orogeny, persist for many kilometers with little observable axial plunge (Brown et al, 1951).

Granite masses, evident in long, linear outcrops in the peninsula, are dated as Late Cretaceous–early Tertiary (Burton and Bignell, 1969). Four major ridges are recognized from west to southeast (Fig. 1): (1) Phuket Ridge, extending through Ranong and Kanchanaburi provinces; (2) Satun Ridge, extending to Ko Samui; (3) Ko Kra Ridge; and (4) Narathiwat Ridge.

The Tertiary orogeny, which formed the present physiographic features, probably closed with Wallachian (late Pliocene) movements, which caused the uplift of the Mesozoic Khorat plateau in the northeast, the compensatory subsidence of the Chao Phraya depression in the central region, and fault-block tectonism in the north with associated basalt intrusions (Brown et al, 1951).

The late Miocene folding and volcanism were responsible for the formation of the three major fault zones: the Klong Marui, Ranong, and the Kwae Noi faults (Fig. 1).

The Chao Phraya depression occupies an area of about 60,200 km^2 (140 by 430 km) and consists of thick Tertiary-Quaternary alluvium and deltaic deposits above an irregular basement. The Chao Phraya, Tha Chin, Mae Klong, and Bang Pakong are the main rivers transporting sediment to the Gulf. The delta progrades an average of 4–5 m/yr (Brown et al, 1951).

Seafloors

Gulf of Thailand—A shallow, flat basin with maximum water depth of 86 m, the Gulf extends from the Chao Phraya deltaic plain near Bangkok southeast about 800 km. Its mouth is defined as a line between the southwestern tip of the Mekong (birdfoot) delta and a small island off the east coast of the Thai-Malay Peninsula near 6°N lat.

[1]Manuscript received, October 21, 1974.

[2]Department of Mineral Resources, Ministry of Industry, Bangkok, Thailand.

The writer is grateful to Saman Buravas, Director-General of Department of Mineral Resources, Ministry of Industry, and Permanent Representative of Thailand to CCOP; and to the petroleum concessionnaires for their special support and permission to publish this paper.

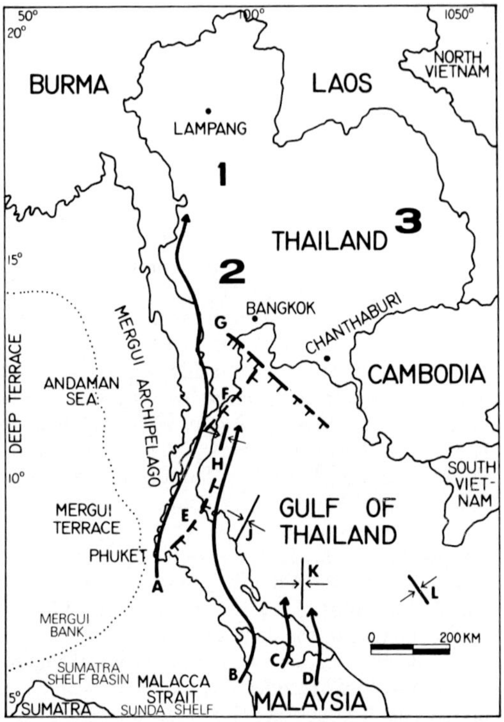

FIG. 1–Location map. *1*, Paleozoic-Mesozoic folded belt. *2*, Chayo Phraya depression. *3*, Khorat plateau. Ridges: *A*, Phuket Ridge; *B*, Satun Ridge; *C*, Ko Kra Ridge; *D*, Narathiwat Ridge. Faults: *E*, Klong Marui fault; *F*, Ranong fault; *G*, Kwae Noi fault. Basins: *H*, Chumphon basin; *J*, Songkhla basin; *K*, Pattani trough; *L*, Malay basin.

In general, the Gulf is floored by modern detrital silts and clays. Along the sides of the Gulf, the seafloor is interrupted by hills, many of which rise above sea level as islands (Emery and Niino, 1963). These islands are composed mainly of granite and/or metasedimentary rocks of late Paleozoic–Mesozoic age and are the source of sand and rock fragments. Some islands lie on the trend of onshore ridges such as the Satun and Ko Kra Ridges (Fig. 1).

The southwest coast of peninsular Thailand is bounded by the Andaman Sea, which is an important tectonic basin extending 1,200 km southward from the Irrawaddy deltaic plain of central Burma to northern Sumatra and the Strait of Malacca. The generalized physiographic features of the Andaman Basin are well defined by Rodolfo (1969). The inner shelf off Phuket Island is about 35 km wide; it increases in width to 130 km at its northern end (the Mergui Archipelago off southern Burma). To the south and east, it merges with the Malacca Strait (Sunda Shelf). The Mergui terrace originates north of 12°N lat. and becomes wider, flatter, and deeper southward. It also becomes concave southward, bifurcating into western (Mergui bank) and eastern arms separated by a deep basin (Sumatra Shelf basin) south of 9°N lat. The Mergui bank crests at 1,000 m below sea level (local pinnacles), and is 70 km wide and 200 km long. The Sumatra Shelf basin is bounded on the east and south by slopes off the inner shelf, the Sunda Shelf, and northern Sumatra. The Sumatra Shelf basin's floor south of the Mergui bank is 1,300 m deep and is open to the continental slope. The continental slope off the Thai-Malay Peninsula is the most persistent physiographic feature within the Andaman Basin. The continental-shelf (outer) break occurs at increasing depths southward, to a maximum of 500 m at Mergui bank. The shelf break, particularly where it forms the western edge of Mergui bank, is marked by pinnacles with 200 m of relief. The continental slope is generally convex upward, has an average gradient of 1.8°, and ends abruptly in a terrace 2,435 m deep (Deep Terrace) between Sewell Seamount and Martaban Canyon.

Geophysical and Drilling Results

Gulf of Thailand—Regional aeromagnetic and seismic data of petroleum concessionaires indicate that granite ridges probably do exist, but are buried at successively greater depths beneath younger deposits toward the center of the Gulf (G.E. Kelley, personal commun.). According to Parke et al (1971), the magnetic anomalies in the Gulf of Thailand are less than 100 γ, with broad gentle gradients in the areas of thick sediment cover. The greater values and steeper gradients occur where acoustic basement is shallower along the sides of the Gulf.

An acoustic basement can be picked readily on the seismic profiles over most of the concession areas. It is characterized by a low-frequency, high-amplitude reflector below which there are no other reliable events.

Four major basins are the Chumphon basin, Songkhla basin, Pattani trough, and Malay basin (Fig. 1). Both basement and basins are overlain by flat-lying rocks of probable Tertiary age. Where the basement becomes deeper, these rocks may be over 6 km thick.

Seismic interval velocities are generally slow for the near-surface sediments, but they increase from less than 2 km/sec at shallow depths to 4.5 km/sec below 2,000 m. Drilling results indicate that strata in the Gulf of Thailand are of Tertiary age and are more than 4 km thick.

Chao Phraya depression—Aeromagnetic data indicate that the Chao Phraya plain (Fig. 1) is floored by basement arches and plutons in association with a diverse assemblage of fault or flexure zones. The plain is located within a broad structural depression filled by strata several thousand feet thick. The basement slopes both toward the Chao Phraya River and southward toward the Gulf of Thailand (Hunting Geophysics Ltd., 1956).

Drilling in the vicinity of Ayudhya, 60 km north of Bangkok, encountered basement—defined as a "leucogranite" plug—at a shallow depth of about 1,158 ft (353 m). However, another Ayudhya well drilled about 6,000 ft (1,800 m) of sedimentary rocks before abandonment (Bunnag, 1959). Basement of metamorphic rock is at 1,434 ft (437 m) near the eastern edge of the head of the Gulf (Phian-Charoen and Buapeng, 1973).

Andaman Basin—Geophysical data from surveys in the Andaman Sea from 1960 to 1967 were reinterpreted by Rodolfo (1969). The following summary is based on his interpretation.

The inner shelf and adjacent Mergui terrace are characterized by offlap and are overlain by unconsolidated sediments about 200 m thick. The irregular surface of the inner shelf is presumably part of a Tertiary terrace cut into acoustic basement. This basement is probably equivalent to the Paleozoic-Mesozoic rocks of the Thai-Malay Peninsula. The Sumatra Shelf basin is interpreted as the result of deformation of the Tertiary erosional surface and overlying strata. In the center of this basin, the sedimentary sequence is more than 1 km thick and was deposited at an estimated rate of 10 cm/1,000 yr. The erosional surface, probably an extension of the Sunda Shelf surrounding the Indonesian islands, was first downwarped to form the shelf basin in late Miocene time. Core samples from the Mergui terrace are shelly quartz sandstone containing Pleistocene and reworked Miocene Foraminifera. Depositional rates and sediment thicknesses indicate that the second erosional surface is late Pliocene.

Paleontologic analysis—Results of micropaleontologic and palynologic analyses of Gulf of Thailand rocks reveal that the deeper wells penetrated sedimentary sequences ranging in age from Holocene to Oligocene. Four to five floral zones are recognized. Interpretation of environments of depositional sequences is based on both microfloral and microfaunal evidence (Table 1).

A marked, late Miocene regional unconformity is recognized in the seismic profile and was intersected in the wells at depths from about 1,000 to 4,500 ft (300–1,370 m). In the northern basins, fluvial and lacustrine deposits are associated with lignite, oil shale, and oil. In the peninsula, terrestrial sediments were deposited in association with lignite and marine materials; especially in the west-coast region.

GEOLOGIC HISTORY

Geologic setting—The emplacement of granites in Late Cretaceous–early Tertiary time in Thailand led to the development of the Chao Phraya depression, the Gulf of Thailand, and the structural basins in the north and in the peninsula. The basement of most of the Tertiary basins, therefore, is composed mostly of metasedimentary rocks of Mesozoic–late Paleozoic age and/or granitic rocks.

Depositional systems and basin tectonics—At the beginning of early Tertiary time, most of Thailand probably was elevated above sea level more than at present, and the Gulf may have formed as a small basin with its head a few hundred kilometers south of the present head. It can be assumed that the present 200-ft (61 m) bathymetric contour marked the edge of the Tertiary gulf basin. A well-integrated drainage should have developed with high gradients. The Tertiary drainage system and sediment-dispersal patterns may have been similar to the present ones.

Oligocene sections in the Gulf of Thailand at about 12,000 ft (3,650 m) were deposited under fluvial and deltaic conditions at the lower paleoslope portion of the fluvial system, probably of the ancestral Chao Phraya River. However, the exact location and boundary of the early Tertiary gulf basin cannot be determined until more detailed study of depositional systems and more drilling have been done.

Drilling results also indicate that the Gulf section consists of clastic rock sequences ranging in age from Holocene to Oligocene that were laid down under alternating sequences of environments (see Table 1). Only one regional unconformity is recognized throughout the section and is defined below the upper Miocene section. These data suggest that the Gulf basin consists of several offlapping sequences of clastic wedges.

Local volcanic and intrusive activity, probably in late Miocene time, resulted in isolated diorite stocks cut by later andesite and rhyolite porphyry intrusions along the eastern margins and the floor of the Chao Phraya basin. Regional uplift of the Gulf and the Andaman Basin, accompanied by faulting (Klong Marui, Ranong, Kwae Noi fault zones) and folding, may have occurred in late Miocene time. After a short erosional period, basalt plugs, dikes, and flows were formed at Lam-

Table 1. Generalized Stratigraphic Section of Gulf of Thailand

<table>
<tr><th>Major Lithologic Units</th><th colspan="2">Floral Zones</th><th colspan="3">Environmental Phases</th><th colspan="2">Age</th></tr>
<tr><td>Light gray to gray-brown clays, silty with lignite interbeds</td><td colspan="2">Podocarpus</td><td colspan="3">Inner sublittoral</td><td colspan="2">Quaternary</td></tr>
<tr><td>Light gray clays with sands and lignite interbeds</td><td colspan="2">Dacrydium</td><td>Coastal swamp</td><td>Littoral</td><td>Sublittoral</td><td>Pliocene</td><td rowspan="5">Tertiary</td></tr>
<tr><td>Gray claystones and shales with sandstone, lignite, and coal interbeds</td><td colspan="2">Florschuetzia meridionalls</td><td>Coastal swamp</td><td>Lagoonal</td><td>Coastal plain</td><td>Late Middle Miocene</td></tr>
<tr><td rowspan="2">Varigated shales with sandstone and coal interbeds</td><td rowspan="2">Florschuetzia levipoli</td><td></td><td>Floodplain</td><td>Coastal plain</td><td>Lagoonal</td><td>Middle Miocene</td></tr>
<tr><td>Alnipollenites verus</td><td>Coastal Plain</td><td>Littoral</td><td>Inner-sublittoral</td><td>Early Miocene</td></tr>
<tr><td>Dark gray shale with sandstone interbeds</td><td colspan="2">Florschuetzia trilobata</td><td>Littoral</td><td colspan="2">Inner sublittoral</td><td>Oligocene</td></tr>
</table>

pang, along the southern edge of the Khorat plateau, southeast of Bangkok, near Chanthaburi, and in Kanchanaburi province (west of Bangkok). Deposition resumed and a clastic wedge of late Miocene–Pliocene age was deposited.

The ancient depositional systems that were active in forming the Tertiary clastic wedges and sequences in the Gulf of Thailand and the Chao Phraya fluvial deltaic plain are, therefore, similar to the modern depositional systems of the Gulf region.

Petroleum Potential

Drilling in the Gulf of Thailand and the Chao Phraya deltaic plain encountered Tertiary strata ranging in age from Holocene to Oligocene. These strata consist of abundant sequences of sandstone beds in association with clay, shale, and lignite.

Oil and gas accumulations found in the Tertiary sandstone sequences of the Gulf suggest that the occurrence of oil and gas is controlled primarily by depositional environments, especially deltaic environments. The results of drilling plus other data prove that the Gulf of Thailand, Chao Phraya depression, and the Andaman Sea are clastic- or terrigenous-filled depositional basins of Tertiary age. It may therefore be concluded that the occurrence of oil and gas in the Gulf of Thailand, Chao Phraya depression, and the Andaman Basin is controlled regionally by depositional environments in association with coincident structures. The potentially producing trends are delta, barrier bar–strandplain, shelf, shelf-edge, and submarine-fan trends. Further drilling results and detailed study of depositional facies and environments will aid in locating and defining these potential petroleum-producing trends.

However, if some of these trends were in areas of abnormally high geothermal heat for long periods of time, only uneconomic dry gas may be produced. In certain areas of the Gulf of Thailand, abnormally high geothermal heat has been detected. The highest known heat-flow value in the Andaman Sea, 5.27 $\mu cal/cm^2/sec$, was measured in the Nicober rift valley at a depth of 4,206 m (Rodolfo, 1969). Therefore, certain petroleum potential trends in the Andaman Basin may be controlled by depth and heat-flow patterns.

References Cited

Brown, G. F., et al, 1951, Geologic reconnaissance of the mineral deposits of Thailand: U.S. Geol. Survey Bull., 984, 183 p.

Bunnag, D., 1959, Geophysical prospecting in the Chao Phraya basin, Thailand. *in* Proceedings of the Symposium on the Development of Petroleum Resources of Asia and the Far East: U.N. ECAFE, Mineral Resources Devel. Ser.. No. 10, p. 216–217.

Burton, C. K., and J. D. Bignell, 1969, Cretaceous–Tertiary events in Southeast Asia: Geol. Soc. America Bull., v. 80, p. 681–688.

Emery, K. O. and H. Niino, 1963, Sediments of the Gulf of Thailand and adjacent continental shelf: Geol. Soc. America Bull., v. 74, p. 541–554.

Hunting Geophysics Limited, 1956, Aerial magnetic survey of the Chao Phraya River basin, Thailand: Final rept. to Dept. Mineral Resources (Dept. Mines).

Parke, M. L., Jr., et al, 1971, Structural framework of continental margin in South China Sea: AAPG Bull., v. 55, no. 5, p. 723–751.

Phian-Chạroen, C. and S. Buapeng, 1973, Hydrogeology of the Lower Chao Phraya delta: Dept. Mineral Resources Ground Water Bull. No. 4 (in prep.)

Rodolfo, K. S., 1969, Bathymetry and marine geology of the Andaman basin, and tectonic implications for Southeast Asia: Geol. Soc. America Bull., v. 80, p. 1203–1230.

Petroliferous Taiwan Basins in Framework of Western Pacific Ocean[1]

C. Y. MENG and J. T. CHOU[2]

Abstract The island of Taiwan was formed as a result of collision of continental and oceanic plates, which may have occurred as recently as post-Miocene time. The epicontinental sea on the autochthonous block includes three sedimentary basins, the Southern basin, south of the Peikang-P'enghu basement high, the Northern basin, between the basement high and the Kuanyin swell, and the Taiwan basin on the shelf of the East China Sea north of the Kuanyin swell. North and west of the island, offshore seismic profiles show that there is acoustic basement above which predeformation and postdeformation sedimentary layers are separated by a widespread unconformity. This unconformity developed after the Miocene. Therefore, we conclude that the predeformation layer was folded under the influence of the Miocene tectonic movement on the margin of the China mainland.

Although the sediments of western Taiwan and those offshore to the north and west were deposited in the same sedimentary basin, the tectonic events on land differed from those offshore. On land the orogeny was intense and late, and its influence was limited to development of structures in western Taiwan. Therefore, the youngest structures on land may be the sites for the accumulation of oil. The offshore part of the basin was not influenced by this late tectonic movement. Thus, the sedimentary deposits and the structures resulting from the Miocene tectonic movement on the margin of the continent should be considered as the possible sites for the generation and accumulation of hydrocarbons offshore.

Autochthonous and Allochthonous Blocks

In northeastern Taiwan the structural trend is eastward toward the Ryukyu arc (Fig. 1). The island of Taiwan sustained a strong westward compression (Ho, 1967) and was folded into high mountains on the east. However, beyond the northeastern bend of Taiwan, the structural trend is a result of extension. and forms the characteristic features of an island arc and a marginal sea (Karig, 1971). The strong contrast between Taiwan and the Ryukyu arc may be interpreted as a result of northwestward movement of the Philippine Sea plate and its collision with the Eurasian plate on the east side of Taiwan (Biq, 1960). This collision caused the buffer zone to be amalgamated with the autochthonous block (Meng, 1962). Additional movement caused this allochthonous block to press west and also to shift toward the north. The Philippine Sea plate met no obstruction until as far north as the Ryukyu Trough, where it underwent subduction (Murphy, 1973).

Formation of the island of Taiwan, which resulted from this collision of the continental and oceanic plates (Fig. 2), took a very short time (Meng, 1967). Sedimentary and structural evidence indicates that the bulging may have started after the Miocene (Meng, 1962) and continued to Holocene time. Although the bulging attained great altitude, the area affected is not wide and is limited mostly to the island or nearby seas.

Coupled with this westward movement and northward shifting of the allochthonous block was a great subsidence of the basement under the South China Sea southwest of Taiwan (Chou, 1970). This created the northeast-southwest hinge-line fault running from Yichu to the Taiwan bank (Yichu fault). The east end of this hinge fault extends inland in southern Taiwan (Hsiao, 1971) to where the east corner of the subsiding South China Sea basement was cut off by the western side of the northward-moving allochthonous block (Central Range). The combined effect brought about a "flipped" subduction zone (Murphy, 1973; Ludwig, 1970).

Thus, the island of Taiwan is composed of an autochthonous block and an allochthonous block which have been compressed together (Meng, 1962). The allochthonous block is the highly metamorphosed part of the Central Range (Yen, 1971) east of the major fault extending from the Ilan plain in the north to the P'ingtung plain in the south (Chiang, 1971). The Coastal Range, considered by some to have formed through "obduction" of the oceanic plate (Biq, 1971), composes a small part of the allochthonous block.

Characteristics of Basins on Autochthonous Block

Three sedimentary basins, two west of Taiwan and one north of Taiwan marginal to the China mainland, are included in the epicontinental sea of the autochthonous block (Chou, 1971). From south to north, these are (1) the Southern basin, south of the northeast-southwest-striking Peikang (Peichiang)-P'enghu basement high; (2) the Northern basin, between the basement high and the Kuanyin swell, protruding eastward from the continental margin; (3) the extensive Taiwan ba-

[1]Manuscript received, November 11, 1974.

[2]Chinese Petroleum Corporation. T'aipei, Taiwan, Republic of China.

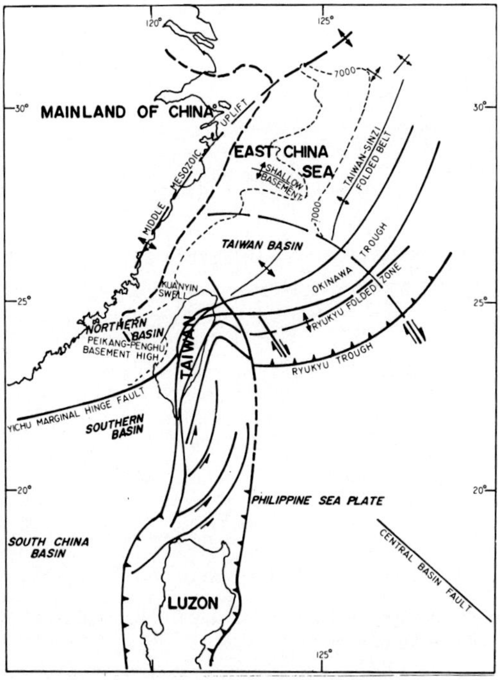

FIG. 1—Structural relation between Philippine Sea plate and China mainland.

sin, on the continental shelf of the East China Sea north of the Kuanyin swell.

Late in the Yenshanian orogeny (late Mesozoic) of mainland China, before the collision of the allochthonous and autochthonous blocks, crustal movements on the autochthonous block caused Cretaceous and older marine deposits to be distorted and indurated. Later, regional emergence, consolidation, and erosion left an uneven basement surface exposed. Not until the Tertiary, after many transgressions and regressions, was this area of the epicontinental sea formed and filled with a sedimentary wedge. The acoustic basement visible in the seismic profile west of Taiwan is equivalent to this unconformity surface. The wells drilled at P'enghu and Peikang in western Taiwan reached this surface, and we may deduce that it is extensive over this area of the epicontinental sea. A land gravity survey in western Taiwan (Hsieh and Hu, 1972) and a marine aeromagnetic survey west of Taiwan (Bosum et al, 1970) generally show the configuration of this surface.

Within the area of the epicontinental sea, from the continental shelf of the East China Sea to the northern edge of the South China Sea, the most noteworthy features are the Taiwan-Sinzi folded belt (Emery et al, 1970), the Kuanyin swell, and the ridge which extends from Peikang through P'enghu to the Taiwan bank, as well as the three sedimentary basins and the thick continental-slope sediments on the outer edges of these basins.

Tectonic Framework of Basins

The three basins differed in their development, and separate tectonic movements were instrumental in the development of each.

1. The framework of orogeny in Fukien and Chekiang on the China mainland affected the formation of the Taiwan-Sinzi folded belt, the Kuanyin swell, and the Peikang–P'enghu–Taiwan bank basement high, resulting in three basins and local distortions of structures involved.

2. The Peikang–P'enghu–Taiwan bank basement high rose and subsided several times. As a result, the sediments on both sides thin or wedge out toward the high; or, in places, sediments have been eroded away or were never deposited (Chiu, 1971; Chou, 1970, 1973).

3. South of the Peikang-P'enghu basement high, a hinge-line growth fault, parallel with the slope of the basement high, caused a thicker sequence of sediments to be deposited in the Southern basin toward the South China Sea. The subsidence of the basement is greatest on land at the southern end of Taiwan.

4. The Peikang-P'enghu basement high subsided in the east, where it intersected a hinge-line growth fault on land. The Miocene sediments of northern Taiwan flowed into the trough resulting from this subsidence; the strata become thicker toward the south (Chang, 1964; Chou, 1973).

5. Between the Peikang-P'enghu basement high and the Kuanyin swell, the cresent-shaped Northern basin continually subsided.

6. In the Taiwan basin north of the Kuanyin swell on the continental shelf, the structures are mostly horsts and grabens resulting from tension.

Petroliferous Prospects

Southern Basin

On the crest of the Peikang-P'enghu basement high, which is the northern boundary of the Southern basin, most of the early and late Miocene sequence is absent. Toward the end of the Miocene, the southern block of the Yichu fault, extending far south of the basement high, subsided greatly. This subsidence resulted from the deep subsidence of the basement to form the South China Sea. Therefore, south of the Yichu fault, the flysch sediments are mostly dark gray mudstone reaching a thickness of 4,000–5,000 m (Chou, 1971, 1973). In many places in this thick

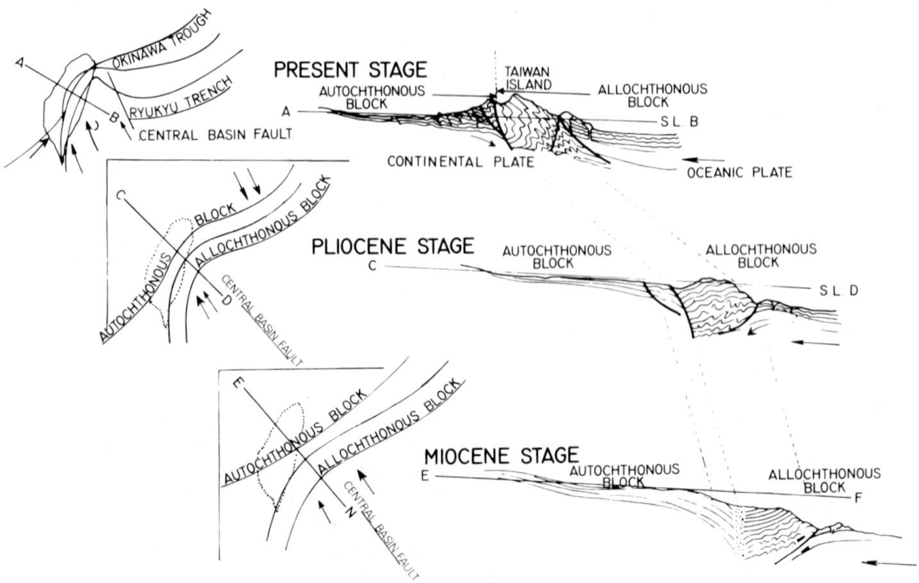

FIG. 2—Sketch showing hypothetical tectonic development of island of Taiwan resulting from collision of autochthonous block with allochthonous block.

mudstone accumulation, sediments are mixed in from submarine slumping. In regions near the Yichu fault, coarse, clean quartz sand washed down from the China mainland and the Peikang-P'enghu basement high and covered the unconformity surface. The thick mudstone cover provides favorable conditions for the accumulation of gas and oil in sandstone reservoirs.

In the foothills east of this mudstone area in southern Taiwan, a thick accumulation of late Miocene and Pliocene fine-grained sandstone and shale resulted from inflow of sediments from the Northern basin through the outlet at the eastern end of the Peikang basement high where the hinge-line growth fault caused great subsidence of the basement. Thus, the basement sank very deep in the basin in which mudstone developed around T'ainan and Kaohsiung and in the trough in which sandstone and shale developed.

Northern Basin

The Northern basin is situated between the Kuanyin swell, protruding east from the margin of the China mainland, and the Peikang-P'enghu basement high. In this basin, which curves toward the west, the basement rises toward the continental margin. The Miocene sedimentary sequence in the Taiwan Strait continues on land in the northwestern part of the island and the depositional environment of the strata becomes more deep sea in character toward the east. The depocenter was near the present-day foothills belt of Taiwan. Owing to rise and subsidence of the basement, the Miocene deposits represent three complete sedimentary cycles of deep- and shallow-marine facies. Westward, on the edge of the continent in the Taiwan Strait, the sandy facies increase, but eastward, near the depocenter, the strata grade to finer grained sandstone and shale. Where the Miocene deposits are thick along the axis of the basin trough, the sequence is complete, indicating continuous subsidence of basement.

Toward the Peikang-P'enghu basement high in the Northern basin, the Miocene strata become thinner, pinch out, or are unconformable, indicating a greater amplitude of rise and subsidence than in the Northern basin. Unconformities on the basement high correspond to regression in the basin, because coarse-grained quartz sandstone and coal seams of littoral facies indicate that the basin also was uplifted, but less than the basement high.

Along the Peikang-P'enghu basement high in the Northern basin, a hinge-line growth fault is detectable as the cause of abrupt thickening of the Miocene sequence in the basin, in contrast to the sequence on the basement high, where strata thin or disappear. This hinge-line fault runs on land in western Taiwan along the south side of the basement high and extends east in an arc.

Between the Central Range in the east and the Northern basin, the crescent-shaped Hsuehshan block bulges toward the west. This block is composed mainly of slightly metamorphosed early Miocene and Paleogene black shale and Eocene coarse-grained quartz sandstone. It may be interpreted as an early extension of the Northern basin which was pushed westward by the allochthonous block into a shape similar to the curved west side of the basin. The longitudinal main fault running west of the Hsuehshan block becomes S-shaped as it continues southward and bulges eastward opposite the Peikang-P'enghu basement high. This indicates that the resistance offered by the Northern basin was not as great as that of the basement high; therefore, the positions of the basin and the basement high before the distortion must be surmised.

The source of sediments before the late Miocene may be determined from the mineral composition of the deposits and the direction of the paleocurrent—that is, from the China mainland. After the late Miocene, the Central Range on the east became the source of sediments. In the northern basin, only a limited area received molasse sediments, and in the Taiwan Strait they are much thinner.

The main gas- and oil-producing area at present is on land, in the northern part of western Taiwan. Production is from folded structures in Miocene sandstone reservoirs, such as the Chuhuangkeng, Chinshui, Tunghsiao, Chingtsaohu, and Chiting anticlines. Under compression from the east, folding in the Northern basin (from on land into the Taiwan Strait) became most intense in the late Pleistocene, so the youngest formations also are folded. The Northern basin manifests the influence of this late crustal movement. Continuous seismic marine profiles outside of this basin all show a postdeformation Pliocene and Pleistocene layer unconformably overlying disturbed Miocene and older formations.

Because of the late deformation of the Northern basin, the migration and accumulation of gas and oil would be later, and the trapping different from, those in the undeformed Pliocene and Pleistocene formations elsewhere.

Taiwan Basin

The Taiwan basin is on the continental shelf north of the Kuanyin swell. It is bounded on the east by the Taiwan-Sinzi folded zone. Seismic surveys found no east-west transverse fault cutting this folded zone near the northern tip of Taiwan, so it should continue all the way to the Peikang-P'enghu basement high. The epicontinental sea extends beyond the Kuanyin swell to include the Northern basin.

The deposits of the Taiwan basin began in the Paleogene. Seismic profiles show an acoustic basement surface and an unconformity between the predeformation and overlying, flat, postdeformation layers.

The structures in this basin resulted from tension. The horsts and grabens present complicated structures. Evidently, this basin—within the Ryukyu arc on the continental shelf—was influenced by the subduction outside the Ryukyu arc. The upper plate expanded southeastward to form the island arc and the Okinawa Trough. Behind the arc, the Taiwan basin was broken into horsts and grabens. Thus, prospecting for accumulations of gas and oil will require somewhat different techniques than in the other two basins.

Petroleum Production

Oil and gas have been produced from onshore fields in northwestern Taiwan since 1904. In 1973, oil was produced at the rate of 4,000 bbl/day and gas at 160 MMcf/day. Gas production is from sandstone reservoirs throughout the Miocene, and oil is from reservoirs of early to middle Miocene age (Chang, 1971).

References Cited

Biq, C. C., 1960, Circum-Pacific tectonics in Taiwan: 21st Internat. Geol. Cong., Nordan, pt. 13, p. 203–214.

——— 1971, Comparison of melange tectonics in Taiwan and in some other mountain belts: Petroleum Geology Taiwan, no. 9, p. 79–104.

Bosum, W., et al, 1970, Aeromagnetic survey of offshore Taiwan: U.N. ECAFE, CCOP Tech. Bull., v. 3, p. 1–34.

Chang, S. S. L., 1964, Regional stratigraphic study of the lower Pliocene and upper Miocene formations in the Chiayi and Hsinying area: Petroleum Geology Taiwan, no. 3, p. 1–10.

——— 1971, Regional stratigraphy and petroleum possibilities of Miocene formations of northwestern Taiwan, China: AAPG Bull., v. 55, no. 10, p. 1838–1865.

Chiang, S. C., 1971, Seismic study of the Chaochou structure, Pingtung, Taiwan: Petroleum Geology Taiwan, no. 8, p. 281.

Chiu, H. T., 1971, Folds in the northern half of western Taiwan: Petroleum Geology Taiwan, no. 8, p. 7–23.

Chou, J. T., 1970, A petrographic study of the Mesozoic rocks underneath the Chiayi plain in western Taiwan: Petroleum Geology Taiwan, no. 7, p. 209–228.

——— 1971, A preliminary study of the stratigraphy

and sedimentation of the mudstone formations in the Tainan area, southern Taiwan: Petroleum Geology Taiwan, no. 8, p. 187–220.

——— 1973, Sedimentology and paleogeography of the upper Cenozoic system of western Taiwan: Geol. Soc. China Proc., no. 16, p. 111–143.

Emery, K. O., et al, 1970, Geological structure and some water characteristics of the East China Sea and the Yellow Sea: Petroleum Geology Taiwan, no. 7, p. 243–280.

Ho, C. S., 1967, Structural evolution and major tectonic forms of Taiwan: Geol. Soc. China Proc., no. 10, p. 3–24.

Hsiao, P. T., 1971, Seismic study of the area between the coastal plain and the foothills, Yunlin, Taiwan: Petroleum Geology Taiwan, no. 8, p. 249–264.

Hsieh, S. H., and C. C. Hu, 1972, Gravimetric and magnetic studies of Taiwan: Petroleum Geology Taiwan, no. 10, p. 283–321.

Karig, D. E., 1971, Origin and development of marginal basins in the western Pacific: Jour. Geophys. Research, v. 76, no. 11, p. 2542–2561.

Ludwig, W. J., 1970, The Manila Trench and West Luzon Trough—III Seismic refraction measurement: Deep Sea Research, v. 17, p. 553–571.

Meng, C. Y., 1962, Wrench-fault tectonics in Taiwan and its relations to the petroleum potentialities, Petroleum Geology Taiwan, no. 1, p. 1–22.

——— 1967, Lateral movement in the northern half of western Taiwan: 3d Symposium on Devel. Petroleum Resources of Asia and Far East Proc., Mineral Resources Development Series, no. 26, v. 1, p. 181–186.

Murphy, Richard, 1973, The Manila trench–West Taiwan foldbelt; flipped subduction zone: Geol. Soc. Malaysia Bull., no. 6, p. 27–42.

Yen, T. P., 1971, Geology of Taiwan—A review, especially on stratigraphy: Sino-American Sci. Coop. Colloquium on Ocean Resources Preprints, v. 1, p. 1–24.

Petroleum Potential of Korean Offshore[1]

CHONG SU KIM[2]

Abstract Areas of major offshore development around the Korean peninsula are in the Yellow Sea off the west coast, in the East China Sea off the south coast, and in narrow belts in the Japan Sea off the east coast. Geological and geophysical surveys in these Korean offshore areas have been carried out since 1966 by the Geological and Mineral Institute of Korea (GMIK) and since 1969 by oil companies that obtained concessions on seven blocks from the Korean government.

The GMIK has conducted reconnaissance seismic and magnetic surveys for general study of submarine geology. Oil companies have conducted preliminary and detailed seismic, magnetic, and gravimetric surveys. Wildcat drilling was started in 1972, and four holes were drilled by three oil companies.

The Yellow Sea area contains a Tertiary sequence thicker than 1,000 m. Several basin structures are separated by uplifted basement or intrusive igneous rocks.

The East China Sea and the southern part of the Japan Sea contain more than 2,000 m of Tertiary strata. The strata are folded and faulted and the structural trend is northeast-southwest.

In the rest of the Japan Sea area, Tertiary strata are present in the Pohang offshore area ($<$1,000 m thick) and extend northward for almost 100 km, whereas, on the continental slope, a 3,000-m-thick sequence is probably of Holocene age.

Petroleum potential in offshore Tertiary rocks is not well known. However, a high organic content increases the prospects in the Yellow Sea area, and the East China Sea and Japan Sea areas have good reservoir rock and structures favorable for hydrocarbon accumulation.

INTRODUCTION

Korea, a country whose geologic features offer little possibility of development of hydrocarbon resources on land, in the early 1960s began an investigation to study the subsea geology and prospects of hydrocarbon resources in the vast Korean continental-shelf areas.

The first attempt was to establish a committee under the Economic Commission for Asia and the Far East (ECAFE) to help in searching for submarine mineral resources. After the Committee for Coordination of Joint Prospecting for Mineral Resources in Asian Offshore Areas (CCOP) was established in February 1966 as a committee of ECAFE, Korea sought assistance from advanced countries through the Committee by submitting a plan for exploring Korean offshore areas.

A general reconnaissance geologic study of the entire Korean continental-shelf areas was conducted (Fig. 1). As a result, foreign oil companies obtained concessions on the whole continental-shelf area of the West Sea (Yellow Sea) and the South Sea (East China Sea), and on the southern part of the East Sea (Sea of Japan) for the purpose of conducting exploration activities. Besides these exploration activities, the Geological and Mineral Institute of Korea, the only national organization conducting exploration of submarine mineral resources, is conducting a detailed geologic survey of nearshore areas and developing its own capability for offshore exploration.

GEOLOGIC CHARACTERISTICS OF LANDMASS OF KOREAN PENINSULA

The Korean peninsula is about 300 km wide and 1,300 km long and has an area of 220,769 km^2. The area of South Korea is about 100,000 km^2.

Topographically, the east coast is uplifted and has a monotonous coastline; in contrast, the south coast and, particularly, the west coast are submerged coasts with irregular coastlines and many large and small islands scattered along them. Dating, by C^{14} methods, of peat from swamps of the west coast indicates that the coast has been submerging at a rate of 0.426 mm/yr for the last 4,000 years and at 1.4 mm/yr during the preceding 2,700 years. The west coast has, therefore, submerged 5.48 m in the last 6,700 years (Park, 1969). It thus appears that the Korean landmass has been uplifted in the east and submerged in the west.

Geologic features of the landmass of South Korea can be grouped into four tectonic units which are, from north to south, the Kyonggi massif, the Okchon geosyncline, the Reinan massif, and the Kyongsang basin (Kim, 1970). These four tectonic units are aligned in a general northeast-southwest direction—the Sinian trend. Thus the tectonics of the Korean landmass can be extended to the China mainland. In fact, the Reinan massif joins with the Fukien massif south of Shanghai, and is called the "Fukien-Reinan massif"; and the Kyonggi massif is related to the Shantung–Laoyeh Ling massif. Actually, the Kyonggi and Reinan massifs can be considered as one on the

[1]Manuscript received, December 26, 1974.

[2]Geological and Mineral Institute of Korea, Seoul, Korea.

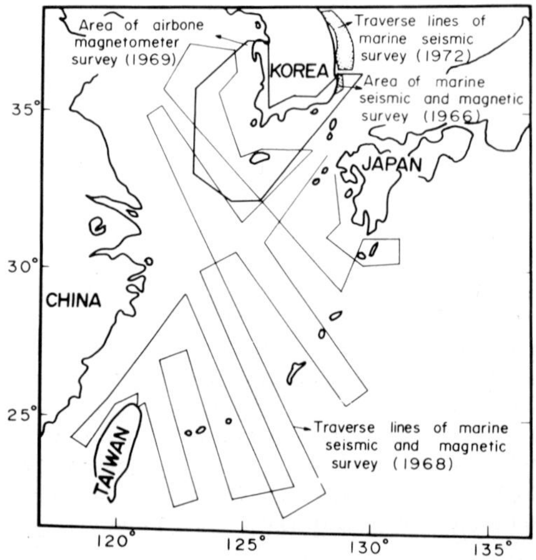

FIG. 1—Map showing traverse lines and area of geophysical surveys conducted in Korean offshore areas since 1966 (work by foreign oil companies is not included).

basis of lithology; they consist of metamorphosed sedimentary rocks and granite gneiss of Precambrian age. The Okchon geosyncline is developed on the Kyonggi and Reinan massifs and is divided into two parts—Paleozoic marine sedimentary formations in the northeastern part and metamorphosed sedimentary formations of unknown age in the southwestern part. These two groups of different ages in the same geosynclinal belt are named the "Okchon neogeosynclinal zone" (northeastern) and the "Okchon paleogeosynclinal zone" (southeastern). They are the result of the migration of the Okchon geosyncline at different times (Kim, 1970).

The Kyongsang basin contains terrestrial sedimentary formations of Mesozoic age. Their correlation with the Inkstone and Wakino Series of western Japan (Kobayashi, 1953) suggests widespread deposition of Mesozoic sediments. The sedimentary formations of the Kyongsang basin show monoclinal structures dipping east; however, near the coast (Pohang and Ulsan areas), shallow marine basins containing Tertiary sedimentary formations are developed on a small scale.

Sedimentary formations of the Kyongsang basin are present in areas of the Okchon geosyncline, all along the south coast and islands, and in the middle part of the west coast. Such distribution indicates that large areas, including areas of the Kyongsang basin and the Okchon geosyncline, were submerged in the Mesozoic and received marine sediments. After uplift and erosion, Mesozoic marine rocks remained on a small scale in the area of the Okchon geosyncline, and the large Kyongsang basin also persisted.

Marine Exploration Activity

On the basis of results of surveys conducted before 1968, the Government of Korea anticipated the possibility of hydrocarbon resources in the sea bottom and claimed seven blocks of concession areas on the continental shelves of the West Sea, the South Sea, and the southern part of the East Sea (Fig. 2). Four foreign companies have made contracts with the government for exploration and exploitation of subsea hydrocarbon resources, and they have carried out marine geophysical surveys since 1969. Three companies have drilled four holes since 1972.

Subsea geologic surveys in the nearshore area of the West Sea have been carried out by the Geological and Mineral Institute of Korea since 1970, and detailed surveys to map subsea geology on a scale of 1:250,000 were begun in 1972. It is planned to cover all the nearshore area of the West Sea by 1976.

Characteristic Features of Continental Shelf

To facilitate discussion of the characteristic features of the continental shelf around the Korean peninsula, I have divided it into three areas—West Sea, South Sea, and East Sea. The West Sea lies between ridges connecting the southwestern part of Korea and the China mainland south of Shanghai (so-called Fukien-Reinan massif), in the south, and the Shantung–Laoyeh Ling ridge in the north. The South Sea lies between the Fukien-Reinan massif, in the west, and a line connecting Pusan in the southeastern part of Korea and Tsushima Island, in the east. The East Sea lies north of a line connecting Pusan and Tsushima (Fig. 3).

West Sea (Yellow Sea)

The Yellow Sea, between the Korean peninsula and the China mainland, in most places has a smooth continental shelf. Along Korea's west coast are many islands having rocky outcrops, and the Han and Kum Rivers flow into the sea. The islands located near the coast consist of the same rocks as does the coastal area, and those far from the coast consist of igneous rocks intruded during the late Tertiary.

The sea bottom has a very smooth topography in general, but small sea-bottom valleys are developed parallel with the coastline by strong bottom currents. Along the east coast of the China mainland, large rivers such as the Yellow and Yangtze

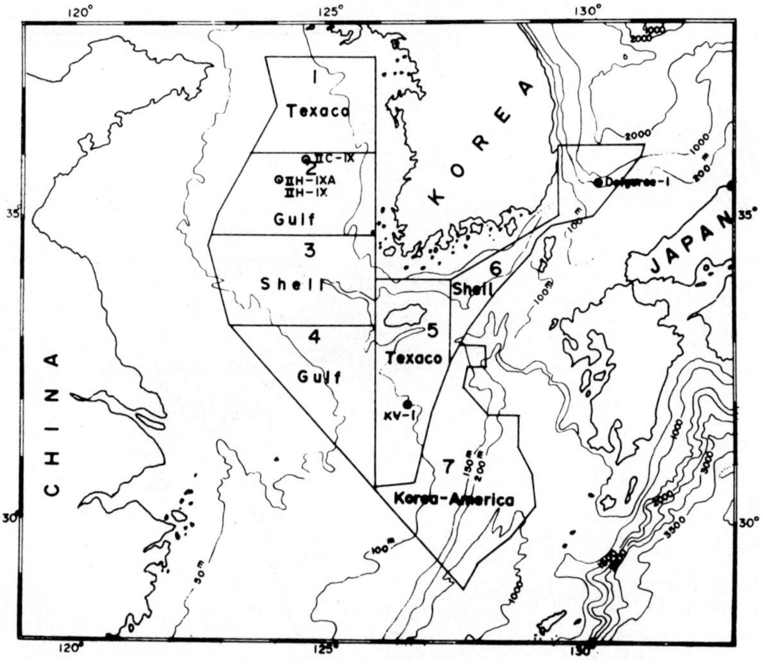

FIG. 2–Map showing concession areas and wildcat drilling sites in Korean offshore area.

Rivers flow into the sea and large deltas are developed. The sea bottom shows a smoother topography than on the Korean side. The bathymetric chart (Fig. 3) shows the maximum depth contour as 40 fm, and depth contour lines are generally aligned parallel with coastal lines of both sides.

South Sea (East China Sea)

Along the southern coast of Korea are many scattered islands consisting of the same rocks as coastal areas. The Naktong River flows into the sea in the eastern part. Sea bottom between the coast and Cheju Island has a rather steep slope, and depth contours are aligned parallel with the coastline. South of Cheju Island, depth contour lines are aligned northwest-southeast, and the sea bottom slopes smoothly eastward into the Okinawa Trough, which is steep near the edge of the continental shelf. Along this edge are several islands that consist of volcanic rocks of Pleistocene age.

In the sea area south from Tsushima Island to the northern end of the Okinawa Trough, contour lines are aligned generally northeast-southwest (Fig. 3).

East Sea (Sea of Japan)

In the sea area between Pohang and Tsushima Island, the continental shelf extends from Korea to Japan. Depth contour lines are aligned parallel with the coastline in a generally northeast direction down to 100-m water depth; deeper contours are aligned east-west. The shelf-slope break occurs on the north, and the continental slope inclines steeply into a small ocean basin at 2,000-m water depth. This basin is called "Ulneung-do (Tsushima) basin." The southern part of the Korean continental borderland, which trends northeast-southwest, can be considered as the eastern extension of the Reinan massif (Fig. 3).

North of Pohang, the continental shelf is only about 25 km wide, and contours parallel the coastline.

Stratigraphy and Structure of Sea Bottom

West Sea

Suspended sediment brought into the Yellow Sea by the Yellow and Yangtze Rivers is deposited as silt and clay on the vast sea bottom covering two thirds of the Yellow Sea. Considering the fact that recent sediments in the deltaic plain north of the Yellow River are more than 960 m thick (Wageman et al, 1970), it is believed that almost 1,000 m of recent sediments is deposited on the sea bottom. These sediments contain organic materials which are derived from land and also some from the sea that are nourished by nutrients supplied by rivers (Emery et al, 1969). On the sea

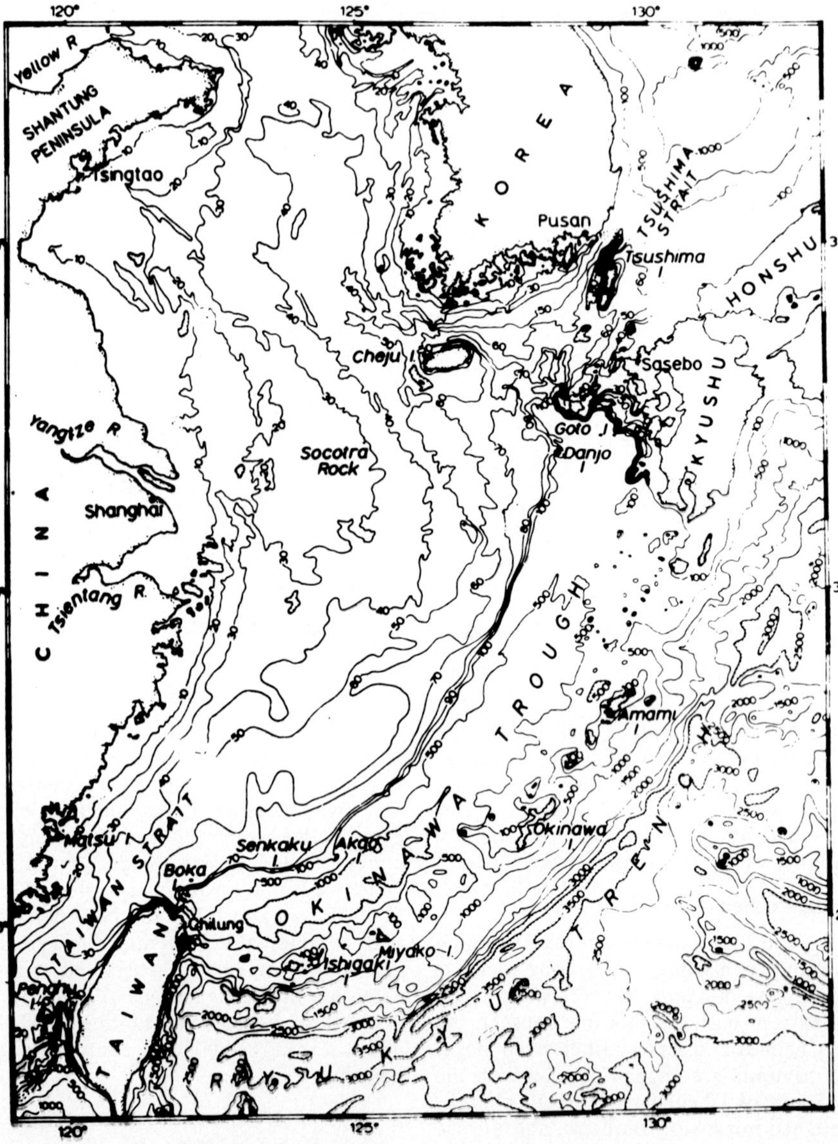

FIG. 3—Bathymetric chart. Contours in fathoms.

bottom near the west coast of Korea, recent sediments consist mainly of sandy materials brought in by the Han and Kum Rivers; their thickness reaches about 250 m at 124°E long.

These thick recent sediments are underlain by Neogene strata and, together, form two basins. The northern basin is elongated northeast-southwest, terminating at the Korean west coast; it contains 1,200–1,300 m of sediments. The southern basin is broad, and its axial trend changes direction from north-south to northwest-southeast. It is near the China mainland and has a sediment thickness of 1,400 m.

Below the Neogene sequence is a third sequence—Paleogene strata and Late Cretaceous shale (Wageman et al, 1970). The lower boundary of this sequence could not be identified, and it is believed that the energy used for the seismic survey was not strong enough to penetrate to the boundary. Considering the 6,000–7,000-m depth to magnetic basement calculated on the basis of data from an airborne-magnetometer survey (Bosum et al, 1971), thickness of this "third layer" is considered to be 4,000–5,000 m.

Development of sedimentary basins in the Yellow Sea can be summarized as follows. In this sedimentary basin limited by the Fukien-Reinan massif on the south and by the Shantung–Laoyeh Ling massif on the north, a succession of sediments was deposited on the pre-Cretaceous basement rocks. The basement surface was sufficiently irregular to cause separation into sedimentary basins, and deposition continued as a result of subsidence of the sea bottom. Also, intrusions of igneous rocks of late Tertiary age and uplift of basement rocks are considered to have separated the sedimentary basins into irregular shapes. In the Korean offshore areas in the Yellow Sea, igneous intrusions are aligned in a nearly north-south direction, which has resulted in a similar orientation of the various subbasins.

South Sea

The pattern of sea-bottom geology in the area between the south coast and Cheju Island differs from that in the area between Cheju Island and the Okinawa Trough. As shown by the results of the airborne-magnetometer survey, subsea geology between the south coast and Cheju Island is considered as an extension of land geology. and the pronounced magnetic anomaly present in narrow zones is interpreted to be caused by intrusion of an igneous body (Fig. 4). Depth to the igneous body is 1.6–3.3 km (Bosum et al, 1971). A remarkable feature is the development of a medium-intensity magnetic anomaly in the sea area west of Cheju Island to the Fukien-Reinan massif. It may be caused by thicker sedimentary layers in this area. The possibility that Paleozoic and Mesozoic sedimentary formations are present here is enhanced by the facts that (1) Paleozoic coal-bearing sedimentary formations extend from the Samchok area, northeastern Korea, to the Hwasun area, southwestern Korea, and (2) Mesozoic shale with oil seeps is present on a small scale in the Haenam area.

In the area between Cheju Island and the Okinawa Trough, sea-bottom geology is characterized by folded sedimentary sequences trending northeast-southwest. Magnetic distribution in the area south of 33°N lat. indicates the presence of a thick sedimentary basin, and a tectonic line developed approximately along the 33rd parallel supports the difference in geologic structure in the south and north. Tertiary sedimentary rocks are present on a small scale in the southern part of Cheju Island and are covered by Quaternary volcanic rocks.

The above-mentioned sedimentary basins extend to Japan and Taiwan, covering large areas. Northeast-southwest folding is represented by the Taiwan-Sinzi folded zone. The Taiwan Basin is developed on the west side, and the Okinawa Trough on the east side. The northern extension covers the northwestern coast of Honshū and the southern extension covers the western half of Taiwan. Neogene units are distributed on Taiwan, Senkoku Island, and Gotō Island near Kyushu, and dredged samples from this sea area have been identified (Emery and Niino, 1967) as Neogene.

The Neogene sequence present in the Taiwan Basin is thought to be rather thick, considering that the thickness of Neogene units in Taiwan is more than 5,000 m (Meng, 1968). Also, a seismic survey indicates a thickness of more than 2,000 m. The Neogene is underlain by Paleogene and Cretaceous rocks, and the total thickness of the sedimentary sequence below the sea bottom of the northwestern part of Taiwan reaches 3,000–11,000 m (Bosum et al, 1970).

Along the edge of the Okinawa Trough and in the Taiwan Basin, vertical faults are developed

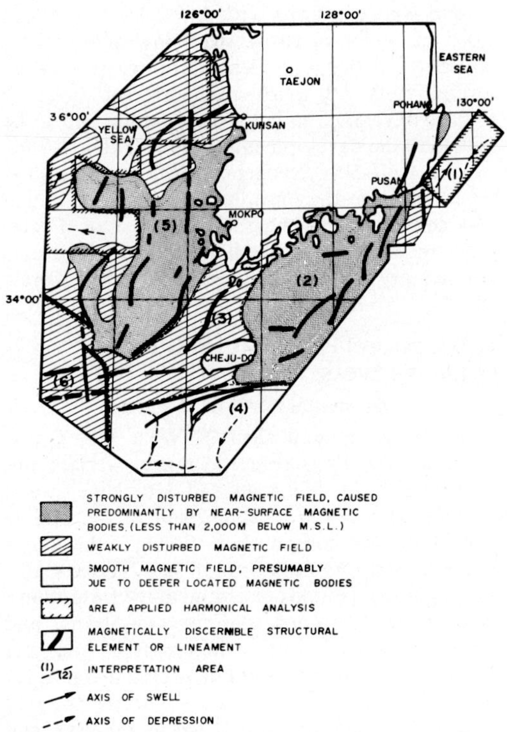

FIG. 4–Structure map of Korean offshore areas, based on aeromagnetic data.

that may have produced the trough (Wageman et al, 1970).

East Sea

Neogene sedimentary formations distributed in the nearshore area around Pohang extend north for 120 km in a zone approximately 25 km wide. This sedimentary basin is cut sharply in the east by an igneous ridge extending north from Kuryong Peninsula, and is bordered on the north by an uplifted metamorphic complex which is considered as an eastward extension of the Reinan massif. North of this complex, Holocene strata lie directly on basement rocks of a granitoid or metamorphic complex and occupy a small basin oriented north-south. The ridge which forms the eastern edge of the southern sedimentary basin consists of metamorphosed volcanic rocks of Paleogene and late Tertiary age. East of this ridge, thick Holocene units (maximum thickness known within the survey area reaches more than 3,000 m) are deposited on the continental slope, which is connected to the small Ulneung-do (Tsushima) basin (Huntec, 1968; Schlüter and Chun, 1973).

The sedimentary basin west of the ridge consists of Neogene units with a thickness of about 1,000 m unconformably overlying older rocks, possibly Paleogene. In this sedimentary basin, several faults with north-south strike are believed to have developed in late Tertiary time and to be related to the development of the recent coastline. Also, the offshore Neogene sedimentary basin is separated from the onshore Pohang basin, which also contains Neogene rocks, by a Late Cretaceous igneous body intruded into the Cretaceous sedimentary formation (Geol. Survey of Korea, 1964).

Hydrocarbon Potential of Korean Offshore Areas

Continental Shelf of West Sea

In the continental shelf of West Sea, Cretaceous and/or Paleogene rocks are overlain unconformably by Neogene and Holocene strata. Cretaceous rocks are also present in the middle part of the west coast and on islands in the southern part. Igneous intrusions of late Tertiary age, causing possible uplift of sedimentary formations, could have developed trapping conditions; and large amounts of organic materials in Neogene and recent sediments could be source material for oil and natural gas. Also, sandy material brought into the sea through rivers such as the Han and Kum is suitable as source material for reservoir rocks.

Thickness of the Neogene and Holocene sequence is less than 2,000 m in the Korean continental-shelf area, but the thickness of the Cretaceous sedimentary formations probably is 4,000–5,000 m. Coal in Cretaceous sedimentary formations in the coastal area is considered as a source material for natural gas. Generally, the deltaic environment is favorable for hydrocarbon generation and trapping. Both the west and east sides of the Yellow Sea contain deltaic environments, and the Fukien-Reinan massif provides conditions for sedimentation under a reducing environment, thus enhancing favorable conditions of hydrocarbon generation (Mainguy, 1970).

Continental Shelf of South Sea

The area east of Cheju Island is considered to contain extensions of the Cretaceous sedimentary formations of the Kyongsang basin, and many igneous intrusions are expected. The part west of Cheju Island is also considered to contain Cretaceous rocks, but igneous intrusions are rather small and Paleozoic sedimentary rocks underlie the Cretaceous. The presence of shale bearing oil seeps on the southwestern coast may indicate the possibility of oil occurrence. According to the data from a geochemical study of Cretaceous rocks of the Kyongsang basin made by Centre de Recherches du Pau of France, Cretaceous sedimentation occurred in a restricted-marine environment.

The sea area south of Cheju Island is underlain by Neogene strata which extend to Taiwan. Considering the occurrence of oil and natural gas in Taiwan and Okinawa, the possibility of finding hydrocarbon resources in this sea area is excellent. Also, oil recently has been discovered west of Gotō Island.

Continental Shelf and Slope of East Sea

The facts that Neogene sediments in the Pohang area were deposited in a reducing environment, and that small amounts of natural gas have been found, offer hope for finding natural gas resources in offfshore areas. In the area between Pohang and Tsushima Island which covers the southern part of the East Sea, thick Neogene deposits are present; they are folded and faulted in a northeast-southwest direction. Below the Neogene are Paleogene strata probably underlain by Cretaceous rocks, as in the Kyongsang basin. Therefore, there is a good possibility that hydrocarbon resources are present.

In the north, adjacent to Pohang, Neogene strata are rather thin in the inner continental-shelf area. Thick Holocene sedimentary rocks are present in the outer continental-slope area, and possibly in the small ocean basin. This deeper sea area may have potential for hydrocarbon accumulations in thick Holocene strata.

CONCLUSIONS

1. The continental shelf of the West Sea was divided into subbasins by igneous intrusions and accompanying uplift of basement during the late Tertiary; the possibility of discovering accumulations of natural gas is good.

2. On the continental shelf of the South Sea, the sea area north of Cheju Island is divided into eastern and western parts on the basis of distintive differences in geologic features. The eastern part is considered to be an extension of the Kyongsang basin. In the western part, where Cretaceous rocks are underlain by Paleozoic sedimentary rocks, the possibilities of accumulation of oil and natural gas cannot be ignored. The sea area south of Cheju Island contains a large basinal area covering Taiwan and the southwestern part of Japan. Thick Neogene rocks are present in it, and the possibility for hydrocarbon accumulation is highly favorable.

3. On the continental shelf of the East Sea, the area between Pohang and Tsushima Island is covered by a thick Neogene section underlain by Cretaceous rocks. The possibility for natural gas accumulation is good. In the area north of Pohang, the continental-shelf area is very narrow and Neogene strata are rather thin; small-scale natural gas accumulations are expected. On the continental slope east of Pohang, a thick Holocene sequence is present, and natural gas accumulation of relatively large scale is expected.

4. To evaluate the possibilities of hydrocarbon accumulation in each of the areas mentioned, detailed and deep seismic surveys and additional drilling are required.

REFERENCES CITED

Bosum, W., E. G. Kind, and J. H. Koo, 1971, Aeromagnetic survey of offshore areas adjoining the Korean Peninsula: U.N. ECAFE, CCOP Tech. Bull., v. 4, p. 1–21.

——— et al, 1970, Aeromagnetic survey of offshore Taiwan: U.N. ECAFE, CCOP Tech. Bull., v. 3, p. 1–34.

Emery, K. O., and Hiroshi Niino, 1967, Stratigraphy and petrôleum prospects of Korea Strait and the East China Sea: Geol. Survey Korea, Rept. Geophys. Explor., v. 1, no. 1, p. 249–263.

——— et al, 1969, Geological structure and some water characteristics of the East China Sea and the Yellow Sea: U.N. ECAFE, CCOP Tech. Bull., v. 2, p. 3–43.

Geological Survey of Korea, 1964, Geological map of Korea, Pohang sheet: scale 1:50,000. Huntec Ltd., Toronto Canada, 1968, Report on the offshore geophysical survey in the Pohang area, Republic of Korea: U.N. ECAFE, CCOP Tech. Bull., v. 1, p. 1–12.

Kim, O. J., 1970, Geology and tectonics of the mid-central region of South Korea: Korean Inst. Mining Geology Jour., v. 2, no. 4, p. 73–90 (in Korean).

Kobayashi, T., 1953. Geology of South Korea: Tokyo Univ. Press.

Mainguy, M., 1970, Regional geology and petroleum prospects of the marine shelves of Eastern Asia: U.N. ECAFE, CCOP Tech. Bull., v. 3, p. 91–107.

Meng, C. Y., 1968, Geologic concepts relating to the petroleum prospects of Taiwan Strait: U.N. ECAFE, CCOP Tech. Bull., v. 1, p. 143–153.

Park, Y. A., 1969, Submergence of the Yellow Sea coast of Korea and stratigraphy of Sinpyeongcheon marsh, Kimje, Korea: Geol. Soc. Korea Jour., v. 5, no. 1, p. 57–66.

Schlüter, H. U., and W. C. Chun, 1973, Seismic surveys off the east coast of South Korea: in press.

Wageman, J. M., T. W. C. Hilde, and K. O. Emery, 1970, Structural framework of East China Sea and Yellow Sea: AAPG Bull., v. 54, no. 9, p. 1611–1643.

Marine Petroleum Exploration of Huksan Platform, Korea[1]

S. B. FRAZIER,[2] S. O. CHOI,[3] B. K. KIM,[4] and D. SCHWARTZ[5]

Abstract Marine geophysical surveys have been performed over the southeastern Yellow Sea, an area composing part of the great North China–Korean platform. A veneer of Plio-Pleistocene sediments covers a broad area of shallow basement called the "Huksan platform." Bordering this positive structural trend, which extends west and southwest of the Korean peninsula, an arcuate series of sedimentary basins is revealed by subcrop patterns below a regional unconformity.

Three exploratory wells that are of significance to a better understanding of the stratigraphy west of Korea have been drilled on the north edge of the study area. Sediments there were deposited in continental environments. High-matrix carbonate in the clastic rocks throughout the section may account for the high interval velocities observed. Although no hydrocarbons were found in rocks penetrated by the wells, additional drilling on the perimeter of the Huksan platform is required before this large area can be fully evaluated.

INTRODUCTION

Figure 1 outlines the mid-1974 configuration of the Blocks I–VII granted for petroleum exploration in the seas surrounding Korea.

Though portions of Blocks II, III, and IV make up the study area of this report, some regional interpretations about Korea and the surrounding seas are expressed on the tectonic map (Fig. 2). There are many published descriptions of the peninsula of Korea; of particular interest for a general discussion of the geologic setting is the paper by O. J. Kim (1971).

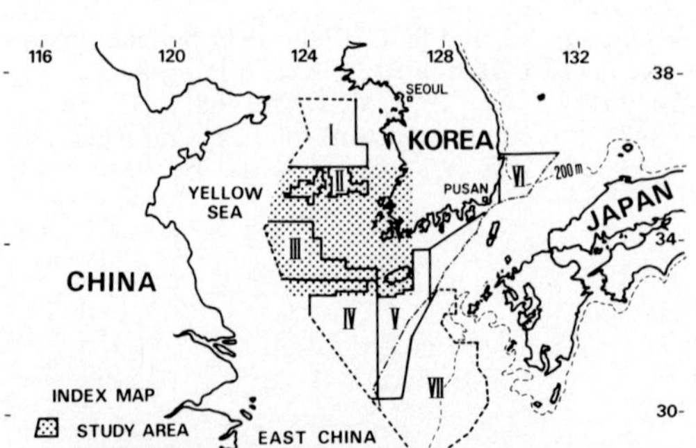

FIG. 1–Index map showing exploration blocks, offshore Korea: I and V, Texaco Korea Inc. and Chevron Oil Co. of Korea; II and IV, Gulf Oil Co. of Korea; III and VI, Korea Shell, N.V.; VII, Korean-American Oil Co., Inc.

The purpose of this paper is to demonstrate the position of the various basins that lie in an arcuate sweep around the Huksan platform—a broad arch located in the Yellow Sea just off the southwestern corner of the Republic of Korea. Information for the maps is derived from seismic, magnetic, and gravity surveys as well as data from three exploratory tests. In addition, geologic, topographic, and hydrographic data have been incorporated into the study.

TECTONIC SETTING

The tectonic map of Korea and surrounding seas (Fig. 2) illustrates the regional tectonic features of the area of concern. Tertiary basins delineated by marine exploration surveys are highlighted, and certain land features are projected out to sea. In trying to extend the tectonic lines, we checked many maps. Some disagreement is found between magnetic, seismic, and topographic interpretations. The following descriptions refer to features not found in onshore Korea. (For an explanation of Korea's onshore framework, see Um, 1974).

[1]Manuscript received, December 26, 1974. Permission to publish was granted by the Bureau of Mines, Ministry of Commerce and Industry, Republic of Korea, and Gulf Oil Company–Asia.

[2]Formerly exploration manager, Gulf Oil Company of Korea, Seoul; presently Gulf Energy and Minerals Co.-U.S., Houston, Texas 77035.

[3]Gulf Oil Company of Korea, Seoul, Korea.

[4]Professor of Geology, Seoul National University; Seoul, Korea.

[5]Formerly geophysicist, Gulf Oil Company of Korea, Seoul; presently Gulf Oil Company of Japan, Tokyo.

Appreciation is extended to C. T. Carlsen and J. M. Cain, who examined well cuttings and logs, and M. S. Oh, for drafting; to R. A. Hodgson, P. E. Howe, J. A. Hafer, E. D. Fleming, and A. Kugler, of Gulf Research & Development Company, who performed geophysical work or supplied data and ideas, and to the staff of Korea Shell, N.V., Seoul, for their contribution of information. Micropaleontologic interpretations of data were made by D. J. Smith and S. J. Stoll; nannofossil studies were done by B. L. Shaffer; palynologic interpretations were made by G. R. Fournier; petrographic interpretations were made by T. D. Altman, E. L. Couch, and T. J. Weismann; and charophytes were studied by Gustavo A. Morales. W. H. Roberts, III and S. J. Stoll critically read the manuscript.

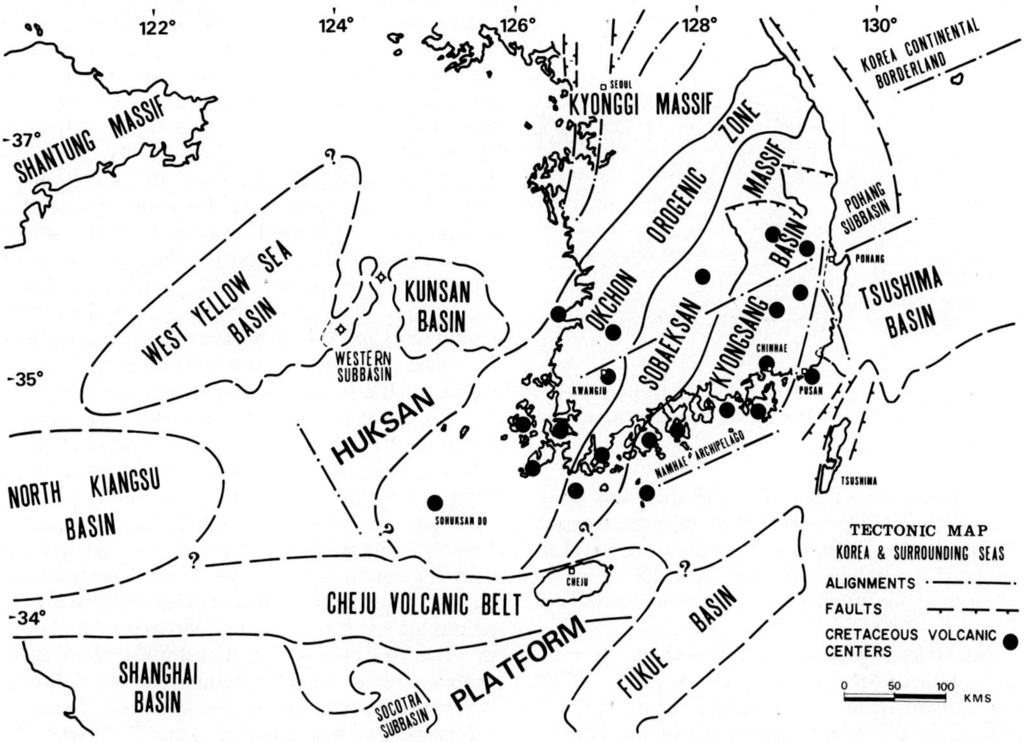

FIG. 2–Tectonic map, Korea and surrounding areas. Alternate spellings of certain names are: Kyonggi (Gyeonggi) massif; Okchon (Ogcheon) orogenic zone; Sobaeksan (Sobaegan or Yongnam, also called "Reinan") massif; Kyongsang (Gyeongsang) basin.

Kunsan basin—The original area of Block II was described by Frazier et al (1972). It contains a gently deformed basin which is named for a nearby Korean port on the Yellow Sea. Seismic lines show a regional unconformity just below a veneer of Plio-Pleistocene strata (Fig. 4). Exploratory well II C-1x was drilled on a basement arch on the west flank of this basin (Fig. 2).

West Yellow Sea, North Kiangsu, and Shanghai basins—These basins in the Yellow Sea were partly outlined by the ECAFE survey (Wageman et al, 1970) and by the tectonic map of Eurasia, (Geol. Inst. Acad. Sci., USSR, 1966).

Western subbasin—Deeply depressed by faulting, this eastern extension of the West Yellow Sea basin contains a sedimentary sequence over 5,000 m thick. Exploratory wells II H-1x and II I-1xA were drilled in this subbasin (Fig. 5).

Socotra subbasin—Named for the only nearby geographic locality—Socotra Rock (U.S. Naval Oceanographic Office, 1959)—this subbasin connects the Shanghai basin with the North Taiwan basin (not shown on Fig. 2) by means of a narrow corridor. In this northwest-southeast corridor or strait can be found about 3,600 m of prospective sedimentary rocks.

Fukue basin—Southeast of the island of Cheju, a Tertiary basin can be outlined in an intermediate position between the Tsushima basin and the North Taiwan basin. On the Coastal Chart (Sanmen Wan to Korea Strait; Defense Mapping Agency, 1973), a seafloor topographic ridge (not shown on Fig. 2) of possible igneous origin marks the southern boundary of this basin and serves to separate it from the North Taiwan basin. In turn, the complex structure of Tsushima Island forms the topographic saddle that separates the Fukue basin from the Tsushima basin.

Huksan platform—This term is proposed for the structurally positive area offshore from southwestern Korea. It composes part of the seaward portions of the tectonic provinces found on land and encompasses a region not readily defined as an extension of an existing feature. Covered by a

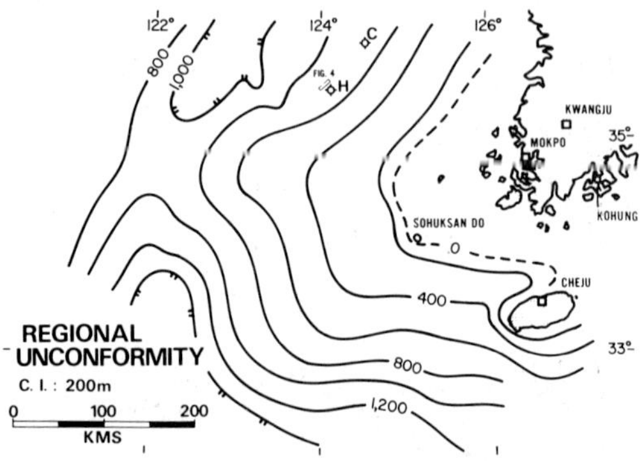

FIG. 3—Structure of regional unconformity surface in study area; datum is base of Plio-Pleistocene (see Fig. 4).

Plio-Pleistocene veneer of sediments, this basement arch is surrounded by an arcuate sweep of shallow basins. Depths to basement on the Huksan platform range from less than 600 m in the north to approximately 1,500 m in the south (Fig. 3).

The Huksan platform is bounded on the south and southeast by the long geosynclinal trough that extends from Taiwan to the Sea of Japan. A portion of this trough just south of the Huksan platform (but not shown on Fig. 2) is called the "North Taiwan basin." Bounded by the trough of the North Taiwan–Fukue–Tsushima basins, all of the other features on the tectonic map, including the Huksan platform, are part of the great North China–Korean platform.

At numerous positions across the Huksan platform, strong magnetic anomalies, sharp bathymetric knolls or spires, and abundant seismic indications give evidence of intrusive or volcanic activity. Although these criteria suggest the presence of igneous rocks, their scattered arrangement could permit other interpretations for the boundaries of the various marine features.

Separated from Korea by the Huksan platform, the Fukien massif of the Cathaysian fold system in China probably does not extend across the Yellow Sea to connect directly with the Sobaeksan (Reinan) massif of the Korean peninsula; nevertheless, this older Sinian trend from China persists to the northeast into southern Korea.

Kwangju-Pohang alignment—An east-northeast lineament (N70°E) stands out clearly on two maps of Korea at scale 1:250,000—as linear topographic segments on the raised relief map (U.S. Army Topographic Command, 1968) and as a series of breaks in the outcrop pattern on the geologic map (Geol. and Mineral Inst. Korea, 1973).

Evidence for the seaward extension of this alignment can be found on the structure map of the Pohang area (Huntec, 1968); the mapped outcrop pattern of shallow igneous rocks indicates that this alignment projects eastward. Also, on the coastal chart (Korea Strait to Mys Nizmennyy; Defense Mapping Agency, 1973), the geometry of the isobaths near the extension of this alignment into the southwest corner of the Sea of Japan offers evidence that this interruption zone could be a wrench fault with right-lateral movement. Age of deformation along this lineament would have to be Late Jurassic (Daebo orogeny), but later movements must have occurred to account for the present topographic alignment.

Late Cretaceous centers of volcanic activity—From their plotted pattern on the map (Fig. 2), it appears that these extrusives could have been triggered by movement along a direction parallel with the strike of the Kwangju-Pohang alignment. Evidence of possible craters is best recognized near the southern coast, where the volcanic centers are distributed in great circles of remnant topographic expression—from Mokp'o to Kohŭng, to Yŏsu, to Chinhae. The distribution of volcanic centers by geologic provinces is as follows: Kyongsang basin, 11 centers; Sobaeksan massif, 3 centers; Okchon orogenic zone, 7 centers (see Fig. 2).

Since these pyroclastic deposits cover a considerable portion of Korea (south of 36°N), there must have been many more volcanic vents than are shown on Figure 2. The most extensive outcrops of acidic-type volcanic rocks occur near the southwest corner of the peninsula, where they can be followed eastward to include the Kohŭng volcanic center. Another area of extensively outcropping volcanic rocks, of a more intermediate type, occurs in southeastern Korea along and west of a north-northeast line parallel with the elongated valleys that form the alignment stretching from Pusan toward Pohang.

Around Chinhae a ring complex stands out on the relief map of Korea (Nahm, 1970). Here, the topographic expression has the appearance of a collapsed caldera with an indented bay stretching far inland to provide the seaport at Masan.

With more study, such lines of volcanic activity could give rise to implications of plate boundaries. (For another view of trend lines in Korea, see the paper by S. M. Lee, 1974.) Also, these centers could have potential for mineral exploration, but such a suggestion is speculative at this time.

Namhae Archipelago alignment—A long line of volcanic centers, from Sohŭksan Do to Pusan, suggests an explanation for the direction of Korea's southern coast. This observation, together with linearity of the sea bottom topography, caused us to place an alignment near the outer reaches of the Namhae Archipelago (Hallyo Sudo islands).

Cheju volcanic belt—A strong east-west trend on the Project Magnet interpretation (Bosum et al, 1971) gave us the first clue to this direction; the change in strike of the contours on the regional unconformity map offered additional evidence (Fig. 3). Confirmation of the pattern was found on seismic profiles near the northwest corner of Block IV, where an anomalous shallow basement area can be mapped. Here, an elongated circular area contains an abundance of intrusive rocks and represents what we interpret to be evidence of late Tertiary volcanism, near 33°05′N, 124° 00′E. Small bumplike protuberances can be seen on seismic sections above the Plio-Pleistocene unconformity surface in that area.

Also, the local configuration of the isobaths on the hydrographic chart adds support to the east-west limits drawn on the map. This direction cuts across the older northeast-southwest trend of the Okchon-Sobaeksan features of the southern Korean peninsula.

A prominent topographic feature on this trend—Mt. Halla on Cheju Island—stands out as a recent volcano. Its last eruption occured in the year 1002 (Suh et al, 1964). Around it a bathymetric depression suggests that the volcanic vent near the center of the island represents a great basalt diapir with a resultant rim syncline.

Pohang subbasin—This area of Tertiary rocks has been described as the "Pohang area" by Um et al (1964), B. K. Kim (1965), and Huntec (1968). Kim dated the section from Miocene to Pliocene. Huntec's seismic surveys indicate that the onshore Neogene sequence of about 800 m thickens eastward into the Tsushima basin. In 1965–67, six dry holes were drilled in this subbasin by the Geological and Mineral Institute of Korea (J. H. Lee et al, 1967).

Tsushima basin and Korea continental borderland—Both of these features are described by Hilde and Wageman (1973).

Geophysical Surveys

Operations

Marine geophysical surveys on Blocks II and IV, made in 1969, 1970, and 1971, resulted in a total of 13,360 km of multisensor control. In

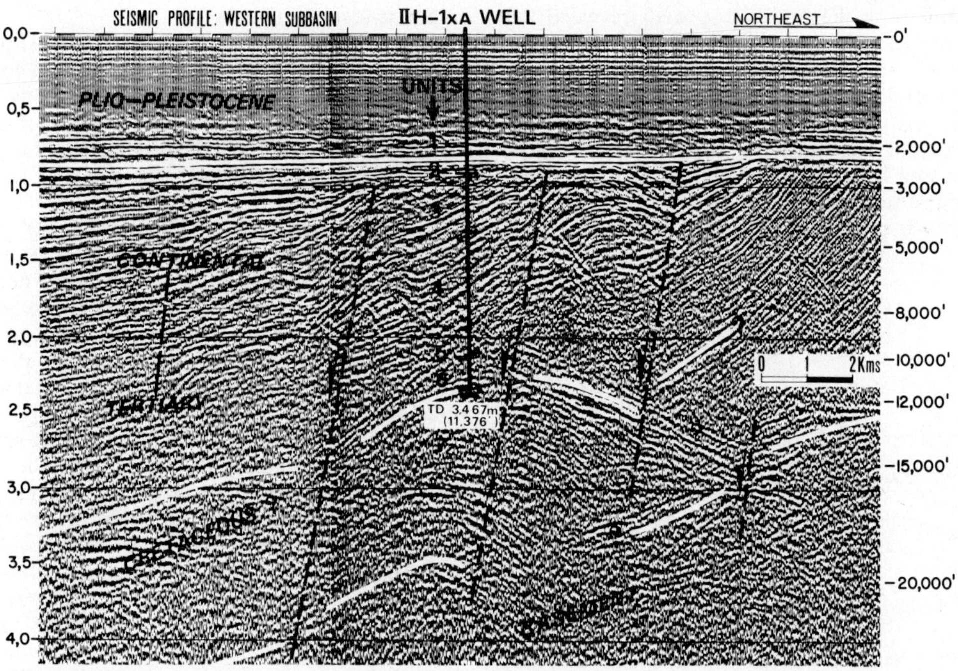

FIG. 4–Seismic profile, Western subbasin. See Figure 5 for location.

Block III, a total of 2,118 km of data was acquired in 1970 and 1971.

Weather proved to be the most serious operational problem and caused most of the lost time. Ships must shut down frequently during the winter gale season from November through March. The typhoon season lasts from June to September, with the highest frequency of storms during August. These typhoons required extensive evasive action by vessels, with resultant loss of time. Experience has demonstrated that the optimum period for geophysical or drilling operations in the Yellow Sea extends from April through July.

Interpretation

Seismic profiles from digital recordings demonstrate a blanket of undeformed, low-velocity strata lying on a marked unconformity (Fig. 4). Below the unconformity, thick sedimentary accumulations are present under portions of the Block II area. Here, the velocities rise abruptly to 3,050–4,575 m/sec, giving an indication of a highly indurated section. Before the wells were drilled, this observation caused us to interpret this section as being of pre-Tertiary age. Partly because of these high velocities, a strong basal reflection is not continuous, but an attempt was made to map the acoustic basement in Block II of the central Yellow Sea (Fig. 5).

Drilling of the exploratory wells revealed that the deeper section (except for the bottom 200 m in II H-1xA) is composed of nonmarine Tertiary rocks, and the abnormal velocities are attributed to a high carbonate content in the calcareous shales and claystones.

In well II C-1x, where the sedimentary section overlies metamorphic rocks, the depth to basement was predicted accurately. However, in well II H-1xA, where the deep section becomes gradually more tightly compacted and indurated, the prediction was more difficult. Reflection seismic information suggested basement velocities at 3,200 m, yet the well was still in sedimentary rock at 3,467 m.

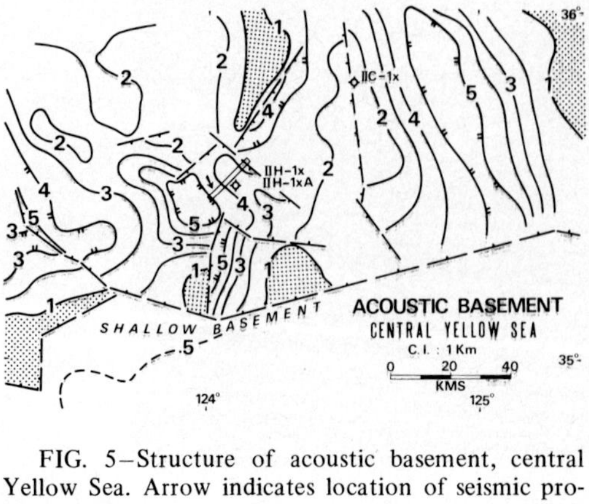

FIG. 5–Structure of acoustic basement, central Yellow Sea. Arrow indicates location of seismic profile of Figure 4.

Gravity control was obtained in conjunction with some of the seismic surveys. The results for Block II have been published previously (Frazier et al, 1972).

Some surface magnetic control was obtained in conjunction with the seismic surveys. The interpretation of these data, combined with those from the aeromagnetic survey—Project Magnet—was carried out at Gulf's research facility in Houston, Texas. The information obtained from magnetics was extremely useful in identifying areas of shallow basement and/or intrusive activity; thus seismic surveys could be concentrated in the more favorable areas.

Exploratory-Well Results

Of several drillable features in Block II prospects II-C and II-H were selected. The three wells drilled (Table 1; Fig. 5) penetrated unconsolidated Plio-Pleistocene sediments above the regional unconformity. Below this break the section consists mainly of continental Tertiary redbeds with minor sandstones (Table 2).

As a result of these wildcat tests, a limited portion of the area has been tested and proved to be unattractive for hydrocarbon accumulations. In our opinion, the wells were "dry" because of lack of source rocks and the limited thickness of reservoir rocks. A considerable area remains to be tested by the drill.

Dipmeter interpretation—Examination of the continuous dipmeter section in well II H-1xA (829–3,464 m) reveals southwest dip as the dominant direction. In the shallower part of the hole, random dips tend to obscure the structural situation until about 1,676 m. There, the rate of dip increases steadily to a maximum of 20° at a depth of 2,134 m.

At 2,170 m, a normal fault occurs with some associated drag folding. Below 2,201 m, the dip decreases slowly to 10° at 2,356 m. An unconformity can be interpreted at 2,406 m; the dip below this depth is less than above it. Below 2,710 m, dip decreases to 5° and the direction of the dip changes to south-southwest. Both this general direction and magnitude are maintained to total depth. Dips in the bottomhole core are difficult to determine, but one contact at 3,465 m indicates approximately 3° of dip.

Table 1. Well Data, Block II

Well	*Location*	*Date Spudded*	*Date Completed*	*Total Depth (m)*
II H-1x	Western subbasin 35°30′32″N, 124°06′12″E	Nov. 12, '72	Suspended Dec. 3, '72, owing to adverse weather conditions; declared lost Mar. 29, '73	828
II H-1xA	Same as above	Mar. 31, '73	Plugged and abandoned, June 8, '73	3,467
II C-1x	Central high area between Western subbasin and Kunsan basin 35°48′50″N, 124°32′38″E	Feb. 11, '73	Plugged and abandoned, Mar. 21, '73	2,017

When the dipmeter is checked against the other electrical logs and the caliper log, the depths of the anomalies tie in with the geologist's interpretation of the induction log. The dips indicated by the dipmeter remain compatible with the geometry as revealed by the seismic reflections (Fig. 4).

STRATIGRAPHY

Miocene to Pleistocene sedimentary deposits occur in Korea in limited outcrops along the east and southeast coasts and on Cheju Island. Along the west coast, Precambrian metasedimentary rocks, Jurassic granites, and a Cretaceous Kyongsang sequence of sedimentary, pyroclastic, and granitic rocks are exposed. No Tertiary deposits are present in that area.

Lithology

On the basis of lithology and well-log information, the section in well II H-1xA is divided into seven units as shown in Figure 4 and Table 2.

Unit 1 (0–719 m) is composed of semiconsolidated gray sandstone and mudstone with intercalated lignites. The contact between units 1 and 2 is marked by a regional unconformity that extends to the west and southwest limits of the study area.

Units 2–6 (719–3,284 m) consist mainly of red-brown mudstone, sandstone, marlstone, and conglomerates. These units can be recognized by their lithologic makeup and by their electric-log characteristics. The seismic and dipmeter data indicate the presence of many unconformities and one acoustic discontinuity between units 2 and 6. The lower part of unit 5 contains a "unique anhydrite-rich shale" that in turn is underlain by an easily recognized sandstone marker bed.

Unit 7 (3,284–3,467 m) is a homogeneous to thinly bedded, red-brown silty shale. The matrix is composed of oxidized and iron-stained clay. The rocks are generally devoid of sedimentary structures except for small-scale bedding and channeling. The unit is similar to the Kyongsang sequence of Cretaceous rocks that can be found in isolated outcrops within the Okchon orogenic zone.

In well II C-1x, units 1, 4, 5, and 6 are correlated with similar units in well II H-1xA, but units 2, 3, and 7 are missing. No metamorphism is noted until, at the bottom of the well (2,017 m), hornfels in a core is taken to be basement. The "unique anhydrite-rich shale" in well II H-1xA is also found in well II C-1x and is the key marker between the wells. The reason for the absence of units 2, 3, and 7 is that the well is located on a structurally high area between two basins (Fig. 5).

Paleontology

The section in well II H-1xA can be broadly divided into four zones on the basis of palynology: (1) 311–497 m, Miocene to Holocene (probably Miocene); (2) 549–716 m, Oligocene to Holocene; (3) 780–3,230 m, Tertiary (characterized by very poor, nondiagnostic palynofloras); and (4) 3,261–3,467 m, indeterminate in age because it is barren. These zones are based on sparse palynofloras and probably are not of regional extent.

Study of the scarce charophytes from this well supports a Tertiary age for the section above 3,261 m.

Scarce, thermally altered *Classopollis* pollen grains and one abraded specimen of the foraminifer *Globotruncana* indicate some reworking of Cretaceous sediments in the Tertiary section.

Table 2. Lithostratigraphic Correlation: Wells II H-1xA and II C-1x[1]

II H-1xA			
2,360		Seafloor–Unit I, veneer	Plio-Pleistocene
	2,360	Regional unconformity	
2,360 - 2,675		Unit 2	
	2,675	2nd unconformity	
2,675 - 4,240		Unit 3	Continental Tertiary
4,240 - 7,895		Unit 4	
	7,895	3rd unconformity	
7,895 - 9,265		Unit 5	
	8,890	Anhydrite-rich shale	Key marker bed
	9,265	4th unconformity	
9,265 - 10,775		Unit 6	Continental Tertiary
	10,775	5th unconformity	
10,775 - 11,376		Unit 7	Cretaceous?
	11,376	Total Depth	
II C-1x			
2,002		Seafloor–Unit 1, veneer	Plio-Pleistocene
	2,002	Regional unconformity	
2,002 - 5,012		Unit 4	Continental Tertiary
	5,012	3rd unconformity	
5,012 - 5,620		Unit 5	
	5,302	Anhydrite-rich shale	Key marker bed
	5,620	4th unconformity	
5,620 - 6,564		Unit 6	Continental Tertiary
	6,564	5th unconformity	
6,564 - 6,619		Basement (hornfels)	Unknown
	6,619	Total Depth	

[1] Depths in feet.

Because of the presence of charophytes and freshwater ostracods and the total absence of nannofossils, all the sediments are thought to have been deposited in continental environments. Floodplain, fluvial, alluvial-fan, lacustrine, and possibly estuarine or lagoonal environments are indicated.

The extremely scarce foraminiferal specimens are most likely reworked but, even if in place, they do not support the existence of open-marine conditions in this area for any significant period of time.

Sedimentation

The sediments are texturally immature, indicating that they were deposited rapidly. They also are immature compositionally, having abundant fragments of volcanic, metamorphic, and sedimentary rocks as detrital components. The exposure of these varied lithologies in the source area and the indicated rapid deposition imply considerable tectonic activity.

Conclusions

1. In the study area, the Kunsan basin, Western subbasin, and Socotra subbasin offer possibilities for additional petroleum exploration in offshore Korea.

2. Certain biostratigraphic and lithologic criteria from the subsurface identify four palynologic zones and seven rocks units in a continental-type facies of Tertiary age.

3. Recognition of an east-northeast trend of the Kwangju-Pohang and Namhae Archipelago alignments adds new directions to the tectonic outline of this area.

4. Our attempt to plot centers of Late Cretaceous volcanic activity points out some caldera-like features not mapped as such before. When verified by more field work, these centers could

be found to mark old plate boundaries. As a speculation, they could highlight areas for mineral exploration.

5. The east-west trend of the Cheju volcanic belt represents a newly determined Tertiary direction of deformation that cuts across the older northeast-southwest Sinian trend; both trends extend from China.

References Cited

Bosum, W., E. G. Kind, and J. H. Koo, 1971, Aeromagnetic survey of offshore areas adjoining the Korean peninsula: U.N. ECAFE, CCOP Tech. Bull., v. 4, p. 1–21.

Defense Mapping Agency, 1973, Coastal charts: N. O. 94028 (San-Men Wan to Korea Strait), 94032 (Northern Part of Yellow Sea including Po-Hai and Liao-Tung Wan), and 95015 (Korea Strait to Mys Nizmennyy), Hydrographic Center, Washington, D.C.

Frazier, S. B., et al, 1972, Korea; petroleum exploration surveys over a portion of the Yellow Sea, *in* Mineral fuels—Combustibles Mineraux, Section 5: Internat. Geol. Cong., Proc.—Cong. Geol. Internat. Prog., no. 24, p. 135–141.

Geological and Mineral Institute of Korea, 1973, Geological maps of Korea: scale 1:250,000, 15 sheets.

Geological Institute of the Academy of Sciences of the USSR, 1966, Tectonic map of Eurasia: scale 1:5,000,000.

Hilde, T. W. C., and J. M. Wageman, 1973, Structure and origin of the Japan Sea, *in* The Western Pacific; island arcs, marginal seas, geochemistry: Nedlands, Univ. Western Australia Press, p. 415–434.

Huntec Ltd., Toronto, Canada, 1968, Report on the offshore geophysical survey in the Pohang area, Republic of Korea: U.N. ECAFE, CCOP Tech. Bull., v. 1, p. 1–12.

Kim, B. K., 1965, The stratigraphic and paleontologic studies on the Tertiary (Miocene) of the Pohang area, Korea: Seoul Univ. Jour., Sci. and Tech. Ser., v. 15, p. 32–121.

Kim, O. J., 1971, Metallogenic epochs and provinces of south Korea: Geol. Soc. Korea Jour., v. 7, no. 1, p. 37–59.

Lee, J. H., S. H. Um, and C. S. Kim, 1967, Report of exploration for natural gas in the Pohang area, Korea: Korea Geol. Survey, Geol. Rept. Coal Fields, no. 8, p. 1–21.

Lee, S. M., 1974, The tectonic setting of Korea, with relation to plate tectonics: Geol. Soc. Korea Jour., v. 10, no. 1, p. 25–36.

Nahm, K. Y., 1970, The tectonics in Korean peninsula shown on the 1:250,000 scale relief map: Geol. Soc. Korea Jour., v. 6, p. 53–60.

Suh, H. G., H. I. Cho and G. Y. Nahm, 1964, Geology and ground-water resources of Cheju island: Korea Geol. Survey, Geol. Ground-Water Resources, no. 1, p. 77–111.

Um, S. H., 1974, The tectonic map of Korea: Submitted by Geol. Min. Inst. Korea, ROK for Inaugural Meeting of Subcomm. for Southeast Asia and . ar East of Comm. for Geol. Map of World.

——— D. W. Lee, and B. S. Bak, 1964, Explanatory text of the geological map of Pohang sheet (Sheet 7022-II): Geol. Survey Korea.

U.S. Army Topographic Command, 1968, Plastic relief map of Korea: scale 1:250,000.

U.S. Naval Oceanographic Office, 1959, Sailing directions for the coast of China: H.O. Pub. 94, sec. 7A-4, p. 340.

Wageman, J. M., T. W. C. Hilde, and K. O. Emery, 1970, Structural framework of East China Sea and Yellow Sea: AAPG Bull., v. 54, no. 9, p. 1611–1643.

Petroleum Fields of Japan with Volcanic-Rock Reservoirs—Summary[1]

T. KATAHIRA and M. UKAI[2]

INTRODUCTION

Oil and gas fields have been found mainly in the Neogene sedimentary basins in the northern half of Japan. Petroleum from the basin along the Japan Sea coast of the northern part of Honshū accounts for about 99% of all Japanese production.

As 5 of 11 major oil and gas fields found during the last 15 years in northeast Honshū island have volcanic-rock reservoirs, volcanic deposits have become an important objective in exploration for petroleum in Japan.

Generally, igneous rocks are inferior to sandstone and carbonate rocks as petroleum reservoir rock; however, igneous rocks can form economic reservoirs in places.

The Neogene sedimentary basin extending along the northern coast of the Sea of Japan began its depression, in association with volcanic activity, in early Miocene time and continued to subside through the Neogene and Quaternary. It is about 700 km long and 80 km wide, and the maximum thickness of the sequence overlying basement is estimated to be about 10,000 m.

Basement rock consists mostly of slightly metamorphosed Paleozoic rocks and granitic rocks which are presumed to have been intruded during the late Mesozoic. Small masses of Mesozoic rocks, but no Paleogene formations, are found in the southern part of the basin.

The basin is divided into two subbasins (Fig. 1): Akita subbasin on the north and Niigata subbasin on the south. They are similar in their geologic development, although differences are seen in the detail of sedimentary facies, amount of sediments, mode and period of volcanic activity, and tectonic movements.

Geologic successions in these subbasins are shown in Figure 2. The formations have been correlated in detail using chiefly planktonic and benthonic foraminifers. Oil and gas in these subbasins occur in the formations ranging in age from middle Miocene to early Pliocene.

Sandstone is the reservoir of many fields. However, the remarkable characteristic in the area is that volcanic rocks and pyroclastic rocks form rather good oil and gas reservoirs in many places. As the sandstone reservoirs are generally tuffaceous, it may be concluded that reservoir rocks in the basin were more or less affected by volcanism.

Oil and gas fields which have reserves of 1 million kl or more are shown in Table 1.[3] Among these, six oil and gas fields occur in volcanic-rock reservoirs.

In the Akita subbasin, only one oil field with a volcanic reservoir has been found, whereas in the Niigata subbasin there are five major oil and gas fields in volcanic reservoirs (Table 1). At least two medium-size gas fields—Kumoide and Fujikawa—can be included. Therefore, the Niigata subbasin may be characterized as a sedimentary basin where development of volcanic-rock reservoirs is remarkable.

This paper is concerned with volcanic-rock reservoirs in the Niigata subbasin only, because more data are available.

CHARACTERISTICS OF VOLCANIC-ROCK RESERVOIRS

Volcanic deposits which form reservoir rocks in this basin range in age from late middle Miocene to early Pliocene. These reservoirs are composed of liparitic and andesitic lava, agglomerate, and tuff breccia. Basaltic reservoirs have not been found. These reservoir rocks (e.g., Yoshii gas field) have many amygdales and, in thin section, show fine-grained andesite with poor phenocrysts. Accordingly, these rocks are considered to have cooled rapidly after eruption. As they are intercalated in marine sediments, they probably represent underwater eruptions. However, there are no data which prove this origin. In thin section, these rocks show no palagonite or chlorite, and only a small amount of celadonite.

The volcanic rocks may have secondary pores such as vugs and/or fractures. These vugs and fractures are believed to originate as the rocks are formed. Because of the rapid cooling, pores from which gas escaped form the vugs, and autobrecciation and/or later tectonic movements may have formed the fractures. Such pores are very important for volcanic rocks to become hydrocarbon-bearing reservoir rocks. Since lavas generally

[1]Manuscript received, November 6, 1974.

[2]Japan Petroleum Exploration Co., Ltd., Tokyo, Japan.

[3]1,000 m^3 of gas is converted to 1 kl of oil.

have such pores, one might expect to find pay zones in the sequence of volcanic and pyroclastic rocks developed largely at the intervals of lava flows.

Oil and Gas Accumulation

Volcanic-rock reservoirs have a general resemblance to carbonate-rock reservoirs. Both have fractures and vugs, relatively high formation resistivities, and reeflike shape. However, the main difference is that carbonate rocks are possible source rocks as well as reservoir rocks, whereas volcanic rocks cannot be source rocks. Therefore, it is important that the source rock be situated near any possible volcanic-rock reservoirs.

At Mitsuke oil field, volcanic and pyroclastic rocks of middle Miocene age (Nanatani Formation, also called "Green Tuff") formed a high in the sedimentary basin, and have remained continuously high through the Neogene and Quaternary because subsidence has been less than in surrounding areas. Petroleum is considered to have migrated into these volcanic bodies from contemporaneous mudstone formations surrounding them.

Development of Oil and Gas Fields

Core data and/or interpretation of geophysical logs indicate that the total effective porosity of pay zones in volcanic-rock reservoirs is generally about 15–20%, whereas secondary porosity, such as fractures and vugs, is estimated to be only about 5%. Effective porosity, therefore, seems to be mainly primary, such as intergranular or intercrystalline porosity. Secondary pores ideally should form connections among primary pores to produce a continuous pore system. Continuity of pore systems varies in the volcanic rocks of different fields.

A typical example of a volcanic-rock reservoir with a limited continuous pore system is the Mitsuke oil field. The reservoir of this field is considered to be mushroomlike bodies of liparite lava. Several small lava bodies form a reeflike mass filled with impervious tuff. In the development of this field, the location and magnitude of the lava bodies presented a serious problem.

In the Katagai and Fujikawa gas fields, characteristics of the volcanic rocks forming the reservoirs are rather uniform. Based on open-flow capacity of producing wells, the potential of the reservoir is a little greater at the top of the structure than at the wing areas. The shapes of these anticlines became distinct and volcanic rocks of the axial area were fractured by tensional stress as these anticlines underwent tectonic movement during the deposition of the overlying formations.

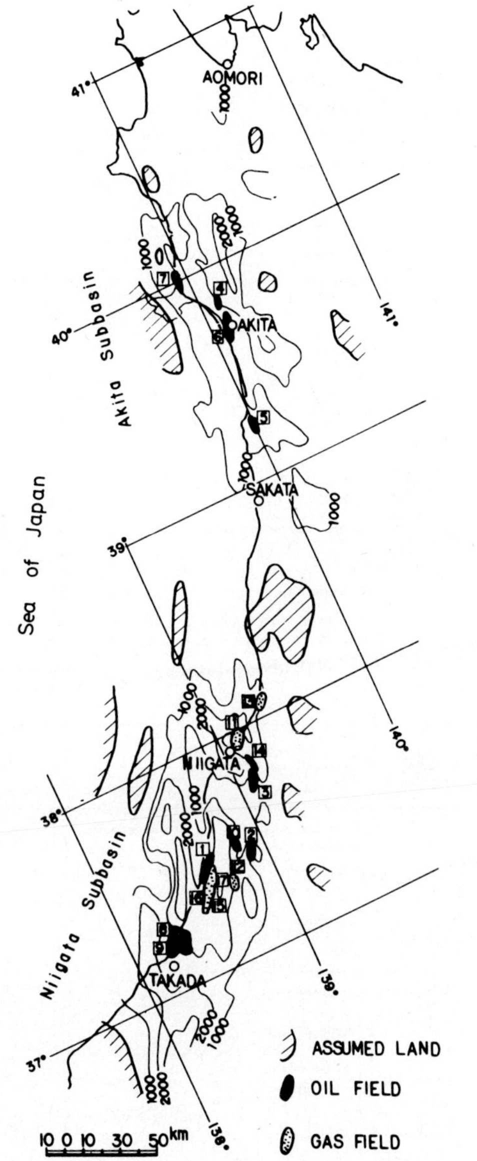

FIG. 1–Neogene sedimentary basin of northeast Japan. Map shows isopachs of upper Miocene sediments and location of major oil and gas fields listed on Table 1. Field numbers are keyed to Table 1.

At Yoshii gas field, now under exploitation, the degree of continuity of the volcanic reservoir is between those of the Mitsuke oil field and the Katagai and Fujikawa gas fields. Figure 3 is an isopore-volume map (drawn by the product of average porosity and net-"pay" thickness of the pay zones) of Yoshii, Myohoji, and Higashi-Kashiwazaki gas fields. The producing capacity of the

Table 1. Major Oil and Gas Fields in Japan[1]

	Name of Field		Sub-basin	Year of Discovery	Producing Horizon	Lithology of Reservoir	Depth of Reservoir
1	Nishiyama	O. F.	N	1883	Sy - Td	Tuffaceous Sandstone	-400 ~ -900 m
2	Higashiyama	O. F.	N	1888	Sy - Td	Sandstone	+400 ~ -700
3	Niitsu	O. F.	N	1896	Sy - Td	Tuffaceous Sandstone	- 50 ~ -850
4	Kurokawa	O. F.	A	1913	On	Andesite	-250 ~ -300
5	Innai	O. F.	A	1923	Kn - Fn	Tuffaceous Sandstone	-600 ~ -1300
6	Yabase	O. F.	A	1933	Kn - On	Tuffaceous Sandstone	-350 ~ -1700
7	Sarukawa	O. F.	A	1958	Kn - On	Tuffaceous Sandstone ~ Tuff	-600 ~ -1500
8	Katamachi	O. F.	N	1958	Td	Tuffaceous Sandstone, Tuff	-400 ~ -1600
9	Kuroi	O. F.	N	1958	Td	Tuffaceous Sandstone, Tuff	-500 ~ -1900
10	Mitsuke	G. F.	N	1958	Nt	Dacite, Dacitic Tuff - breccia	-1500 ~ -2000
11	Higashi - Niigata	G. F.	N	1959	Ny	Sandstone, Tuffaceous Sandstone	-1200 ~ -3000
12	Katagai	G. F.	N	1960	Ny	Andesitic Agglomerate	- 800
13	Nakajo	G. F.	N	1961	Ny - Sy	Sandstone	-800 ~ -2000
14	Minami - Aga	O. F.	N	1964	Sy	Tuffaceous Sandstone	-2250
15	Yoshii	G. F.	N	1968	Nt	Andesite, Andesitic Tuff - breccia	-2600
16	Higashi-Kashiwazaki	G. F.	N	1969	Nt	Andesite, Andesitic Tuff-breccia	-2600
17	Myohoji	G. F.	N	1969	Nt	Andesite, Andesitic Tuff - breccia	-2600

A : Akita Sub-basin; N : Niigata Sub-basin; Kn : Katsurane F.; Fu : Funakawa F.; On : Onnagawa F.; Ny : Nishiyama F.; Sy : Shiiya; Td : Teradomari; Nt : Nanatani

[1] Oil and gas fields with reserves of 1 million kl or more (1,000 m^3 of gas is converted to 1 kl of oil).

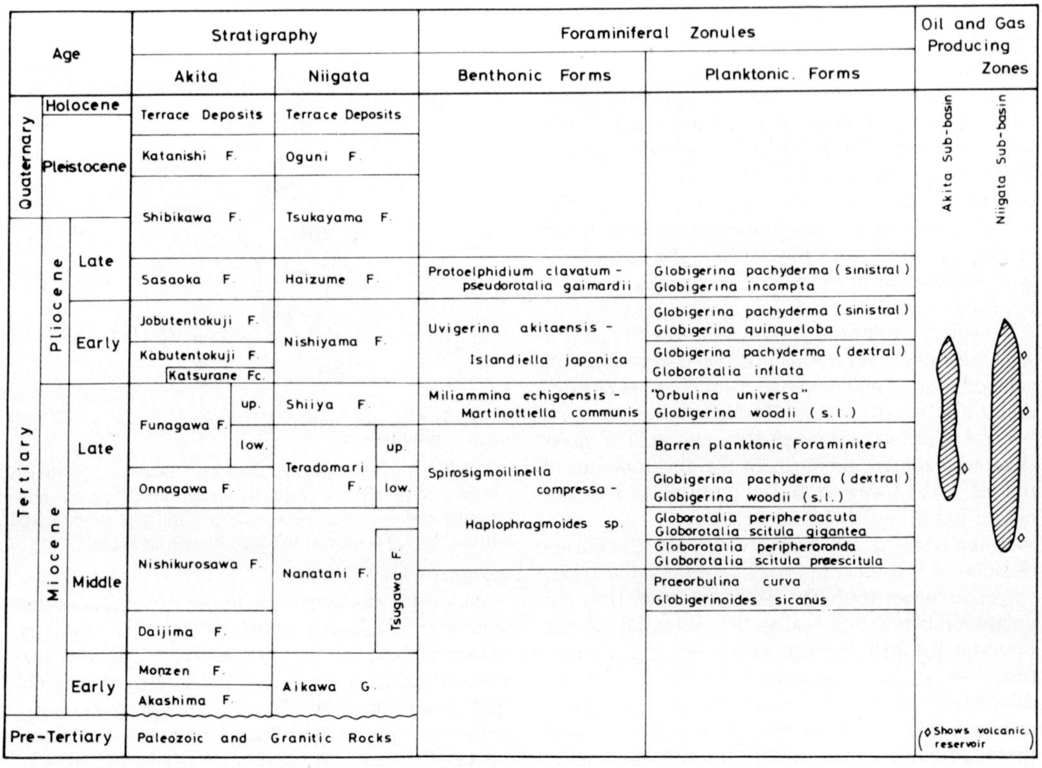

FIG. 2–Cenozoic stratigraphy of oil fields in northeastern Japan.

wells drilled on the five anomalies shown on this map is rather high compared with that in less anomalous areas. These oval anomalous areas are 2–4 km in diameter. Andesite lava is recognized to be developed on these anomalies.

Characteristics of Oil and Gas Production

Oil and gas reservoirs will have certain pressure distributions, according to their permeabilities, distribution of wells, amount of production of hydrocarbons, and encroachment of water. Data from Katagai and Fujikawa gas fields show that there is uniformity of reservoir pressures in both fields, even for varied cumulative production. The Katagai gas field is a dry gas reservoir with no water drive, and the Fujikawa gas field is a gas-condensate reservoir with an active water drive. The standard deviations of pressure distributions are 0.38 kg/cm^2 and 1.7 kg/cm^2 for the Katagai and Fujikawa gas fields, respectively. These fields are good examples of homogeneous reservoirs.

In the Yoshii gas field, the standard deviation of reservoir pressure, obtained from a survey made in October 1971, is 4.72 kg/cm^2. Because this gas-condensate field has been producing for only 4 years, neither the extent of water encroachment nor its future performance can be predicted. However, reservoir pressures of the wells are slightly scattered positively owing to the heterogeneity of the reservoir rock. This heterogeneity of reservoir rock is indicative of the local distribution of lava bodies (Fig. 3). The standard deviation of the pressure of the wells drilled at the anomaly of the central part of the field (Fig. 3) is 2.35 kg/cm^2, and the standard deviation of the pressure of the wells drilled at the southern anomaly is 0.98 kg/cm^2.

The standard deviation of reservoir pressures in Mitsuke oil field (observed in October 1964) is a relatively high value of 4.05 kg/cm^2, which indicates the poor continuity of the volcanic-rock reservoir there.

Although both the Fujikawa gas field and the Mitsuke oil field have fairly strong water drives, the manner of water encroachment is different for each field. In the Fujikawa gas field, water seems to encroach from the northeast side of the field and the gas-water contact is suspected to incline from north to south. Fujikawa SK-9, which was completed in the northeast part of the reservoir with the bottom near the gas-water contact, is an example of high water cut which has risen rapidly. Fujikawa SK-6 produced water at the time when it was completed and the water cut rose gradually, but the water cut seems to depend on the gas production rate. This well is located on

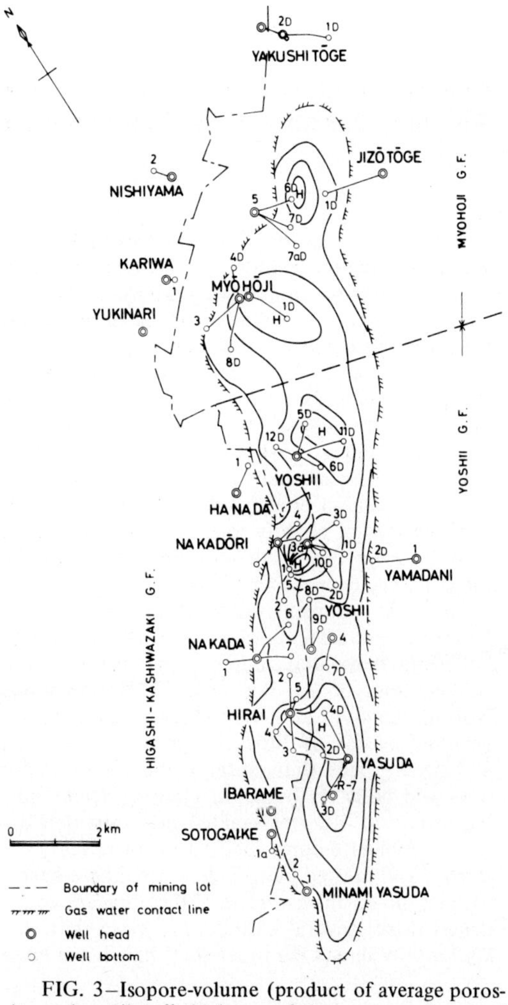

FIG. 3–Isopore-volume (product of average porosity and net-"pay" thickness of the pay zones) map of Yoshii, Myohoji, and Higashi-Kashiwazaki gas fields.

the east flank of the structure and was completed over a long interval from top to bottom of the reservoir.

At Mitsuke oil field, water enters the reservoir vertically from the bottom, so that "drowned" wells are located structurally higher, irrespective of their areal position. Once water is produced, the water cut rises rapidly and oil production declines sharply. As a result, the reduction of the production rate has little effect in reducing the water cut. However, in some wells the water cut rises gradually and is very sensitive to oil production rate. Generally, the higher initial potentials indicate the likelihood of rapid water encroachment. This behavior is believed to be related to the state of fracture development.

Cenozoic and Mesozoic Petroleum Prospects, Aleutian–Bering Sea Region—Summary[1]

D. W. SCHOLL, M. S. MARLOW, and E. C. BUFFINGTON[2]

Petroleum prospects of the extensive Aleutian–Bering Sea region include: (1) thick sections (basins) of Cenozoic and, in some areas, Cretaceous strata underlying broad areas of the Bering shelf; (2) deformed Mesozoic rocks unconformably underlying these basins; (3) domal and diapiric structures associated with the more deeply submerged (water depth 2,000 m) Aleutian and Bowers basins and Umnak plateau areas; (4) thick masses of Neogene beds in summit basins along the crestal region of the Aleutian Ridge; and (5) thick wedges of slightly to moderately deformed beds flanking the north and south sides of the ridge in very deep water (4,000 m).

The most promising prospects are the thick (2–10 km) accumulations of early Tertiary through Holocene beds that underlie the shallow submerged floor of the Bering shelf. For example, Norton, Anadyr', and Bristol basins are large inner-shelf basins of little-deformed coastal-plain and neritic beds that underlie the shelf's major bays and gulfs. Pribilof, St. George, Zhemchug, Navarin, and other associated but unnamed basins are elongate outer-shelf basins of broadly deformed and faulted marine deposits. These basins parallel the northwest trend of the adjacent continental slope; several of them are more than 200 km long. Whereas the inner-shelf basins are large structural sags, many of the outer-shelf basins appear to be fault-controlled grabens or half-grabens. Some of the shelf basins (e.g., Anadyr' and probably Navarin) may include a basal sequence of Late Cretaceous strata. However, many of the outer-shelf basins are underlain by folded Cretaceous and Jurassic(?) beds—stratigraphic units that not only are prospects in themselves but may have supplied hydrocarbons to overlying Cenozoic structures.

Other promising prospects are the fairly large (as much as 30 × 80 km) summit basins of the 2,200-km-long Aleutian Ridge. Roughly rectangular in shape, these structures are physiographic as well as geologic basins. The floors of two of them, the neighboring Amukta (171.7°W) and Amlia (173°W) basins, are overlain by about 1,000 m of water; they are underlaain by a 3–4-km section of sedimentary beds, chiefly of late Miocene and younger age. These basins, elongate parallel with the ridge, are bordered by major normal faults. The infilling section is broadly folded and disrupted along high-angle growth faults.

[1]From Circum-Pacific Energy and Mineral Resources Conference program booklet.

[2]U.S. Geological Survey, Menlo Park, California 94025.

Structural Evolution of Late Mesozoic and Cenozoic Basins in Western North America[1]

EUGENE R. ORWIG, JR.[2]

Abstract The North American segment of the Circum-Pacific, east of the Aleutian Trench, is distinguished from adjoining segments by the absence of an oceanic trench and the lack of recorded seismicity with focal depths exceeding 70 km. The continental margin seaward of the North American Cordillera incorporates a series of irregularly distributed sedimentary basins of late Mesozoic and Cenozoic age. The structural development of these basins was determined by progressive, episodic subsidence accompanied by interbasin uplift.

The tectonic history, structural style, and distribution of the basins demonstrate the presence of a composite system of mobile vertical-stress fields which has endured over the past 140 m.y. The primary forces which created these basins were expressed by increments of episodic and randomly distributed vertical movements, in the same manner in which this fundamental tectonic process is occurring today. A superimposed tangential stress field, also active today, is expressed by strike-slip faults and related structures. These data are mechanically incompatible with hypotheses of overriding crustal plates as currently proposed.

Twenty-one billion barrels of oil and 37 Tcf of gas have been discovered in seven basins. A practical understanding of the structural style and framework of these prime target areas is critical to the realization of the energy potential of the relatively untested basins of this province.

Introduction

The North American segment of the Circum-Pacific province occupies the northeastern one sixth of its perimeter. In contrast to adjoining segments and most of the western Pacific, the area is distinguished by the absence of a deep-sea trench along its seaward flank. Correspondingly, intermediate and deep-focus earthquakes, which generally are associated with trenches, have not been recorded. These distinctions imply that arc-trench tectonic processes are absent in North America, which in turn serves to emphasize the more universal basin-producing processes which are, perhaps, also more fundamental.

The format of this paper is to describe the tectonic history, structural style, and morphology of the upper Mesozoic–Cenozoic marginal basins in this province as the product of a composite system of vertical stresses which has endured over the past 140 m.y. Its objective is to present a frame of reference for the role of basic tectonic processes and dynamic structural style in the occurrence of hydrocarbons.

Tectonic Setting

Metamorphic and plutonic rocks of the Cordilleran fold belt form the stable foreland of the province. Eugeosynclinal deposits of Late Jurassic and Cretaceous age, coeval in part with shelf-basin sediments, are exposed in interbasin uplifts. These rocks also comprise the floor of parts of younger Tertiary basins. Major strike-slip faults are aligned predominantly northwestward.

Thirty-six well-defined basins occupy the marginal belt between the North American Cordillera and the submerged edge of the continental shelf (Fig. 1). They are disposed in an irregular manner lacking linear continuity, and range greatly in areal dimensions; however, there is a general conformity in elongation parallel with the mountain front. Each of these sediment-filled depressions had a distinct and independent history, but all share a common mode of origin and tectonic expression.

Although descriptions of the Alaskan and Canadian portions of this province are not included in this paper, it is my belief that the same genetic elements apply throughout.

Incipient Basin Development

Major uplift of the North American Cordilleran fold belt occurred during the waning stages of Late Jurassic diastrophism, terminating the geosynclinal-orogenic cycle which had its beginning in early Paleozoic time. A new cycle of deformation and sedimentation then came into being on the denuded western flank of the older terrane. The initial stage of Pacific-margin basin formation is conceived as a narrow shelf separating the emergent cordillera from the deep, open ocean of the Pacific. Owing to large gaps in the geologic

[1]Manuscript received, November 11, 1974.

[2]Mobil Oil Corporation, Denver, Colorado 80217.

Appreciation is expressed to Mobil Oil Corporation for permission to publish this paper, and for support including drafting and manuscript preparation. Thanks are due A. G. Alpha, C. E. Tranter, and J. A. Forman of Mobil and Parke Snavely, Jr., E. H. Lathram, and D. L. Jones of the U.S. Geological Survey for helpful critique; however, the writer is solely responsible for the opinions expressed.

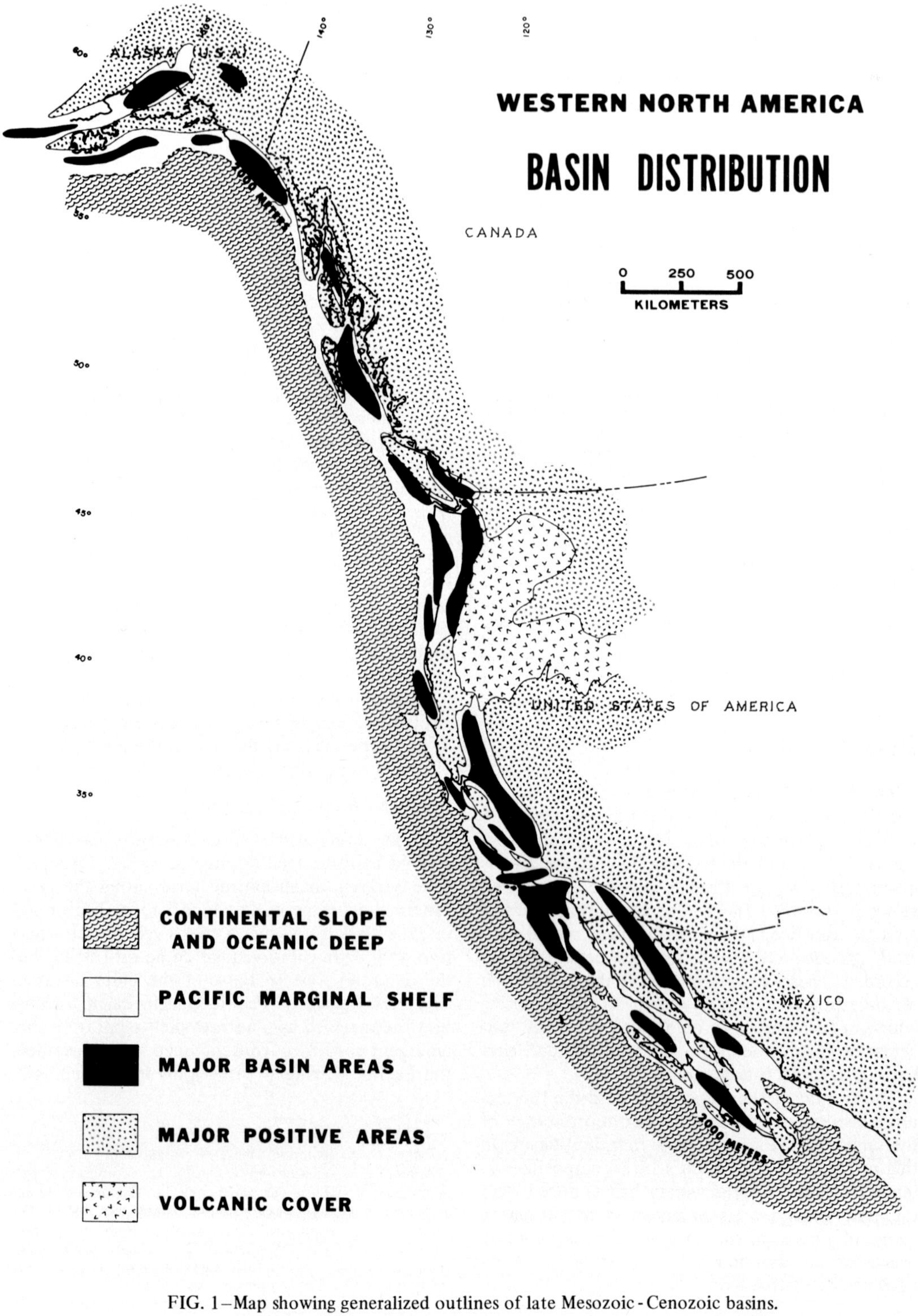

FIG. 1–Map showing generalized outlines of late Mesozoic - Cenozoic basins.

record and serious conflicts between well-documented sets of criteria, the history and amounts of movement on the San Andreas fault are highly controversial. For the purposes of this paper, with its emphasis on basin development, the crustal blocks on either side will be treated independently.

Sediment derived from the highland filled the incipient shelf depressions, and the excess was carried beyond the shelf edge and deposited in the form of coalescing deep-sea fans. This concept is similar to present-day processes.

Aside from physiography, much remains to be learned from detailed study of the internal structure and stratigraphy of modern deep-sea fans. From the sparse data available, it can be stated that the present fans postdate demonstrable diastrophic events in which older fan deposits have been deformed and incorporated in the adjacent continental shelf. No material has been lost; it simply has been added to the continental margin. In at least one area, evidenced by core data from DSDP sites 32 and 34, the present fan dates to Oligocene time. In order to define these coalescing deep-sea-fan deposits—both modern and ancient—the term "continental-slope piedmont" is herein proposed for the collective suite of sediments deposited at the base and impinging upon the continental slope. A new term is considered justified in preference to the use of such terms as "flysch," "trench," or "continental rise," which are inapplicable as originally defined. These deposits are clearly distinguishable genetically, but require recognition, from the shelf-basin sediments with which this paper is primarily concerned.

Cretaceous Evolution

Through the remainder of the Early Cretaceous interval, the seas progressively advanced across the widening shelf left by erosion of the cordilleran mountain front. The beginning of this interval was marked by a localized tectonic event, in which the gently curving, northerly trend of the cordillera was broken by the westward thrusting of a large crustal block known as the "Klamath salient." The relatively brief period during which this thrusting took place is well documented by paleontology and field relations. It occurred near the end of Valanginian time, as determined by a 90-km horizontal offset of the Valanginian shoreline reported by Jones and Irwin (1971), and before submergence of the salient beneath the seas of the succeeding Hauterivian Stage. Following the relatively rapid transgression of the Hauterivian seas across this westward projection, subsequent Early Cretaceous shorelines were realigned in the previous northerly trend.

Widening of the shelf during the Hauterivian to Albian interval permitted the development in Oregon and California of distinct localized depositional troughs separated by the shallowly submerged Klamath salient. Basin subsidence was more than compensated by continued uplift of the cordillera. A superabundance of detrital material was carried from the rising mountains to fill the shelf depressions to thicknesses of 5–10 km and to build the previously formed deep-sea fans off the continental slope outward into a broad and thick continental apron. South of the San Andreas fault, evidence for the clear distinction of individual basins has been obliterated by erosion or later metamorphism, as in Baja California.

Continued transgression of Late Cretaceous seas was accompanied by further differentiation of the earlier basins by intervening uplift. In Baja California, late-phase cordilleran diastrophism metamorphosed the Lower Cretaceous deposits and the seas regressed in this area. Elsewhere, renewed uplift of the Klamath salient and uplift of isolated positive areas in southern California and southwestern Canada separated the shelf into distinct basinal trends. Individual subsiding troughs received sediments from the cordilleran highlands and the isolated positive areas in thicknesses up to 7 or 8 km.

The tectonic setting at the beginning of the Tertiary Period (Fig. 2) summarizes the preceding stages of basin development. North of the San Andreas fault the shelf seas had receded, leaving in their wake broad alluvial plains which, in a relative sense, were still basinal. Major uplifts occurred with maximum intensity along the axes shown, accompanied by a lateral shifting of the axes of maximum subsidence and deposition in the basins. Juvenile uplift of formerly subsident areas, along the site of the present California Coast Ranges north of the San Andreas fault, and in the central coastal belt southwest of the fault, introduced new tectonic elements which strongly influenced subsequent basin development. To the north, along the unstable continental margin, extensive basaltic extrusives began to erupt from a chain of submarine vents.

A model for the tectonic development up to this stage is presented by a series of profiles on Figure 3. In the early formative stage, the rising continental mass is compensated by a slight downwarp on the seaward margin, differentiated into a narrow shelf on the east and the continental-slope piedmont deposited beyond the shelf

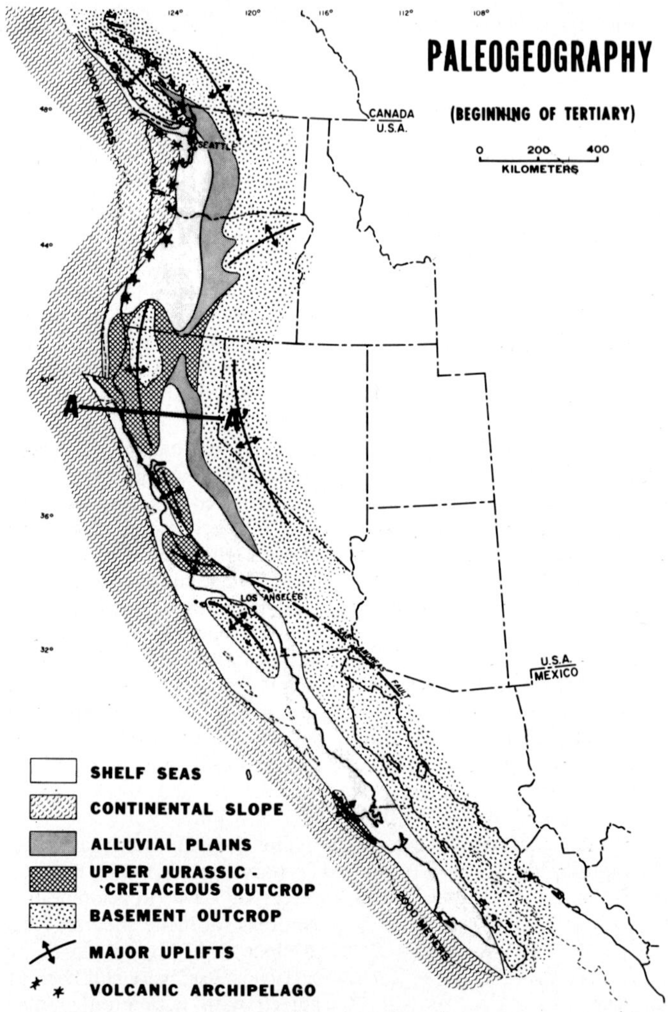

FIG. 2–Generalized paleogeographic map at beginning of Tertiary.

margin. The upper diagram illustrates the situation at the end of this incipient basin stage about 130 m.y. ago. As erosion of the cordilleran front proceeded (second profile), even greater uplift in the foreland was accompanied by distinct basinal subsidence on the widened shelf and the developing foredeep at the foot of the continental slope. The stage of maximum subsidence was reached near the end of Cretaceous time, as shown in the third diagram. The profound depression of the older deposits of the continental-slope piedmont had reached the depths of burial at which the indicated low-temperature metamorphic minerals were developed (Bailey et al, 1964; Coleman, 1967). The stress system which formed the subsiding shelf basin and continental foredeep resulted from strong negative vertical forces, notwithstanding any lateral compressive forces which may or may not have existed. As these negative forces eventually weakened, the ever-present force of gravity became dominant and the thick prism of foredeep sediments was expelled from its isostatically highly unstable former position. The lower diagram (Fig. 3) was drawn from an actual cross section as seen today, but it is representative of the relations at the end of the Cretaceous. The

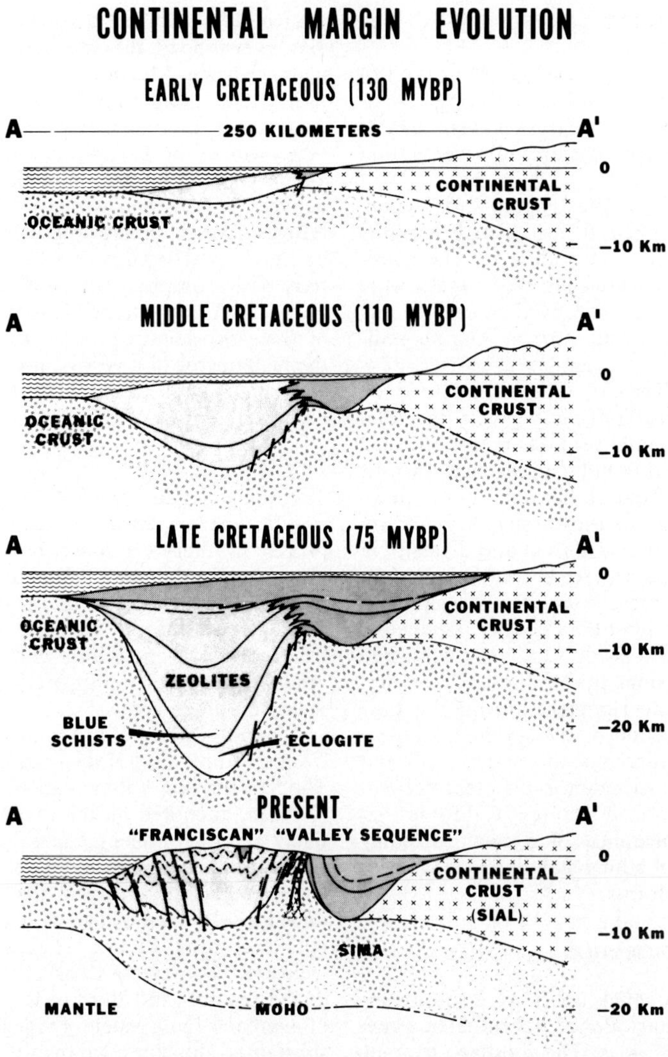

FIG. 3–Structural evolution of Cretaceous basins depicted through section *A-A'* of Fig. 2. Continental-slope piedmont deposits are distinguished from shelf-basin sediments by lighter shading.

emergence of the continental piedmont deposits, which are called "Franciscan" in California (Berkland et al, 1972), was vigorous and was accompanied by intense folding, high-angle reverse faulting, local thrusting, diapiric structures, and emplacement of chaotically disturbed mélange. Serpentinized ultramafic intrusives occur as sills and dikes, perhaps best expressed at the present contact between the Franciscan facies and the coeval shelf-basin sediments known as the "Valley sequence" on the west flank of California's Great Valley. In places oceanic crust has reached the surface, exposing an ophiolite sequence, as would be expected, overlain by a relatively thin pelagic layer. There are no traces, to my knowledge, of foreign-derived material anywhere on the present continent, even in those areas where oceanic basement is exposed. It is concluded that this entire sequence, including its basement, is 100% endogenetic, having been deposited and later deformed in a regime of vertically directed tectonic stress. A corollary conclusion from this model is that no crustal shortening is required to explain the observed structure of this area.

Tertiary Development

The basins continued to develop during the Eocene in the patterns established during Cretaceous time; the primary differences were greater fractionization of earlier basins and accentuation of structural relief.

The marine basalts in Washington and Oregon, as previously mentioned, formed a high volcanic ridge along the lower and middle Eocene continental border. These Hawaiian-type basalts were formed *in situ* as evidenced by their gradational undeformed contact with normal marine sediments to the east.

During middle Tertiary time, represented by Oligocene to middle Miocene deposits, the total extent of the continental seas was more restricted owing to widespread eustatic emergence, and the loci of thick sedimentation became more numerous. Although many of the former elements remained, the irregular distribution and diversity in size of individual basins were becoming manifest.

The outlines of present-day basins were well established by the end of the Pliocene. Thicknesses of sediments in excess of 3 km were deposited in localized depressions during the late Tertiary. The principal new developments during this time were the fractionization of the southern California continental borderland into many small and very deep basins of subsidence, the great deepening of previously formed southern California basins, and the appearance of a new subsiding trough on the site of a long-standing positive area in the Gulf of California.

Quaternary Modification

Continued differential uplift and subsidence during the Quaternary Period accentuated structural relief in most areas. The resultant tectonic pattern of today is shown on Figure 4 in context with the major faults in the area. Many of these strike-slip or wrench faults have significant bearing on hydrocarbon accumulations where they have exerted modifications of internal basin structure.

Basin development is summarized on the profiles of Figure 5 (keyed to lines A–D on Fig. 4). The profiles illustrate maximum subsidence by the age groups indicated, as recorded in thicknesses of marine sediments; each diagram is terminated on the left at the continental margin for corresponding ages. Depths shown have a common datum at the base of the Quaternary.

On profile A, the axis of maximum subsidence in the easterly basin is seen to have migrated westward with time, whereas the coastal-basin axis overstepped the seaward-prograding continental margin. Maximum development of the interior basin along profile B was attained in Cretaceous time; the coastal basin had its greatest rates of growth in the Late Cretaceous and in the late Tertiary. The opposing directions of asymmetry between eastern and western basins in profile C reflects growth of the intervening Coast Range uplift. On profile D, the superposition of late Tertiary basins on preexisting emergent areas as well as older subsident trends is shown. The net effect of these subsidence profiles is the illustration of the high degree of diversity, nondirectional lateral asymmetry, and irregular subsidence rates which characterize the evolution of the basins.

Producing Basins

To illustrate the principles described and their significance to petroleum accumulation, the oil-profilic southern California basins provide examples. Six oil-producing basins are included in the relatively small area—San Joaquin, Salinas, Cuyama, Santa Maria, Ventura, and Los Angeles basins. Twenty-one billion U.S. barrels (2.84 billion MT) of oil are ultimately recoverable from these combined areas.

A variety of structural configurations evolved from a common set of tectonic forces in this area. The San Joaquin Valley is an example of a primary basin, anchored on the east flank by the relatively stable continental mass of the Cordillera. Discontinuous but progressive subsidence from Cretaceous time through the Pleistocene produced a deep axial trough with a composite thickness exceeding 10 km. The San Andreas fault truncated the western flank of the basin and produced second- and third-order structures typical of a wrench-fault system. Evidence of a history of substantial dip-slip components of movement on the fault is also recorded in the strata on the west side. Entrapment of hydrocarbons ranges from sandstone pinchout and truncation along unconformities to complex folds and fault traps on the flanks, and broad open folds in the mildly deformed center of the basin.

The Ventura basin is an example of a trough which once was much broader but has been dissected by intrabasin upthrust blocks. The present compressed Ventura basin, with its thick upper Miocene and Pliocene deposits, is a Neogene development. The basin is flanked by reverse faults which curve downward with depth. These structural data are mechanically compatible with the upthrust concept developed by Proucha et al (1965) for the fracture response of crustal rocks to base-

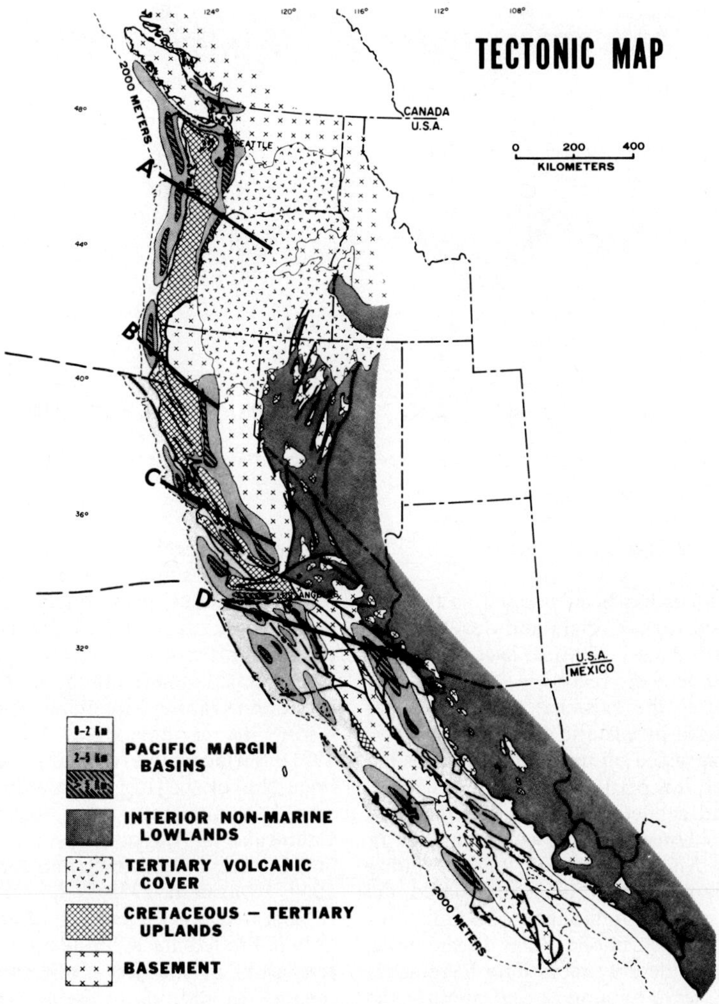

FIG. 4—Map showing present-day composite late Mesozoic - Cenozoic basins in tectonic setting.

ment uplift. Although other interpretations have been applied to this area, including large-scale crustal shortening and block rotation, the rocks simply do not show the strain effects which would inevitably result from such stresses. It is of interest to note that geodetic data show recent uplift on the north flank of the basin in the vicinity of the coast (Buchanan, 1973), and comparative elevation studies (Castle et al, 1974) have established continuing upthrusting of the structurally related San Gabriel Mountains to the east. Owing to the profound relief produced by these late Tertiary events, most of the oil accumulations are directly or indirectly related to the resulting structure.

The Los Angeles basin is a composite of the three primary tectonic elements described: (1) major progressive subsidence generated in a localized trough on the continental shelf, (2) upthrust flanks which, in concert with downwarping, produced high structural relief, and (3) internal basin modification of structure and development of drag folds by strike-slip faulting. The favorable relationship between multiple sandstone reservoirs and interbedded rich source material are largely responsible for the Christmas-tree configuration of many of the oil fields in this area, which are widely known for high petroleum yield—up to 550,000 bbl per surface acre.

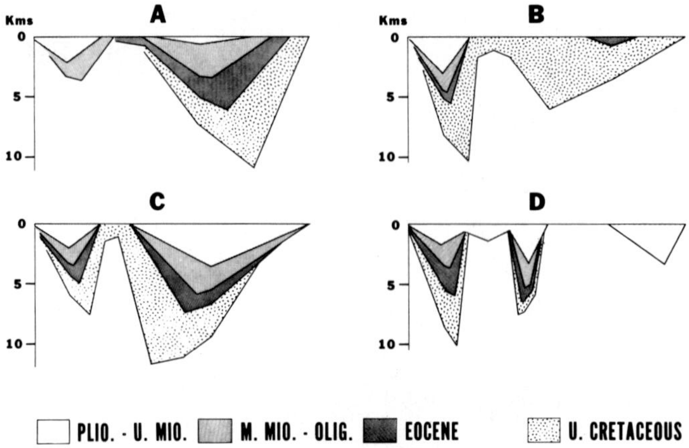

FIG. 5—Basin mobility diagrams showing thicknesses of marine sediments by age groups. Profiles drawn through lines *A-D* on Figure 4.

Geomorphic and Geodetic Evidence

Vertical tectonics has been stressed as the primary mechanism for the origin and development of western North America basins, based on interpretation of the geologic record.

No one doubts the existence of the vertical movements caused by eustatic changes in sea level, so vividly expressed on many coastlines of the world. However, it is pertinent to reemphasize the geomorphic and geodetic evidence of *differential* uplift and subsidence which is commonplace in western North America. In 1949, Gilluly, citing the data available at that time, elucidated the principle of episodicity in mountain-building processes with worldwide implications. In the ensuing 26 years, considerably more data have accumulated which serve to support and amplify the basic truth of this concept.

In general, geomorphic expression of local episodic uplift is supported by geodetic data, earthquake-producing uplifts with or without surface rupture, and older preserved fault scarps. Multiple elevated marine terraces and incised streams with raised lateral benches, truncated tributary fans, and related phenomena are particularly abundant throughout the province. Antecedent streams commonly cross rising structures on valley floors, and "instant" terraces, such as those which appeared during the great Alaska earthquake of 1964 (Plafker, 1965), probably have been formed in many instances in the past.

Fault scarps of historic and prehistoric age, recording net vertical slips up to 10 m, relate to older geomorphic expression of uplifts. Recent investigations by Morton and Yerkes (1974) reveal a recent, but prehistoric, scarp on the pediment surface of the south flank of the eastern San Gabriel Mountains in southern California. Other better-known examples are the scarp produced by the 1872 earthquake at Owens Valley flanking the Sierra Nevada uplift and those produced by uplift of the San Gabriel Mountains near San Fernando during the earthquake of 1971 (Grantz et al, 1971). Surface rupture, recording dip slip or oblique slip of varying magnitudes, has been observed in numerous other localities from Baja California to Alaska (Carson, 1973; Dickinson and Grantz, 1968; Minch and James, 1974; Page, 1969; Stout et al, 1973).

Repeated precise leveling surveys are available only in limited areas, but tectonic growth of positive areas in southern California is well documented. In addition to the data previously cited, the U.S. National Ocean Survey data cover the major highways between San Francisco and Los Angeles. Even if the subsidence attributable to subsurface fluid removal is discounted, differential tectonic uplift or depression is evident in many places (Holdahl, 1974; Lofgren, 1967; Parkin, 1948). Differential subsidence at rates of 15 cm/yr is recorded by tide gauges in southeastern Alaska (S. R. Holdahl, personal commun.), and the adjoining fiord region of western Canada is tectonically rising (Bostrom, 1968). Elsewhere along the North American coast, differential uplift and deformation of marine terraces are well documented by Palmer (1967), Plafker (1965), Plafker and Rubin (1967), and many others.

Although this brief sketch does not do justice to the research done in these investigative fields, a host of geodetic and geomorphic data leaves little doubt that mountains are rising and basins are sinking throughout this dynamic belt. The episodic and randomly distributed nature of the increments of vertical movements observed today fits the geologic record in precise application of the time-honored uniformitarian principle.

SEISMICITY

All recorded seismic events in western North America are of shallow origin—that is, less than 70 km, other than those associated with San Andreas fault movement. These events are scattered, and there does not appear to be a pattern to their distribution. It is clear that the scattered events are attributable to strain release by fracturing of cohesive, brittle rock under conditions of vertical stress. First-motion studies, strain measurements, geodetic data, and other instrumentally produced facts, along with fault-scarp observations and geologic setting, present cogent testimony to this conclusion. Of the 174 recorded severe earthquakes in the western United States between 1769 and 1972 (Mercali intensity VII, Richter magnitude 5.5), 100 are located outside the San Andreas system. A similar proportion, or about 60% vertical movements to 40% lateral, holds for all intensity groupings, large or small.

Of the 57 major recorded earthquakes of intensity IX or magnitude 7.2 and larger in continental western North America, 37 events involved chiefly vertical movement.

The random distribution of seismic epicenters recorded on the North American continent contrasts strongly with other segments of the Circum-Pacific in which linear belts of shallow, intermediate, and deep-focus earthquakes occur. However, the scattered loci of vertical movements in North America are identified with structural elements which, through geodetic, geomorphic, and geologic data, are known to be rising. The gross difference in seismicity is interpreted as being due to the absence in North America of the stress system which has produced Benioff zones elsewhere. Therefore, the stress system which is more fundamental to the creation of the basins in western North America is more clearly revealed.

CONCLUSIONS

Dynamic, vertically directed stress fields have generated episodic increments of crustal movements which evolved, in the course of the 140 m. y. under this regime, into the present basins. These movements represent a fundamental geotectonic process which is overt, perceptible, and measurable. Whereas these vertical forces are seen to be the governing factor in basin evolution, it would be contrary to the purposes of this paper to imply that these are the *only* tectonic stresses in operation. Lateral movements along wrench faults play an important role, albeit a modifying role, and extensional and compressional forces are present as well.

It is well known that geology is a discipline of inductive rather than deductive reasoning, dependent upon accreting knowledge through the observation of the complex interplay of a multitude of natural processes displayed in infinite variety. Efforts to apply an Einsteinian formula or universal law governing all tectonic expression are contrary to the perversity of nature, which does not always behave as we will that it should. Without intending to dispute the concept of plate tectonics, or any other causative mechanism as a working hypothesis where tenable, I strongly suggest that the doctrine of universal application to areas where geologic support is lacking be seriously questioned, or at most retained with an open mind and full awareness of contradictory data.

There lies a great promise of finding important new hydrocarbon reserves in the 36 sedimentary basins defined in western North America, only 7 of of which are currently productive. Imaginative, constructive thinking, constrained by sound principles and inspired by scientific observation, will be required to realize the energy potential of this province.

REFERENCES CITED

Bailey, E. H., et al, 1964, Franciscan and related rocks, and their significance in the geology of western California: California Div. Mines and Geology Bull. 83, p. 89–112.

Berkland, J. O., et al, 1972, What is Franciscan?: AAPG Bull., v. 56, no. 12, p. 2295–2302.

Bostrom, R. C., 1968, Oceanward spreading of the continental crust in the fiord region of northwest America: Trend Engineering, v. 20, no. 2, p. 4–9.

Buchanan, J. M., J. I. Ziony, and R. O. Castle, 1973, Recent elevation changes across part of Transverse Ranges near Ventura, California: Geol. Soc. America Abs. with Programs, v. 5, no. 1, p. 17.

Carson, R. J., 1973, First known active fault in Washington: Washington Div. Mines and Geology Geol. Newsletter, v. 1, no. 3.

Castle, R. O., et al, 1974, Elevation changes preceding the San Fernando earthquake of February 9, 1971: Geology, v. 2, no. 2, p. 61–66.

Coleman, R. G., 1967, Low temperature reaction zones and ultramafic rocks of California, Oregon and Washington: U.S. Geol. Survey Bull. 1247, 49 p.

Dickinson, W. R., and A. Grantz, eds., 1968, Proceedings of conference on geologic problems of San Andreas fault system: Stanford, California, Stanford Univ. Press, 473 p.

Gilluly, James, 1949, Distribution of mountain building in geologic time: Geol. Soc. America Bull., v. 60, p. 561–590.

Grantz, Arthur, et al, 1971, The San Fernando, California, earthquake of February 9, 1971: U.S. Geol. Survey Prof. Paper 733, p. 1–253.

Holdahl, S. R., 1974, Unpublished comments, illustrations, graphics, tables and formulas, relevant to a discussion of sea-slope on the west coast of the United States: National Geodetic Survey, NOS/NOAA, Rockville, Maryland.

Jones, D. L., and W. P. Irwin, 1971, Structural implications of an offset Early Cretaceous shoreline in northern California: Geol. Soc. America Bull, v. 82, p. 815–822.

Lofgren, B. E., 1967, Tectonic movement in the Grapevine area, Kern County, California: U.S. Geol. Survey Prof. Paper 550-B, p. B6–B11.

Minch, J. A., and A. H. James, 1974, Evidence for a major fault parallel to the Gulf of California: Geol. Soc. America, Abs. with Programs, v. 6, no. 3, p. 221–222.

Morton, D. M., and R. F. Yerkes, 1974, Spectacular scarps of the frontal fault system, eastern San Gabriel Mountains, southern California: Geol. Soc. America, Abs. with Programs, v. 6, no. 3, p. 223–224.

Page, R., 1969, Late Cenozoic movement on the Fairweather fault in southeastern Alaska: Geol. Soc. America Bull., v. 80, p. 1873–1877.

Palmer, L. A., 1967, Marine terrace deformation in Pacific coastal United States: Osaka City Univ. Jour. Geosciences, v. 10, p. 69–81.

Parkin, E. J., 1948, Vertical movement in the Los Angeles region, 1906–1946: Am. Geophys. Union Trans., v. 29, no. 1, p. 17–26.

Plafker, G., 1965, Tectonic deformation associated with the 1964 Alaska earthquake: Science, v. 148, no. 3678, p. 1675–1687.

——— and M. Rubin, 1967, Vertical tectonic displacements in south-central Alaska during and prior to the great 1964 earthquake: Osaka City Univ. Jour. Geosciences, v. 10, p. 53–66.

Proucha, J. J., et al, 1965, Basement controlled deformation in Wyoming province of Rocky Mountain Foreland: AAPG Bull., v. 49, no. 7, p. 966–992.

Stout, J. H., et al, 1973, Evidence for Quaternary movement on the McKinley strand of the Denali fault in the Delta River area, Alaska: Geol. Soc. America Bull., v. 84, no. 3, p. 939–947.

Tectonic Framework of Petroliferous Rocks in Alaska[1]

ARTHUR GRANTZ[2] and C. E. KIRSCHNER[3]

Abstract Alaska, which contains about 28% of the land and continental shelf of the United States, is estimated by the U.S. Geological Survey to contain about one third of the nation's undiscovered oil and about one sixth of its undiscovered natural gas. The Survey estimates that fields discovered in Alaska through 1972 ultimately may produce about 26 billion bbl of oil and 68 Tcf of natural gas.

In northern Alaska, Paleozoic and Mesozoic shelf and slope carbonate and clastic rocks of the Brooks Range orogen were thrust relatively northward over the depressed south margin of the Paleozoic and Mesozoic Arctic platform. A foredeep, the Colville geosyncline, developed across the depressed margin of the platform in earliest Cretaceous time. Detritus from the Brooks Range filled the foredeep and prograded northward to fill the Cretaceous and Tertiary North Chukchi and Umiat-Camden basins and form the progradational Beaufort shelf. The largest petroleum reserves (Prudhoe Bay and associated fields) and the best prospects for additional large discoveries in Alaska lie in the areally extensive upper Paleozoic to Tertiary carbonate and clastic rocks of northern Alaska.

In southern Alaska, a series of arc-trench systems developed on oceanic rocks during Jurassic and Cretaceous time. Between these arcs and the metamorphic (continental) terranes of east-central and northern Alaska, large back-arc and arc–trench gap basins received thick volcanic and detrital deposits. These deposits were extensively, and commonly intensely, deformed and disrupted by mid-Jurassic to Tertiary plutonism, Laramide oroclinal bending, wrench faulting, and arc-related compression. This deformation, coupled with low porosity (in part produced by diagenetic mobilization of labile constituents), has left these rocks with only modest, local prospects for petroleum.

Laramide events compressed and consolidated ("continentalized") the late Mesozoic back-arc basin deposits and welded them to the older continental terranes on the north and east. Subsequent sedimentation was localized and nonmarine except in onshore and offshore coastal basins, where thick sections of mixed marine and nonmarine Tertiary sediments accumulated. The Aleutian arc and the associated Queen Charlotte transform-fault system have dominated structural and depositional patterns in southern Alaska, including many of the Tertiary coastal basins, since the early Cenozoic.

The Tertiary coastal basins are areally extensive, and in some areas contain many large folds. They are known to be petroliferous in Bristol Bay and the Gulf of Alaska, and to contain major accumulations of oil and gas at Cook Inlet, but they are relatively little explored.

INTRODUCTION

This paper relates the occurrence of petroleum in Alaska to the tectonic history. We consider that plate-tectonics theory and the ideas of Karig (1971) and others on marginal ocean basins, migrating arcs, and back-arc sedimentation provide the most fruitful available concepts for understanding Alaskan tectonics, and our discussion is colored by these ideas. We present a simplified outline of the tectonics of Alaska to provide a conceptual basis for evaluating its petroleum resources. For the sake of brevity, we use an assertive style that may imply a greater confidence than we actually have regarding details of the tectonic history; and we have not attempted to reconcile the many recently proposed worthy, but commonly conflicting, ideas on the subject. For our purposes, Alaska includes both its land areas and bordering continental terraces. These terraces constitute more than half of the total continental mass in and adjacent to Alaska and offer large tracts that appear to be attractive for exploration.

TECTONIC SETTING

Alaska lies in the northwestern part of the late Mesozoic and Cenozoic North American crustal plate. Seafloor spreading in the North Atlantic has produced a continent-to-continent compressional boundary between the North American and Eurasian plates along the west front of the Cherskiy fold belt of northeastern USSR (Churkin, 1972), as well as extensive east-west megascale compression in western and northern Alaska. Similarly, spreading at the East Pacific Rise has produced a compressional boundary between oceanic crust of the Pacific plate and its predecessors and continental rocks of the evolving North American plate near the present Aleutian arc and Gulf of Alaska. These two interplate collisions have dominated the late Mesozoic and Cenozoic tectonic evolution of Alaska.

Viewed broadly (Fig. 1), Alaska is an isthmus between Eurasia and North America. Its topography and structure are dominated by the Cordilleran fold belt. Collisions with the Eurasian continental plate and the Pacific oceanic plate

[1]Manuscript received, November 21, 1974. Published with permission of U.S. Geological Survey and Standard Oil Company of California.

[2]U.S. Geological Survey, Menlo Park, California 94025.

[3]Amoseas Indonesia Inc., Jakarta, Indonesia.

We are indebted to many of our colleagues for helpful information and comments, and to W. P. Brosgé, E. H. Lathram, and I. L. Tailleur for reviewing the manuscript.

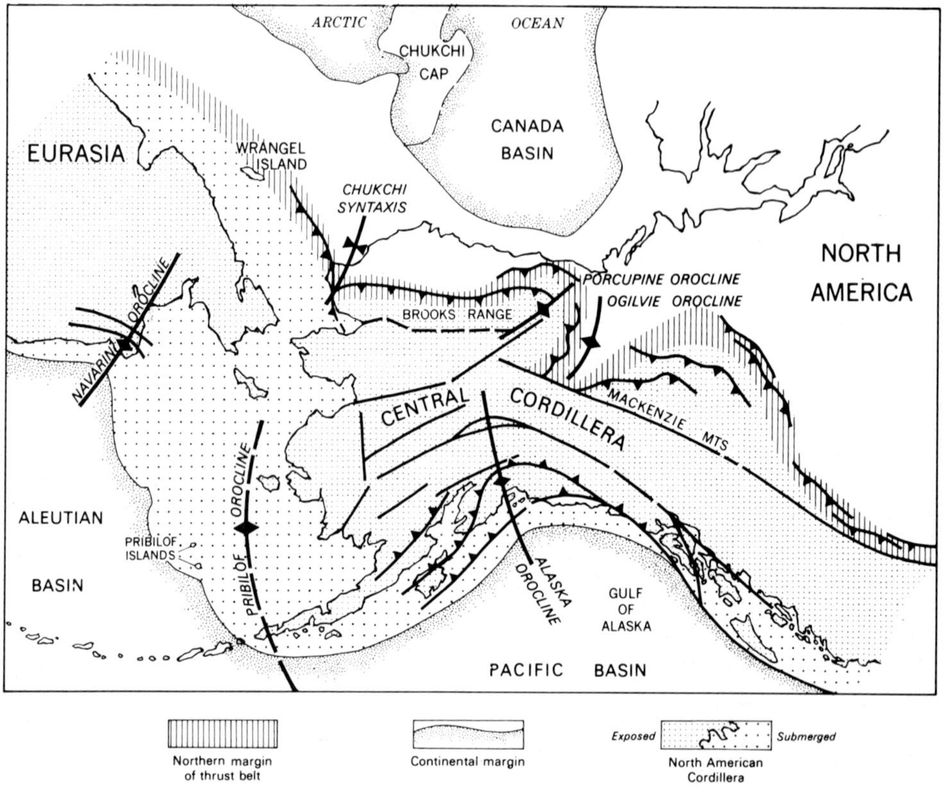

FIG. 1—Broad patterns of Cordilleran belt in Alaska.

produced Laramide megascale oroclinal deformation in the fold belt and enlarged it by the accretion of oceanic rocks. The collision with Eurasia is postulated to have produced the genetically linked Alaska, Pribilof, and Navarin oroclines in southern Alaska and the Bering shelf, and the Chukchi syntaxis and Porcupine and Ogilvie oroclines in northern Alaska.

Physiography and Structural Setting

The major physiographic provinces of Alaska and some of the major faults are shown in Figure 2. The Southern and Interior provinces and the Brooks Range of the Northern province make up the Cordilleran system in Alaska. The Northern province is dominated by a late Paleozoic–Mesozoic shelf and a Cretaceous-Tertiary thrust-fault system, whereas the Interior and Southern provinces are dominated, respectively, by deformed late Mesozoic back-arc basins and Mesozoic and Tertiary arc-trench systems.

Northern Province

The Northern province consists of the tectonically related Arctic coastal plain, foothills, and Brooks Range subprovinces. The coastal plain and foothills are underlain by a stable shelf, the Arctic platform, containing Upper Devonian to lowest Cretaceous rocks of northern provenance. The shelf is apparently truncated on the north by the linear Beaufort scarp, a steep continental slope, and is disrupted on the south and southwest by Cretaceous and early Tertiary thrust faults in the Brooks Range and central Chukchi Sea. To the east it is partly uplifted in the northeastern Brooks Range and partly underlies Cretaceous and Tertiary beds beneath the Arctic coastal plain, Beaufort shelf, and Mackenzie delta. The delta was localized at the intersection of the Beaufort scarp and the Canadian platform subsequent to development of the Canada basin.

Proposed origins of the Beaufort scarp must be compatible with those for the Canada basin and Alpha Ridge of the Arctic basin (Figs. 1, 2). If the Alpha Ridge was an active spreading axis from 200 to 40 m.y. ago (see Ostenso, 1973, for a recent summary of this hypothesis), then the Beaufort scarp could be an extinct transform fault that extended east from the western end of the ridge. However, the presence of latest Cretaceous rocks

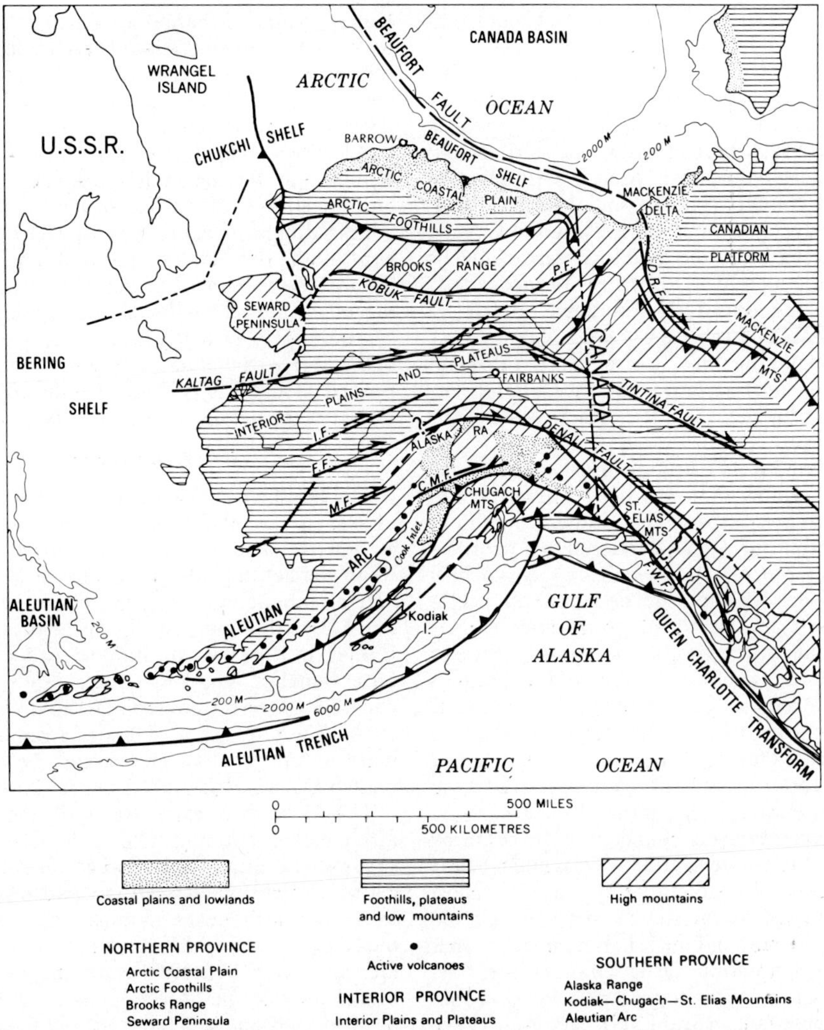

FIG. 2—Physiography and selected major faults of Alaska, after Wahrhaftig (1965), Grantz (1966), and King (1969). *P.F.*, Porcupine fault; *I.F.*, Iditarod fault; *F.F.*, Farewell fault; *M.F.*, Mulchatna fault; *C.M.F.*, Castle Mountain fault; *F.W.F.*, Fairweather fault; *D.R.F.*, Donna River fault zone.

on the eastern end of the ridge (Ling et al, 1973) suggests that any spreading on the ridge ended during the Late Cretaceous. The Canada basin may thus, in part, have opened by relative southeast movement of the Canadian platform away from the Alpha Ridge, along the Beaufort scarp. Right slip on such a Beaufort transform fault might have dissipated into oblique slip along the Jurassic and Early Cretaceous Cordilleran front via an ancestral Donna River right-slip fault, much as the present Donna River fault zone of northwestern Canada (Fig. 2) appears to transfer its right slip to right-oblique slip faults in the northwestern Mackenzie Mountains (see Geol. Survey Canada, 1970a, b). The present curvilinear trace of the postulated Beaufort–ancestral Donna River alignment resulted from oroclinal bending of the eastern Brooks Range by relative northeastward displacement of northern Alaska along the Kaltag-Porcupine and Tintina faults (Figs. 1, 2). On the other hand, Pitman and Herron (1974) and Herron et al (1974) suggest that the Alpha Ridge is an incipient subduction zone–island arc. They view it as a compressional feature complementary to early opening of the North Atlantic in Laramide time (81–63 m.y. ago) about a pivot in northern Greenland. In this case, the Canada basin would be older than, and unrelated in origin to, the Alpha Ridge. Among models compatible with this hypothesis is the one proposing that the Beaufort scarp is a left-lateral fault along which, in Jurassic time, the Kolymski block drifted out of the western Arctic basin to its present position in eastern Siberia (Pitman and Herron, 1974; Herron et al, 1974), thereby opening the Canada basin. Alternatively, the Beaufort scarp could be a tensional fault which, together with the Canada basin, was produced by rotational rifting of northern Alaska away from the Canadian Arctic Islands beginning in Jurassic time (Tailleur, 1973). Tailleur postulates that the rotation occurred about a pivot in the northern Mackenzie River valley.

Latest Jurassic to Tertiary thrust faults in the northern foothills of the Brooks Range carried Brooks Range rocks over the southern margin of the Arctic platform and depressed it to form a foredeep, the Cretaceous Colville geosyncline. The foredeep deposits thin and buttress northward against the Arctic platform, and become coarser southward toward the thrust belt and Brooks Range. The Brooks Range and foothills rocks and structures correspond in time and style to those of the Rocky Mountains province of Canada, and they have been tentatively traced into northeastern USSR and Wrangel Island via the Seward Peninsula (Churkin, 1973, p. 47ff).

The Brooks Range is bounded on the south by the Kobuk fault zone (Patton, 1973), much as the Mackenzie–Rocky Mountains system is, in most places, bounded on the southwest by the Tintina and Rocky Mountain faults. In both areas, Upper Cretaceous nonmarine clastic sediments were deposited in and near the topographic trenches that lie along these faults (Brabb, 1969; Patton, 1973). Accordingly, we postulate that the Kobuk fault zone may be a deformed and offset extension of a pre-Tertiary Tintina fault zone.

Interior Province

The Interior province lies between the Kobuk-Tintina fault system on the north and the Denali fault system, southern Alaska Range, and Aleutian arc on the south. Its commonly metamorphosed Paleozoic rocks and infolded, complexly deformed Mesozoic geosynclinal clastic rocks extend southeastward into Canada and westward beneath patchy Tertiary basins on the Bering shelf.

The Interior province contains a mainly Late Cretaceous to middle Tertiary en-echelon system of northeast-striking fractures consisting of the Kaltag, Iditarod, Farewell, Mulchatna, and perhaps other right-slip faults. These faults may have formed in response to rotational stresses about vertical axes within the active western limb of the Alaska orocline (Grantz, 1966). On the other hand, preexisting structural features such as the Kobuk and Tintina fault zones (Roddick, 1967) and the strands of the Denali fault in the central Alaska Range appear to have been either offset or bent by the deformation and rotation in the west limb of the orocline. The northeast-trending Kaltag and Farewell faults of the echelon fracture system appear in part to offset—in part to splay or trail into—the older, northwest-trending Tintina and Denali faults. They do not appear to be folded segments of the older faults.

Southern Province

The Southern or Pacific province is a complex of arcuate mountain ranges, linear lowlands, and a narrow continental shelf whose structural trends reflect the influence of the Alaska orocline and of subduction along the Cenozoic Aleutian arc. Most outcropping rocks in the province were formed in a system of Jurassic and Cretaceous arcs, trenches, and arc–trench gap basins and back-arc basins (terminology after Dickinson, 1974). The province is separated from the Pacific basin by the Aleutian Trench and the genetically related Queen Charlotte transform-fault system.

The Aleutian volcanic arc enters the continent at Unimak Island and follows the Alaska Peninsula to the central Alaska Range. There, the associated Tertiary igneous activity and historic Benioff-zone earthquakes terminate near the axis of the Alaska orocline. Subparallel with the volcanic arc, and lying between it and the Pacific, is a broad nonvolcanic or tectonic arc that is defined by Kodiak Island and the Kenai-Chugach Mountains. An actively growing outer arc is present beneath the outer continental shelf along the Aleutian Trench.

Summary

The Kaltag-Porcupine fault appears as an important break north and south of which deformation of the Alaskan intercontinental "isthmus" followed somewhat different geometries. To the north, a box fold of syntaxes and oroclines related to right slip along the fault dominates the megastructure of the Northern province; to the south, an S-fold pattern of complementary oroclines dominates the structure of the Interior and Southern provinces and their offshore extensions. On both sides of the fault, however, the causative stress was east-west shortening in response to relative westward drift of North America. The oroclinal deformations may have been localized in western and northern Alaska, because there the isthmus between Eurasia and North America was narrowest.

STRATIGRAPHY

Alaska's stratigraphy resembles that of the Canadian platform and the Pacific Cordillera. In general, Paleozoic to Early Cretaceous sedimentation was miogeosynclinal to the north or east and eugeosynclinal to the south or west. However, early Paleozoic eugeosynclinal rocks underlie much of northernmost Alaska. Large areas in the western Interior and Southern provinces remained oceanic until late Mesozoic time, when they were finally compressed and consolidated ("continentalized") and were accreted to North America.

Paleozoic and Triassic

The Paleozoic stratigraphy of Alaska has been summarized by Brosgé and Dutro (1973) and Churkin (1973), and some data on Paleozoic and Triassic rocks were given by King (1969). The late Paleozoic and Triassic rocks of northern Alaska are shelf deposits of northerly provenance that lie unconformably on both eugeosynclinal and miogeosynclinal rocks of early Paleozoic age (see Figs. 3, 4). The unconformity marks a major Late Devonian orogenic event that consolidated the early Paleozoic geosynclinal rocks into a stable shelf, the Arctic platform, upon which the late Paleozoic and Triassic rocks were deposited. These rocks are cut off on the north as a result of earliest Cretaceous and preceding erosion and by the Beaufort scarp, which we speculate was a major late Mesozoic fault. Displacement along this fault zone removed the northern sourceland of the shelf deposits of the Arctic platform, and Cretaceous and Tertiary clastic sediments prograded northward across the platform to the Beaufort shelf and slope. Therefore, the postulated Beaufort fault zone was probably active in Jurassic and/or earliest Cretaceous time.

During early Paleozoic to Triassic time, most of southern Alaska was oceanic and contained volcanic arcs, but the boundary between these ensimatic rocks and the shelf deposits to the north was apparently mobile. Note that the western two thirds of the Carboniferous to Triassic miogeosynclinal-eugeosynclinal boundary in Alaska (Fig. 4; the specific position of the boundary is an interpretation) must lie north of the present position of the early Paleozoic miogeosynclinal-eugeosynclinal boundary (Fig. 3). These relations suggest that the southern margin of the early Paleozoic continental terrace split away from the protocontinent to the north and drifted southward, much as the Lomonosov ridge of the Arctic Ocean appears to have broken away from the Siberian continental terrace. The rifting may have occurred during the Late Devonian orogeny, which strongly affected northern Alaska, or during Carboniferous time. We postulate that a late Paleozoic marginal ocean basin, with ensimatic sedimentation, then formed in the gap created in the older shelf. The distribution of ophiolitic rocks in the same region led Patton (1970; personal commun., 1974) to postulate that the Cretaceous volcanic and sedimentary rocks of the Yukon-Koyukuk province (Fig. 6) filled a Permian, Triassic, or Jurassic rift in this same gap. Rifting and marginal-ocean-basin expansion may thus have occurred from the Late Devonian or Carboniferous to the Jurassic or Early Cretaceous in west-central Alaska. Episodes of expansion may have alternated with periods of basin contraction, as suggested by Churkin (1974) for the Cordilleran fold belt to the southeast, but the chronology of such events in the two areas may have differed. Broad bands of pelitic schists, marble, serpentine, and greenstone lie along much of the boundary between the miogeosynclinal and eugeosynclinal rocks in central Alaska (Fig. 5). We speculate that these schists, where they lie near the late Paleozoic facies boundaries or along the rifted ridge of early Paleozoic miogeosynclinal rocks, include metamorphosed continental-rise deposits. The local presence of serpentines and blueschists (Forbes et al, 1971; Patton, 1973) suggests that in places these schists also mark ancient, but now disrupted and obducted, subduction zones.

Jurassic and Cretaceous

The Jurassic Period brought changes to Alaskan depositional patterns and tectonics that were comparable in scope to those created by the Late Devonian orogeny. These changes heralded a period of crustal mobility and compression that, by Tertiary time, had created a continent out of the back-arc and arc–trench gap terranes of the western Interior and Southern provinces. The Jurassic and Cretaceous sedimentary basins of Alaska are shown in Figure 6.

In northern Alaska, clastic sediments of northern provenance and platform facies were deposited from the Late Devonian through the Jurassic. Uplift and northward-directed thrust faulting in the Brooks Range began in the latest Jurassic, and a genetically related foredeep, the Colville

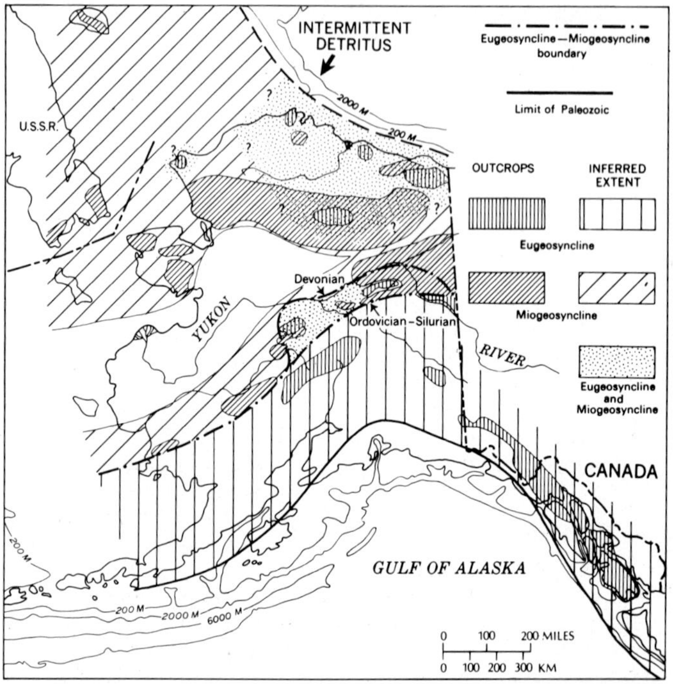

FIG. 3–Present distribution of Cambrian to Middle Devonian facies, after Brosgé and Dutro (1973) and Churkin (1973).

geosyncline, formed to the north. The rising thrust belt shed voluminous orogenic and post-orogenic detritus northward; the detritus over-filled the foredeep during the Cretaceous and prograded northward, across the Barrow arch, to form the North Chukchi and Umiat delta systems and basins. The Umiat basin underlies the eastern part of the Arctic coastal plain and adjacent continental shelf.

Jurassic mafic rocks in the southern Brooks Range mark a zone of northward-directed obduction and indicate the proximity of the southern edge of the continent at that time. Late Paleozoic and early Mesozoic ophiolites in the southern Brooks Range, and at many localities in the Interior and Southern provinces, are presumed to represent deep seafloor deposits formed in marginal ocean basins. Remnant arcs (Karig, 1972) of Paleozoic (particularly late Paleozoic) and Triassic age and at least one large ridge of rifted Precambrian and Paleozoic continental shelf were scattered across this marginal ocean basin, which must have resembled parts of the modern southwest Pacific. Upon this complex foundation between the southern Brooks Range and a frontal arc-trench system in and north of the present Kodiak-Chugach-St. Elias Mountains (see Fig. 6), an assemblage of Jurassic and Cretaceous volcanic arcs and back-arc basins developed. These basins, some apparently formed on oceanic crust and some on continental crust, were filled with volcanic and clastic deposits of diverse facies. The earliest deposits were typically marine volcanic material and turbidites of local volcanic provenance; they were succeeded by shallow-marine, deltaic, and finally coal-bearing sequences of increasingly continental provenance.

The back-arc basins formed behind (north of) a Middle Jurassic to Late Cretaceous arc–trench gap basin in southern Alaska (the Matanuska geosyncline; see Fig. 6) and a Jurassic and Cretaceous magmatic arc and sourceland contiguous with the basin on the north. The arc–trench gap basin contains shelf and slope clastic rocks much like those of similar age in the Sacramento Valley of California. A tectonically related eugeosyncline of Late (and perhaps older) Jurassic and Cretaceous age parallels the arc–trench gap basin on the south. It lies across the regional Border Range fault (MacKevett and Plafker, 1974), which is thought to represent a Laramide subduction zone. The eugeosyncline contains continental-rise and trench turbidites and submarine basalts that have been tectonically disordered and

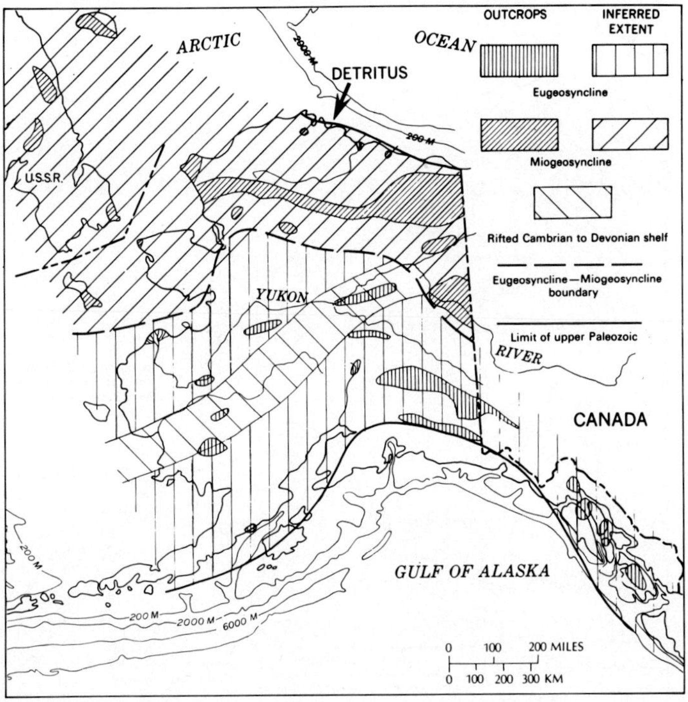

FIG. 4—Present distribution of Carboniferous to Triassic facies, after Brosgé and Dutro (1973), Churkin (1973), and King (1969).

collapsed to a fraction of their former extent by underflow of the Pacific basin.

Tertiary

Tertiary deposition in interior Alaska (Fig. 7) was restricted to nonmarine, intermontane basins, a legacy of Cretaceous "continentalization" in the Interior and Southern provinces. Thick marine and nonmarine clastic basins, however, are present semicontinuously around Alaska's continental margins and constitute some of its most attractive exploration targets.

The interior basins contain modest thicknesses of clastic sedimentary rocks that are generally coal-bearing. Intermontane, half-graben basins form the principal sediment traps. Deformation is typically slight to moderate.

The Tertiary coastal basins are larger and deeper than the interior basins and in many places are more intensely deformed. In addition to coal-bearing rocks, they contain paralic, shallow-marine shelf, and deltaic rocks, deep-marine turbidites, and organic shale. All Tertiary epochs are represented. Deformation ranges from gentle to intense, even metamorphic.

The Aleutian volcanic-arc and trench system, which formed in latest Cretaceous or earliest Tertiary time, and the tectonically linked Queen Charlotte transform-fault system controlled or strongly influenced Cenozoic sedimentation in the St. Elias, Kodiak, Cook Inlet, and Bristol Bay coastal basins.

The Cook Inlet and Kodiak basins are parts of a compound Tertiary arc–trench gap basin lying between the Aleutian volcanic arc and trench (Fig. 8). The Cook Inlet basin and its extensions into the Susitna lowland and Shelikof Strait occupy a tectonic trough between the Aleutian volcanic arc and the Kodiak-Chugach tectonic arc. This trough contains nonmarine clastic rocks and some first-cycle volcanic materials. Deformation is progressively greater in older units and may result in part from the impressed rotational strains developed in the Alaska orocline. The Kodiak basin lies between the tectonic arc and the Aleutian Trench and contains both marine and nonmarine rocks. Its eastern extension, the St. Elias basin, lies between the Chugach–St. Elias tectonic arc and a poorly developed trench in the Gulf of Alaska, and is bounded by splays of the Queen Charlotte transform-fault system.

The Tertiary basins along the outer Bering shelf are half-grabens formed as a result of slumping and collapse of a late Mesozoic conti-

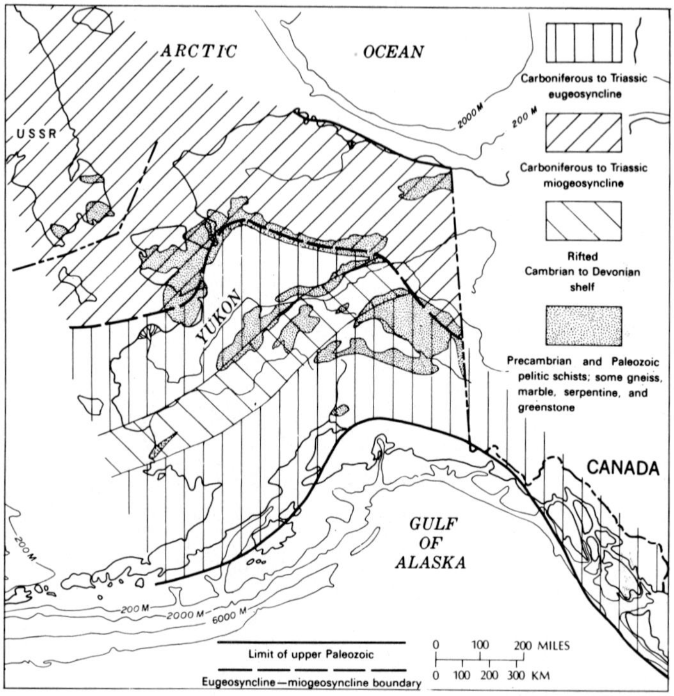

FIG. 5—Present distribution of Precambrian and Paleozoic schists, after Churkin (1973).

nental margin. Their structural position and depth, as well as limited sampling, suggest that they include both marine and nonmarine facies. The Tertiary basins of the inner Bering shelf and the southern Chukchi Sea are thinner, and may be predominantly nonmarine.

Tertiary beds in the North Chukchi and Camden basins and the Beaufort shelf in the Northern province probably consist chiefly of deltaic and progradational continental-shelf sequences. They presumably contain marine as well as nonmarine beds offshore as they do onshore in northeastern Alaska.

Clastic sediments that were swept across the Alaska continental shelves have formed thick Cretaceous and Cenozoic continental-rise sedimentary prisms in the Beaufort and Bering Seas and have contributed to thick sedimentary prisms overlying oceanic crust in the Canada and Aleutian basins. They have also partly filled the Cenozoic Aleutian Trench.

Petroleum Prospects

The sedimentary basins in Alaska that appear, on the basis of publicly available data, to have significant potential for oil and gas are shown and classified in Figure 9. Their potential can be related to Alaska's stratigraphic and tectonic history.

In the extensive Northern province (Fig. 10), petroliferous late Paleozoic and early Mesozoic marine shelf sediments on the Arctic platform contain broad structures that, at Prudhoe Bay near the crest of the Barrow arch, have stratigraphically and structurally trapped giant oil and gas deposits (Morgridge and Smith, 1972; Brosgé and Tailleur, 1971). Significant, but much smaller, deposits have also been found in the overlying Cretaceous beds of the Colville geosyncline. We anticipate that additional substantial accumulations will be found in both suites of rocks, which are virtually untested.

Significant prospects may also occur in the Cretaceous and Tertiary rocks of the coastal and offshore Umiat-Camden and North Chukchi basins and the Beaufort shelf, but older prospective rocks may be thin or absent north of the Barrow arch. The Camden basin and Beaufort shelf contain large folds eastward from Camden Bay (146°W). In the North Chukchi basin, an extensive offshore wedge of Cretaceous and Tertiary deltaic and progradational shelf clastic rocks contains many diapirs, perhaps of Cretaceous shale.

The foothills belt of the Northern province consists of overthrusts in Paleozoic shelf deposits on the south and detachment folds in thick wedges of late Mesozoic orogenic clastic rocks on the

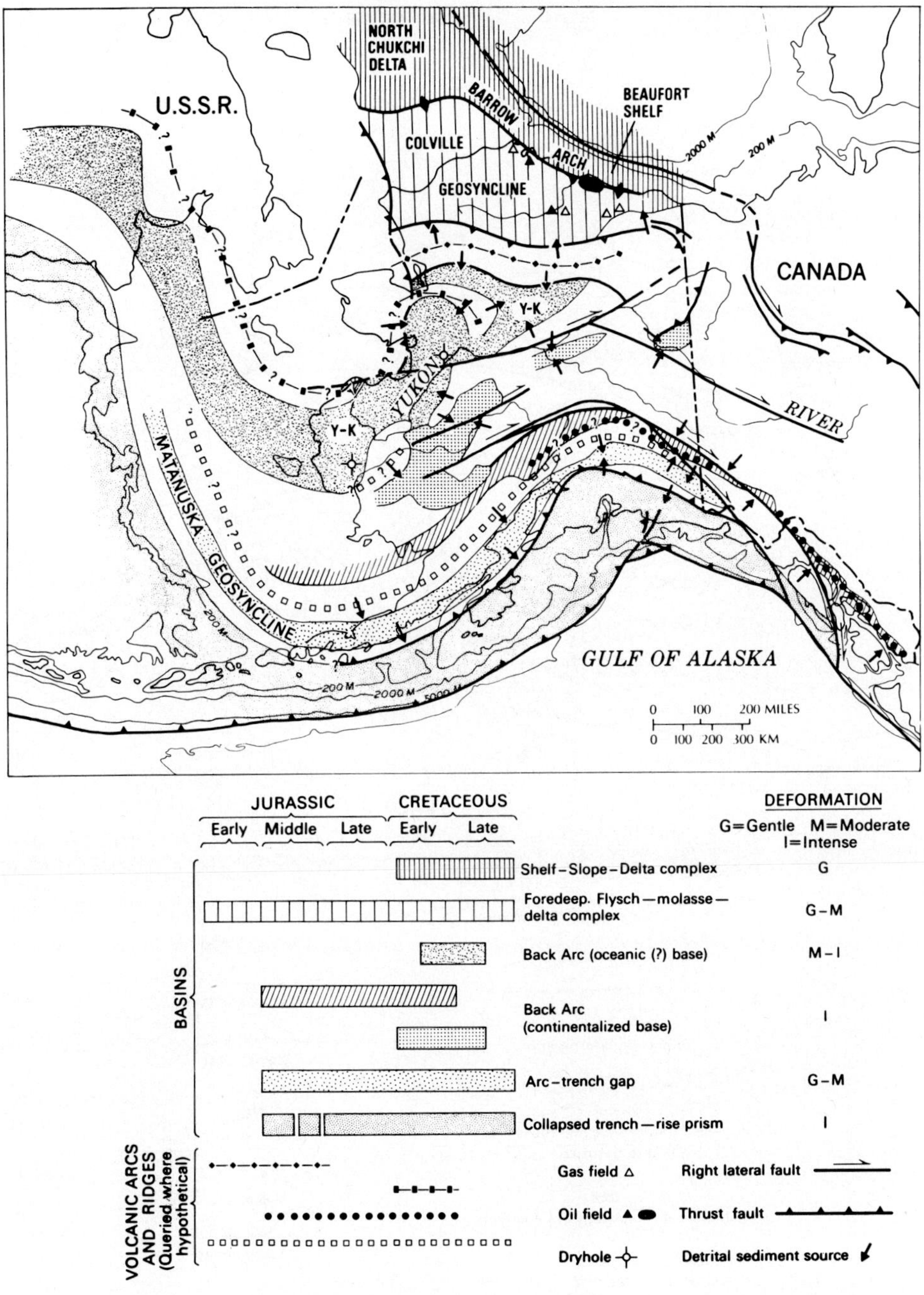

FIG. 6–Present distribution of Jurassic and Cretaceous sedimentary basins. *Y-K*, Yukon-Koyukuk province. Sources for this figure and related text are Berg et al (1972), Brabb (1969), Brosgé and Tailleur (1971), J. M. Hoare (personal commun., 1974), King (1969), Lerand (1973), Marlow et al (1976), Morgridge and Smith (1972), Patton (1973), Scholl et al (1975), Tailleur and Brosgé (1970), and our own studies.

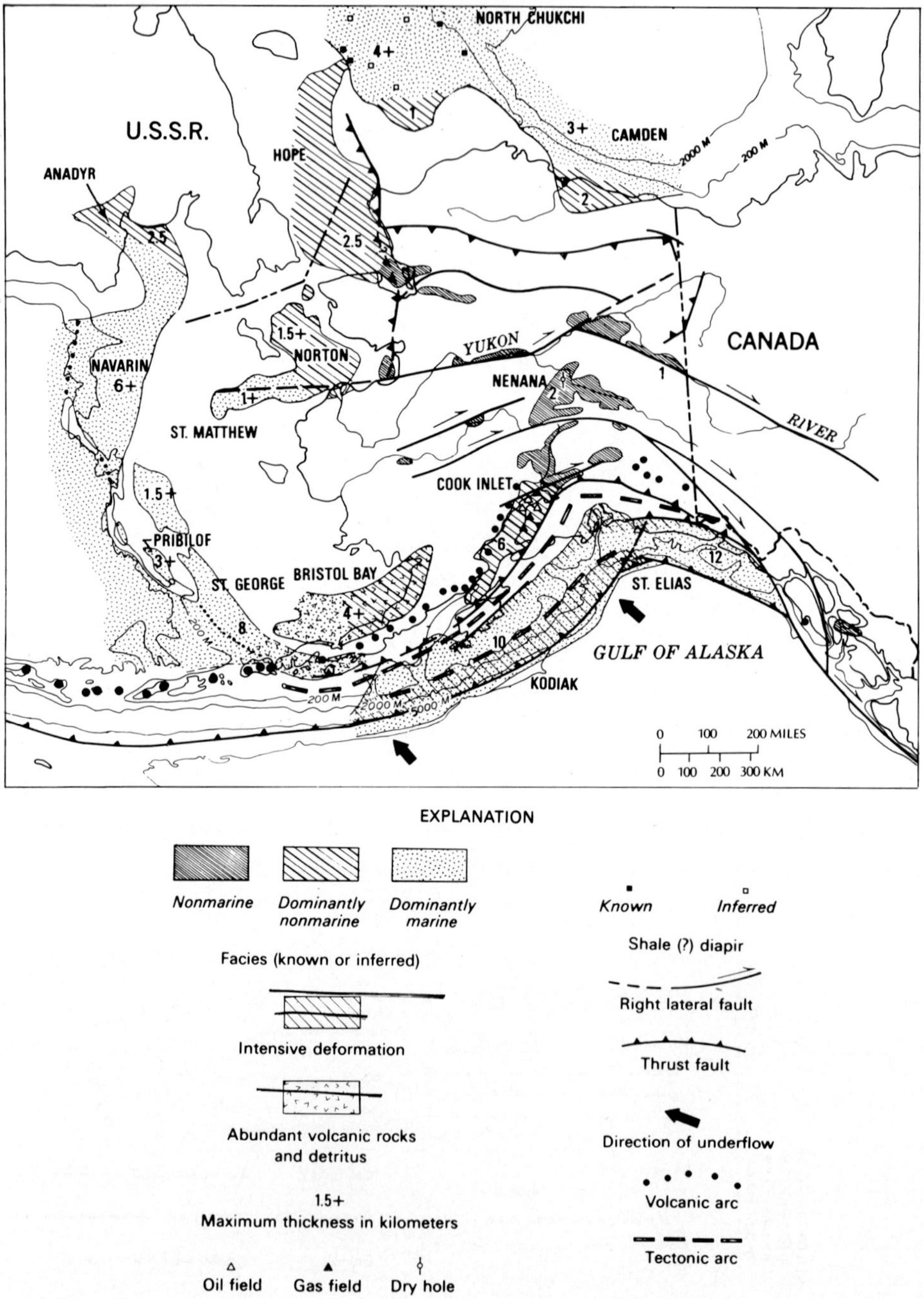

FIG. 7–Tertiary sedimentary basins. Sources for this figure and related text are Barnes (1969, 1971), Grantz et al (1970), Hatten (1971), King (1969), Kirschner and Lyon (1973), Marlow et al (1976), Plafker (1971), von Huene et al (1971), M. S. Marlow and D. S. Scholl (personal commun., 1974), and our own studies.

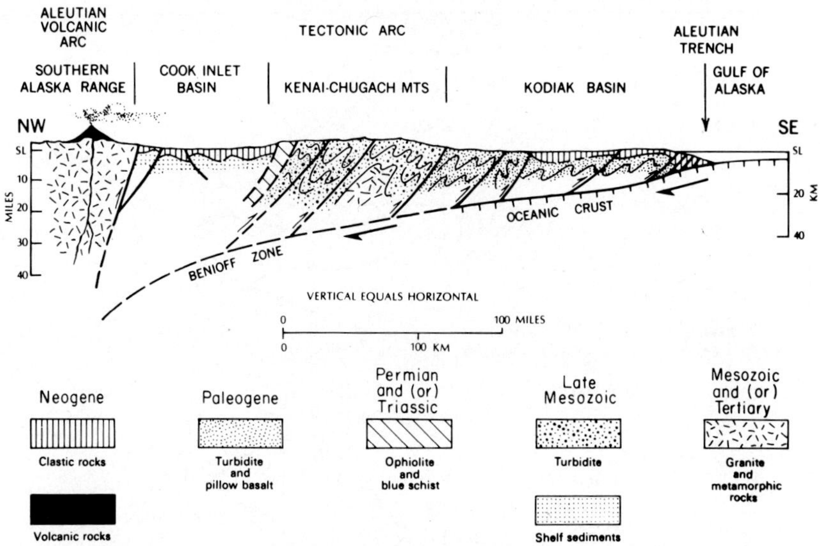

FIG. 8—Generalized cross section showing tectonic setting of Cook Inlet and Kodiak basins between Mount Spurr volcano and Gulf of Alaska near Middleton Island, after Plafker (1969, Fig. 29).

north. Many of the sedimentary units are petroliferous, but the structural definition of viable prospects will be difficult.

In the Interior province, excluding Bristol Bay, only three exploratory tests have been drilled, all barren. Two of these, near the Yukon River in western Alaska (Fig. 6), penetrated diagenetically compacted and intensely deformed clastic rocks typical of the Jurassic and Cretaceous back-arc basins of the Interior province. We conclude, therefore, that these basins have little potential.

The third well, in the Nenana basin (Fig. 7), penetrated Neogene nonmarine conglomerate, sandstone, shale, and coal and bottomed in Paleozoic schist. The sequence is probably typical of the small, relatively shallow Tertiary basins of the Interior province. Viable prospects in this setting, which would be for dry gas, may require greater thicknesses and volumes of sedimentary rocks than are generally present.

The Bristol Bay basin (Fig. 11; Hatten, 1971) is a large back-arc, half-graben basin that lies on continental crust and contains a thick section of moderately to gently folded Tertiary marine and nonmarine clastic and volcanic rocks. The basin becomes more marine and deeper toward the southwest. Shows of oil and gas were logged in several onshore test wells, but there has been no production. Broad structures on the northwest shelf may provide attractive exploratory prospects. The southeastern margin of the basin overlies locally petroliferous late Mesozoic beds.

In the Southern province, thick Middle Jurassic to Late Cretaceous marine shelf and slope clastic rocks of the Matanuska geosyncline trend more than 1,600 km from Canada to the outer Bering shelf (Figs. 6, 9). These arc–trench gap, basin deposits are generally moderately folded, but in places they are intensely deformed or intruded by dikes and small plutons. Despite large oil seeps and common oil or gas shows in exploratory tests, these rocks have been discouraging owing to both originally low and diagenetically diminished porosity. However, they offer the only significant, albeit modest, prospects for Mesozoic oil and gas in Alaska south of the Colville geosyncline.

The best prospects for large oil or gas accumulations in the Southern province lie in the coastal Tertiary basins. The Cook Inlet basin (Kirschner and Lyon, 1973) is an intermontane half-graben containing a thick section of Oligocene to Pliocene deltaic, estuarine, and continental clastic deposits—the Kenai Group of Calderwood and Fackler (1972; Fig. 12). These rocks overlie Paleocene sandstone, shale, and coal and Jurassic and Cretaceous marine strata with oil shows. Estimated proved reserves in the basin exceed 2.6 billion bbl of oil and 5 Tcf of gas (after Crick, 1971). The deposits are in the Kenai Group and cluster in the north-central part of the basin, apparently

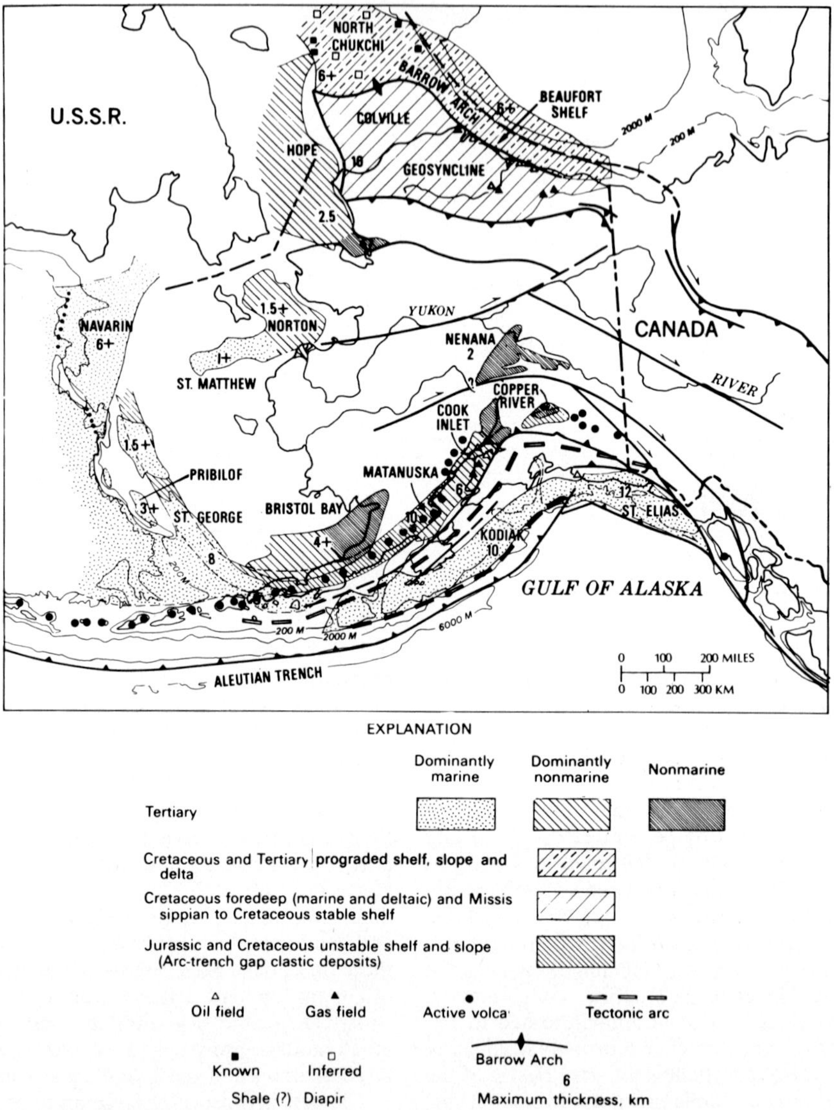

FIG. 9–Known and possible significant oil and gas basins in Alaska.

at an optimum stratigraphic position. Exploratory tests elsewhere in the basin have been noncommercial, and the potential for additional large reserves is speculative.

The Kodiak and St. Elias basins (Figs. 13, 14) lie largely beneath the continental terrace between the Aleutian Trench and the tectonic arc represented by the Kodiak–Chugach–St. Elias Mountains. The boundary between these stratigraphically and structurally related basins is the easternmost of the northeast-striking faults and folds that are related to subduction at the Aleutian Trench.

The Kodiak basin (von Huene et al, 1971) overlies the Aleutian Trench Benioff zone and is an arc–trench gap basin. It contains a linear, gently continentward-dipping half-graben wedge of upper Tertiary clastic rocks bounded by large faults along the outer coast of Kodiak Island and by a semicontinuous arch near the seaward margin of the continental shelf. Tertiary and Quaternary strata in the arch and beneath the continental

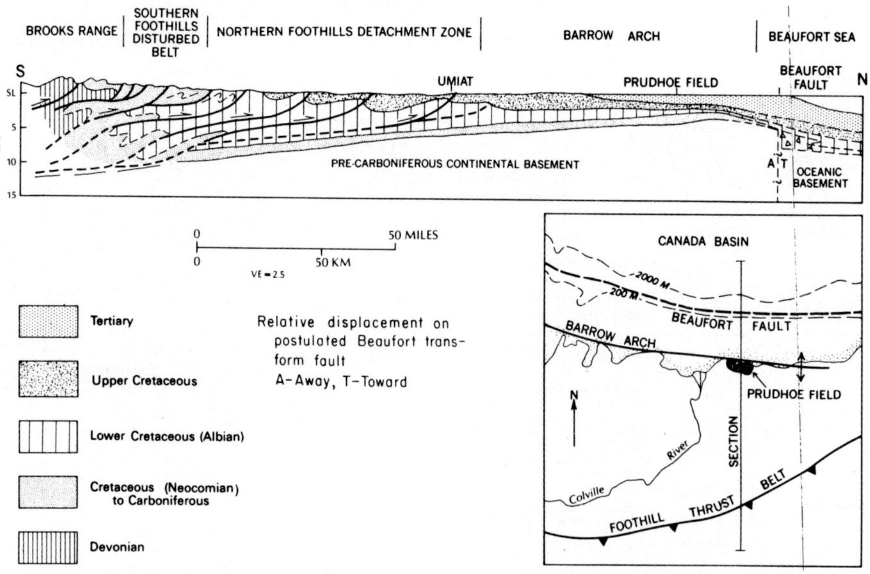

FIG. 10–Generalized cross section, North Slope (Northern province).

slope are being tectonically churned by slumping, folding, and imbricate thrusting as a result of underflow of the Pacific plate. No exploratory wells have been drilled in this mainly offshore basin, and little is known of its stratigraphy. On Kodiak and adjacent islands, Oligocene continental and paralic strata and Miocene and Pliocene marine sandstone and siltstone overlie complexly deformed and mildly metamorphosed turbidites and mafic volcanic rocks of Paleocene and Eocene age. The Kodiak basin offers a large area of modest potential for exploration, but the attractiveness of its sedimentary volume is offset by uncertainty about reservoir rocks and by locally complex deformation.

The St. Elias basin (Plafker, 1971; Stoneley, 1967) has been structurally telescoped by slip along bordering right-reverse faults which are

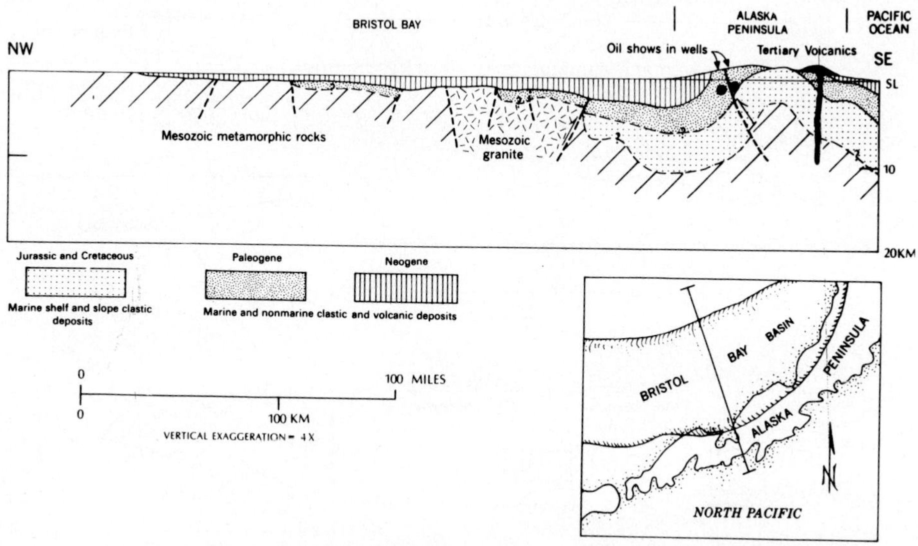

FIG. 11–Generalized cross section, Bristol Bay basin, after Burk (1965) and Hatten (1971).

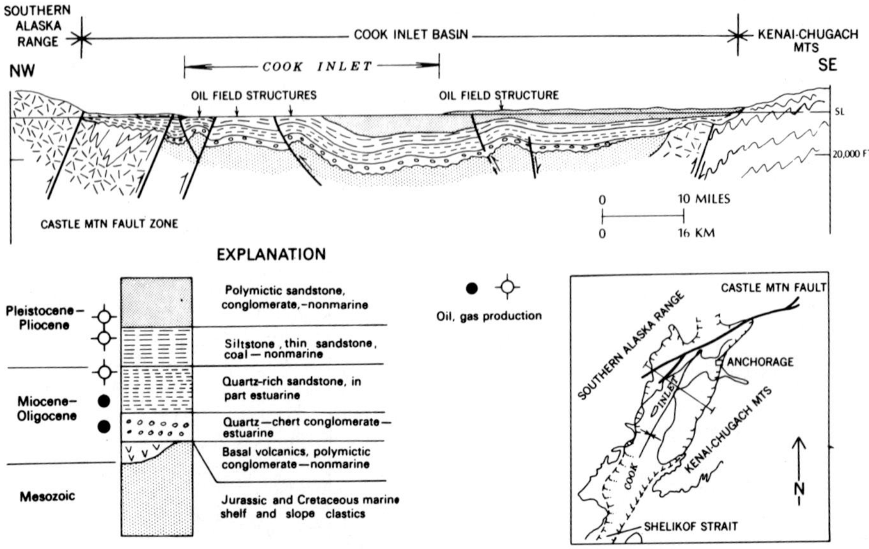

FIG. 12–Generalized cross section, Cook Inlet basin, after Kirschner and Lyon (1973).

splays of the active Queen Charlotte–Fairweather transform-fault system. These faults appear to have absorbed large amounts of right slip from the Queen Charlotte fault system. As a result the basin has a complex stratigraphic and structural history. In the west, where its strata are very thick, imbricate thrusting in the St. Elias thrust system is succeeded southward by tight east-west-striking thrust folds near the coast and broad asymmetric folds offshore. Eastward, the basin thins and its structure is dominated by a broad southeast-striking syncline.

Paleocene and Eocene turbidites and mafic volcanic rocks at the base of the section in the St. Elias basin are intensely deformed or mildly metamorphosed and offer no potential for petro-

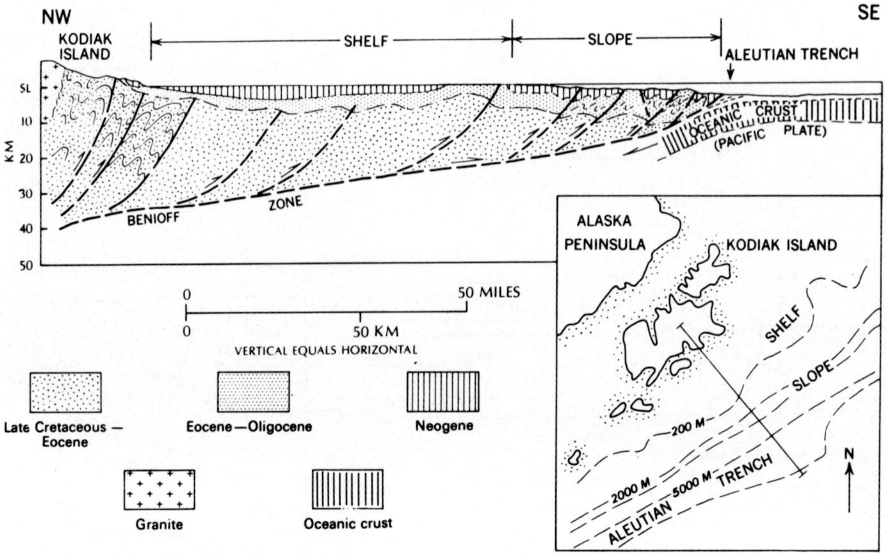

FIG. 13–Generalized cross section, Kodiak basin, after von Huene et al (1971).

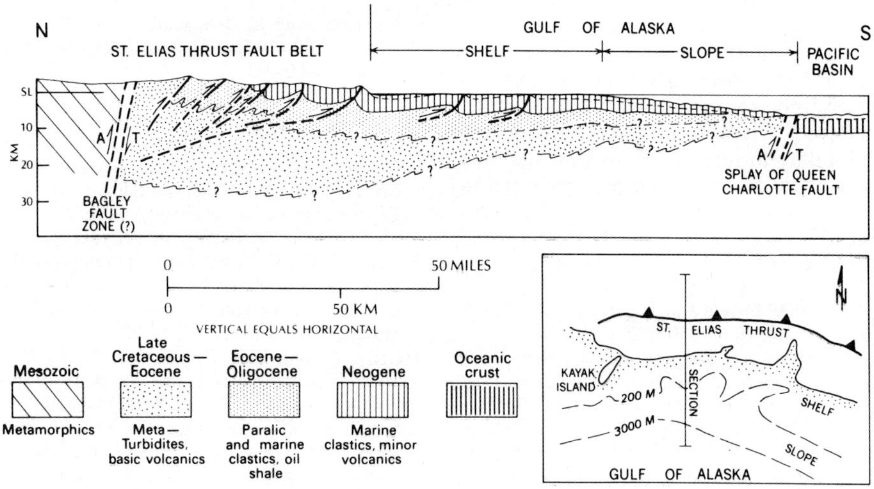

FIG. 14–Generalized cross section, St. Elias basin, after Plafker (1971) and Stoneley (1967).

leum. They are succeeded by Eocene to lower Miocene deltaic sandstone, coastal-plain shale and coal, and thick transgressive Oligocene and Miocene petroliferous siltstone and oil shale. These rocks are overlain, in turn, by thousands of feet of Neogene marine sandstone and tillite. The petroliferous rocks, oil seeps, large structures, and thick sedimentary section spurred the drilling of 25 onshore exploratory tests in the basin. All were unsuccessful owing to complex structure and diagenetically diminished porosity. The offshore offers more attractive prospects in large, and perhaps less complex, structures, but the presence of adequate reservoir rocks will be critical.

The recent discovery of several Tertiary sedimentary basins beneath the Bering and Chukchi shelves (Figs. 7, 9) has greatly increased the known prospective area in the Alaskan offshore (see Marlow et al, 1976; Grantz et al, 1970). The inner basins of the Bering and southern Chukchi shelves—Norton, St. Matthew, and Hope—contain moderate thicknesses of strata and may have formed by crustal extension and subsidence behind the large terrane in northern Alaska that was displaced relatively northeastward by right slip on the Kaltag fault during the Laramide orogeny. Broad folding and faulting have provided possible structural traps in these basins, but the stratigraphy is almost unknown. Marginal outcrops and a few seafloor samples suggest that both nonmarine and marine beds of Late Cretaceous(?) and both Paleogene and Neogene ages may be present. Natural gas has been found in the partly analogous Anadyr' basin in northeastern Siberia (Meyerhoff, 1972). The deeper parts of the three basins appear to offer prospects for natural gas, and perhaps for some oil as well.

The outer basins of the Bering shelf, which include the large and deep Navarin and St. George Tertiary half-graben basins (6 and 8 km, respectively), formed by collapse of a continental margin that has been tectonically passive since formation of the Aleutian arc during latest Cretaceous or earliest Tertiary time. These basins thus escaped the intense tectonism that underflow of the Pacific plate produced in the Kodiak and St. Elias basins. Tilting, faulting, and subordinate folding in the outer basins may have provided structural traps for petroleum, but the adequacy of source and reservoir rocks is not known. Seafloor sampling by Hopkins et al (1969) has shown that Oligocene to Pliocene shallow-marine clastic rocks are present at least locally. Their tectonic setting, together with facies relations in the adjacent Anadyr' and Bristol Bay basins, suggests that the outer basins contain predominantly marine rocks. The great extent and volume of possibly favorable rocks, their silled character, and a stable tectonic environment make the outer Bering basins attractive for oil and gas exploration.

About 11 billion bbl of oil and 32 Tcf of natural gas had been discovered in Alaska through 1972 (Miller et al, 1975). It is estimated that the ultimate production from the fields containing these deposits will approximate 26 billion bbl of oil and 68 Tcf of natural gas (Mast and Dingler, 1975). In addition, the U.S. Geological Survey estimates that Alaska contains about one third of the undiscovered oil and about one sixth of the undiscovered natural gas in the United States (Miller et al, 1975). Although this optimistic estimate is certainly speculative, it is obvious that Alaska offers for petroleum exploration a wide variety of mildly to intensely deformed sedimentary basins of diverse ages and tectonic environments.

References Cited

Barnes, D. F., 1969, Progress on a gravity map of Alaska: Eos, v. 50, no. 10, p. 550–552.

——— 1971, Preliminary Bouguer anomaly and specific gravity maps of Seward Peninsula and Yukon Flats, Alaska: U.S. Geol. Survey Open-File Rept., 9 p.

Berg, H. C., D. L. Jones, and D. H. Richter, 1972, Gravina-Nutzotin belt—tectonic significance of an upper Mesozoic sedimentary and volcanic sequence in southern and southeastern Alaska: U.S. Geol. Survey Prof. Paper 800-D, p. D1–D24.

Brabb, E. E., 1969, Geologic map of the Charley River quadrangle, east-central Alaska: U.S. Geol. Survey Misc. Geol. Inv. Map I-573, scale 1:250,000.

Brosgé, W. P., and J. T. Dutro, Jr., 1973, Paleozoic rocks of northern and central Alaska, *in* M. G. Pitcher, ed., Arctic geology: AAPG Mem. 19, p. 361–375.

——— and I. L. Tailleur, 1971, Northern Alaska petroleum province, *in* I. H. Cram, ed., Future petroleum provinces of the United States—their geology and potential: AAPG Mem. 15, p. 68–99.

Burk, C. A., 1965, Geology of the Alaska Peninsula—island arc and continental margin: Geol. Soc. America Mem. 99, pts. 1–3, 250 p.

Calderwood, K. W., and W. C. Fackler, 1972, Proposed stratigraphic nomenclature for Kenai Group: AAPG Bull., v. 56, no. 4, p. 739–754.

Churkin, Michael, Jr., 1972, Western boundary of the North American continental plate in Asia: Geol. Soc. America Bull., v. 83, p. 1027–1036.

——— 1973, Paleozoic and Precambrian rocks of Alaska and their role in its structural evolution: U.S. Geol. Survey Prof. Paper 740, 64 p.

——— 1974, Paleozoic marginal ocean basin–volcanic arc systems in the Cordilleran foldbelt, *in* R. H. Dott, Jr., and R. H. Shaver, eds., Modern and ancient geosynclinal sedimentation: Soc. Econ. Paleontologists and Mineralogists Spec. Pub. 19, p. 174–192.

Crick, R. W., 1971, Potential petroleum reserves, Cook Inlet, Alaska, *in* I. H. Cram, ed., Future petroleum provinces of the United States—their geology and potential: AAPG Mem. 15, p. 109–119.

Dickinson, W. R., 1974, Sedimentation within and beside ancient and modern magmatic arcs, *in* R. H. Dott, Jr., and R. H. Shaver, eds., Modern and ancient geosynclinal sedimentation: Soc. Econ. Paleontologists and Mineralogists Spec. Pub. 19, p. 230–239.

Forbes, R. B., et al, 1971, Tectonic implications of blueschist facies metamorphic terranes in Alaska: Nature; Phys. Sci., v. 234, no. 49, p. 106–108.

Geological Survey of Canada, 1970a, Geological map of Canada: Map 1250-A, scale 1:5,000,000.

——— 1970b, Tectonic map of Canada: Map 1251-A, scale 1:5,000,000.

Grantz, Arthur, 1966, Strike-slip faults in Alaska: U.S. Geol. Survey Open-File Rept. 267, 82 p.

——— et al, 1970, Reconnaissance geology of the Chukchi Sea as determined by acoustic and magnetic profiling, *in* W. L. Adkison and M. M. Brosgé, eds., Proceedings of the geological seminar on the North Slope of Alaska: Pacific Sec. AAPG, p. F1–F28.

Hatten, C. W., 1971, Petroleum potential of Bristol Bay basin, Alaska, *in* I. H. Cram, ed., Future petroleum provinces of the United States—their geology and potential: AAPG Mem. 15, p. 105–108.

Herron, E. M., J. F. Dewey, and W. C. Pitman, III, 1974, Plate tectonics model for the evolution of the Arctic: Geology, v. 2, no. 8, p. 377–380.

Hopkins, D. M., et al, 1969, Cretaceous, Tertiary and early Pleistocene rocks from the continental margin in the Bering Sea: Geol. Soc. America Bull., v. 80, p. 1471–1480.

Karig, D. E., 1971, Origin and development of marginal basins in the western Pacific: Jour. Geophys. Research, v. 76, p. 2542–2561.

——— 1972, Remnant arcs: Geol. Soc. America Bull., v. 83, p. 1057–1068.

King, P. B., compiler, 1969, Tectonic map of North America: U.S. Geol. Survey, scale 1:5,000,000.

Kirschner, C. E., and C. A. Lyon, 1973, Stratigraphic and tectonic development of Cook Inlet petroleum province, *in* M. G. Pitcher, ed., Arctic geology: AAPG Mem. 19, p. 396–407.

Lerand, Monti, 1973, Beaufort Sea, *in* R. G. McCrossan, ed.. Future petroleum provinces of Canada—their geology and potential: Canadian Soc. Petroleum Geologists Mem. 1, p. 315–386.

Ling, H. Y., L. M. McPherson, and D. L. Clark, 1973, Late Cretaceous (Maestrichtian?) silicoflagellates from the Alpha Cordillera of the Arctic Ocean: Science, v. 180, p. 1360–1361.

MacKevett, E. M., Jr., and George Plafker, 1974, The border ranges fault in south-central Alaska: U.S. Geol. Survey Jour. Research, v. 2, no. 3, p. 323–329.

Marlow, M. S., et al, 1976, Structure and evolution of Bering Sea shelf south of St. Lawrence Island: AAPG Bull., v. 60, no. 2, p. 161–183.

Mast, R. F., and Janet Dingler, 1975, Estimates of inferred plus indicated reserves for the United States by states, *in* B. M. Miller et al, 1975, Geological estimates of undiscovered recoverable oil and gas resources in the United States: U.S. Geol. Survey Circ. 725, p. 3–4.

Meyerhoff, A. A., 1972, Russians look hard at the Anadyr basin: Oil and Gas Jour., Oct. 23, p. 124–129; Oct. 30, p. 84–89.

Miller, B. M., et al, 1975, Geological estimates of undiscovered recoverable oil and gas resources in the United States: U.S. Geol. Survey Circ. 725, p. 3–4.

Morgridge, D. L., and W. B. Smith, 1972, Geology and discovery of Prudhoe Bay field, eastern Arctic Slope, Alaska, *in* R. E. King, ed., 1972, Stratigraphic oil and gas fields: AAPG Mem. 16, p. 489–501.

Ostenso, N. A., 1973, Seafloor spreading and the origin of the Arctic Ocean basin, *in* D. H. Tarling, and S. K. Runcorn, eds., Implications of continental drift to the earth sciences: London and New York, Academic Press, p. 165–173.

Patton, W. W., Jr., 1970, Discussion of Tailleur, I. L., and Brosgé, W. P., Tectonic history of northern Alaska, *in* W. L. Adkison and M. M. Brosgé, eds., Proceedings of the geological seminar on the North Slope of Alaska: Pacific Sec. AAPG, p. E20.

——— 1973, Reconnaissance geology of the northern Yukon-Koyukuk Province, Alaska: U.S. Geol. Survey Prof. Paper 774-A, 16 p.

Pitman, W. C., III, and E. M. Herron, 1974, Continental drift in the Atlantic and the Arctic, *in* L. Kristjansson, ed., Geodynamics of Iceland and the North Atlantic area: NATO Adv. Study Inst. Ser. C-11, 15 p.

Plafker, George, 1969, Tectonics of the March 27, 1964, Alaska earthquake: U.S. Geol. Survey Prof. Paper 543-I, p. I1–I74.

——— 1971, Possible future petroleum resources of Pacific-margin Tertiary basin, Alaska, *in* I. H. Cram,, ed., Future petroleum provinces of the United States—their geology and potential: AAPG Mem. 15, p. 120–135.

Roddick, J. A., 1967, Tintina Trench: Jour. Geology, v. 75, p. 23–32.

Scholl, D. W., E. C. Buffington, and M. S. Marlow, 1975, Plate tectonics and the structural evolution of the Aleutian-Bering Sea region: Geol. Soc. America Spec. Paper 151, p. 1–31.

Stoneley, R., 1967, The structural development of the Gulf of Alaska sedimentary province in southern Alaska: Geol. Soc. London Quart. Jour., v. 123, p. 25–57.

Tailleur, I. L., 1973, Probable rift origin of Canada basin, Arctic Ocean, *in* M. G. Pitcher, ed., Arctic geology: AAPG Mem. 19, p. 526–535.

——— and W. P. Brosgé, 1970, Tectonic history of northern Alaska, *in* W. L. Adkison and M. M. Brosgé, eds., Proceedings of the geological seminar on the North Slope of Alaska: Pacific Sec. AAPG, p. E1–18.

Von Huene, Roland, E. H. Lathram, and Erk Reimnitz, 1971, Possible petroleum resources of offshore Pacific-margin Tertiary basin, Alaska, *in* I. H. Cram, ed., Future petroleum provinces of the United States—their geology and potential: AAPG Mem. 15, p. 136–151.

Wahrhaftig, C. A., 1965, Physiographic divisions of Alaska: U.S. Geol. Survey Prof. Paper 482, 52 p.

Sedimentation and Tectonics of Pacific Continental Margin of British Columbia—Summary[1]

D. L. TIFFIN and B. E. B. CAMERON[2]

A triple junction of three lithospheric plates at the British Columbia continental margin has had considerable influence on structure and sedimentation in marginal basins. South of the triple junction, compressional forces from subduction of the oceanic plate beneath the continent have resulted in deformation mainly by folding and block faulting; however, north of the junction, stresses have been relieved by strike-slip movement along transform faults and only broad folding of Tertiary strata has occurred.

Tofino basin, south of the triple junction, has undergone major uplift, linear echelon folding, and elongate diapirism on the outer shelf. More than 12,000 ft (3,700 m) of Tertiary mudstone and siltstone has been drilled adjacent to such structures by Shell Canada Ltd. The fine clastic material in these distal facies is not conducive to the accumulation of petroleum. However, potentially productive reservoir beds may exist in the proximal turbidite sequences nearer shore. Subsequent uplift and erosion of the latter also may have resulted in clean second-cycle wedges west of the uplift boundary. In the Tofino basin, as in other areas of the west coast, hydrocarbon prospects in stratigraphic traps appear to have the highest potential.

North of Brooks Peninsula, structural style is dominated by shelf-edge faulting which, west of Queen Charlotte Islands, represents transform movement between the Pacific and North American plates. The Queen Charlotte basin has undergone net subsidence of several thousand feet. Late Tertiary nonmarine strata overlie Tertiary and Mesozoic volcanic basement and Paleozoic intrusives in the north, and late Tertiary marine sedimentary rocks overlie Tertiary volcanic rocks in the south. The sequence overlying the basement is as thick as 15,000 ft (4,600 m). Permeabilities are reduced by siltstones and claystones, but facies changes between interfingering marine and nonmarine depositional sites should make good stratigraphic traps.

Winona basin, at the base of the slope, is folded only gently and broadly in the north but is more highly deformed into prominent sedimentary ridges in the south. The sedimentary sequence at the deepest point under the base of the slope is 3–6 km. The oldest sediments at the flank of the basin are Pliocene. The present deep-water basin undoubtedly has received a high proportion of second-cycle clastic deposits from the uplifted, older Tertiary belt.

[1]From Circum-Pacific Energy and Mineral Resources Conference program booklet.

[2]Geological Survey of Canada, Vancouver, British Columbia.

Oil Resources of the Pacific for Peace and Security—Summary[1]

J. P. TRUNZ[2]

Although the idea of establishing nationally owned petroleum reserves on public land was seriously considered over 70 years ago during the administration of Theodore Roosevelt, it remained for his successor, William Howard Taft, to inaugurate the naval-petroleum-reserve system. The first of these Reserves was initiated in 1909 at Elk Hills in Kern County, California. Elk Hills and the nearby Buena Vista Reserve were officially designated Naval Petroleum Reserves by a Presidential order in 1912. A small area, Teapot Dome in Wyoming, was added in 1915, and the inclusion in 1923 of the Alaskan Reserve—Naval Petroleum Reserve No. 4 (NPR-4)—completed the holdings. When initially set aside, none of the Reserves contained known oil production.

Buena Vista and Teapot Dome did not develop into viable Reserves. Elk Hills, on the other hand, is the largest remaining oil field in the lower 48 states in terms of production. Its 46,000 acres (18,630 ha) contain 1,000 wells tapping reserves that approach 1.5 billion bbl of oil and 1 trillion cu ft of natural gas. Using current facilities, daily production of 160,000 bbl of oil and 100 million cu ft of gas could be attained. During thc history of the reserve, Elk Hills has produced a total of 300 million bbl of oil, even though it has been in a reserve status at all times except during World War II. At present, remedial and offset production is about 3,500 bbl/day.

NPR-4, in the northwest corner of Alaska, covers over 6% of the land surface of Alaska—about 37,000 sq mi (95,830 km^2). Estimates of reserves range from 10 to 33 billion bbl. This Reserve may well prove to have more oil than Algeria or Venezuela or, in fact, the majority of the producing nations associated in OPEC (The Organization of Petroleum Exporting Countries). The first important survey of NPR-4, from 1944 to 1953, was sponsored by the Navy in cooperation with the U.S. Geological Survey. Eighty wells were drilled, and significant oil deposits were found at Umiat, Simpson, and Fish Creek, and natural gas at Barrow, Gubic, Wolf Creek, Oumalik, Meade, and Square Lake. It now seems that the reserves of NPR-4 and those of Prudhoe Bay (50 mi or 80 km east of NPR-4) are part of a system of oil and gas deposits that stretch along the Arctic coasts and hinterlands of Canada and the Soviet Union as well as Alaska.

Although the naval petroleum reserves have not been used, except for remedial production, since World War II, their importance for the future has become increasingly clear. The growing dependence of the United States on foreign energy sources has made the need for large domestic supplies for use in emergencies ever more urgent. There will be neither time nor capital to develop new resources once an emergency is upon us. Unless the wells, the pumps, and the pipelines are already in place, deliverability will be slow in coming no matter how desperately it may be needed by our armed forces or industry or anyone else.

Optimum exploitation of NPR-1 at Elk Hills, California, will involve full pressure maintenance under full production. Our present 5-year development program will increase the well density to a total of about 2,000. This should add 250,000 bbl/day to present capacity, making the total about 400,000 bbl/day. In the early stages of production, all gas will be reinjected to stimulate oil production. During a national emergency, the Elk Hills Reserve, producing at maximum level, would supply a substantial portion of the needs of the U.S. Armed Forces. Put in another context, it would replace some of the oil imports to the United States denied by enemy action.

The plans for NPR-4 call for 10-year programs of assessment and development. Test wells would be drilled to depths ranging from 2,000 to 25,000 ft (610–7,620 m), and about 80 coreholes would provide geologic samples. By this means the extent of the reserve would be established, and the huge gap (20+ billion bbl) between the high and low estimates would be narrowed considerably. The 10-year development program, which could begin almost simultaneously with the assessment program, would be expected to result in a daily production of about 3,000,000 bbl. Although the cost of developing resources in Alaska is high, the cost of alternative domestic sources to fulfill the

[1]Manuscript received, November 6, 1974.

[2]Captain, CEC, USN; Director, Naval Petroleum and Oil Shale Reserves, Washington, D. C. 20360.

same needs is even higher—probably many times higher.

For more than 50 years, Congress has kept the NPRs and their use as a matter of special and detailed concern. Whatever policy changes may take place,[3] it is unlikely that Congress will delegate its responsibility to any part of the executive, be it the Navy, the Department of the Interior, or anyone else. To have, intact and protected, such a vast resource lends an assurance to long-term planning for energy development and national security that will become increasingly valuable.

Reference Cited

Oil and Gas Journal, 1976, Elk Hills, other NPR's to flow soon: v. 74, no. 15, p. 50.

[3]Legislation passed in April 1976 has authorized commercial production from NPRs 1–3 to begin within 90 days (Oil and Gas Jour., 1976).

Hydrocarbon Potential of Offshore California—Summary[1]

ANTHONY E. L. MORRIS[2]

Seaward from the tideline of the California coast to the base of the continental slope is an area of 100,000 km^2 containing 15 sedimentary basins covering 60,000 km^2 with a volume of 156,000 km^3.

The shelf is less than 10% explored, and the activity has been largely confined to the coastal fringe of the Santa Barbara Channel and to the area near Los Angeles. Nevertheless, 1.5 billion bbl of oil and 39.5 billion m^3 of gas have been produced from 3,100 exploratory and development wells. Reserves are estimated at 2.8 billion bbl of oil and 94 billion m^3 of gas.

The region has three distinct tectonic provinces. Clastic sedimentary units with minor pyroclastic and extrusive igneous rocks are common to all, and cherts are abundant in the Santa Maria basin of the northern province. There are no carbonate rocks.

The southern province of 46,600 km^2 is a region of tectonic extension with northwest-trending horst and graben development that is clearly reflected in the bathymetry. Nine sedimentary ba-

[1]Manuscript received, February 10, 1975.

[2]Petroleum Consultant, Los Angeles, California 90064.

I am indebted to Union Oil Company of California for drafting assistance and slide preparation for the original presentation. Reports and summaries published by the California Division of Oil and Gas, the Conservation Committee of California Oil Producers, and the American Petroleum Institute provided the basic statistical data.

Table 1. Summary of Hydrocarbon Potential

Basin	Area (km^2)	Volume (km^3)	Oil Factor (bbl/km^3)	Hydrocarbon Potential: Oil ($bbl \times 10^9$)	Hydrocarbon Potential: Gas ($m^3 \times 10^9$)
Southern Province[2]					
Santa Monica	2,600	5,400	300,000	1.62	38.88
San Pedro	3,000	5,600	500,000	2.80	67.20
Santa Cruz	3,900	12,300	240,000	2.95	70.80
San Nicolas	3,600	14,000	200,000	2.80	67.20
Catalina	2,100	2,300	130,000	.30	7.20
San Diego	2,200	5,800	160,000	.93	22.32
San Clemente	1,000	2,100	110,000	.23	5.52
Tanner	2,800	4,600	160,000	.74	17.76
Others[1]	2,600	4,600	100,000	.46	11.04
Subtotal	23,800	56,700		12.83	307.92
Santa Barbara Channel[3]	5,200	31,700	260,000	8.24	329.60
Northern Province[2]					
Santa Maria	8,500	13,800	50,000	.69	16.56
Santa Cruz	2,600	3,400	10,000	.03	.72
Bodega	3,600	5,400	10,000	.06	1.44
Point Arena	7,000	15,000	10,000	.15	3.60
Eel River	9,300	30,000	40,000	1.20	28.80
Subtotal	31,000	67,600		2.13	51.12
TOTAL	60,000	156,000		23.20	688.64

[1]Includes the relatively small, isolated East and West Cortes and Long basins.
[2]Average gas:oil ratio for southern and northern provinces is estimated as 24 m^3/bbl.
[3]Average gas:oil ratio for Santa Barbara Channel is estimated as 40 m^3/bbl.

sins occupy the deep-water areas, and islands and shallow banks are underlain by thin or older deposits and basement rocks. Maximum sedimentary thickness is probably about 4,000 m and volume is about 56,700 km^3.

The Santa Barbara Channel province trends east-west. Compressional forces have formed several lines of sharp folds along the northern edge of the basin. Many large faults on the north and south borders show left-lateral movement. Rocks from Cretaceous to Pleistocene age aggregate up to 20,000 m in thickness. Basement is estimated at 12,000 m at the eastern end and rises to the seafloor at the western end. The basin has an area of 5,200 km^2 and a volume of 31,700 km^3.

The northern province has five basins covering 31,000 km^2 with a volume of 67,600 km^3. Structural trends are northwest, and compressional forces are indicated by marginal thrust faulting and folding. Erosional remnants of Upper Cretaceous and lower Cenozoic rocks are common, but prospective sedimentary units are of Miocene and Pliocene age. Except for the Eel River basin, which may be as deep as 4,500 m, all of the basins are 3,000 m deep or less.

Hydrocarbon potential of offshore California is estimated as 23.2 billion bbl of oil and 689 billion m^3 of gas (Table 1).[3]

[3] *Note added in proof*: Information developed in the 2 years since this paper was written, including that from a 3,328-m corehole on the western edge of San Nicolas basin, suggests that the petroleum potential listed in Table 1 for the southern province is too high. A more realistic value is 3.75 billion bbl of oil and 80 billion m^3 of gas.

Petroleum Potential of Continental Rise off Central California—Summary[1]

P. WILDE,[2] W. R. NORMARK,[3] and T. E. CHASE[4]

INTRODUCTION

The Deep Sea Drilling Program (DSDP) has demonstrated the technical feasibility of drilling in the deep ocean. However, except for petroleum shows in the Sigsbee Knolls in the Gulf of Mexico, there is no firm evidence of petroleum reservoirs in the deep ocean beyond the limits of the continental slope. Emery (1973) and Beck and Lehner (1974) suggested that vast petroleum resources may lie in deep water adjacent to the continents in areas of hemipelagic sedimentation. One such area adjacent to the United States is the continental rise off California between Point Conception and Cape Mendocino.

Based on available seismic reflection profiles, concentrated on the Monterey fan, this paper evaluates the possibilities for economic petroleum deposits for the region off the central California coast (Fig. 1).

CONTINENTAL RISE

The continental rise off central California (from Point Conception to Cape Mendocino) covers about 200,000 km^2 of the seafloor (Fig. 1), and is composed of three submarine fans—from south to north: (1) Arguello fan heading in Arguello Canyon; (2) Monterey fan heading in Monterey Canyon with tributary canyons Sur-Partington-Lucia, Carmel, Ascension, Pioneer, and Farallon; and (3) Delgada fan heading in Delgada Canyon with tributary canyons Bodega, Noyo, and Spanish (Shepard and Dill, 1966). The sediments of the fans are chiefly hemipelagic green muds, but coarser deposits are found in submarine channels and fan valleys extending from the mouths of the various canyons (Wilde, 1965). The green color implies reducing conditions for the bulk of the fan sediments. The provenance of the sand-size particles, the depositional fabric, the location of coarse deposits in the axes of submarine channels (Wilde, 1965), and the presence of levees on the upper fan valleys (Shepard, 1966) all support the contention of Dill et al (1954) and Menard (1960) that the bulk of the sediment on the fans is deposited and shaped from spent turbidity currents transporting continental-shelf sediment to the deep sea.

SOURCE-BED POTENTIAL

The hemipelagic green muds of the California continental rise, which are derived from the nearshore areas, could be excellent source beds for petroleum if sufficient organic content were maintained during transportation to the fans and subsequent deposition. The only detailed analyses of the organic content of the sediments of the California deep-sea fans are for samples from DSDP sites 32, 33, and 34, on the toe of the Delgada fan (Table 1). The rate of deposition increases toward the apex of the fan or toward the source. The rate of sedimentation for the toes of both the Monterey and Delgada fans is on the order of 30 m/m.y., and the rate increases shoreward to 430 m/m.y. on the upper Monterey fan near the mouth of Ascension Canyon. The organic content in the borderland basins (Table 1) is much higher than that of the DSDP samples, which suggests that (1) the preserved organic content increases with an increase in the rate of deposition, and (2) the average value of organic content for the green muds on the fan would be intermediate between the DSDP and borderland values.

It is difficult to determine whether the green muds on the fans have been buried or heated sufficiently to convert the organic material into migrating hydrocarbons. However, Cordell (1972) has shown that some reservoirs containing petroleum could form under conditions of shallow burial and low temperature.

RESERVOIR POTENTIAL

If conditions within the source-bed area are favorable, the next requirement for an economic

[1]Manuscript received, October 21, 1974.

[2]Institute of Marine Resources, University of California, Berkeley, California 94720. Present address: Lawrence Berkeley Laboratory, Berkeley, California 94720.

[3]Department of Geology and Geophysics, University of Minnesota. Present address: Pacific-Arctic Branch, U.S. Geological Survey, Menlo Park, California 94025.

[4]Geologic Data Center, Scripps Institution of Oceanography, La Jolla, California 92093. Present address: Pacific-Arctic Branch, U.S. Geological Survey, Menlo Park, California 94025.

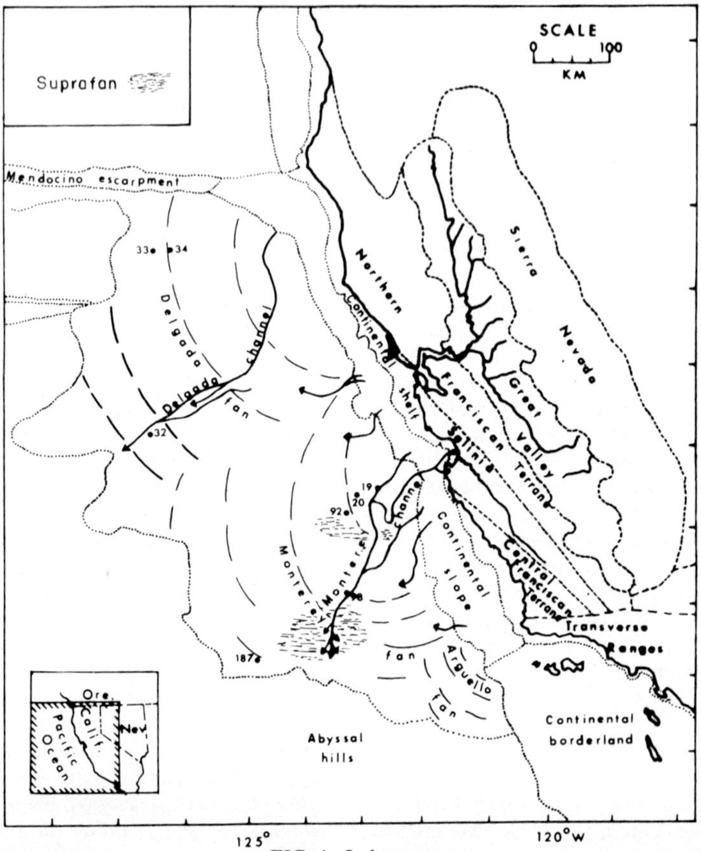

FIG. 1—Index map.

deposit is a suitable reservoir for the hydrocarbons. In general, the sandy sediments deposited by turbidity currents on levees and on the fan surface are poorly sorted with a muddy matrix, and would make poor reservoir rocks with low porosity and permeability. However, the sands deposited in a channel are coarser grained and better sorted, analogous to stream channel deposits. Wilde (1965, p. 138–139) found in a piston core taken in the Monterey Channel, at 4,100 m water depth, pebble lenses and medium- to fine-grained, well-sorted sandstone layers tens of centimeters thick. MacManus et al (1970, p. 16–17, 109) reported moderately well-sorted sandstones up to 40 cm thick at DSDP site 32, and 200 cm of well-sorted friable sandstones in beds 5–30 cm thick from core 6 of DSDP site 34.

Stratigraphic Traps

The geometry of the reservoir-potential sand bodies on a deep-sea fan depends on the local depositional environments during channel development. As Wilde (1965) and Normark (1970a, b; 1974) have shown, the development of a fan is not entirely constructional, and the sedimentary regime depends on the rate of supply and grain-size distribution of the material transported to the fan by turbidity currents. Three valley-channel types (Normark, 1970b)—termed "depositional," "erosional," and "compound"—can be found in the Monterey fan.

Where the fan valley is depositional—generally on the upper fan—prominent levees are constructed, and both the valley floor and levee crests build above the level of the adjacent fan. The Ascension fan valley on the upper Monterey fan is a well-developed constructional valley whose floor lies 300 m above the open fan. Sand units in the Ascension system are thicker, coarser, and cleaner than those on the adjacent levee crest (Hess, 1974). Deep-tow 3.8-kHz reflection profiles suggest that sands become less common down the backside of the levee (Normark, 1971). In depositional fan valleys, the turbidity currents lose sediments and the gradient of the channel is reduced.

If the gradient becomes too low and the valley floor is prominently elevated, the channel may abandon its old position and shift to a new area with a steeper gradient. Shepard (1966) has shown dramatic evidence for shifting of deposition channels where the Monterey fan valley goes through a tight meander on the upper fan. Normark (1970a) interpreted the meander as a westward shift, into the Ascension fan valley, of the depositional valley which extended from the Monterey Canyon (Monterey East valley); the result was that (1) the lower portion of the Monterey East valley was beheaded and (2) the upper portion of the Ascension fan valley was left as a hanging tributary to the present Monterey canyon/valley system. In general, this and similar forms of channel piracy and abandonment on all fans must occur rather frequently in order to produce the arcuate shape of the fans.

In erosional fan valleys, levees are absent or poorly developed, as in the present lower Monterey fan valley. Previously deposited turbidite and pelagic sediments on the fan are incorporated into the eroding flow, and the local gradient may be steepened. Because overflow deposition is minimal in erosional channels, the sandy sediments in the valley do not interfinger with levee muds, but rather have very abrupt lateral terminations against steep, terraced valley walls such as in the Miocene fan valleys of the Capistrano Formation of southern California (Bartow, 1966; Piper and Normark, 1971; Hess, 1974). These ancient fan valleys contain thick (several meters), massive sandstones and conglomerates which commonly become thinner and finer grained during later stages of channel filling. This change in grain size is probably a result of a gradual loss of sediment supply as the valley cuts through its levee upstream and begins to feed sediment to a new section of the fan.

On the Monterey fan, compound channels are formed as a result of a marked change in (1) channel gradient or (2) turbidity-current regime (i.e., the volume and grain size of material in the flow).

After channel abandonment, the formerly active channel probably is partially filled by slumps from the channel walls, overflow from nearby active distributaries, material reworked by contour currents, and eventually by pelagic sedimentation. As noted, the coarsest and best-sorted material is in the axis of the channel, whereas the poorly sorted, finer grained material is on the levees and the fan surface. Thus, the filling of the abandoned channel would produce a reasonably impermeable cap of silts and clays over the better

Table 1. Organic Content[1]

Site	*Water Depth (m)*	*No. Samples*	*% Organic Carbon*
DSDP			
32	4,758	12	1.54[2]
33	4,284	30	0.86[2]
34	4,322	35	0.90[2]
California Borderland			
Santa Barbara basin		11	5.8
Santa Catalina basin		32	8.6
Continental slope		15	5.4
Blue and green mud		52	7.66
Shale		78	1.25

[1] After Weser (1970, p. 586-588) and Emery (1960, p. 235).

[2] Using Trask and Patnode's (1942, p. 43) conclusion that percent organic matter is approximately two times percent organic carbon.

sorted sands of the channel axis. For example, a buried channel from the Monterey fan is covered by an estimated 500 m of sediment, and the uppermost 300 m is composed of regularly bedded muds deposited by overflow from the nearby levee of the present Monterey fan valley.

On many fans where the turbidity currents carry appreciable amounts of sand-size sediments, the valley on the upper fan terminates in a depositional lobe called a "suprafan" (Normark, 1970b). The suprafan is characterized by an anastomosing set of distributary channels of low relief, generally without levees. The suprafan area, usually on the mid-fan, appears as a convex-upward bulge on a longitudinal fan profile. The distributary channels are floored by sand, but sandy lenses may compose much of the suprafan section (Piper and Normark, 1971).

After burial by green mud, the distributary channels and sandy sections within the suprafan would provide a network of entry points for the migration and concentration of petroleum upslope into the main channel. A continuous gradient from the mouth of the submarine canyon to the suprafan areas must be maintained during fan growth. Accordingly, the potential petroleum traps (the sands of the channels and the suprafans) are deposited with a tilt in the up-channel direction. On the Monterey fan, the present slope or gradient decreases from 28′ at the apex of the fan to 07′ at the outer edge in a water depth of 4,500 m (Wilde, 1965, p. 1). Locally, steeper slopes reaching 2° exist on the backsides of the levees and along the leading edges of suprafan lobes on some fans. Thus, the abandoned channels would make excellent stratigraphic traps; the petroleum could migrate up the channel gradient to the coarse and more permeable sands of the channel at the apex of the fan.

To estimate the number and size of potential petroleum reservoirs within the California continental rise, we attempted to define both (1) the surface and (2) the buried valley systems of the Monterey fan, the only area for which we have a reasonable amount of seismic reflection data. The surface channel system consists of: (a) presently active channels, which receive major turbidity-current flows directly from the submarine canyons; (b) overflow channels, which include recently abandoned major channels and channels receiving occasional overflows; and (c) suprafan areas.

Three suprafans are identified on the Monterey fan (Fig. 1). The largest and southernmost one is associated with the present termination of the Monterey fan valley and has many distributary valleys. The central suprafan probably has been abandoned. It appears to have been cut off from the Monterey fan valley, which has been deepened by headward erosion from its present area of termination (Normark, 1970a). This suprafan marks the previous termination of the Monterey fan valley before the valley breached the northwest-trending basement ridge (now seen as a line of seamounts protruding through the fan surface). The smallest suprafan (36°N, 123°W), probably the product of the beheaded Monterey East valley, is relatively inactive.

The active channels of the Delgada and Monterey fans are in the southern sector of each fan. Accordingly, the network of abandoned channels, now buried by subsequent deposition, should extend in radiating patterns from the apex of the fans as suggested by the arcuate outline of the fans. Several westward-trending buried valleys can be identified on the Monterey fan. One of the buried channels lies underneath the large levees of the present Monterey fan valley.

Prospects

The best prospects for petroleum accumulation in stratigraphic traps would be near the apices of the various fans near the mouths of the major submarine canyons. The primary prospects would be the buried channels (1) west of the present Monterey channels; (2) west of the present Delgada channels (not seen but postulated by analogy with the Monterey fan); and (3) west of the mouth of Arguello Canyon. Secondary prospects would be seaward of the mouths of Farallon and Pioneer canyons and south of Davidson Seamount.

With the limited amount of subbottom-profiling coverage, it would be conjectural to postulate the existence of tectonically produced structures modifying the buried channels. However, tectonic activity is suggested along (1) the eastward extension of the Murray fracture zone and (2) the northwest-trending line of seamounts and escarpment which lies approximately perpendicular to the long axis of the Davidson Seamount.

The immense size of these channel systems and their proximity to refining facilities and markets on the West Coast, combined with our developing capacity to operate in the deep ocean, make buried channels on the central California continental rise attractive targets for petroleum exploration in the future.

References Cited

Bartow, J. A., 1966, Deep submarine channel in upper Miocene, Orange County, California: Jour. Sed. Petrology, v. 36, p. 700–705.

Beck, R. H., and P. Lehner, 1974, Oceans, new frontier in exploration: AAPG Bull., v. 58, no. 3, p. 376–695.

Cordell, R. J., 1972, Depths of oil origin and primary migration; a review and critique: AAPG Bull., v. 56, no. 10, p. 2029–2067.

Dill, R. F., R. S. Dietz, and H. B. Stewart, 1954, Deep-sea channels and delta of the Monterey submarine canyon: Geol. Soc. America Bull., v. 65, p. 191–194.

Emery, K. O., 1960, The sea off southern California: New York, John Wiley, 366 p.

——— 1973, Oil on the shelf: Oceanus, v. 17, Spring, p. 11–17.

Hess, G. R., 1974, Submarine fanfare; a comparison of modern and Miocene deep-sea fans: M.S. thesis, Univ. Minnesota, 118 p.

McManus, D. A., et al, 1970, Regional aspects of deep-sea drilling in the Northwest Pacific, *in* D. A. McManus, et al, Initial reports of the Deep-Sea Drilling Project, v. 5: Washington D. C., U.S. Govt. Printing Office, p. 621–636.

Menard, H. W., 1960, Possible pre-Pleistocene deep-sea fans off central California: Geol. Soc. America Bull., v. 71, p. 1271–1278.

Normark, W. R., 1970a, Growth patterns of deep-sea fans: AAPG Bull., v. 54, p. 2170–2195.

——— 1970b, Channel piracy on Monterey deep-sea fan: Deep-Sea Research, v. 17, p. 837–846.

——— 1971, Mini-topography of deep-sea fans; geometric considerations for facies interpretations in turbidites: Pacific Sec. Soc. Econ. Paleontologists and Mineralogists Field Trip Guidebook, p. 22–36.

——— 1974, Submarine canyons and fan valleys: factors affecting growth patterns of deep-sea fans, *in* R. H. Dott, Jr., and R. M. Shaver, eds., Modern and ancient geosynclinal sedimentation: Soc. Econ. Paleon-

tologists and Mineralogists Spec. Pub. 19, p. 56–68.
Piper, D. J. W., and W. R. Normark, 1971, Re-examination of a Miocene deep-sea fan and fan-valley, southern California: Geol. Soc. America Bull., v. 82, p. 1823–1830.
Shepard, F. P., 1966, Meander in valley crossing a deep-ocean fan: Science, v. 154, p. 385–386.
——— and R. F. Dill, 1966, Submarine canyons and other sea valleys: Chicago, Rand McNally, 381 p.
Trask, P. D., and H. W. Patnode, 1942, Source beds of petroleum: Tulsa, AAPG, 556 p.
Weser, O. E., 1970, Lithologic summary, *in* D. A. McManus, et al, Initial reports of the Deep-Sea Drilling Project, V. 5: Washington D. C., U.S. Govt. Printing Office, p. 569–620.
Wilde, P., 1965, Recent sediments of the Monterey deep-sea fan: California Univ. Berkeley Hydrol. Eng. Lab. Rept. HEL 2-13, 153 p.

Potential for Exploration and Development of Hydrocarbons in Atrato Valley and Pacific Coastal and Shelf Basins of Colombia[1]

RAFAEL BUENO S. and CARLOS GOVEA R.[2]

Abstract The Pacific coastal basin of Colombia covers an area of approximately 56,000 km² between the Pacific Ocean and the Western Cordillera. Another basin of deposition—the Pacific shelf basin—lies offshore and west of the coastal basin and comprises an even larger area. These basins are the least-explored, potentially petroliferous areas in Colombia. Only 6 wells (2 offshore, 4 inland) have been drilled; 3 are in the northern Atrato area of the basin. A partial reason for lack of exploration is the relative inaccessibility of the area, one of the most humid and unhealthy tropical rainforest zones in the western hemisphere.

Unsuccessful results may also be traced to the lack of a concentrated effort at deciphering the regional geology. Surface geology does not everywhere accurately reflect the subsurface picture; in many places it masks the true underground structure.

The Pacific coastal basin is currently regarded as one of promising potential for a country that badly needs to find new oil to meet ever-rising demands. Seismic investigations and photogeologic work permitted the mapping of a body of igneous and metamorphic rocks that crop out between the Docampadó and Baudó Rivers in the central or San Juan area of the basin (San Juan paleohigh).

The petroleum potential of the area also appears clear, considering that sediments in excess of 20,000 ft (6,100 m) have accumulated in the basin, and that possible source rocks of both Tertiary and Late Cretaceous age occur within the sedimentary area. Furthermore, numerous oil and gas seeps have been reported throughout, and previously drilled wells, though not commercial, had encouraging oil and gas shows.

INTRODUCTION

The Pacific coastal basin of Colombia has an area of 56,000 km². It is bordered on the east by the mafic and ultramafic complex of the Western Cordillera. The western border is marked by the Baudó Range, which is submerged beneath the Pacific Ocean from a point near Cabo Corrientes; the range reenters the continent in the subsurface south of Buenaventura. On the north and northeast, an important tectonic feature—the Sautatá arch—separates the basin from the Sinú and San Jorge basins.

The Pacific coastal basin has been considered a part of the Bolivar geosyncline (Schuchert, 1935; Nygren, 1950), which was thought to extend along the west coast of South America and probably to be deepest in Peru (Burgl, 1961). However, Duque Caro (1972), Bandy (1967), Case et al (1969), and others have expressed doubts about the existence of the postulated landmass on the west (Nygren, 1950; Jacobs, 1963), which could have supplied clastic material to fill the geosyncline. Their doubts are based mainly on investigations indicating that the Pacific (Nazca) oceanic plate is subducted under the South American continental margin. Also, the character of planktonic foraminifers and radiolarians and the presence of ophiolites indicate conditions of deep-marine sedimentation.

West of the onshore Pacific coastal basin, a trend of lows, filled with great thicknesses of sedimentary rocks, is indicated by seismic profiles. It is apparent that another basinal area is present on the continental shelf, but is separated from the coastal basin by a trend of basement highs. This basin is herein referred to as the "Pacific shelf basin" and, although some information is available about its structures, practically nothing is known about its stratigraphy.

Both of these basins have many structural features favorable for the entrapment of hydrocarbons.

STRUCTURE

Pacific Coastal Basin

Structurally, the Pacific coastal basin can be divided into three subbasins: north or Atrato area, central or San Juan area, and south or Tumaco area.

[1]Manuscript received, November 6, 1974.

[2]Empresa Colombiana de Petróleos, Bogotá, Colombia.

The writers are indebted to Mario Galán Gómez, President, and Numael Nieto Gracia, Director of Production, Empresa Colombiana de Petróleos (Ecopetrol), for authorization to publish this paper. Bernardo Taborda, Assistant Director of Production, and Alonso Quintero, Exploration Manager, kindly reviewed the manuscript and illustrations and offered valuable suggestions. Roberto Martínez, head of the Exploitation Division, was also very encouraging. Special thanks are expressed to Hernam Duque-Caro, Regional Biostratigrapher of the Instituto de Investigaciones Geológico-Mineras (Ingeominas) of Colombia, who assisted the writers in establishing the regional stratigraphic picture of the basin and its geologic evolution.

Several oil companies currently operating in Colombia kindly allowed the writers to consult key reports.

Atrato Area

The northern subbasin—the Atrato area—extending between Cabo Corrientes and the Panama border, comprises the area of the Atrato River valley. The basin limits can be seen at the surface, where a fault zone marks the western margin (Figs. 1, 2). West of the fault zone lies the Baudó Range, a complex of igneous and metamorphic rocks of various ages. Faulting is also present on the eastern margin, at the contact between the Tertiary sedimentary rocks of the basin and the Cretaceous metamorphic Dagua Group, which crops out in the western foothills of the Western Cordillera. At some localities, the relations at the eastern margin are not clear, but could be unconformable.

In this northern region, the structures nearly parallel the strike of the Western Cordillera (approximately N15°W). Farther north, toward Panama, regional strike becomes more pronouncedly northwest. In the lower Atrato River valley, the Sautatá arch (a regional basement high) strikes N45°W and becomes the major structural feature in the area. This arch separates the Pacific coastal basin from the Sinú basin on the northeast, and links the Darien highland of Panama with the Western Cordillera. Several anticlinal features of possible petroleum importance occur within this northern portion of the basin. Some of them have been mapped for several kilometers in length and show good horizontal and vertical closures. Most are symmetrical and fairly gentle, having dips rarely exceeding 10°, whereas others show asymmetry of the flanks and have dips in the 20–30° range.

Salaquí anticline—The structure lies near the Panama border, between the Truandó and Salaquí Rivers. Its northeastern plunge is marked geomorphologically by the Salaquí River. Lower Miocene and Oligocene rocks crop out along its crest. The structure has not been tested for hydrocarbons.

Chintado anticline—This fold is located near the Chintado River not far from the hamlet of La Nueva. Its structural closure seems to be determined mainly by faulting. The main drilling objectives probably would be the Oligocene and upper Eocene in fault traps.

Opogadó anticline—This structure lies between the Domingodó and Opogadó Rivers west of the hamlets of La Isla and Opogadó. It strikes north-northwest with dips difficult to measure owing to poor bedding of the strata capping the structure. Seismic work indicates good structural closure. The main objectives are the sandstones and limestones of the upper Eocene.

Napipí anticline—The Napipí is a small structure which lies southeast of Opogadó. It strikes northeast-southwest. Very little information is available regarding its structural and stratigraphic conditions owing to accessibility difficulties.

Buchadó anticline—This fold is in the middle Atrato valley and is over 30 km long. It is a gentle, symmetrical fold with dips not exceeding 5° on both flanks at the surface. It covers an area of about 20,000 ha. and vertical closure is about 300 m.

Tagachi anticline—Located between the Tagachi and Beté Rivers, this fold has been established mostly by photogeologic work, since surface expression is rather poor for field-work definition. Some recent seismic work better defined the structure, which is regarded as a good oil prospect.

Urodó anticline—This feature is located on the west side of the basin. Dips are much steeper than at Buchadó, ranging in the order of 40–50° at the surface. Seismic studies indicate possible overturning to the west at depth.

Berreberre anticline—Located a short distance east of the Baudó River, this anticline is nearly 20 km long; dips are 20–25°. The area of closure is in the order of 12,000 ha, and a vertical closure of 1,300 m has been reported. An attempt was made to drill this highly prospective feature, but the project was abandoned as a result of inaccessibility.

San Juan Area

The central part of the basin, lying in the general area of the Baudó and San Juan River valleys, is characterized by an interruption of the regional structural pattern of the north (Fig. 1). A fairly complex series of faults and folds strikes N45°E. Dips commonly exceed 30°, and the zone shows evidence of tectonic compression. This intensely folded and faulted zone has been referred to as the "shear zone." Jacobs (1963) interpreted it as the result of rotational strain induced by east-west compression of blocks to the northwest and southeast, representing the possible interaction of oceanic and continental tectonic plates. This zone, however, corresponds to the general area where the existence of the San Juan paleohigh (Fig. 3) is postulated on the basis of photogeologic and seismic evidence. The change in structural grain (different here from the prevalent structural grain to the north and south) could also be related to the tectonic relation caused by an

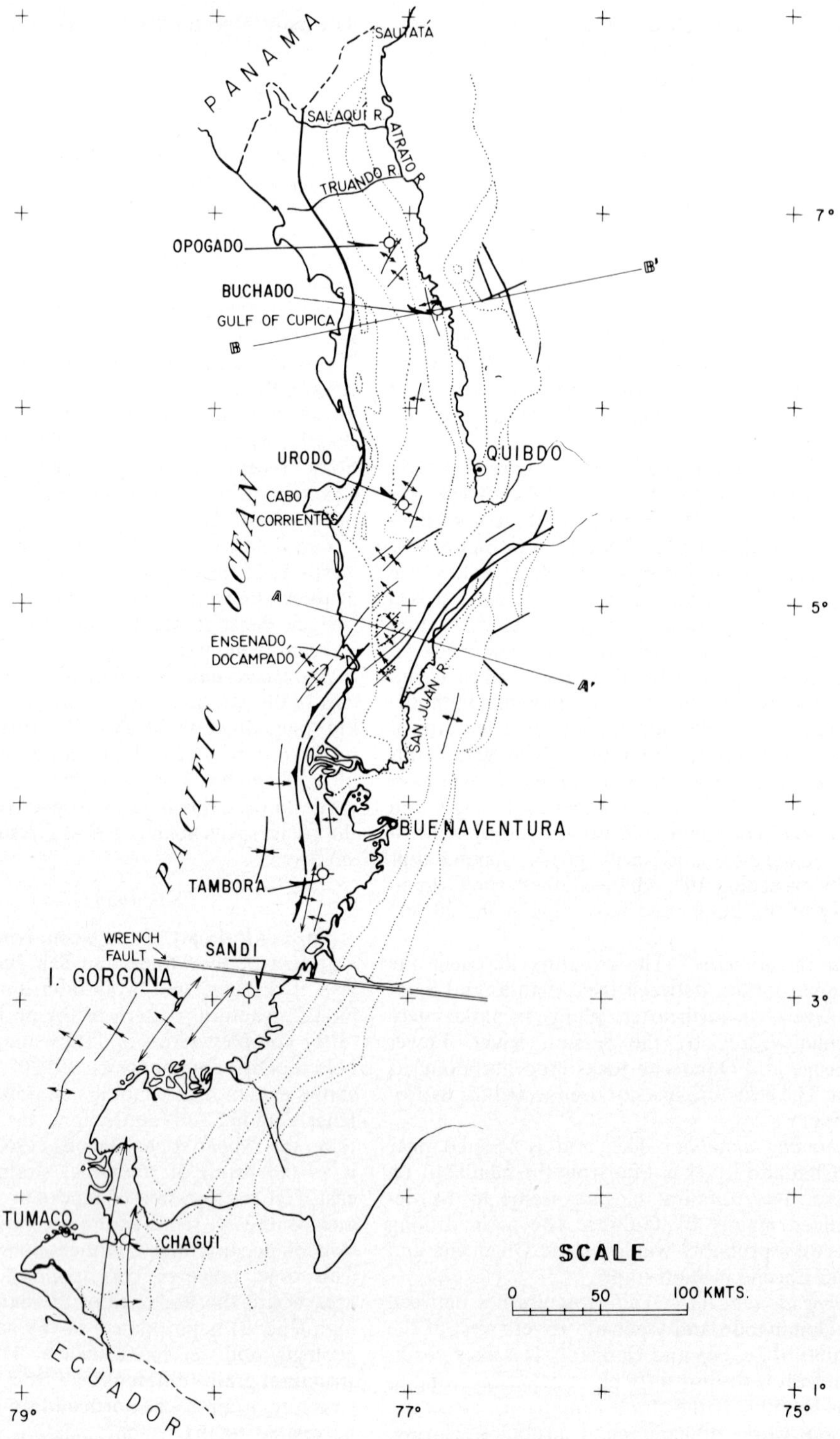

FIG. 1–Generalized geologic map of Pacific coastal and shelf basins.

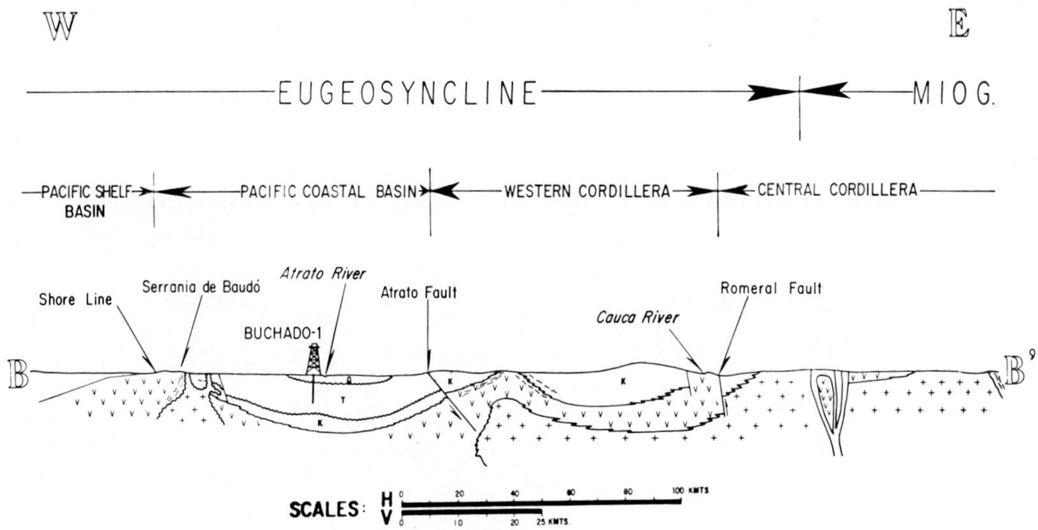

FIG. 2–Generalized regional cross section *B-B'*.

archlike spur of possible continental crust. Seismic interpretation indicates that this basement high, possibly a spur of the Central Cordillera, extends into the Pacific Ocean and reenters the continent in the subsurface farther south. If this interpretation is correct, the petroleum possibilities of the central part of the basin are considerably improved, since the existence of an old landmass during Tertiary cycles of deposition should have created the conditions for marginal sedimentation, permitting biohermal limestones and sandstones to be deposited either during late Eocene or early to middle Miocene times.

Several anticlinal folds occur within this part of the basin, but none have been tested by the drill.

Siguirisua anticline—This fold lies in south-central Chocó Department, north of the Docampadó River. The structure strikes northeast-southwest and has been mapped for over 35 km. Dips are 25–40°, and areal closure is about 7,000 ha.

Copoma anticline—This fold lies in the general area of the Copoma and Cucurrupí Rivers. The structure is well established by photogeology, but surface work has not determined its exact characteristics because of the presence of coarse clastic rocks of possible Pliocene to Holocene age which give poor indications of dip.

Calima anticline—This feature has been recognized at the east end of Malaga Bay, about 25 km northeast of Buenaventura. Well-developed sea cliffs suggest that the area was recently uplifted.

Tumaco Area

An east-west-trending wrench fault, possibly corresponding to a geofracture, separates the central and southern subbasins (Fig. 1). The Tumaco subbasin is less complicated structurally than the San Juan subbasin. Regional structure returns to a more northerly strike with some convergence of the structures in the extreme south of the basin, possibly owing to a general shallowing of the area of sedimentation. The following are the two major anticlinal folds found in this southern area.

Remolino Grande anticline—The structure lies north-northeast of the town of Tumaco, a lumber and fishing industry port in Nariño Department. The northern plunge of the anticline lies in the lowermost Patia River valley. The Patia River indicates structural closure to the north; to the south the closure is caused by transverse faults, well established by surface geology (G. F. Velasco, personal commun., 1974). A gas seep is present in the area of this structure, which is regarded as very promising for petroleum.

Chagüi anticline—This feature is also in Nariño Department, east of Tumaco Bay. It is one of the largest surface features in the basin—nearly 45 km long and 15 km wide.

Pacific Shelf Basin

South of Cabo Corrientes, the Pacific shelf basin roughly parallels the shoreline until about lat. 3°N, where it also is displaced by the east-west wrench fault. The basin continues south into offshore Ecuador. The strata within this basin are intensely folded and several normal faults are present. Most of the faults dip west and are arranged in an echelon manner causing basement blocks to be raised progressively from west to east.

Domelike intrusions that pierce the strata within the basin are common. The nature of these in-

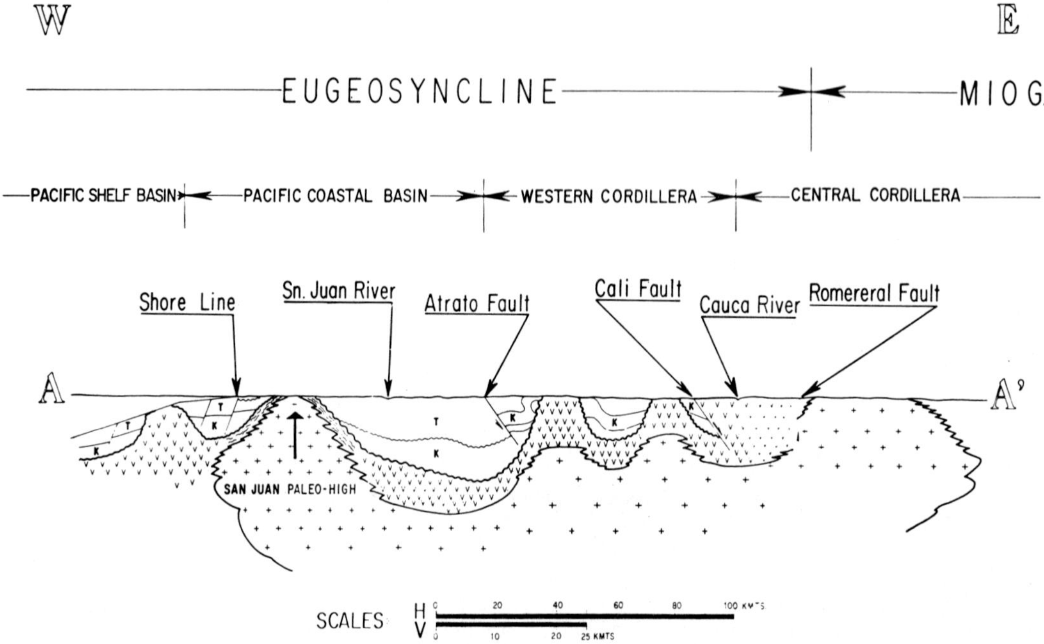

FIG. 3–Generalized regional cross section *A-A'*.

trusions is not known. They do not show typical salt-dome characteristics; more probably they are shale plugs.

Thickness of the sedimentary section has not been established since no wells have been drilled in this basin. Two wells—Sandí-1 and Tambora-1—were drilled offshore in the western margin of the Pacific coastal basin. Anticlinal and synclinal folds of varying lengths can be mapped within the shelf basin. For the most part, they nearly parallel structural trends inland. Wedging, lensing, and other stratigraphic-type traps are also shown by the seismic records, and the basin appears to have hydrocarbon potential.

Stratigraphy

Stratigraphic studies in most of Colombia are hampered by the lack of a uniform code of stratigraphic nomenclature; different nomenclature is used by almost every independent geologist and each oil company. The Pacific coastal basin, which has been studied mainly by oil companies, is poorly covered by published literature. The problem is further complicated by the complex sedimentary history of the area, where abrupt vertical and horizontal facies changes are the rule rather than the exception. The close association of sediments of marine or nearshore origin with volcanic rocks and igneous intrusions, together with the erratic distribution of such deposits, creates additional difficulties.

For these reasons, it seems more appropriate to discuss rock units in terms of their ages, rather than to group them under poorly understood rock-stratigraphic units. Possible correlations or name equivalencies are provided simply as an identification guide.

Basement

Basement is represented by outcrops related to the Sautatá arch, the Western Cordillera, the Baudó Range, and the San Juan paleohigh. Metamorphic rocks are mostly greenschists and phyllites, and the igneous body which crops out at Pavarandocito has been identified as granodiorite (Haffer, 1967).

The eastern margin of the Pacific coastal basin is marked by the sheer cliffs of the Western Cordillera, where the "Diabase Porphyry Group" crops out extensively. The lower portion of the group is a series of hard dark shales and slates with altered limestones and cherts interbedded with flows of diabase and diabase porphyry. Phyllites, quartzites, and graphite schists also occur. Most writers assign a Cretaceous age to this group.

The geology of the Baudó Range is not well known, and published literature is scarce. Morales (1963) mentioned that, in the headwaters of the Salaquí and Truandó Rivers, basement is composed mostly of a brownish basalt which might have been the source for the platinum found in western Colombia. Elsewhere, mainly at Cabo Corrientes and the Gulf of Cupica, a thick section of pillow basalts and gneisses crops out. Farther south, at Gorgona Island, gabbros and peridotites have been described by Oppenheim (1949). The geologic picture has been complicated by the presence of Tertiary basic intrusives and extrusives and by late folding and faulting. According to Jacobs (1963) and G. F. Velasco (personal commun., 1974), the age of this complex is middle to Late Cretaceous. Other oil-company geologists believe that the Baudó Range represents an outer arc of volcanic islands which foundered to the south of Cabo Corrientes, where it is now submerged under the waters of the Pacific Ocean.

The San Juan paleohigh has been postulated on the basis of photogeologic and seismic evidence. Unfortunately, field evidence related to this igneous-metamorphic complex was not available at the time of writing.

Cretaceous

Very little is known about the Cretaceous of the area. Investigation of rocks of this age has been hampered by difficult access and scarcity of outcrops.

Lithology and occurrence—Hubach (1930a) reported Cretaceous ammonite remains from dark gray shales in Mequera Creek, a tributary of one of the western headwaters of the Napipí River in the lower Atrato valley. Oppenheim (1949) reported Cretaceous outcrops in the Iró River area. These rocks are mostly dark blue and black, slightly metamorphosed claystones interbedded with dark gray limestones and dark cherts. This section is intensely affected by intrusives, flows of dark igneous rocks, diorites, and other lavas. On the west side of the basin, near Cabo Corrientes, a series of interbedded dark cherts and dark gray siliceous shales with varying amounts of conglomerate, quartzite, and sandstone crops out at the eastern flank of the Baudó Range.

Cretaceous rocks were also identified, on the basis of fossil evidence, in the Domingodó River area, where they are faulted against pillow basalts and the section is affected by several minor gravity faults. From base to top the section consists of black shales and sandstones, black limestones interbedded with claystones, and green siliceous limestones. A section of black and reddish chert marks the contact with the Eocene.

Conglomerate, sandstone, and siltstone of Cretaceous age were reported in the Nercua graben by Bandy (1967). These conglomerates are composed of clasts of basalt, ranging from boulders to pebbles, cemented in a sandy calcareous matrix. They are tan to light gray, thin- to medium-bedded units which are highly fractured and jointed. Siltstone is tuffaceous, white to gray, and highly fractured. Sandstone is tuffaceous, white to gray, very fine to coarse grained, and slightly to highly weathered.

Petters (1955) reported Cretaceous foraminifers in rocks cropping out at Cerro Pulga, 7 km south of Sautatá, near the point where the Sautatá arch crosses the basin.

Southward, in the San Juan River valley and close to the eastern border, a sequence of black, slaty shales is present; it includes some beds rich in Foraminifera.

The scarce lithologic and stratigraphic information does not permit a complete definition of the lithofacies relations within this sequence. In the north-northwestern part of the basin, the facies is predominantly ophiolitic, whereas on the northeast and south shallower water facies predominate. Farther south, the Cretaceous is associated with igneous and volcanic activity, as in the Iró River section. South of the San Juan area, Cretaceous rocks are mostly black, slaty shales with no evidence of metamorphism or association with igneous or volcanic activity. In the Tumaco area, no Cretaceous rocks are present.

Thickness—In Chocó, Cretaceous rocks are approximately 700 m thick on the west side of the Pacific coastal basin and 400–600 m on the east side. In the San Juan river area, thickness seems to be reduced to about 300 m.

Contacts—The absence of Paleocene and lower Eocene strata indicates an unconformable contact between the Cretaceous and younger rocks. The lower contact is clearly unconformable on the metasedimentary and igneous rocks of the basement complex.

Petroleum possibilities—There are several oil and gas seeps related to Cretaceous outcrops, especially in the central part of the basin. Where present, Cretaceous rocks are probably a good source of petroleum.

Correlation—The Cretaceous rocks described are partly equivalent to the Nercua Group (Bandy, 1967) and the Iró Formation (Oppenheim, 1949).

Tertiary

Tertiary rocks in the Pacific coastal basin range in age from middle Eocene to Pliocene. Duque Caro (1972) recognized a regional hiatus between the Upper Cretaceous and the middle Eocene on the basis of the similarity of faunal successions in Colombia, Ecuador, and Peru and the considerable difference in depositional environments between the two.

Middle Eocene

Lithology and occurrence—In the Atrato area, the middle Miocene consists of dark gray to dark brown, siliceous, locally calcareous shales. Interbeds of soft claystone, foraminiferal limestone, and siltstone are common. The unit also contains several thin, dark gray to black, locally conglomeratic sandstones with a high content of pyroclastic material. At outcrop, these rocks are commonly associated with mafic igneous intrusions.

In the Napipí River section, gray and dark gray shales form the base of the unit. In the San Juan area, two facies (volcanic and nonvolcanic) are believed to be of middle Eocene age. The rocks of volcanic origin, or associated with igneous activity, consist of olive-green cherts associated with ophiolites; the nonvolcanic rocks are mostly greenish-gray, calcareous claystones and sandstones which locally are partly tuffaceous.

No Eocene outcrops have been reported in the Tumaco area. The Chagüi-1 well (Fig. 1) had not reached Eocene strata at a depth of 13,107 ft (3,995 m).

Thickness—The maximum thickness of the middle Eocene rocks is estimated at 1,000 m in the northern part and 500 m in the San Juan area. The rocks wedge out toward the south and are totally absent in the Tumaco area.

Contacts—The middle Eocene overlies the Cretaceous with pronounced angular unconformity. Its relation to the overlying upper Eocene also seems to be unconformable.

Petroleum possibilities—These units appear to have been deposited under conditions unfavorable for the accumulation of oil.

Correlation—The middle Eocene is correlative with the "Lower Suruco" formation as used by Richmond Petroleum Co.

Upper Eocene

Lithology and occurrence—In the Atrato area, the upper Eocene rocks consist mainly of siliceous limestones with interbeds of black chert, calcareous sandstones, and dark gray, hard shale. At the Murrí River section, a coarse conglomerate of igneous clastics is developed toward the base. In the subsurface, the limestone is light gray, hard, and crystalline; some thin-bedded glauconitic sandstones with a high content of apatite, tourmaline, etc, are present, as well as well-indurated, calcareous, foraminiferal shales.

Although no upper Eocene outcrops have been reported from the San Juan area, it is believed that biohermal limestones (related to the postulated San Juan paleohigh) could be present in the subsurface. Upper Eocene rocks seem to be absent in the southern part of the basin.

Thickness—According to subsurface information, the thickness of the upper Eocene rocks does not exceed 200 m. They could be thicker, but outcrop measurements are difficult because of folding and faulting. Thicknesses of up to 700 m have been reported at Salaquí River.

Contacts—The contact with the overlying Oligocene is unconformable.

Petroleum possibilities—The development of limestones and sandstones makes this section highly favorable as a reservoir for petroleum accumulation, especially in the central part of both the Pacific coastal and Pacific shelf basins.

Correlation—These rocks may be equivalent to the "Upper Suruco" formation (Standard) and the Salaquí formation (Mobil).

Oligocene

Lithology and occurrence—In the northern part of the basin, the Oligocene consists predominantly of marlstones, calcareous claystones, and hard calcareous sandstones. In this area, the differentiation between the Oligocene and lower Miocene rocks is not clear. Toward the south the presence of possible Oligocene rocks is more clearly established. At Nauritá River, the facies changes to calcareous sandstones, and limestones are developed at Mumbaradó River.

In the Tumaco area, most of the basal portion of the Oligocene consists of tightly cemented, coarse conglomerates interbedded with claystones and siltstones; the top part is tuffaceous sandstones.

Thickness—Thickness of the Oligocene section seems to vary. In the north it is 1,600 m thick, in the central part it thickens to 1,900 m, and it may reach 3,000 m in the Tumaco area (Guapí River).

Contacts—The contact with the upper Eocene is unconformable, but the relation with the overlying lower Miocene is conformable in the Atrato and San Juan areas and seems to be disconformable and transgressive toward the basin edges.

Petroleum possibilities—The Oligocene seems to have good petroleum potential, owing to the development of sandstones and limestones. Shows of tar and oil were reported from Oligocene rocks at the Buchadó-1 well.

Correlation—The Oligocene sequence may be partly correlative with the Uva formation (Mobil), the Tutunendo Formation of Oppenheim (1949), and part of the Truandó Group in the sense of Bandy (1967).

Lower and Middle Miocene

Lithology and occurrence—The lower Miocene sequence marks the beginning of the second Tertiary marine transgression. In the Atrato area, the lower and middle Miocene sequence consists of dark greenish-gray and greenish-black, hard, locally silty and calcareous shales interbedded with brownish-gray, fine- to medium-grained, slightly shaly, hard sandstones (Napipí and Salaquí River sections). Bands of silty limestone concretions are common. Thick-bedded limestones appear toward the base. Shell fragments and both macrofossils and microfossils are abundant. Toward the San Juan area, the facies becomes more sandy, and in the Tumaco area it consists of olive-green and gray siltstones with interbeds of tuffaceous sandstones and fine conglomerates. In the subsurface, a section of nearly 200 m of tuffaceous sandstones and siltstones was found in the Chagüi-1 well.

Thickness—The thickness of the lower and middle Miocene rocks has been estimated at 1,400 m in the northern part of the basin. The sequence thins toward the south.

Contacts—The upper boundary with the upper Miocene is unconformable and may mark a regional hiatus. The lower boundary could be conformable in certain areas (Atrato and San Juan) where the Oligocene is present, but it is unconformable toward the basin edges and also where Oligocene rocks pinch out and rocks of early Miocene age directly overlie rocks older than Oligocene.

Petroleum possibilities—The sandstones and fine conglomerates of the lower and middle Miocene sequence make it a drilling objective in the Tumaco area. The basal limestones in the Atrato area and sandstones in the San Juan area are attractive as potential reservoirs.

Correlation—This unit is correlative with the Bojayá formation (Richmond), the Napipí shale (Mobil), and, partially, with the Naya Formation of Oppenheim (1949) and the Rio Salado Group in the sense of Bandy (1967).

Upper Miocene–Pliocene

Lithology and occurrence—The lithology of the upper Miocene is fairly constant throughout the basin, from the Atrato to the Tumaco area. Dark bluish-gray claystones and siltstones with thin interbeds of brownish sandstones make up the bulk of the unit. The bluish-gray claystones and shales are micaceous and carbonaceous; they commonly show leaf imprints. Claystones are calcareous in only a few places, and bands of red mudstone concretions are found locally. Sandstone percentage increases toward the top, and red to grayish colors are common. Conglomerates occur at various levels within the unit. Pelecypods and gastropods are common, as are diagnostic microfossils.

The base of the Pliocene in the Atrato area is marked by a thick-bedded conglomerate formed by pebbles and cobbles of basaltic extrusives and a few granitic intrusive rocks with very little sandy matrix. This basal conglomerate is overlain by blue-gray, micaceous, massive claystone commonly containing thin silty to sandy laminae not more than 2–5 mm thick. In the Tumaco area the Pliocene is represented by three facies—marine, continental, and sedimentary-volcanic. The marine facies is composed of bluish-gray sandstones, siltstones, and conglomerates; the latter are restricted to the upper part and have pebbles of mafic igneous rocks in a quartzose matrix. The continental facies, apparently contemporaneous in deposition with the marine facies, is interfingered with it and consists of dark bluish-gray claystones and siltstones with abundant carbonized wood. This facies is well exposed along the Mira River. The volcanic facies is contemporaneous with at least the upper sections of the continental and marine facies; it is composed of tuff, ash, and volcanic breccia. It is present only in the southwest part of Nariño Department. The pyroclastic facies is well exposed along the Caunapí River.

Thickness—The thickness of the upper Miocene and Pliocene rocks has been estimated at between 1,300 and 1,500 m. Approximately 1,600 m of this sequence was penetrated by the Buchadó-1 well.

Contacts—The upper Miocene strata are transgressive, especially in the Atrato area, and cover the underlying units with slight angular unconformity. The Pliocene-Quaternary contact also is unconformable.

Petroleum possibilities—The basal portion of the upper Miocene could be a potential reservoir in the deeper parts of the basin, especially where late tectonic activity may have provided migrational channels for oil which originated in older rocks.

Correlation—Partial correlation may be established between this sequence and the Chocó Istmina Formation of Oppenheim (1949), the Quibdo formation (Mobil), and the Pató formation (Richmond).

Quaternary

Lithology and occurrence—Quaternary deposits cover the lowlands and the low coastline up to a distance 10 km inland. Along the coast there are beach sands and lagoonal clays and silts. Fluvial deposits range from boulder and gravel conglomerates along the mountain front to sandstones and finer grained units in the lower valleys of most rivers. There are also vast amounts of recent tuffs and volcanic agglomerates near the Western Cordillera, toward the central and southern parts of the basin.

Thickness—The thickness of Quaternary deposits ranges from 100 to 150 m.

Contacts—The Quaternary deposits unconformably overlie older rocks, mostly with angular unconformity.

Petroleum possibilities—There are no possibilities of hydrocarbons in Quaternary deposits.

History of Exploration

The history of exploration dates back to the middle 1940s, when both government and private-company geologists initiated surface mapping in the basin. Their efforts (Jacobs, 1963) were hampered by difficult field conditions and the lack of base maps. Between 1945 and 1952, surface mapping was performed by Richmond Petroleum Company (now Chevron—a subsidiary of Standard Oil of California) and Socony-Vacuum Oil Company of Colombia (Mobil). Tropical Oil Company secured good gravity reconnaissance coverage of the basin. Richmond and others did some seismic shooting in the early years of exploration. The first exploratory test (Buchadó-1) was drilled in 1953.

New activity in the area was spurred recently by the need to discover new fields to replace rapidly declining production from established provinces. Intercol (Exxon), Gulf, Phillips, Conoco, and Superior, among others, did surface geology and seismic work, and five more wells were drilled (Chagüi, Sandí, Tambora, Urodó, and Opogadó). In August 1971, United Geophysical Corp. shot a 700-km marine seismic profile for Ecopetrol on the continental shelf of the coastal and shelf basins. In January 1973, Geophysical Service Inc. performed more detailed seismic work between Cabo Corrientes and Tumaco in an approximate area of 10,800 km^2 with about 3,350 km of seismic profile.

Photogeologic mapping traditionally had been difficult owing to the almost permanent cloud cover. About 70% of the area has been covered by conventional vertical air photographs. The combination of radar imagery and stereoscopic photography led to the first attempt at photogeologic mapping of the entire basin. Such a project is currently being performed by a Colombian consulting firm, and its preliminary results were the basis for the postulation of the ancient landmass—the San Juan paleohigh.

The following exploratory wells have been drilled within the basin (see Fig. 1).

Buchadó-1—This rank wildcat was the first test drilled in the basin; the operator was Richmond Petroleum Co. The well was spudded October 30, 1953, and completed August 4, 1954, at a total depth of 15,539 ft (4,736 m)—one of the deepest wells ever drilled in Colombia. The well was abandoned, but indicative shows of oil were reported at 9,130 ft (2,783 m) and tar at 11,650 ft (3,551 m). Some gas shows were reported down to approximately 8,000 ft (2,440 m), and a blowout of dry gas occurred at 5,800 ft (1,768 m); high gas pressures were encountered at many levels. The well was a satisfactory test of the structure, but its basin position was poor.

Chagüi-1—Drilled by Mobil, this well was spudded September 13, 1955, and completed June 24, 1956. Shows of light oil were reported from 5,930–5,950 ft (1,807–1,814 m) and 6,190–6,280 ft (1,887–1,914 m). The lower zone yielded gas on testing. High-pressure gas was encountered from 4,230 ft (1,289 m) to total depth of 13,107 ft (3,995 m). The basin position of this well also was poor.

Tambora-1—This test, drilled by Intercol, was the first to be drilled offshore, although within the limits of the coastal basin. It was drilled in approximately 170 ft (52 m) of water to a final depth of 11,365 ft (3,464 m). Spudded February 18, 1967, it was abandoned May 18, 1967, with only minor shows of gas. Stratigraphically, the well penetrated a thick (9,000 ft or 2,743 m) sequence of turbidites, and the bottom 2,300 ft (700 m) consisted entirely of conglomerates. The presence of such a thick section of conglomerates indicates a position close to a source of clastics. This situation tends to support the existence of the San Juan paleohigh.

Sandí-1—Spudded May 20, 1967, and abandoned July 8, 1967, the well was located southwest of the Tambora-1. It reached a total depth of 12,161 ft (3,707 m) and was drilled in approximately 200 ft (60 m) of water. No shows were reported from this well, which apparently was located in an unfavorable basin position.

Urodó-1—Drilled in 1973 by Superior Oil Company, this well tested the west flank of the Urodó anticline. The well reached a total depth of 15,000 ft (4,572 m) and, according to biostratigraphic studies (Duque Caro, personal commun., 1974), it did not reach the Eocene. Owing to steep dips and structural complications, the well apparently penetrated only about 9,000 ft (2,743 m) of sedimentary rocks. High-pressure gas was encountered throughout the drilling, but there apparently were few reservoir strata (sandstone and limestone) and no significant hydrocarbon shows were reported. The well was dry and abandoned.

Opogadó-1—This test, located in the northern part of the coastal basin, on the Opogadó structure, reached a total depth of 11,372 ft (3,466 m) without reaching Eocene deposits, the main objective for the test. It was abandoned by Gulf Oil Company in June 1974.

Petroleum Potential

The petroleum potential of an area depends on the existence of source and reservoir rocks as well as traps and trapping mechanisms suitable for the accumulation of commercially exploitable quantities of hydrocarbons. These three main controlling factors are present within the limits of the Pacific coastal and Pacific shelf basins.

Source Beds

Marine rocks of Cretaceous age are believed to be important sources of petroleum. The section described by Hubach (1930b) at Mequera Creek is undoubtedly of marine origin and rich in organic matter. South of the San Juan area, the facies also seem to be favorable. Generally, rocks of Cretaceous age can be considered a good source of petroleum where they have not been affected by severe igneous or metamorphic activity. Some oil seeps have been reported from Cretaceous outcrops in the east-central part of the basin. One of the best known is that at Iró River, where oil comes both from marine sedimentary and fractured volcanic rocks.

Late Eocene and late Oligocene to early Miocene limestones also could have acted as source beds of petroleum.

Reservoir Rocks

The best potential reservoir rocks occur mainly in the upper Eocene and lower Miocene sections, but the upper Oligocene also has good possibilities.

The existence of the San Juan paleohigh and the possible development of an associated shelf and conditions of tectonic stability lead us to believe that the central or San Juan area of both the Pacific coastal and shelf basins offers the best petroleum potential. These circumstances substantially modified the conditions of abyssal sedimentation prevalent on the north and south during long periods of geologic time.

Traps

Only structural traps have been tested by the drill, but it is apparent that more thorough considerations should be given to stratigraphic and combination traps in the Pacific coastal and shelf basins. Lithofacies, as related to paleogeography, could represent a very important type of trapping mechanism.

Structural and paleogeographic conditions as anticipated for the San Juan area should have created the characteristics needed for the development of reefs, offshore bars, channels, and other features related to neritic to bathyal environments as opposed to abyssal sedimentation.

References Cited

Bandy, Orville, 1967, Atlantic-Pacific Interoceanic canal; Route 25 geology data: Panama, Office Interoceanic Canal Studies, 20 p.

Burgl Hans, 1961, Historia Geológica de Colombia: Rev. Acad. Sci. Fisic. Nat., v. 11, no. 43, p. 137–191.

Case, J. E., et al, 1969, Investigación gravimétrica y magnética a través de Istmo de Panama (noroeste Colombiano): Geología Colombiana, v. 6, p. 5–16.

Duque Caro, H., 1972, Relaciones entre la bioestratigraf ía y la cronoestratigrafía en el Ilamado geosinclinal de Bolívar: Colombia, Serv. Geol. Nac. Bol. Geol., v. 19, no. 3, p. 25–68.

Jacobs, Cyril, 1963, Geology and oil possibilities in the Pacific coast basin of Colombia, South America: Bogotá, Private Company Rept.

Haffer, J. H., 1967, On the geology of the Uraba and Northern Chocó regions, northwestern Colombia: Bogotá, Private Company Rept.

Hubach, E., 1930a, Apreciación de los proyectos del canal interoceánico por el Napipí y por el Truandó, según puntos de vista geológicos: Colombia, Bol. Min. Petr., v. 3, no. 13, p. 15–34.

——— 1930b, Manifestaciones petrolíferas en la Intendencia del Chocó: Colombia, Bol. Min. Petr., v. 4, nos. 21–22, p. 336–337.

Morales, L. G., 1963, Regional geology and oil possibilities of the Urabá Gulf area and the Pacific geosyncline of western Colombia: Bogotá, Private Company Rept.

Nygren, W. E., 1950, Bolívar geosyncline of northwestern South America: AAPG Bull., v. 34, no. 10, p. 1998–2006.

Oppenheim, V., 1949, Geología de la costa sur del Pací fico de Colombia: Inst. Geofís. Andes Colombianos, Bol. 1, 23 p.

Petters, V., 1955, Developments of Upper Cretaceous foraminiferal faunas in Colombia, S. A.: Jour. Paleontology, v. 29, no. 2, p. 7–35.

Schuchert, C., 1935, Historical geology of the Antillean region: New York, John Wiley and Sons, v. 1, 811 p.

Exploration and Development of Hydrocarbon Resources in Pacific Basins of Ecuador—Summary[1]

ALFONSO JARRIN[2]

INTRODUCTION

The Pacific basins of Ecuador have excellent geologic characteristics for the accumulation of hydrocarbons. Geologic, seismic, gravity, and magnetic surveys provided a basis for exploratory drilling in the provinces of Esmeraldas, Manabí, and Guayas.

Exploratory drilling in the Santa Elena Peninsula has resulted in the discovery of the Ancon, Petropolis, Santa Paula, Concepción, Certeza, and other oil fields. In 1970, the Amistad gas field was discovered in the Gulf of Guayaquil. Shows of gas and oil were obtained in stratigraphic units penetrated in wildcats elsewhere in Guayas Province.

The hydrocarbons are found in the Atlanta Formation (Atlanta sandstone) and the Socorro Formation. Shales of the San José Formation are considered to be the source rocks.

The best petroleum prospects of the Ecuadorian littoral zone are placed offshore in the Gulf of Guayaquil. Important gas accumulations are present at depths between 9,000 and 14,000 ft (2,740–4,270 m), principally in the Miocene Subibaja Formation (Fig. 1).

Cumulative production of oil in this area to June 1974 was 106,300,000 bbl. Although data on gas reserves are scarce, proved reserves in the Gulf of Guayaquil are estimated at 145,111 MMcf, and probable reserves at 373,536 MMcf.

STRATIGRAPHY

Cretaceous

The sedimentary pile of the Ecuadorian Pacific basins is up to 13,500 m thick in the Guayas depression. Approximately a fourth of these rocks is of Late Cretaceous (Maestrichtian) age. The rocks unconformably overlie a volcanic series—the Piñon Group—which is a complex of pyroclastic rocks, basaltic sills, and interstratified sandstone and gray tuffaceous shale. The base of the sedimentary pile is the Upper Cretaceous Cayo Formation, which has a thickness of 3,500–4,000 m. This formation has been divided into three members, from the base upward—the Calentura, Cayo *sensu stricto,* and Guayaquil.

The Cretaceous is overlain by a Tertiary sequence that ranges from Paleocene to Pliocene. The Tertiary is of most interest for petroleum production and prospects.

Azúcar Group (Paleocene–Middle Miocene)

The Tertiary sequence begins with the Estancia, Chanduy, and Engabao Formations of the Azúcar Group; they represent sublittoral and brackish-water deposits. Their lateral equivalents are marine facies which, in the Santa Elena Peninsula subsurface, are known as the San José Formation, the Atlanta sandstone, and the Atlanta shale.

The stratigraphic relation of these strata with the underlying Guayaquil Formation is disconformable. The pebbles which constitute conglomerates of the Chanduy Formation are lithologically similar to the rocks of the Cretaceous Piñon, Calentura, Cayo, and Guayaquil.

The San José and Atlanta Formations are important because they are oil productive in the Santa Elena Peninsula fields.

To the north in the Chongón-Colonche Hills (in Manabí and Esmeraldas Provinces), the Paleocene–lower Eocene is absent, and the Tertiary section begins with the middle Eocene.

San José Formation—The San José Formation is composed of sandstone at the base and shale at the top. The sandstone units are micaceous, hard, dark gray, somewhat clayey, and oil bearing. As a group, the sandstones are 1,000 m thick. The overlying shale is dark gray to black, considerably fractured, and is considered to be the source rock for hydrocarbons in this area. The shale is 100–150 m thick. The San José has been assigned to the Danian.

Atlanta Formation—The Atlanta Formation includes gray, hard, medium- to coarse-grained, fragmented sandstones which make up the principal oil-producing zone in the Santa Paula field on Santa Elena Peninsula. The sandstones are 800 m thick. Overlying them is the Atlanta shale, which is composed of black to dark gray shales with gray sandstone intercalations. The Atlanta shale ranges in thickness from 0 to 800 m. The thick-

[1]Manuscript received, October 21, 1974.

[2]Corporacion Estatal Petrolera Ecuatoriana, Quito, Ecuador.

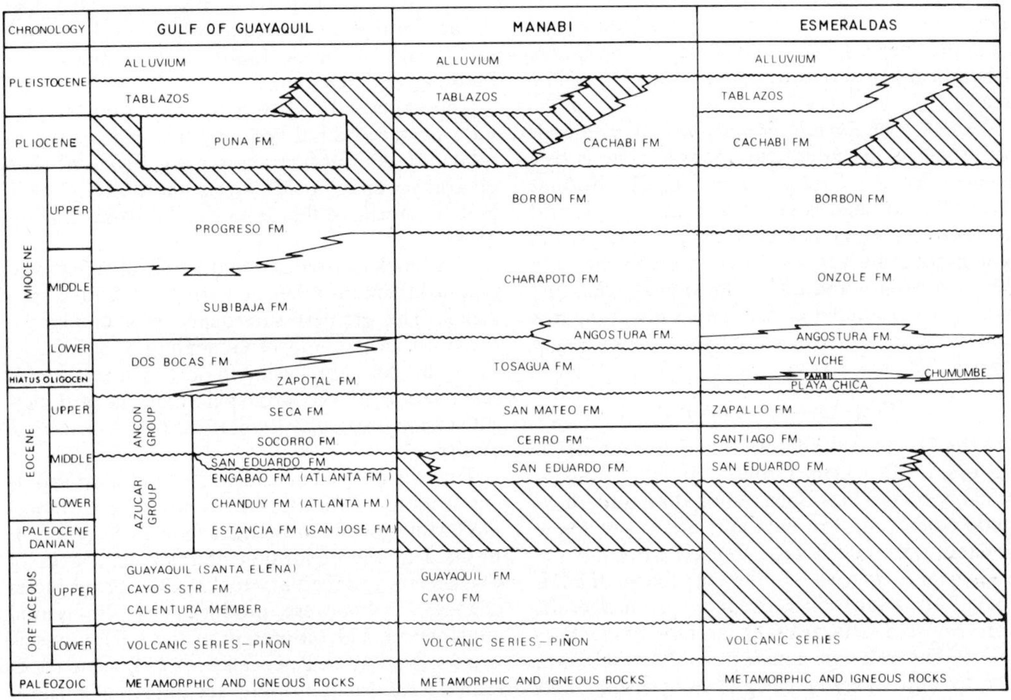

FIG. 1–Stratigraphic chart of Pacific Ecuadorian littoral rocks.

ness variations are due to a disconformity at the top. On the basis of the fossil content, the Atlanta ranges in age from early Eocene to early middle Eocene.

San Eduardo Formation—Placed chronologically in the lower part of the middle Eocene, the San Eduardo overlies the upper part of the Azúcar Group or, farther north, elements of the Cayo or Piñon. This formation, a sandy limestone, is widespread in Ecuador.

Ancon Group (Middle-Upper Eocene)

Overlying the San Eduardo, Cayo, or Azúcar disconformably is the Ancon Group. Lithologically, it can be divided into two units which are composed of sandstones, siltstones, and shales with a total thickness of approximately 2,000 m. The formations which compose this group are the Socorro, below, and the Seca, above.

Socorro Formation—The Socorro is divided locally into three members—the Santo Tomas, Clay Pebble Beds, and Socorro *sensu stricto.* The Santo Tomas member, at the base of the formation, ranges in thickness between 0 and 220 m, and produces economic amounts of oil. It is composed of poorly sorted, light gray, siliceous and/or calcareous, conglomeratic sandstone. Silty and quartzitic pebbles and, in the upper part, well-rounded quartz clasts are interstratified.

The Clay Pebble member ranges in thickness from 0 to 750 m and has varied lithology of siltstones and silty sandstones.

The Socorro Formation *sensu stricto* consists of alternating sandstones, shales, and siltstones; the sandstones predominate. The unit is gray to greenish gray, and the shales are gray to dark gray. The unit is middle Eocene and is more than 700 m thick. The sandy zones are the reservoirs for about 20% of the oil produced in Ancon field.

Seca Formation—The Seca Formation lies on, and is concordant with, the Socorro Formation. The Seca Formation has a maximum thickness of 1,200 m and terminates upward in sandy, conglomeratic strata of the Ancon Point member. The Seca Formation is late Eocene; it contains foraminifers, radiolarians, and mollusks.

Lateral equivalents of the Ancon Group are the Cerro and San Mateo Formations in Manabí, and the Santiago and Zapallo Formations in Esmeraldas.

San Mateo Formation—The upper part of this formation contains conglomerates and well-sorted sandstones which, because of their porosity and permeability, should be good reservoir rocks for hydrocarbons. Thickness ranges from less than 100 m to 1,500 m. A black shale facies present in the deeper basin sections may be the source of hydrocarbon shows detected in drilling in this area.

Santiago and Zapallo Formations—In Esmeraldas Province, the middle to upper Eocene is represented by the Santiago and Zapallo Formations. The Santiago is sandy and calcareous, and the overlying Zapallo is dark gray to buff shale, rich in foraminifers, which is intercalated with some sandstones and tuffs. This organic-rich formation is considered as the hydrocarbon source rock in the Esmeraldas area. Its thickness ranges between 200 and 300 m.

Post-Ancon Formations

In the Progreso subbasin of the Gulf of Guayaquil basin, the Ancon Group is overlain unconformably by the Dos Bocas Formation and/or its lateral equivalent, the Zapotal Formation. These formations are late Eocene to early Miocene.

Subibaja Formation—In the same basin, the Subibaja Formation, which produces gas in the offshore Amistad field, overlies the Dos Bocas Formation. The Subibaja consists of sublittoral, dark gray to buff, slightly calcareous, resistant siltstones with intercalated thin beds of fine-grained, compact sandstones. These beds are rich in foraminifers and mollusks. The lower part of the formation is characterized by greenish-gray siltstones grading to silty shales; they bear carbonaceous fragments and megafossils. Toward the center of the basin, the formation has characteristics of source rocks, and may be the source of the gas produced from the sandstones in the Amistad field.

The post-Ancon formations in Manabí and Esmeraldas Provinces are shown in Figure 1. At present, they are not considered prospective for petroleum.

Structure

Tectonically, the Pacific littoral zone of Ecuador—onshore and offshore—is a platform related to the Cordillera Occidental, which has formed as the result of block faulting. The blocks are bounded by high-angle, normal faults.

The structural style was impressed on the pre–Late Cretaceous basement by the principal faults related to the Cordilleran orogeny. From the Gulf of Guayaquil to 0°35′N lat., these faults strike N30°E; north of that latitude, the strike is N70–75°E.

The blocks subsided rapidly and differentially, and sedimentation began during the Late Cretaceous. The greatest subsidence was in Guayas Province, in the Gulf of Guayaquil.

In the Manabí basin, in late Miocene time, reactivation of the principal faults caused new subsidence and deposition of the San Mateo Formation.

That differential subsidence occurred in different sections of the Pacific littoral zone is important with respect to hydrocarbon occurrence. It resulted in the deposition of thick sequences of Cretaceous and Tertiary rocks with abrupt facies changes. Folds were produced by differential compaction. The oil present in the western reservoirs originated in Paleocene–lower Eocene rocks (San José Formation) and accumulated in sandstones of the Atlanta and Socorro Formations.

Conclusions

1. Accumulation and trapping factors of major significance in the western Ecuadorian geologic environment are related to block faults that have had significant movement.

2. There are no anticlinal folds, and all the structural slopes are the result of differential compaction.

3. Several stratigraphic zones have good source-rock and good reservoir characteristics.

4. The best prospects for hydrocarbons are offshore in the Gulf of Guayaquil, as proved by the drilling of three gas-producing wells.

5. Exploration should continue in this area in order to determine an accurate estimate of reserves.

Hydrocarbon Potential of Coastal Basins of Peru[1]

RUSSELL B. TRAVIS, GRÓVER GONZALES, and ALFREDO PARDO[2]

Abstract Along the coast of Peru, the shelf between the Andes mountains and the 100-fm isobath is narrow and reaches a width of 100 km in few places. Between lats. 6 and 14°S, nearly all the shelf lies offshore. This shelf was the site of marine deposition throughout the Tertiary. Basement highs and irregularities define six subbasins. Beginning in the north, the mainly offshore Progreso basin extends into Peru from the Gulf of Guayaquil. It is filled with upper Tertiary shales, sandstones, and conglomerates with a maximum thickness of 6,000 m. The small, abandoned Zorritos field is located in this basin, and some oil and gas have been discovered recently. The basin has an estimated potential of 350 million bbl.

The Talara basin, which has produced over 800 million bbl of oil, is both onshore and offshore. The strata are Upper Cretaceous and lower Tertiary silicate clastic rocks with a maximum thickness of 8,000 m; nearly all production comes from Eocene deltaic, fluvial, and turbidite sandstones. Intense block faulting, gravity sliding, and submarine slumping complicate development operations. The onshore part does not have significant potential for future discoveries, but the offshore is estimated to have a potential of 1 billion bbl. The Sechura basin is situated between the Andes and a discontinuous coastal chain of low mountains which separates it from the Talara basin. It is mostly onshore but extends offshore to the south. Up to 3,000 m of marine, brackish, and nonmarine deposits, including diatomite and phosphatic and tuffaceous elements, fills the basin. The most widespread sedimentary rocks are upper Tertiary, but lower Tertiary and Cretaceous rocks are also present. In the 1950s, 28 wildcat wells were drilled and moderate reserves of gas were discovered. Total potential for the basin is estimated to be 100 million bbl.

The Salaverry basin, the largest of the coastal basins, has a length of 500 km and a width up to 100 km. Entirely offshore, it extends southward to about 100 km north of Lima. It contains up to 3,000 m of Tertiary marine shales, siltstones, and calcareous rocks with sandstones at the base. Two wells have been drilled in this basin. The estimated potential for the basin is 500 million bbl. The Pisco basin begins about 100 km south of Lima in the offshore, but splits into onshore and offshore parts southward. Up to 2,000 m of lower and upper Tertiary rocks are present. The lower Tertiary is composed of conglomerates, sandstones, and calcareous shale. The upper Tertiary consists of strata similar to those of the upper Tertiary in the Sechura basin. One well has been drilled in the basin. The estimated potential for the basin is 100 million bbl. The Moquegua basin is a narrow onshore basin between the Andes and the coastal chain of mountains. Marine strata are found only in the northern part and are of insufficient volume to have significant potential. The total potential of the coastal basins is estimated to be about 2 billion bbl.

INTRODUCTION

Peru is made up of three distinct physiographic and climatic regions—a narrow coastal belt; a high complex Cordillera of the Andes; and a broad, low, jungle-covered plain of the Amazon. The coastal belt is a narrow strip of desert which receives almost no rain, but is watered in the subsurface along river courses by runoff from the high Andes. The Andes Cordillera is made up of spectacular ranges of mountains reaching elevations of nearly 7,000 m. The largest physiographic region is the upper Amazon drainage area. It is vast, low, and wet. Its 600,000-km^2 area is covered by heavy rain forest. Only 10% of Peru's population lives there.

The shelf, both onshore and offshore to the 100-fm isobath, is narrow along the southern two thirds of the coast. The mountains extend literally to the water's edge. Along the northern third, the shelf is about 100 km wide, mostly offshore in the southern part and onshore in the northern. The offshore shelf has an area of about 80,000 km^2. The Peru-Chile Trench parallels the trend of the Andes 120–200 km to the west, shallowing toward the north.

REGIONAL GEOLOGIC SETTING

The general geology of Peru and the location of the coastal basins are shown in Figure 1. The term "basin" is used here not only in the conventional sense, but also to designate areas of deposition.

During Jurassic and Triassic times, geosynclinal conditions existed over large parts of the present Andes area. During the Cretaceous, the geosyncline increased in extent to cover most of Peru. Along the present coast, a vigorous submarine volcanic regime gave way to a nonvolcanic marine regime a short distance to the east. The geosyncline was floored by Precambrian and Paleozoic metamorphic and sedimentary rocks. In Late Cretaceous and early Cenozoic time, the entire area of the geosyncline rose above sea level, volcanic activity essentially ceased, and the Andes area became a source of nonmarine deposits on the east side and marine deposits on the west. These western marine sediments, deposited

[1]Manuscript received, October 21, 1974.

[2]Petróleos del Perú, Lima, Peru.

Most of the information in this paper came from the geological files of Petróleos del Perú. The writers are grateful to Petróleos del Perú for permission to use the data and to publish this paper.

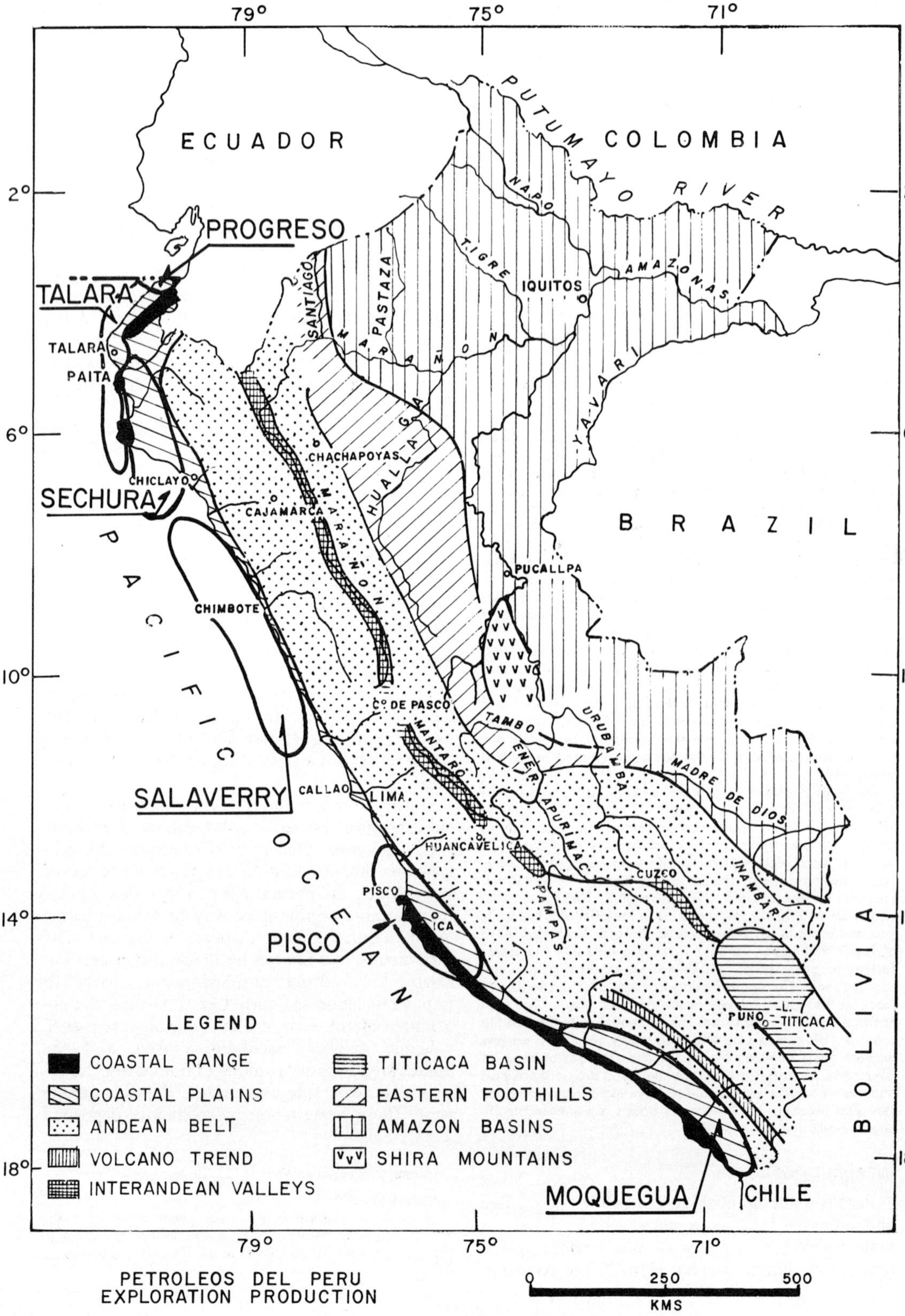

FIG. 1–Morpho-structural map of Peru.

from Late Cretaceous time until the present, are the ones whose hydrocarbon potential will be considered in this paper.

The main geomorphic features of the coast from east to west are the Andes mountains, the coastal shelf, the Coastal Range, the continental slope, and the Peru-Chile Trench. The principal elements of the western part of the Andes mountains are the coastal batholith, Paleozoic metamorphic rocks, Mesozoic volcanic and nonvolcanic strata, and a discontinuous mantle of late Cenozoic volcanic rocks of generally andesitic composition. The coastal batholith is a complex of intrusions of various granitic rock types, the most abundant being granodiorite. Paleozoic metamorphic rocks are of various ages and are metamorphosed to different degrees. For example, they include phyllite, slate, schist, quartzite, and amphibolite. Triassic, Jurassic, and Cretaceous submarine volcanic rocks are found along the western foot of the Andes. An incomplete composite thickness in excess of 10,000 m consists of pillow lavas, pyroclastic rocks, ignimbrite, and minor amounts of fossiliferous marine shale, sandstone, and chert. The correlative nonvolcanic rocks are marine limestone, shale, and sandstone or quartzite.

The name "Coastal Range" refers to a discontinuous string of low mountains and offshore subsurface basement[3] highs, subparallel with and west of the Andes mountains. The northern part is formed by a chain of low mountains along the coast (Fig. 1). The rocks consist of late Paleozoic argillite, quartzite, slate, and phyllite intruded by granitic rock of pre-Cretaceous (probably Jurassic) age. The southern onshore part of this range is formed by a series of hills and low mountains along the coast south of Pisco (Fig. 1). The range here consists of a complex of igneous and metamorphic rocks of late Precambrian and possibly early Paleozoic age, and isolated blocks of sedimentary rocks of late Paleozoic age. Along certain stretches of the offshore, the range is represented by a persistent trend of subsurface basement highs. Much of the coastal shelf lies between the Andes and the Coastal Range. Where this situation exists, a structural basin is formed on the shelf. The general architecture of the coastal elements associated with such a basin is shown in Figure 2. In places where the coastal range is not present, no basin structure is found and shelf sediments form a continuous blanket with upper-slope sediments. The outlines of the basins shown in Figure 1 are for location purposes only and do not necessarily indicate the limits of the sediment blanket on the west side.

[3]The term "basement," applied to the offshore, refers to acoustic basement; the basement no doubt is composed of different lithologies in different places.

General Stratigraphy

Very little drilling has been done outside the Talara area, so offshore stratigraphy is poorly known. Seismic data show considerable thicknesses of sedimentary rock on both the shelf and the upper slope, which is as far as detailed seismic lines go. Thickness is controlled locally by faulting and varies erratically from place to place. Some troughs and grabens have as much as 5,000–6,000 m of sedimentary strata, and in the Talara basin there is up to 8,000 m. No generalizations can be made about where the thicker sections are. Along the upper slope in water depths from 500 to 1,000 fm, the thickness ranges from 1,000 to 4,000 m, but not uniformly. Composite drilled sections offshore and onshore show up to 2,500 m of Campanian-Maestrichian, 10,000 m of Paleocene-Eocene, and 5,000 m of post-Eocene rocks. The Paleocene-Eocene section here is one of the most complete in the world.

In general, the lithology is relatively uniform, consisting mainly of shale, siltstone, and sandstone. Conglomerate is locally abundant, and thin limestone beds form a minor part of the section in some places. The sediments are almost entirely marine and represent environments from abyssal to fluvial. The oldest deposits of the coastal-basin sequences are Campanian-Maestrichtian shale and sandstone. These strata overlie rocks of the Andes or Coastal Range and are known mainly in the Talara basin. The sequence thickens toward the offshore, and subbasins containing Campanian-Maestrichtian rocks may be expected along the coast. However, good reservoir beds are uncommon in known sections, so, although shales with good source potential are present, there is no significant oil production.

Paleocene rocks are restricted to the Talara basin and consist mainly of shale and only minor sandstone. Conglomerate has been reported on the coast of the Sechura area and may extend seaward, but it is probably of local extent. There is only minor oil production from the Paleocene.

Lower and middle Eocene strata are also restricted to the Talara basin. These rocks account for nearly 90% of the oil that has been produced in Peru. Some sandstones on the coast of the Sechura area have been referred to the early Eocene, so it is possible that sediments of this age may be found seaward in the offshore areas.

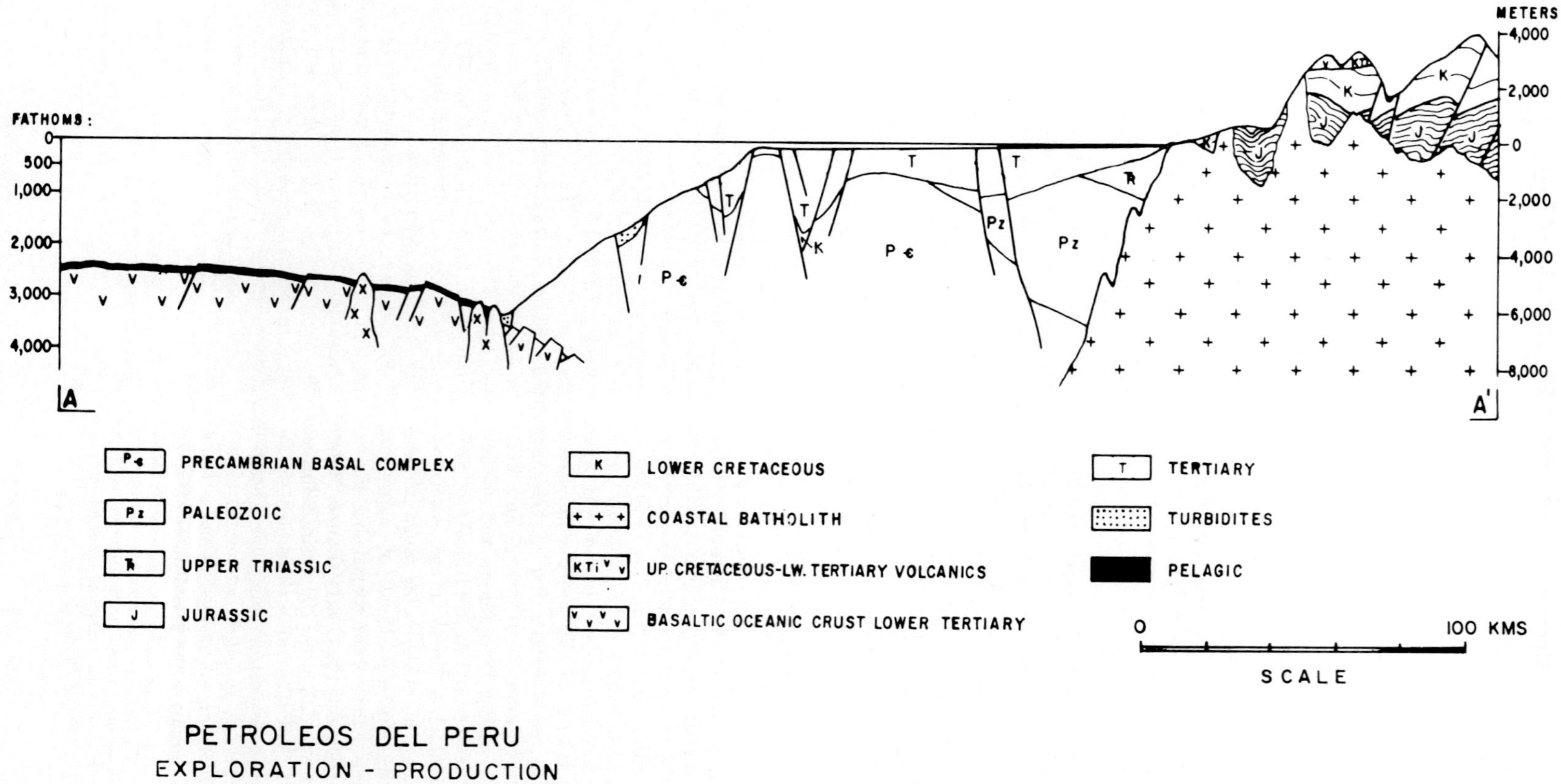

FIG. 2–Cross section at Salaverry parallel (10°S), based on seismic profiles.

Upper Eocene rocks are present in all the basins except Progreso, although they are limited to the west side of the Pisco basin and the west side and north end of the Sechura basin. They consist mainly of shale and sandstone. Upper Eocene rocks have produced 8% of the oil from the Talara basin and have tested gas in the Sechura basin. Some shales have rated high in analysis for source potential. The upper Eocene beds are prospective in all the basins.

Post-Eocene[4] formations are thickest in the Progreso basin where Oligocene, Miocene, and Pliocene rocks are present. However, post-Eocene units are present in all the basins except Talara, where they probably have been removed by erosion. Shale, siltstone, and conglomerate make up the sections, but proportions of these lithologies range widely from place to place.

In summary, the most promising rocks, because of their established productivity, are the lower and middle Eocene. These rocks are known to be present only in the Talara basin. If their presence in the offshore can be established in some definable patterns, the attraction of the offshore basins would be greatly enhanced. The potential of strata of other ages depends largely on the presence of adequate reservoir beds.

Structures Within Basins

The structural style affecting the prospective strata onshore is extensive normal faulting ranging from intense to moderate. Sizes of blocks and of faults range widely, and displacements are as great as 2,000–3,000 m. The largest faults delineate high trends with which oil accumulations are associated in the Talara basin. These trends are themselves made up of numerous smaller blocks. Normal faulting has been active since at least Late Cretaceous time, and probably earlier. It is likely that block faulting created considerable relief during deposition.

Geophysical data indicate that the structural style in the offshore is similar. Relief of the interpreted basement is considerable though erratic, and has the form of fault blocks. Although horsts and grabens are irregularly distributed, there are high and low trends. Younger strata are less affected than older rocks, and there are flexures without much faulting. In many places, the younger rocks are barely disturbed, but onlap, offlap, and truncation are common features. Gravity slumping and sliding, in places on a large scale, and shale flowage have been interpreted on seismic sections. There seems to be an ample variety of trapping mechanisms for hydrocarbons.

[4]In some areas, there are questions about ages of these strata, so the inclusive term "post-Eocene" has been adopted.

Hydrocarbon Potential

The only oil production from the Pacific Coast of Central and South America comes from southwestern Ecuador and northwestern Peru, which together are estimated to have an ultimate recovery of 1.5–1.8 billion bbl. In Peru, the only important prospective area for discovering new coastal accumulations is the offshore. Private companies may enter into operating contracts with the state oil company, Petróleos del Perú, for the exploration and exploitation of 400,000-ha. offshore blocks. Marine seismic work has been conducted over the entire area.

Progreso Basin

The Progreso basin extends from the Gulf of Guayaquil, Ecuador, into northwestern Peru (Fig. 1). In Peru, the basin has an area of 2,500 km^2 and lies onshore or in water less than 100 fm deep. It is bounded on the east by mountains of the Coastal Range and on the south and southwest by a large prominent structure which lies transverse to the regional structural grain of coastal Peru (Fig. 3) but has the same trend as major structures to the north and northwest (Santa Elena Peninsula, Ecuador). In the central part of the basin, the sedimentary section is up to 6,000 m thick and consists of shale, siltstone, and sandstone. Most of the section is of Miocene age, but the Oligocene, Pliocene, and Pleistocene are also represented. The section thins abruptly toward the coast and eastern limit of the basin with notable facies change from marine to fluvial-deltaic deposits. No pre-Oligocene has been found, but no offshore wells have reached basement.

Some idea of structure in the Progreso and northern Talara basins is given by the seismic map in Figure 3. In addition to numerous normal faults, gravity-slide features have been identified on seismic sections. Domal and large anticlinal structures can be observed on this map.

Although relatively small, the Progreso basin has substantial potential. The small, abandoned Zorritos oil field, discovered in 1870, produced only 4 million bbl, but had a yield of 26,500 bbl/surface acre from depths of 305–610 m. An offshore well found several productive zones, 3–7.5 m thick, with a producing potential of 9,000 BOPD and a few million cubic feet of gas. One 3-m-thick sandstone produced at a potential flow rate of 2,500 BOPD. A 2.5-km offset failed to confirm the extension of such producing capacity, but additional drilling is planned. Undeveloped

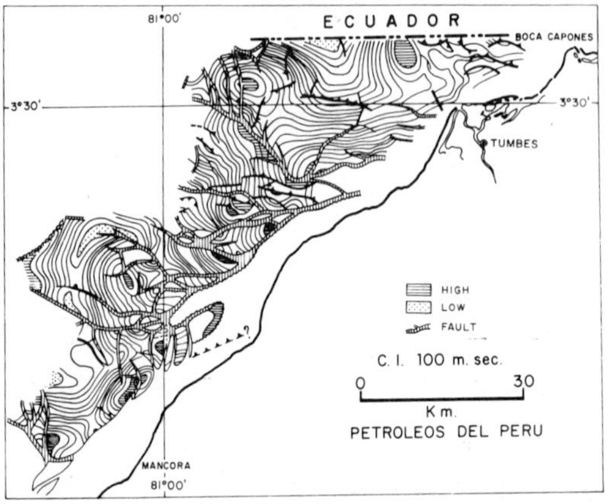

FIG. 3—Seismic map of offshore parts of Progreso and Talara basins.

gas accumulations have been discovered in the Ecuador part of the basin. A future potential[5] of 350 million bbl is estimated for the Progreso basin in Peru.

Talara Basin

The Talara basin, which has yielded 97% of Peru's oil production to date, has an area of 16,500 km^2 out to the 500-fm isobath. More than half of the basin is offshore. It is bounded on the east and southeast by the Coastal Range. The southern limit is arbitrarily taken near 6°30'S lat. where the 500-fm isobath crosses the western limit of the Sechura basin, although it may extend farther south.

Hydrocarbons of the Talara basin were exploited before the Spanish conquest and have been produced by conventional methods since 1874. Of the total 867 million bbl produced to year-end 1973, about 800 million bbl came from onshore. Sedimentary rocks in the basin consist of shale, sandstone, and conglomerate with a maximum thickness of 8,000 m. They range in age from Campanian to late Eocene, and represent deep-marine to fluvial environments. Nearly all production comes from Eocene deltaic, fluvial, and turbidite sandstones representing three main cycles of deposition. Some production comes from fractured Paleozoic (Coastal Range) rocks which otherwise constitute effective basement.

Structure in the Talara basin is complicated (Fig. 4). Although production is generally related to high trends or areas, intense block faulting causes exceptions and makes development of new pools difficult. Movement of large masses of rocks on low-angle surfaces, probably representing gravity slides, creates chaotic zones and causes repetition of section in several places. Submarine slumping on a smaller scale, and possibly on as large a scale as post-consolidation sliding, is also recognized.

Onshore, the basin is in a late stage of development with little prospect for the discovery of large new reserves. Structure and stratigraphy in the offshore are essentially the same as onshore. It is estimated that the total potential of the Talara basin offshore, including the reserves already discovered, is on the order of 1 billion bbl.

Sechura Basin

The Sechura basin (Fig. 1) has a total area of 29,000 km^2 out to the 500-fm isobath. The basin is mainly onshore (17,000 km^2), but it extends southward into the offshore. It is bounded on the west by the Coastal Range, on the east by the Andes, and is limited on the north by exposed Cretaceous sedimentary rocks bearing volcanic material. In the south, it is separated from the Salaverry basin (Fig. 1) by a north-south trend of subsurface basement highs. On the southwest, it is arbitrarily limited by the 500-fm depth, although 1,000–2,500 m of sedimentary strata are believed to be present.

The onshore part of the basin for the most part is filled with post-Eocene sedimentary rocks up to 3,000 m thick. These rocks represent shallow-marine and brackish-water conditions and are made up of silty shale and soft argillaceous sandstones. Glauconite, tuffaceous material, fish remains, diatomite, and phosphate beds of commercial grade characterize the section. Eocene and Late Cretaceous sedimentary rocks are present, but are locally restricted. The contrast between the Sechura and Talara sections is significant in considering the potential of the Sechura basin. For all practical purposes, none of the producing sandstones of Talara are present in the onshore part of the Sechura basin. The Coastal Range marks the eastern limit of these sandstones.

Numerous fault structures are present in the onshore Sechura basin. Seismic data indicate that faulting is more intense offshore, especially on the west flank of the Coastal Range. Several structures with closure are present. Oil seeps are pres-

[5]"Potential," as used in this paper, refers to the maximum amount of recoverable oil and/or gas that could be reasonably expected in terms of known or inferred geologic factors including hydrocarbon indications. It is assumed that operations in water depths up to 500 fm will be technologically feasible in the near future. "Total potential" includes both discovered and estimated undiscovered hydrocarbons; "future potential" includes only undiscovered hydrocarbons.

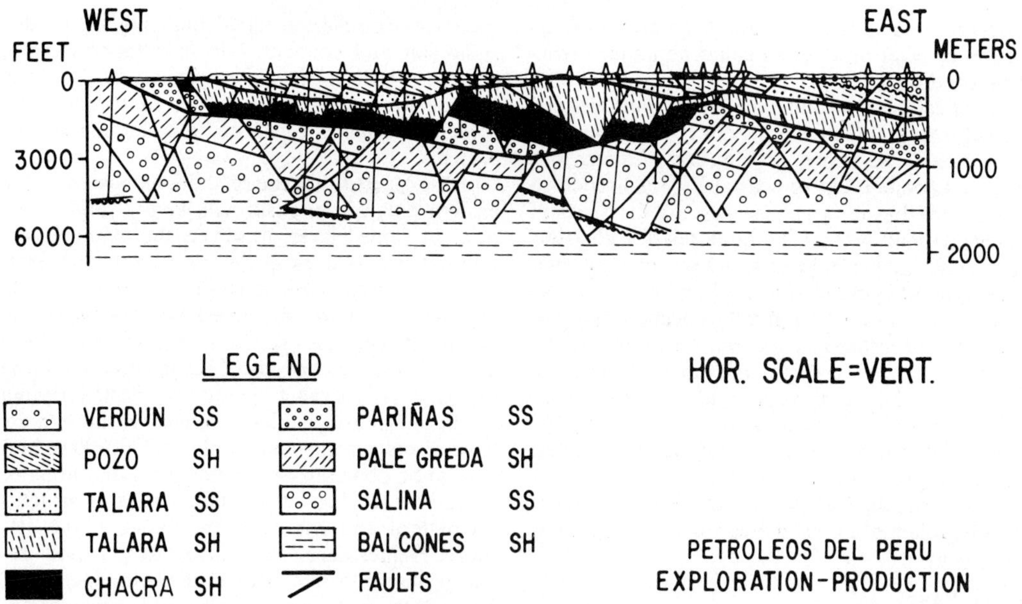

FIG. 4—Structure section of onshore Talara basin.

ent around exposures of the Coastal Range including the larger islands on the west border of the basin. In the 1950s, 28 wildcat wells were drilled onshore with only the discovery of some gas which has not been developed. Based on the results of drilling onshore and the apparent limitations of distribution of the Talara productive section, it is unlikely that potential of the Sechura basin could exceed 100 million bbl.

Salaverry Basin

The Salaverry basin (Fig. 1) is entirely offshore and is the largest of the coastal basins, having an area of 54,000 km² to the 500-fm isobath and 38,000 km² to the 100-fm isobath. The northern part, to about 10°30′S lat., is an elongate basin between the Andes mountains and subsurface basement highs of the Coastal Range (Fig. 2). Near lat. 10°30′S, the 100-fm isobath shifts abruptly to the east, and the continental slope between 100 and 500 fm is much less steep than it is elsewhere. In this area there is no structural border on the west. The southern end of the basin is marked by a thinning of the section over a series of basement highs which give way to a large basement platform extending from about 100 km north of Lima to about 50 km south of Lima. This feature is expressed by the westward bulge of the 100-fm isobath and the steep upper continental slope.

The maximum thickness of sedimentary rocks in the northern Salaverry basin is about 3,000 m; the section thins toward the sides and southwest, but thickens to 4,000 m west of the Coastal Range trend. In the southern part, strata generally thicken westward, reaching a maximum of about 3,000 m along the western boundary of the survey near the 500-fm isobath.

The only direct information about lithology and age of these strata comes from two wells drilled in 1971. The section consists of shale, siltstone, marlstone, minor thin limestone beds and, in the basal part, sandstone. The lower part is dated as late Eocene and the overlying rocks, post-Eocene. The section is somewhat similar to that in the Sechura and Pisco basins.

Although seafloor topography is relatively smooth, basement topography is irregular owing to block faulting. Faults pass into the overlying strata with diminishing displacement upward. Although widespread, faulting is not as intense as in the Talara basin. In the axial part of the basin, some of the basement structures were active during deposition of the overlying sediments. In addition to numerous closed structures related to block faulting, broad flexures or low folds are present.

The two wells drilled by Occidental Petroleum in the Salaverry basin were not very encouraging, but the area is large, and a variety of stratigraphic and structural features is present. Tests of post-

Eocene strata in the Progreso basin give an indication of what production could be under favorable conditions.

Practically all production from southwestern Ecuador and northwestern Peru comes from Eocene rocks. The known distribution of lower and middle Eocene rocks of coastal Peru is limited to the Talara basin, which is west of the Coastal Range. It may be that the same relative distribution exists in the Salaverry basin and that such rocks will be found west of the trend of the Coastal Range. If lower and middle Eocene strata are present, the prospects for the basin would be greatly enhanced, even though much of the prospective area might be in water depths greater than 100 fm. Because of uncertainties about the presence of adequate reservoir rocks and the fact that the more promising strata might be in water of excessive depth, a somewhat conservative figure of 500 million bbl is estimated for the potential of the basin.

Pisco Basin

The Pisco basin has an area of 25,000 km^2, of which 18,000 km^2 is onshore or in waters of less than 100 fm. It is bounded on the east by the Andes mountains and limited arbitrarily on the west by the 500-fm isobath. The northern end is marked by basement highs and the basement platform in front of Lima; the basin is divided into two parts by the Coastal Range. This range extends northward in the subsurface and also divides the offshore portion of the basin into two parts. The eastern part is limited at the south by the convergence of the Andes and the Coastal Range. The western part is all offshore.

The maximum thickness of sedimentary rocks in the eastern part is 2,500 m and in the western part, 3,000 m. Much of this western basin is in water greater than 500 fm deep. The strata onshore east of the Coastal Range are of Miocene age and are very similar to the Miocene rocks in the Sechura basin; diatomite, fossiliferous shale, fine tuffaceous sandstones, and phosphate beds are present. A well drilled in 1955 penetrated 500 m of diatomite. These Miocene strata generally overlie Coastal Range basement rocks or, on the east, Cretaceous rocks of varied lithology.

In a few places on the west flank of the Coastal Range, Miocene sedimentary rocks are underlain by upper Eocene fine tuffaceous sandstone and shale interbedded with thin limestone beds. A coarse basal conglomerate of erratic thickness is present at the contact with the Coastal Range basement complex. Isolated fault blocks of Mississippian and younger Paleozoic rocks are also present. These rocks are not abundant, but they include some adequate reservoir sandstones, and some of the shales appear to have source-rock potential.

It is expected that the offshore extension of the eastern part of the basin will have a similar section. In some places, however, the offshore section is twice as thick as the onshore. In the western part, seismic sections show a considerable thickness of older, deformed beds between the basement and the generally undisturbed beds above. Lower and middle Eocene strata could be included. As elsewhere along the coast, this area has been extensively block faulted. Onshore faults have displacements commonly of 200–300 m and tend to be downthrown to the west. No folding of Tertiary strata has been noted. The same structural pattern and character are found in the offshore. Higher basement blocks have created domal structures with closure in the overlying strata. Some of these show growth during sedimentation. Because of the unattractive Tertiary sedimentary section onshore, the relatively thin sections over the structures, and the excessive depths of water over much of the thicker sections, the potential of the Pisco basin is estimated at only 100 million bbl.

Moquegua Basin

The Moquegua basin (Fig. 1), another sedimentary basin between the Andes and the Coastal Range, is almost entirely onshore and has an area of about 19,000 km^2. The only marine strata are in the northern part, and their thickness does not exceed 1,500 m. As in the other coastal basins, block faulting is the principal structural feature in the Tertiary rocks. The volume of sedimentary rocks in this basin is insufficient to have significant potential.

Conclusion

The coastal basins of Peru have a very large area and volume of marine strata. Full evaluation of some of the more prospective parts will require deep-water drilling. Productive lower Eocene fluvial and deltaic sandstones in the Talara area are overlain by units interpreted to have been deposited in water depths near 1,000 fm, indicating the possibility that such shallow-water prospective rocks may be found in the subsurface deep water. The Peruvian offshore will be a highly prospective area for many years.

Hydrocarbon Potential of Amazon Basins of Colombia, Ecuador, and Peru[1]

F. ZUÑIGA y RIVERO, ALFREDO PARDO, HUGO VALDIVIA, and PEDRO VELARDE[2]

Abstract The Oriente, Ucayali, and Madre de Dios basins, located in the Amazon drainage of Colombia, Ecuador, and Peru, are part of a series of large asymmetric depressions between the Andes Cordillera and the Guyana and Brazilian shields. They are separated by basement arches and have areas of 458,000, 200,000, and 95,000 km^2, respectively. The whole area is topographically low, covered by heavy rain forest, traversed by numerous huge tributaries of the Amazon, and very sparsely populated.

From early Paleozoic time until the Maestrichtian, seas repeatedly invaded the area, depositing a variety of sediments, but mostly calcareous and silicate clastic material, over wide areas. At the beginning of the Tertiary, predominantly marine deposition gave way to nonmarine deposition, reflecting the Andean orogeny and topographic development of the Andes mountains. The depositional cycle of major importance for hydrocarbons took place in the Cretaceous. A complete marine cycle of miogeosynclinal sedimentation is represented; the maximum thickness is 2,500 m, but the sequence thins consistently and becomes sandier toward the east. Although the sequence consists mainly of sandstones and shales, limestones and sandy limestones are important potential reservoirs. Most prospective structures in the basin are anticlines, generally fault-bounded and somewhat steeper on the east. Salt domes and other diapiric structures are also present. Amplitude of structures and intensity of deformation decrease eastward. The formation of structures and the migration and entrapment of hydrocarbons in them appear to have occurred at various times in the Tertiary.

The Colombian part of the Oriente basin is in an advanced state of exploration and has little potential for future significant discoveries. In Ecuador, although the peak of exploration activity has been passed, the future potential is thought to be substantial. In Peru, exploratory drilling in the Oriente basin has already discovered reserves on the order of 800 million bbl of oil. Based on these facts and on information from Colombia and Ecuador, the total potential of the Oriente basin is estimated to be 25–35 billion bbl. About 20 wildcats have been drilled in the Ucayali basin; two small oil fields and an undeveloped, but potentially large, gas field have been discovered.

On the basis of these results and estimates for the Oriente basin, the Ucayali basin has an estimated potential of 5–10 billion bbl. In addition, Paleozoic sedimentary rocks have some potential. The Madre de Dios is the least explored of the three basins. Favorable conditions are known to exist and oil seeps are present. A potential of 6–12 billion bbl is estimated. The total estimated potential for the three basins is about 45 billion bbl.

INTRODUCTION

The Oriente, Ucayali, and Madre de Dios basins, located in the Amazon drainage of Colombia, Ecuador, and Peru, form part of a series of large asymmetric depressions between the Andes Cordillera and the Guyana and Brazilian shields. They are structurally separated by basement arches projecting from the shields. They have a total area of about 750,000 km^2, if shallower parts in Brazil are included. The area is low, flat, covered by heavy rain forest, and traversed by huge tributaries of the Amazon. In the rainy season from October to March, large areas are under water. Although the Amazon drainage constitutes 60% of the area of Peru, only 10% of the population lives there.

REGIONAL TECTONIC SETTING

Figure 1 shows the locations of the basins and the major geologic elements of the region: (a) the Andes Cordillera, shown as Cretaceous and pre-Cretaceous rocks on the west; (b) the Amazon plains, shown as Cenozoic rocks, and; (c) the shields, shown as pre-Cretaceous on the east. These regional elements are the result of movements and deformation during the Andean orogeny that took place in Late Cretaceous and Cenozoic time. The present geomorphology and drainage are influenced by these major structural features. For example, the topographic divide between the Amazon and Orinoco drainages in Colombia reflects the Vaupés swell, a broad basement feature projecting westward from the Guyana shield, which separates the Oriente and Llanos basins. Some of the deep subsurface features in the Amazon plains resulted from tectonic movements as late as the Pliocene.

Figure 2 shows in cross section the structural character of these major elements and their relationships. Both of these sections are from Peru, but the basic features in Ecuador and Colombia are the same. Three main physiographic regions are apparent: the Cordillera Oriental of the Andes, the Subandean zone or foothills belt, and the Amazon plains.

The Cordillera Oriental is composed of Precambrian and Paleozoic metamorphic and plutonic igneous rocks and folded Mesozoic sedimentary rocks. Folds are rather tight, oriented

[1]Manuscript received, November 6, 1974

[2]Petróleos del Perú, Lima, Peru.

The writers are indebted to the management of Petróleos del Perú for permission to make full use of company files and reports and to publish this paper. We also acknowledge the assistance of various geologists, engineers, draftsmen, and secretaries.

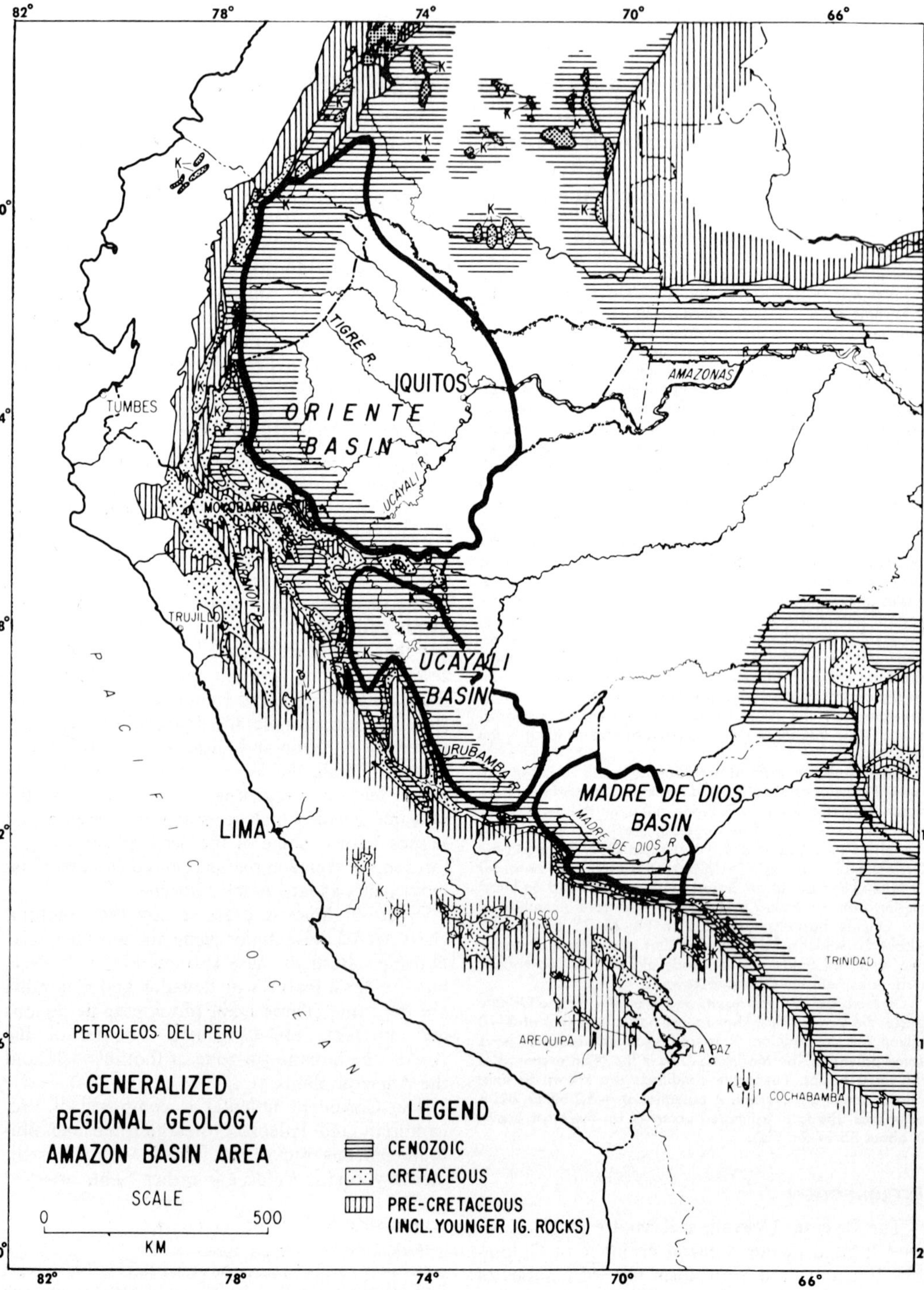

FIG. 1—Generalized regional geology of Amazon basin area.

northwest-southeast, and asymmetric; they are steeper or overturned on the east.

In the foothills belt, composed of folded Mesozoic and Cenozoic rocks, structures are similar and are associated with high-angle reverse faults. In some areas, such as the middle Huallaga valley, salt diapirism accompanied deformation. The structural boundary with the Amazon plains is abrupt, and is expressed by escarpments formed by reverse faults or large monoclinal structures.

The Amazon plains are underlain by a thick sequence of Cretaceous and Tertiary rocks, gently folded in the Oriente basin and more intensely folded in the Ucayali and Madre de Dios basins. A break in the slope of the basement at the Corrientes field represents the basin hinge line (Fig. 2, top section). East of this line is the platform or stable shelf; west of it is the mobile part—the "deep" basin. The slope of the basement is greater than the slope of the base of the Cretaceous, which in turn is greater than the slope of the top of the Cretaceous.

Some of the important tectonic features which affected Cretaceous deposition in the region are shown in Figure 3. In addition to the shields, they are the Vilcanota geanticline or arch and its extension, the Shira uplift, and the Marañon geanticline or arch. The Vilcanota and Marañon geanticlines were formed during the late Paleozoic orogeny.

A miogeosyncline bounded by these elements was formed toward the end of the Triassic and reached maximum extent during the Cretaceous. Secondary tectonic forces active during the deposition of the Cretaceous formed the basement arches such as Vaupés, Contaya, Fitzcarrald, and Jatunari, which delimit the basins in a north-south direction. Regional variations in thickness of Cretaceous rocks are shown in Figure 3.

Regional Stratigraphy

Ordovician

Lack of information on Paleozoic, Triassic, and Jurassic rocks prevents delineation of the principal tectonic elements which limited the deposition of these sediments. Apparently the Ordovician basin was deep and extensive and bordered the Guyana and Brazilian shields. The Contaya Group (the oldest Ordovician rocks) consists of argillites and black slates with graptolites (Fig. 4). Because of intensive deformation, it is impossible to measure thickness in the Cordillera Oriental, but to the southeast the correlative San José and Sandia Formations exceed 7,000 m in thickness. Ordovician rocks must have covered a large part of eastern Peru, but they are limited at present to the Brazilian border area; exposures occur in the Cerros de Contaya and in the Madre de Dios basin. Owing to extreme induration and metamorphism, these rocks are not prospective for petroleum.

Silurian

Epeirogenic movements at the end of the Ordovician and the beginning of the Silurian interrupted deposition for a short time. The older strata are discordantly overlain by the San Gaban Formation (Fig. 4), which consists of turbidite sequences; fragments of igneous rocks are interbedded with conglomerates resembling tillites. Exposures are limited to the southern part of the Madre de Dios basin, and measured thickness is only 160 m. These rocks are not considered prospective.

Devonian

The Devonian Cabanillas Group consists of marine, fossiliferous black shales interbedded with gray sandstones and lenses of limestone. Incomplete sections measure up to 800 m. These rocks are found in the Madre de Dios, Urubamba, and upper Ucayali valleys and in the Cutucú area of Ecuador, where they are metamorphosed. The Devonian section has good potential source rocks and fair reservoir rocks; it is productive in Bolivia. It could be an important prospective zone but, at present, little is known about it in the subsurface.

Mississippian

After a prolonged period of erosion, reflecting the Hercynian movements, the Mississippian Ambo Formation (Fig. 4) was deposited. It consists of 300–800 m of marine and deltaic gray sandstones, shales, and coal beds. Like the Devonian, it has good potential source rocks and fair reservoir rocks.

Pennsylvanian

After deposition of the Ambo Formation, uplift and erosion occurred, followed by Middle and Late Pennsylvanian marine transgression in a basin that was oriented generally north-south but also extended between the Guyana and Brazilian shields. In the axial part of the basin the strata make up the Tarma Formation and consist principally of greenish-gray sandstones and subordinate limestone. In the Oriente basin, basal sandstones are overlain by gray micritic limestone with nodules of chert and dolomite in the upper part and intercalations of black and greenish-gray

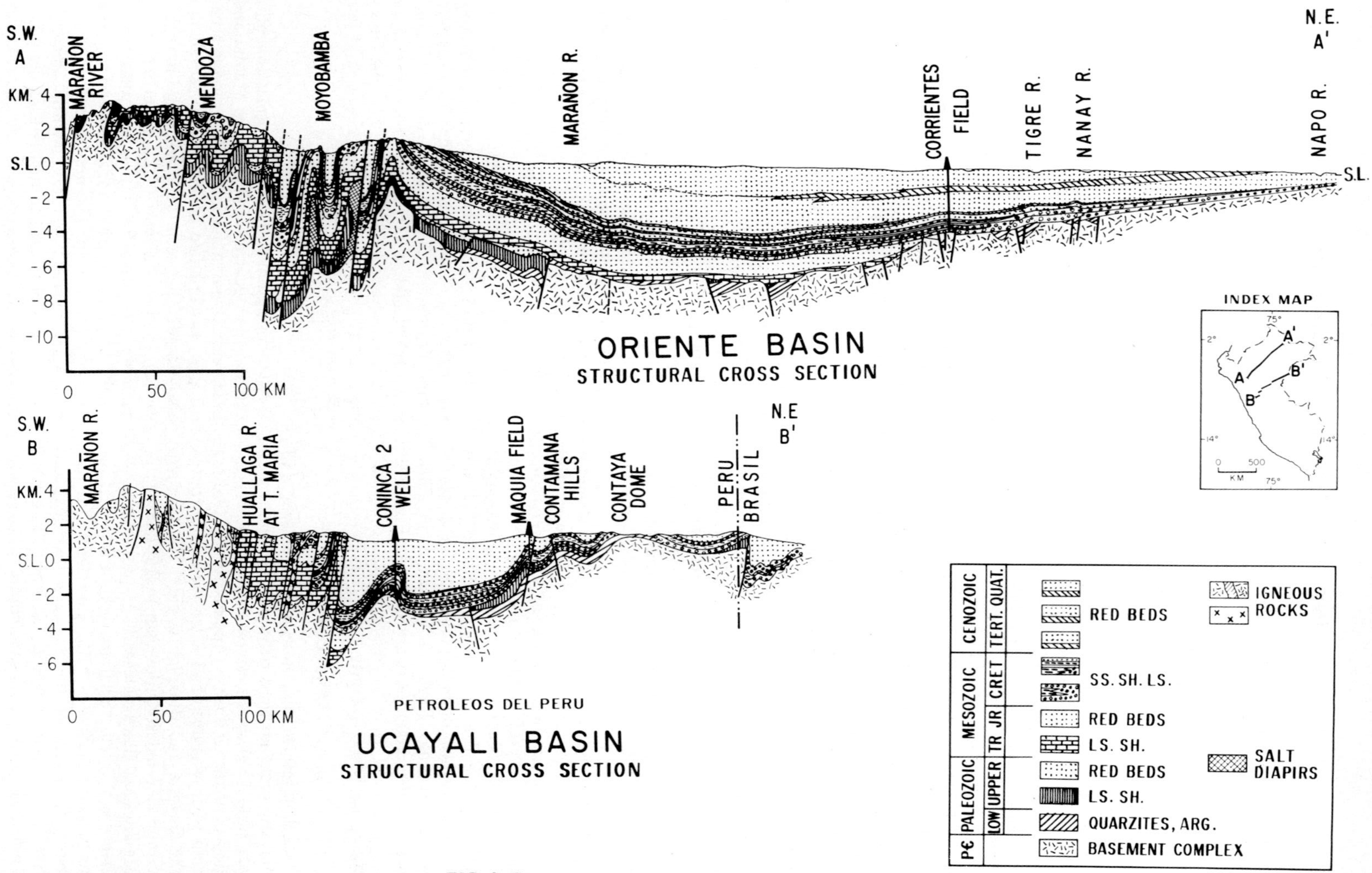

FIG. 2–Structural cross sections of Oriente and Ucayali basins.

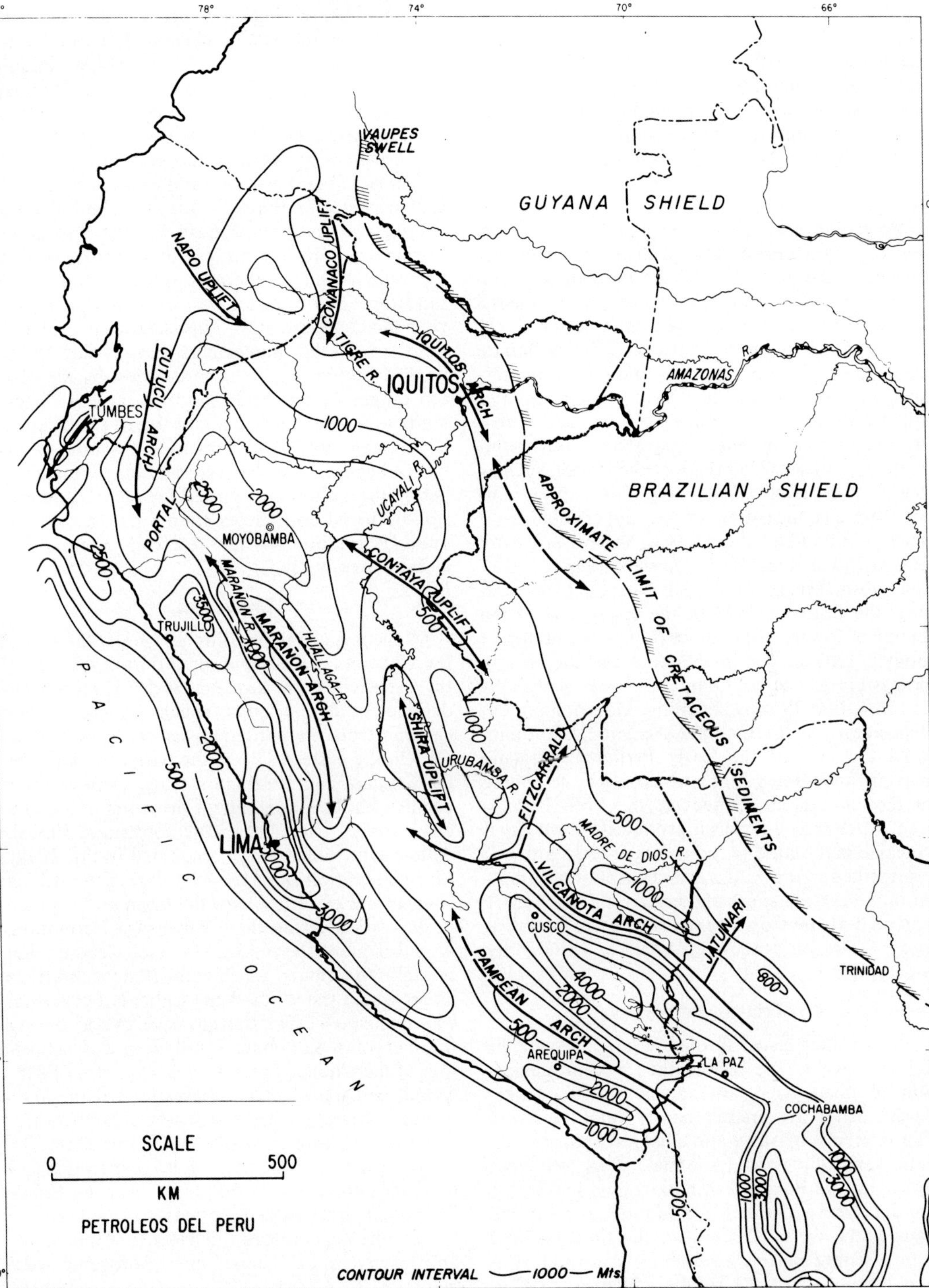

FIG. 3–Cretaceous isopach map showing main tectonic features.

shales. These strata are limited to the east flank and foothills of the Andes in the Madre de Dios, Manú, upper Urubamba, and upper Ucayali River sections; they are up to 300 m thick. Source and reservoir potential are rated as good and fair, respectively, for the Pennsylvanian. Carboniferous rocks are productive in Bolivia.

Permian

After another epierogenic uplift and erosion, an important marine transgression occurred, resulting in the deposition of the Permian Copacabana Formation (Fig. 4). This formation consists of limestone, dolomite, shale, and lesser thicknesses of sandstone and siltstone. The carbonate rocks are dark gray, micritic, and very fossiliferous; in many places they have vugular cavities filled with oil. The dolomite exhibits good porosity. Exposures of the Copacabana are found mainly in the eastern ranges of the Andes, such as the Cordilleras of Vilcabamba, Shira, and Oriental. Sections measured in the upper Madre de Dios, Urubamba, and Ucayali valleys are from 200 to 700 m thick. The Copacabana is found in the subsurface in the upper Ucayali basin and near the Contaya arch. In the upper part of the Permian, toward the southeast between Peru and Bolivia, redbeds are intercalated and the section grades upward to continental deposits of the Mitú Formation. Northward in the Marañon region, deposition was marine until the Middle Permian.

At the end of the Early Permian, orogenic movements created highs which were the source of the continental redbeds of the Mitú Formation. Attendant volcanism produced interbedded volcanic material. The sedimentary rocks are well represented in the Andes, but distribution is limited in the eastern spurs and eastern plains. Source potential of the Copacabana is rated as excellent and its reservoir potential is fair. The Mitú is not prospective.

Triassic-Jurassic

Erosion that lasted through a large part of the Triassic leveled the region, then marine deposition resumed over large areas of the present Andes. Strata are predominantly limestones with thin interbeds of shale and siltstone and subordinate sandstone and dolomite. This sequence makes up the Pucará Formation. The limestones are dark gray, fossiliferous, micritic, and bituminous. Some are bioclastic with chert nodules and emit a strong petroleum odor. The formation is best developed in the northern Andes. Toward the plains, to the east and southeast, the sequence thins abruptly and facies change to mainly sandstone and shale. Exposures are found in the eastern spurs of the Andes (Cerros de Cushabatay, upper Ucayali, Pachitea, Tingo María, middle Huallaga, Moyobamba, Rioja, and Río Santiago valleys). The sequence is not present in the Madre de Dios basin. The Pucará has excellent potential and excellent source and reservoir beds.

It is possible that uplift and erosion of the Pucará occurred in the early Liassic. The Pucará is overlain by nonmarine strata of the Sarayaquillo Formation, which is present in a narrow strip in the central and northern Andes and consists of sandstone, red mudstone and, in the upper part, volcanic agglomerates. This sequence probably reflects Nevadan movements. East of the Andes, its distribution is limited on the west by the eastern ranges of the Andes and, in the east, it extends to a line running along the rivers Ucayali-Urubamba and Tigre-Corrientes. Tectonic movements beginning in the Jurassic, but terminating before the Tithonian, uplifted the region, and erosion occurred until Neocomian time. The Sarayaquillo Formation has fair reservoir potential but lacks source beds.

Cretaceous

Miogeosynclinal deposition began at the end of the Neocomian and continued during the rest of the Cretaceous. The sequence is divided into well-defined lithologic units which undergo facies changes from the borders to the center of the basin. On the eastern side of the basin, bordered by the Guyana and Brazilian shields, sedimentation began with the Cushabatay Formation of the Oriente Group. The Cushabatay consists of fluvial-deltaic sandstones which originated in the shield. Marine invasion from the west during the Albian reached the eastern side of the basin and resulted in deposition of the shaly Esperanza Formation. A subsequent regression in the Cenomanian brought the return of fluvial-deltaic conditions under which the sandy Agua Caliente Formation was deposited. Two transgressive cycles during the Turonian-Santonian resulted in the deposition of the Chonta Formation, a sequence of shaly and calcareous rocks interbedded with sandstones. Toward the northwest, the Chonta contains a greater proportion of limestone. On the eastern side of the basin, it is composed largely of sandstone, and in the south, near the Bolivian border, interbedded redbeds are common.

With the beginning of movements of the Andean orogeny in the Campanian, synorogenic sediments of the Vivian and Cachiyacu Formations were deposited. These formations are overlain by the continental redbeds of the Huchpayacu For-

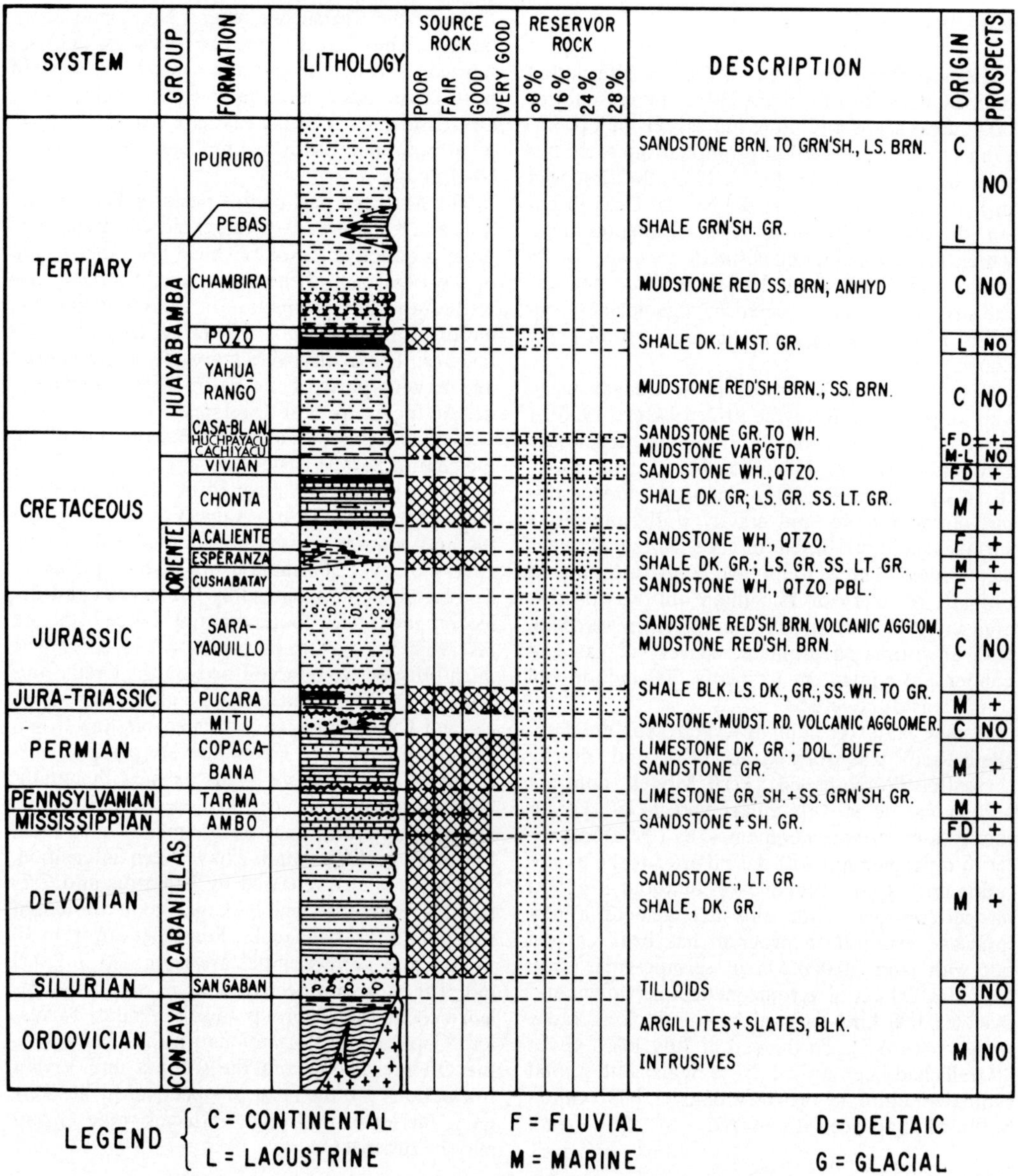

FIG. 4—Composite stratigraphic column, eastern Peru.

mation of Campanian and/or Maestrichtian age.

Cretaceous rocks in Peru are the reservoir rocks in presently productive fields; therefore, they are considered to have potential for future hydrocarbon discoveries. All the sandstones have good to excellent porosity, and the interbedded shales are considered to be good source rocks. The relative uniformity of the section over wide areas makes all these areas prospective.

Tertiary

During the Tertiary, a continental environment predominated. As the basin continued to subside, it was filled with continental and lacustrine sediments of great thickness. The Tertiary sequence consists of sandstone, conglomerate, siltstone, mudstone, claystone, evaporites, and volcanic products.

Exploration History

Eastern Peru has been subject to alternating periods of exploration since 1919. The presence of oil and gas seeps and other indications of a promising sedimentary section prompted the early but sporadic interest (Fig. 5). In 1939, the first well drilled in eastern Peru, in the central Ucayali basin, discovered the small Aguas Calientes field, which has an estimated ultimate recovery of 18 million bbl. This event has some fame because the structure was discovered by a geologist (Robert B. Moran) flying as a passenger on an army plane.

Only four other wells, all in the Ucayali basin and all unsuccessful, were drilled before 1953. A surge of activity ensued with the granting under Law 11780 of over 11 million ha. in concessions to 14 oil companies. This activity included seismic, aeromagnetic, and gravity work, as well as surface geology. Based on this work, 21 wells were drilled. The Maquía field, with an estimated ultimate recovery of 15 million bbl, was discovered in 1957 and, in 1959, the Aguatia gas field, with an estimated ultimate recovery of gas and condensate equivalent to nearly 500 million bbl of oil, was discovered.

A new state oil company, Petróleos del Perú (PetroPerú) was formed in 1970, and shortly thereafter the so-called "Perú Model" contract system was devised to replace Law 11780. Under this system, private companies, as operators, enter into partnership with PetroPerú for the exploration and exploitation of areas of about 1 million ha. per contract. Under this arrangement, a very intensive exploration program has been carried out with over 30,000 km of seismic lines shot, over 100,000 km of aeromagnetic lines flown, and over 600,000 km^2 covered by side-looking radar imagery (SLAR). To the end of July 1974, about 40 wells had been drilled. Nine fields, with a total estimated ultimate recovery of over 800 million bbl of oil, had been discovered.

Hydrocarbon Potential

Stratigraphic Factors

The stratigraphy of the Amazon basins within Colombia, Ecuador, and Peru is essentially uniform in regional terms for the Cretaceous and Tertiary. The subdivision into the Oriente, Ucayali, and Madre de Dios basins is based on structural features.

Oriente basin—This basin is characterized by a thick sequence ($\pm$10,000 m) of Mesozoic and Cenozoic rocks in the western part, but marked thinning occurs eastward. The prospective rocks are those of the Cretaceous, which have wide distribution. The Triassic-Jurassic rocks, which are limited in distribution to the central and western parts of the basin, and the Paleozoic rocks, which are found mainly in the east, are thought to have great potential, but it has not been corroborated by drilling.

The northern part of this basin in Peru, Ecuador, and Colombia is being explored intensively. Nine productive structures have been discovered in the Peru part of the basin, and it already has well-established production in Colombia and Ecuador of about 300,000 BOPD, all from the Cretaceous. These discoveries are all in the central and eastern parts of the basin where the fluvio-deltaic facies are best developed.

Ucayali basin—This basin is located mainly in the central part of eastern Peru. During the years 1930–60, exploration resulted in the discovery of three Cretaceous fields—Aguas Calientes, Aguaytía, and Maquía—with an estimated ultimate recovery of over 500 million bbl of oil equivalent. The basin is characterized by Cretaceous and Paleozoic rocks with good potential source beds and reservoir beds with porosities of 15–25%. The most prospective reservoir beds in the Cretaceous are the Vivian, Chonta, Agua Caliente, Esperanza, and Cushabatay. In the Paleozoic, the Copacabana, Tarma, and Cabanillas are prospective.

Madre de Dios basin—This basin is the southernmost one and the least explored because of its remoteness and difficult accessibility. Since 1968, however, geologic studies have been intensified. The basin is characterized by Paleozoic and Cretaceous sections which have good potential source and reservoir rocks. Porosities are from 10 to 20% and permeabilities are from 400 to 2,000 md. The most prospective sections are in the Paleozoic Cabanillas Group and the Ambo, Tarma, and Copacabana Formations. In the Cretaceous, the Oriente Group and the Chonta and Vivian Formations are the most prospective. In the Tertiary, the lower part of the Huayabamba Group may be prospective.

Structural Factors

Most of the structures in the Amazon basin are asymmetric anticlines oriented north-south or northwest-southeast in an echelon arrangement. The east flanks are normally steeper and faulted. However, there are symmetrical structures, some of which are productive along the Ecuadorian border in Peru. In the basins, anticlines are broader and of lower amplitude than along the foothills of the Cordillera Oriental or in the Subandean zone. Structures situated on the east

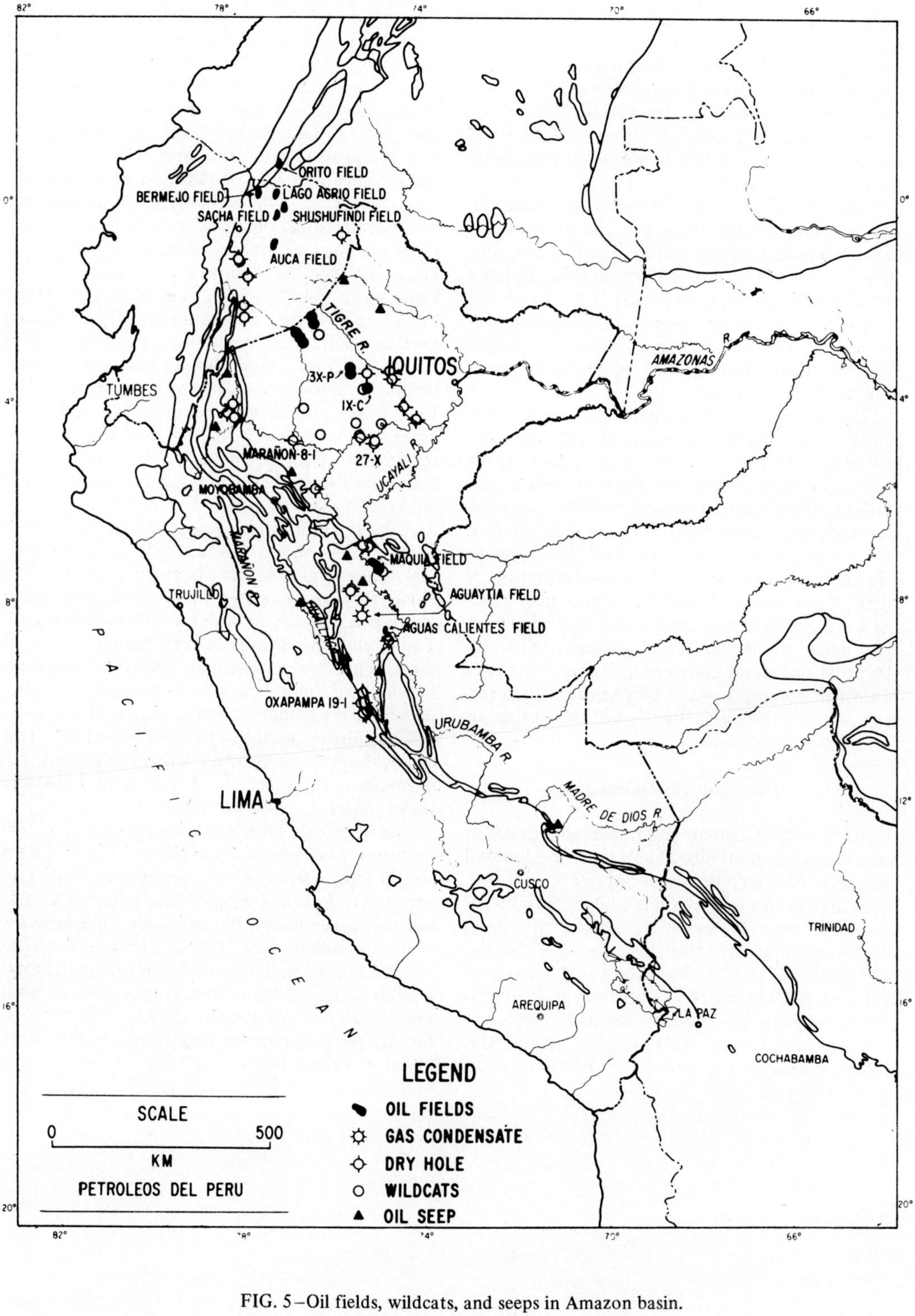

FIG. 5–Oil fields, wildcats, and seeps in Amazon basin.

flanks of the basins, especially those over the platform, probably originated from flexuring caused by fault movements in the basement at this level; response to the regional compressive forces was apparently reactivation and reversal of movements on faults produced by earlier tensional forces. This interpretation is borne out by seismic data in several places.

Structures in the Ucayali basin have greater relief than those in the Oriente basin, as indicated in the lower section of Figure 2. In other respects, such as asymmetry, they are very similar. In parts of the Oriente basin in Peru, east of the hinge line where the pre-Cretaceous–post-Precambrian section is thicker, nearly equidimensional domal structures have been detected which have greater areal extent and closure.

There are numerous adequate structures in the Amazon basins, although many of them are rather young. The final shaping of structures in the region occurred during the Pliocene Quichuana orogeny. This phase, characterized by compressive movements directed toward the east, affected pre-Cretaceous basement and modified previous structures. Under these conditions, formation of some of the more obvious structures may postdate hydrocarbon migrations, and thus the structures would be barren. It will be important to reconstruct structural events in order to locate sites of former structures which may have retained reservoired hydrocarbons by virtue of faulting or some other situation.

Estimates of Potential

The Oriente basin in Colombia and Ecuador has an area of about 80,000 km^2 and an estimated volume of Cretaceous strata of about 45,000 km^3. Allowing for future revisions and extensions of discovered reserves, we estimate that 8 billion bbl of hydrocarbons have been discovered. These figures would indicate a yield factor for Cretaceous rocks of 180,000 bbl/km^3. Cretaceous stratigraphy in the Peruvian Amazon basins is very similar to that in Ecuador and Colombia, and producible oil has been discovered in Cretaceous rocks. The volume of Cretaceous strata in the Peruvian part of the Oriente basin is about 355,000 km^3. If this volume had a yield factor equal to that in the Colombia-Ecuador part, the potential would be 64 billion bbl. Although a relatively small part of the Peruvian Oriente basin has been drilled, several of the structures are barren. On the basis of the number and size of detected structures, the fact that many will probably be unproductive, and the statistical observation that oil accumulations are not uniformly distributed in basins, it is estimated that the Peruvian part of the Oriente basin has a good chance of containing 20 billion bbl of recoverable hydrocarbons. It is believed that the Colombian portion does not have large undiscovered potential, though this does not preclude the discovery of substantial reserves. The Ecuadorian portion very definitely has additional potential. An estimate of 2–5 billion bbl would be reasonable. With the present discovered reserves, the total estimated potential for the entire Oriente basin is 25–35 billion bbl.

Although areally only about half of the Ucayali basin is in Peru, the Brazilian part is rather shallow and several wells have been unsuccessful. The Peruvian part has an area of 100,000 km^2 and a volume of Cretaceous rocks of about 85,000 km^3. The Cretaceous section is very similar to that in the Oriente basin; potential reservoir sandstones are of equal quality. Structures also are adequate. Based on these factors, the potential of the Ucayali basin is estimated to be 5–10 billion bbl. This estimate applies only to the Cretaceous rocks. No estimate is made for the potential of Paleozoic rocks, which are also prospective.

The Madre de Dios basin has an area of 95,000 km^2 and a volume of Cretaceous rocks of about 70,000 km^3. Details of the geology of this basin are poorly known owing to the remoteness and difficult accessibility, but full-scale exploration is currently under way. Taking these and stratigraphic factors into account, we estimate the total potential at 6–12 billion bbl, which includes both the Cretaceous and Paleozoic rocks. The total estimated potential for the three basins is, therefore, about 45 billion bbl.

Hydrocarbon Potential of Bolivian Sedimentary Basins—Summary[1]

J. OBLITAS and C. SALINAS-E.[2]

Bolivia, in the center of the South American continent, has an area of 1,098,581 km^2, of which 450,000 km^2 (41% of the territory) constitutes sedimentary basins favorable for the search for hydrocarbons.

The sedimentary basins have undergone a constant and very marked evolution controlled by positive cratons or shields, and dorsals or highs that were active during different geologic periods.

During the early Paleozoic, positive tectonic elements were the Brazilian shield in the northeast, the Pampean massif in the southwest, and the Asunción massif in the southeast.

The reactivation of the Pampean massif and two newly developed highs—the Michicola high in the extreme south of the country and the Izozog high in the southeast—produced regression of the seas during the late Paleozoic that climaxed at the end of the Paleozoic with complete withdrawal from Bolivian territory.

During most of the Triassic and Jurassic Periods, the whole country was a positive area.

The transgression of the Cretaceous seas did not reach Bolivia until the Late Cretaceous and formed the Eastern and Western sedimentary basins. The two basins were separated by the Aiquile dorsal high, which constitutes an extension of the Marañón high in Peru. At this time the Eastern basin was controlled by tectonic transverse highs such as the Apolo high and the Isarsama high, whereas the Western basin was affected by the transverse highs of Puerto Acosta, Andamarca, and Esmoraca. During the Tertiary Period, normal sedimentation of the seas in a rapidly sinking basin covered a great extension of the country in which the Aiquile dorsal high was a prominent feature.

The present distribution of the sedimentary basins shows a very extensive Eastern basin between the Brazilian shield on the northeast and the Andean Mountains on the west, and a Western or Altiplano basin between the two chains of the Andean Mountains.

Until a few years ago, hydrocarbon exploration in Bolivia was limited to the Santa Cruz and Chaco subbasins in the southern part of the Eastern basin.

Recent regional studies have established the presence of 5,000 m of dark gray and black shales of early Paleozoic age. Late Paleozoic sedimentary units have a thickness of 500 m.

The Ordovician and Silurian shales, together with Devonian strata, have proved to be the most important sources of hydrocarbons in Bolivia. In the Altiplano basin, a shaly sequence of 1,100 m in the Cretaceous also has been proved by geochemical analysis to be a source bed for hydrocarbons.

The structural setting of the northern area of the Oriental basin, where geologic, gravity, magnetic, and some seismic information is available, shows the development of Devonian and Cretaceous folding with structures of larger dimensions than their equivalents in the productive areas in south-central Bolivia.

Geologic, gravity, and seismic data in the Altiplano basin have been used to locate anticlines, diapirs, and salt domes that appear to be attractive for oil exploration.

Geologic studies indicate that *in-situ* oil reserves in the Bolivian basins are 6 billion bbl.

Since exploration to find this oil and the development of these reserves demand capital investments which the national oil entity (Yacimientos Petrolíferos Fiscales Bolivianos) cannot undertake, the government promulgated, in March 1972, a very advanced petroleum legislation—the "General Law of Hydrocarbons"—to enable participation of private capital investments in an equitable relation with the national oil entity.

Under the new legislation, YPFB to this date has signed 10 service contracts with different oil corporations, each covering an area of approximately 1 million ha.

[1]Manuscript received, November 11, 1974.

[2]Yacimientos Petrolíferos Fiscales Bolivianos, La Paz, Bolivia.

HYDROGEOLOGY

Hydrogeology of the Pacific—Summary[1]

JOHN F. MINK[2]

Plate tectonics has become a unifying concept in the earth sciences, and within its framework rational explanations of heretofore diverse and apparently singular geologic phenomena have followed. In the Pacific, hydrogeologic provinces fit naturally within the plate-tectonics framework.

Hydrogeologic provinces in the Pacific basin are classified as: (1) continental coasts and islands at the margins of continental plates, (2) arc islands associated with the margins of ocean plates, and (3) ocean islands thought to have formed as the result of "hot spots" underlying ocean plates.

The hydrogeology of continental coasts and islands is usually an extension of continental-type hydrogeology, except that coastal hydrology commonly is overwhelmingly influenced by the relation between fresh water and seawater in aquifers. Arc-island hydrogeology is dominated by andesitic volcanic rocks, commonly acting as aquicludes, and in many places by fossil reefs, which are normally suitable as freshwater aquifers if extensive enough. The hydrogeology of ocean islands is characterized by permeable basalts, relatively impermeable sedimentary units, and perhaps fossil reefs. Atolls are emerged fossil reefs overlying preexisting arcs or ocean islands.

The hydrogeology of islands is particularly crucial to successful economic development.

1Manuscript received, November 11, 1974.

2Consultant, Honolulu, Hawaii 96813.

Developing Dike-Complex and Basal-Lens Water Supplies on Basaltic Islands—Summary[1]

D. LUM[2]

Rainfall, the primary source of water resources of basaltic islands such as Hawaii, is distributed contrastingly between low and elevated areas. Typically, in mountain areas rainfall averages 200 to 300 in. (5,100–7,600 mm) a year, whereas in coastal areas rainfall averages only 20 to 30 in. (510–760 mm) a year. In spite of the prolific rainfall in mountain areas, direct runoff in some watersheds amounts to as little as 20% of rainfall. A permeable basaltic terrane accounts for the great absorption of rainfall and the consequent creation of large bodies of groundwater in basaltic rocks. High-level groundwater bodies, present at elevations of as much as 1,000 ft (305 m) in volcanic-dike complexes of wet mountain areas, are called "dike water." Unseen below the surface, groundwater presumably moves from these high-level bodies into even greater and more extensive bodies of groundwater standing basally a few feet to several tens of feet above sea level as buoyant bodies of fresh water on salt water. Salt water saturates the island at depth. The problems of developing dike-complex water supplies relate primarily to hydrologic yield and utilization of natural storage, but the problems of developing basal-lens supplies involve the added constraint of the quality of water produced under pumping conditions. Infiltration tunnels, Maui-type shafts, and drilled wells are utilized in developing dike-complex and basal-lens water supplies. The method of development chosen depends upon the hydrology of the area, water-supply design requirements, and economics.

[1]From Circum-Pacific Energy and Mineral Resources Conference program booklet.

[2]Hydrologist-Geologist, Department of Land and Natural Resources, Honolulu, Hawaii.

Control of Earthquakes at Rangely, Colorado—Summary[1]

J. D. BREDEHOEFT,[2] C. B. RALEIGH,[3] and J. H. HEALY[3]

Following the experience with earthquakes at the Rocky Mountain Arsenal, it was recognized that there might be other places where earthquakes were triggered by fluid injection. In fact, it was suggested that it might be possible to control earthquakes by controlling subsurface fluid pressures. A seismic observatory at Vernal, Utah, reported many earthquakes near the Rangely oil field in northwest Colorado. The United States Geological Survey observed earthquakes in the field for 1 month in 1967 and established two active zones, both on the periphery of the field and both areas of waterflooding and high reservoir pressure.

In 1969, the Geological Survey began a 4-year experiment to control earthquakes along one active fault. Through an arrangement with Chevron Oil Company, the operator of the Rangely unit, the Survey gained operating control of four waterflood injection wells in one of the zones of earthquake activity. The experiment was conducted in three phases: (1) an initial phase of reducing fluid pressure—an attempt to stop the earthquakes; (2) a second phase of reinjection and increased pressure—an attempt to reactivate the fault; and (3) a fluid phase of reducing pressure—an attempt to terminate earthquake activity along the fault. The experiment was completed successfully during the summer of 1973.

The results demonstrate conclusively that, by controlling the effective stress through the injection and withdrawal of fluids, it is possible to control earthquakes, at least in favorable geologic environments.

[1]From Circum-Pacific Energy and Mineral Resources Conference program booklet.

[2]U.S. Geological Survey, Lakewood, Colorado 80225.

[3]U.S. Geological Survey, Menlo Park, California 94025.

Development of Australia's Groundwater Resources—Summary[1]

N. O. JONES and F. A. HATFIELD[2]

Most of Australia has no perennial streams; hence groundwater has been important in meeting the demand for water. Groundwater was used chiefly for stock and domestic supplies until about 1950, when extraction was greatly increased for irrigation. Groundwater also has been significant in urban water supply in Australia. Usage of groundwater in Australia in 1971 was about 2.5×10^9 m^3 and, although some aquifer systems already are fully developed or overdeveloped, total groundwater resources are well in excess of usage.

Half of the groundwater used in Australia comes from aquifers in the larger sedimentary basins. Most of the remainder is drawn from alluvium and other unconsolidated aquifers. Fractured and weathered zones in hard rocks are important locally.

Groundwater in Australia is used by both the public and private sectors within the overall control, planning, and coordination of the Australian Water Resources Council. Several cases illustrate the interplay of hydrogeologic setting, economics, technologic change, and social attitudes in the pattern of groundwater use.

Few records are available on the use of groundwater prior to the discovery of the Great Artesian basin in 1878. Rapid exploration and development of groundwater with consequent declining yields and inefficient water use resulted in increased investigations. Groundwater in coastal sands and the alluvium of rivers was used during early settlement, but the main growth in usage for irrigation has occurred in the last 2 decades. The Burdekin delta and the Hunter River valley are examples. In the Murray-Darling River system, interest in groundwater was intensified by waterlogging and soil salinization following intensive irrigation of the riverine plains. Water is supplied from the surface, although recent investigations have shown that many areas contain valuable groundwater resources. In the Perth basin, groundwater and surface water are being used for expanding urban and industrial developments. Groundwater sources now are incorporated into the existing system. Development of new mineral fields in northwest Australia has depended on availability of large quantities of water. Only groundwater has been available in the quantities needed.

To meet the increasing demands for water in Australia from new groundwater sources, the main requirement is that groundwater resources be economically, socially, environmentally, and regionally managed.

[1]From Circum-Pacific Energy and Mineral Resources Conference program booklet.

[2]Canberra College of Advanced Education, Canberra, Australia.

Artificial Recharge of Groundwater in Burdekin Delta, Australia—Summary[1]

N. O. JONES, R. E. VOLKER, S. E. JAMES, K. K. WATSON, C. BURDEN-JONES, W. T. SPRAGG, and F. A. HATFIELD[2]

Aquifers underlying the 500-km^2 onshore part of the Burdekin delta are among the most prolific in Australia. Annual production of groundwater is mainly for irrigation and is in excess of 300×10^6 m^3. Aquifer thickness ranges from zero, where the river crosses a bedrock bar, to 100 m. Mean annual natural recharge has been estimated at 210×10^6 m^3, part of which is supplied by seasonal rains and part by the Burdekin River. This river has a mean annual flow of 10×10^9 m^3 and yet is ephemeral. The seasonal pattern of recharge and pronounced variations in recharge year by year cause major fluctuations in the water table. Intensive pumping has accentuated water-level fluctuations and, in 1971, at the end of a dry period, the water table at one point was 5 m below sea level in an aquifer with a mean transmissivity of the order of 5,000 m^2/day. The short-term problem in the delta is severe fluctuation of groundwater levels, and the long-term danger is saline intrusion.

Local water boards have been established to build and operate recharge works—the first such substantive program in Australia. Water is pumped from the river into a system of natural and artificial channels; thereby, increasing rates of recharge have been achieved. Pumpage for artificial recharge now is approaching 100×10^6 m^3/acre. Suspended clay in the recharge water precludes recharge through bores without expensive pretreatment.

Since commencement of artificial recharge the system has had sharply declining recharge rates, particularly in the constructed facilities, and in 1971 the Australian Water Resources Council initiated a group of research projects: (1) numerical modeling of the groundwater system, which will provide a basis for design and management of the expanding artificial-recharge program; (2) detailed study of unsaturated flow below the recharge pits, including the effects of low-permeability layers on and below the floor of the pits; (3) Study of the role of the sediment load in the recharge waters in the reduction of recharge rates; and (4) study of the biota of the warm recharge waters and their effect in limiting recharge.

Other studies are in progress on tritium dating of the groundwater—to assist in definition of the flow system—and on the heterogeneity of the materials around the recharge pits.

[1]From Circum-Pacific Energy and Mineral Resources Conference program booklet.

[2]Canberra College of Advanced Education, Canberra, Australia.

Saline Groundwater Inflows to Murray River in South Australia—Summary[1]

W. R. P. BOUCAUT[2]

The Murray River, with a total length of 2,600 km, flows for about 650 km through South Australia. There, it meanders in an incised channel across a floodplain 1–10 km wide. The floodplain is underlain by about 500 m of predominantly bryozoan limestones and sandstones of Tertiary age forming the Murray basin, which is about 250,000 km^2 in area. Natural groundwater in the basin commonly has a salinity of 30,000 mg/l.

Groundwater gradients are relatively flat (0.25 m/km), and flow is generally toward the river, reflecting the recharge from the higher country forming the margins of the basin. The Tertiary strata form aquifers of moderate transmissivity (200–500 m^3/day/m). Some inflows of saline groundwater occur naturally through the river bed, but studies have shown that most inflows are due to man-made structures and human activities.

The waters of the Murray River are vitally important to South Australia for domestic supplies and for irrigation of citrus orchards along the river banks. This irrigation has resulted in the build-up of groundwater mounds which are relieved by a tile drainage system. Drainage water, now of increased salinity, is pumped to evaporation basins on the floodplain close to the river. The raised water levels in these basins and also in the river weir system have increased inflows of saline groundwater to the river. In drought years these salt accessions can render the water unsuitable for irrigation.

Piezometers have been established to monitor the regional groundwater regime, the irrigation groundwater mounds, and the effect of raised water levels in storage basins. The mechanisms of leakage from these basins have been studied to determine remedial measures necessary to intercept and remove saline groundwater underflow. Alternative evaporation basins, sited on clay for water tightness, have been located away from the river floodplain. Pumping schemes for the transport of drainage water to the basins are proposed.

[1]Manuscript received, October 21, 1974.

[2]Supervising Geologist, Engineering Division, South Australian Department of Mines, Adelaide, South Australia 5000.

Groundwater Potential of Limestone Terranes: Examples from Indonesia—Summary[1]

M. M. PURBO-HADIWIDJOJO and R. SOEKARDI[2]

Although limestone constitutes only a few percent of Indonesia's land area, it ranks third among the best water-bearing rock types—after volcaniclastic rocks and alluvial deposits. In Indonesia, limestones range in age from Permo-Carboniferous to Holocene but most are Tertiary.

An indication of the groundwater potential of limestones in Indonesia is given by examples of two typical limestone-terrane areas—Gunung Kidul in central Java and Lenteng on Madura Island.

In the Gunung Kidul area, southward-dipping middle Miocene limestone about 200 m thick is deeply dissected by solution channels which probably follow fractures. The whole area exhibits a typical karst morphology. In fact, the topography in the Gunung Sewu ("Thousand Hills") part of Gunung Kidul is called "cockpit karst" because of the "cockpits" between the hemispheroidal hills.

Baron Spring, on the south coast of Java, has a dry- and rainy-season discharge that fluctuates between 6 and 20 m^3/sec. This spring is thought to be the main outlet through which the area is drained.

Rainfall in the area is about 2,000 mm/yr, and the mean daily temperature is 26°C. About 45% of the rainfall is thought to infiltrate into the ground through vertical solution channels and cracks. The water table is not well defined. Groundwater occupies the fractures, solution channels, and interstices in the massive limestone. Interconnected caves and other passages greatly affect the flow.

There are two groups of aquifers in the Lenteng area, one between 8 and 70 m below ground surface and one below 228 m. In the proximity of Lenteng village, the groundwater potential of the upper group is 0.33 m^3/sec. The upper aquifers—about 200 m thick—consist of reef limestone with intercalated marlstone and gravel conglomerate; they are of Pliocene age. A Miocene foraminiferal limestone with marlstone intercalations forms the lower aquifers. It has not been penetrated completely, and the thickness is unknown.

The Lenteng area is drained by the Saroka River.

Mean annual rainfall in the Lenteng area is 1,500 mm, and the mean daily temperature is thought to be about 26°C. Infiltration rate in the area is estimated as about 25%. The water table in the eastern part is about 3 m below ground surface, but it gradually deepens to the west.

Both the Gunung Kidul and Lenteng areas have a relatively dense population (about 400 persons/km^2), but the groundwater potential at present is practically untapped.

[1]Manuscript received, November 6, 1974.

[2]Geological Survey of Indonesia, Bandung, Java, Indonesia.

Groundwater Potential of Areas Underlain by Volcaniclastic Rocks: Examples from Indonesia—Summary[1]

M. M. PURBO-HADIWIDJOJO and R. SOEKARDI[2]

Introduction

Indonesia is one of the largest volcanic areas in the world. Situated on a volcanic arc, the country has more than 100 volcanoes which are considered active, that is, having erupted in historical time. (Because of incomplete records for many Indonesian volcanoes, "historical time" dates back to 1600 A.D.)

For the most part, Indonesian volcanoes are built up of andesitic or basaltic-andesitic rocks. The volcanic material, combined with a high annual rainfall of 3,000–4,000 mm, has caused the development of fertile soils over large parts of the country. As a result, agriculture flourishes.

For centuries wet-rice culture has made use of irrigation water from springs at the foot of the volcanoes (Purbo-Hadiwidjojo, 1967) or from streams which originate on the volcano slopes. Groundwater is becoming increasingly important to increase agricultural yields.

In discussing the groundwater potential of volcanic terrains, three examples are put forward. Each of the terrains is underlain by volcaniclastic rocks and each is important from the point of view of agriculture or urban development. They may serve as "type areas." The areas are: Yogyakarta in Central Java, Bandung in West Java, and Nganjuk-Kertosono in East Java (Fig. 1). Each area has different characteristics.

Yogyakarta Area

The Yogyakarta area has a gently southward-sloping terrain with an elevation between 25 and 300 m above sea level. It is dominated by Mt. Merapi (2,911 m), one of the most active volcanoes in the world. The area is covered by products of this volcano, which erupts nearly every year, with larger eruptions every 5–6 years. With each eruption new material is accumulated around the summit. During the rainy season this material is washed away or slides down in the form of an avalanche which later turns into lahars or mudflows. Lahars are common in streams which originate on the higher slopes of Mt. Merapi. Where those streams enter the plain, the lahar is spread out and large tracts of land may thus be covered by debris (Purbo-Hadiwidjojo and Suryo, 1968).

The results of drilling at several localities and of geophysical surveys indicate that the volcaniclastic rocks in the Yogyakarta area are about 100 m thick. The material consists of ash, sand, gravel, and coarser aggregate. The material, especially in the northern part of Yogyakarta, is poorly sorted, and is composed mainly of medium-grained sand and pebbles. Southward the grain size diminishes.

The high rainfall, together with the nature of the soil, results in rapid infiltration. A transmissivity figure of 520 m^2/day has been obtained. The wet-rice fields north of Yogyakarta contribute even more to this infiltration.

In this unsorted material, well-defined aquifers are not present. Storage capacity of the material varies from place to place and with depth. However, there are indications that certain zones transmit water better than other parts of the volcanic rocks. Hence, it has been suggested that the material behaves like a leaky aquifer.

Saturation takes place to a depth of about 70 m, where the material is more indurated. It is possible that the lower part of this material is derived from another period of eruption of Mt. Merapi.

The groundwater table in the plain ranges between 3 and 4 m and becomes higher southward. Owing to water withdrawal for use in Yogyakarta, the water table has dropped to 10 m or more.

Pumping tests at 25 localities indicate that specific capacities are largest at depths less than 10 m, then decrease, but increase again at greater depths ($>$30 m). The specific capacity of the material as a whole averages 0.5–5 l/sec per meter of drawdown. At depths greater than 70 m the specific capacity decreases even more.

As regards the quality of the groundwater, a change in chemistry with depth has been observed. Whereas the uppermost zones contain fresh water which is dominated by weak acids, the alkaline earth (Ca^{+2}) and strong acids (SO_4^{-2}) dominate in deeper layers. In a recent well, drilled for a hotel in Yogyakarta, water ob-

[1]Manuscript received, November 6, 1974.

[2]Geological Survey of Indonesia, Bandung, Java, Indonesia.
The cooperation of hydrogeologists with Sir M. Macdonald and Partners, Hunting Technical Services, and Nielsen and Rauschenberger–Cowiconsult is gratefully acknowledged.

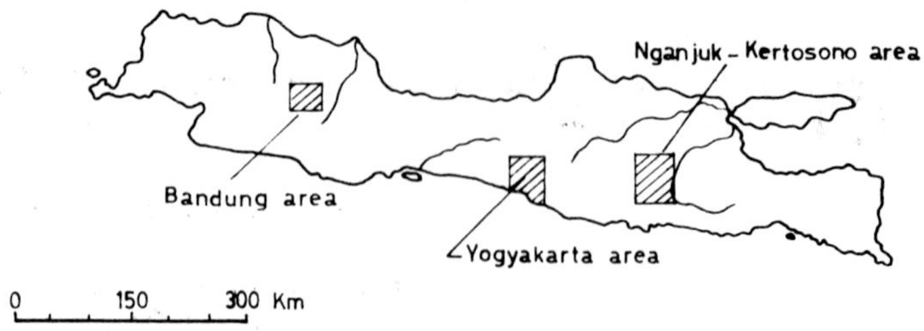

FIG. 1–Location map.

tained from a depth of 60 m contained iron in the amount of 0.8 ppm and manganese, 0.2 ppm. These figures are above the limits of public water-supply standard.

The use of groundwater in the Yogyakarta area at present is still very limited. In addition to dug wells which are widespread in the area, groundwater is tapped by pumped wells for various purposes ranging from small-scale industries to the city water-supply system, which obtains daily 12,960 m^3 from its wells north of the city.

The feasibility of the use of groundwater for agriculture is being studied. Also under study is the possibility of extracting larger amounts of groundwater for extension of the city water supply. With industry increasing and modern hotels being built to meet the inflow of tourists, it is expected that in the near future the demand for groundwater will increase considerably.

Bandung Area

In the Bandung area (elevation between 700 and 1,100 m above sea level), material is from two sources. The oldest rocks cropping out in this area are those from the hypothetical Sunda volcano of presumably late to middle Pleistocene age. The rocks are breccias and lavas, some of which are very dense. Remnants of the Sunda volcano can be seen as scattered mountain blocks around Lembang, north of Bandung.

Part of the Sunda material was covered by younger material from Tangkubanparahu volcano (2,093 m), at a time when its activity was severe. According to Van Bemmelen (1937), this volcanic activity can be divided into three periods, which he called "Eruption Phases A, B, and C." Phase A was the most violent and resulted in the production of huge amounts of pumiceous tuffs which covered extensive areas. Bandung is built on the periphery of this tuff fan. During this period of activity, Lake Bandung was formed as a result of the damming of the Citarum River, the stream that drained the intermontane basin of Bandung. Although no carbon dating has been done, artifacts from an ancient culture and a well-known legend relating to that lake indicate that those mudflows most probably took place in the not-so-distant past, perhaps less than 10,000 years ago. The highest level attained by the lake water was 722 m above sea level. When spilling took place, the lake was gradually drained until ultimately the present-day Bandung plateau was formed.

During phase B, basalt flows descended from Tangkubanparahu volcano. They followed several valleys—the farthest reaching a distance of 12 km—and in places formed waterfalls. Phase C produced pyroclastics, though in lesser amounts than during phase A.

This discussion is concerned mainly with Bandung and its surroundings. From drilling records it is known that the thickness of the Tangkubanparahu material is between 120 and 250 m.

As can be expected in such unsorted material, there are no well-defined aquifers and no layers of uniform composition which are spread over large areas. Consequently, there are great variations in hydraulic properties from place to place and with depth.

The groundwater table in Bandung ranges between 5 and 6 m and becomes shallower southward. In the hilly country north of the city, it may be deeper than 10 m.

The specific capacity of the Tangkubanparahu material ranges between 1 and 5 l/sec per meter of drawdown; that of the underlying Sunda material is between 0.5 and 1 l/sec per meter of drawdown, although there are exceptions. Recent drilling east of the city, for example, penetrated through lava and encountered a water-bearing zone with a specific capacity of 5 l/sec per meter of drawdown. West of Bandung, at approximately the center of the tuff fan, the piezometric level is about 10 m above ground surface, although a piezometric level far above this figure has been recorded.

In general the quality of the groundwater is good and meets public water-supply standards. As in Yogyakarta, the deeper layers may yield water rather high in iron and manganese.

Unlike Yogyakarta, the use of groundwater in Bandung has greatly accelerated. The city water-supply system includes spring water (125 l/sec), purified river water (750 l/sec), and 40 l/sec of water from 12 artesian boreholes. Industries, primarily textile manufacturers, consume an estimated additional 300–400 l/sec of water from private wells.

A recent study on meeting the demand for water in the next decades favors, in the first stage, dependence on groundwater; an additional amount of 1,000 l/sec may be extracted at reasonable cost.

Nganjuk-Kertosono Area

The Nganjuk-Kertosono area is practically flat, and bounded by gently sloping terrains. In the central part, the elevation is 40 m above sea level. In the course of its history, this basin has received clastic material from many sources, mostly volcanic. Subsequently, the Wilis, Anjasmoro, Pandan, and Kelut volcanoes have contributed to the shaping of the basin. Among those volcanoes only Mt. Kelut is still active, with frequency of eruption ranging between 3 and 15 years. Tertiary to lower Quaternary marly rocks of the Kendeng Mountains to the north have contributed nonvolcanic components.

The sand and silt from the Kelut eruptions have made the Brantas River—the main stream which drains this area—heavily laden with this material. This stream has meandered toward the mountains around Mt. Kelut. As a result, the drainage of the Nganjuk-Kertosono area has become very poor so that large swamps have formed. It is possible that at one time this area was occupied by a lake. A layer of heavy clay, which at places has a thickness of 90 m and occupies a large part of the flat country, may have been deposited in this situation. At present, thanks to the control of the main rivers—Brantas and Widas—much has been converted into arable land.

The maximum thickness of the basin fill is about 300 m. The clay layer that tops the basin fill prevents easy infiltration. This fact, combined with the flatness of the country, makes flooding a common phenomenon during the rainy season. However, because of the poor water-retention capacity of the soil, during the dry season a water shortage is felt nearly everywhere. Exploratory deep drilling for water was conducted by the Dutch before the end of the last century in an effort aimed at giving relief to the drought-stricken areas. Local people now know deeper seated water-bearing layers exist, and several holes have been made, by very simple means, for domestic use and for the cultivation of secondary crops.

As this area is an intermontane basin, the aquifers are better defined than in the areas discussed previously. Statistical analyses of data obtained from pumping tests at 109 drilled wells indicate that specific capacities are largest at depths between 75 and 125 m, with averages of about 2.2 l/sec per meter drawdown.

As a rule, the quality of the groundwater at the foot of volcanoes is good. Where the basin borders the Tertiary mountains, the unconfined groundwater may have a rather high chloride content—>250 mg/l. In this part of the basin, the artesian groundwater also may have a high chloride content locally. For example, the water from a well at Lestari sugar mill, about 2 km from the Tertiary hills, has a chloride content of 350 mg/l.

Conclusions

1. In Indonesia, terrains that are underlain by volcaniclastic rocks are the most promising for groundwater development.

2. Although capacities may vary from place to place and from depth to depth, as a whole they are higher than the capacities of other rock formations. Capabilities range between 0.2 and 10 l/sec per meter drawdown.

3. The water of the uppermost layers is best. Deeper water may contain higher amounts of iron and manganese which may exceed the limit as set forth in public water-supply standard.

References Cited

Purbo-Hadiwidjojo, M. M., 1967, Hydrogeology of strato-volcanoes: Hannover, 1965, Internat. Assoc. Hydrog. Cong., Mem., v. 7, p. 293–298.

——— and I. Suryo, 1968, Volcanic activity and its implications on surface drainage; the case of the Kelut Volcano, East Java, as an example: Malang, East Java, 1968, ECAFE Irrig. Seminar.

Van Bemmelen, R. W., 1937, Geologische Kaart van Java (Geologic Map of Java), Toelichting by Blad 36 (Explanatory Note to Sheet 36): Bandung, Dienst van den Mijnb.

Groundwater in Volcanic Areas in Japan—Summary[1]

SOKI YAMAMOTO[2]

INTRODUCTION

To determine the nature of volcano springs in Japan, and to clarify the mode of occurrence of groundwater on and around volcanoes, the writer carried on hydrogeologic and hydrologic studies of several volcanoes, including Mt. Fuji. The amount of discharge and water temperature were measured and some chemical analyses, including tritium content, were made of the spring and well waters on and around the volcanoes.

Many volcanoes are large reservoirs which absorb and store rainwater as groundwater and release it over a long period. Hydrologic knowledge involving mode of occurrence of water, storage capacity, and water balance on a volcano makes it possible to evaluate the water potential of such a reservoir.

GEOLOGY AND GEOMORPHOLOGY OF MT. FUJI

Mt. Fuji is the highest (3,776 m) and most typical stratovolcano in Japan. Its surfaces are sculptured by numerous radial valleys, none with a perennial stream.

The existence and the shape of a valley on the mountain slope have a close relation to the potential for furnishing groundwater. Despite its simple conical, gentle slope, Mt. Fuji has no single structure but is composed of a group of superposed cones. Two contemporaneous volcanoes—Mt. Ashidaka and Mt. Komitake—underlie the Miocene Misaka volcanic rocks (Nishiyatsushiro Group) of Mt. Fuji. Rocks of Mt. Fuji are divided into an older stage—Kofuji—and the present Fuji.

Mt. Komitake is composed of lava sheets with interstratified agglomerate and other pyroclastic material. The lavas of this volcano, and of Mt. Ashidaka, are olivine-bearing two-pyroxene andesite. Kofuji is represented principally by agglomeratic mudflow deposits, scoriaceous lapilli, and many kinds of basalt (aphyric, olivine, augite-olivine, hypersthene-augite-olivine, and hypersthene-olivine). These agglomeratic mudflows are present directly beneath ejecta of the present Fuji, which has a thin black soil. The mudflows act as a hydrologic boundary for the occurrence and movement of groundwater in this area.

After a short period of quiescence, eruptive activity of the younger Fuji began. The oldest lava flows are near the foot of the mountain. These oldest flows and most of the lavas erupted in the early stage of activity of the present Fuji are olivine basalt, but lavas of succeeding stages of activity include basalts of almost every petrographic type. They alternate with pumiceous sand and gravel or scoriaceous lapilli. The lavas, sands, and gravels all form aquifers, but there are great differences in porosity. Even in the same lava, porosity can range from 10 to 90%.

SPRINGS AND WELLS

On and around Mt. Fuji there are about 60 springs. Most of the larger springs are distributed at its base. Springs generally decrease in number at higher levels of the mountain. Total discharge in this area is about 55 m^3/sec. The discharge of springs in the period from August to October varies from 0.0004 l/sec to 10,000 l/sec.

Many wells have been drilled around the base of Mt. Fuji. In the southern Gakunan area, well density is 34/km^2. Discharge of 1.4 million m^3/day from wells for industrial use has been recorded. Owing to overpumping of this area, considerable seawater intrusion has occurred along Suruga Bay.

GEOPHYSICAL PROSPECTING AND AQUIFER TEST

Considerable electrical resistivity prospecting and seismic refraction exploration have been carried on in this area. Each volcanic layer has similar velocities on all sides of Mt. Fuji. The electrical and seismic data show the feasibility of distinguishing each layer as a group and the possibility of locating buried valleys present on the surface of a Kofuji aquitard and/or aquifuge.

Permeability coefficients and specific discharges obtained from aquifer tests are shown in Table 1.

DISCUSSION AND CONCLUSION

On Mt. Fuji, older lavas composed of olivine basalt are generally more productive than others; volcanic sands and gravels are also good aquifers.

If these sand and gravel layers or olivine-basalt lava sheets are deposited in a buried valley that is

[1]Manuscript received, November 11, 1974.

[2]Department of Geography, Tokyo Kyoiku University, Tokyo, Japan.

Table 1. Results of Aquifer Tests

Rock	*K (cm/sec)*	*Sp. Yield (m^2/day)*
New scoria	$1.1 - 7.0 \times 10^{-2}$	
New mudflow	$1.4 - 9.8 \times 10^{-4}$	
New scoria (lower)	1.0×10	100-600
Old lava	7.0×10	250
(Yoshiwara)	$1.0 - 3.3 \times 10^{-1}$	
(mean of Mishima)	2.0×10	2,000-3,000
Kofuji scoria	$3.1 - 10.3 \times 10^{-2}$	3,450
Kofuji mudflow	4.0×10^{-5}	
Ashidaka Tuff breccia	1.6×10^{-2}	

formed on a Kofuji impermeable layer, the valley provides a good subterranean groundwater passage. Large springs gush out at the outlet of these buried valleys, and good wells are found in the lower course of such valleys.

The role of dikes, flank openings (lava tunnels), and parasitic cones in groundwater occurrence on this mountain was studied, but no good relation could be shown. We still have no good explanation of the mechanism of artesian flow by each well.

Mt. Fuji can be divided into three hydrologic zones: (1) an upper recharging zone with no groundwater; (2) a middle recharging zone with scanty groundwater, and (3) a spring zone with abundant groundwater (Fig. 1).

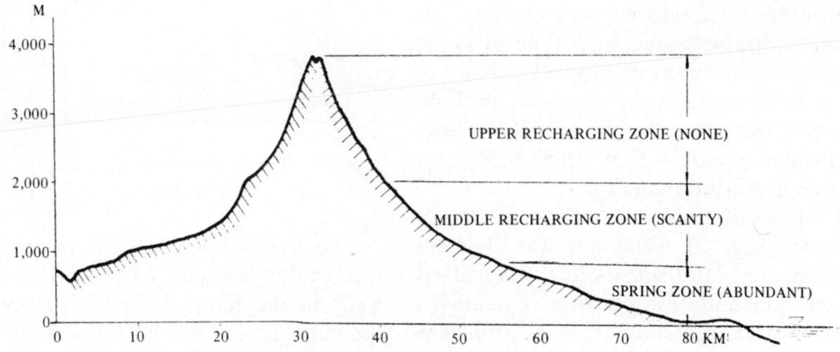

FIG. 1–Diagrammatic division of Mt. Fuji into three hydrologic zones.

Land Subsidence in Japan Resulting from Fluid Extraction—Summary[1]

SHIGERU AOKI[2]

CHARACTERISTICS OF LAND SUBSIDENCE

The characteristics of land subsidence can be summarized as follows:

1. The ground surface settles in a short period, at a rate of 1–10 cm/yr, or at times more than 20 cm/yr.
2. The ground surface generally is sinking rather than rising.
3. The area of subsidence is not large, the maximum being several hundred square kilometers.
4. Continuous subsidence of the ground surface commonly is caused by the compaction of the subsurface layers.
5. More often than not, land subsidence is a result of excessive groundwater withdrawal or exploitation of underground resources.

The movement of the land surface here considered is considerably different from crustal movement. Most instances of land subsidence are caused by human activities, such as pumping of groundwater or the exploitation of underground mineral resources. Therefore, they are distinguished from natural disasters as a type of *kogai* (environmental destruction) in Japan.

Land subsidence in Japan is not new; in Tokyo, for example, the ground surface of Kōtō lowland has subsided about 4–5 m since 1890, and the subsidence is continuing (Fig. 1).

Recently, areas of land subsidence in Japan have increased (Fig. 2). Wadachi and Hirono (1942) theorized that land subsidence was caused by accelerated consolidation because of a lowering of groundwater pressure. This theory was proved by the fact that the subsidence stopped in Tokyo and Osaka during the last stage of World War II when groundwater usage was curtailed.

Accordingly, we should examine the reason why subsidence continues even though the cause has been clarified.

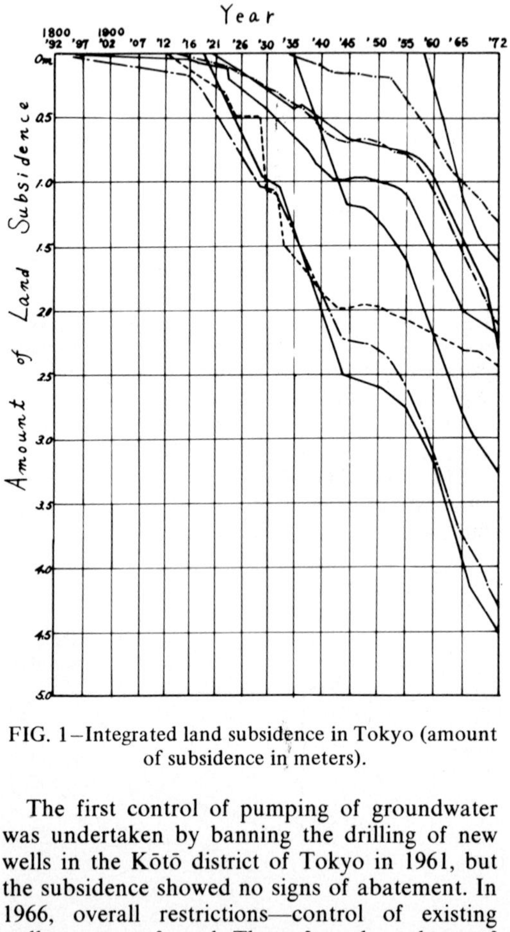

FIG. 1—Integrated land subsidence in Tokyo (amount of subsidence in meters).

The first control of pumping of groundwater was undertaken by banning the drilling of new wells in the Kōtō district of Tokyo in 1961, but the subsidence showed no signs of abatement. In 1966, overall restrictions—control of existing wells—were enforced. Thereafter, the volume of groundwater withdrawal began to show a remarkable decrease, the groundwater level showed a rapid rise, and the rate of subsidence decreased.

LAND SUBSIDENCE IN SOUTHERN KWANTO REGION

The land-subsidence area was restricted within the lowland of the city of Tokyo during the 10 years after the end of World War II. Since 1960, when rapid economic growth began in Japan, the area of land subsidence has been extending to the outlying districts.

[1]Manuscript received, November 6, 1974.

[2]Igarashi, Niigata, Japan.

I express my sincere appreciation to the late Naomi Miyabe for his supervision, and to Soki Yamamoto, Tatsuo Shibasaki, Isamu Kayane, and Shizuo Shindo for their discussions and encouragement. The materials used herein are from published and unpublished data of the Landsubsidence Research Group of Southern Kwanto region, which was organized by the Tokyo Metropolitan, Chiba, Saitama, and Kanagawa prefectural governments.

However, the following year the rates of recovery of the water level and of land subsidence were limited.

These phenomena were considered to be caused by the continuous pumping of the numerous wells in the neighboring district which remained unrestricted. Therefore, it is apparent that it is necessary to restrict excessive water withdrawal, not only in the area which is affected by subsidence, but also in other areas that belong to the same groundwater system.

In the southern Kwanto region, the main aquifers of the confined water belong to the Pleistocene Narita Group, which is widely developed in basins. This group consists mainly of gravel layers intercalated with sand in the lower part, and alternating sand, gravel, and clay in the upper part. On the basis of stratigraphy and hydrogeology of the Narito Group, the groundwater system of this region can be considered to be one hydrologic system of a large groundwater basin.

Countermeasures to land subsidence include determining the so-called "safe yield" of a groundwater basin which should be considered with relation to the water balance. In the first phase of control of groundwater withdrawal in Tokyo, the geologists of the Geological Survey of Japan emphasized the existence of the so-called "Urawa permeation zone or water vein zone" under the subsided area. An amount of groundwater estimated at 1,000,000–1,500,000 m^3/day was believed to have been recharged into the zone from rainfall; and this amount of recharge was said to be the "safe yield." However, the existence of such a permeation zone has been denied hydrogeologically, and the concept of "safe yield" had to be reexamined in view of the fact that the control of groundwater withdrawal, which had been enforced on the basis of this theory, had failed as a countermeasure to land subsidence.

The amount of the groundwater pumping has increased rapidly in the southern Kwanto region, especially in the district surrounding Tokyo City. The volume of water pumped was estimated as 1,300,000 m^3/day in 1959, and 2,340,000 m^3/day in 1969. Uses of groundwater in this region are: industrial, 42.2%; municipal, 41.9%; building, 11.4%; and agricultural, 4.5%. The widespread and rapid lowering of groundwater level (Fig. 3), resulting from withdrawal of water and dissolved natural gas in part of the Tokyo and Chiba Prefectures, caused widespread land subsidence in this region.

FUTURE UTILIZATION OF GROUNDWATER

From the analysis of tritium content of groundwater, the recycling rate of the Kwanto confined groundwater basin is considered to be very slow, and there is no natural flow or water vein. The amount of groundwater pumping obviously exceeds recharge into the basin, which Kayane et al (1973) estimated to be 1 mm/day.

The regional control enforced by the local governments from 1970 to 1973 was unsatisfactory as a countermeasure to land subsidence in the Kwanto region. However, it must be realized that people have been unconcerned in their use of groundwater resources, and that water has been supplied too cheaply.

In view of the enormous sums of money spent for construction of industrial water-channel and antiflood projects in the land-subsidence areas,

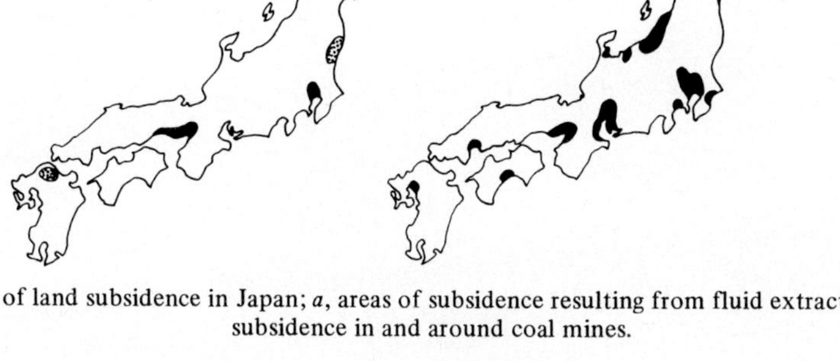

FIG. 2–Areas of land subsidence in Japan; *a*, areas of subsidence resulting from fluid extraction; *b*, areas of subsidence in and around coal mines.

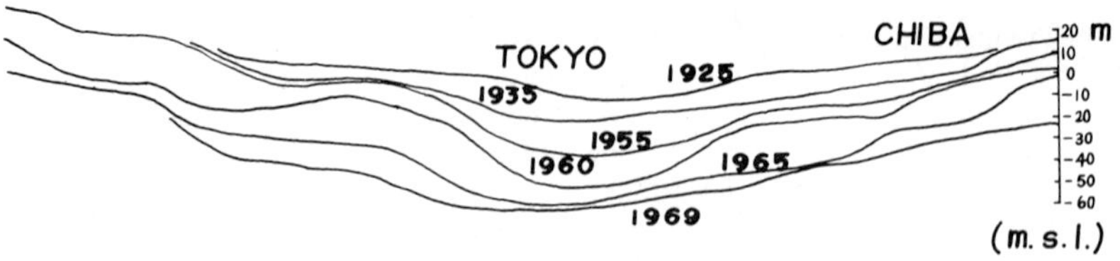

FIG. 3–Change in profile of groundwater level in Tokyo area (Shindo, 1972).

there seem to be only three corrective measures applicable to future utilization of groundwater resources:

1. Stop groundwater use entirely. This strategy should be undertaken in the "zero-meter area" and its close neighboring areas as soon as possible. In the other areas, however, this strategy may not be practical, because groundwater use has greater merit than the demerit of land subsidence.

2. Limit the amount of water pumped to the amount of natural recharge in the groundwater basin. This method requires more study, as it is difficult to estimate accurately the amount of natural recharge.

3. Enforce artificial recharge. This method has been successful in the land-subsidence area in and around Niigata City. In this district, about 190,000 m^3 of water per day has been pumped into several underground aquifers. (Natural gas was first separated from the water.) Since October 1973, the groundwater level has recovered as a result of this injection.

From study of land subsidence in the southern Kwanto region and the Niigata region, it appears that the first method—stopping the use of groundwater entirely—is effective as a countermeasure to land subsidence. In order to apply the third method—artificial recharge—a large amount of water for injection must be easily obtainable, and groundwater must not be used as a drinking-water supply. The second method is transitional between the other two.

Groundwater must be preserved as a future water resource; accordingly, the first and third methods are recommended for use in the subsidence areas in Japan.

References Cited

Kayane, I., et al, 1973, Development and management of groundwater resources (in Japanese): Water Use Inst., p. 1–418.

Shindo, S., 1972, Groundwater in the southern Kwanto region (in Japanese): Soil Mechanics and Found. Eng., v. 20, no. 5, p. 25–36.

Wadachi, K., and T. Hirono, 1942, On the landsubsidence in west of Osaka, pt. 3 (in Japanese): Rept. Nat. Disasters Research Inst., p. 1–33.

Impact of Economic Development on Demands for Groundwater and Wastewater Disposal in Honolulu—Summary[1]

GEORGE YUEN[2]

Like many large cities, Honolulu, the capitol of Hawaii, is suffering from the strains imposed by an ever-increasing population. Honolulu has experienced a rapid economic growth and has become the center of trade in the Pacific region. The relentless demand for an adequate supply of water and an environmentally acceptable sewage-disposal system are just two of the many crises facing Honolulu today.

Honolulu—city and county—encompasses the whole island of Oahu and contains about 82% of the state's population. This concentration imposes a heavy burden on the available groundwater supply. The bulk of recent increases in groundwater use is attributable to population growth.

The groundwater resources of Oahu are of three different types. The most extensive is the so-called basal fresh water that floats on seawater under much of the southern and northern portions of the island. Basal water bodies exist because of the difference in density between fresh water and seawater. The fresh water, being slightly lighter than seawater, floats on the latter.

Less widespread is water restrained between impermeable vertical rock structures—dikes—located in the cores of mountain ranges. These high-level bodies of water, although somewhat limited in volume, are important, particularly from an operational standpoint, because they supply areas of need through gravity flow.

The third type of groundwater is water held up, or "perched," on horizontal layers of impermeable beds such as volcanic ash. This type is of minor significance on Oahu.

Rainfall is the basis of the water supply. Of the water which becomes a part of the groundwater resources, about 550 million gal/day is developable. With a present groundwater draft of 440 million gal/day, only 110 million gal/day is available for future use. It has become apparent that we must direct our efforts toward a two-pronged program of conservation and development of alternative sources of supply.

Three alternative sources have been under consideration. They are (1) impounding of surface water, (2) reclamation of wastewater, and (3) desalination. In planning for future needs on a long-term basis, particularly when it is necessary to weigh the cost of alternatives, the need for conservation and limitation of water use must be considered. We have advocated conservation, but we now find that it may be necessary to enforce regulatory measures in order to realize more meaningful results.

Waste-disposal systems play a critical role in protecting groundwater supplies. During the past several years, the problem of sewage disposal for the Island of Oahu has become increasingly acute. Sewage systems have not kept pace with the city's development, and preventing the widespread indiscriminate disposal of sewage by use of cesspools, septic tanks, and other similar facilities has not been an easy task. Rules and regulations by governmental agencies have been implemented to control waste-disposal facilities and protect the groundwater supplies. The uncertainty associated with viral detection and destruction is one of the major reasons for the conservative measures applied to waste-disposal facilities.

As we look to the future, further economic development will accentuate the belief that from now on the whole proposition of providing potable water and the disposal of wastewater will be so closely interrelated that each will merely be a part of an integrated function with a common objective—to preserve and maintain the general health, well-being, and economic stability of the community.

[1]Manuscript received, November 11, 1974.

[2]Manager, Board of Water Supply, Honolulu, Hawaii 96843.

Simulation Model of Honolulu Basal Aquifer—Summary[1]

R. H. DALE[2]

History of Development of Honolulu Basin

The first well drilled in the Honolulu basin was completed in 1880. In the following decade, about 50 wells were drilled. Many of the wells were for irrigation, but some were used as an energy source for rice mills. The energy was available because the initial freshwater head in the basin was 42 ft (12.8 m) above sea level and many of the wells were at or near sea level.

The development of the groundwater basin was haphazard and unplanned, and by the year 1900 wells were beginning to become salty and there was a gradual reduction in freshwater head. In 1925, the Territorial government initiated a conservation policy. The strategy was to seal unused wells and to prohibit the use of artesian wells as an energy source. This strategy has remained unchanged for almost 50 years. With the establishment of the conservation policy, the freshwater head increased to a quasi-equilibrium head of about 26 ft (7.9 m).

In addition to the conservation policy, an operational policy of maintaining a relatively constant annual pumping rate developed. Thus, for the past half century the basin has been operated under approximately steady-state conditions.

Purpose and Scope

As the city of Honolulu has grown, the demand for water has risen accordingly. This demand has been met by importing water from outside the Honolulu basin; more than half of the water supplied in 1973 was imported from the Pearl Harbor area.

Because developable water of the Pearl Harbor basin must have a finite limit, water must be imported to the Honolulu area from much greater distances or the Honolulu basin must be redeveloped.

In order to redevelop the Honolulu basin, a good understanding of the mechanics of groundwater flow in the basin is needed so that the amount of developable freshwater can be defined. Therefore, the U.S. Geological Survey in 1970 initiated a program with the Honolulu Board of Water Supply to determine the feasibility of developing a digital simulation model of the Honolulu basin.

The first modeling attempt—to modify the method used by Pinder and Bredehoeft (1968, p. 1069) to fit the insular hydrologic conditions—was unsuccessful. Therefore, an insular model had to be developed, starting from basic physical laws.

Geohydrologic Framework

With respect to reservoir description, the geologic framework of the Honolulu area can be mapped in terms of three geologic units (Fig. 1). Two of the units are of volcanic origin and the third is a sedimentary sequence.

Most of the lava flows that formed the basal-water aquifer of the Honolulu area were erupted from volcanic feeder dikes that are oriented along long narrow rift zones. The feeder dikes cut preexisting lava flows, but away from the rift zone the lava flows are essentially dike-free. The presence or absence of dikes changes the hydraulic properties of the lava flows. Therefore, the dike-intruded lava flows are delineated from the dike-free lava flows.

The sedimentary coastal-plain and valley-fill deposits act as low-permeability confining units that restrict the flow of groundwater both toward the ocean and parallel with it.

Reservoir Rocks

Dike-intruded lava flows—The principal rift zone of the Koolau Range runs almost parallel with the crest of the range. The Kaau rift zone is a secondary rift zone that passes through Diamond Head and Kaau crater.

The feeder dikes range from 1 to 10 ft (0.3–3 m) thick. The dikes generally have a steep dip and are aligned parallel with the rift zone. They are extremely dense, and have low porosity and permeability values in comparison to the flows. The dikes cut preexisting lava flows that dip toward the ocean at about 5°.

The lava flows sandwiched between the dikes have high permeability and porosity, and these flows form dike-compartment groundwater reservoirs. However, the reservoirs are small, and the

[1]Manuscript received, October 21, 1974. Publication authorized by the Director, U.S. Geological Survey.

[2]Hydrologist, U.S. Geological Survey, Honolulu, Hawaii 96815.

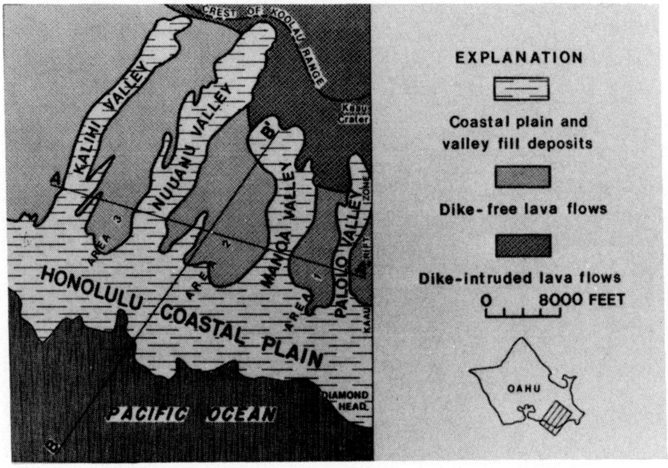

FIG. 1–Generalized geologic map of Honolulu area.

net effect of the intrusion of dikes into the lava flows is to reduce substantially the overall permeability of the reservoir rocks.

Dike-free lava flows—The dike-free lava flows are flows that erupted from the rift zones. These flows are thinly bedded, contain numerous clinker beds, and dip at about 5° away from the rift zone. The slope shown on Figure 2 is essentially a dip slope.

These dike-free lava flows form the reservoir rocks for the basal-water system. The necessary conditions for a basal-water system are that the reservoir rocks be highly permeable and be hydraulically connected with the ocean. The dike-free lava flows satisfy these conditions (Fig. 2).

Coastal-plain and valley-fill deposits—Deep valleys were cut into the lava flows during the Pleistocene. Nearshore deposition of the sediments contributed to the building of the thick sedimentary wedge of the Honolulu coastal plain. Sea-level rise relative to the land surface in late Pleistocene time caused a backfilling of the deep valleys, which generated the valley-fill deposits (Fig. 3). In plan view, it is difficult to delineate the coastal-plain deposits from the valley-fill deposits, but in section view the two units are easily differentiated. Figures 2 and 3 indicate the unit with a typical sand and gravel pattern. The deposits are, in part, sand and gravel but, because of extreme lateritic weathering, these deposits have a very low value of hydraulic conductivity in comparison to the dike-free lava flows.

The coastal-plain deposits (Fig. 2) restrict the flow of fresh groundwater to the ocean and are locally termed "caprock." The deep valley-fill deposits (Fig. 3) restrict groundwater flow parallel

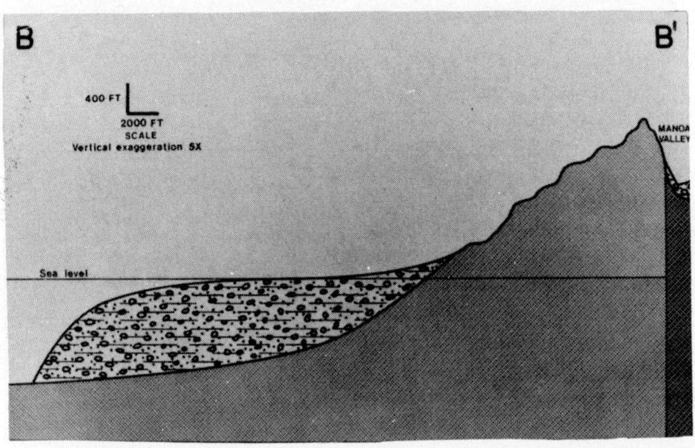

FIG. 2–Geologic section *B-B'*. Location is shown on Figure 1.

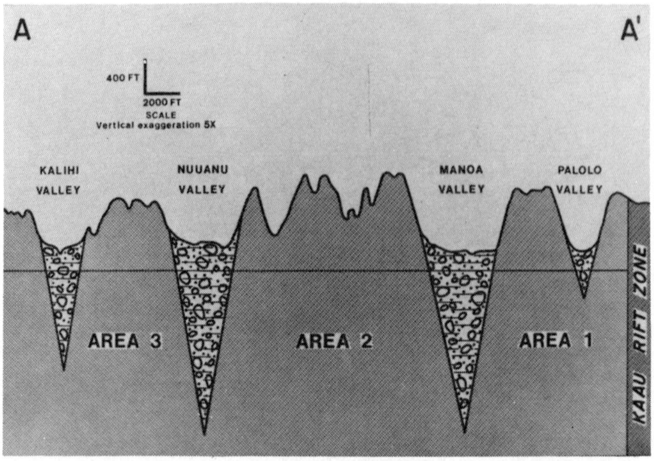

FIG. 3–Geologic section *A-A'*. Location is shown on Figure 1.

with the coast and divide the Honolulu area into three smaller basins which locally are termed "areas 1, 2, and 3."

Hydrologic Properties of Reservoir Rocks

Definition of the flow within these units depends, in part, on the hydraulic conductivity and the storage parameter for each unit. Ideally, these parameters should be known for all points in the reservoir rocks, but the existing data do not permit this.

Hydraulic conductivity—The definition of permeability varies between groundwater analysts and petroleum analysts. In groundwater problems, a single fluid is involved that is of relatively constant viscosity. In petroleum problems, several fluids are involved, each with a different viscosity. "Hydraulic conductivity," the term used in groundwater studies, is the rate of flow of groundwater of existing viscosity through a unit cross-sectional area, as measured perpendicular to the flow under a unit hydraulic gradient.

Storage parameters—There are two ways by which groundwater storage may change—confined and unconfined modes. In the confined mode, the control volume is fixed by means of an overlying aquiclude, and change in storage is related to the elastic properties of both the groundwater and the aquifer. In the unconfined mode, the control volume is not fixed by an overlying aquiclude, and the change in storage is related to the displacement of a movable boundary. In continental groundwater problems, both modes are important. In petroleum problems, usually only

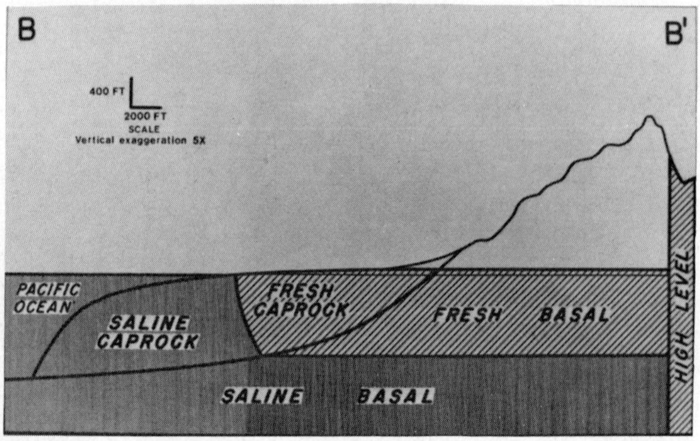

FIG. 4–Groundwater bodies. Flow in freshwater system is from high-level water body toward ocean; flow in saline-water system may be toward or from ocean, and over long periods of time, net saline-water flow is zero. Location is shown on Figure 1.

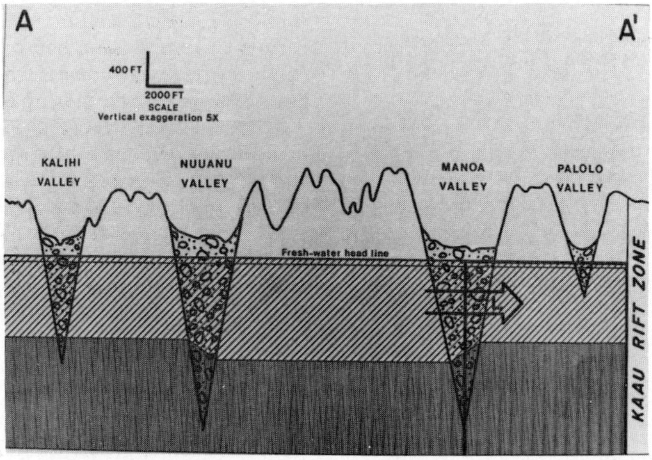

FIG. 5–Hydrologic section. Location is shown on Figure 1.

the confined mode is considered; in insular basal groundwater problems, only the unconfined mode is important.

There are two unconfined boundaries for the basal groundwater problem (Fig. 4). If only the fresh basal-water body is considered as being within the control volume, then both the top and bottom of this water body are unconfined, and the storage is referred to as top or bottom storage, depending on which free surface moves.

Insular Hydrologic Concepts

The groundwater beneath Oahu consists of a single fluid of variable density; however, from the standpoint of descriptive hydrology, it is convenient to subdivide the fluid into five water bodies (Fig. 4). These water bodies depend on the geologic unit and the fluid density. A high-level water body occurs in the dike-intruded lava flows; a fresh basal-water body and a saline basal-water body occur in the dike-free lava flows, and a fresh caprock-water body and a saline caprock-water body occur in the coastal-plain deposits. The principal concern of this report is the fresh basal-water body.

Coastal-Plain Leakage Model

The total freshwater flux is the amount that is being pumped plus the leakage through the coastal-plain deposits. Thus, there is considerable justification and motivation to model the coastal-plain leakage.

Within each artesian area, there is little variation in geologic conditions perpendicular to the section. Therefore, it is possible to determine the leakage per unit width of section, and then to determine the area leakage as the product of the

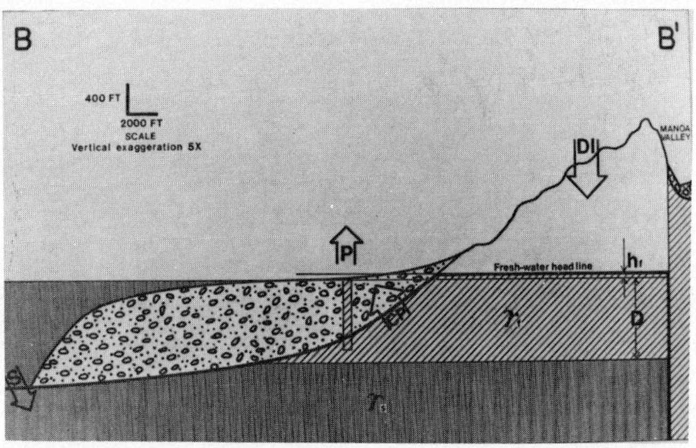

FIG. 6–Hydrologic section. Location is shown on Figure 1.

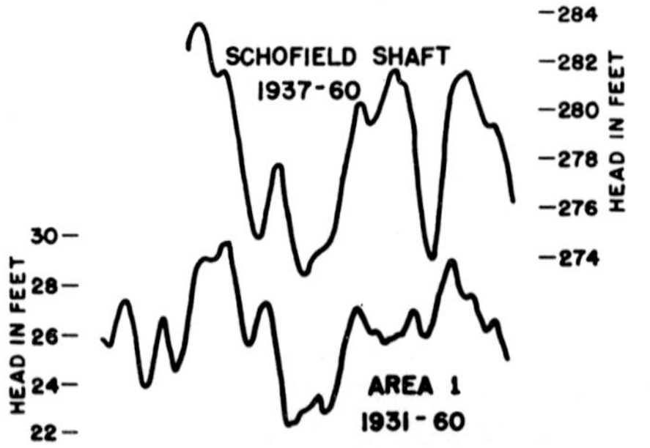

FIG. 7–Groundwater head trends for island of Oahu.

unit width leakage and the total width of the artesian area.

Basin-Storage Model

Maintaining adequate groundwater reserves that can be used during periods of drought is the principal motivation for development of a basin-storage model.

For this model, the control surface is placed around the fresh and the saline basal-water bodies (Figs. 5, 6). Flow across the control surface consists of deep infiltration, saltwater inflow, lateral inflow, coastal-plain leakage, and pumpage. Of the types of inflow and outflow for the Honolulu aquifer area, pumpage and deep infiltration are dominant. The lateral flow is probably small because the driving freshwater head has not been very large, the summation of salt water over long periods of time is zero, and coastal-plain leakage has been relatively constant because of the relative constancy of the fresh basal-water head.

Schofield shaft, located at Wahiawa in the high-level water body, best shows the climate-related trends in groundwater head, because this is a large, relatively undeveloped groundwater reservoir. The filtered heads for Schofield shaft and area 1 illustrate that the major long-term variations in head are caused by variations in deep infiltration (Fig. 7).

Conclusions

Insular groundwater problems are different from continental problems. In Honolulu, wells do not run dry but, rather, the water becomes salty. At this time, construction of a mathematical model of the Honolulu area that will satisfy management needs appears possible. Modeling success will depend primarily on the collection of saltwater-head data, because it is the combination of the freshwater head and the saltwater head that determines the volume of fresh groundwater in storage.

Reference Cited

Pinder, G. F., and J. D. Bredehoeft, 1968, Application of the digital computer for aquifer evaluation: Water Resources Research, v. 4, no. 5, p. 1069–1093.

Effect of Subsurface Waste-Disposal Practice on Groundwater Resources in Hawaiian Islands[1]

F. L. PETERSON[2] and K. J. TAKASAKI[3]

Abstract Groundwater in Hawaii is present as basal water, as dike-impounded water, and as perched water. Basal water, the fresher part of it forming a lens-shaped body floating on saline groundwater, is in dike-free volcanic rocks and in sedimentary rocks. Dike-impounded water is confined to volcanic rocks in eruptive zones. Perched water is present in all rocks and at all altitudes.

Most recharge to groundwater is in the wet interior mountains. Therefore, the main areas of recharge are upslope from developed areas where most wastes are disposed of in the subsurface. This natural deterrent, by position of the recharge area, so far has kept much of the groundwater in its pristine state. Deterioration of the groundwater will increase as land development encroaches toward the recharge areas—the degree will depend greatly on the waste-disposal practices.

Most natural discharge of groundwater is along the shore, downslope from disposal wells. Contamination of the beaches, rather than the water supply, is the main concern regarding subsurface disposal of wastes in the low areas under present conditions.

The success of injection operations depends primarily on injection capacity and fate of the injected waste. To evaluate these factors properly, an understanding must be obtained of local hydrogeologic conditions, hydrodynamics of injection under Ghyben-Herzberg lens conditions, and possible chemical and biologic effects. The hydrogeology of Hawaii is understood fairly well and, where adequate information is not available, it usually is possible to collect these data by careful field investigation. Considerable information is available from other parts of the world on the hydrodynamics of wastewater injection. However, much of this information is not directly applicable to injection in the Hawaiian environment. Particularly troublesome are the complications caused by the extreme heterogeneity of receiving formations and Ghyben-Herzberg lens effects. Likewise, because chemical and geologic reactions depend on the nature of the injected waste, the receiving waters, and the receiving formations, many of the data collected elsewhere are not applicable in Hawaii.

Geologic Framework and Hydrologic Conditions

In Hawaii, liquid waste injected into the subsurface and the leachate from waste deposited on the land surface behave nearly the same as freshwater recharge from rainfall. The waste moves through and saturates the same rocks under the same conditions, mostly because the dissolved-solids content of the liquid waste is low compared to that of seawater, which saturates most of the rocks below sea level. Therefore, the geologic and hydrologic environments controlling the movement of water must be understood if hydrologic problems related to the disposal of waste are to be identified.

The Hawaiian Islands are the tops of shield volcanoes in the Pacific Ocean. Each of the islands consists of one or more volcanic domes, most of which are composed of thousands of thin-bedded, highly permeable, basaltic lava flows. Associated with each volcano are lavas extruded in repeated eruptions from narrow zones of fissures. When eruption ceased, lava remaining in the fissures was quickly chilled to form low-permeability dikes. This rock assemblage of highly permeable lavas, intruded by dikes of low permeability, makes up the principal aquifer in the Hawaiian Islands.

In most of the island volcanoes, these highly permeable lava flows were capped subsequently by less permeable lava flows or by weathered rock. The interior mass of the volcano, however, mostly remained unweathered and retained much of its original high permeability. The resultant geologic framework consists of a highly permeable interior mass, intruded in places by dikes of low permeability and generally capped by a layer of less permeable rock. Because of the high permeability of the interior, most of the rocks below the level of the ocean are saturated with seawater.

Groundwater occurs in this framework as basal water, as dike-impounded water, and as perched water. In dike-free rocks, fresh water floats on, and displaces part of, the saline water and is known in Hawaii as *basal groundwater.* Fresh water which is impounded and separated from the saline water by dikes is known as *dike-impounded* or *dike-confined groundwater,* and that separated from other bodies of water by nearly horizontal beds of low permeability is known as *perched groundwater.* Dike-impounded or perched groundwater which occurs at high altitudes is

[1]Manuscript received, October 21, 1974. Publication authorized by the Director, U.S. Geological Survey.

The part of this paper dealing with the geologic framework and hydrologic conditions as they relate to subsurface waste disposal was written by Takasaki. That part discussing the various aspects of liquid waste injection was written by Peterson.

[2]Associate Professor, Department of Geology and Geophysics, University of Hawaii, Honolulu, Hawaii 96814.

[3]Hydrologist, U.S. Geological Survey, Honolulu, Hawaii 96814.

commonly referred to as "high-level groundwater," in contrast to "basal groundwater," which occurs at or near sea level.

Rainfall is the principal source of the recharge to groundwater and is heaviest in Hawaii (1) near the crest of mountains with an altitude of less than 6,000 ft (1,829 m) and (2) between altitudes of 2,000 and 4,000 ft (609–1,219 m) on the windward slopes of mountains higher than 6,000 ft. Most of the rain results from cooling of moist trade-wind air as it is orographically lifted. On the major islands, rainfall is more than 300 in. (7,620 mm) per year over the mountainous areas, but it decreases rapidly to less than 20 in. (508 mm) per year at the crest of the higher mountains and along some of the leeward shores.

Land development is heaviest in the coastal-plain areas and on the lower slopes of mountains. During the last 10 years, tourist resorts and rural residential subdivisions have developed rapidly in dry leeward areas heretofore relatively undeveloped.

Most recharge, therefore, is generally upslope from the low-lying developed areas where waste disposal is more likely to occur. This natural deterrent, by position of the recharge area, has kept much of Hawaii's groundwater supply from being polluted. However, as land development encroaches toward and into the undeveloped recharge areas, some deterioration of the groundwater supply is to be expected; the degree will depend greatly on future waste-disposal practices.

The resort areas and rural residential complexes in the coastal areas are presently in various stages of development. Being relatively new developments, these areas are not yet serviced by municipal sewer facilities. Most, however, have some kind of sewage collection, treatment, and disposal facilities. A major portion of this sewage effluent is presently injected into the subsurface. In the coastal plain, where the underlying basal water in the volcanic aquifer is confined and under artesian conditions, liquid waste injected at or near the surface is intercepted by permeable beds in the confining coastal-plain sediments. Some developers are utilizing sewage effluent for golf-course and other lawn-type irrigation.

Impending danger of contamination and deterioration of the potable groundwater supply is small, owing to the relatively small volume of liquid waste being injected and to the generally favorable location of these injection sites. Nevertheless, waste disposal is a growing environmental problem in Hawaii because of rapid and intensive urban and resort growth and a lag in the installation of satisfactory disposal measures to cope with the increasing volume of waste. The increase in subsurface injection of wastes, especially of inadequately treated sewage, is of primary concern to health officials and water developers. In inland areas, the concern is the possible future contamination of potable water supplies; in nearshore disposal areas, the concern is the possible pollution of beaches and the presently unknown effects on the offshore environment.

Hydraulic, Geochemical, and Monitoring Aspects

In Hawaii, most waste injection occurs in shallow coastal aquifers where the ultimate fate of the injected waste is to discharge, often quite rapidly, into shallow, offshore coastal waters. Furthermore, most liquid waste injected into the subsurface in Hawaii is treated sewage effluent. Consequently, because of its low density compared to seawater, it can be treated, from a fluid-flow standpoint, essentially as freshwater recharge.

The success of any injection operation in Hawaii depends ultimately on two critical factors. First, the injection capacity of the receiving formation must be sufficiently large to allow a suitable rate of injection which can be maintained throughout the life of the facility. Second, the ultimate fate of the injected waste must be known. In particular, the pattern of movement of the waste in the subsurface environment, as well as any changes the waste may undergo as it moves through the subsurface, must be understood.

In the Hawaiian environment, two areas are of concern: possible contamination of potable groundwater bodies and contamination of shallow nearshore coastal waters. Fresh potable groundwater primarily is restricted to lava flows, and usually is not present in developable quantities in the immediate coastal areas where most waste injection occurs. Thus, with present waste-disposal practices, in most areas contamination of fresh groundwater supplies is not a major problem. However, the quality of nearshore coastal waters, both from health and esthetic standpoints, is of considerable concern, as these waters are the ultimate sink for most injected waste.

To evaluate properly the injection capacity and the possible fate of injected wastes in any area, an understanding must be obtained of the local geology and hydrology, the regional groundwater flow pattern, the effects of injection on the regional groundwater flow pattern, and possible chemical and biological effects.

Geologic considerations are exceedingly important, as they exert the primary control over rates of subsurface waste injection, total waste storage, and waste movement. Rock permeability and porosity, storage coefficient, and geologic structures and boundaries are of primary importance. The permeability and porosity of Hawaiian rocks, especially the lava flows and coastal-plain sediments into which most wastes are injected, are characterized by extreme heterogeneity and anisotropy. Consequently, although the hydrogeology of the Hawaiian Islands is fairly well understood, and average values for the aquifer parameters can be obtained on a regional basis, the details must be evaluated for each individual injection site.

An understanding of the site geology, as well as the regional groundwater flow pattern and the anticipated distribution of injection-head buildup, is essential. The details necessary to evaluate local groundwater flow patterns can be obtained by careful field investigation. A knowledge of the injection capacity and injection-head distribution can be obtained by well-test analysis. Ideally, the well testing should include both injection and pump testing and, wherever possible, the use of water for injection of approximately the same density as the expected waste. This will allow a better evaluation of possible Ghyben-Herzberg lens effects.

With the regional groundwater flow and injection-head information, it is possible to compute mathematically the position and movement of a theoretical injection plume, provided a simple flow model is assumed. For example, if waste effluent of approximately the same density as fresh water is injected into the fresh portion of the basal lens, initially the effluent will be diluted by mixing with the aquifer water. After sufficient time, however, the aquifer water will be displaced by a plume of waste effluent which eventually discharges into nearshore coastal waters. Figure 1 shows the mathematical solution for frontal positions of the waste plume for a single well in a uniform regional groundwater flow. In this model (Bear and Jacobs, 1964), a water divide ultimately will be established upstream at a distance of

$$X_s = Q/2\pi TJ. \tag{1}$$

The curve of the water divide is given by

$$X = Y \operatorname{cotg} Y/X_s, \tag{2}$$

and is asymptotic to the width of

$$W = 2\pi X_s. \tag{3}$$

In these expressions, Q is the discharge or recharge rate, T the transmissivity, and J the natural slope of the water table.

The frontal positions will be given by

$$e^{\bar{x} - \bar{t}} = \cos\bar{y} + \frac{\bar{x}}{\bar{y}} \sin\bar{y}\ , \tag{4}$$

where $\bar{x}$, $\bar{y}$, and $\bar{t}$ are dimensionless parameters given by

$$\bar{x} = \frac{2\pi TJ}{Q} x\ ; \bar{y} = \frac{2\pi TJ}{Q} y\ ; \bar{t} = \frac{2\pi (TJ)^2}{bn\,Q} t\ , \tag{5}$$

in which b is the bulk thickness of the aquifer, n is the porosity, and t is the time.

All available mathematical models, however, assume uniform regional groundwater flow patterns and homogeneous and isotropic media. None of the models take into account two very serious departures from these assumptions in the Hawaiian environment: (1) extreme heterogeneity of aquifer properties such as porosity and permeability, and (2) Ghyben-Herzberg lens effects. It has been well documented that heterogeneity increases dispersion; thus, the actual injection plume in the Hawaiian environment probably is greatly distorted from the ideal plume predicted from theory. Field measurements of injection in Israel (Mercado, 1967) into sandstone aquifers, which are much more homogeneous than Hawaiian lavas and coastal-plain sediments, show dispersivity values three to four times greater than predicted from theory. In Hawaii the departures probably are even greater. Furthermore, in Hawaiian wastewater-injection practices, the effluent most commonly is not injected into the fresh basal lens but, rather, into underlying brackish water. Although the general effects of injecting low-density liquids into higher density receiving waters are recognized (e.g., if domestic sewage effluent is injected into higher density saline water, it is well understood that the sewage effluent tends to migrate upward in order to achieve density equilibrium), detailed effects of the Ghyben-Herzberg lens on injection dynamics are not well understood.

Chemical and biological effects, likewise, are extremely important in determining the ultimate fate of injected wastes. In particular, chemical and biological clogging and changes in the chemistry of the waste as it migrates through subsurface formations must be evaluated.

Owing to the siting of most injection facilities in coastal areas, some variation of the plume flow model described probably best defines most injected effluent movement in Hawaii. In this model, the main body of waste effluent arrives at the coast with little dilution and discharges into coastal waters through a zone of seepage along the ocean bottom. Hence, any effects on the quality of coastal waters depend on dilution of the effluent seepage by seawater, which in turn depends on the area of the seepage zone and the oceanographic conditions over the area. Mink (1973) calculated that, if waste injected into a

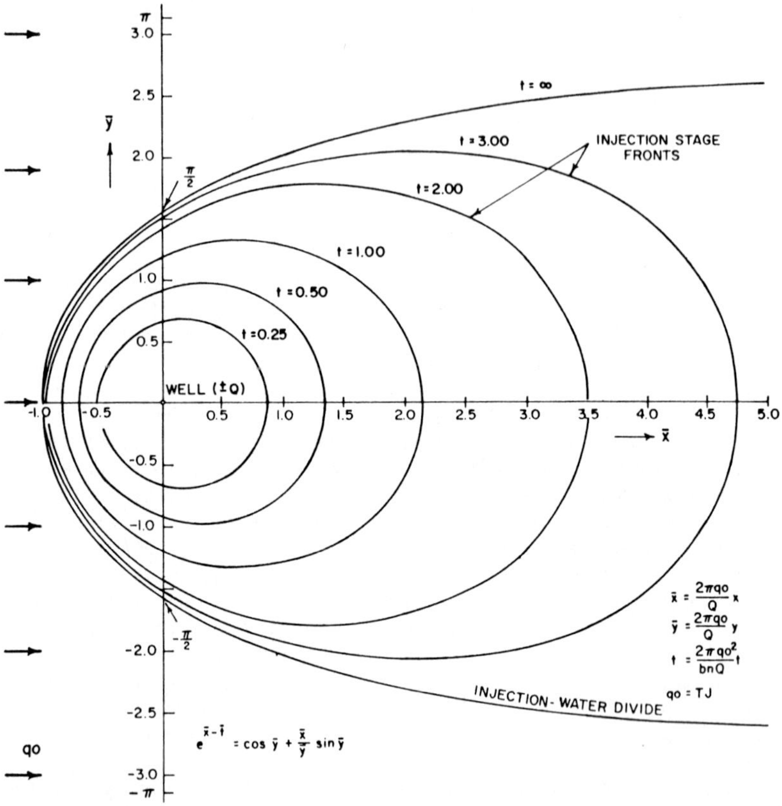

FIG. 1—Front positions for a single well in uniform regional groundwater flow (after Bear and Jacobs, 1964).

fresh basal lens discharges along a uniformly distributed seepage zone on the ocean bottom, the seepage rate per unit, Q/A, is given by

$$Q/A = 2\,ak, \tag{6}$$

where a is the difference in density between the seawater and the effluent divided by the density of the effluent, and k is the permeability. However, if the waste seepage occurs as point discharges, such as might be the case if lava tubes channelize flow, or if waste effluent is injected into salt water below the basal lens, mathematical treatment becomes virtually impossible.

Although the main body of effluent probably will travel as a plume and discharge at the coast, undoubtedly chemical and biological reactions will occur in the immediate vicinity of the injection well and may pose potential clogging problems. Little detailed information is available on chemical and biological aspects of waste injection in the Hawaiian environment. However, in virtually every reported instance where treated sewage effluent is injected, the injection capacity has decreased owing to chemical and biological clogging. For example, at Bay Park, New York, injection of tertiary-treated domestic sewage effluent which meets drinking-water standards resulted in substantial chemical and microbiological clogging, despite the use of fiberglass conductor pipe. Water recovered after injection at the Bay Park site showed concentrations of iron, phosphate, volatile solids, and bacterial content many times greater than in the injected wastewater (Vecchioli and Ku, 1972). One favorable factor in Hawaii, however, is that most studies report considerably less clogging in lava flows and limestones than in sandstones and other sedimentary formations.

The limited information available comes primarily from work at Mililani, Oahu, where treated sewage effluent is being spread on the ground surface and percolates into the soil cover. Lau et al (1972) reported that 5-ft-thick (1.5 m) lysimeters at Mililani are effective in removing BOD_5, TOC, phosphates, potassium, boron, coliforms, and viruses, but that nitrogen removal is not nearly as efficient. These results were obtained from unsaturated soil columns, however, and may have limited applicability to saturated rock formations.

After an injection facility is completed and waste injection begins, it is extremely desirable to monitor both the movement of the waste plume and any possible quality changes which might occur in the waste. Owing to a very poor understanding of these processes in the Hawaiian environment, it is not easy to decide where to monitor, or even what to monitor for. However, several things seem apparent in regard to waste-injection monitoring in Hawaii. Based on observations elsewhere, most pronounced chemical and biological reactions probably occur in the immediate vicinity of the injection wells; consequently, monitoring of these effects must occur close to the injection wells. Owing to the possibility of channelized flow in both lava flows and limestone formations, it will be virtually impossible to detect all possible waste flow patterns. To do so would require almost an infinite number of wells. Since the ocean is the ultimate sink for virtually all the injected waste, it probably will be necessary to provide a nearshore coastal-water monitoring program in addition to inland monitoring sites.

In Hawaii, the hydrogeology, regional groundwater flow patterns, injection capacity, and distribution of injection-head buildup generally are fairly well understood. If any of these factors is not adequately known for a specific area, it usually is possible to collect these data by careful field investigation. The detailed effects of rock heterogeneity, Ghyben-Herzberg density stratification, and possible chemical and biological reactions are not well understood for the Hawaiian environment. Furthermore, it is not readily possible to collect these kinds of data from field investigations prior to actual injection operations. In fact, it may require years of waste injection before the effects of these factors are fully understood. Furthermore, techniques for monitoring waste injection in Hawaii are not well developed owing to the lack of understanding of the detailed effects of these factors. Consequently, in order to evaluate properly the effects of current and anticipated future injection practices in Hawaii, a considerable amount of additional basic research is required.

References Cited

Bear, J., and M. Jacobs, 1964, The movement of injected water bodies in confined aquifers, underground storage and mixing of water: Technion-Israel Inst. Technology Prog. Rept. 2, 45 p.

Lau, L. S., et al, 1972, Water recycling of sewage effluent by irrigation—a field study on Oahu: Hawaii Univ. Water Resources Research Center Tech. Rept. 62, 89 p.

Mercado, A., 1967, The spreading pattern of injected water in a permeability stratified aquifer, *in* Artificial recharge and aquifer management symposium, Haifa, Israel: UNESCO Pub. 72, p. 23–36.

Mink, J., 1973, Kauai Water Quality Management Plan, 1973: Rept. by Sunn, Low, Tom, and Hara, Inc., Honolulu, Hawaii.

Vecchioli, J., and H. Ku, 1972, Preliminary results of injecting highly treated sewage-plant effluent into a deep sand aquifer at Bay Park, New York: U.S. Geol. Survey Prof. Paper 751-A, 14 p.

MINERALS

Ferromanganese Nodules of the Pacific[1]

P. L. BEZRUKOV and N. S. SKORNYAKOVA[2]

Abstract The distribution, morphology, internal structure, and chemical composition of ferromanganese nodules in the Pacific Ocean reflect the circumcontinental and latitudinal zonality of sedimentation processes and such local factors as depth of the ocean, composition of substratum, intensity of diagenetic processes, intensity of halmyrolysis of volcanic rocks, hydrothermal activity, and changes of near-bottom current velocity. The age of formation of ferromanganese nodules and crusts on the ocean floor varies in different regions. In the pelagic regions the nodules are mainly of Pleistocene or, locally, pre-Quaternary age. Detailed research on selected geologic polygons gives important information for the study of the irregularity of ferromanganese distribution and composition within main ore fields.

INTRODUCTION

Soviet investigations of ferromanganese nodules and bottom sediments of the Pacific Ocean began about 20 years ago and were carried out in oceanographic and geologic expeditions, mainly on R/V *Vityaz* (1954–55, 1957–61, 1965–70, 1973), R/V *Akademik Kurchatov* (1968), and R/V *Dmitry Mendeleev* (1972). These investigations covered, in different degrees, almost all the main areas of ferromanganese deposits on the ocean floor with the exception of the Antarctic area. During two cruises of the *Vityaz* (1968, 1970), study of ferromanganese nodules was carried out on polygons—small bottom areas, on the average 24 × 24 km—chosen for more detailed geologic research. Samples of nodules and bottom sediments were recovered from about 1,500 stations, and more than 2,000 photographs of the ocean floor were taken at 150 stations.

Results of the study of manganese nodules have been published in many papers (Bezrukov, 1960, 1969, 1971, 1973; Glagoleva, 1972; Skornyakova and Zenkevich, 1961; Skornyakova et al, 1962; Skornyakova and Andruschenko, 1964, 1970; Strakhov et al, 1968).

DISTRIBUTION

Ferromanganese nodules are distributed in the Pacific Ocean in different tectonic and facies environments, at various depths and stratigraphic levels of the sedimentary strata. They cover the largest areas of the floor in the deep (4–6 km) basins with hilly bottom topography and very low sedimentation rates—<3 mm/1,000 yr (Menard, 1964; Bezrukov, 1970; Lisitzin, 1972). They lie mostly in pelagic red clay or in siliceous diatom-radiolarian oozes. At a depth less than 4.0–4.5 km they are commonly found in calcareous pelagic oozes. Ferromanganese nodules and encrustations at the surface of the volcanic and sedimentary rocks are found commonly on large rises at depths of 2 km and less. Outcrops of hard rocks at great depths also are ore encrusted.

Study of the processes of sedimentation in the Pacific Ocean permits distinction of two closely associated, major types of ocean environments—mainly terrigenous continental-margin and pelagic environments.

In the peripheral parts of the ocean, at the continental margins and abyssal plains covered with hemipelagic muds and characterized by high sedimentation rates, the nodules are sparsely scattered or entirely absent.

In the pelagic areas of the ocean, there are three large latitudinal zones where nodules commonly are highly concentrated on the ocean floor. Their position is determined by the latitudinal zonality of sedimentation processes and by some other factors. All three zones are better developed in the eastern Pacific than in the western part of the Pacific.

The northern zone is situated between 6–8°N and 30°N lat. The nodules are distributed in pelagic red clay and, south of 15°N, in clayey-siliceous and siliceous sediments. The nodule coverage on the ocean floor here ranges from 0 to 100%, but more commonly is between 10 and 50%; the average is about 35%. The highest density of nodules is more than 20 kg/m². The nodule coverage generally increases on tops and slopes of abyssal hills.

On the Mid-Pacific guyots and some other mountains where the thickness of ferromanganese

[1]Manuscript received, November 11, 1974.

[2]P. P. Shirshov Institute of Oceanology, USSR Academy of Sciences, Moscow, USSR.

crusts reaches 10–15 cm, the quantity of ore material locally increases up to 200–300 kg/m^2.

The second large zone of ferromanganese deposits is separated from the first by the equatorial belt of high biological productivity and rapid sedimentation rates (up to 30 mm/1,000 yr). In this zone, situated between 7–8°S and 40–45°S, the ferromanganese nodules are associated mostly with red clay and partly with calcareous sediments. Their coverage of the floor ranges from 0 to 100%, but on the average is higher than in the first zone, particularly near the Cook Islands, where it reaches about 50%. The maximum density of nodules is about 70 kg/m^2. Ferromanganese nodules are scarce on the East Pacific Rise.

The third narrow band of ferromanganese deposits, discovered by the USNS *Eltanin*, is situated between 55 and 60°S under the Antarctic convergence zone, for which high flow rates of near-bottom currents and very low rates of sedimentation are typical. The nodule coverage on the bottom here commonly reaches 50–100% (Goodell et al, 1970).

Between these zones and at their periphery, high concentrations of nodules and thick crusts are found only on separate mountains and hills where there is little or no sedimentation.

Beneath the ocean floor the ferromanganese nodules were found, in some places, at different levels in the upper 3 m of sedimentary strata and at the base of the sedimentary sequence, above basalts, according to the Deep-Sea Drilling Project results. These nodules are probably of geothermal origin.

The change in nodule quantity on the ocean floor is accompanied by a change in size, form, and internal structure of the nodules. In the areas of high nodule coverage, large spheroidal, ellipsoidal, and discoidal varieties with relatively small cores predominate. In the marginal parts of the ocean, nodule size decreases and their form generally is very irregular. These changes, as well as the changes in quantitative distribution of nodules, reflect the circumcontinental zonality of sedimentation in the ocean (Bezrukov, 1970). Within the limits of the main ore fields, the morphology of nodules changes markedly with different forms of bottom topography.

There is a wide range in the age of formation of the ferromanganese nodules and crusts in different regions. In the marginal regions of high sedimentation rates, where nodules are generally small, they are everywhere of Holocene or late Quaternary age. In the pelagic areas of very low sedimentation and nodule growth, their age corresponds to most or all of the Pleistocene. Paleontologic data indicate that the underlying red clay and siliceous oozes are, in many places, also of pre-Quaternary age, as are the ferromanganese crusts on the surfaces of Late Cretaceous and Paleogene rocks of the Mid-Pacific Mountains. This variety of ages of the ferromanganese deposits must be taken into account in regional considerations of their quantitative distribution and composition. The study of laminated textures of spheroidal nodules, especially with several cores, shows that the intermittent precipitation of ore material probably is partly connected with changes in deep-sea current velocity.

The chemical composition of ferromanganese nodules also undergoes regional changes. These changes are most evident far away from continents and between different latitudinal zones. The geochemistry of ferromanganese nodules and encrustations depends on many local factors (e.g., depth of ocean, composition of substratum, intensity of diagenetic processes, intensity of halmyrolysis of volcanic rocks, and hydrothermal activity).

The metal content changes from the periphery of a nodule to its core and from one concentric layer to another. For this reason, regional characteristics of the chemical composition are based on average composition of nodule crusts (or whole nodules).

In the marginal regions of the ocean, as was established long ago, the upward migration of mobile iron and manganese from the reducing into the oxidizing sediment zone is the main process leading to nodule formation. This process occurs under conditions of high organic content and is a wholly diagenetic process. Nodules of the peripheral regions have a variable content of Fe and Mn, but more commonly they are relatively enriched in Fe (Mn/Fe ratio is <1), and almost everywhere they have a low content of Ni, Cu, and Co ($<$0.1–0.2%).

In the pelagic area, where the content of the labile organic matter is very low and conditions are highly oxidizing, the diagenetic processes are less intensive, and the precipitation of manganese and iron from solution in the water is the main factor in nodule formation. The greater thickness of ore crust on the upper surfaces of tabular nodules and slabs, as compared to the lower surfaces, confirms this. However, the spotty coloration of the siliceous and clayey-siliceous oozes and the local occurrence of FeO traces in these sediments show that diagenetic processes may take place in pelagic sediments. For example, some lower ore crusts of tabular nodules and slabs are enriched in manganese, and some varia-

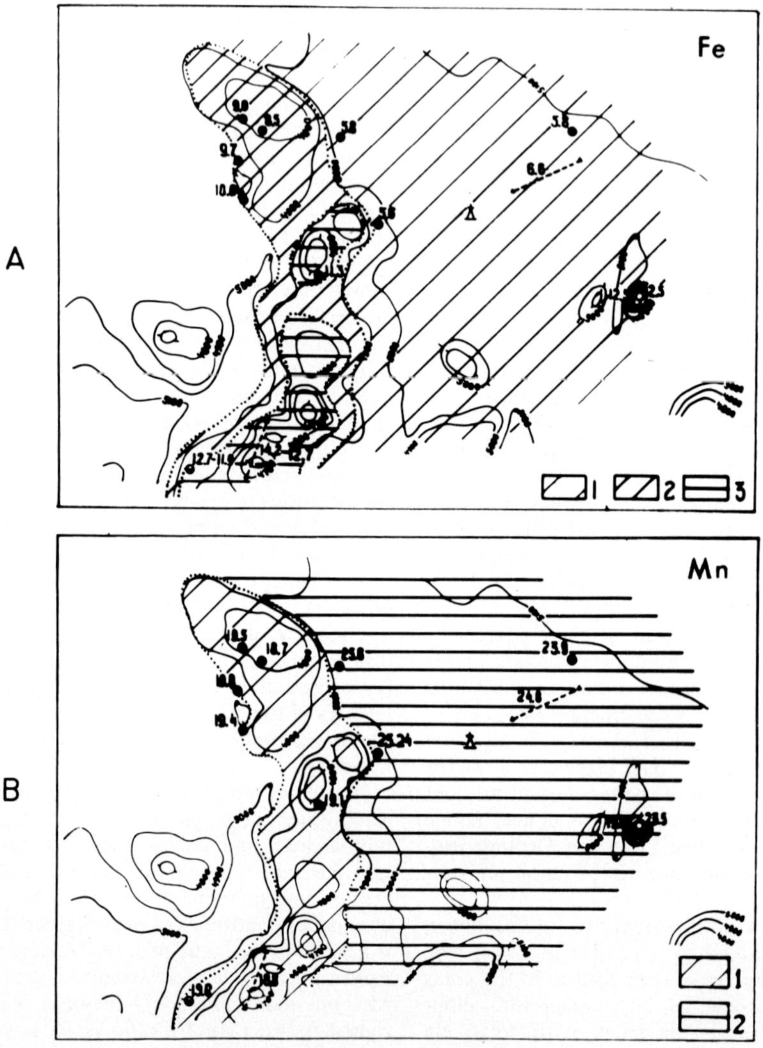

FIG. 1–Distribution of Fe and Mn in nodules, polygon station 5996. **A.** Fe: *1*, <6%; *2*, 6 - 12%; *3*, >12%. **B.** Mn: *1*, <20%; *2*, >20%.

tion in the chemical composition of nodules occurs in different zones of the ocean.

The distribution of all metals in nodules on the ocean floor displays some latitudinal zonality. For example, the highest manganese content (25–35%) occurs mainly in the siliceous sediments of the northern tropical zone. The distribution of nickel and copper repeats the distribution of manganese, which acts as their host in absorption processes. The highest contents of nickel (>1.5%) and copper (>1.5%) occur mainly in the siliceous sediment of the same belt. An analogous pattern was found in the Indian Ocean (Bezrukov and Andruschenko, 1973). The distribution of iron is contrary to that of manganese. The highest iron content (20–25%) in nodules occurs in red clay in the South Pacific basin. The cobalt distribution is similar to that of iron, but the highest values of cobalt (>0.5%) occur mainly on the tops of submarine mountains.

All the above data are in good agreement with the data of other scientists who have studied the regional changes of nodule chemical composition in the Pacific (Cronan and Tooms, 1969; Horn et al, 1972, 1973; Menard, 1964; Price and Calvert, 1970).

A

B

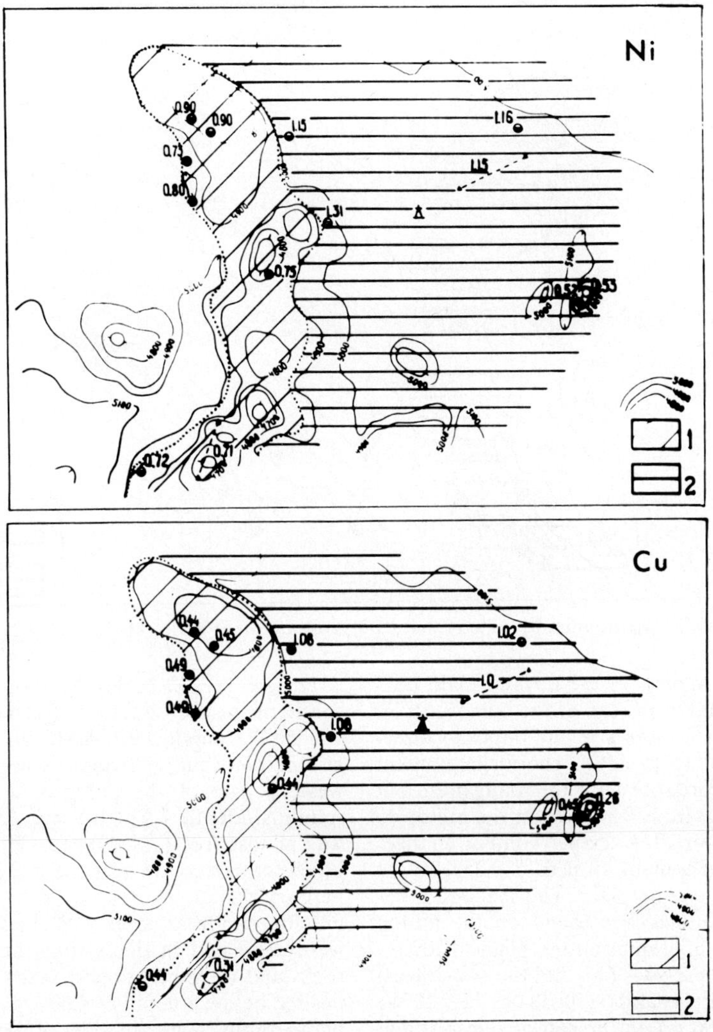

FIG. 2–Distribution of Ni and Cu in nodules, polygon station 5996. **A.** Ni: *1*, 0.5 - 1.0%; *2*, >1.0%. **B.** Cu: *1*, 0.25 - 0.5%; *2*, >1.0%.

Polygon Surveys

The important information for the study of the irregularities of the ferromanganese-nodule distribution and composition within main ore fields was obtained by polygon surveys. The geologic work included precise echo soundings; seismic profiling; bottom photography; and collection of sediment samples, volcanic rocks, and ferromanganese nodules by means of corers, grabs, trawls, and dredges. At each polygon, 15–62 samples and several long series of photographs were obtained from different forms of bottom relief.

Polygon station 5988—This station is in the southern part of the Central Pacific basin (11° 04′S, 156°12′W; depth 4,776–5,350 m), in the area of pelagic and zeolite clay. The greatest coverage of nodules (50–100%) and greatest density (up to 32 kg²) were found on the large abyssal hill in the eastern part of the polygon. Coverage was a little less on the gentle swell in the western part. Large spheroidal nodules predominate here. On the ocean floor around the hill, the nodules are sparsely scattered and are smaller.

The chemical composition of nodules shows large variation within the polygon. The manga-

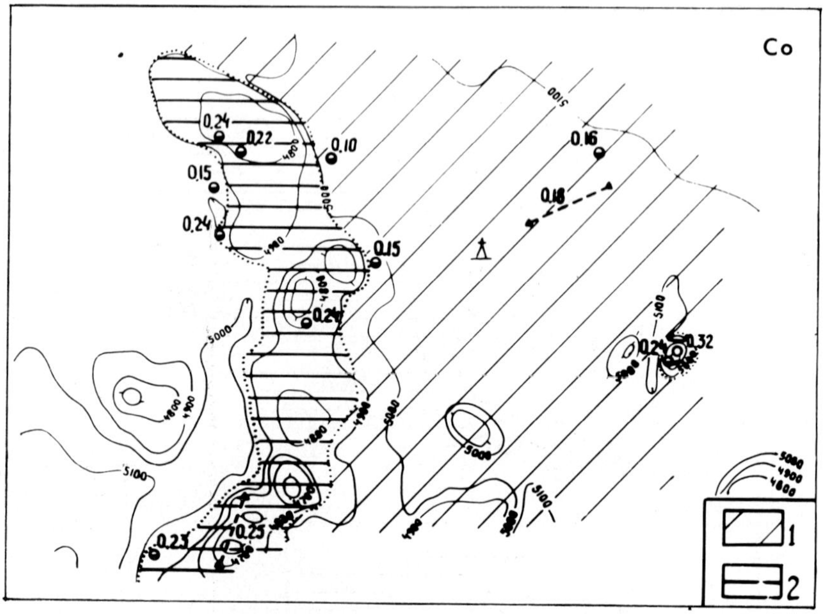

FIG. 3–Distribution of Co in nodules, polygon station 5996. *1*, <0.2%; *2*, >0.2%.

nese content is highest at the abyssal plain (20–21%) and minimal at the top of the hill (16–18%); the iron content increases in the opposite direction (from 10–11 to 15–16%). The nickel content increases in accordance with the manganese but more intensively, from 0.3–0.4% on the hilltop to 0.85% on the plain. The copper content changes along with nickel content, from 0.15–0.25% on the hill to 0.4–0.6% on the plain. The maximum cobalt value in nodules was found on the hilltop (0.38%) and the minimum on the plain (0.20%).

Polygon station 5996—This station is southeast of the Hawaiian Islands (10°58′N, 152°23′W; depth 4,700–5,100 m), in the belt of siliceous and calcareous oozes. Here, too, the nodules are irregularly spaced. The maximum coverage (more than 50%) and quantities were found on abyssal hills. On the plain, where Oligocene rocks crop out, the large tabular and discoid nodules have a more scattered distribution.

The metal content of the nodules changes in the same mode as in the previous polygon, but values of Mn, Ni, and Cu here are higher, and Fe and Co are lower.

On the tops of abyssal hills the manganese content of nodules is about 19%, and on the plain, about 24–26%. The iron content changes in the opposite direction—from 14–19% to 5–6%. The nickel and copper values are highest on the plain (each about 1.0%), and the cobalt content reaches its maximum on the abyssal hills (up to 0.25%; Figs. 1–3).

Polygon station 6298—At station 6298, in the southern basin, south of Rarotonga (22°41′S, 160°50′W; depth 2,970–4,915 m), the pattern of nodule distribution is more complicated. In the western part of the polygon, a volcanic ridge (minimal depth 2,970 m) was identified which was almost free of sediments. Pillow basalt and blocks of palagonite tuffs covered by ferromanganese encrustations crop out in the area. Calcareous foraminiferal oozes with manganese nodules were deposited in depressions between volcanic rocks and on gentle slopes; marly clays were deposited below a depth of 4,600 m. As observed in photographs, there are slides of rocks and nodules on steep slopes. According to seismic-profiling data, east of the volcanic ridge, in the submeridional depression at depths of 4,800–4,900 m, the sedimentary strata thicken to 60–100 m. In this area, sediments are mainly pelagic clays partly replaced by zeolitic clays. The northeastern part of the polygon has a hilly relief and is covered chiefly by marly clays. Two zones of hydrothermally altered sedimentary and volcanic rocks, ascribed to a fracture zone, have been discovered in the central depression. Lenses of mineralized clays have been found in cores of pelagic clays. The mineralized clays contain up to 17% iron and 13% manganese (Bezrukov, 1973; Skornyakova et al, 1973).

The ferromanganese nodules are spaced rather irregularly within the polygon. Over short distances on the slopes and tops of the volcanic ridge, their coverage ranges from 0 to 100%; in the central depression it ranges from 40 to 70%; and in the hilly rises in the eastern part of the polygon it ranges from 60 to 100%. The highest density of nodules here reaches 50–70 kg/m^2.

The chemical composition of nodules shows more irregular changes. The highest content of manganese (18.6%) was found near the top of the volcanic ridge, where high values of iron and cobalt also were noted. At the places where hydrothermally altered rocks crop out, great variations in the Mn/Fe ratio (0.2–50) and relatively high contents of nickel and cobalt were found.

This polygon represents an example of the influence of hydrothermal activity on the process of oceanic ore formation outside the limits of mid-oceanic-rise systems where the wide distribution of metalliferous sediments has been discovered recently.

References Cited

Bezrukov, P. L., 1960, Sedimentation in the north-western part of the Pacific Ocean: Copenhagen, 21st Internat. Geol. Cong. Rept., pt. 10, p. 39–49.

——— 1969, The 43rd cruise of the R/V *Vityaz* to the central Pacific Ocean: Oceanology, v. 9, p. 153–160.

——— ed., 1970, The Pacific Ocean; sedimentation in the Pacific Ocean: Akad. Nauk SSSR, v. 6, pt. 2, 419 p. (in Russian).

——— 1971, Geological structure and mineral resources of the Pacific Ocean: Akad. Nauk SSSR Vestnik, no. 5, p. 86–94 (in Russian).

——— 1973, On the sedimentation in the northern part of the South Pacific, *in* Oceanography of the South Pacific 1972: Wellington, New Zealand Comm. UNESCO, p. 217–219.

——— and P. F. Andruschenko, 1973, Geochemistry of ferromanganese nodules from the Indian Ocean: Akad. Nauk SSSR Izv., Ser. Geol., no. 9, p. 18–37 (in Russian).

Cronan, D. S., and J. S. Tooms, 1969, The geochemistry of manganese nodules and associated pelagic deposits from the Pacific and Indian Oceans: Deep-Sea Research, v. 16, no. 4, p. 335–359.

Glagoleva, M. A., 1972, Characteristics of the changes in the chemical composition of iron-manganese concretions in sediments of the northwestern part of the Pacific Ocean: Litol. Polez. Iskop., no. 4, p. 40–49 (in Russian).

Goodell, N. G., and M. A. Meylan, and B. Grant, 1970, Ferromanganese deposits of the South Pacific Ocean, Drake Passage, and Scotia Sea, *in* Antarctic oceanology, I: Antarctic Research Ser., v. 15, p. 27–92.

Horn, D. R., B. M. Horn, and M. N. Delach, 1972, Ferromanganese deposits of the North Pacific: Lamont-Doherty Geol. Observ. Tech. Rept., no. 1, 78 p.

——— ——— and ——— 1973, Ocean manganese nodules; metal values and mining sites: Lamont-Doherty Geol. Observ. Tech. Rept., no. 4, 57 p.

Lisitzin, A. P., 1972, Sedimentation in the world ocean: Soc. Econ. Paleontologists and Mineralogists Spec. Pub. No. 17, 218 p.

Menard, H. W., 1964, Marine geology of the Pacific: New York, McGraw-Hill, 271 p.

Price, H. B., and S. E. Calvert, 1970, Compositional variation in Pacific Ocean ferromanganese nodules and its relations to sediment accumulation rates: Marine Geology, v. 9, p. 145–171.

Skornyakova, N. S., and P. F. Andruschenko, 1964, Ferromanganese nodules of the Pacific Ocean: Litol. Polez. Iskop., no. 5, p. 21–36 (in Russian).

——— and ——— 1970, Ferromanganese nodules in the Pacific, *in* P. L. Bezrukov, ed., The Pacific Ocean: Akad. Nauk SSSR, v. 6, pt. 2, p. 202–268 (in Russian).

——— and N. L. Zenkevich, 1961, The distribution of ferromanganese concretions in the surface of sedimentary layers of the Pacific Ocean: Oceanology, v. 1, no. 1, p. 86–94.

——— P. F. Andruschenko, and L. S. Fomina, 1962, Chemical composition of ferromanganese nodules in the Pacific Ocean: Oceanology, v. 2, no. 2, p. 264–277.

——— et al, 1973, Bottom sediments and volcanic rocks of a polygon in the southern basin of the Pacific Ocean: Litol. Polez. Iskop., no. 1, p. 17–28 (in Russian).

Strakhov, N. M., et al, 1968, Geochemistry of a sedimentary manganese ore-forming process. Akad. Nauk SSSR Geol. Inst. Trudy, v. 185, p. 7–178 (in Russian).

Manganese-Nodule Structures: Relation to Nodule Origin—Summary[1]

J. ANDREWS and S. MARGOLIS[2]

Manganese nodules, as seen on the seafloor, generally occur in fairly discrete and recognizable populations both locally and regionally. Nodules from the North Pacific are distinct in most cases from the Pacific-Antarctic cannonball forms, and these in turn differ from the discoidal to ellipsoidal rough-surfaced forms common to the copper- and nickel-enriched belt north of the equator in the eastern Pacific. On a smaller scale, local nodule populations may be either distinct and abruptly separated or mixed in a single deposit. Figure 1 shows a mixed deposit with two populations of strikingly different sizes and moderately distinct morphologies. If slow linear rates and constant growth are assumed, then these nodules were seeded by two widely separated events. This photo shows seafloor formed at the crest of the East Pacific Rise approximately 60 m.y. ago. If nominal growth rates (1–10 mm/m.y.; Ku and Broecker, 1967) are assumed, the larger nodules might have originated at that time to reach their present size of 8–16 cm diameter, whereas the smaller nodules may have appeared first about 20–30 m.y. ago. During growth of the nodules, this portion of seafloor was moving westward from the crest of the East Pacific Rise and simultaneously moving north with the entire East Pacific spreading system. These motions would have carried the nodules away from volcanic sources at the rise crest, and across the zone of high biologic productivity in the equatorial upwelling zones.

Major changes in the environment of manganese-nodule growth were produced by (1) shifts in spreading patterns in the western Pacific, (2) overriding of the East Pacific Rise by the North American continent, (3) adjustments along nearby fracture zones, and (4) the appearance and introduction of Antarctic Bottom Waters into the Pacific basin. Therefore, it is apparent that the internal structure patterns, which reflect nodule-growth history in general, need to be examined.

Nodule structure as seen in polished sections, X-radiographs, and SEM photographs reveals a variety of growth patterns which reflect the history of a particular nodule population. These structures range in size from less than 0.25μ to several millimeters. The smallest laminar structures routinely seen by the scanning electron microscope range from 0.25 to 10μ, and average 3μ. These may represent discrete growth units, or possibly diagenetic recrystallization as the nodule ages (Margolis and Glasby, 1973). Figure 2 illustrates these structures on an etched section of a nodule, and Figure 3 shows the same structures exposed on fracture surfaces of micronodules from deep-sea sediments. The fact that these structures appear to be common to all nodules examined to date (deep-sea, freshwater, and land nodules and micronodules) suggests that their origin most likely is due to growth, and that they represent some fundamental processes in nodule formation.

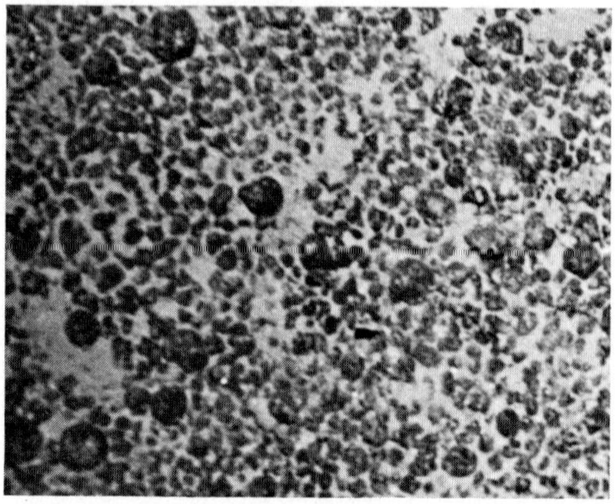

FIG. 1—Bottom photograph at 13°N, 141°W showing two manganese-nodule populations.

As structural forms, these microbands are fairly continuous around the nodule within the limits of broader structural units, and they form the outer botryoidal morphology of the nodules (Fig. 2). Figure 4 shows, however, that laminar layers are also accompanied by massive zones lacking distinctive layering and columnar structures within individual nodules. Indeed, growth habits appear to change radically with time. In Figure 5a, a massive or perlitic zone surrounds a small off-center seed. This zone is succeeded by a manganese-rich band (up to 45% Mn and enriched in Cu and Ni) of finely laminated structure; this band is succeeded in turn by a poorly laminated columnar zone. Substantiation of a relation between these structures and environmental factors influ-

[1]Manuscript received, December 26, 1974.

[2]Department of Oceanography, University of Hawaii at Manoa, Honolulu, Hawaii 96822.

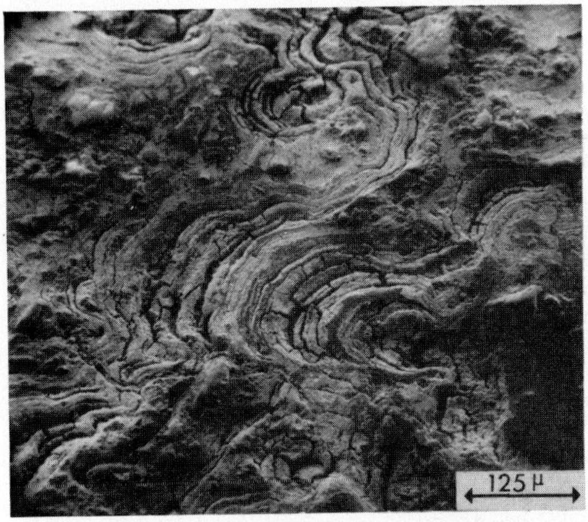

FIG. 2—Nodule microstructures photographed via SEM. Etched to show laminar structure.

encing growth can be seen in Figure 5b, which shows an intergrown double nodule from the same site as the nodule in Figure 5a. The same structural zones are present here. Nodule contact and intergrowth occurred during the third growth interval. These nodules are from the base of a secondary abyssal hill near the Mendocino fracture zone. Seafloor in the area is approximately 70 m.y. old, and the abyssal hill, which is perhaps a third that age, formed along a feather fracture developed by motion on the fracture zone (Andrews, 1971). It is possible that the inner zone in these nodules was the product of growth during movement of the crust away from the crest of the East Pacific Rise; that the fine laminar zone was directly related to some intermediate event, perhaps the volcanism involved in the formation of the hill; and that the outer layer is the product of

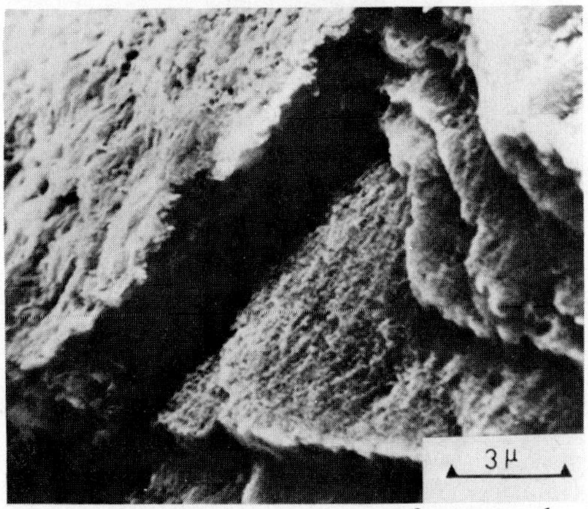

FIG. 3—Nodule microstructures on fractures and surfaces in micronodules; SEM photo.

recent growth. Greenslate (1974) and Dudley and Margolis (1974) have described benthic foraminifers on the surface of nodules. Many of the foraminifers enrich their tests with particular materials which may markedly influence nodule growth. The structures within the massive zone in Figure 5a and b suggest that many such biogenic structures may have been incorporated into the nodules at this site during the long initial growth phase; thus they point to particulate growth of nodules rather than physicochemical growth from solution. When manganese oxides are leached from the nodule thin section (by 3N HCl), it becomes clear that the basic structures are formed of iron oxides and have been infilled by manganese oxides. This fact supports a biogenic particulate ori-

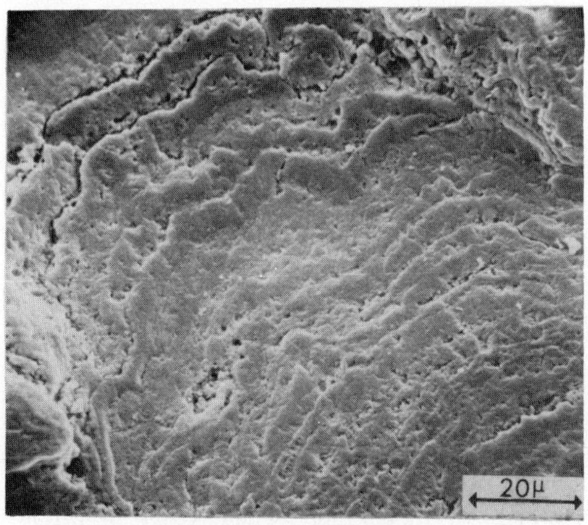

FIG. 4—Massive and columnar zones in nodule structure; SEM photo.

gin, as many organisms produce iron chelates in test building and by excretion or through metabolic processes (Dudley and Margolis, 1974).

Abrupt changes in growth pattern point to a rapid and marked alteration of the local environment which affected the entire nodule population. In this case, the change may well have been the tectonic and volcanic development of a small abyssal hill at the site. Nodules which were apparently seeded by this volcanic episode are cobalt enriched (Hubred, 1970), and microprobe studies are currently being pursued to determine whether this same enrichment pattern occurs in the growth zone attributed to this same time period for all nodules in the population. Finely laminated bands have high Mn, Cu, and Ni contents (and low Fe), whereas Fe-rich layers are low in Mn, Cu, and Ni and high in Co. Chemical inhomogeneities also are observed in horizontal and ver-

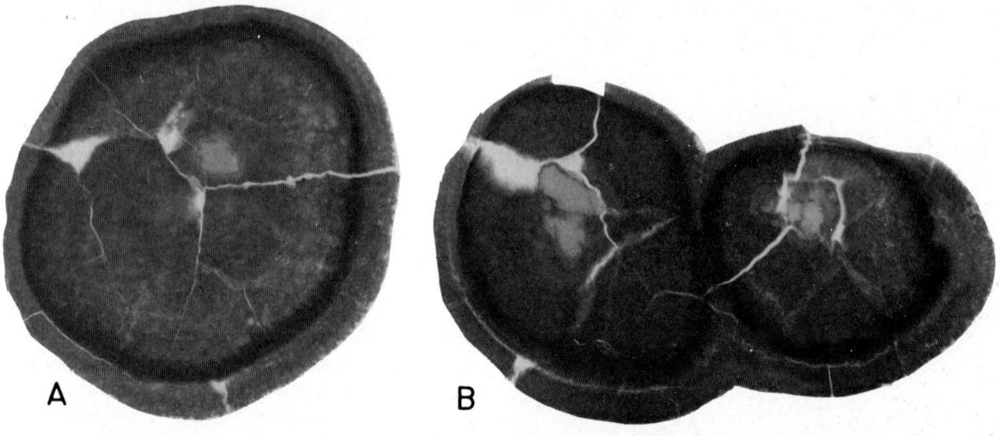

FIG. 5–X-radiographs of nodules from site in North Pacific near Mendocino fracture zone: **A,** single nodule, three growth episodes; **B,** double nodule, intergrown during third episode.

tical microprobe traverses within a single, small growth layer. However, compositions are usually within a given range within each layer packet of fine laminations. This suggests a particulate mode of accretion rather than chemical precipitation from solution.

Figure 6 illustrates a more complex nodule history. This nodule is from the belt of copper- and nickel-enriched nodules north of the equator. Alternate iron- and manganese-rich bands are seen in the X-radiograph as density contrasts. At least eight growth intervals are exhibited in these structures. The darker manganese-rich bands are more laminar in appearance, whereas the iron-rich zones are massive to columnar. Not all of the zones completely circumscribe the nodule. It is possible, therefore, to suggest a history for the orientation of the nodule on the seafloor. Although buried nodules are not uncommon, most nodules are exposed on the seafloor, and a history of rolling is certainly the simplest explanation for this nodule's structure. Growth occurred at the nodule/water interface, but only rarely on that part of the nodule embedded in the sediment. For growth to have occurred only on the buried nodule surface would require the nodule to have been almost completely buried, as the discontinuous bands generally circumscribe more than three quarters of the nodule. Reorientation would then have required a sequence of severe erosional events to uncover, move, and rebury the nodule.

The growth pattern suggests that episodes of growth followed events which reoriented the nodule and/or influenced the geochemical environment. The specific events cannot be identified because of the problems of dating nodule layers over the time interval in question and the lack of detailed knowledge of local seafloor events. Microprobe studies of Southwest Pacific (Cook Islands) nodules reveal localized thin layers enriched in copper and nickel to the level of the equatorial nodule deposits, although bulk analysis of these nodules does not indicate ore-grade concentrations. These observations suggest periodic similarities in depositional events in the South Pacific and the equatorial zone although the metallogenic events occur more commonly in the equatorial zone.

It is becoming increasingly apparent that growth episodes exist. Many workers have demonstrated rapid growth of ferromanganese oxides on artificial seeds (e.g., Morgenstein and Felsher, 1971). Rates may be as high as several centimeters per million years. Morgenstein (1974) has

FIG. 6–X-radiograph of nodule from 13°N, 141°W (near site of Fig. 1).

demonstrated periodic reseeding of the crust-style deposits on terraces in the Kauai Channel (Hawaiian Islands), with a resultant rapid stratigraphic development of intervals of ferromanganese crusts alternating with volcanogenic sediments.

The problem of long-term and short-term stability in the deep-sea environment is being investigated by many workers in a variety of disciplines and research programs. Understanding of these factors and their relation to the periodic and possibly rapid growth of manganese nodules may prove to be an extremely useful tool for future assessment of the nodule resource as well as for identification of smaller areas of economic potential without an extensive and random exploration program. It is hoped that the International Decade of Ocean Exploration Ferromanganese Research Program will contribute substantially to the solution of these intriguing and important scientific and economic problems.

References Cited

Andrews, J. E., 1971, Abyssal hills as evidence of transcurrent faulting on North Pacific fracture zones: Geol. Soc. America Bull., v. 82, p. 463–470.

Dudley, W. C., and S. V. Margolis, 1974, Iron and trace element concentration in marine manganese nodules by benthic agglutinated Foraminifera (abs): Geol. Soc. America Abs. with Programs, v. 6, no. 7, p. 716.

Greenslate, J., 1974, Microorganisms participate in construction of manganese nodules: Nature, v. 249, p. 181–183.

Hubred, G. L., 1970, Relationship of morphology and transition metal content of manganese nodules to an abyssal hill: Hawaii Inst. Geophysics Rept. HIG-70-18, 38 p.

Ku, T. L., and W. S. Broecker, 1967, Growth rates of deep-sea manganese nodules (abs): Am. Geophys. Union Trans., v. 48, p. 239.

——— and ——— 1969, Radiochemical studies on manganese nodules of deep sea origin: Deep-Sea Research, v. 16, p. 625–637.

Margolis, S., and G. Glasby, 1973, Microlaminations in marine manganese nodules as revealed by scanning electron microscopy: Geol. Soc. America Bull., v. 84, p. 3601–3610.

Morgenstein, M., 1974, Sedimentary diagenesis and manganese accretion on submarine platforms, Kauai Channel, Hawaii: PhD dissert., Univ. Hawaii.

——— and M. Felsher, 1971, The origin of manganese nodules—a combined theory with special reference to palagonitization: Pacific Science, v. 25, p. 301–307.

Exploitation of Manganese Nodules in South Pacific—Summary[1]

G. P. GLASBY[2]

Introduction

To date, comparatively little work has been carried out on the distribution and metal contents of manganese nodules in the South Pacific. Certain regions of the South Pacific have a potential for the commercial recovery of manganese nodules, and their exploitation would have a beneficial effect on the economies of adjacent islands.

Nodule Distribution

A recently published chart of the New Zealand Oceanographic Institute (Glasby and Lawrence, 1974a) shows the distribution of manganese nodules in the South Pacific Ocean. These data are plotted on the Soviet bathymetric chart of the region (Udintsev, 1963) in order to show the relation between nodule distribution and bathymetry. The chart shows only the occurrence of manganese nodules in individual dredge hauls and gives no indication of nodule abundance or morphology. Because of the nature of oceanographic sampling, it is expected that some bias may occur in the data. For example, the relatively high occurrence of nodules in the South Pacific Ocean may reflect the substantial work of the USNS *Eltanin* in that region, whereas the relative absence of nodules in other regions may reflect the lack of sampling in deep-water areas far from land. Because these data are compiled from various published sources, it is not easy to verify whether nodules discussed by different authors represent the same dredge haul. This may have resulted in assigning more than one symbol for a given dredge haul, and hence in an apparently higher nodule abundance at a site.

From the data presented by Glasby and Lawrence (1974a), it is apparent that manganese-nodule distribution in the South Pacific can be classified into three major categories: (1) basin-type environments such as the Southwest Pacific, Peru, Chile, and South Tasman basins and the basin north of Tahiti (predominantly red-clay areas); (2) the circumpolar belt, approximately 1,000 km wide (predominantly a siliceous-ooze area); and (3) the mountain region bounded by the Manihiki Plateau, Society Islands, Tahiti, and the Tuamotu Archipelago. In addition, limited, noncommercial quantities of nodules are found in plateaus and seamount environments.

The sedimentation rate probably is the dominant factor controlling nodule distribution throughout the South Pacific. Nodules form preferentially in regions of low sedimentation such as the submarine basins, where red clay is the dominant sediment type, and in the siliceous-ooze provinces around Antarctica. The paucity of nodules in the Tasman basin may, in fact, reflect its shallower depth and relative proximity to sources of terrigenous sediment compared with other basin environments. Nodules are also far less abundant in elevated regions where the accumulation of calcareous sediments increases the sedimentation rate and prevents the formation of conditions favorable to manganese-nodule growth. The occurrence of manganese deposits in elevated regions such as the Campbell Plateau probably reflects the fact that the phosphorite nuclei on which the nodules form are a lag deposit which has been swept clear of sediment by bottom currents. This has created local conditions favorable for nodule formation and is analogous to the situation on the Blake Plateau (Glasby and Summerhayes, 1975).

The occurrence of extensive deposits in the circumpolar region of Antarctica also reflects other factors controlling nodule distribution—bottom-current velocity, sediment substrate, and availability of potential nucleating agents on which the manganese nodules can accrete. It is probable that the siliceous-ooze substrate is conducive to nodule formation, as in the siliceous-ooze province south of Hawaii. The efficacy of siliceous ooze as a substrate for nodule formation is probably a function of the open-pore structure of the ooze, which permits facile migration of trace metals through the interstitial waters of the sediment (Horn et al, 1973). Finally, the presence of glacial erratics to serve as nodule nuclei is an important factor controlling nodule morphology at high latitudes (Glasby, 1972). Watkins and Self (1972)

[1]Revised manuscript received, October 16, 1975.

[2]New Zealand Oceanographic Institute, Department of Scientific and Industrial Research, Wellington, New Zealand.

I thank M. A. Meylan and Mrs. J. Z. Frazer, who generously made unpublished geochemical analyses available for incorporation in data charts, and T. Crayford, K. Dean, R. H. Frisby, P. H. Brown, and P. Lawrence of Cartographic Section, D.S.I.R., who prepared the data charts for publication.

noted a steady decrease in the percentage of manganese nodules in dredge hauls with increasing latitude toward the Antarctic continent. This reflects a dilution of nodule abundance by glacial erratics.

The origins of the manganese nodules in the mountain region of the central South Pacific are complex. Bezrukov (1973), for example, pointed out that nodule distribution is very heterogeneous in the Manihiki Plateau region. Sediments in this region are predominantly calcareous above the carbonate compensation depth and chiefly red clays at greater depths; sediment content is greatly influenced by volcanic processes. Availability of potential nuclei probably plays some role in controlling the distribution of nodules in this region, but other factors such as sediment type, the influence of bottom currents, and the considerable geologic age of the region may also be important. According to Horn et al (1972), nodules in this region occur on the flanks and in the deeps of intermountain basins. Associated sediments are detrital carbonate on submarine slopes and red clays in adjacent deeps. High sedimentation rates associated with sediment slumping may restrict nodule formation in the immediate vicinity of some island groups (e.g., the Samoan Islands; Lonsdale, 1975).

Of interest is the fact that the limited nodules so far recovered at depth in JOIDES deep-sea drill cores in the South Pacific commonly are found away from the main areas of nodule occurrence. Manganese nodules and pebbles recovered from a lag deposit at a depth of 91.7 m from JOIDES drillhole 274 (68°59.81′S, 173°25.64′E; 3,326 m), directly north of the Antarctic continent (P. J. Barrett, personal commun.), suggest that this deposit was formed at the sediment surface farther north from its present position and subsequently was buried and migrated south during the separation of Antarctica from the Australian continent. Since the nodule zone is dated as approximately early Pliocene to middle or late Miocene (5–12 m.y.), the nodules would have formed 130–300 km farther north (assuming a half-spreading rate of Antarctica from Australia of 2.6 cm/yr; Weissel and Hayes 1971, 1972). If this hypothesis is correct, buried nodule provinces do not necessarily lie beneath present-day nodule provinces but may be displaced from them by a considerable distance. Such a trend has been shown for the North Pacific (Horn et al, 1973a).

Metal Contents of Nodules

The contents of certain metals of economic significance, notably Mn, Fe, Cu, Ni, and Co, in South Pacific nodules are shown in a series of recently published charts (Glasby and Lawrence, 1974b–f). Data for each chart are compilations drawn from many publications; most of these data are stored in the Scripps Data Bank (Frazer and Arrhenius, 1972; J. Z. Frazer, personal commun.). Each data point represents the mean of up to 12 analyses carried out by different analysts on different samples in different laboratories using different sampling and analytical techniques. Some of the analyses were carried out almost a century ago (Murray and Reynard, 1891). The data are therefore uneven in quality. Variations in metal content within a single dredge haul can vary by more than an order of magnitude. In addition, interpretation of regional trends for individual elements is complicated by the fact that nodules from adjacent stations commonly show marked differences in composition. Despite these limitations, it is thought that the charts give a useful generalized picture of regional variations in nodule composition and, in particular, permit areas of obviously high metal contents to be selected for further investigation as sites of potential economic importance.

Manganese—Nodules with high manganese contents (>25%) generally are found in oceanic basins such as the Peru basin and the oceanic basin north of Tahiti. Lesser concentrations occur in elevated regions such as the mountain region between the Cook Islands and Tahiti and the Nazca and Chile Ridges. Lower contents of manganese are also found in the siliceous-ooze province around Antarctica, particularly in the Drake Passage–Scotia Sea region, where the lowest manganese contents are recorded (generally <10%). Excluding nodules from seamounts, banks, or similar topographic highs, there appears to be a decrease in manganese content of basin-type nodules away from the eastern margin of the Pacific toward the southwestern Pacific.

Iron—The iron content of nodules is, on the average, lower than that of manganese and shows almost the opposite pattern of distribution. Highest values occur on elevated regions and in the siliceous-ooze province around Antarctica.

Nickel—Nickel is the most abundant of the three trace metals studied. Its distribution closely follows that of manganese. Highest values (>1.5%) are found mainly in the oceanic basins, particularly in the Peru basin and the oceanic basin north of Tahiti. In elevated regions, the nickel contents are lower (generally <0.5%).

Copper—Copper occurs in much lower concentrations than nickel, but follows the same pattern of distribution. Concentrations of 0.7% are common for oceanic basins, although values as high as 1.3% are observed. In elevated regions, values of <0.2% are common.

Cobalt—Cobalt has the lowest concentration of the elements studied and generally follows the distribution of iron. High concentrations (2.23%) are found in the mountain region between the Cook Islands and Tahiti. These are, however, rare. Most values are <0.4%.

Metalliferous Sediments

Moderately to highly metalliferous sediments are concentrated almost exclusively in the vicinity of the East Pacific Rise, where manganese nodules are largely absent as a result of the high rates of carbonate sedimentation. This absence of nodules in the region of metalliferous-sediment occurrence suggests that the formation of manganese nodules and metalliferous sediments is mutually exclusive and that different processes contribute to their formation (Bonatti et al, 1972; Glasby, 1973, in press). It is possible that volcanogenic processes dominate the formation of metalliferous sediments, whereas manganese nodules form largely from seawater. This conclusion is in agreement with the observations of Blissenbach and Fellerer (1973) that the formation of these mineral deposits reflects stages of continental drift; metalliferous muds are formed during the early stages of rifting, whereas manganese nodules are generated on abyssal plains formed in the advanced stages of continental drift.

Metal contents of these sediments are at least an order of magnitude lower than those of manganese nodules of economic grade, and the sediments are not thought to be commercially exploitable.

Economic Potential of Nodules

Several criteria must be met if manganese nodules are to be exploited commercially: (1) the nodules on the seafloor must be abundant over a considerable area; (2) the metal content of the nodules must be high (nodules containing a combined Cu + Ni content of 3% have been defined as being of economic interest); (3) the nodules must be found within a reasonable distance of port facilities, in regions of smooth topography, and preferably where reasonable weather conditions prevail; and (4) they should not be diluted by other rock fragments on the seafloor.

Probably the most extensive manganese deposits in the South Pacific are found in the circumpolar siliceous-ooze province, but they are not of economic interest because of their relatively low metal contents, their dilution by glacial erratics, their great distance from land, and the unfavorable weather conditions. More promising are the basin-type manganese deposits. These occur, at least locally, in high abundance, have high metal contents in certain regions, are generally within reasonable distance of land, are found in regions of smooth topography, and are not subject to prolonged periods of adverse weather conditions.

From the trace-metal data so far available (Glasby and Lawrence, 1974–f), few nodules satisfy the basic economic criterion of having a copper plus nickel content exceeding 3%. Although nodules having nickel contents in excess of 1.5% are relatively common in basin environments, no nodules have a copper content approaching this value. In fact, only three nodules, all from the ocean basin north of Tahiti, have been found to have a copper content in excess of 0.8%. Since copper is the major element of interest from a commercial standpoint, this lack severely limits the South Pacific as a potential source of economic-grade nodules. Nodules having cobalt contents in excess of 1.5% are found in the intermountain region between the Cook Islands and Tahiti. These nodules are deficient in copper and nickel and occur in regions of steep topography; thus they are of no economic interest. In the deep ocean basins, the content of cobalt is much lower (generally <0.4%) and, therefore, only of marginal interest from a commercial standpoint.

On the basis of the presently available metal contents, only the deep oceanic basin north of Tahiti appears to have definite potential as a source of economic-grade manganese nodules. The Peru, Chile, and Southwest Pacific basins all appear to be of marginal interest. However, extremely high surface concentrations of manganese nodules, comparable to those of the red-clay–siliceous-ooze province south of Hawaii (Glasby et al, 1974), have been discovered in the Southwest Pacific basin; also, many of the deep-water areas farther from land (where Cu-Ni–rich nodules may be found) are largely unexplored. Thus, there is potential for nodule exploitation in the deep ocean basins.

References Cited

Bezrukov, P. L., 1973, On the sedimentation in the northern part of the South Pacific, *in* Oceanography of the South Pacific, 1972: New Zealand Natl. Commission for UNESCO, Wellington, p. 217–219.

Blissenbach, E., and R. Fellerer, 1973, Continental drift and the origin of certain mineral deposits: Geol. Rundschau, v. 62, p. 812–840.

Bonatti, E., T. Kraemer, and H. Rydell, 1972, Classification and genesis of submarine iron-manganese deposits, *in* D. R. Horn, ed., Ferromanganese deposits on the ocean floor: Natl. Sci. Found., Washington, D.C., p. 149–166.

Frazer, J. Z., and G. Arrhenius, 1972, World-wide distribution of ferromanganese nodules and element concentrations in selected Pacific Ocean nodules: Natl. Sci. Found., Office Internat. Decade Ocean Explor., v. 2, 51 p.

Glasby, G. P., 1972, Manganese deposits in the Southwest Pacific: Hawaii Inst. Geophysics Rept. HIG-72-23, p. 59–82.

——— 1973, The role of submarine volcanism in controlling the genesis of marine manganese nodules: Oceanography and Marine Biology, Annual Rev., v. 11, p. 27–44.

——— in press, Deep-sea manganese deposits, *in* K. H. Wolf, ed., Sedimentary and volcanic ores: New York, Elsevier.

——— and P. Lawrence, 1974a, Manganese deposits in the South Pacific Ocean—distribution: New Zealand Oceanog. Inst. Chart, Misc. Ser. 33.

——— and ——— 1974b, Manganese deposits in the South Pacific Ocean—manganese content: New Zealand Oceanog. Inst. Chart, Misc. Ser. 34.

——— and ——— 1974c, Manganese deposits in the South Pacific Ocean—iron content: New Zealand Oceanog. Inst. Chart, Misc. Ser. 35.

——— and ——— 1974d, Manganese deposits in the South Pacific Ocean—copper content: New Zealand Oceanog. Inst. Chart, Misc. Ser. 36.

——— and ——— 1974e, Manganese deposits in the South Pacific Ocean—nickel content: New Zealand Oceanog. Inst. Chart, Misc. Ser. 37.

——— and ——— 1974f, Manganese deposits in the South Pacific Ocean—cobalt content: New Zealand Oceanog. Inst. Chart, Misc. Ser. 38.

——— and C. P. Summerhayes, 1975, Sequential deposition of authigenic marine minerals around New Zealand—paleoenvironmental significance: New Zealand Jour. Geology and Geophysics, v. 18, p. 477–490.

——— et al, 1974, Extensive manganese nodules discovered in the Southwest Pacific near New Zealand: Meerestechnik Marine Technology, v. 5, p. 145–147.

Horn, D. R., B. M. Horn, and M. N. Delach, 1973, Ocean manganese nodules metal values and mining sites: Natl. Sci. Found., Office Internat. Decade Ocean Explor., v. 4, 57 p.

——— et al, 1972, World-wide distribution of manganese nodules: Ocean. Ind., v. 7, no. 1, p. 26–29.

Lonsdale, P. F., 1975, Sedimentation and tectonic modification of Samoan archipelagic apron: AAPG Bull., v. 59, p. 780–789.

Murray, J., and A. F. Reynard, 1891, Deep sea deposits: Rept. Sci. Results Explor. Voyage Challenger, 525 p.

Udintsev, G. B., ed., 1963, Pacific Ocean: Map prepared by Scientific Editorial Board of Cartographic Department of G.U.G.K., Moscow, using Urmaev pseudocylindic projection (scale 1:25,000,000).

Watkins, N. D., and R. Self, 1972, Examination of the *Eltanin* dredged rocks from high latitudes of the South Pacific Ocean, *in* R. J. Adie, ed., Antarctic geology and geophysics: Oslo, Universitetsforlaget, p. 61–70.

Weissel, J. K., and D. E. Hayes, 1971, Asymmetric seafloor spreading south of Australia: Nature, v. 231, p. 518–522.

——— and ——— 1972, Magnetic anomalies in the Southeast Indian Ocean, *in* D. E. Hayes, ed., Antarctic oceanology II: Baltimore, Australian–New Zealand Sec. Am. Geophys. Union, 364 p.

Winterer, E. L., et al, 1974, Structure and acoustic stratigraphy of the Manihiki Plateau: Deep-Sea Research, v. 21, p. 793–814.

Nickel- and Copper-Rich Nodules of Equatorial North Pacific—Summary[1]

D. R. HORN, B. M. HORN, and M. N. DELACH[2]

Two years ago, the authors compiled all available data on the distribution and metal content of ferromanganese deposits throughout the oceans. Resulting maps revealed only one geographic area where nodules were abundant and consistently rich in copper and nickel. It is an east-west band in the equatorial North Pacific defined by coordinates 8°00′N to 10°00′N and 125°00′W to 154°00′W. Nodular deposits are most abundant at the eastern and western ends of this rectangular province and sparser in the central part. Close inspection of analytical data reveals that the deposits are at least twice as rich in copper and nickel as the world average for nodules. Moreover, metal values suggest east-west axes of maximum values of both copper and nickel at the northern and southern limits of the province. There, copper and nickel constitute 2.5–3.0% (dry weight) of the nodules.

In order to explain the unusually rich province of valuable metals, the east-west trends of maximum metal values, and variations in abundance of the nodules, a comprehensive study of the nodules and their substrate is being conducted. Moreover, to quantify the data, parameters have been established, and the controlling influence of each on the distribution and metal content of the deposits has been determined. These parameters include submarine physiography, topographic grain of seabed, distribution of seamounts, distance from major fracture zones, substrate type, porosity and moisture content of substrate, and age of sediment. In addition, characteristics of the nodules themselves—such as average size and size range, percent of seafloor covered by nodules, and concentration of nodules—were combined with the preceding measures.

Our study has led us to the following conclusions:

1. The geographic limits of equatorial North Pacific nodules rich in copper and nickel reflect the distribution of exposures of Tertiary sediments on the ocean floor.

2. The east-west trend of the province of enriched nodules is a function of exposures of Miocene, Oligocene, and Eocene strata.

3. Copper and nickel contents of nodules appear to increase with increasing age of substrate.

4. The regional patterns of major nodule fields are related directly to locations of seamounts and fracture zones. Presumably, subsea weathering and explosive submarine volcanism release and disperse seed grains of basalt from vents and steep flanks and cliffs associated with volcanoes and fracture zones. Nodules therefore are most abundant in the eastern and western thirds of the enriched zone—where submarine volcanoes are common—and along a narrow band which follows the Clarion fracture zone. The central section has considerably fewer nodules because of a limited number of volcanoes and the fact that the Clipperton fracture is a site of rapid sedimentation, which eliminates this feature as a source of basaltic debris.

5. Further evaluation of this resource will be best served if future surveys are directed toward definition of Eocene, Oligocene, and Miocene exposures close to sources of basaltic seed grains which serve as surfaces for ferromanganese accretion.

6. Porosity and moisture content of radiolarian deposits are higher than for any other ocean sediment, and they may play a role in upward diffusion of nickel and copper within the substrate and their subsequent incorporation in nodules.

7. It is speculated that differential erosion and deposition within the area of Tertiary exposures may have imposed a secondary pattern of nodule concentration over crests and flanks of abyssal hills and dilution by burial in adjacent valleys.

[1]From Circum-Pacific Energy and Mineral Resources Conference program booklet.

[2]Lamont-Doherty Geological Observatory of Columbia University, Palisades, New York 10964.

Iron Ore, Coal, and Steel Production in New Zealand—Summary[1]

THOMAS MARSHALL[2]

Iron Ore

Titaniferous iron sands—derived from weathering, coastal transport, and concentration of volcanic ash, lava, and andesite—constitute the largest known iron-ore resource in New Zealand. For 500 km along the west coast of the North Island, the presence of iron sands is indicated by black beaches, coastal iron-sand dunes, and bedded deposits extending several kilometers inland. Offshore, iron sands extend out to a water depth of about 27 m (15 fm). Deposits of commercial size and grade occur sporadically along this 500-km coastal zone. New Zealand iron sands are of Pleistocene-Holocene age; transport and concentration on beaches are evident at present. Most of the harbors along this coastline are bar-bound by drifting iron sands.

Commercial ore bodies take the form of layered deposits of free-flowing or poorly cemented sand, up to about 40 m thick and a few kilometers in lateral extent, which are easily mined by use of dredges or scrapers. Titanomagnetite concentrates are readily extracted by magnetic and/or gravity methods. The titanomagnetite content varies widely; the average content in deposits of commercial interest is in the range of 10–47%. Offshore, bottom sampling has indicated grades of 3–30% in the areas shown in Figure 1.

The deposit at Waikato North Head (*A*, Fig. 1) has been thoroughly drilled to delineate reserves of about 100 million MT of concentrate (further substantial reserves occur directly to the north) in sands with an average grade of 18% titanomagnetite. This deposit is being scraper-mined to supply the nearby plant of New Zealand Steel Ltd. A second major deposit at Taharoa (*B*, Fig. 1) has also been drilled to establish reserves of about 300 million MT in dune sands with an average grade of 35% titanomagnetite. This deposit is being dredged for export of concentrates to a limited and specialized market in Japan (Goodfellow, 1974); offshore shiploading by slurry-pipeline techniques is being used. In the Wanganui district (*C*, Fig. 1), Luke (1955) reported approximate titanomagnetite quantities exceeding 200 million MT, but much of this is under high-quality farmland. One deposit at Waverley is being dredged to produce concentrates for export to Japan via offshore loading; the concentrate reserves exceed 80 million MT.

Numerous smaller deposits in the 500-km coastal zone have been explored and drilled to various extents. The total onshore reserves (above about 7% cut-off grade) are presently assessed at about 757 million MT of concentrates (Buist, 1974). The inclusion of offshore reserves would increase the total resources substantially.

A 9–million MT limonitic iron-ore deposit (Luke, 1955) in the South Island was used for a few years, up to 1935, to produce pig iron in a small blast furnace.

Recently a gabbroic magnetite-ilmenite deposit of potentially substantial size was discovered in the southern end of the South Island (Alexander, 1972). Milling and magnetic concentration of surface samples produced high-quality vanadium-bearing magnetite concentrates. However, exploration of this deposit, which is in extremely rugged country, is not yet beyond the prospecting stage; grades and quantities of the deposit are not reliably known.

Coal

Locations of the main New Zealand coalfields are indicated in Figure 1. The Waikato coalfield (*1*, Fig. 1), at present the most important in the country, may possibly extend as a Tertiary coal basin for at least 40 × 20 km beneath undulating farmlands and swamps. This field produces low-sulfur subbituminous coal noted for its high reactivity and for the high reactivity of char produced from it. Seams are undulating with dips up to 20°; some coal occurs in multiple seams. The seams may be faulted, and their depths in the center of the "basin" are not known but probably exceed 500 m.

Present production is about 1.5 million MT/yr from opencast and bord-and-pillar underground

[1]Manuscript received, November 6, 1974.

[2]Deputy Director, Chemistry Division, Department of Scientific and Industrial Research, Lower Hutt, New Zealand. Formerly metallurgist, New Zealand Steel Investigating Company Ltd.

The writer is greatly indebted to D. R. Buist, N. T. Evans, and J. R. Tulloch of New Zealand Steel Ltd. for assistance in the preparation of this paper.

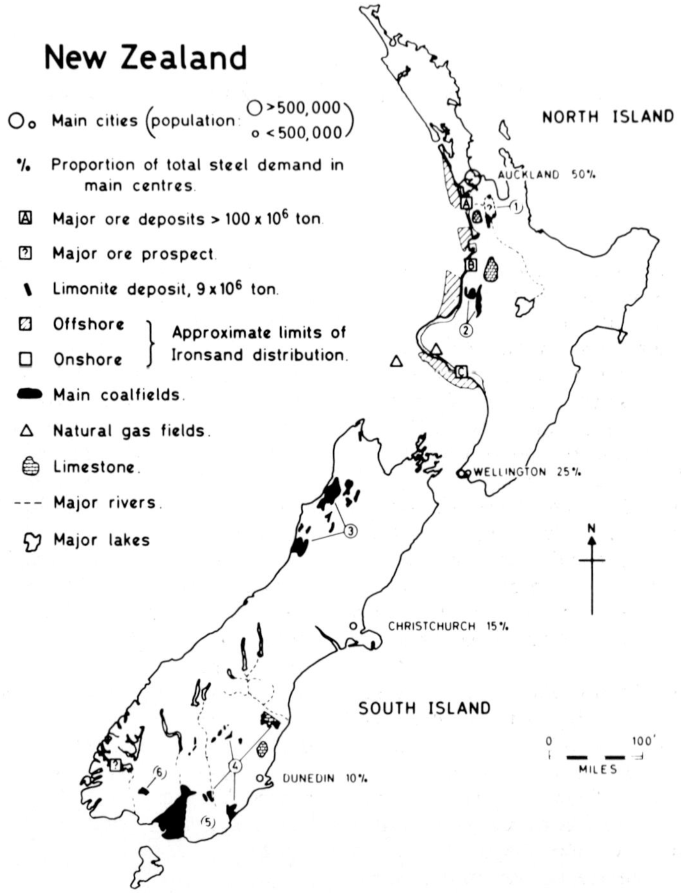

FIG. 1—Locations of steel-making raw materials in New Zealand.

mines, where experimental application of continuous miners is in progress. Total (measured + indicated + inferred) recoverable reserves are estimated to be 313 million MT. Only the northern and southern ends of the field have been explored by drilling and mining, but exploration toward the center is in progress.

The Taranaki coalfield (*2*, Fig. 1), in very rugged, largely inaccessible country, contains flat-lying seams of subbituminous coal. Present estimates of total recoverable reserves are 89 million MT; the field has not been delineated in depth and extent by drilling, as much of it is virtually unexplored.

In the South Island, the Cretaceous-Tertiary Westland coalfields (*3*, Fig. 1) are located in rugged, mountainous country, mostly at high altitudes. Coal seams commonly are contorted and steeply dipping (some are vertical), and Westland coals are highly variable in rank and quality from subbituminous to high-rank bituminous. Sulfur and ash contents range from very low to very high. Production is about 469,000 MT/yr, mainly from bord-and-pillar underground operations but with some opencast and hydraulic mining. Present estimates of total recoverable reserves are 157 million MT. Further exploration for coking coal is in progress.

Otago lignite and subbituminous coal deposits (*4*, Fig. 1) occur in Tertiary depressions within an extensive schist terrane, and they crop out near the boundaries of the depressions. Depths are usually shallow (<150 m), though some seams dip beneath thick overburden to depths greater than 1,000 m. Deposits near the main city have been delineated by drilling, but more remote deposits are virtually unexplored; hence total reserves are not precisely known. They have been estimated at 148–250 million MT, mostly in the inferred category.

The Southland coalfields also occur in Tertiary depressions. Ohai field (*6*, Fig. 1) is of limited extent (about 10 × 16 km), whereas Mataura Valley (*5*, Fig. 1) extends, possibly discontinuously, within an area about 60 × 30 km. Coal seams are usually shallow, but in some localities they dip to depths of at least 1,000 m. Ohai coal is subbituminous with low ash and sulfur contents; Mataura Valley coals are lignites with mainly low sulfur and ash contents. Present production, which supplies nearby towns and factories, is about 362,000 MT/yr, mainly from underground operations in the Ohai field plus localized opencast mining in the Mataura Valley field. The Ohai field has been explored by drilling, establishing total recoverable reserves of 106 million MT. Although more than 200 scattered drill intersections, outcrops, and pit exposures have indicated the presence of lignite throughout the Mataura Valley, it has not been drilled systematically. Estimates of total recoverable reserves range from 257 to 1,239 million MT (mostly in the inferred category); however, because this could be the largest and most accessible coal deposit in New Zealand, further exploratory drilling is in progress.

Other Energy and Material Resources

Deposits of limestone are widely distributed in New Zealand. Some of these, and some major rivers which provide process water supply, are shown in Figure 1.

A small onshore gas-condensate field (see Fig. 1) with pipeline-quality gas reserves estimated at 11.4×10^9 m^3 is in production, and a larger offshore gas-condensate field with estimated recoverable reserves of 151×10^9 m^3 has been discovered recently (Scott, 1974). The present and planned development of these fields is for electric power generation and city gas supplies; further oil and gas exploration is in progress.

Electric power is supplied from a national grid with installed capacity of 4,258 Mw in 1972 and further capacity under construction.

Steel Production

The major steel-making raw materials of New Zealand (titaniferous ore, noncoking coal) are unconventional in character and generally unsuited to conventional blast-furnace processing. The New Zealand market for steel products is relatively small and highly diversified; thus it is unsuited to large-scale steel production. These factors have defeated numerous attempts since 1848 (Martin, 1955) to develop steel production from iron sands. However, in the 1960s, a government-financed company was set up to investigate steel production from iron sands.

This company established that adequate coal and iron-sand resources were available to meet raw-material requirements for at least 50 years, and that projected markets were sufficient to support small-scale steel production. This stage of the investigations revealed a highly favorable logistic situation for local steel production: an iron-sand deposit and a major coalfield located close to the main market center of the country, potentially low costs for concentrate production, good road and rail communications, and good water and limestone supplies, all within a radius of 48 km.

In the search for a steel-production method capable of utilizing the unconventional ore and coal, many techniques were investigated, and the SL/RN kiln–arc furnace system (Fraser and Grigg, 1965) was selected as being most suitable.

A plant was constructed, and operations were commenced in 1969 under the name "New Zealand Steel Ltd." The plant design was based on separation of primary sand-size titanomagnetite concentrates, followed by grinding and reconcentration to produce secondary concentrates containing about 60% Fe, pelletizing the secondary concentrates with bentonite, and feeding the green pellets to the SL/RN kiln for metallization.

Because several major operational difficulties were encountered, the company decided to simplify the steel-making system by changing the kiln operation to direct feeding of raw coal plus primary iron-sand concentrates.

The company is operating a remarkably simple system for steel production: magnetic concentration of iron sands; direct feeding of concentrates (5% Fe, 8% TiO_2), raw subbituminous coal, and recycled char to the rotary reduction kiln; roof feeding of metallized iron sand (about 72% metallic iron, 20% gangue) to arc furnaces charged with some scrap; followed by continuous casting of steel billets for rod and wire production. This system started operation in 1975, and, although optimum operating conditions have not yet been established, it is working very well. Good market conditions and improved production in all plants led to a substantial profit the year that this system has been in operation.

Future plans include production of plate and sheet steel when the market becomes large enough to support a flat-products mill.

This very simple production system has only one inherent disadvantage—a relatively high electric-power consumption of about 1,200 kwhr/MT of steel produced from the metallized iron-sand component of the arc-furnace charge. Research is in progress in government laboratories to develop single-stage production of essentially gangue-free metallics from iron sand plus noncoking coal. The main objectives are minimizing electricity consumption, costs, and environmental pollution in the production of steel from iron sands.

References Cited

Alexander, D., 1972, Possible implications of a magnetite-ilmenite-vanadium bearing gabbroic differentiate: New Zealand Eng., v. 27, no. 1, p. 40–47.

Buist, D. R., 1974, The exploration and continuing operations of New Zealand Steel Mining Ltd. at Taharoa, New Zealand: New Zealand Inst. Engineers Trans., v. 1, no. 2, p. 155–164.

Fraser, M. J., and C. R. Grigg, 1965, The SL/RN process—its place in the iron and steel industry: Eng. Jour. Canada, v. 48, no. 6, p. 29–34.

Goodfellow, T. B., 1974, The Taharoa project—a general description: New Zealand Inst. Engineers Trans., v. 1, no. 2, p. 137–154.

Luke, J., 1955, Iron ore resources of New Zealand: New Zealand Dept. Sci. and Indus. Research, Inf. Ser., Bull. 6, p. 28.

Martin, W. R. B., 1955, The iron and titanium ores of New Zealand: New Zealand Eng., v. 10, no. 10, p. 317–336.

Scott, W. E., 1974, Maui gas will fire N.Z. power stations: Energy Internat., v. 11, no. 4, p. 34–35.

Submarine Phosphorite Deposits of Chatham Rise Near New Zealand—Summary[1]

A. WESLEY KARNS[2]

INTRODUCTION

Phosphate nodules occur several places on the Chatham Rise, a broad submarine platform about 130 km wide and 960 km long, which extends eastward from Banks Peninsula on the east coast of the South Island of New Zealand to slightly beyond the Chatham Islands. The platform of Chatham Rise is outlined by the 500-m isobath (Fig. 1). Average elevation of the platform is approximately 400 m below sea level. The rise is terminated abruptly on the north where the seafloor descends steeply into the Hikurangi Trench; the southern and eastern slopes are more gentle. Pukaki Gap, a narrow saddle with a maximum known depth of 575 m, separates the rise from the shelf at Banks Peninsula; the Pegasus submarine canyon cuts the island shelf and trends northward into the south end of the Hikurangi Trench.

Several shallow banks are present on the Chatham Rise. In the western section, Mernoo, Reserve, and Veryan Banks shoal at 51, 220, and 154 m, respectively. Fleming and Reed (1951) reported channels and ridges on Mernoo Bank which they suggested indicate a period of subaerial erosion. Steep slopes, anomalous geographic location, and volcanic-rock recoveries suggest that Veryan Bank is of volcanic origin, possibly a guyot. Matheson Bank, in the central region of the rise, extends upward in stairstep fashion to a depth of 219 m.

On a smaller scale, the bottom topography is characterized by rather smooth, gently sloping profiles and a few jagged profiles of up to 10 m of relief.

PHOSPHORITE OCCURRENCE

Sampling of Chatham Rise sediments in 1950 first revealed phosphorite on the central region of the rise (Reed and Hornibrook, 1952). In describing the phosphorite material of one sample, Reed observed that the nodules consist of highly concentrated *Globigerina* ooze and contain glauconite and some quartz and schist fragments. The degree of phosphatization varied in the different nodules examined; in some it was complete but, in the majority, large unaltered cores remained. The main phosphate mineral was collophane.

Norris (1964) stated that all the nodules he examined consisted of phosphatized Miocene foraminiferal ooze. Two types of nodules were observed: (1) an abundant, irregular, slabby type with a dull black crust showing a greenish cast enclosing a buff to pale brown calcareous material; and (2) a less abundant type that has a mottled brown surface and is mollusk bored. The blackish crust was found to be mainly glauconite. In considering economic aspects of Chatham Rise, Norris drew attention to the very widespread occurrence of phosphorite of reasonable grade, the very considerable amount of associated calcium carbonate, and the high reactivity of the glauconite in terms of potash release. Summerhayes (1967, p. 275) speculated that 90 million tons of nodules could be present in one relatively small area and considered the economics of recovering materials from the rise for fertilizer use.

The first economic investigation of the phosphate deposits was undertaken in 1967–70 by the Oceanic Division of Global Marine Inc. About 300 bottom samples were dredged from 289 m minimum to 468 m maximum depth. Both frequency of phosphorite occurrence and reserves are concentrated at around 390–400 m depth. Samples were assessed for phosphate content and size distribution, and were analyzed for K, F, U, V, Se, and other trace elements.

ORIGIN OF PHOSPHATE

Several hypotheses have been advanced to account for the origin of the Chatham Rise phosphorite nodules. Reed and Hornibrook (1952) suggested that the Chatham Rise deposits were derived from Miocene sediments (mainly *Globigerina* ooze). During deposition or immediately following the post-Miocene uplift responsible for the formation of the rise, physicochemical and biological conditions were altered, bringing about saturation or supersaturation with respect to phosphate. This provided a favorable phosphati-

[1]Manuscript received, October 21, 1974.

[2]Managing Director of Earth Scientists Pty. Ltd., Aukland, New Zealand.

The writer is greatly indebted to the Directors of J.B.L. Minerals, who gave their permission for the use of confidential reports.

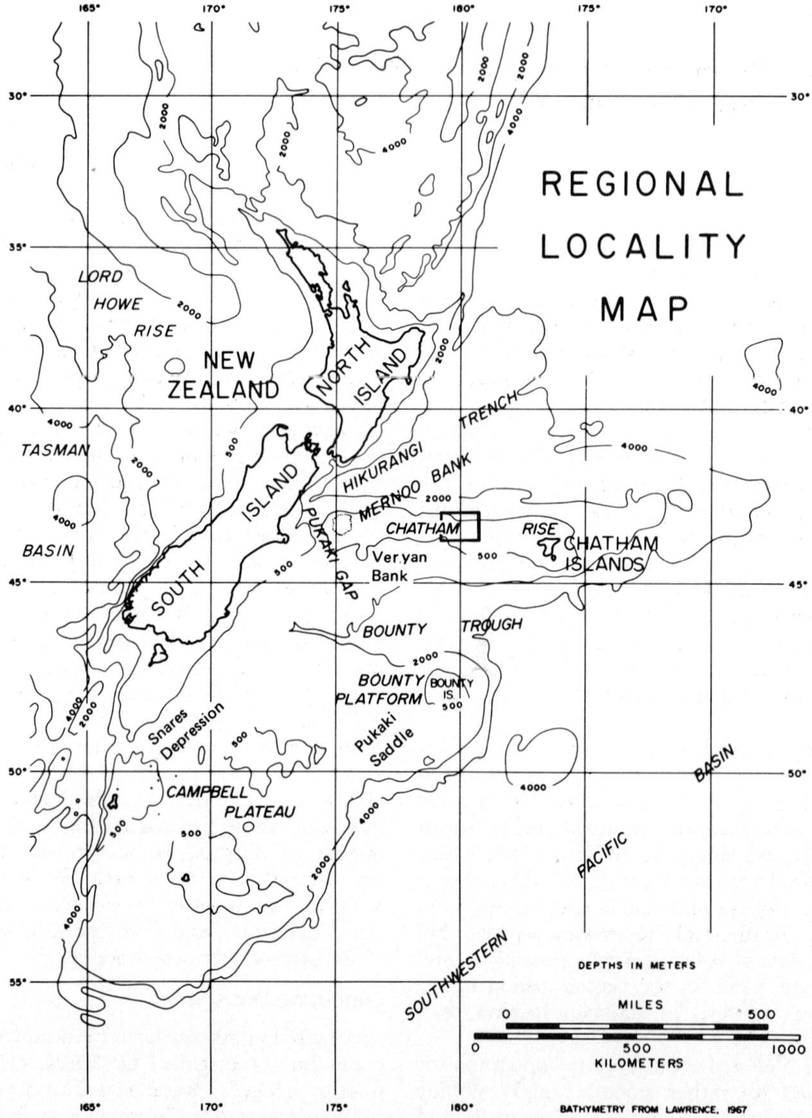

FIG. 1–Regional locality map, reproduced from Austin et al (1973). Bathymetry after Lawrence (1967).

zation environment. Because foraminifers contained in the nodules are Miocene, Reed suggested that no phosphate has been deposited since the Miocene. Norris (1964), however, stated that some phosphate deposition is taking place at present, and that conditions favorable for deposition may have existed since the Miocene. Watters (1968), from work in the Chatham Islands, suggested that conditions favorable for phosphorite formation may have lasted for a long period and would be expected particularly in Tertiary time.

The phosphorite is found in an oxidizing environment affected by tidal and other currents typical of the open sea, particularly those upwelling from a great depth. However, the various forms of glauconite present indicate a microenvironment that was neither as oxidizing as the open sea nor as reducing as stagnant basins.

Mero (1965, p. 61) speculated that the predominant occurrences of phosphorite nodules along coasts and on the crests of submarine ridges reflect Eh changes in the waters at these points. It is postulated that rapid temperature changes due to currents rising into warmer waters could kill large numbers of pelagic or deep-water organisms and thus form considerable layers of decomposing, phosphate-rich organic matter on the nearby ocean bottoms. In areas where there are large amounts of decaying organic matter, an environment is created that allows the phosphate to dissolve. As this dissolved phosphate migrates from this area to the prevailing oxidizing atmosphere of the ocean and, in particular, onto ridges like Chatham Rise, it precipitates.

Another clue to the origin of the Chatham Rise phosphorites is suggested by Smirnov (1957), who showed that the solubility of phosphate has a marked dependence on the carbon dioxide content of the seawater. Thus, when the bottom layers of water are forced up over Chatham Rise, the resultant lowering of pressure and rising of temperature with the concomitant loss of carbon dioxide would cause the phosphates to precipitate in colloidal form or to replace the calcium carbonate in the oozes. Usually, colloidal particles of this nature possess an electric charge which causes the particles to be preferentially drawn to hard reactive surfaces. Nodules already forming could provide an ideal surface for the further accretion of colloidal precipitates. The abundance of replacement evidence suggests this to be the main mode of phosphate concentration on Chatham Rise.

Economic Aspects

The Chatham Rise phosphorite is economically interesting largely because the closest deposits are only 320 km from New Zealand, which is the largest per-capita consumer of phosphate in the world and has no significant indigenous deposits. The recent price rise of phosphate rock in New Zealand to as high as NZ$40.75 (US$62.76) has focused attention on exploitation of the Chatham Rise deposits.

Tests showed that low-grade Chatham Rise nodules ranging from 17.4 to 19.2% P_2O_5 could be beneficiated by flotation or calcination to 28% P_2O_5, or to 32.5% by acetic acid extraction and calcination at 875°C (Gallaher, 1973).

It had been noted in earlier tests that Chatham Rise phosphate ranks very high—10—on the Hoffman and Breen scale of phosphate-ore activity. Nauru, Christmas Island, and Florida pebble phosphates rank only 7 on this scale. This high reactivity suggests that the Chatham Rise phosphate may also have value as a direct-application fertilizer. Experiments showed the Chatham Rise phosphorite to be 166% more effective as a direct-application fertilizer than Christmas Island phosphate, 171% more effective than Nauru phosphate, and only 26% less effective than superphosphate. This type of fertilizer would have application largely where a useful long-term effect is desired, as in forestry. It also could have use in grassland farming once sulfur is added. The price, however, must be kept at about 45% of the cost of superphosphate to offset the larger quantity needed and the higher distribution costs.

Recent calculations by Earth Scientists (1974), based on data obtained by a suction dredge with submerged centrifugal pumps, indicate landed costs per metric ton as NZ$6.35 (US$9.79) for 326,000 MT annually. This estimate is based on a plant capital cost of NZ$5,000,000 (US$7,700,000) and includes 10% depreciation, finance charges, and estimates of all other overhead. More detailed mining-engineering feasibility studies are needed, however, before definitive estimates of mining costs can be made.

Phosphate Nodule Reserves

Published estimates of minimum concentrations (surface density) of nodules necessary for ocean mining vary somewhat less than estimates for mining costs; they range from 25.2 kg/m^2 (J. L. Mero, personal commun., 1971) to 10.1 kg/m^2 (Archer, 1974). The average specific gravity of the phosphate nodules is 2.65, so the weight of nodules completely covering 1 m^2 to a depth of 15 cm is 411 kg.

Evaluation of dredge samples reported by Global Marine revealed that the majority of P_2O_5 occurs in the +7-mesh fraction (>0.32 cm diameter). It is also noteworthy that the $CaCO_3$ content is at a minimum for this size fraction.

Global Marine (1970) delimited five areas having nodules making up 34–94% of the +7 material (i.e., 133–389 kg/m^2). Potential reserves averaging 21.3% P_2O_5 were estimated to be 59 $\times$ 10^6 MT) over 341 km^2. From all available dredging data, Earth Scientists (1973) constructed phosphorite isocon maps from which it was similarly calculated that the same area contained 70 $\times$ 10^6 MT of phosphorite nodules at the same tenor.

If potential recoverable reserves are extended to include areas of concentration of 35.6 kg/m^2, about 8% concentration, then the potential reserves are approximately 200 $\times$ 10^6 MT over an area of about 2,590 km^2.

In the foregoing calculations, it was assumed that the nodule deposit is superficial. Should further work reveal a layered deposit, then the reserves could be doubled for every 15 cm of additional depth.

Conclusions

The Chatham Rise phosphorite deposits seem to meet all the criteria for economic exploitation put forth by proponents of marine mining.

1. The phosphorite deposits are of sufficient size and grade for an economic operation.
2. They are close to a large market which is open for penetration.
3. Cost of delivered, beneficiated phosphorite seems favorable compared to the landed value of the product.
4. No adverse governmental regulations or restrictions are anticipated.
5. No significant environmental problems are expected.
6. There is the possibility of a high return on capital invested.

Although detailed engineering studies are necessary to determine exact figures, it would seem that mining and landing costs to New Zealand for the Chatham Rise phosphorite are unlikely to exceed NZ$7.20/MT (US$11.09). With the estimated value per metric ton of the nodules ranging from NZ$14.76 (US$22.73) to NZ$23.06 (US$35.51), depending on use, and a large market assured, the Chatham Rise deposits appear to represent a very attractive mining venture.

References Cited

Archer, A. A., 1974, Progress of marine mining: Mining Mag., v. 130, no. 3, p. 150–163.

Earth Scientists Pty. Ltd., 1973, The economic potential of the Chatham Rise phosphorite: Unpub. Rept. for JBL Consolidated (NZ) Ltd., 86 p.

——— 1974, Notes on current estimated value of Chatham Rise phosphorite: Unpub. Rept. for JBL Consolidated (NZ) Ltd., 5 p.

Fleming, C. A., and J. J. Reed, 1951, Mernoo Bank, east of Canterbury, New Zealand: New Zealand Jour. Sci. Technology, sec. B, v. 32, no. 6, p. 17–30.

Gallaher, P. J., 1973, Economic evaluation of Chatham Rise phosphorite: Unpub. Rept. by N. Z. Fertiliser Manufacturers' Research Assoc. Inc. for JBL Consolidated (NZ) Ltd., 10 p.

Global Marine, 1970, Compendium of the geology and phosphorite deposits of Chatham Rise, east of New Zealand: Unpub. Rept. to the N. Z. Mines Dept., 40 p.

Mero, J. L., 1965, Mineral resources of the sea, 1st ed.: New York, Elsevier Publishing Co., 312 p.

Norris, R., 1964, Sediments of Chatham Rise: New Zealand Dept. Sci. and Indus. Research Bull. 159, 39 p.

Reed, J. J., and N. de B. Hornibrook, 1952, Sediments from the Chatham Rise: New Zealand Jour. Sci. Technology, sec. B, v. 34, p. 173–188.

Smirnov, A. I., 1957, The problem of the genesis of phosphorite: Akad. Nauk SSSR Doklady, 119, p. 53–56.

Summerhayes, C. P., 1967, Marine environments of economic mineral deposition around New Zealand—a review: New Zealand Jour. Marine and Freshwater Research, v. 1, no. 3, p. 267–282.

Watters, W. A., 1968, Phosphorite and apatite occurrences and possible reserves in New Zealand and outlying islands: New Zealand Geol. Survey Rept. 33, 32 p.

Nickel Resources of Australia—Summary[1]

R. WOODALL[2]

INTRODUCTION

Rocks of every geologic age are present in Australia, including large areas of Precambrian rock. The country contains a wealth of mineral deposits and has had an active mining industry for 120 years.

The small population is concentrated in coastal cities, and the interior, much of which is semiarid to arid, is sparsely settled. Much of the interior is flat, deeply weathered, and soil covered. It is only partly explored in terms of near-surface mineral deposits. Extensive surface concealment, leached outcrops, and electrically conductive overburden combine to make exploration in many areas difficult and costly. The fact that Australia's nickel potential remained unknown until 1966 can be attributed in part to these factors. Rising world demand for metals over the past 20 years has stimulated mineral exploration in Australia, and the nickel discoveries have resulted from this new era of exploration.

HISTORY OF NICKEL EXPLORATION IN AUSTRALIA

Gossans representing the surface outcrops of nickel sulfide deposits were discovered by gold prospectors as early as 1898, but surface diggings did not extend into the underlying sulfides.

The nickel silicate garnierite was reported in the northwest of South Australia in 1953. The area subsequently was explored by International Nickel Company of Canada, and a deposit of laterite nickel was discovered at Wingellina in Western Australia (Fig. 1).

Although these activities received considerable publicity, they did little to stimulate exploration for nickel in the more accessible, mineralized Archean areas of Western Australia because, at that time, Archean rocks were not considered to have a high potential for nickel sulfide occurrence.

In 1964, following indications from prospectors, geologists employed by Western Mining Corporation Limited recognized nickel-bearing gossans at Kambalda in Western Australia (Figs. 2–4), and in January 1966, drilling intersected high-grade sulfide nickel ore beneath these outcrops. The developing world shortage of nickel led to one of the most vigorous periods of mineral exploration in the history of Australia. As a result

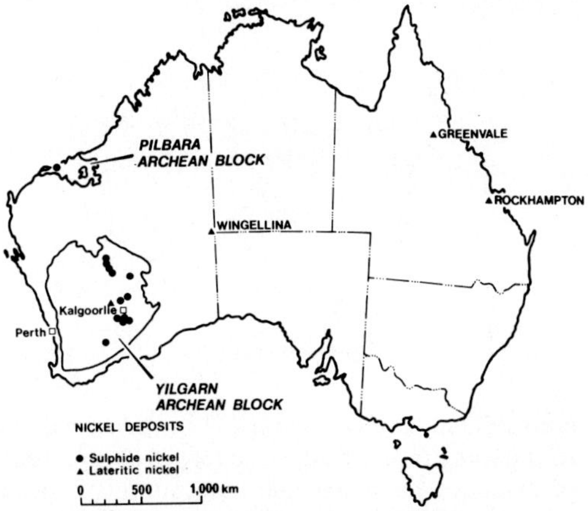

FIG. 1—Nickel deposits in Australia.

of numerous nickel sulfide discoveries made during the period 1966–71, Western Australia has emerged as a potential major world source of nickel. Reserves as of June 1973 are given in Tables 1 and 2.

Nickel-enriched laterite was discovered at Greenvale (Fig. 1) in Queensland in 1957. Several occurrences of laterite nickel were also discovered in Western Australia during exploration for sulfide nickel; the most important deposits are at Ora Banda (Fig. 5), 65 km north of Kalgoorlie. The locations of other important deposits of laterite nickel are listed in Table 3 and shown in Figure 1.

NICKEL SULFIDE DEPOSITS

Australian nickel sulfide deposits occur in or adjacent to magnesium-rich ultramafic rocks in Archean volcanic belts (Figs. 1–4). These belts are linear and generally are elongated in a north-northwest direction, paralleling major strike

[1]Manuscript received, October 21, 1974.

[2]Western Mining Corporation Ltd., Kalgoorlie, Western Australia.

The writer was assisted by many members of the staff of Western Mining Corporation Ltd., and geologists from various companies also made available unpublished articles on the individual deposits. This invaluable help is gratefully acknowledged. The illustrations were prepared by N. W. Keddy and the drafting staff of the Exploration Division of Western Mining Corporation Ltd. Their contribution is also sincerely appreciated.

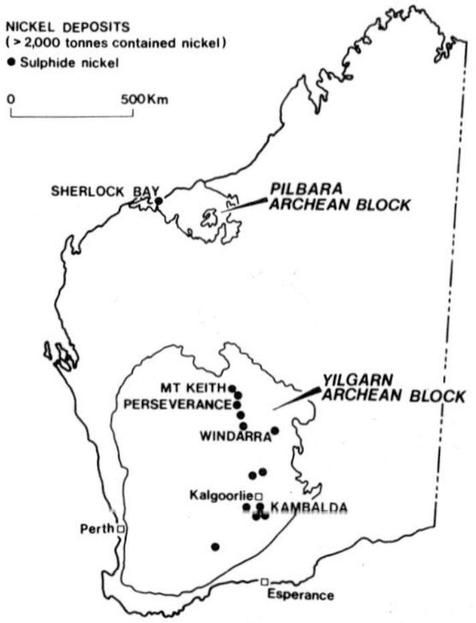

FIG. 2–Nickel sulfide deposits in Western Australia.

faults. Gemuts and Theron (1975) described the stratigraphy in the Kalgoorlie region and presented evidence for a regional association of nickel mineralization with a stratigraphic marker bed.

The ultramafic rocks, which are usually the host rocks for the mineralization, are serpentinite or talc-carbonate derivatives of olivine-rich extrusives and intrusives. The nickel mineralization is typically found in the lower, locally thickened

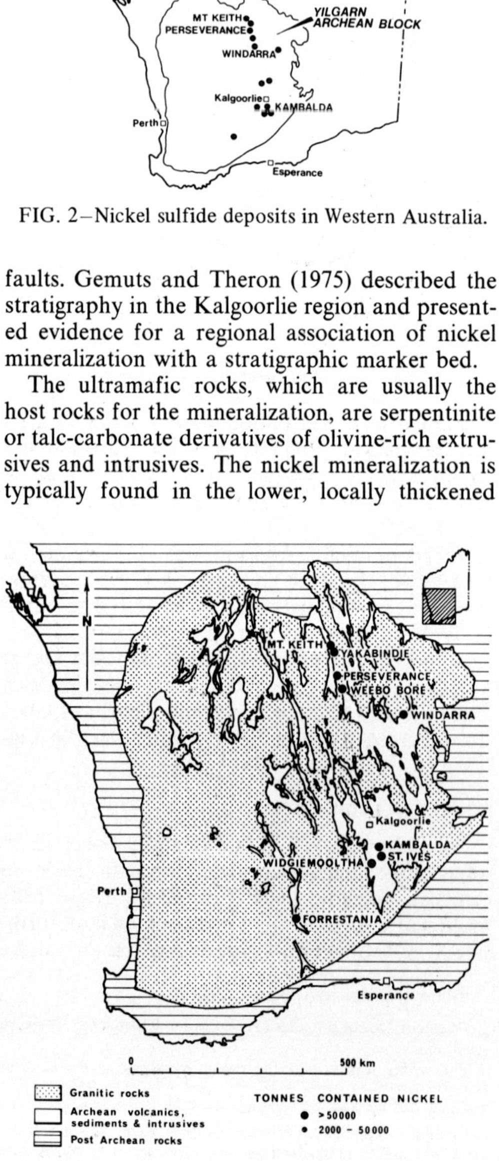

FIG. 3–Nickel sulfide deposits in Yilgarn block, Western Australia.

part of the ultramafic host rocks, and is commonly most marked at the basal contact in a flexure or embayment.

O'Driscoll (1971a, b) has developed a tectonic pattern for the nickel sulfide province related to northeast- and northwest-trending intracontinental lineaments. The thickened ultramafic sequences and embayments associated with ore occurrence may relate to this tectonic model, which is a concept consistent with the affinity of ore concentrations for those parts of the environment that have been tectonically prepared.

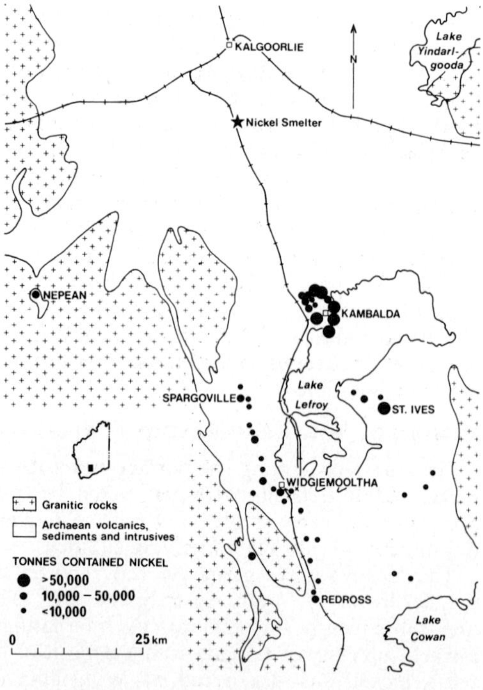

FIG. 4–Nickel sulfide deposits in Kambalda-Widgiemooltha area, Western Australia.

The dominant primary sulfides in the ores are pentlandite and pyrrhotite; the minor sulfides are pyrite and chalcopyrite.

Most deposits are deeply weathered and, as a result, exhibit leached outcrops with a well-developed *oxide zone* in which all sulfides have been altered to limonite. The oxide zone may extend as deep as 50 m below surface. It passes into a secondary sulfide *pyrite-violarite zone*, which in turn gives way to primary sulfides at depths of 150–200 m (Woodall and Travis, 1969; Nickel et al, 1974).

Table 1. High-Grade Nickel Sulfide Deposits in the Archean of Western Australia[1]

Location	Million MT	Grade % Ni	Contained Nickel (MT)	Remarks
Kambalda and St. Ives	24.1	3.2	778,000	
Agnew	40.0	2.2	880,000	
Windarra - Mt. Windarra	5.4	2.2	117,700	
- So. Windarra	3.5	2.0	70,700	
Widgiemooltha Area	7.0	2.5	175,000	Unofficial Est.
Others (including Forrestania Area, Scotia, Nepean, Carr Boyd)	7.4	2.5	184,300	Unofficial Est.
Total	87.4	2.5	2,205,700	

[1] Cutoff 1% Ni; reserves as of June 1973.

Kambalda deposits—Nickel sulfide mineralization occurs at the base and within a sequence of magnesium-rich ultramafic extrusives which are overlain and underlain by metabasalts (Ross and Hopkins, 1975). The ultramafic rocks associated with the nickel sulfides are altered olivine-peridotite flows and possibly shallow intrusives which are now represented by serpentinites and talc-carbonate and serpentine-talc-carbonate rocks. Twelve separate deposits have been discovered around a domal flexure, principally in structural depressions or flexures in the footwall contact of the ultramafic complex.

Typical basal-contact ore consists of a layer of massive sulfides overlain by a well-defined layer of disseminated sulfides. Sulfides within the ultramafic succession are mostly disseminated. The average nickel:copper ratio is 13:1 and the nickel: cobalt ratio is 54:1.

Kalgoorlie region—Most nickel sulfide deposits in the region (Figs. 4, 5) are similar to the Kambalda deposits, although the relative amounts of massive and disseminated mineralization varies, and in some deposits the sulfides have been redistributed by deformation. The Carr Boyd deposit is an exception. There, sulfides are associated with brecciated, pipelike, bronzite pegmatoid intrusions in a mafic to ultramafic complex (Purvis et al, 1972). The nickel:copper ratio is 3:1.

Windarra deposits—Two deposits, Mt. Windarra and South Windarra, occur 14 km apart (Fig. 3) in approximately the same stratigraphic position, near the base of an ultramafic body which overlies an iron formation (Roberts, 1975). The

Table 2. Low-Grade Nickel Sulfide Deposits in the Archean of Western Australia[1]

Location	Million MT	Grade % Ni	Contained Nickel (MT)	Remarks
Mt. Keith	290	0.6	1,740,000	Burt & Sheppy, 1975
Yakabindie	60	0.6	360,000	Unofficial Estimate
Sherlock Bay	75	0.5	375,000	Miller & Smith, 1975
Others (including Kambalda, Agnew, Mt. Windarra)	110	0.7	770,000	Unofficial Estimate
Total	535	0.6	3,245,000	

[1] Cutoff 0.5% Ni; reserves as of June 1973.

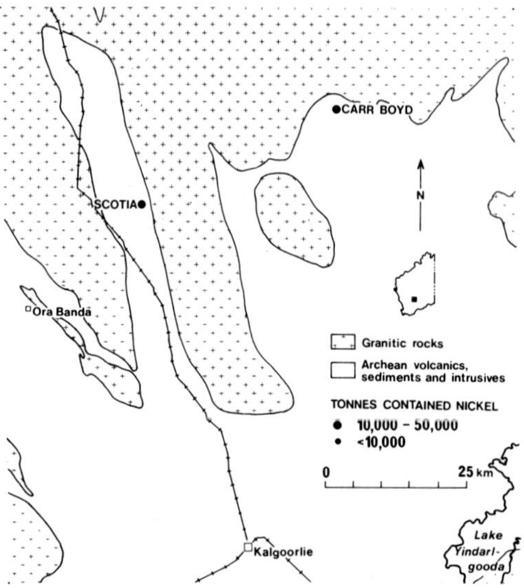

FIG. 5—Nickel sulfide deposits in Scotia - Carr Boyd area, Western Australia.

ultramafic body is serpentine-talc-carbonate rock which probably represents metadunite or metaolivine peridotite. The ore is mainly disseminated in type, although some massive sulfides underlie the disseminated zone and some breccia ore is associated with the footwall iron formation. The South Windarra ore contains mainly secondary sulfides (pyrite-violarite) with some millerite. The nickel: copper ratio is 11:1 at Mt. Windarra and 22:1 at South Windarra.

Perseverance deposit—Nickel sulfide mineralization occurs in an embayment along one contact of a steeply dipping dunite-serpentinite body intruded into metasedimentary rocks (Martin and Allchurch, 1975; Figs. 2, 3). The mineralization consists of disseminated sulfides grading into massive sulfides at the contact. The nickel:copper ratio is approximately 25:1.

Mt. Keith deposit—Disseminated sulfides occur in the central section of a steeply dipping, serpentinized, dunitic intrusive (Burt and Sheppy, 1975; Figs. 2, 3). The sulfides are interstitial to serpentinized olivine, and commonly are intergrown with magnetite. The mineralized zones have a uniform composition and sharp boundaries. The nonsulfide nickel content of mineralization containing 0.4–0.8% nickel is 0.11% Ni. The nickel: copper ratio is 60:1 and the nickel:cobalt ratio is 34:1.

Sherlock Bay deposit—The host rock for the Sherlock Bay deposit (Fig. 2) is a banded quartz-magnetite-amphibole-sulfide schist within a volcanic sequence. The deposit is unusual because the host is not an ultramafic rock, although ultramafic rocks are present nearby. The mineralized zone, which Miller and Smith (1975) consider to be a mineralized fracture zone, has a nickel:copper ratio of 5:1.

Laterite Nickel Deposits

Nickel may concentrate during weathering of nickel-rich ultramafic rocks and give rise to laterite nickel deposits. Four occurrences of possible economic significance have been found in Australia—at Greenvale and Rockhampton in Queensland, and at Wingellina and Ora Banda in Western Australia.

The most intense lateritic weathering developed during the Tertiary. The resultant weathered zone over ultramafic rocks varies greatly in mineralogy but, in general, the profile is: (a) soil, (b) hard-cap or pisolitic limonite zone, (c) ochrous limonite zone, (d) secondary silicate or "saprolite" zone (montmorillonite, nontronite, serpentine, chlorite), and (e) unweathered ultramafic bedrock. Secondary silica occurs irregularly in the profile, but is most common within the secondary silicate zone near the boundary with the limonite zone.

Nickel may concentrate in either the limonite zones or the secondary silicate zone, giving rise to two general ore types which are typified by their Fe, MgO, and SiO_2 content. Both silicate and limonite ore may occur in the same deposit. The nickel in limonite ore is probably adsorbed onto the limonite whereas, in silicate ore, it is held in various ways in the secondary silicate minerals and associated manganese and iron oxides.

Greenvale and Rockhampton—Nickel-enriched laterites formed on Paleozoic intrusive serpentinites during the Tertiary (Fletcher and Couper, 1975; INAL Staff, 1975). The ore reserve at Greenvale (Fig. 1) is made up of approximately 50% limonite ore, 25% silicate ore, and 25% siliceous silicate ore. The Rockhampton deposit (Fig. 1) is predominantly silicate ore.

Wingellina—Ochrous limonite containing 1.3% nickel occurs in laterite overlying Proterozoic peridotites (Sprigg and Rochow, 1975; Fig. 1). The nickel distribution is extremely erratic both laterally and vertically.

Ora Banda—Nickel is concentrated in a clay-rich zone developed over Archean dunite (Loftus-Hills, 1975; Fig. 5). The nickel is mainly contained in clays in proportion to their iron content. The nickel-rich material has the following general

Table 3. Laterite Nickel Depoits of Australia

Location	*Deposit*	*Million MT*	*Grade % Ni*	*Contained Nickel (MT)*	*Remraks*
Eastern Australia	Rockhampton Area	29	1.28	371,000	INAL Staff, 1975
	Greenvale	40	1.57	628,000	1973 Reserves
Western Australia	Ora Banda	30	1.3	390,000	Unofficial Estimate
	Wingellina	56	1.2	672,000	Sprigg & Rochow, 1975
Total		155	1.3	2,061,000	

composition: 16% Fe; 8% MgO; 44% SiO_2; 1.4% Ni; and 0.08% Co.

Summary and Conclusions

Australia is third in world nickel production, behind Canada and New Caledonia. It has the potential to become an even more important world source of nickel. Nickel production in 1973 was 40,000 MT, and projected production for 1976 is 90,000 MT. The deposits on which this production estimate is based have all been discovered since 1966.

The largest reserves of nickel are in Western Australia, where nickel sulfide deposits have been found associated with ultramafic rocks in Archean volcanic belts. Although widespread in the region, they remained undiscovered until recently despite 70 years of exploration associated with gold mining. The environment has the potential for further major discoveries. Concentrations of laterite nickel occur overlying ultramafic rocks in Queensland and Western Australia.

The size of the nickel market, the price of nickel in Australian currency, and the cost structure in Australia relative to other producing countries are critical factors which will determine the extent to which Australia's nickel resources are exploited.

References Cited

Burt, D. R. L., and N. R. Sheppy, 1975, Mt. Keith nickel sulphide deposit, *in* Economic geology of Australia and Papua New Guinea—metals: Melbourne, Australasian Inst. Mining and Metallurgy, p. 159–168.

Fletcher, K., and J. Couper, 1975, Greenvale nickel laterite, *in* Economic geology of Australia and Papua New Guinea—metals: Melbourne, Australasian Inst. Mining and Metallurgy, p. 995–1001.

Gemuts, I., and A. Theron, 1975, Stratigraphy and mineralization of the Archean between Norseman and Coolgardie, Western Australia, *in* Economic geology of Australia and Papua New Guinea—metals: Melbourne, Australasian Inst. Mining and Metallurgy, p.66–74.

INAL Staff, 1975, The nickeliferous laterite deposits of the Rockhampton area, Queensland, *in* Economic geology of Australia and Papua New Guinea—metals: Melbourne, Australasian Inst. Mining and Metallurgy, p. 1001–1006.

Loftus-Hills, G. D., 1975, Ora Banda lateritic nickel deposits, Western Australia, *in* Economic geology of Australia and Papua New Guinea—metals: Melbourne, Australasian Inst. Mining and Metallurgy, p. 1010–1011.

Martin, J. E., and P. D. Allchurch, 1975, Perseverance nickel deposit, Agnew, *in* Economic geology of Australia and Papua New Guinea—metals: Melbourne, Australasian Inst. Mining and Metallurgy, p. 149–155.

Miller, L. J., and M. E. Smith, 1975, Sherlock Bay nickel-copper deposit, *in* Economic geology of Australia and Papua New Guinea—metals: Melbourne, Australasian Inst. Mining and Metallurgy, p. 168–174.

Nickel, E. H., J. R. Ross, and M. R. Thornber, 1974, The supergene alteration of pyrrhotite-pentlandite ore at Kambalda, Western Australia: Econ. Geology, v. 69, no. 1, p. 93–107.

O'Driscoll, E. S. T., 1971a, Deformational concepts in relation to some ultramafic rocks in Western Australia: Geol. Soc. Australia Spec. Pub. No. 3, p. 351–366.

——— 1971b, The tectonic pattern of the nickel province of southwestern Australia: Soc. Mining Geology Japan, Spec. Issue, p. 23–33.

Purvis, A. C., R. W. Nesbitt, and J. A. Hallberg, 1972, The geology of part of the Carr Boyd Rocks complex and its associated nickel mineralization, Western Australia: Econ. Geology, v. 67, no. 8, p. 1093–1113.

Roberts, J. B., 1975, Nickel deposits at Windarra, Western Australia, *in* Economic geology of Australia and Papua New Guinea—metals: Melbourne, Australasian Inst. Mining and Metallurgy, p. 129–143.

Ross, J. R., and G. M. F. Hopkins, 1975, The Kambalda nickel sulphide deposits, Western Australia, *in* Economic geology of Australia and Papua New Guinea—metals: Melbourne, Australasian Inst. Mining and Metallurgy, p. 100–121.

Sprigg, R. C., and K. Rochow, 1975, Nickeliferous ochres in the Hinckley (W. A.) and Tomkinson (S. A.) Ranges, *in* Economic geology of Australia and Papua New Guinea—metals: Melbourne, Australasian Inst. Mining and Metallurgy, p. 1008–1009.

Woodall, R., and G. A. Travis, 1969, The Kambalda nickel deposits, Western Australia: London, 9th Commonwealth Mining and Metallurgy Cong. Proc., Paper 26, p. 17.

Australian Bauxite Deposits—Discovery and Development[1]

H. J. EVANS[2]

Abstract With the discovery in 1955 of the large reserves of bauxite at Weipa, world attention became focused on Australia as a potential source of raw material for the aluminum industry, and several aluminum companies commenced exploration activities. Australia has joined the Caribbean area and West Africa as a major world source of bauxite; total reserves are in the order of 5,800 million MT. Of this total, 3,250 million MT is contained in the three main deposits—Weipa, Darling Range, and Gove—in which mining operations are presently being carried out. The bauxite deposits occur as residuals of Tertiary laterite and are predominantly pisolitic in texture. The bauxites are developed on several rock types including sedimentary rocks, basalt, and schist. Since 1955 the growth rate of the bauxite mining industry has been rapid—from less than 1 million MT in 1966 to over 19 million MT in 1973. The integrated aluminum-industry growth has been no less spectacular. Alumina production has risen from 30,279 MT in 1960 to 4 million MT in 1973 and, in the same period, aluminum metal production has grown from 11,842 MT to 203,210 MT. The Australian reserves are strategically located to serve the long-term needs of the Pacific area, particularly the developing countries of Southeast Asia.

INTRODUCTION

Australia now ranks as one of the major bauxite countries in the world with total reserves in the order of 5,800 million MT. Three large-scale mining operations have been established with a total annual capacity of 22 million MT. The main bauxite deposits occur in the north and west coastal regions as residuals of a widespread Tertiary laterite.

The discovery of major deposits of bauxite in Australia has resulted in the establishment of a vigorous aluminum industry and enabled these resources in the form of bauxite and alumina to be phased into the pattern of world trade. In 10 years (1963–73), Australia not only has become self-sufficient in bauxite, alumina, and aluminum, but has also become a major supplier to the world aluminum industry. Australia has surpassed Jamaica as the leading world producer of bauxite and is second to the United States in alumina production in the Western world.

HISTORY OF DISCOVERY

The occurrence of bauxite in Australia had been known from at least the late 1800s. By 1945 several minor deposits were known, and small tonnages of bauxite were mined for local use in the chemical and steel industries.

With the discovery of Weipa in 1955, and its initial evaluation, the Australian bauxite reserves increased to 2,743 million MT. Following the Weipa discovery, several major aluminum companies commenced exploration activities. This resulted in the discovery of the Mitchell Plateau deposits, the reassessment of the Darling Range deposits in Western Australia, and the exploration and evaluation of deposits around the original Weipa discovery.

BAUXITE RESERVES AND GRADES

Of the 21 known deposits, at least 12 have sufficient reserves to support large-scale mining operations (Table 1). Of the 5,800-million MT total reserves, about 58% is contained in three major deposits—Weipa, 44%; Darling Range, 9.6%; and Gove, 4.4% (Fig. 1). The grade of bauxite varies considerably. Alumina content ranges from 30 to 60% and silica from 1.4 to 15%. Although published reserves are not available for some deposits, the known reserves of commercial-grade bauxite are more than sufficient to maintain the expanding Australian aluminum industry for the foreseeable future.

GEOLOGIC ENVIRONMENT

Bauxite deposits occur in different environments, and the origin of some deposits is still in dispute. However, it is generally agreed that bauxite forms by weathering conditions suitable for the retention of alumina and iron and for leaching of the other constituents of the parent rock. The process, which is essentially desilication, proceeds more rapidly under tropical climatic conditions. The main controlling factors in the development of bauxite are: (1) high rainfall, ideally with alternating dry periods; (2) porosity and permeability of the source rocks sufficient to permit access and free circulation of groundwater; (3) a well-established water table; (4) the presence

[1]Manuscript received, October 21, 1974.

[2]C.R.A. Exploration Pty. Ltd., Melbourne 3000, Victoria, Australia.

The writer thanks the Geological Survey of Queensland for permission to use published tables, and Comalco for much of the basic information that made possible the presentation of this paper.

Table 1. Australian Bauxite Reserves and Grades

MAJOR DEPOSITS:

	COMPANY	LOCALITY	RESERVES (Million tons)	GRADES OF BAUXITE Al_2O_3	SiO_2
QUEENSLAND	COMALCO	WEIPA *	2500	45 - 60%	2·0 - +10% total silica
		ESCAPE RIVER	not available	not available	
CAPE YORK PENINSULA	ALCAN	WENLOCK RIVER (Adjoining Comalco leases)	296	+45%	-15% total silica
	TIPPERARY / HOLLAND AL. / PECHINEY	ARCHER RIVER (South of Comalco leases)	600	not available	not available
	AUSTRAL PACIFIC / FALKIRK	South of Comalco leases	? 92	"	"
	C.R.A. / N.B.H.C.	WENLOCK RIVER	45	"	"
WESTERN AUSTRALIA	ALCOA	JARRAHDALE / DEL PARK *	500	25 - 38% available alumina	0·5 - 2·5% (reactive silica)
DARLING RANGE	ALWEST	MT. SADDLEBACK	200	30 - 35% available alumina	not available
	PACMINEX	CHITTERING	150	31%	"
	PROJECT MINING	BUNBURY	34	not available	"
	WESTERN MINING	WALPOLE	"substantial"	"	"
KIMBERLEY AREA	AMAX	MITCHELL PLATEAU	230	40 - 57%	Low
		CAPE BOUGAINVILLE	980	36%	not available
NORTHERN TERRITORY	NABALCO	GOVE *	250	50%	4·2% (total silica)
		TOTAL OF MAJOR RESERVES	+5877	* — operating mines	

MINOR DEPOSITS:

QUEENSLAND	—	TAMBOURINE MOUNTAINS	2	35 - 41%	4·0 - 7·6%
		HAMPTON	<1		
NEW SOUTH WALES	—	INVERELL	15	35 - 41%	2·7 - 6·6%
		MOSSVALE	4	31 - 54%	4·0 - 7·5%
		TRUNDLE	<1	43 - 51%	5·0 - 8·0%
VICTORIA	—	BOOLARRA	1	44 - 58%	3·7 - 11·5%
TASMANIA	—	OUSE / ST. LEONARDS / MYALLA	1	38 - 42%	1·4 - 8·0%
NORTHERN TERRITORY	GOVERNMENT RESERVE	MARCHINBAR ISLAND	10	48 - 51%	4·1 - 6·5%
		TOTAL OF MINOR RESERVES	34		

of carbon dioxide in the circulating water; (5) low to moderate topography; (6) a warm to tropical climate; (7) tectonic stability over a long period of time; and (8) suitable source rocks. Of these, the source-rock type is the least important. The major factors would seem to be the duration and intensity of weathering under the right combination of rainfall, topography, and drainage.

Type of Bauxite

The Australian deposits are almost exclusively the "blanket" type, the only exceptions being several minor deposits in Victoria and New South Wales. The texture is predominantly pisolitic, but some development of massive to tabular and concretionary texture occurs in parts of the Gove, Darling Range, and Mitchell Plateau deposits. All of the deposits are lateritic in origin and are mainly the result of in-situ weathering of the source rocks during the widespread Tertiary laterization. The predominant aluminous mineral in the bauxites is gibbsite, but significant amounts of boehmite occur in the Weipa deposit and, to a lesser degree, at Gove and Mitchell Plateau. Details of the elevation, source rocks, mineral composition, and type and grade of bauxite of the four principal deposits in Australia are shown in Table 2.

Exploration and Evaluation

Because the Australian bauxite deposits are confined to Tertiary laterites, exploration has been relatively simple. All of the major deposits occur as residual plateaus which generally have distinctive vegetation types. Thus, potential areas of bauxite are easily selected by aerial reconnaissance or study of aerial photos, photo-mosaics, or geologic maps.

The deposits are superficial blanket deposits, normally no more than 12 m thick. Evaluation is carried out either by pitting or by auger drilling. Spacing of auger holes or pits depends on lateral variation in grade in individual deposits. At Weipa, the initial reconnaissance was carried out with

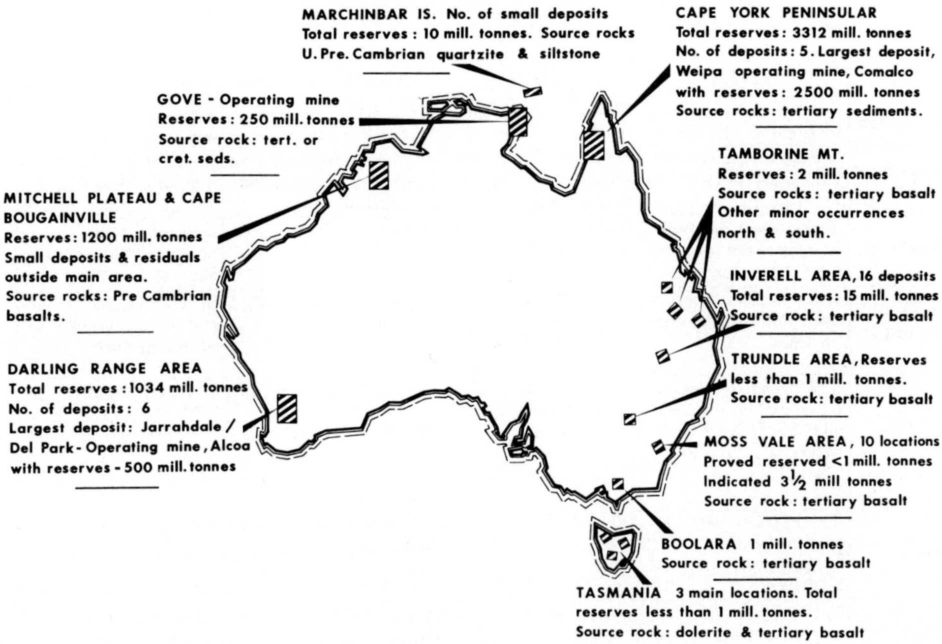

FIG. 1 – Locations of Australian bauxite deposits and known reserves.

pits and auger holes spaced at 610-m intervals and line spacing from 610 to 2,438 m. Selected areas with potential of the order of 200 million MT were filled in at 305-m centers. Final evaluation was made by drilling auger holes at 152-m centers in areas selected for initial mining.

Industry Development and Integration

The Australian aluminum industry started in 1945 with the formation of the Australian Aluminium Production Commission. The twofold objective of the Commission was: (1) to establish an aluminum smelter in Australia, and (2) to assess the bauxite potential of the country. The infant industry was founded on the basis of cheap hydroelectric power in Tasmania, and in 1955 a small smelter and alumina refinery was commissioned at Bell Bay in Tasmania. The smelter had a rated capacity of 11,176 MT of aluminum per year, and production was based on imported bauxite.

Since 1955, with the discovery of major bauxite deposits, the growth rates of bauxite mining and of the integrated aluminum industry in Australia have been phenomenal. Alumina production increased from 2,540 MT in 1955 to 4.02 million MT in 1973, and aluminum production increased from 1,268 MT in 1955 to 203,210 MT in 1973.

In this period of 2 decades, three major bauxite mines have been established and a fully integrated aluminum industry has been developed, making Australia more than self-sufficient in all forms of raw, processed, semifabricated, and fabricated aluminum.

Export Markets

Although the bauxite-alumina-aluminum industry in Australia is relatively young, it is firmly established in the world market. The main market outlets are (1) bauxite: Japan, E.E.C. countries, Taiwan, and North and South America; (2) alumina: E.E.C. countries, Japan, Norway, New Zealand, United States, and Canada; and (3) aluminum: Hong Kong, Japan, East Africa, and New Zealand.

Australia has now joined the Caribbean area and West Africa as a major world source of bauxite. The Australian reserves, particularly the large deposits at Weipa, are strategically located to serve the long-term need of the Pacific area including Japan and the developing countries of Southeast Asia.

Social Environment

With the exception of the Darling Range deposits, the bauxite mining operations have been essentially "grass-roots" developments. Both Gove and Weipa were in isolated tropical regions

Table 2. Topography, Mineral Composition, and Grade of Principal Australian Bauxite Deposits[1]

	WEIPA	GOVE	DARLING RANGE	MITCHELL PLATEAU
Elevation	Gradual slope from coast to height of 30m inland	10-100m above sea level	250-500m in plateau areas & slopes of ridges	300-550m gentle northerly slope to coast
Underlying rocks	Tertiary arkose & sandy clay, dip 4° west	Flat-lying Cretaceous sandstone & shale on Archaean basement	Archaean granite & some metamorphics intruded by dolerite dykes	Precambrian basalt (Carson Volcanics)
Mineral	Gibbsite with boehmite important near surface; 4·1 gibbsite: boehmite	Gibbsite with hematite & a little boemite near surface	Gibbsite with iron oxides & quartz	Gibbsite, some boehmite & hematite near surface
Type of Bauxite	Loose pisolites, some cemented; surface cap 1m thick	Upper base pisolites; intermediate cemented pisolite; lower tabular or vesicular	Hardcap underlain by friable zone	Ferruginous, pisolitic & gibbsitic zones with increasing depth
Grade	52% available Al_2O_3, beneficiable to 58% with 5% SiO_2, 7% Fe_2O_3, 2·5% TiO_2	52% available Al_2O_3, 2-6% reactive SiO_2, 13-18% Fe_2O_3	30-35% available Al_2O_3 1-2% reactive SiO_2 15-20% Fe_2O_3	40-45% available Al_2O_3 2-5% reactive SiO_2 20% Fe_2O_3
Thickness of Bauxite	1-10 m below surface cap 1 m thick	Up to 10 m, average 3-4 m	Up to 12 m, averages 4 m overburden 0·6 m	1-10 m, average 3·2 m
Area	500-800 km² within area of laterite residuals more than 1300 km²	120 km²	Orebodies discontinuous within zone extending 300 km north-south & 50 km wide	210 km²
Reference	Evans, 1965	Somm, 1973	Baker, 1973 Kirke, 1973	Joklik et al., 1973 Sofoulis, 1966

[1] From Queensland Government Mining Journal (March, 1974).

with no established ports, townships, or reliable communication services; the only access before developments was by sea or air. The establishment of an industry in these areas required a great deal of predevelopment research to obtain the basic data required for the planning of mining and treatment facilities, port and harbor installations, and townships. The studies included detailed hydrographic and land surveys, tidal measurements, establishment of standard meteorologic stations, water-resource and pest-control surveys, and local material-resource surveys.

For townships, a major consideration in design philosophy was the predictably long life of the orebodies, and thus the permanence of the community dependent upon them.

Regeneration of Mined Areas

The three operating mines are all in relatively well-timbered areas. Opencut mining methods involve the removal of large surface areas to shallow depths. Under mining legislation, the surface soil must be replaced and the original natural contours restored where possible.

Although there is variation in the regeneration procedure in each area, the practices are similar. Areas in which mining has been completed are surveyed to determine the drainage pattern. The compacted mine floor is ripped to allow root penetration and then contoured by grading. Topsoil is then reclaimed from storage dumps and spread over the graded surface and raked ready for the application of fertilizer.

The bauxite mining companies have carried out a great deal of research to determine the best methods of restoring the mined-out areas. Nurseries have been established and trial plantings of a wide variety of tree types have been made to determine the species best suited for each locality. Particular attention has been paid to those types that may prove to be a useful source of timber in the future.

Some alternate land usage is being considered. For example, at Weipa an experimental area of 42 acres (8.5 ha.) of pasture has been developed and is at present stocked with 21 head of cattle.

The regeneration work at Weipa and Darling Range has shown that the mined areas can be effectively reforested, but it will be some time before it is known if an economically viable timber crop can be obtained.

Social Benefits

In today's atmosphere of active conservation and environment philosophies, mining is a common target for criticism, and too often only the debit side is widely publicized. Because the opencut operations of bauxite mining are choice targets for such criticism, it is timely to present some of the advantages that have accrued to Australia from the development of the aluminum industry.

1. The establishment of a fully integrated aluminum industry has made the country more than self-sufficient.

2. A strong export market has been developed and phased into the consumption pattern of the world industry.

3. Major communities and ports have been established in remote areas, providing springboards for northern development.

4. Valuable research into timber growth and usage, pastures, soils, meteorology, and community living in tropical environments has been undertaken and is continuing.

5. At least $1.5 billion has been invested in Australia, and direct employment for more than 15,000 people has been established. In addition, it is conservatively estimated that 30,000 jobs have been created indirectly.

References Cited

Baker, G. F. U., 1973, Bauxite deposits of the Darling Range, Western Australia, *in* Metallogenic provinces and mineral deposits in the south western Pacific: Australia Bur. Mineral Resources Geology and Geophysics Bull. 142, p. 223.

Evans, H. J., 1965, Bauxite deposits of Weipa, *in* Geology of Australian ore deposits, 2d ed.: 8th Commonwealth Mining and Metallurgy Cong., v. 1, p. 396–401.

Geological Survey of Queensland, 1974, Aluminium: Queensland Govt. Mining Jour., v. 75, no. 869, p. 94–102.

Joklik, G. F., W. D. Jackson, and J. A. Zani, 1973, Bauxite deposits of the Kimberley, Western Australia: Australasian Inst. Mining and Metallurgy Conf. Proc., 1973, p. 141.

Kirke, E. A., 1973, Alumina from Darling Range bauxite: Australasian Inst. Mining and Metallurgy Conf. Proc., 1973, p. 553–564.

Sofoulis, J., 1966, Bauxite deposits of north Kimberley region, Western Australia: Western Australia Dept. Mines Rept. 1965, p. 117–118.

Somm, A., 1973, Bauxite deposits of Gove Peninsula, Northern Territory, Australia, *in* Metallogenic provinces and mineral deposits in south western Pacific: Australia Bur. Mineral Resources Geology and Geophysics Bull. 142, p. 223.

Iron-Ore Deposits of Western Australia—Geology and Development[1]

J. H. LORD and A. F. TRENDALL[2]

Abstract Although abundant iron ore was reported in Western Australia in the 1890s, only a few easily accessible deposits attracted interest before 1960. Intensive exploration since 1960 has increased the indicated and inferred reserves from a few hundred million to 24 × 10^9 MT. Most of the ore is high grade, and 95% of it is in the northwest of the state. Export began in 1965, and in 1974 six mines were operating solely for export, one for domestic use, and two for both purposes. Total production in 1973 was 74 million MT, and the total export tonnage (1965–73), including pellets, was 281.5 million MT.

Geologically, there are three distinct ore types: hematite enrichment ore, pisolitic limonite ore, and sedimentary ore. Hematite enrichment ore developed during Proterozoic time within both Archean and Proterozoic banded iron formations (BIFs). Structurally controlled deep-water circulation removed Si, Ca, Mg, Na, K, and CO_2, and introduced Fe, to effect metasomatic conversion of the BIFs to hematite and develop stratiform orebodies, a few of which exceed 1 × 10^9 MT. Secondary development of goethite has lowered the grade to about 63.5% iron. Pisolitic limonite formed along Tertiary drainages off the Hamersley basin. This material now caps sinuous mesas; it contains 40–60% iron. Sedimentary ore is quantitatively minor, and is restricted to the small high-grade Proterozoic coastal deposits of Yampi Sound.

A total capital investment of A$1,570 million since 1965, the extent and quality of Western Australian ore reserves, and proximity to the Japanese market—all justify confidence that the present export status will be maintained; however, several problems must be solved before export and domestic-processing growth rates can increase.

HISTORICAL BACKGROUND

Early geologists, during exploratory reconnaissances of the vast State of Western Australia (nearly 2.6 million km^2), noted the abundance of iron ore. Woodward (1890) reported, "This is essentially an iron country. . . There is enough to supply the whole world should the present sources be worked out." However, he later added (Woodward, 1894), ". . . but as iron ores are of no value, it is useless to trouble about them." Except for some interest in deposits near shipping facilities, nothing was done to examine, test, or develop iron-ore deposits until after 1960.

Deposits on Cockatoo and Koolan Islands in Yampi Sound were known to early pearlers, who commonly used iron ore from the coastal deposits for ballast. In 1920, the State of Queensland financed drilling on Cockatoo Island for a proposal to ship iron ore to Queensland for smelting with local coal. The proposal was eventually abandoned, and the leases passed to a company controlled by Broken Hill Pty. Ltd. (BHP), Australia's only iron and steel producer.

Leases on Koolan Island were purchased in 1936 by Brasserts Ltd., of England, who began to test and develop the deposit for the export of ore to Japan. However, in 1938, before production began, the Commonwealth Government imposed an embargo on export of all iron ore from Australia because of the inadequacy of known iron-ore resources for national needs. The Commonwealth immediately commenced an exploration program to evaluate known deposits, but it was discontinued early in 1939. The Koolan Island leases later reverted to the Crown.

The next major decision affecting the development of iron ore came in 1952 when the State Government granted BHP leases on all the Yampi Sound deposits, as well as a temporary reserve over the Koolyanobbing Range deposit. All other deposits were reserved to the Crown. Whereas iron ore previously could not be exported, now mining tenements for iron ore were not available to prospectors or companies.

In 1957 and again in 1959 the State Government made representations to the Commonwealth for a permit to export iron ore, because between 1952 and 1959 they had drilled the deposits at Koolyanobbing, Tallering Peak, and Mt. Goldsworthy with promising results. Finally, in December 1960, the Commonwealth Government announced that it would give limited approval for the export of iron ore from deposits other than those held by BHP—which recently had been granted the Koolyanobbing deposit—provided that the company establish an Integrated Steel Works at Kwinana.

In 1961 the State Government called for applications to develop the Mt. Goldsworthy deposit and the small deposit at Tallering Peak. Then it partly lifted the prohibition on mining titles for iron ore by inviting applications in other areas for temporary reserves, each with a maximum area of 50 sq mi (12.95 km^2).

An active period of exploration followed, and exploration soon became confined to the north-

[1]Manuscript received, November 6, 1974.

[2]Geological Survey of Western Australia, Perth 6000, Australia.

west of the state and the Hamersley Range area, in particular, where iron ore was known to occur. Several large, high-grade deposits were identified within a short time in this area. Export permits later were granted by the Commonwealth Government, which, owing to the amount of iron ore found, no longer restricted export permits to a portion of the ore proved, as was originally proposed.

Although exploration commenced late in 1961, export did not begin until 1966 because of the arid and isolated nature of the area concerned. After the location and testing of a deposit, it was necessary to negotiate a long-term sales contract with Japan (the only country interested) in order to raise the huge capital required for development. Once all this was achieved, the large task of constructing modern townships, a railway, and a new harbor started.

Six mines now are producing ore exclusively for export. Two of these have tripled their originally planned production rate and all are continuing to expand. A seventh small mine at Koolanooka has closed because of the depletion of the deposit.

Although it was known for many years that a large potential for iron ore existed in Western Australia, there was no incentive to prospect for it as there was no market. Later, when a world demand developed for iron, there were two political decisions which prevented prospecting. If these decisions had been reversed any later than 1961, significant Australian entry into the world iron-ore market might not have been possible, in spite of the available resources.

Geology and Distribution of Ore Types

The three main types of Western Australian iron ore—*hematite enrichment ore, pisolitic limonite ore,* and *sedimentary ore*—are presently (1974) being mined in the approximate proportions 20:2:1 (Trendall, 1974). Detrital ferruginous material, variously referred to as "scree ore," "conglomerate ore," "detrital ore," or "canga," is commonly associated with hematite enrichment ore, and may be mined in the near future; reserves of potentially *beneficiable iron formation*, or *taconite*, are very large, but this material does not have the status of ore.

Although the genetic term "hematite enrichment ore" is preferred for geologic classification, this ore type generally contains a variable amount of goethite, and the synonymous term "hematite-goethite ore," introduced by MacLeod (1966), is also used in this paper.

The distribution of these three ore types in relation to the broad geologic structures of Western Australia is shown in Figure 1.

Hematite Enrichment Ore

Hematite enrichment ore occurs stratigraphically within Archean (>2,500 m.y.) banded iron formations (BIFs) of the Pilbara and Yilgarn blocks (Fig. 1) and within Proterozoic BIFs of the Hamersley basin.

As in the Archean "nuclei" of most other continents, the Pilbara and Yilgarn blocks consist essentially of sinuous "greenstone belts" of tightly folded metasedimentary and metavolcanic rocks which enclose extensive areas of granite. The greenstone belts occupy about a fifth of the total outcrop area and, within them, BIFs form less than 1% of the total stratigraphic thickness; most individual BIF units are probably no thicker than 10 m, and the thickest is about 65 m. So far as can be seen from the normally tightly folded and steeply dipping stratigraphic sections, strike continuity may reach 75 km, but is mainly much less. Lithologically, the BIFs consist of thin, alternating laminae of red or white chert (quartz) and either magnetite or hematite, although other minerals are present in very minor amounts outside areas of high metamorphism.

The intracratonic, ovoid Hamersley basin, with an area of about 150,000 km^2, developed over an eroded surface of these Archean rocks about 2,400–1,800 m.y. ago (Trendall and Blockley, 1970). After accumulation of the thick lavas and mainly volcaniclastic sediments of the basal Fortescue Group, the BIFs of the central Hamersley Group were laid down in a quiet environment which was terminated by the deposition of the largely terrigenous clastic sediments of the uppermost Wyloo Group. Hematite enrichment ore of economic interest occurs in the basal Marra Mamba Iron Formation and in the Dales Gorge and Joffre Members of the Brockman Iron Formation, all within the Hamersley Group (Fig. 2). Later uplift and erosion have provided excellent exposure of relatively undisturbed and virtually unmetamorphosed Hamersley Group rocks in the northern part of the present outcrop area, but in the central and southern parts there is a steady increase in the intensity of folding, until in the extreme southeast there is local overturning.

By contrast with the Archean BIFs, those of the Hamersley Group differ in being thicker (Fig. 2) and more persistent laterally. The three economically important Hamersley Group BIFs have usual thicknesses of 180 m (Marra Mamba), 150 m (Dales Gorge Member), and 260 m (Joffre

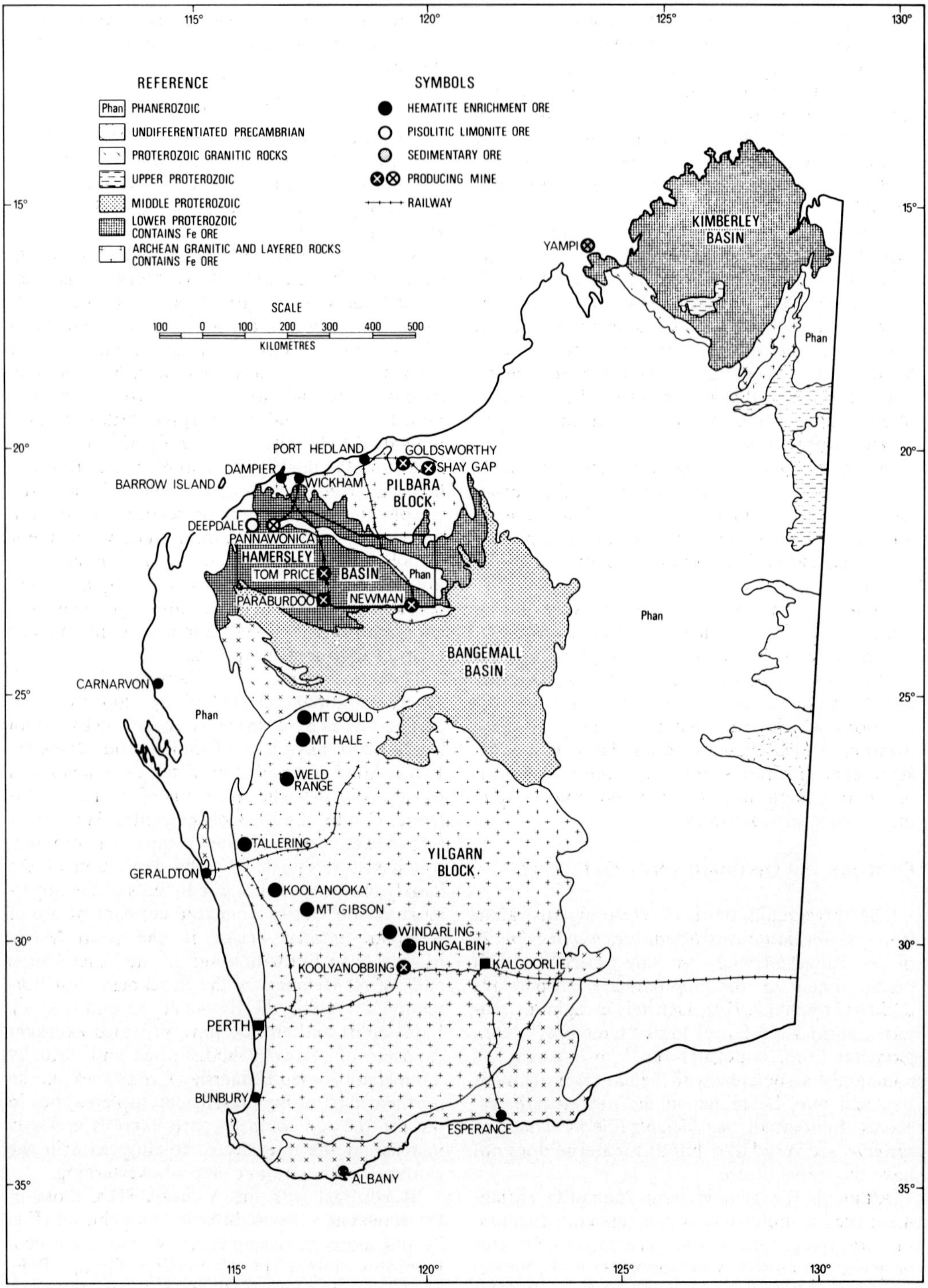

FIG. 1–Simplified map of Western Australia showing distribution of principal iron-ore deposits in relation to main tectonic units.

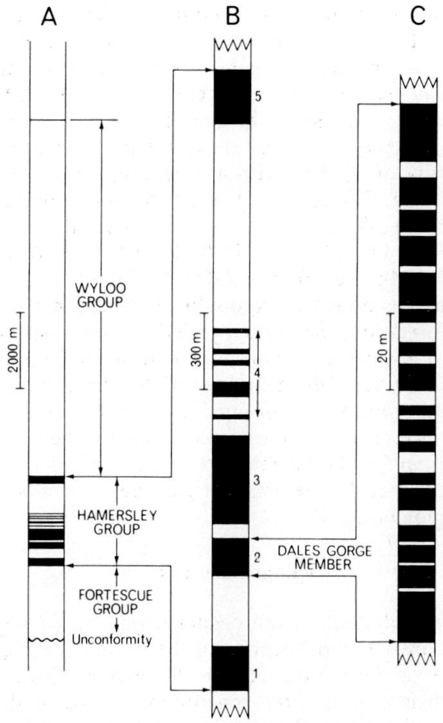

FIG. 2–Stratigraphic columns showing various features of distribution of BIFs in Hamersley basin; BIFs are shown as solid black. **A.** Position of Hamersley Group in total basin stratigraphy. **B.** Five major BIF units of Hamersley Group: *1*, Marra Mamba Iron Formation; *2*, Dales Gorge Member of Brockman Iron Formation; *3*, Joffre Member of Brockman Iron Formation; *4*, Weeli Wolli Formation; *5*, Boolgeeda Iron Formation. **C.** Thin shale bands (16) within Dales Gorge Member which have basin-wide persistence.

Member). All can be followed confidently over the outcrop area, and laminae less than 1 mm thick have been correlated over a distance of 300 km. Although the Proterozoic BIFs of the Hamersley Group are basically similar to the Archean BIFs in that they have regularly alternating laminae of chert (quartz) and an iron mineral, they differ lithologically in the following important respects: (1) the chert bands tend to be thicker and to have a finer internal lamination (microbanding) defined by iron-bearing minerals, so that the cherts themselves have a significant iron content; (2) carbonate minerals including ankerite, members of the siderite-magnesite series, and minor dolomite and calcite are important constituents of the BIFs, and stilpnomelane, minnesotaite, and riebeckite are common.

Deposits of hematite enrichment ore occur as sharply defined bodies within both the Archean and Proterozoic BIFs; in neither do they constitute a large proportion of the total volume of the BIFs down to minable depths, and they certainly form less than 2% (but possibly as much as 1%) of the BIF outcrop area.

Most commonly, individual orebodies occupy the full stratigraphic thickness of their host BIF, so that two of the bounding surfaces are subparallel stratigraphic boundaries, commonly against shale or similar fine-grained impermeable rocks. The shapes of orebodies in sections parallel with the stratigraphy are poorly known, but most appear to have simple closed shapes elongate along the major regional structural axis. A major difficulty in determining the original shape of the bodies is that they crop out at the existing ground surface, which developed much later than the hematite and truncates the orebodies to an unknown extent. The present shapes of bodies vary widely and correspond to the variety of structures possessed by the host BIF prior to enrichment. Thus orebodies in synclines are bowl shaped, bodies in areas of gentle dip are stratiform and monoclinal, bodies in isoclinally folded areas are dikelike, etc.

The larger individual orebodies occur in the BIFs of the Hamersley basin, where mineralization extends to depths of 400 m, and strike lengths of several kilometers are known. Individual orebodies exceed 1,000 megatonnes, below which there is continuous gradation to the limits of economic interest. The stratigraphic thickness of such bodies is normally about 50% of that of the enclosing BIF. For example, bodies within the Dales Gorge Member of the Brockman Iron Formation have a stratigraphic thickness of 60–80 m, compared with an unenriched thickness of 150 m. In this stratigraphic unit, because of the presence of 16 thin shale bands (Fig. 2), all the internal lithologic subdivisions are clearly traceable from the unaltered BIF into and through the hematite orebody. This evidence is critical for an explanation of the genesis of the ore—that is, the hematite results from secondary replacement of BIF similar to the adjacent unenriched host rock. Simple calculations from compositions, densities, and stratigraphic thickness relations show that, for the formation of hematite enrichment ore from a BIF, the initial iron of the BIF (about 30%) remains in place while the silica and most minor constituents (Mg, Ca, K, Na, CO_2) are removed. Simultaneously, additional iron constituting up to 50% of that initially present in the BIF is introduced. Paleomagnetic evidence suggests that this process took place during the approximate interval 1,600–1,300 m.y. ago, immediately after the main folding in the Hamersley basin at about 1,700 m.y. ago. Surprisingly, this evidence suggests that the ore in both the Archean and Proterozoic BIFs formed simultaneously.

Such a replacement origin is consistent with the texture of the ore. The more solid hematite ore, especially at lower levels of the orebodies, normally has a banded structure defined by alternate massive and porous layers of unstrained hematite mosaic of about the same thickness as the banding of the adjacent unenriched BIF. Such ore normally has a specific gravity in the range 3.5–5.0, from which a porosity in the range 5–35% is implied. In fact, goethite is a common component of the ore, and is more abundant near the surface; it is believed to represent a later, probably Tertiary, modification of the ore. In addition to the microporosity of the solid hematite, much of the ore is well jointed and also has larger channels and cavities. Goethite is more abundant in the more open-structured ore. The bulk in-situ specific gravity of much ore is probably not greater than 3.6, corresponding, for exploration purposes, to a density factor of 10 cu ft/LT (0.2783 m^3/LT).

In composition, most of the mined Hamersley Range ores correspond closely to a theoretical mix of 70% hematite, 23% goethite, 5% kaolin, and 2% quartz, to yield a composition of about 63.5% Fe, 4.3% SiO_2, 2.0% Al_2O_3 and 3.0% L.O.I. (loss on ignition). Phosphorus is the most commercially important trace constituent; most mined deposits contain about 0.05%. The ore associated with Archean BIFs is characteristically of higher grade than that of the Hamersley Range; it generally has lower Al_2O_3 and SiO_2 and a very low phosphorus content.

Pisolitic Limonite Ore

In contrast to hematite enrichment ore, pisolitic limonite ore is of Tertiary age and occurs along drainage lines with catchments in the Hamersley Range area. Typically, pisolitic limonite ore forms caps up to about 60 m thick (but usually $<$30 m) on elongate discontinuous mesas whose flat summits stand up to 75 m above the level of the present rivers. Such mesas, which lie close to and along the line of present rivers, are particularly well developed within the drainage basin of the Robe River, where the ore has been formally named the "Robe Pisolite." They have a wide distribution within and around the northern margins of the Hamersley iron province.

The ore consists of pisoliths mostly 1–3 mm in diameter, and is composed of varying proportions of amorphous limonite, goethite, maghemite, and hematite, closely packed in a variable matrix of the same minerals, together with minor amounts of clay, calcite, and chalcedony. The pisoliths may be structureless or concentrically layered. The ore may be earthy or very hard, and has a porosity of 20–30% with a specific-gravity range of 2.5–3.5. The material has the following composition: Fe, 40–60%; H_2O^+, 7.5–11.0%; SiO_2, 3.7–7.0%; Al_2O_3, 1.8–4.5%; P, 0.015–0.05%; S, 0.02–0.07%; Mn, 0.05–0.15%; TiO_2, 0.04–0.30%.

There is no reason to doubt the common-sense interpretation, clearly suggested by the disposition of the ore, that it accumulated by the fixation of iron in the beds of sluggish, flat-floored, meandering rivers which drained the Hamersley iron province prior to a fairly recent rejuvenation. An age limit for the pisolitic ore is given by its clear discordance to the older Cretaceous Yarraloola Conglomerate.

Sedimentary Ore

The sedimentary ore occurs within the locally developed Yampi Member of the Pentecost Sandstone, a formation of the Kimberley Group, which is one of three constituent groups of the Kimberley basin. This basin occupied an area of at least 120,000 km^2 of the northernmost part of Western Australia about 1,800 m.y. ago, at about which time it was filled relatively quickly by mainly clastic sediments.

The Yampi Member is best developed on Koolan and Cockatoo Islands, which are the only places where it contains economically extractable iron ore. On both islands the member is tightly folded about subhorizontal axes lying within axial planes which dip steeply southwestward. A thickness of a least 450 m consists of quartzite, sandstone, conglomerate, and shale in beds less than 1 m to a few meters thick; the beds are arranged in cycles 10–45 m thick. Hematite is a common component of each lithologic cycle, but tends to predominate in the uppermost bed of each cycle. It occurs both as clasts and matrix in the coarser sediments, and as finer grains, some with euhedral octahedral form, in the pelites; magnetite cores and martitic lamellae are commonly present. The orebodies are those stratigraphic units thick enough, and rich enough in hematite, to be minable. Reid (1965) gave the following approximate compositional range of ore: Fe, 57–69%; SiO_2, 0.4–6.0%; Al_2O_3, 0.2–6.0%; TiO_2, 0.4%; P, 0.02–0.05%; H_2O^+, 0.4–6.0%. The mined orebodies are up to 40 m wide, but generally are about 25 m, and reach over 2 km in length. They lie in steeply dipping beds. The material is friable below the surface, but a hard crust is locally present.

All authors concerned with the Yampi sedimentary ores have concluded that they originated by the detrital accumulation of hematite as one component of normal clastic sedimentation.

Ore Reserves

Australia has become recognized in the last decade as one of the few countries with major iron-ore resources. Ore reserves have increased from a few hundred million to an estimated 24 billion MT of indicated and inferred ore. Most of the ore is the hematite-goethite type containing more than 60% iron. About 95% of these reserves occur in, or are developed alongside, the Hamersley basin. Throughout the Pilbara and Yilgarn blocks and the Hamersley basin, huge areas of BIFs contain 30% or more iron, which may in the future be a source of ore.

To illustrate the immense resources involved, it is estimated that the Brockman Iron Formation alone has 6.4×10^{12} MT in outcrop section available for extraction by opencut mining (BIF to overburden ratio of 6:1, allowing 45° batter). In addition, within the Hamersley basin there are the resources of the Marra Mamba and Boolgeeda Iron Formations and the possibility of deeper mining.

Exploration

Following identification of major orebodies, company and government geologists quickly established the stratigraphy of the rocks; and, by use of air photographs, it was relatively easy to follow the favorable BIFs either on foot, by vehicle, or by helicopter. Although it was possible to identify BIFs from the air, it was necessary to identify the enriched portions on the ground. Prospects were tested by rotary percussion drilling and, for control, some diamond-drill coreholes were used.

The early prospecting located the more obvious orebodies, but later prospecting has been aided by a proper appreciation of the detailed stratigraphy involved. Gamma-ray logging is now used extensively to determine the stratigraphy (Jones et al, 1973), both in exploration and in mine-development drilling. During the last 3 years, geophysics has been used more and more for exploration, in particular for the Marra Mamba Iron Formation, which is commonly soil covered. Airborne and surface magnetic surveys are used and, in some places, gravity surveys.

Production

The first regular production of iron ore began in 1948 near Wundowie, 67 km east of Perth, from small lateritic deposits. The ore was used in a small, local charcoal-iron plant. In 1951, production commenced at Yampi Sound for consumption by the Australian iron and steel industry. However, it was not until 1966 that the dramatic upsurge in production began (Fig. 3). That year, three mines commenced production and export: Koolanooka, Mt. Goldsworthy, and Mt. Tom Price.

In 1967 the mine at Koolyanobbing started production of ore to feed a new (BHP) blast furnace at Kwinana near Perth. The Mt. Whaleback (Newman) deposit commenced production in 1969, followed by Pannawonica in 1972, and Paraburdoo and Shay Gap in 1973.

The companies exporting ore are Goldsworthy Mining Ltd. from Mt. Goldsworthy and Shay Gap, Hamersley Iron Pty. Ltd. from Mt. Tom Price and Paraburdoo, Mt. Newman Mining Co. Pty. Ltd. from Mt. Whaleback, and Cliffs W.A. Mining Co. Pty. Ltd. from Pannawonica (Fig. 4).

In 1973 Hamersley Iron produced 27.1 million MT and Mt. Newman Mining produced 26.1 million MT, representing over 70% of the total production of the state. These two companies are continuing to expand. Mt. Newman, which was due to complete an expansion to a production rate of 35.6 million MT/yr (costing A$100 million) by the end of 1974, announced a further expansion to 40.6 million MT/yr (costing A$85 million) by early 1976.

This rapid rate of expansion of existing mines apparently is keeping pace with the increase in demand of the major customer, Japan, and is making it very difficult for other companies to obtain the new long-term contracts necessary to raise the capital required to develop new mines.

Economics

It is estimated that about A$1,570 million had been invested by the companies with producing mines by the end of 1973. Hamersley has the highest capital investment because it has developed two mines—Tom Price and Paraburdoo.

The pattern of establishing a mine has been to locate and test an orebody to the satisfaction of a consumer (the Japanese steel industry), who would be prepared to enter into a long-term contract for the supply of ore. The company then had to raise the capital—some from Australia and the remainder from North American, European, and Japanese sources. Hamersley raised 54% of its

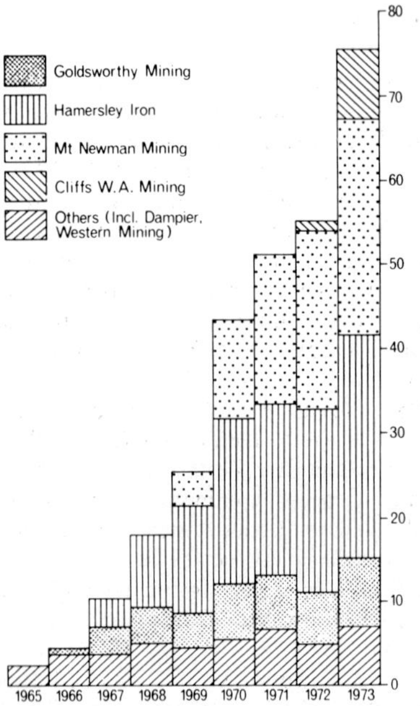

FIG. 3—Iron-ore production by companies since commencement of export in Western Australia.

capital in Australia and the remainder from overseas. Overseas investors own a 78% interest in Hamersley and 49.6% in Mt. Newman. A difficulty for any new company attempting to raise capital has been the fairly recent requirement of the Australian Government that at least 25% of any overseas loan be deposited with the Reserve Bank on an interest-free basis. This amount was reduced to 5% in August 1974.

The profitability of the iron mines has decreased rapidly in recent years owing to revaluation of currencies, high inflation rates that brought rises in production costs, and industrial troubles. As all contracts have been written in U.S. dollars, currency revaluations alone have caused a loss of 24–29% to the mining companies, of which only 15–18% has been recovered by negotiated price increase.

Iron-ore production and export are of great value to Western Australia. In 1973, 73 million MT, valued at $454 million and providing A$28 million in royalties to the state, were produced. About 85% of the exported iron ore goes to Japan to supply over 45% of its requirements; the rest goes mainly to European countries.

Australia is thus providing about 10% of the world's iron-ore requirements and, with the reserves available, will continue to be one of the world's leading producers.

Future Outlook

Western Australia will remain a major exporter of iron ore with every chance of retaining its 45% share of the Japanese market because of its proximity (Port Hedland to Japan is 6,700 km). However, because of the distance to European and North American markets, it will be difficult to export in that direction except by processing ore locally and exporting crude steel. A major feasibility study, by a consortium of Australian, Japanese, British, and American companies, has commenced on a proposal to establish a "jumbo" steel plant with an ultimate capacity of 10 million MT/yr of crude steel.

There are deposits, each with indicated reserves of hundreds of millions of metric tons, awaiting development at Jimblebar, Rhodes Ridge, Marillana, Mining Area "C," Koodaiderie, and Marrandoo.

Among the problems to be overcome—mostly economic—are the winning of large, long-term export contracts which will justify the establishment of new mines; the maintaining of profitability against inflation and currency revaluation; and the settling of industrial unrest. The established mines will continue to increase their production rate, making it difficult for the development of new mines.

Another problem is the high phosphorus content of some of the high-grade ore of the Hamersley province. Before this ore can be mined, steel mills will have to agree to accept it, or the ores for export will have to be a blend of high- and low-phosphorus ores.

Because huge reserves of ore containing over 50% iron are yet to be developed, it is too early to speculate on whether the colossal reserves of the BIFs will be developed.

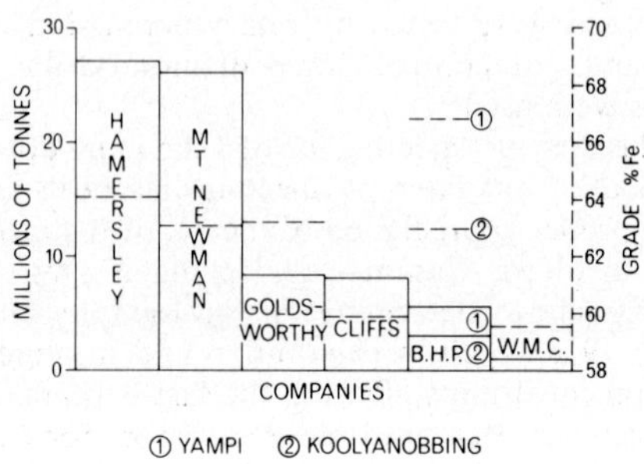

FIG. 4—Total production and grade by each company in 1973. Note that Hamersley and Cliffs include pellets.

References Cited

Jones, H., F. Walraven, and G. G. Knott, 1973, Natural gamma logging as an aid to iron ore exploration in the Pilbara region of Western Australia: Australasian Inst. Mining and Metallurgy Conf., ser. 2, p. 53–60.

MacLeod, W. N., 1966, The geology and iron deposits of the Hamersley Range area, Western Australia: Western Australia Geol. Survey Bull. 117, 170 p.

Reid, I. W., 1965, Iron ore deposits of Yampi Sound, *in* John McAndrew, ed., Geology of Australian ore deposits, 2d ed.: 8th Commonwealth Cong. Mining and Metallurgy Pub., v. 1, p. 126–131.

Trendall, A. F., 1974, Geology of Western Australian iron ore, *in* C. L. Knight, ed., Economic geology of Australia and Papua–New Guinea; metallic minerals: Australasian Inst. Mining and Metallurgy.

——— and J. G. Blockley, 1970, The iron formations of the Precambrian Hamersley Group, Western Australia: Western Australia Geol. Survey Bull. 119, 366 p.

Woodward, H. P., 1890, Annual general report for 1888–1889: Perth, Govt. Printer, p. 25–26.

——— 1894, Mining handbook to the colony of Western Australia: Perth, Govt. Printer, p. 48.

Mining Heavy-Mineral Sand in Australia—Summary[1]

J. PINTER[2]

Introduction

The mining of mineral sands of rutile, zircon, ilmenite, and monazite is linked with a variety of new and developing technologies.

Extraction of mineral sands commenced on the Australian East Coast 40 years ago at Byron Bay, 800 km north of Sydney. In its first year, the operation produced only a few hundred tons of zircon. Since then, the industry has expanded rapidly, and is now active on both sides of the continent. Total production in 1973 amounted to 325,000 MT of rutile, 357,000 MT of zircon, and 700,000 MT of ilmenite. The total value of the production was over US$100 million; more than 95% of this production was exported to about 40 countries.

The individual minerals are briefly described as follows:

Rutile is nearly pure titanium oxide (TiO_2) and is used for the manufacture of titanium pigments via the chloride route and for welding electrodes, titanium metal, and various titanium alloys and chemicals.

Zircon is zirconium silicate ($ZrSiO_4$). It has excellent refractory properties and is used extensively as a special molding sand for high-quality steel castings. Increasing use is being made of zircon in the manufacture of refractory bricks for the glass and steel industries. Zircon is also used for ceramic glazes, for the manufacture of zirconium metal to sheath fuel elements in atomic reactors, and for zirconium chemicals and cosmetics.

Ilmenite is a titanium iron oxide ($FeTiO_3$) and is traditionally used for the manufacture of titanium pigments via the sulfate route. Ilmenite is now also being upgraded, prior to shipment, to 92 or 95% TiO_2, thereby making it more acceptable for use in pigment manufacture by the chloride process.

Monazite is basically a mixed rare-earth phosphate, (Ce, La)PO_4, which also may contain thorium. It is used in the production of polishing powders and as a source of europium and yttrium for use in color television tubes.

The industry also produces some minor products such as leucoxene, garnet, and kyanite. Small quantities of fine gold are also recovered. The average grain size of all minerals is 100–150 mesh.

History

Australian mineral sands are mainly produced for export; thus the rate of production over the years tends to follow world economic and political changes.

The first commercial development of the black sands began in 1934 at Byron Bay, N.S.W. The first commercial production of pure rutile in Australia was in Melbourne in 1941. The rutile was produced from Byron Bay concentrates and was shipped to England under wartime convoy. As the demand for rutile continued to increase, high-grade seams were gradually depleted and lower grade deposits were worked.

The first titanium boom occurred in the mid-1950s. Many new rutile producers entered the field, and the price for rutile trebled and peaked at $300 per ton. With the coming of *Sputnik* in 1957, the titanium boom collapsed rapidly. Strategic thinking turned away from manned aircraft and thus from titanium metal. Rutile prices and production dropped sharply. Soon after (1960–61), the titanium pigment producers became interested in rutile as a feed material for the newly developed chloride process, and rutile production increased to meet the demand.

It was during this period that mineral-sand mining commenced in Western Australia. Ilmenite was mined from very high-grade and good-quality deposits near Bunbury, south of Perth. By 1965, ilmenite production in that area was about 400,000 MT/yr, and relatively small tonnages of zircon, rutile, and monazite were produced.

Methods of concentration and separation were improved, and new methods were developed, so that the East Coast low-grade deposits could be mined economically.

On the West Coast, new rutile- and zircon-rich deposits were discovered at Eneabba, north of Perth. Western Australia, with its large output of ilmenite, is now set to become an important producer of rutile and zircon also.

The relationship between operators has changed greatly. In the early days, individual pro-

[1]Manuscript received, December 26, 1974.

[2]Managing Director, Associated Minerals Consolidated Ltd., Southport, Queensland, Australia.

ducers were jealous of each other and seldom cooperated. A classic example was the necessity to construct two parallel roads—several miles in length through bush country—in order to service two competing separation plants built within half a mile of each other.

Competition for mining areas also tended to transcend normal legal argumentation. At one stage, a battle raged for several hours between two D8 bulldozers. One was trying to dig a hole across the access road of another company, while that company's bulldozer, of equal make and size, was trying to push the dirt back so as to make the road trafficable again. It sounded and looked like a modern-day version of a battle of roaring dinosaurs.

These events have long been forgotten. Smaller operators have been absorbed into larger units which are mostly backed by substantial, publicly listed companies. Thus, there are now only seven well-established operators on the East Coast and six on the West Coast.

GEOLOGY

Eastern Australia

Mineral-sand deposits have been found at various locations from Broken Bay, north of Sydney, to Rockhampton in Queensland—a distance of about 1,600 km. The mass of sand is silica and, of the heavy minerals present, the rutile content ranges between 50% in the south and 4% in the north. Over the same area, zircon content ranges from 35 to 10% of the heavy-mineral fraction, and ilmenite content ranges from 10 to 70%. In general, these three minerals make up 80–90% of the heavy minerals. The balance consists mainly of various amounts of garnet, leucoxene, tourmaline, monazite, and cassiterite.

Despite this unique heavy-mineral assemblage, most of the coastal sand mass is uneconomic, except where natural forces have combined to form and preserve heavy-mineral concentrations. These are best formed during storms, when wave action separates the "light" silica fraction and carries it seaward, leaving a heavy-mineral concentration on the beach. Usually, these concentrations are dissipated by subsequent wind and tidal action, but some of the seams have been covered quickly with wind-blown sand which eventually is built up into a foredune behind the beach.

These "fossil" deposits, which can contain up to 80% heavy minerals, are commonly 1 m thick, 10 m wide, and several kilometers long. They make up the high-grade rutile and zircon orebodies. This process has been repeated many times in some areas, resulting in the preservation of a series of orebodies beneath a system of parallel dunes as the beach migrated eastward with the buildup of sand. Fossil deposits occur up to 8 km from the present coastline.

These beach concentrations formed during the interglacial periods. One group, consisting of unconsolidated sand, appears to have been formed in the last interglacial period, within the last 10,000 years. The second main group, consisting of sand cemented with colloidal organic material, resembles a soft brown to black sandstone; it was formed over the period from 20,000 to 50,000 years ago.

In addition to the beach concentrations, there are considerable low-grade reserves contained in remnants of a system of high sand dunes formed during the glacial period, when large amounts of sand were exposed on the continental shelf as the sea level fell. Not all of these high sand dunes are economic, but in some sections the winnowing action of the wind has been sufficient to form large low-grade orebodies.

Large remnants of the high dune system form Fraser, Moreton, and Stradbroke Islands off the southeast coast of Queensland. On these islands, sand dunes exceed a height of 250 m.

It is generally agreed that the mineral sands are the end products of erosion and that the Precambrian shield was the initial source. These minerals have persisted through the formation and erosion of younger sedimentary rocks, the last of which are the Mesozoic sandstones of the Sydney, Clarence, and Moreton basins.

Western Australia

The main Western Australian deposits occur along 400 km of coastline north and south of Perth. The deposits are associated with several raised shoreline systems, and are as much as 30 km from the present coastline. All are essentially beach and foredune concentrations formed by wave and wind action.

The best-known deposits are located south of Perth near Bunbury. Of the heavy minerals present, ilmenite can constitute from 70 to 85%, zircon ranges between 3 and 15%, rutile between 1 and 5%, and monazite from 1 to 4%. These orebodies are more massive than those of the East Coast. Seams can be up to 4 m thick, 200 m wide, and several kilometers long.

The deposits have been partially lateritized by precipitation of iron, which is released with the alteration of ilmenite to leucoxene. Up to 20% of the ilmenite may be altered to leucoxene; thus there is wide variation in the titanium dioxide content of the ilmenite product.

The deposits north of Perth, between Jurien Bay and Eneabba, have been discovered over the last 5 years and are situated on northern extensions of the raised shoreline systems. In this area, the rutile and zircon content is significantly higher than in the Bunbury area. Rutile composes about 10% of the heavy-mineral fraction, and zircon composes up to 30%. High clay content is common, and very hard sandstone occurs toward the base of mineralization in some areas as a result of lateritization. Igneous and metamorphic basement rocks exposed in the area are the source of the heavy minerals.

Exploration

The presence of "black sands" on East Coast beaches has been known since the last century. The obvious deposits were worked for their gold and platinum content, and some buried fossil deposits were traced inland. These were first thought to be alluvial leads. However, as the rutile and zircon content became important, further study of the deposits showed them to be fossil beach deposits.

The unconsolidated-sand areas were prospected initially by hand methods such as shovel, auger, and sludge pump below water. It was seen early that, because of the long narrow orebodies, drillholes must be closely spaced across the strike of the beaches and parallel dune systems.

Later, aerial photographs were used to show the full extent of these dune systems, and machine drilling was introduced to prospect the older consolidated-sand formation.

Geophysics has not played a major role because most of the heavy minerals are only weakly conductive and magnetic susceptibility is low. In Western Australia, where the ilmenite deposits are massive, magnetometers have had a limited application; and, where the monazite content is high (up to 4% of the heavy-mineral fraction), scintillometers were used in regional surveys.

However, the low cost of drilling shallow sand deposits favors photointerpretation followed by drilling. Initially, the prospecting grid has frequent holes on widely spaced lines across the dune strike. Subsequently, short lines across the strike are used to delineate particular orebodies.

The need to prospect high dune formations has led to the development of several methods of drilling uncased holes to 100 m in unconsolidated sand.

Methods of Production

Mining

The economic deposits of mineral sands along the east and west coasts occur in a large number of orebodies of different sizes and grades, and with varying physical characteristics. The equipment used to mine them is equally varied.

Suction cutter dredges are used wherever possible, as they are the most economical method of mining. Dredge capacities range from 100 to 1,500 MT/hr. When mining unconsolidated sand, the cutter is used to pulp the feed and direct it toward the suction pipe. In indurated deposits, the hard-faced teeth on the arms of the cutter "shave" the face continuously.

Where dredging is not feasible, standard earth-moving equipment is used. Bulldozers, front-end loaders, mechanical shovels, draglines, and bucket-wheel excavators have all been used. The mineral feed is then supplied to the primary concentrating plant by means of conveyor belts or transfer bins in which the dry feed is pulped with water and then pumped to the concentrating plants.

Hydraulic sluicing was also used in deposits containing a high proportion of cemented sand or "coffee rock." The sand was sluiced from the face to a sump, from which it was pumped to the concentrating plant by a submerged gravel pump.

In general, mining costs represent about 75% of the total direct operating costs of the industry.

Concentrating Plants

These plants either float directly behind the dredge or are land based, commonly mounted on frames for rapid repositioning close to the mining unit. After removal of the heavy mineral, the plant tailings are returned to the mined area.

Modern concentrating plants vary in design and circuit used, but the basic form consists of three stages of upgrading. Final concentrates of 80–90% heavy mineral are produced from feed grades which may be well below 1%.

The concentrating equipment consists of three major types—Reichert cones, pinched sluices or trays, and spirals. Cones or trays are used in the first and second stages as they have a high capacity per unit, require little operator attention, and give good recovery. With a feed-pulp density maintained at 56–60% solids by weight, recovery of heavy minerals exceeds 90%. Spirals are used mainly in the final stage, where capacity requirements are low but more operator control may be required. Most concentrators are constructed of fiberglass with rubber-lined wearing surfaces; the light-weight units provide long, trouble-free service.

Rehabilitation of Mined Lands

Mineral-sand mining takes place on both frontal and hind dunes and restoration methods vary to suit local conditions. At all times, emphasis is placed on the use of native plant species.

Frontal Dunes

Erosion is an important feature of the eastern Australian coastline. It commonly is started by grazing animals or by recreational activities. Once erosion has commenced, the prevailing southeast winds accelerate the process and move large masses of barren sand westward—often engulfing existing vegetation.

Revegetation of the frontal sands subsequent to mining will stop the flow of additional sands from the beach and thus halt this destructive process. The restored and planted areas are usually better able to withstand the erosion forces of wind and wave action than they were prior to mining.

Hind Dunes

The hind dunes are usually covered with heath or forest vegetation. Prior to mining, a thorough vegetation survey is carried out to define plant communities and to list the species they contain.

When mining is completed, the tailings are contoured to resemble the original topography and the topsoil is replaced. This is done as rapidly as possible so as to retain the maximum vitality of the seeds it contains. A seed mixture consisting of a cover crop, various grasses, and the seed of suitable native trees and shrubs is sown to stabilize the topsoil and assist the regeneration of the native species.

Competition for Land Use

Most mining in Australia is undertaken in remote inland areas where there is little competition from other land users. However, the coastal region of the East Coast, where the various mineral-sand operations are situated, is heavily populated. About 70% of Australia's population lives in this area. The population is increasing at a rate which promotes rapid development of the region.

In localities where minerals occur, the mining interests are obliged to compete with other interests. This competition may be considered under three broad headings.

Total development—Urban development and intensive agriculture belong in this category. Since suitably restored mined land is ideal for such developments, the conflict with mining interests is one of timing. The mining interests require time to obtain titles, explore, develop mines, extract the minerals, and restore the land. This lengthy process must precede urban or agricultural development if the valuable mining resources are not to be locked in.

Partial development—The second group of interests which can conflict with orderly mining retain the general features of the natural environment while providing access roads, power, telephones, water supplies, and limited building development. Recreational areas and grazing developments are examples. Again, the chief problem is one of timing, but the conflicts are capable of resolution through thoughtful planning.

Nondevelopment—The third group of conflicting interests wants to preserve the wilderness areas. The conflict can be resolved only if alternate, nonmineralized wilderness areas are available.

In essence, the only real solution to the competition for land is advance planning of all developments. This requires a thorough knowledge of all the natural resources, not just the mineral resources, of the land in question.

Marketing

Australia produces 95% of the world's rutile. Most of the rutile goes to the United States, and lesser quantities go to Europe, Japan, and elsewhere. Prices obtained today for the little rutile that is available are very high.

Australia produces 75% of the world's zircon; most of it goes to Japan and Western Europe. Zircon was a by-product of both rutile and ilmenite production. Zircon was used mainly as a molding sand in steel foundries; however, because of its excellent refractory properties, zircon is used increasingly as a high-grade refractory, particularly in the glass and steel industries. As a result, prices have risen steeply.

Australia produces 20% of the ilmenite (or slag) used in the sulfate pigment process. Australian ore competes with ilmenites from the United States, Canada, Norway, Finland, India, and elsewhere. However, the advent of the chloride process has made it economical to use ilmenite which has been upgraded to a 90–95% TiO_2 product—an artificial rutile. The first such operation has commenced in Western Australia.

Future for Heavy-Minerals Industry

Australian East Coast production of rutile and zircon is not likely to rise above its present level. However, Australia's total production will increase as a result of the exploitation of the Eneabba resources in Western Australia.

The combined earnings from rutile, ilmenite, and zircon should be sufficient to enable the continued working of low-grade deposits. The mineral-sands industry of Australia will thus remain one of the world's major suppliers of high-grade titanium and zirconium ores for many years to come.

Occurrence and Development of Sedimentary Manganese Ore, Groote Eylandt, Northern Australia[1]

W. M. LONIE[2] and J. L. McINTOSH[3]

Abstract One of the world's major deposits of high-grade sedimentary manganese ore is on Groote Eylandt in the western section of the Gulf of Carpentaria in northern Australia.

Production and marketing of the ore commenced in 1966 after establishment of mining and treatment equipment together with ancillary facilities such as housing, roads, port facilities, water, and power supply. Subsequently, further detailed exploration, development, and metallurgical work resulted in the construction of a comprehensive ore-treatment and beneficiation plant and additional handling facilities.

Current production capacity is of the order of 1.25 million MT/yr of manganese ore, of which 80% is exported. An expansion program will provide for a production capacity of the order of 2 million MT/yr by the end of 1975, bringing the total capacity expenditure on the developments to approximately A$70 million.

The manganese ore is a tabular bed of approximately 30 km². It crops out adjacent to the western coastline of Groote Eylandt, and is covered by soft Tertiary and recent sediments.

Marine arenaceous Foraminifera of Early Cretaceous age have been identified from within the ore zone, and no apparent tectonic deformation of the zone has occurred.

The deposit appears to have been formed in a shallow-water environment, as it is associated with typical shallow-water sediments including unconsolidated clays and sands containing typical shallow-marine (inner sublittoral) faunal assemblages.

INTRODUCTION

Groote Eylandt, an island 2,330 km² in area, is located near the western margin of the Gulf of Carpentaria (lat. 14° S, long. 136° E) in northern Australia (Fig. 1). Although named in 1644 by a Dutch explorer, the island remained virtually unknown and undeveloped until 1962, when its potential for mining development was first recognized.

Until that time the population comprised aboriginals and missionaries, living under subsistence conditions and rarely visited except by supply vessels and an occasional aircraft.

An occurrence of manganese oxide on Groote Eylandt was recorded during a visit by the South Australian government geologist in 1907. However, it was not until 1960 that a regional geologic survey by the Australian Bureau of Mineral Resources revealed additional outcrops on the island, and some investigation of the nature of the oxide occurrences was undertaken.

Preliminary investigations and sampling were rather inconclusive, and it remained for W. C. Smith, of The Broken Hill Proprietary Company Limited, to recognize that the relatively small, flat, inconspicuous pavement-type outcrops were surface exposures of a continuous and major sedimentary deposit of manganese. The orebody subsequently was proved to be over 12,000 ha. in area.

Preliminary investigations into the economic possibilities of the deposit involved mapping, shallow drilling, and, subsequently, shaft sinking and costeaning in selected strategic locations to define the boundaries of the orebody and to investigate the nature and characteristics of the ore.

This early work confirmed the tabular nature of the deposit and showed that, although basically pisolitic and lateritic in character, the naturally occurring high-grade manganese oxides are subject not only to wide variation in physical characteristics, but also to contamination from the associated, loosely consolidated sediments.

Because of these characteristics, it was deemed necessary at an early stage to develop sampling and testing techniques for drillhole samples, taking into account likely future treatments for the production of a marketable product. Basically, these techniques involved small-scale standardized ore-dressing investigations on all samples obtained during exploration activities, in lieu of the more conventional sampling and chemical-analysis techniques applied to orebodies with tabular characteristics.

These ore-dressing techniques proved successful in assessment of the likely ore characteristics, and modified techniques have been continued in conjunction with some blast-hole drilling, on a grid pattern, as a regular feature for preproduction planning and quality-control purposes.

[1]Manuscript received, October 21, 1974.

[2]General Manager, Minerals Production and Marketing, The Broken Hill Proprietary Company Ltd., Melbourne 3000, Victoria, Australia.

[3]Coordinator of Development, Groote Eylandt Mining Company Ltd., Groote Eylandt, Northern Territory, Australia.

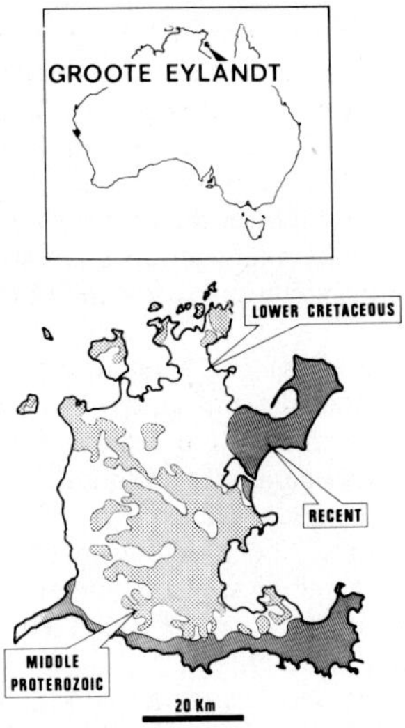

FIG. 1—Location and geology of Groote Eylandt.

Larger samples from shafts and costeans were tested in a pilot plant, established in 1965. Samples for trial smelting purposes were prepared and tested in electro-smelting furnaces in Tasmania and Norway in order to assess the suitability of the ore for production of ferromanganese.

Mining, treatment, and ore-handling equipment, together with the necessary port facilities, township, roads, and services, were established by July 1966 for a projected production of 200,000 MT/yr of metallurgical lump ore.

Additional mining equipment and shift production introduced in 1968 doubled the capacity to 400,000 MT/yr. Improvements to the original crushing and screening plant continued until 1972, when a larger beneficiation plant incorporating crushing, screening, scrubbing, and heavy-media separation processes was commissioned. This plant has a design capacity of 1.25 million MT/yr.

Facilities installed at the same time as the beneficiation plant included rubber-tired mining equipment, a 150,000-MT stockpile and shiploading plant, workshops, stores, power-house extensions, additional houses and township amenities, and expanded quality-control facilities.

Total ore products shipped for the calender year 1973 amounted to 1.4 million MT, bringing the total tonnage shipped to over 5 million MT.

Further expansion activities to provide for sales up to 2 million MT/yr are expected to be completed early in 1976. On completion, a total investment of the order of A$70 million will have been undertaken in the development of the Groote Eylandt ore deposits.

Stratigraphy

Precambrian—Dolerite and metamorphic rocks on nearby islands are the only indication of the Archean–lower Proterozoic igneous-metamorphic complex of Arnhem Land in the vicinity of Groote Eylandt. The middle Proterozoic Groote Eylandt Beds, comprising about 600 m of massive, blocky, well-jointed and crossbedded sandstones, unconformably overlie this complex and form the core of the island.

Lower Cretaceous—Unconformably overlying the Groote Eylandt Beds and outcropping as fringing coastal plains are the Lower Cretaceous Mullaman Beds of claystones, quartz sandstones, quartzite-pebble conglomerates, manganiferous marlstones, and the manganese-ore zone. These beds do not occur at elevations higher than 60 m above sea level on Groote Eylandt. However, they do reach a maximum thickness of more than 100 m. Apart from localized claystone ridges, the basal zone commonly is a quartz sandstone. In the vicinity of sandstone inliers, the manganese-ore zone normally lies directly on the Groote Eylandt Beds.

Tertiary—Despite the lack of fossil evidence, a Tertiary age is ascribed to a lateritic unit consisting of cemented limonitic sandy clays which, in many places—and particularly where they truncate the ore—contain manganese oxide pebbles and pisolites. This zone forms the outcropping manganiferous pavements where recent soil cover is sparse or absent.

Evaluation of Resources

Three evaluation stages are involved in the overall assessment and proving of the resources in preparation for development.

Reserve evaluation is based on drilling of areas underlain by manganese oxide zones on a 240-m^2 grid and sampling every 0.3 m in the ore zones. Costeans are programmed to ensure correlation between drillholes and to provide bulk samples. Drill samples are bulked on site by visual examination into specific minable ore zones and these samples are then wet-screened, beneficiated by a float/sink technique, and analyzed for the various elements of significance in future blending.

Mining and quality-control evaluation phases consist of regular infill drilling of the 240-m grid at 120- and 60-m intervals, respectively, to obtain detailed information required for mine development planning and the programming of production activities for the various marketable ore grades.

Ore Characteristics

Physically, the ore can be grouped into either massive or pisolitic types. However, gradations occur within short horizontal and vertical distances and, in field mapping, local standard descriptive terms are employed to indicate anticipated behavior of the ore type in the beneficiation process and its ultimate utilization as a raw material.

The manganese-ore zone ranges in thickness from 0.5 to 15 m, although in existing quarry areas the ore characteristically is of the order of 3–4 m thick. It consists generally of boulders, cobbles, and pebbles of manganese oxides in a matrix containing fine-grained particles of manganese oxide and sandy clay ranging between 10 and 50% volume.

There is no evidence of tectonic deformation of the ore zone but, in places, the ore exhibits features indicative of weathering, remobilization, and redeposition of the pisolites. In other places, where the ore is in the form of boulders or pebbles, cementing and reworking have apparently taken place.

The major constituent minerals are cryptomelane and pyrolusite, but psilomelane, todorokite, lithiophorite, and nsutite also have been identified in the main orebody. Rhodochrosite and pyrite have been reported near the base of the Mullaman Beds, although there is no evidence of these minerals where the manganese ore is currently being worked.

Cryptomelane occurs as both a primary product (in pisolitic layers and concretions) and an epigenetic product (in fracture-filling veins of the concretions and as a cement for the pisolites). Pyrolusite occurs as a replacement of cryptomelane in pisolites, as discrete oolites of wad, and as a pyrolusite powder.

Pisolites are commonly composed of concentric shells of pyrolusite and cryptomelane, as alternate or random layers or as partly discordant replacements of one another. Abrasion of pisolites is common, and some cryptomelane pisolites exhibit plastic deformation. Syneresis cracks filled with cryptomelane or pyrolusite are common, and the only visible impurities are a few sand-grain nuclei. Examples of partial replacement of claystone and clay pisolites have been noted.

Some nodular concretions or possibly mud balls of manganese oxides occur in a small claystone bed directly underlying the ore zone.

In the deeper sections of the deposit, the manganese oxide zone becomes less continuous and the ore consists of thin seams or layers of small pebbles in a sandy clay or of a succession of pebbles dispersed through sandy clay.

Adjacent to highs in the underlying sandstones, the ore normally exists in more massive form and exhibits remobilization and cementing characteristics.

The ore zone is characteristically overlain by unconsolidated silts and sands; in a few areas it is overlain by a manganese-bearing ferruginous laterite. It is also generally underlain by similar unconsolidated sands, silts, clays, claystones, and soft sandstones. The interstices between the boulders, cobbles, pebbles, pisolites, oolites, etc, comprising the ore zone proper are filled with similar fine-grained sediments.

Despite this close association with gangue materials, the manganiferous boulders, pisolites, etc, show little or no contamination from the surrounding sediments except in limited areas where sand grains have obviously formed nuclei for pisolite formation. These boulders, pebbles, and other forms can be separated from the adjacent gangue materials into high-grade marketable products by relatively simple ore-dressing operations.

Ore Production and Treatment

Mining operations are carried out by clearing the trees and scrub by bulldozers, then stripping the overburden with 25-m^3-capacity twin-engine scrapers, drilling and blasting where necessary, and stockpiling the ore with bulldozers before loading into 50-MT dump trucks with rubber-tired front-end loaders equipped with 6-m^3 buckets. Mine faces are maintained and worked in several areas simultaneously in order to ensure consistent product grades. Special grades are also produced from specially selected areas.

The ore is crushed to <130 mm and stockpiled prior to treatment in the beneficiation plant. Treatment consists of four major steps: (1) feed preparation involving secondary crushing and wet-screening of the coarse feed and classification of the finer sizings; (2) scrubbing to remove the contaminant coating from the ore; (3) heavy-media separation of the coarser fractions to remove gangue sandstone and low-grade ore; and (4) treatment using jigs to upgrade the finer fractions.

After beneficiation the ore is hauled 16 km in 90-MT loads by road-train transport to a stockpile adjacent to the loading wharf.

Eight grades of ore are produced to rigid specifications based on market requirements. Both physical and chemical properties are important specification factors, and a feature of the Groote Eylandt activities is the close quality control exercised throughout each operation.

Currently the ore is sold to 17 countries. The average size of shipments is of the order of 30,000 MT.

Throughout the activities on Groote Eylandt, particular attention is given to environmental factors. Land restoration of worked-out areas is carried out to predetermined plans. Rehabilitation of the restored land commences with replacement of topsoil and reestablishment of crop, pasture, and forestry species to suit the local environment.

Genesis of Manganese Deposits

An apparently simple marine origin can be ascribed to the Groote Eylandt manganese deposits in view of the presence of marine arenaceous Foraminifera beneath, within, and above the ore zone.

The faunal assemblages indicate a shallow-marine (inner sublittoral) environment of deposition (Smith and Gebert, 1969). However, the source of the manganese and its deposition without the general presence of iron are more difficult to postulate.

Algal concentration, photosynthesis, and simple chemical reactions in a warm, shallow-marine environment have been suggested by several authors as suitable conditions and mechanisms for achieving concentrations of soluble manganese compounds and their subsequent precipitation as oxides in a euxinic environment.

Possible original source rocks for the Groote Eylandt deposits are present west of Groote Eylandt in the Arnhem Land area, where they contain relatively low concentrations of manganese. They include: the Archean–lower Proterozoic mafic igneous rocks and upper Proterozoic dolerites; minor manganiferous beds in the lower Proterozoic McArthur Group—which includes dolomites, siltstones, sandstones, and cherts; and the Grindall Metamorphics, which crop out in the islands northwest of Groote Eylandt.

However, it is difficult to postulate the mechanism of solution and precipitation of the high-purity oxides in their present concentrated form, even if the possibility of these sources is accepted.

Evidence points to possible preconcentration of manganese solutions before deposition as solid oxides, and to reworking—and possible resolutions—along the shoreline and around sandstone inliers in a shallow embayment on the western margin of Groote Eylandt. This area of deposition appears to have been relatively undisturbed since late Proterozoic time.

Major sea-level changes or tectonic movements appear to have been unlikely in the area since Cretaceous time. The fact that the orebody generally occurs within a zone from 15 m below to only 30 m above present sea level may be significant. Furthermore, the relatively unconsolidated nature of the directly underlying and overlying units would tend to indicate a relatively recent origin for both the sediments and the ore, were it not for the fossil evidence and the scattered areas of truncation of the ore zone by laterite.

The regular concentric nature of the pisolites may result from growth or from a steady accumulation in a gently agitated environment.

Remobilization of the oxides during diagenesis or lateritization, or perhaps during deposition, has led to cementation of pisolitic zones and, in some areas, to almost total dissolution of the pisolite structure.

Reserves

Evidence of in-situ reserves of the order of 300 million MT indicates the significance of the deposit in terms of resources in the Pacific basin and the world in general.

Reference Cited

Smith, W. C., and H. W. Gebert, 1969, Manganese at Groote Eylandt, Australia: Mining and Petroleum Geology, 9th Commonwealth Mining and Metallurgy Cong., p. 585–604.

Stratiform and Strata-Bound Metal Concentrations in Australia—Summary[1]

HADDON F. KING[2]

Introduction

The stratiform and strata-bound ore deposits of the world include some of the most famous—for their size or richness, or both: the Rand, the Copperbelt of Zambia and Katanga, the Rio Tinto mines, the lead and zinc deposits of the middle United States, the Noranda field, the Sullivan mine, the iron ores, and the phosphates of the western United States. In Australia, ore deposits of this type include concentrations of Pb and Zn, Cu, Sn, W, Au, Ni, U, Fe, and phosphate.

Most of the major Australian stratiform orebodies were found between 1870 and 1885. Significant gold discoveries had been made up to 25 years earlier, but an appreciation of the importance of gossans came later. Mount Morgan, Mount Lyell, Cobar, and Broken Hill all had prominent gossans, some of which had been well-known features of the landscape for up to 20 years before their metal content was realized. Mount Isa was found in 1923, 50 years after the development of the nearby Cloncurry copper field; no other discoveries of large, high-grade outcropping stratiform base-metal orebodies have been made since. All orebodies found later have been in fields which had been known and worked for many years.

Discoveries of stratiform ore deposits have contributed greatly to the recent expansion of the Australian mining industry. Whereas, 30 years ago the only stratiform ore type known to be plentiful was Pb-Zn, there are now large resources of Cu, Ni, Sn, U, Mn, and P, and very large, newly discovered resources of Fe.

All of the longer known and many of the newer deposits have contributed greatly to the development of their geographical region, to the mining industry, and to the national economy. One mining center, Broken Hill, is outstanding in this respect. In addition to initiating secondary processing in the form of smelting, refining, and metal manufacturing, the mining companies at Broken Hill have, over the last 60 years, been directly responsible for, or identified with: (1) the existing Australian iron and steel industry; (2) discovery and development of the Weipa bauxite deposit and its associated industry ranging from bauxite export to metal fabrication; (3) revival of copper mining at Cobar; (4) discovery and development of the Mount Tom Price iron-ore deposit and its associated productive and secondary-processing facilities; (5) development of similar iron-ore export production at Mount Newman, in the eastern part of the Hamersley iron-ore province; (6) development of the Groote Eylandt manganese operation; (7) discovery and development of Bass Strait oil and gas; (8) discovery and development of the Bougainville copper deposit; and (9) discovery of the phosphate deposits of northwestern Queensland.

Stratiform Deposits

The stratiform and strata-bound deposits represent all the principal Australian base-metal deposits except bauxite and titanium minerals. The important individual deposits (including figures on production plus reserves) are the following:

Mount Lyell, Tasmania. Copper, in Cambrian volcanic rocks. Production plus reserves, ~1 million ST Cu plus some Ag and Au. Operating mine.

Renison Bell, Tasmania. Tin, in Cambrian shales, ~90,000 ST metal. Operating mine.

Rosebery, Tasmania. Lead-zinc-copper, in Cambrian volcanic rocks. 4 million ST Pb plus Zn, with minor Cu, Ag, and Au. Operating mine.

Mount Bischoff, Tasmania. Tin, in Cambrian sedimentary rocks. 50,000 ST metal produced mainly between 1873 and 1929.

Mount Cleveland, Tasmania. Tin, in Cambrian shales in volcanic rocks. ~30,000 ST metal. Operating mine.

Savage River, Tasmania. Magnetite-pyrite in Cambrian rocks. Operating mine.

King Island Scheelite, Tasmania. Tungsten, as scheelite, in Cambrian limestone. ~60,000 ST WO_3. Operating mine.

Captains Flat, New South Wales. Lead-zinc-copper in Silurian volcanic rocks. ~600,000 ST Pb plus Zn, with minor Cu, Ag, and Au. Mined out.

Woodlawn, New South Wales. Copper-lead-zinc, in Silurian volcanic rocks. ~150,000 ST Cu and 1 million ST Pb plus Zn; also some Ag and Au. Not yet in production.

Cobar, New South Wales. Seven principal copper, copper-zinc, copper-gold, and gold orebodies in Silurian volcanic rocks. Production ~200,000 MT Cu, 38,000 kg Au, and 120,000 kg Ag. One copper mine in operation.

[1]Manuscript received, February 10, 1975. Summary prepared by AAPG staff.

[2]Consultant, Melbourne, Australia.

Broken Hill, New South Wales. Lead-silver-zinc. Major deposit with dimensions 7.2 km × 610 m × 150 m. Grade to date is about 25% Pb and Zn plus 180 g/ST Ag. Known deposit perhaps 50 million ST combined Pb and Zn plus ~30 million kg Ag. Also contains ~20 million ST S, 8 million ST Mn, and 1 million ST F. Operating mines.

Dawsonvale, Queensland. Iron ore of Triassic-Jurassic age. 200 million ST, 37% Fe.

Mount Morgan, Queensland. Gold-copper, in Devonian volcanic rocks. ~500,000 MT Cu and 300,000 kg Au. Operating mine.

Mount Isa, Queensland. Lead-silver-zinc-copper, in Proterozoic dolomitic shales. Perhaps the most perfectly stratiform lead-zinc deposit known. ~9 million ST Pb and Zn, 5 million ST Cu, and 10 million kg Ag. Operating mine.

Hilton deposit, Mount Isa, Queensland. Lead-zinc-silver, in same stratigraphic zone as Mount Isa. >6 million ST Pb and Zn plus 6 million kg Ag.

Dugald River, Queensland. Zinc-lead, as small oreshoot in Proterozoic black shales. Grade around 12% Zn plus minor Pb. Undeveloped.

Mary Kathleen, Queensland. Uranium–rare earths, in calcsilicate rocks of Proterozoic age. ~10,000 ST U_3O_8. Not now in production.

Lady Loretta, Queensland. Newly discovered lead-zinc-silver in Proterozoic shales; reserves 2 million ST Pb and Zn.

Lady Annie, Queensland. Oxidized copper in Proterozoic dolomitic shales. 90,000 ST Cu.

Duchess and others, Queensland. Phosphates, in Middle Cambrian rocks. >2,000 million ST 17% P_2O_5.

Constance Range, Queensland. Iron ore, in Proterozoic sedimentary rocks. 300 million ST sideritic and oxidized. Sideritic ore, 51% Fe; oxidized ore, 57% Fe.

H.Y.C. deposit, McArthur River, Northern Territory. Lead-zinc, in bituminous dolomitic Proterozoic shales, undeformed and unmetamorphosed; exceedingly fine grained. >28 million ST Pb and Zn.

Sherwin Ironstone, Roper River, Northern Territory. Iron ore, in Proterozoic sedimentary rocks. 400 million ST 32–52% Fe.

Groote Eylandt, Northern Territory. Manganese, in Cretaceous rocks. >200 million ST 40–50% Mn. Operating mines.

Rum Jungle, Northern Territory. Uranium, uranium-copper, copper, and lead concentrations, in Proterozoic black shales. Production 5,000 ST U_3O_8 and 26,000 ST Cu. 1 million ST Pb in low-grade body; undeveloped.

East Alligator, Northern Territory. Uranium; seven recently discovered deposits. Reserves ~100,000 ST. U_3O_8.

Yampi Sound, Western Australia. Iron ore; hematite occurs in Proterozoic rocks. Operating mine.

Hamersley Iron Province, Western Australia. About 30 significant to major iron concentrations in banded iron formations. Operating mine.

Archean Iron Ore, Western Australia. Several deposits in banded iron formations, two north of Hamersley. Operating mines.

West Australian Nickel Sulfide, Western Australia. About 20 individual deposits. ~2 million ST Ni. Operating mines.

Middleback Ranges, South Australia. Iron ore, in Proterozoic iron formations. Main source of iron ore in Australia since 1915, and still the chief source for domestic needs. Operating mines.

Nairne, South Australia. Pyrite, 10% S, in Cambrian sedimentary rocks. Operating mine.

Beverley, South Australia. Uranium, in Tertiary sedimentary rocks. 11,000 MT U_3O_8.

Two examples of modern stratiform deposition are the very extensive pisolitic limonite ore of the Hamersley Range and the Yeelirrie uranium deposits in Western Australia. The limonite ore in the Hamersley Range originally was deposited in, and commonly filled, broad valleys. It now occurs mainly as caps on mesas. It is up to 60 m thick and contains intercalated shales and many fossil wood fragments. The Yeelirrie type of uranium concentration, with major uranium content, also originated from deposits in old drainage channels. If left alone for geologically long periods, both of these deposits would appear as stratiform concentrations deposited on an unconformity.

Discovery and Potential

One of the earliest discoveries of stratiform base-metal bodies was Great Cobar in New South Wales (1870), where secondary copper minerals were conspicuous, but interest in gossans was delayed another decade or two. Thus, the Broken Hill and Mount Morgan outcrops were untouched during 15–20 years of settlement and prospecting. By 1885, however, the prominently outcropping copper-lead-zinc deposits of Mount Lyell, Rosebery, Mount Bischoff, Cobar, Broken Hill, and Mount Morgan had been found. Most of these are still in production, and some of them produce at a greater rate and have a greater assured ore supply than in earlier years.

Thereafter, no important base-metal deposits were discovered for nearly 40 years. Mount Isa, discovered in 1923, also was a prominent outcrop. Then no discovery of importance was made for another 20 years, until exploration reached the poorly exposed or concealed orebodies.

There are several reasons for the gap of years between the first prospecting or exploration efforts in the vicinity and the discovery of some of the less conspicuous deposits: (1) poor or unrepresentative surface expression; (2) occurrence as "blind" bodies, not reaching the surface; and (3) lack of exploratory effort.

Geology of Deposits

The discovery of base-metal deposits is related to exploration and expertise, but potential is dependent on geology, in particular, the understanding of those facets of geology that produce the concentrations that we call "ore."

Twenty years ago, these stratiform deposits had not been identified as a group; nor had they been distinguished genetically from vein or other transgressive metal concentrations from which traditional ideas of ore genesis had been developed. Since that time, opinions have changed from a firm belief in the genetic importance of mineral textures to a realization that these are transient; from an unquestioned acceptance of emplacement as a late event of deep-seated igneous activity to a recognition of emplacement as an early and commonly contemporaneous result of sedimentary processes; and from a belief in the exotic introduction of minerals to one of their occurrence as the logical, if unusual, results of particular kinds of environments.

These changes in opinion were initiated mainly, but not entirely, in the course of exploratory and mining geologic studies of Broken Hill in 1946–60. It is interesting to note some earlier exposition of these ideas which, with more attention or more inherent vigor, could have led to earlier development of present thinking. The newer thinking was foreshadowed, but unfortunately not led, by Crook (1933). The modern statement of environmental concepts is given by Stanton (1972).

Thus we have now recognized an environmental pattern of occurrence of stratiform ores. Whereas the emphasis was once on local features of the geology such as faulting, folding, or proximity to intrusives, the emphasis now is on lithologic, basinal, and broad tectonic associations, as well as on geologic age.

There are certain periods of geologic time when deposition of specific types of ore minerals was most active. For example, in Australia, the following associations are recognized:

Archean—iron formations and iron ores, nickel-copper bodies, and some gold bodies;
Proterozoic—iron formations and iron ores, lead-zinc and lead-zinc-copper ores, and uranium;
Paleozoic—lead-zinc-copper pyritic ores, tin ores and tungsten ores (especially in Cambrian and Silurian);
Mesozoic—manganese; and
Tertiary—uranium.

This pattern has been ascribed to geochemical and geobiological factors (Pereira and Dixon, 1965). It is clearly not merely a feature of the distribution of those materials we call "ore" but, rather, it is a feature of earth history. This pattern is basically the same on all continents. The deposits long antedate the present disposition of continents and oceans and are therefore not related to the present features of the Pacific margin.

The stratiform deposits also show persistent compositional differences. The various metals occurring in stratiform deposits seem to have distinctive concentration patterns, and in places form almost monomineralic rocks.

In the lead-zinc-copper group, copper and zinc may occur in any proportions, as may lead and zinc; but lead-zinc ore normally contains less than 1% copper, and copper-rich bands or bodies contain little or no lead (or lead and zinc). These relations no doubt are due to unexplained chemical or biochemical affinities and antipathies.

Another important characteristic is grain size, which is all-important in the mill. Grain size evidently is related to metamorphism and to the size of the rock-forming minerals. Copper minerals are usually coarser than the lead and zinc minerals.

Though much has been learned about the mode of formation of stratiform sulfide bodies (Trudinger et al, 1972), the questions become more numerous and more complex as knowledge increases. An interesting aspect now being explored is the relation of biostratigraphy to the metal concentrations. In recent years, microfossils have been recognized in the Precambrian sulfide ore of the McArthur deposit (Muir, 1974), and a study is being made elsewhere of Precambrian mineralized sequences for possible biogenic material.

The source of the metals is another unanswered question. In a few deposits, the source is identifiable or plausible, but in most it is entirely conjectural. Deep-seated magmatic "emanations" are the first choice as a source, followed by seawater (or special currents) and leaching from an old weathered surface.

On the basis of the foregoing considerations, we now see stratiform and strata-bound ore deposits as normal, if sparse, features of principally Precambrian and Paleozoic rocks (also Mesozoic and Tertiary elsewhere in the world) of certain types. They include concentrations of Fe, Cu, Pb, Zn, Ni, Sn, W, U, and Ag (with Mn in a Mesozoic concentration) in several distinctive types of associations. The lead-zinc deposits usually carry valuable amounts of silver and cadmium.

The stratiform and strata-bound deposits have world importance in size or grade, or both. The shapes of the deposits may vary from elongated bladelike or pipelike bodies with dimensional ratios ranging from $100 \times 10 \times 1$ to $5 \times 1 \times 1$, to disklike bodies of perhaps $10 \times 10 \times 1$. The stratigraphic and lateral boundaries of ore are usually sharp, though in some places mild mineralization may be extensive.

Because of possible (and some actual) coincidence of elongation with flat net plunge and/or sharp ore boundaries, all subject to sparse occurrence, only a portion of the concentrations can be well enough exposed at the present surface to form prominent outcrops. Australian experience to date is that half of the 33 occurrences listed represent mineralization with a poor or unrepresentative surface expression or none at all. In addition, there probably are some deposits without surface expression and not findable through proximity to known deposits, but still within practicable and economic mining depth. Such deposits represent the next, and perhaps the principal, source of new discoveries of certain metals, especially lead and zinc.

The Australian Resource

The Australian continent contains more than 1.5 million km^2 of Precambrian and Paleozoic basins in six principal regions, *all* of which have produced major metal deposits other than iron ore. Exploration experience and patterns of ore occurrence suggest that these basins may be expected to contain other stratiform deposits in particular stratigraphic layers within localized suitable environments.

The layered rocks of these basins will probably be found to have flat net dips despite locally steep dips at outcrop; thus, a steep-dipping favorable layer may not be lost forever. The present surface gives us only a random horizontal section and a very imperfect view of the rock assemblage. The unseen parts are not necessarily without promise; they are merely unknown.

Early prospecting and subsequent exploration have probably located all the near-surface deposits that have some surface expression recognizable as being due to mineralization. The search will now be for deposits which are not self-revealing. It will have to employ—and largely develop—exploration criteria other than mineralization.

One lesson that has been learned is that, in these ore environments, the possibilities of mineralization are beyond our imagining. We should therefore be very careful about assigning limits to the potential, even in long-known environments.

References Cited

Crook, T., 1933, History of the theory of ore deposits: London, Thomas Murby, 163 p.

Muir, M. D., 1974, Microfossils from the middle Precambrian McArthur Group, Northern Territory, Australia, *in* Cosmochemical evolution and the origins of life: Dordrecht, Netherlands, D. Reidel, v. 1, p. 105–118.

Pereira, J., and C. J. Dixon, 1965, Evolutionary trends in ore deposition: Inst. Mining and Metallurgy Trans., v. 74, p. 505–527.

Stanton, R. L., 1972, Ore petrology: New York, McGraw-Hill Book Co., 713 p.

Trudinger, P. A., I. B. Lambert, and G. W. Skyring, 1972, Biogenic sulfide ores—a feasibility study: Econ. Geology, v. 67, no. 8, p. 1114–1127.

Background to Present Mineral Search in British Solomon Islands and Fiji—Summary[1]

J. C. GROVER[2]

The effects of earlier mining policy have inhibited exploration in the British Solomon Islands and Fiji. In both countries, government geological surveys were established amid hopes that government geologists might supplant prospecting activity by private enterprise. Long-range ground exploration, with continued support and technical collaboration from London, produced geologic maps. Modest finances demanded a basic philosophy differing from the advice from the private sector, but outstanding results were produced which complemented and supported the role of private enterprise in resources development.

In the Solomons, the situation demanded the evolution of a dynamic role for government. Unprecedented support from many sources and international collaboration in a series of ventures led to a buildup in fundamental exploration which deployed scientists, engineers, army personnel, research vessels, and ships of the Royal Navy. Nickel and other minerals were found and drilled. Activity culminated in a regional, pre-investment, airborne geophysical survey with shipborne and ground follow-up teams of competent indigenous staff, each team led by a scientist. The survey involved an expenditure in excess of $1 million. Copper and bauxite were the main discoveries. The knowledge was applied later to Fiji, where a different approach was needed.

Mapping progress and observations in Fiji suggested the need for complementary activity by the private sector. Copper mineralization indicating porphyry-type potential was described in an official letter to more than 80 major mining companies of the world; press advertisements also were used. The mining companies were not interested. A government decision to continue exploration led to the discovery of anomalies and large copper deposits now being test-drilled. Other minerals have also been found.

The remaining 13,000 km^2 of the main Fiji Islands was the subject of a decision in the same year. Within well-informed mining-policy guidelines, agreement was reached with a highly skilled research group for the expenditure of $2 million.

Hopes of financial independence now are held where formerly there had been none. However, the need for a well-informed government role in resources development is emphasized as a means to national self-realization.

[1]From Circum-Pacific Energy and Mineral Resources Conference program booklet.

[2]Consultant, Australia. Address unavailable.

Fluorite Resources of Thailand—Summary[1]

MANAS VEERABURUS[2]

Introduction

Beautiful cubic crystals of fluorite have been cut as gemstones in Thailand for a long time. Local fluorite was first introduced for domestic uses in the country by the Royal Thai Air Force Engineering Department during World War II.

The first export of fluorite from Thailand, in 1960, was 922 MT, from a total production that year of 3,460 MT. A record 427,498 MT was produced in 1972, of which 346,782 MT was exported. About 60% of the total production comes from the northern provinces of Lamphun, Chiengmai, Lampang, and Mae Hong Son. From 1960 until 1973, total fluorite production has been about 2,350,605 MT of metallurgical grade and 76,404 MT of acid grade.

Most of the commercial fluorite has been easily won from shallow surface outcrops in granite and in adjacent country rocks. It occurs in the forms of fissure veins, fault-fissure fillings, and replacement deposits. In the future, microscopic techniques, geophysical methods, intensive exploration for fluorite in depth, underground mining, and more efficient beneficiation facilities will be needed to maintain high production levels.

Geologic Background

Rocks ranging in age from Cambrian to Holocene have been recognized in Thailand. Most of the Paleozoic rocks—a mixture of schist, slate, argillite, limestone, sandstone, and shale—are marine deposits formed in the Thai-Malaya geosyncline. Mesozoic rocks, mostly of nonmarine origin, include sandstone, shale, and some limestone. Sedimentary rocks of marine, lacustrine, and fluvial origin include limestone, marlstone, sandstone, conglomerate, shale, oil shale, and lignite. They were deposited in various basins during Tertiary time. Quaternary deposits include unconsolidated silt, sand, and gravels which in many places contain placer tin.

Granites, the major type of igneous rocks in the country, are widespread. They occur as elongated masses with north-south trends in the northern and western mountain ranges, and continue both southward into the Malay Peninsula and northward into Burma. At least three ages of granites have been recognized in the country. Mineralization probably took place in Early Carboniferous, Early Triassic, and Tertiary times.

Major orogenic events which produced the present structural configuration occurred in Carboniferous, Triassic, and post-Jurassic times. Upper Tertiary to Pleistocene sediments have been affected by later and very recent activities.

Fluorite Deposits

Most of the fluorite deposits are in the western half of the country, in or near the granite mountain ranges and their contacts with sedimentary rocks. The most important factors in ore localization were the major faults, shear zones, fault breccias, and fault fissures, which acted as channelways for the mineralizing hydrothermal solutions. Fluorite deposits have been found along many major faults, and some deposits occur close to hot springs, although no definite evidence has been noted which would prove that fluorite was deposited by solutions from the hot springs. However, fluorite has been found commonly as thin films or fillings in voids in various fluorite deposits.

Minerals associated with fluorite are quartz, chalcedony, calcite, barite, galena, stibnite, pyrite, chert, and siderite.

On the basis of present knowledge, the fluorite deposits in Thailand can be classified as follows:

1. *Fissure-vein type*—Fluorite occurs as epithermal veins in fractures and fault fissures chiefly in granite and in the host sedimentary rocks close to the contact. This type of vein deposit is of high grade and has a massive texture. It is seen chiefly in Chiengmai, Ratchaburi, and Petchaburi Provinces.

2. *Pegmatite type*—Fluorite has been noted in pegmatite dikes consisting of coarse-grained quartz, feldspar, and muscovite. The mineralization, which occurred under conditions of high pressure and temperature, is related to hydrothermal solutions. Fluorite in these pegmatites is hard to recover by hand sorting. These deposits occur in granites and in their host sedimentary rocks in Ratchaburi and Petchaburi Provinces of west-central Thailand.

[1]Manuscript received, October 21, 1974.

[2]Department of Mineral Resources, Bangkok, Thailand.

3. *Cavity-filling type*—This type of fluorite deposit occurs in open fractures or open cavities, mostly in limestone. Deposits of this type show colloform banding or crustified textures. Mineralization seems to have been under conditions of very low temperature and pressure. Good crystal forms are seen in all such deposits.

4. *Replacement type*—The grade of fluorite in this type of deposit is generally low, but reserves are large. The mineral occurs in limestone, sandstone, conglomerate, and shale. Breccia zones seem to afford the most important structural control. Ore textures show relic fragments and textures of the original rocks. In some stratified sandstones and shales, fluorite tends to select sandstone, particularly arkosic sandstone, for replacement. Deposits of this type can be seen in Chiengmai, Kanchanaburi, and Ratchaburi Provinces. The replacement-type ores and the unmineralized country rocks have very similar visual characteristics, and they can be differentiated only under the microscope. Microscopic study of thin sections of replacement ores should be carried out in order to assist in the beneficiation of crude fluorite ores.

Geophysical Exploration

According to field observations, some fluorite in the vicinity of Lamphun Province shows relic fragments of the country rocks, and mineralization can be traced for more than 10 km. The deposits have a linear arrangement, suggesting that they lie on a fault line trending north-south. Experiments using geophysical methods for locating fluorite deposits along the fault line have been introduced into exploration work in the area 40 km south of Lamphun City. The LaCoste-Romberg gravimeter and the Askania magnetometer have been applied indirectly in the search for fluorite deposits—they are used to detect geologic structures.

Reading stations 10 m apart along traverse lines 25 m apart have been used. Terrain, drift, and diurnal corrections were applied to all readings, and then gravity and magnetic contour maps were prepared. Both geophysical interpretations and geologic data have indicated areas of high potential for fluorite deposits along the fault structure. Although several thousand feet of diamond drilling has been done, only the fault structures and traces of fluorite have been discovered.

Geophysical methods have been found useful only in a broad way for outlining structures; fluorite cannot be detected directly by these methods. Further study and more experiments are required.

Mining and Processing

Presently, 73 fluorite deposits are being mined in Thailand. Most deposits are shallow surface workings or open pits. Because of the nature of the deposits and the technical ability of the operators, only five deposits are being mined underground by shafts and drifts.

Fluorite is sold with little processing after mining—the usual means being by hand sorting, washing, and screening. The quality of the fluorite shipping product can hardly be controlled in this way; consequently, grade control is one of the largest problems in marketing Thai fluorite. The recovery of fluorite by hand sorting is only about 60%. Beneficiation facilities are needed in order to improve efficiency and recovery of fluorite.

In 1972, the first 50,000 MT/yr of acid-grade fluorite was produced by flotation at the Thai Fluorite Processing Company plant constructed in Petchaburi Province. Also, a heavy-media plant of the Universal Mining Company, with a capacity of 60,000 MT/yr, produced metallurgical-grade fluorite. This operation is located in Lamphun Province in the northern region. These two processing plants can maintain both an adequate-quality fluorite product and planned production levels for the country as a whole. It is expected that similar plants will be constructed in Thailand in the near future.

Reserves

The largest fluorite deposits known in Thailand are situated in the northern region, which now accounts for about two thirds of the total production of the country. The largest reserves are in Chiengmai and Lamphun Provinces. Other provinces having important fluorite resources are Kanchanaburi, Ratchaburi, and Petchaburi. More than 100 deposits are known in the northern region and about 50 deposits in the central region; only 9 deposits have been reported in the southern region. At present, no fluorite deposits are known in the east and northeastern regions, but the potential for discovering fluorite deposits in those areas is fair.

A complete study of fluorite reserves is not available. Some deposits have been drilled extensively by the owners, and some diamond drilling has been done by the Department of Mineral Resources. However, according to the last compilation of mineral reserves, made in July 1973, the estimated potential reserves of metallurgical-grade fluorite in Thailand are thought to exceed 11,500,000 MT.

Porphyry Copper Deposits of Southwest Pacific: Discovery and Development—Summary[1]

D. S. CARRUTHERS[2]

In the Southwest Pacific, extending from the Philippines south, economic porphyry copper deposits are known in the Philippines, Sabah, Irian Jaya, Papua, and Bougainville. Other deposits, either uneconomic under present conditions or not yet proved to be economic, are known in Sulawesi, the New Guinea highlands, New Britain, New Ireland, Manus, Guadalcanal, Fiji, eastern Queensland, New South Wales, and New Zealand.

The economic deposits are associated with Tertiary to late Quaternary intrusives of the western rim of the Circum-Pacific mobile belt, including the major west-northwest offset trending through the Solomons and New Guinea to Sulawesi.

The first significant mining opportunity was recognized in the Philippines in the mid-1950s. Since the early 1960s the rate of discovery and development has increased rapidly as a result of a deliberate search in favorable geologic environments.

Favorable conditions of terrain and rainfall, together with the size of the deposits, have made geochemical stream-sediment sampling an ideal technique for exploration and delineation of porphyry coppers.

To varying extents in different parts of the region, remoteness, rugged terrain, high rainfall, shortage of skilled workers, shortage of local capital, and rapidly evolving political institutions have influenced the evaluation and development of these deposits.

Because of the large scale of the projects, their location, and the generally low-grade ore, economic evaluation has been expensive and has been carried out with great care. Porphyry copper projects are capital intensive, and therefore investors require confidence in the political future.

Such projects can produce quick economic benefits to a developing country, such as capital inflow, export income, government revenue, and improved opportunities for employment and training. They also can lead to economic distortions such as local inflation and diversion of the few skilled workers from other parts of the country.

The social and political impact of establishing a porphyry copper project in an underdeveloped community can be immense because of the requirement for access roads, the highly technical nature of the operations, the introduction of expatriate workers, the requirement for significant land areas, and the very fact that extraction of natural resources is involved.

The long-term success of any project will depend on the understanding with which such problems are handled by both the operating company and the host government.

[1]From Circum-Pacific Energy and Mineral Resources Conference program booklet.

[2]C.R.A. Exploration Pty. Ltd., Melbourne 3000, Victoria, Australia.

Geologic Environment and Economic Possibilities of Porphyry Copper Deposits in the Philippines—Summary[1]

DOMINADOR H. ALMOGELA[2]

Introduction

A porphyry copper deposit, as used in this paper, is a hypogene copper sulfide deposit in which the ore minerals occur as disseminated grains and/or networks of veinlets within acid to intermediate intrusive bodies and the rocks they intrude. This definition emphasizes the occurrence and genetic significance without consideration of the economic aspects, and avoids the use of descriptive relative terms such as low grade and large.

Porphyry copper deposits now constitute the main source of copper production in the Philippines. The total ore reserve of this type of copper deposit is about 3 billion MT, ranging between 0.30 and 0.64% Cu. The total daily production stands at 137,700 MT. There are eight producing porphyry copper mines and eight deposits which are expected to be on stream within 3 years. Eighteen other major and promising prospects are presently under extensive drilling exploration.

General Geology of The Philippines

The Philippines is an archipelago of roughly triangular form, located at the border of the Western Pacific between 5 and 20°N lat. and 115 and 127°E long. It includes more than 7,000 islands, the largest of which are Luzon, Mindoro, Masbate, Samar, Panay, Negros, Cebu, Leyte, Bohol, Palawan, and Mindanao.

The eastern part of the Philippine archipelago, from Luzon through the central Philippines and Mindanao, constitutes the "mobile zone" in contrast to the "stable region" of the Palawan and Sulu Sea areas (Gervasio, 1966a). The mobile zone includes all the copper-mineralized belts. To date, no porphyry or disseminated copper deposits have been reported in the stable region. A very pronounced feature of the mobile zone, the Mindanao Deep, is one of the deepest oceanic trenches; it formed during the Miocene orogeny. It extends from southern Mindanao to a point just east of Catanduanes in southern Luzon, where it terminates abruptly.

In regard to plate tectonics, the Philippine archipelago lies at the junction of the Eurasian and Philippine Sea plates; the southeastern islands lie near the junction of these two plates and the Australian plate. The Philippine Sea plate is underthrust and subducted along the Mindanao Trench east of the archipelago. At the northwestern border of the archipelago, another subduction zone appears along the Manila Trench. The dip of this underthrust is eastward and opposite that of the Mindanao Trench subduction zone. The activity along these subduction zones during the late Cenozoic is manifested by volcanic arcs parallel with the Manila and Mindanao Trenches. The recent activity along the Philippine fault (a left-lateral wrench fault comparable in magnitude to the San Andreas fault) is very likely related to the recent activities along these subduction zones.

Tectonism related to plate movement prior to the late Cenozoic is not clear. However, a clearly defined tectonic regime began at about late Oligocene–early Miocene time and ended at about late Miocene–Pliocene time. This regime is characterized by the development of narrow geanticlinal cordilleras and geosynclinal troughs that remain as the major structural elements and control the present physiography.

The period from Late Jurassic to Oligocene time embraces another tectonic regime characterized by oceanic basaltic magmatism and emplacement of ultramafic masses; general northeast-southwest fold patterns developed during a Paleocene orogenic episode. The general structural grain developed in the Paleocene (NE-SW) is in marked contrast to that developed during the middle Cenozoic (N-S to NNW-SSE); thus a sharp change in the general stress-field orientation is indicated. Plate movements during the Paleocene must, therefore, be practically unrelated to present movements.

Fragments of continental crust are present in northern Palawan, Mindoro, and Zamboanga, as manifested by the occurrence of masses of metamorphosed upper Paleozoic quartzites, pelitic schists, and small granitic intrusives. The structural grain of these metamorphic masses generally follows that of the succeeding tectonic stage.

[1]Manuscript received, October 21, 1974.

[2]Supervising Geologist, Philippine Bureau of Mines, Manila, Philippines.

Table 1. Potassium-Argon Dating of Philippines Rocks

Locality	*Millions of Years*
Cebu (Lutopan Diorite)	59.7–close of Paleocene
Guimaras Island, Panay (diorite)	59.0
Rapu Rapu (schist)	42.0
Rizal (Sta. Ines Diorite)	36.0
Bantayan	28.7
Bantayan (andesite in well)	22.3
Bantayan	20.3
Bantayan	20.0
Baguio District (Antamok Diorite)	14.8
Paracale, Camarines Norte (diorite)	14.8
Baguio District (Itogon Quartz Diorite)	9.7–close of Miocene
Bantayan (diorite)	9.4
Batangas (San Juan Diorite)	9.2
Mati, Davao (diorite)	6.7

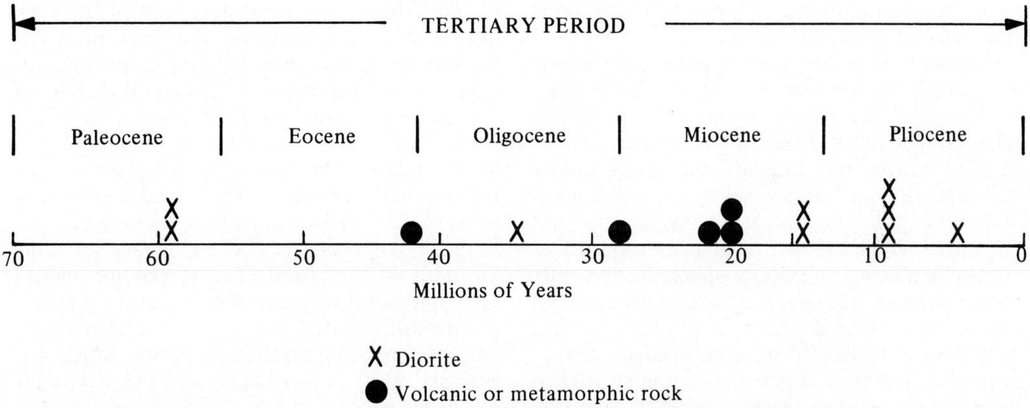

Mineral distribution and geologic setting in the Philippines show that there are certain geologic periods when deposition of specific types of ore minerals was most active. Regular patterns of ore occurrences are bounded in space and time: they occur within particular igneous complexes and are dated as middle to Late Cretaceous to Holocene. Several episodes of igneous activity occurred from late Paleocene to Holocene time.

Potassium-argon dating of Philippines rocks reported by Wolfe (1970) is shown in Table 1. From this table, it appears that the diorites in Cebu and Guimaras Island, Panay, were intruded near the end of Paleocene time, whereas the diorites clustered in the Baguio district, northern Luzon, to which the other porphyry copper deposits are related, were intruded during the middle to late Miocene.

Geologic Environment of Porphyry Copper Deposits

All of the porphyry copper deposits are near the axis of Tertiary geanticlines and within or adjacent to Tertiary dioritic plutons. Most of the deposits occur near pronounced fault zones, especially where the faults branch out or intersect other faults. Mineralization took place essentially along intensely fractured apophyses of the intrusives and along roofs or cupolas of quartz diorite stocks or batholiths that extend into the intruded, regionally metamorphosed volcanic and sedimentary rocks.

Baguio Mineral District, Northern Luzon

In the Baguio mineral district, northern Luzon, the porphyry copper deposits are confined within crumbled zones composed of volcanic wackes and lavas intruded by hornblende-biotite diorite. They are sporadically distributed close to the western margin of the Agno batholith, which constitutes the backbone of the Cordillera Central and Central Sierra Madre mountain ranges. This batholith intrudes a highly faulted and broadly folded sequence of undifferentiated Cretaceous to Paleogene metamorphosed volcanic and sedimentary rocks.

The producing porphyry copper mines and prospects in this region appear to be related genetically to the hypabyssal and felsic stocks of dioritic porphyries and their related facies disposed along a major structural line generally trending north-south. The mineralized crumbled zones possess at least three mappable alteration haloes that are distinct but somewhat overlapping—an outer argillic alteration zone; a middle, narrow, partly silicified and partly argillic-pyritic zone; and an inner biotitized-chloritized-silicified zone where most of the copper mineralization is concentrated (Gervasio, 1966a).

Chalcopyrite is the main copper sulfide mineral; it occurs with minor bornite, chalcocite and/or covellite, and traces of molybdenite. These minerals occur as disseminated specks and as veinlets across the gangue and along quartz stringers. In most places, the veinlets assume a stockwork pattern. Associated metals are gold and silver, which are recovered as by-products of copper beneficiation. Locally, sphalerite and galena are present. Magnetite occurs as veinlets and minute disseminated grains. In the Philex Mine, magnetite is recovered as a by-product. Supergene enrichment is apparent but, unlike that in the porphyry copper deposits in other countries, is not significant.

Breccia pipes have been recognized very recently in the Philex and Sto. Niño Mines. The breccia pipes are composed of chloritized metavolcanic rocks, diorite, and clastic fragments. In the Philex Mine, the breccia pipe is within the ore zone containing 3.0% copper. It measures about 20 m in diameter, and its upper dimension is larger than the lower. The breccia pipe at Sto. Niño is barren of copper. However, less than 0.5 km southeast of the barren breccia pipe, a fault breccia with an associated vein system contains 1.0% copper.

Central and Southern Luzon

Central Luzon is represented by one explored porphyry copper deposit—the Dizon copper prospect—localized at the eastern fringe of the Zambales Range, within the intersection of two major northeast- and northwest-trending fractures. Zambales Range is principally cored with ultramafic rocks which constitute one of the world's greatest sources of refractory chromite. This prospect is believed to be associated with a dioritic intrusive of probably Miocene age. The diorite is mainly covered with andesitic flows and pyroclastic rocks. About 1 km west of the Dizon copper orebody, dioritic bodies are exposed which are now the object of extensive prospecting and exploration by various mining companies.

In Marinduque Island, south of Luzon, the mineralization in the Marcopper Mine, and in another porphyry copper prospect that will eventually be in production, is localized within intensely fractured and silicified margins and/or roofs of quartz diorite bodies. Mineralization generally extends a few tens of meters into the intrusive mass as well as into the intruded rocks. Values commonly decrease as the coarser granitoidal portions of the intrusives are reached. The Island of Marinduque is considerably west of the Philippine rift system. However, dominant fault systems in the island generally trend northwest, parallel with and related to the structural element of the island that controlled the disposition of the diorite stocks in the region. One of these northwest-trending structures is an irregular contact between early Tertiary volcanic and sedimentary rocks which are interrupted by a series of Neogene (Tertiary) dioritic stocks cropping out along the crest of the island. The stocks are aligned northwest-southeast, and the producing Marcopper deposit lies near the center. The Ino-Capayang property of Consolidated Mines, which was expected to be in production by 1975, lies at the northwest end.

Cebu and Negros District, Central Philippines

The Atlas Mine, in the island province of Cebu, is associated with Paleocene intrusives. The Atlas diorite masses with K-Ar dates of 50.7 ± 1.2 m.y. (early Paleocene; Table 1) show an extensive low-grade copper sulfide mineralization primarily of chalcopyrite, occurring as encrustations and fillings along closely spaced fracture systems which were developed in the intrusives and surrounding rocks. Three large orebodies at Toledo, Cebu, are clustered within a 3-km-wide, northeast-southwest-trending intrusive belt. The disposition of the three orebodies and the intrusive belt is approximately parallel with the general structural grain of central Cebu.

The orebodies of Atlas Mine are entirely confined to the diorite; they correspond to the shattered zones that parallel the contacts between the diorite and metavolcanic rocks. Ore controls are both lithologic and structural. The stockwork character of the orebodies and their association with prominent northeast-trending faults indicate the role of structures that were superimposed on lithologic control. It could also be that structures guided the intrusives into place (Bryner, 1968).

Hydrothermal alteration is clearly indicated. Pyritization, argillization, and silicification are very common, and they characterize the outer halo of the orebody, which extends about 300 m away from the ore zones. Silicification in the outer fringes is generally in the form of translucent, relatively impure quartz. In the orebodies, the quartz is typically transparent and in fine aggregates and/or micro-honeycombs, or locally in drusy cavities. Very recently, two breccia pipes about 300 m apart were recognized in one of the orebodies. Breccia pipes in this region may serve as good markers for porphyry copper deposits.

Deposits similar to those at the Atlas Mine are found at Sipalay Mine and three other major prospects in southwestern Negros. These deposits lie along a northwest-trending anticline which is 60 km long and 25 km wide. This belt is underlain by broadly folded and faulted Cretaceous-Paleogene volcanic rocks with spilitic lava flows at the base and fine to coarse pyroclastics above. The volcanic rocks are intruded by a large northwest-trending batholith consisting mainly of diorite, quartz diorite, and hornblende diorite porphyry.

The Sipalay Mine consists of three major orebodies localized along the contact of the metavolcanic rocks and the diorite. The diorite occurs as a plutonic stock, believed to be an apophysis of the dioritic batholith in the region, the main body of which is at the northernmost tip of the porphyry copper belt. In one of the orebodies, a mineralized breccia pipe developed by fracturing, brecciation of the spilite, and intrusion of andesite is associated with pebble dikes (Fernandez, 1966). These pebble dikes are being used as a prospecting guide to locate potential ore in breccia pipes.

Other similarly occurring deposits within the porphyry copper belt are the Hinobaan project of Lepanto, the Basay project of CDCP Mining, and the Aya-Aya copper prospect. They are now being extensively explored by diamond drilling. In these properties, copper mineralization occurs mainly in the pyroclastics and, to a lesser degree, in the diorites and quartz diorites and in their aphanitic-porphyritic facies along the eastern margin of the batholith. However, in the Basay and Aya-Aya properties, the copper sulfide mineralization is greater in the diorite. The main orebody in the Hinobaan project of Lepanto is localized at the crest of a broad northeast-trending nose of the eastern flank of the batholith at its intersection with the northwest-trending major fault zone of the region. The orebodies of the Basay and Aya-Aya properties are localized at the intersections of the northwest-trending major fault zone with an east-dipping contact zone between the pyroclastics and the batholith.

Davao District, Mindanao

Mindanao, the second largest island in the Philippines, has one producing porphyry copper mine and several good prospects. These deposits are within the vicinity of the terminus of the Philippine fault zone of the mobile belt, in southeastern Mindanao. Studies by Malicdem et al (1966) disclosed that the mineralization in the area is not sporadic but, rather, shows a close affinity to small bodies of quartz diorite dikes, sills, and stocks of possible middle to late Miocene age, and to post–quartz diorite hypabyssal intrusives, flows, and pyroclastic rocks of late Miocene to Pleistocene age. The porphyry deposits in the region are related to the middle to late Miocene main stage of orogenic development accompanied by intrusion of quartz diorite, which probably supplied the mineralizing solutions. Northeast-trending longitudinal fractures representing the splays of the Philippine rift system in the fine-grained quartz diorite and pre-Tertiary volcanic flows are the controlling structures of the porphyry-type copper mineralization in the area. They are confined to the eroded nose of a northwest-plunging anticline.

Summary

Most of the major prospects in the country occur within the vicinities of operating mines or proved areas of porphyry copper deposits. All the porphyry copper deposits in the country are related to diorite intrusives disposed along specific orogenic belts. The Philippine rift system—the possible surface manifestation of the subduction zone—is believed to be the major controlling structure of the disposition of the dioritic masses. The Atlas and Sipalay deposits in the central Philippines are found within the Paleogene orogenic belt of the archipelago, whereas the Philex, Marcopper, Batong Buhay, Dizon, Sto. Niño, and Black Mountain deposits in Luzon, and the lower Masara deposit in Mindanao, are within the Neogene orogenic belt (Gervasio, 1966b). They are localized along the cracked and brecciated margins or at the contact between the intrusive masses and the intruded rocks. Generally, major structural lines are within or near the axis of Tertiary geanticlines, all within the mobile belt of the archipelago. Dioritic and quartz diorite stocks and sills, believed to be apophyses, roofs, or cupolas of batholithic masses controlled by north-trending regional faults, are the favored habitats of these deposits.

Breccia pipes have received a great deal of attention recently as a phenomenon that has some relation to the localization of porphyry copper mineralization. They are present in most deposits in the Philippines and are found within or near the ore zones. Consequently, the breccia pipes are being used as prospecting guides in the search for potential porphyry copper deposits.

Economic Possibilities

The gross mineral production values by 1973 amounted to 3.5 billion pesos, up 1.3 billion pesos (57%) from the 2.2 billion registered in 1972. Of this increase in production values, copper accounted for 72.61%, 80% of which came from porphyry copper deposits. The increased value of copper production was attributed partly to the increase of copper prices over the previous years and the record increase in production capacities by most of the operating mines.

Mineral production for the ensuing years can be expected to increase at an accelerated pace, largely as a result of the groundwork of the past years. Copper will remain as the major mining industry. Copper production for 1974 can be expected to be higher by about 8% with the entry of Western Minolco as a producing company, the complete shift of Philippine Iron Mines Co. to copper production, and the increases of the production capacities of most of the producing porphyry copper mines.

During the last 5 years, seven properties with porphyry copper deposits were proved to contain commercial ore reserves. They are Batong Buhay Mines in Kalinga-Apayao; Tawi-Tawi property of BCI in Bokod, Benguet; Dizon copper property in San Marcelino, Zambales; Taysan Copper Mines of BCI in Batangas; Ino-Capayang-Bintakay property of Consolidated Mines in Mogpog, Marinduque; the Hinobaan copper project of Lepanto Consolidated; and Basay copper deposit of CDCP Mining in southern Negros.

Under the global-lithosphere, plate-tectonics concept, the Philippines is an island-arc system at the terminus and boundary of the Pacific Ocean. The oceanic plate is underthrust westward into the Philippine Trench, in the direction of the cratonic China mainland. The Philippines lies within the Circum-Pacific "Ring of Fire," the site of deep-focus earthquakes, faulting and rifting, calc-alkaline volcanism, and consanguineous felsic plutonic intrusions. The porphyry copper deposits may be related to those tectonic events and associated igneous intrusions. If this relation holds true, potential localities of copper mineralization should occur in many places not yet explored, from the southernmost part of Cotabato, Mindanao, to the Central Cordillera and Sierra Madre in Luzon—i.e., throughout the Philippine mobile belt.

References Cited

Bryner, L., 1968, Notes on the geology of the porphyry copper deposits of the Philippines: Mineral Eng. Mag., v. 19, no. 3 p. 12–23.

Fernandez, H. E., 1966, Origin of the mineralized breccia pipe of Baclao Mine, Sipalay, Negros Oriental: Inst. Filipino Geologists Bull., v. 1, no. 1, p. 47–54.

Gervasio, F. C., 1966a, A study of the tectonics of the Philippine archipelago: Philippine Geologist, v. 20, no. 2, p. 51–75.

——— 1966b, The age and nature of orogenesis of the Philippines: Philippine Geologist, v. 20, no. 4, p. 121–140.

Malicdem, D. G., et al, 1966, Analysis of structures in the Masara Mine area, Mabini, Davao: Inst. Filipino Geologists Bull., v. 1, no. 1, p. 14–21.

Wolfe, J. A., 1970, Interpretations of potassium-argon (K/Ar) dating in the Philippines (abs.): Geol. Soc. Philippines Jour., v. 24, no. 4, p. 9.

Geologic Controls of Mineral Deposits in Taiwan—Summary[1]

T. P. YEN[2]

Geologic controls of mineral deposits can be classified into stratigraphic, magmatic, and structural, and each control may be divided into several kinds. Generally the three controls are so closely related in time and space that most mineral deposits are associated with at least two of them, or with all three.

The economically workable mineral deposits of Taiwan include bedded cupriferous iron sulfide deposits, gold-silver-copper deposits, carbonate deposits, sulfur deposits, and geotherms. The bedded cupriferous iron sulfide deposits are related to stratigraphic (primary geosynclinal to basinal), magmatic (basic volcanic), and structural (secondary macroscopic to microscopic) controls; the gold-silver-copper deposits to stratigraphic (secondary local), magmatic (intermediate volcanic), and structural (primary mesoscopic) controls; the carbonate deposits to stratigraphic (primary geosynclinal to basinal) and structural (secondary macroscopic to microscopic) controls; and the sulfur deposits and geotherms to magmatic (intermediate volcanic) and structural (primary mesoscopic) controls.

The geologic controls of mineral deposits which have been studied in Taiwan may be applicable in locating and prospecting for mineral deposits in other regions of the northwestern Pacific island arcs where geologic features are similar.

[1]From Circum-Pacific Energy and Mineral Resources Conference program booklet.

[2]Geological Survey of Taiwan, Taichung, Taiwan, Republic of China.

Mineral Resources of Korea[1]

OK JOON KIM[2]

Abstract The most economic mineral resources, other than fuel, in South Korea are gold, silver, lead, zinc, copper, tungsten, molybdenum, iron, fluorite, graphite, kaolin, talc, and pyrophyllite. These mineral deposits are related closely to the geologic settings and tectonic patterns of the peninsula.

South Korea is divided geologically and tectonically into four segments. The Kyonggi-Ryongnam massif, composed of Precambrian schists and gneisses, constitutes a base for the succeeding formations. The Okchon geosynclinal zone stretches from southwest to northeast diagonally across the peninsula in a direction known as the "Sinian direction." Its northeastern part is composed primarily of Paleozoic to early Mesozoic sedimentary formations and the southwestern part is composed primarily of the late Precambrian Okchon metamorphic series. The Kyongsang sedimentary basin occupies the southeastern part of the peninsula and is made up of a thick series of Cretaceous terrestrial sedimentary and andesitic rocks. A few small Tertiary basins are scattered in the eastern coastal area and in Cheju Island; they are composed of marine sedimentary and basaltic rocks.

Jurassic Daebo granites intrude the Kyonggi-Ryongnam massif and the Okchon zone, and Late Cretaceous Bulkuksa granites are randomly scattered in the Kyongsang basin.

Most of the mineral deposits are related to the acidic intrusives and are classified chiefly as hypothermal to mesothermal vein and metasomatic-replacement deposits, depending on the predominance of structural or lithologic control. In Precambrian terranes, gold-silver, lead-zinc, tungsten, molybdenum, and fluorite are present as vein deposits; and tungsten and iron deposits occur as metasomatic replacements.

In the Paleozoic terranes, gold-silver, lead-zinc, tungsten, and iron deposits are of metasomatic-replacement type, although some gold-silver deposits may be of vein type. In the Mesozoic terranes, iron, copper, and lead-zinc veins are common in the andesitic rocks and in cherty slate, whereas pyrophyllite deposits occur in acidic tuffs as a result of postigneous hydrothermal activity.

These mineralized areas, except the Kyongsang basin, are grouped into belts trending subparallel with the Sinian direction, and they generally show a rough zoning defined by the general gradient of genetic temperatures. The metallogenic epochs of these deposits are classified into four stages: Precambrian, Paleozoic, Jurassic to Early Cretaceous, and Late Cretaceous to early Tertiary.

Geologic Setting

Separated from North Korea by the north-northeast–trending Chugaryong rift valley, South Korea is divided geologically and tectonically into four segments: the Ryongnam-Kyonggi massif, the Okchon geosynclinal zone, the Kyongsang sedimentary basin, and the Tertiary basins. As shown in Table 1, the geology of South Korea includes all geologic sequences from early Precambrian to Quaternary, except for a break extending from the Late Ordovician to the Early Carboniferous.

Stratigraphy

The writer, in collaboration with colleagues and assistants, has been studying the Precambrian stratigraphy and the Mesozoic granites, and the results of this study are summarized briefly below. The major differences between the sequences summarized by Cheong (1970) and the writer are in the Precambrian geology.

Mesozoic granites—Up to 1970, these rocks were called "younger granites" and were thought to be of Cretaceous age. The writer (Kim, 1973) has differentiated the "younger granites" into the Jurassic Daebo granite and the Cretaceous Bulkuksa granite. The Daebo granite in the Precambrian terrane is aligned in the Sinian direction (northeast-southwest) across the Korean peninsula, but the Bulkuksa granite is randomly distributed in the Mesozoic Kyongsang sedimentary basin.

Precambrian geology—All the middle Precambrian rocks had been collectively named "granite gneiss system" by earlier workers. The writer (Kim, 1969a, 1972; Kim et al, 1963a, b) differentiated the "granite gneiss system" into orthogranites, gneisses, and schists, and divided the latter into formations, tentatively correlated as shown in Table 1. In the Kyonggi massif, the early Precambrian Yonchon system is divided into three metamorphic complexes—Puchon, Sihung, and Yangpyong. The middle Precambrian is divided into two newly established groups—the Jangrak and Chunsong. In the Ryongnam massif, the Ryongnam and Yulri systems are divided into series and tentatively correlated with the metamorphic complexes in the Kyonggi massif (Kim, 1974).

Unconformably overlying the metamorphic complexes is the Proterozoic Okchon system, which stretches diagonally in the Sinian direction in the central part of South Korea. The Okchon system, the most controversial among Korean geologists as to its age and sequence, was formerly believed to be metamorphosed Paleozoic and Mesozoic formations (Kobayashi, 1953;

[1]Manuscript received, November 6, 1974.

[2]Geology Department, Yonsei University, Seoul, Korea.

Table 1. Generalized Geologic Sequence in South Korea

		C. H. Cheong (1970)		O. J. Kim (1973)		
Age		*System*	*Series*	*System*	*Series*	
Cenozoic		Quaternary		Quaternary	basalts	
		Tertiary	Yonil Janggi	Tertiary	Yonil Janggi	
Mesozoic	Cretaceous	Kyongsang	Bulkuksa (granite & porphyry intrusion) Silla Naktong	Kyongsang	granites (Bulkuksa granites) volcanics Silla Naktong granites (Daebo granites)	
	Jurassic	Daedong	Ryukyong Sonhyon granites (in N. Korea)	Daedong	undifferentiated in S. Korea granite?	
	Triassic	Pyongan	Nokam	Pyongan	Nokam	
Paleozoic	Permian		Kobangsan		Kobangsan	
	Carboniferous		Sadong Hongjom		Sadong Hongjom granite?	
	Devonian Silurian		Chonsongri?		absent	
	Ordovician	Chosun	Great Limestone	Chosun	Great Limestone	
	Cambrian		Yangduk		Yangduk granite?	
Proterozoic		Sangwon	Kuhyon Sadangwu (in N. Korea) Jikhyon	Okchon	Kunjasan Hwanggangri Changri Munjuri Hyangsanri Kemyongsan — granite gneiss Taebacksan	Upper Precambrian
Archeozoic		granite gneiss		Yulri	Kosonri Kakhwasa Chunchon: Chunsong, Jangrak granite gneiss	Middle Precambrian
		Yonchon		Ryongnam	Wonnam Kisong Pyonghae Yonchon: Yangpyong, Sihung, Puchon	Lower Precambrian

Son, 1969). The writer, however, has concluded that the Okchon is late Precambrian (Kim, 1968) and that the Hwanggangri Formation is of tillitic origin (Kim, 1969a). These conclusions were supported by Reedman et al (1973).

Geologic Provinces

South Korea is divided geologically and tectonically into four provinces: Kyonggi-Ryongnam massif, Okchon geosynclinal zone, Kyongsang sedimentary basin, and the Tertiary basins (Fig. 1).

Kyonggi massif—Rocks of the Precambrian Yonchon system, as well as the younger Jangrak and Chunsong groups and associated granite gneisses, are widely distributed in the Kyonggi massif. The Mesozoic Daebo granite is widespread across the peninsula in the Sinian direction.

Ryongnam[3] *massif*—The Precambrian Ryongnam and Yulri systems, granite gneisses, and Mesozoic Daebo granite are widespread. The Daebo granite is aligned in the Sinian direction, but not so clearly as in the Kyonggi massif.

Okchon geosynclinal zone—This geosynclinal zone is within the Kyonggi-Ryongnam massif. Paleozoic to Mesozoic sedimentary formations are distributed in the northeastern neogeosynclinal zone, and the Precambrian Okchon system is present in the southwestern paleogeosynclinal zone. The Jurassic Daebo and Cretaceous Bulkuksa granites are scattered in both zones.

Kyongsang sedimentary basin—Cretaceous terrestrial sedimentary formations and associated volcanic flows and tuffs are present, and the Cretaceous Bulkuksa granite intrudes randomly into the sedimentary rocks of the basin.

Tertiary basins—Neogene sedimentary formations and associated basaltic flows are present in

[3]"Ryongnam" is the correct name of the district in Korea, although it was pronounced "Reinan" by the Japanese. Thus the name "Fukien-Reinan massif" is common in the older geologic literature. Recently, Korean geologists have used the name "Sobaeksan" instead of "Ryongnam." However, "Ryongnam" has precedence and is the correct name.

the small Tertiary basins and in Cheju Island south of the peninsula. Granites also crop out there, but they have not been dated.

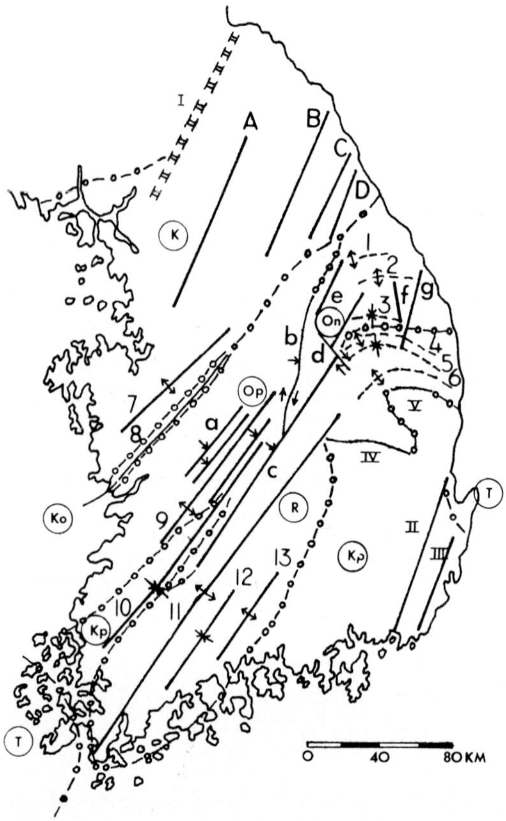

FIG. 1–Geologic provinces and tectonic map of South Korea.

Provinces (circled letters): *K*, Kyonggi land; *R*, Ryongnam land; *Op*, Okchon paleogeosynclinal zone; *On*, Okchon neogeosynclinal zone; *Kp*, Kyongsang proper basin; *Ko*, Kyongsang trough in Okchon zone; *T*, Tertiary basins.

Pre-Triassic faults: *A*, Pyunggang fault; *B*, Inje fault; *C*, Hyonri fault; *D*, Changchon fault.

Triassic deformation: *1*, Jongson syncline; *2*, Jungbongsan anticline; *3*, Hambaek syncline; *4*, Sobaeksan anticline; *5*, Yulri syncline; *6*, Andong anticline.

Jurassic deformation: *7*, Charyong anticline; *8*, Kongju syncline; *9*, Okchon anticline; *10*, Yongdong syncline; *11*, Dukyusan anticline; *12*, Kure syncline; *13*, Jirisan anticline; *a*, Okchon thrust; *b*, Bonghwajae thrust; *c*, Jomchon thrust; *d*, Danyang fault; *e*, Pyongchang fault; *f*, Samchok fault; *g*, Osipchon fault.

Cretaceous-Tertiary deformation: *I*, Chugaryong rift; *II*, Yangsan fault; *III*, Tongrae fault; *IV*, Andong thrust; *V*, Ilwolsan thrust.

Tectonism and Related Granites

The character and episodes of metamorphism and deformation that took place in the Precambrian terranes have become clearer in recent years. In the area of the Kyonggi massif, two major unconformities have been recognized (Kim, 1973); three unconformities have been identified in the area of the Ryongnam massif (Kim et al, 1963a, b). The metamorphic rocks separated by these unconformities show different degrees of metamorphism and different attitudes of deformation. Thus, the metamorphism and deformation can be categorized in three periods (Kim, 1972), although the ages of episodes and their correlation between areas are uncertain. Therefore, the geologic structure of Precambrian terranes in both areas is considered as pre-Triassic; structural breaks in Korea are definitely known to have occurred at the end of the Triassic (Songrim disturbance), the Late Jurassic (Daebo orogeny), the Late Cretaceous (Bulkuksa disturbance), and the mid-Tertiary (Yonil disturbance). The generalized tectonic pattern is shown in Figure 1; Tertiary structure is omitted because of its insignificance.

Pre-Triassic Deformation

In the Kyonggi massif, the foliation of the metamorphic complex is so diverse, even within the same system, that no generalization can be made. However, the prevailing trend is NNW-SSE in the west and NNE-SSW in the central to eastern parts of the massif. Four major faults that trend NNE-SSW are cut by the Daebo granite at their southwestern ends. There is no evidence that the foliation of the complex was affected by the intrusion of the Daebo granite. Although the Charyong anticline was rejuvenated at the time of the Daebo orogeny, it probably originated in Precambrian time.

The foliation in the Ryongnam massif is also very diverse, but it becomes nearly northeast-southwest toward the southwestern part and is cut slightly obliquely by the Jomchon thrust. The Sobaeksan anticline, Yulri syncline, and Andong anticline are classified as Triassic, and the Dukyusan anticline, Kure syncline, and Jirisan anticline are considered as Jurassic; however, all these features are thought to have originated in Precambrian time and to have been rejuvenated during a later orogeny.

Triassic Deformation (Songrim Disturbance)

In the Okchon neogeosynclinal zone in east-central South Korea, Paleozoic and Trissic sedimentary formations are folded and the axes of folds trend west-northwestward. This deformation is thought to have been caused by the Songrim disturbance at the end of the Triassic, because the Jurassic sedimentary rocks in the area were not affected by this deformation. The western ends of these folds were bent to the Sinian direction during the Jurassic. In the northeastern portion of the Ryongnam massif, the Sobaeksan anticline, Yulri syncline, and Andong anticline trend generally west-northwest, but it is surmised that they were rejuvenated by the Triassic deformation, although they might have been originally Precambrian structures.

Jurassic Deformation (Daebo Orogeny)

The Jurassic deformation, caused by the Daebo orogeny, took place and continued from Early Jurassic to Early Cretaceous time (as determined from age-dating of the Daebo granites). This orogeny was the most intense one in Korea, and some formations were severely folded and faulted. The nature of the Daebo orogeny is manifested in the distribution of the Jurassic Daedong sedimentary formations and the alignment of the Daebo granites, which are well exposed in the Okchon geosynclinal zone and adjacent Precambrian terranes.

The Jurassic Daedong sedimentary rocks are scattered in both the Okchon paleogeosynclinal and neogeosynclinal zones and are aligned in a northeast-southwest direction, whereas the Cretaceous Kyongsang rocks in these geosynclinal zones are not aligned or deformed in the same direction. The granites in the central region of South Korea, exclusive of the southeastern Kyongsang basin, crop out in the core of the four major mountain ranges, which are composed mostly of Precambrian formations. The trend of granite distribution is also northeast-southwest, parallel with the axis of the Okchon geosynclinal zone. The Precambrian formations adjacent to the Okchon zone, in both the Kyonggi and Ryongnam massifs, have been rejuvenated as synclines and anticlines which were originally deformed in Precambrian time. The time of tectonic development in the area is determined, on the basis of age-dating of the granites, to be Early Jurassic to Early Cretaceous. The foliation of the Okchon system strikes northeast-southwest (Sinian direction), coinciding with the axis of the Okchon geosynclinal zone, and dips mostly northwestward. These are mostly overturned isoclinal folds in which at least three overthrusts are apparent. Consequently, the same formations appear repeatedly in a narrow strip of the Okchon zone.

As shown in Figure 1, four anticlinoriums and three synclinoriums (nos. 7–13) run alternately from the southern border of the Kyonggi massif to the Ryongnam massif through the Okchon zone. The Okchon thrusts are in the Okchon zone; the Bonghwajae thrust bounds the Okchon paleogeosynclinal and neogeosynclinal zones; and the Jomchon thrust, which is intersected by the Bonghwajae thrust, bounds the Okchon zone and the Ryongnam massif toward the southwest. These anticlinoriums constitute major mountain ranges, and the younger Jurassic and Cretaceous strata are scattered in a few isolated locations within the synclinorium areas. As mentioned, the western ends of Triassic folds turn toward the Sinian direction, which indicates that they were modified by the Jurassic Daebo orogeny.

Late Cretaceous to Tertiary Deformation

Only minor folds are observed in the Cretaceous Kyongsang sedimentary basin, although the sedimentary formations in the basin show a general homoclinal structure to the southeast. The fragmentation of the basin caused by the post-Bulkuksa disturbance resulted in formation of the upthrusts at Andong and Ilwolsan, which bound the Precambrian basement and the Cretaceous strata. Along the Chugaryong rift valley, the Cenozoic basalt flows which cover the old river beds were extruded. High heat flows were noted along the Yangsan and Tongrae faults (Chang, 1970). These facts seem to indicate that these three fault zones might be related to recent northwestern Pacific tectonism.

Associated Granites

The granites of various geologic times are closely associated with the orogenies in Korea. The ages of the Precambrian granites are not certain, although they are estimated as shown in Table 1. It is also unknown whether these granites have any relation to metallogenesis.

The Jurassic Daebo granites—syntectonic plutons of the Daebo orogeny—were intruded along the Sinian direction in the cores of the Okchon folded mountain belts and in the Precambrian Kyonggi-Ryongnam landmass. These acidic intrusives are related to the genesis of most of the mineral deposits.

At the end of the Cretaceous and probably extending into early Tertiary time, the Bulkuksa granites and the associated acidic intrusives were intruded in the Kyongsang basin area and the adjacent Okchon zone as small stocks. They did not cause pronounced deformation. The granites accompanying the Bulkuksa disturbance caused induration of the surrounding rocks and produced various metalliferous lodes in them.

Granites of the post-Chosun (mid-Paleozoic) and Songrim disturbances (Late Triassic) have been reported in North Korea recently, but so far have not been discovered in South Korea. The granites of the Yonil disturbance (mid-Tertiary) are also expected to exist, but not enough age-dating has been done to locate them.

Productive Minerals and Classification of Ore Deposits

The minerals that are produced in great quantities in Korea are gold-silver, lead, zinc, copper, tungsten, molybdenum, iron, fluorite, graphite, kaolin, talc, and pyrophyllite. Except for the last four, these minerals are associated with the acidic igneous intrusives.

The types of mineral deposits in South Korea which are genetically related to the acidic intrusives have certain characteristics as observed in Precambrian, Paleozoic, and Mesozoic terranes.

In Precambrian terranes, gold-silver, lead-zinc, tungsten, molybdenum, and fluorite deposits are mostly associated with the Daebo granites and occur as hypothermal to mesothermal vein types, in pegmatites, or as epithermal deposits. Some of the tungsten and iron deposits occur as metasomatic-replacement types in intercalated limestone beds.

In Paleozoic terranes, gold-silver, lead-zinc, tungsten, and iron deposits are mostly associated with the Daebo granites. They occur as metasomatic-replacement types in the Cambro-Ordovician limestones, but the majority of gold-silver deposits are hypothermal to mesothermal vein-type deposits.

In Mesozoic terranes, copper, lead-zinc, and a few iron and gold deposits are associated with the Bulkuksa granites and are implanted in the fissures in either andesite or cherty slate of Late Cretaceous age. These are more or less mesothermal to epithermal deposits.

Graphite deposits are distributed in the Precambrian graphite schist; kaolin deposits occur as weathered products of anorthositic rocks; talc deposits are present in the late Precambrian dolomite beds or serpentine rocks; and pyrophyllite deposits occur in acidic rhyolitic tuffs as the products of post-igneous hydrothermal activity. Uranium deposits, nonproductive to date, occur widely in the carbonaceous beds of the Okchon system in central South Korea.

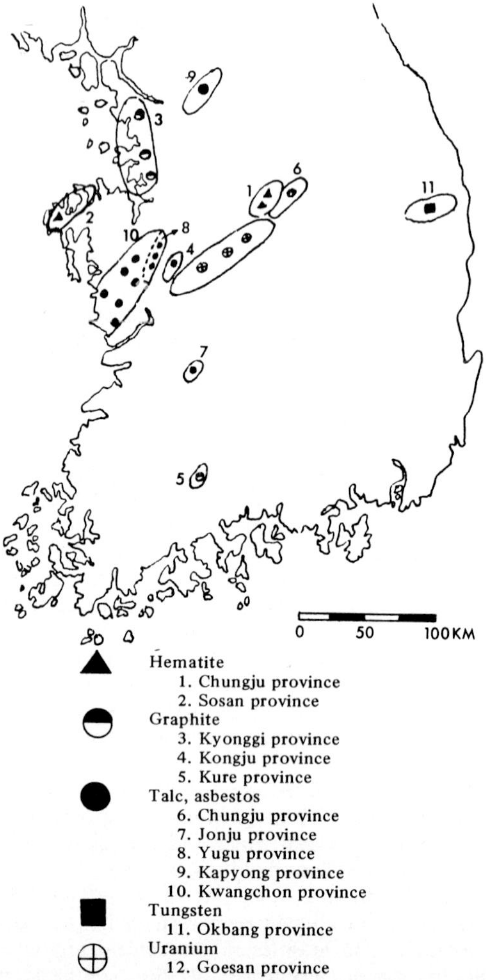

FIG. 2–Precambrian metallogenic province.

Gold deposits are distributed in practically all parts of Korea, particularly in the Precambrian terranes, and some of them are rather large. Lead-zinc deposits embedded in the Cambro-Ordovician limestone beds are of contact-replacement type; the Ulchin, Yeonhwa, and Shinyemi deposits are being operated on a large scale. The Sangdong scheelite deposit is embedded in Cambrian slate and limestone. This deposit was believed by earlier workers to be a contact-replacement deposit (Klepper, 1949; John, 1963), but the writer postulated that it was a segregation deposit (Kim, 1971a). The Okbang scheelite deposit is also believed by the writer to be a segregation type (Kim, 1969b), although others have classified it as a pegmatite deposit (Lee, 1967).

The copper lodes in the Kyongsang basin are of fissure-filling vein type in the andesite and cherty slate of Late Cretaceous age; they are concentrated in the southern coastal region. A copper deposit similar to porphyry copper deposits was discovered recently and is under intensive investigation in the Kyongsang basin. Fluorite deposits are concentrated in the Precambrian and Paleozoic terranes, and high-grade talc deposits occur in the late Precambrian dolomite beds (Tongyang talc mine) and in serpentine.

In general, the abundance or scarcity of mineral commodities is closely related to the geology of Korea. Abundance of gold, silver, molybdenum, tungsten, and fluorite is associated with predominance of granitic rocks and gneisses. Scarcity of chromite, nickel, and cobalt is due to the lack of mafic to ultramafic intrusives. Epithermal deposits of mercury, antimony, and sulfur are entirely absent because of the absence of recent volcanic rocks. The post-Ordovician sedimentary formations are mostly terrigenous deposits, and no petroleum or natural gas deposits have been found thus far except in the continental-shelf areas, where oil prospecting is currently under way.

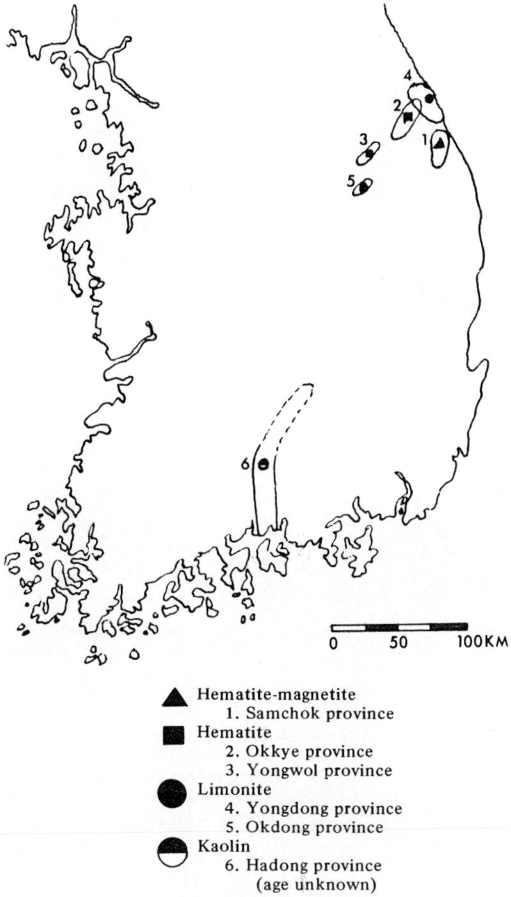

FIG. 3–Paleozoic and age-unknown metallogenic province.

METALLOGENIC EPOCHS

Metallogenesis of mineral deposits associated with igneous rocks is related to the igneous activity which accompanied orogenies and disturbances. The metallogenic epochs may coincide with the periods of syntectonic or subsequent igneous activity that accompanied the orogenies and disturbances. Metallogenic epochs, including those where metallogenesis was of sedimentary origin, that are certain in Korea are classified as (1) Precambrian, (2) Paleozoic, (3) Jurassic to Early Cretaceous, and (4) Late Cretaceous to early Tertiary (Kim, 1971b). However, mineral deposits in Korea have been studied in too little detail to permit differentiation of their metallogenic epochs; it is even difficult to discriminate the Precambrian from the Jurassic metallogenic epoch. Most mineral deposits which seem to be related to the Daebo granites are classified in the Jurassic to Early Cretaceous metallogenic epoch, although some probably are of Precambrian origin. Thus, the spatial distribution of the mineral deposits belonging to the Jurassic to Early Cretaceous epoch is widespread throughout Korea, except in the Kyongsang basin.

Asbestos, talc, some scheelite, hematite, and graphite may belong to the Precambrian metallogenic epoch. Talc is associated with Precambrian dolomite, asbestos with serpentine, and scheelite with amphibolite. However, their metallogenic epochs are uncertain.

Precambrian iron deposits of sedimentary origin are known in a few places. These deposits are mainly hematite accompanied by some magnetite. Graphite occurs in the graphite schist formations, and hematite in quartzite beds of Precambrian age.

Hematite beds are known to occur in the Dongjom quartzite of Early Ordovician age and at the base of the Late Carboniferous Hongjom series; limonite beds are known to occur in the Sadong formation of Permian age. These deposits are grouped together as representing the Paleozoic metallogenic epoch.

The major mineral commodities that are associated with the Jurassic to Early Cretaceous metallogenic epoch are gold, silver, lead, zinc, mo-

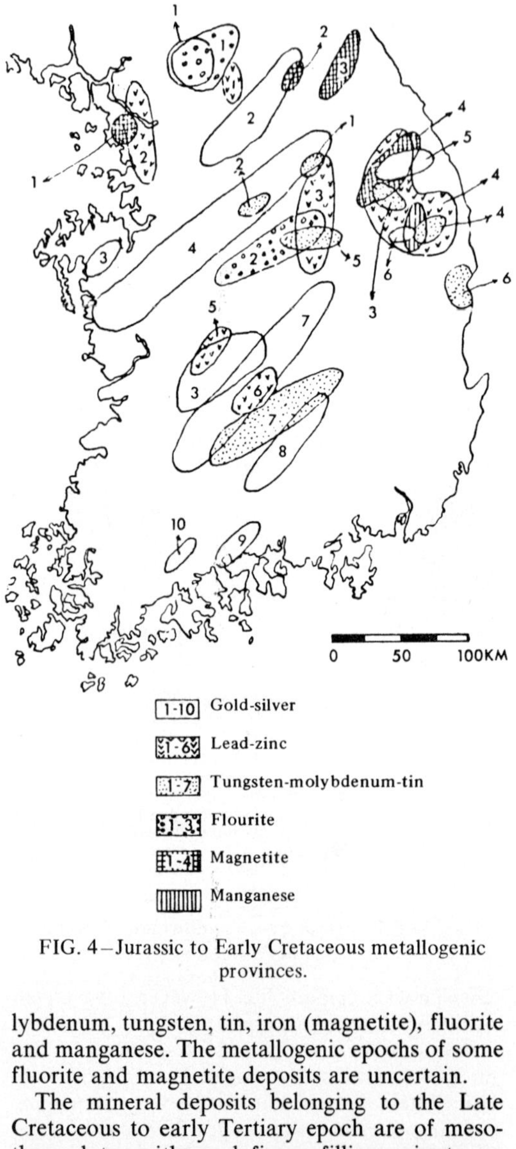

FIG. 4—Jurassic to Early Cretaceous metallogenic provinces.

lybdenum, tungsten, tin, iron (magnetite), fluorite and manganese. The metallogenic epochs of some fluorite and magnetite deposits are uncertain.

The mineral deposits belonging to the Late Cretaceous to early Tertiary epoch are of mesothermal to epithermal fissure-filling vein types. They are widely distributed in the Kyongsang basin and in minor portions of the Okchon geosynclinal zone where Bulkuksa granite crops out.

Minerals of the Late Cretaceous to early Tertiary metallogenic epoch are nearly identical to those associated with the Jurassic epoch, except that copper and pyrophyllite prevail and fluorite and tin are nearly or entirely absent.

Metallogenic Provinces

Of 952 deposits plotted, 40 deposits—12 Precambrian hematite-magnetite deposits, 17 Paleozoic hematite and limonite deposits, and 11 Precambrian graphite deposits—are of sedimentary origin but were deformed and/or metamorphosed in the later periods. The remaining 912 deposits are genetically related to igneous activity and are closely distributed within certain directions and areal limits; however, many deposits have random distribution and no spatial relation with delineated mineral zones. The mineral provinces and epochs are shown in Figures 2–5.

All the Precambrian metallogenic provinces are aligned along the foliation of the metamorphic complexes in which they are embedded. The Paleozoic metallogenic provinces coincide exactly with the distribution of the formation in which they are intercalated. The Jurassic to Early Cretaceous metallogenic provinces are distributed in the areas covered by the Daebo granite and its environs; hence the provinces are aligned chiefly

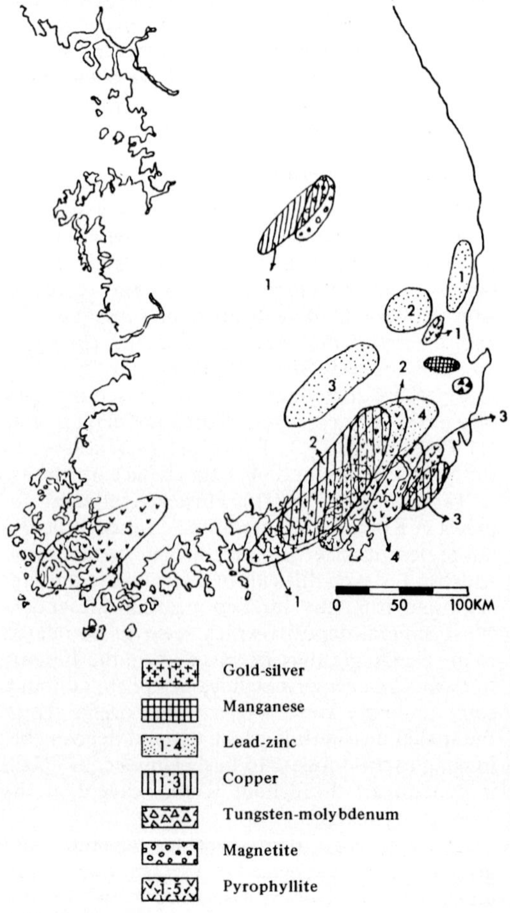

FIG. 5—Late Cretaceous to early Tertiary metallogenic provinces.

in the northeast-southwest Sinian direction (Fig. 4). The Late Cretaceous to early Tertiary metallogenic provinces, except the Jungwon copper belt, are distributed exclusively in the Kyongsang basin; their trends are more or less NNE-SSW, which coincides with the general trend of the Late Cretaceous to Tertiary faults.

Zonal Distribution of Deposits

The mineral deposits in the Kyongsang sedimentary basin are embedded in cracks and fissures in either andesite and related volcanic rocks or indurated black slate of the upper Silla series (Cretaceous). It is evident from this relation that lithologic control predominates over structural control in the basin area.

A rough zonal distribution of the mineral zones in the Kyongsang basin area can be determined (see Fig. 5). The copper deposits occur in the southernmost region of the area, and the lead-zinc deposits occur in the surrounding inland areas north of the basin. It is supposed that magnetite, pyrophyllite, and epithermal gold deposits are younger than other mineral deposits, although they overlap the copper zones. In other regions, the zonal distribution of the mineral deposits is not apparent even in the area of the same metallogenic epoch. It is the writer's opinion that, in general, metallogenic epochs become progressively younger toward the southeast in the Korean peninsula, as do the granites.

References Cited

Cheong, C. H., 1970, New introductory geology: Seoul, Korea, Tamgudang, 374 p.

John, Y. W., (Yong, Won John), 1963, Geology and origin of Sangdong Tungsten Mine, Republic of Korea: Econ. Geology, v. 58, no. 8, p. 1285–1300.

Kim, O. J., 1968, Stratigraphy and tectonics of Okchon system in the area between Chungju and Munkyong: Korean Inst. Mining Geol. Jour., v. 1, no. 1.

——— 1969a, Geology and tectonics of the mid-central region of South Korea; Korean Inst. Mining Geol. Jour., v. 2, no. 4, p. 73–90.

——— 1969b, On the genesis of Okbang scheelite deposit: Korean Inst. Mining Geol. Jour., v. 2, no. 2, p. 73–75.

——— 1971a, Geologic structure and ore deposits of Sangdong scheelite mine, *in* International association of the genesis of ore deposits: Tokyo-Kyoto, Internat. Mineral Assoc. Mtg. Proc. and Papers, p. 144–149; 1971, Japan Mining Geol. Spec. Issue No. 3.

——— 1971b, Metallogenic epochs and provinces of South Korea: Geol. Soc. Korea Jour., v. 7, no. 1, p. 37–59.

——— 1972, Precambrian geology and structure of the central region of South Korea: Korean Inst. Mining Geology Jour., v. 5, no. 4, p. 231–240.

——— 1973, The stratigraphy and geologic structure of the metamorphic complex in the northwestern area of the Kyonggi massif: Korean Inst. Mining Geology Jour., v. 6, no. 4, p. 201–218.

——— 1974, Geology and Tectonics of South Korea: CCOP Tech. Bull., v. 8, p.17–37.

——— et al, 1963a, Explanatory text of the geological map of Pyong Hae sheet (sheet 7024-II): Korea Geol. Survey, 28 p.

——— et al, 1963b, Explanatory text of the geological map of Samgun sheet (sheet 7024-IV): Korea Geol. Survey, 36 p.

Klepper, M. R., 1949, The Sangdong tungsten deposits, southern Korea: Econ. Geology, v. 42.

Kobayashi, T., 1953, Geology of South Korea: Tokyo, Japan, Tokyo Univ. Press.

Lee, S. M., 1967, Geology and ore deposits at the Okbang Scheelite Mine, North Gyeongsang Province, Korea: Geol. Soc. Korea Jour., v. 3, no. 1, p. 1–19.

Reedman, A. J., et al, 1973, Geology of the Hwanggan-gri mining district, Republic of Korea: Anglo-Korean Mineral Exploration Group, Geol. Survey Korea.

Son, C. M., 1969, Geology of Korea, Kwangjin, No. 3: Seoul, Korea, Korea Mineral Promotion Corp.

Relation Between Tectonics and Metallogenesis in Peripheral-Sea–Island-Arc Complex, Japan—Summary[1]

CHIKAO NISHIWAKI, HOKUICHIRO OMACHI, and ISAMU KOBAYASHI[2]

Introduction

This study is an attempt to correlate some geophysical and geological features of the presently active subduction zone with the mineral concentration which has taken place on the island arcs of Japan.

A reverse chronological extrapolation was attempted for the mineral zoning and distribution during Quaternary-Holocene, Miocene, late Mesozoic, and late Paleozoic times. It is based on a hypothesis of plate tectonics which includes the Pacific-ward shifting of the Japan arcs as the Japan Sea developed behind the arcs.

Recent Sulfur and Pyrite Ore Deposits

Many volcanogenic sulfur and pyrite deposits are present near the volcanic front of the Japan arcs. The shapes and types of these deposits are highly variable. They include solfataric sublimation deposits, sedimentary deposits, shallow-replacement deposits, and a large sulfur flow. Replacement deposits commonly include a large amount of iron sulfide and variable amounts of antimony and bismuth sulfides (Mukaiyuma, 1970).

Carr et al (1973) delineated distinct and abrupt discontinuities on the deep seismic zones under the Japanese arcs by careful use of the foci of intermediate and deep-focus earthquakes. Their work pointed out that these discontinuities generally resemble transverse faults, and these faults and segments are related to surface geology. However, as is seen on geologic or tectonic maps of Japan, no pronounced crosscutting or transverse structures have been recognized except the Fossa Magna, which roughly divides northeast and southwest Japan.

In northeast Japan, an extraordinarily close relation exists between the clusters of sulfur deposits and the intersection of the volcanic front with many of the transverse discontinuities in the deep subducting plate beneath the Japanese arcs. Ninety percent of the Pleistocene and recent sulfur deposits are located near the intersection of the present volcanic front with the surface trace of the transverse discontinuities.

The scarcity of expressions of the transverse discontinuities in the present topographic, geologic, or structural trends can be explained by the great depth of the subducting plate (100–300 km) beneath the Japanese arcs, and also by the presumed short time (<40–30 m.y.) since these discontinuities were initiated (Uyeda and Miyashiro, 1973). At the present rate of subduction (8–9 cm/yr), a portion of a subducting plate will take about 5 m.y. to move from the trench to a position under the western coast of northern Japan.

Neogene Mineral Distribution

Miocene mineral deposits in Japan are well represented by the kuroko (black ore) deposits, which are submarine-volcano exhalative deposits with mainly lead, zinc, copper, and iron sulfides associated with barium and calcium sulfates. They commonly occur in Miocene submarine basins in association with rhyolitic domes. The kuroko deposits are in the Miocene submarine volcanic area ("Green Tuff area") at the western (continental) side of the present volcanic front and are associated with acidic to intermediate volcanic rocks of the tholeiitic suite. These deposits are in the area of presently high rates of heat flow (>2μcal/cm^2/sec).

The locations of the clusters, as well as each deposit within a cluster, seem to be controlled not by younger, shallow, north-south crustal structural trends, but by an older northwest to north-northwest structural trend. This older trend must have been present in the basement rocks underlying the Miocene volcanic pile in which the kurokos were emplaced. This trend is the same as the direction of plate movement prior to 40–30 m.y. ago (Jackson et al, 1973; Morgan, 1971; Uyeda and Miyashiro, 1973).

The coincidence of the present high rate of heat flow and the Miocene submarine volcanic area—

[1]Manuscript received, December 26, 1974.

[2]Institute for International Mineral Resources Development, Fujinomiya, Japan.

The writers express their gratitude for discussions with and suggestions by A. Sasaki and S. Ishihara of the Geological Survey of Japan; T. Tatsumi and S. Uyeda of the University of Tokyo; and M. Yoshii of the Meteorological Bureau of Japan. We also are indebted to Charles C. Pineo III, of the Institute for International Mineral Resources Development, for the English editing of the manuscript.

consequently, the kuroko area as a whole—can be explained well if the present rate of heat flow is controlled by magmatic activity which occurred at a deep subducting plate in long past geologic time. Considering the low efficiency of heat transmission in the earth's crust, even if the mechanism of heat transmission was by a voluminous uprise of magma from along the subducting plate, present heat-flow rates could be explained by this model.

LATE MESOZOIC GRANITIC BATHOLITHS AND MINERAL ZONING

The Japanese arcs were not excluded from the worldwide intrusions of late Mesozoic batholiths of acidic magma. Acid magmatic activity began in the Jurassic and seems to have reached its peak in Late Cretaceous time. Granitic batholiths are associated with many metal concentrations, such as copper-iron and lead-zinc massive skarn deposits and tungsten and molybdenum deposits ranging from vein to massive disseminated types.

The asymmetric zonal distribution of the tungsten and molybdenum deposits is well documented by Ishihara (1971; 1973a, b); it shows close parallelism with the petrochemical zoning of acidic intrusives as a whole. The correlation of metal concentration and petrochemical zoning thus inferred suggests a deep and fundamental relation between the metallogenesis and geochemistry of the magma generated above the subduction zone under the present island arcs.

It is not possible to identify any deep tectonic lineament of Mesozoic age which might have controlled the local concentration of these metals.

PERMIAN GEOSYNCLINAL SEDIMENTATION AND MINERAL CONCENTRATION

The late Paleozoic Era in Japan was characterized by extensive geosynclinal sedimentation. Some facies are terrestrial, but sedimentation occurred mainly under shallow to intermediate marine conditions, most probably on the Precambrian bedrock (Kanmera, 1971; Nishiwaki, 1973). These late Paleozoic strata are intensely folded and faulted. They have been metamorphosed in several zones—some to high-pressure, low-temperature blueschist facies and others to low-pressure, high-temperature facies.

Strata-bound, massive sulfide deposits in metamorphosed eugeosynclinal sedimentary rocks associated with a predominantly alkali-basaltic volcanic pile occur principally in the Sambagawa system. They are characteristically very poor in zinc and lead. Also, there are many small, bedded manganese deposits in the so-called Chichibu geosynclinal sedimentary units of late Paleozoic age. These are associated with well-developed chert, slate, and graywacke and with lesser amounts of basic tuff or lava flows and limestone.

We think that the majority, if not all, of these two types of deposits are of submarine-volcano exhalative origin, having been formed at a stage of volcanic activity in a eugeosyncline situated at the eastern periphery of the Asian continent. They were deformed and metamorphosed to various grades and were moved out toward the Pacific (Kimura, 1973; Kimura and Tokuyama, 1971).

There are similar, but smaller and less abundant, massive sulfide and manganese deposits of Jurassic age in the less extensive eugeosynclinal pile in the outer zone of the Japanese arcs. Similar small, bedded manganese deposits of Neogene age are present in the Fossa Magna zone.

So far, there are no directly observable relations between the distribution of the kuroko deposits and the Paleozoic and Jurassic submarine-volcano extrusive deposits. However, their geologic features and genetic environments do not fundamentally contradict the assumed plate-tectonic model.

Most of the sulfur and pyrite deposits, with very minor amounts of chalcophile metal sulfides, occur on or near the volcanic front, which obviously is the shallowest point of magma generation above the subduction zone. So far, we have no evidence of present deposition of base-metal sulfides west of the front, even along these discontinuities, which represent the most probable channelways to carry metal elements toward the shallower crust.

We can well imagine that base-metal concentration may be taking place relatively far to the west of the present front in fissures above the discontinuities and somewhat deeper under the present surface, if concentration is occurring at all. The intersection of the discontinuities and deep tectonic fissures approximately 30–50 km west (continental side) of the volcanic front could be the most attractive area for base-metal concentration.

There is a very distinct compatibility between the distribution of kuroko and other Neogene mineral deposits and the present rate of heat flow. Furthermore, genetic relations between magmatic mineral concentration and plate subduction are strongly suggested by the facts that (1) kuroko deposits are almost everywhere associated with volcanic rocks of the tholeiitic suite or rocks at the boundary of the calc-alkalic suite in the petrologic zoning above the subduction zone, and (2) they are generally parallel with the direction of the arc-trench system.

There is an overall spatial parallelism between the distribution of the late Paleozoic massive sulfide zone and the present arc-trench system. This parallelism suggests a close spatial and genetic relation between the deposition of minerals and later tectonic movements in the tectonic history of this island-arc system from the time of deposition to the present.

References Cited

Carr, M. J., R. E. Stoiber, and C. L. Drake, 1973, Discontinuities in the deep seismic zones under the Japanese arcs: Geol. Soc. America Bull., v. 84, p. 2917–2930.

Ishihara, S., 1971, Major molybdenum deposits and related granitic rocks in Japan: Geol. Survey Japan Rept. 239 (in Japanese with English resume).

——— 1973a, Molybdenum and tungsten provinces in the Japanese islands and North American Cordillera—an example of asymmetrical zoning in Pacific type orogeny: Australia Bur. Mineral Resources Geology and Geophysics Bull., no. 141, p. 173–189.

——— 1973b, The Mo-W metallogenic provinces and the related granitic provinces: Mining Geology, v. 23, p. 13–32 (in Japanese with English resume).

Jackson, E. D., E. A. Silver, and G. B. Dalrymple, 1973, Hawaiian-Emperor chain and its relation to Cenozoic circum-Pacific tectonics: Geol. Soc. America Bull., v. 83, p. 601–608.

Kanmera, K., 1971, Paleozoic and Mesozoic geosynclinal volcanicity in Japan: Geol. Soc. Japan Mem., no. 6, p. 97–110.

Kimura, T., 1973, The old "Inner" arc and its deformation in Japan, *in* P. J. Coleman, ed., The western Pacific—island arcs, marginal seas, geochemistry: Univ. Western Australia Press, p. 255–273.

——— and A. Tokuyama, 1971, Geosynclinal prisms and tectonics in Japan: Geol. Soc. Japan Mem., no. 6, p. 9–20.

Morgan, W. J., 1971, Convection plumes in the lower mantle: Nature, v. 230, p. 42–43.

Mukaiyama, H., 1970, Volcanic sulphur deposits in Japan, *in* T. Tatsumi, ed., Volcanism and ore genesis: Tokyo, Japan, Tokyo Univ. Press, p. 285–294.

Nishiwaki, C., 1973, Metallogenic provinces in Japan: Australia Bur. Mineral Resources Geology and Geophysics Bull., no. 141, p. 81–94.

Uyeda, S., and A. Miyashiro, 1973, Geologic history of Japanese islands, *in* R. Suguyama, ed., Median tectonic line: Tokyo, Japan, Tokai Univ. Press, p. 289–334.

Geologic Aspects of Uranium Resources of Japan—Summary[1]

TEIJI KAMIYAMA,[2] YOSHIHIKO SHIMAZAKI,[3] and KYOSUKE KUBO[2]

INTRODUCTION

Systematic investigation of uranium resources in Japan was started by the Geological Survey of Japan in 1954. Until then, uranium minerals were only known to occur in specimen quantity from pegmatites and placers derived therefrom.

During the course of the next 10 years, two significant uranium concentrations—in the Ningyo-toge and Tono areas—were discovered by carborne radiometry, and 10 areas of the country were delineated as warranting detailed work. Contrary to the expectation of the geologists when the project first started, these were all areas of Tertiary sedimentary formations overlying granitic basement; neither pegmatites nor metallic deposits were included.

Despite strong efforts by the staff of the Geological Survey of Japan and by the Power Reactor and Nuclear Fuel Development Corporation to investigate domestic uranium resources, it appears that the uranium reserves in Japan are far from sufficient to meet demands. Although we are in the midst of an unprecedented energy crisis, we are also at the threshold of immense technical innovation. For example, the development of fast breeder reactors will not only cut down the consumption of nuclear fuel, but it will lower the cost of the fuel in proportion to the electricity produced, so that low-grade materials which are discarded today may become commercially feasible.

These factors show that future exploration must be planned with insight and that we must be prepared to cope with the rapidly changing needs of society. Thus it is necessary to evaluate carefully the various elements involved and to continue basic investigation.

DISTRIBUTION AND OCCURRENCE OF URANIUM IN JAPAN

Numerous radioactive anomalies are found in various geologic environments throughout Japan. The distribution of major uranium occurrences and acidic intrusive masses is shown in Figure 1. All significant uranium concentrations are related to granitic masses, and their occurrences are restricted to areas where large granitic bodies are present.

The reserves of major uranium deposits (see Table 1) increased 39% between 1969 and 1974 as a result of active exploration. The cutoff grade of this tabulation is 0.02%; the total reserves of ores of higher grade than 0.099% U_3O_8 are 3,955 MT. More than 97% of the uranium is concentrated in the "inner zone" of southwest Japan, which includes the Tono, Ningyo-toge, Sanin, and Okutango areas. Approximately 93% is concentrated in the two large deposits at Tono and Ningyo-toge.

These uranium deposits occur in host rocks of Neogene fluvial conglomerates and sandstones which unconformably overlie basement granitic rocks. Generally, the deposits have formed at the marginal parts of small sedimentary basins with channel structures, which are troughs carved out of the surface of the basement granite by pre-Miocene streams.

Organic material such as lignite is common in the sedimentary rocks. Although small patches of marine sedimentary strata are found in the Tono area, most of the host rocks were deposited under nonmarine conditions.

Basement is mostly granitic, late Mesozoic to early Tertiary rock, except at southern Kyushu and Ashizuri, where the basement is Miocene granite.

Ningyo-toge deposits—The deposits of the Ningyo-toge area are located in the vicinity of the boundary of Tottori and Okayama Prefectures. More than 12 deposits have been found in the approximately 400-km² area.

The uranium is concentrated in the basal part of Miocene conglomerate, arkose, and sandstone which unconformably overlie the Mesozoic-Paleogene granitic basement. The highest concen-

[1]Manuscript received, December 26, 1974.

[2]Power Reactor and Nuclear Fuel Development Corporation (PNC), Tokyo 162, Japan.

[3]Geological Survey of Japan (GSJ), Tokyo, Japan.

We acknowledge with pleasure the kind assistance of the staffs of PNC and GSJ during the preparation of this paper. A. Sato of the Chubu Exploration Office of PNC offered many valuable data and spent long hours discussing various problems. S. Ishihara of GSJ gave the writers much valuable advice, and K. Doi of PNC and Y. Sakamaki of GSJ kindly permitted use of their unpublished data.

FIG. 1–Distribution of major granitic bodies and uranium concentration in Japan.

tration of uranium occurs at the marginal parts of the channels rather than at the bottom.

A representative deposit consists of several channels with the total dimensions in the order of 1,000 m long, 100 m wide, and 1–2 m thick. The average grade of ore is 0.05–0.1% U_3O_8.

Tono deposits—The Tono area comprises four major uranium deposits. The host rocks consist exclusively of the Miocene Toki lignite-bearing formation. The major lithologies of this formation are conglomerate, sandstone, and shale. Large amounts of volcanic ash are present also.

Table 1. Uranium Ore Reserves in Japan, April 1, 1974

Area		*Ore (10^3 MT)*	*Grade (% U_3O_8)*	*U_3O_8 (MT U_3O_8)*
Tono		10,863	0.054	5,944
Ningyo-toge		4,957	0.050	2,471
Sanin		650	0.054	345
Tarumizu		235	0.047	110
Oguni		295	0.030	88
Okutango		143	0.043	62
Okujiri		54	0.063	34
Total	1974	17,197	0.052	9,054
	(1969)	(12,600)	(0.052)	(6,530)

The basement of the area consists of late Mesozoic to early Tertiary granitic rocks and Paleozoic formations. The uranium concentration occurs exclusively over the granitic basement. The element is concentrated at the basal part of the sedimentary rocks in the channels cut into the basement granite. The ores occur in the smaller channels in the upstream parts rather than in the larger main channels and the major basin. Also characteristic of this area is the concentration of the ores along the major fault at Tsukiyoshi, forming the largest deposit of the area. It is believed that the damming effect of the clayey material associated with the fault created the concentration of the mineralizing solution.

The orebodies usually have dimensions in the order of several hundred meters in width, several kilometers in length, and 1–5 m in thickness.

Geochemistry of Uranium Concentration

Although these occurrences of uranium in the usual sandstone-type deposits are found in many parts of the world, the physicochemical states of the concentration of both the Ningyo-toge and Tono areas are unique.

The major ore mineral of the deposits in the Ningyo-toge area is ningyoite. This is a new phosphate mineral discovered by Muto et al (1959). It occurs as minute friable grains, several microns or less in diameter, on the surface of pyrite and carbonaceous material. Thus, it is very difficult to identify, and only two analyses have been made to date. These analyses were made on minute amounts of the mineral separated from the ore, and the results are far from satisfactory owing to incomplete separation. The nature of the mineral was established, however, by comparison with synthetic material. The formation of this mineral is interpreted by Muto (1962) as the result of reaction between uranium-bearing solutions and apatite at temperatures between 35 and 100°C. Ando et al (1969) raised doubts as to the validity of this hypothesis on the ground that the amount of apatite in many parts of the conglomerate was insufficient to account for the phosphorus and rare earth elements contained in the mineral. They did clarify that the abundance pattern of the rare earth elements of the mineral was similar to that of granites; thus a genetic relation between the two is strongly indicated. The known occurrence of this mineral is unique to this area. Although it is probable that ningyoite may be found in uranium deposits of other parts of the world with the development of methods (such as the electron microprobe) for analyzing fine-grained particles, it cannot be denied that these deposits were formed under uncommon conditions. Other primary ore minerals in the unoxidized zone are coffinite and uraninite. They are gray to dark gray and commonly are associated with organic material. The ores of the oxidized zone commonly are associated with limonite, and the major ore mineral is autunite; also, B-uranophane and carnotite have been identified.

A rather unusual uranium concentration is found at the deposits of the Tono area. Here, coffinite and uraninite are the major primary ore minerals, and autunite is found in the oxidized zone. However, a very large part of the uranium occurs adsorbed on zeolites. The tuffaceous material of the Miocene sedimentary formation was strongly montmorillonitized and also zeolitized by diagenesis. The zeolites are of the heulandite-clinoptilolite group (Katayama et al, 1974).

Noting that uranium is adsorbed under pH conditions between 4 and 9, and also that gypsum is associated, Hirono (*in* Katayama et al, 1974) maintained that adsorption on and leaching from

Table 2. Uranium Content of Groundwaters at Mitake, Tono Area, Japan (ppb U)[1]

Uranium Content	*Granite (203 samples)*	*Contact (130 samples)*	*Paleozoic (22 samples)*
Average	0.089	0.026	0.009
Median	0.01	0.01	0.00
Maximum	10.00	1.82	0.09
Minimum	0.00	0.00	0.00

[1] From unpublished data of K. Doi.

zeolites occurred repeatedly owing to the fluctuation of the pH of the mineralizing solution: the pH decreases as a result of the production of sulfuric acid from the oxidation of pyrite and increases as a result of the formation of gypsum. The maximum amount of uranium which can be adsorbed on zeolite is about 0.9%, and these ores show almost maximum adsorption. Hirono also demonstrated that water in equilibrium with this zeolite must contain about 10 ppm uranium. Such a high concentration of uranium in groundwater is normally not attained, and Hirono believes that it possibly is due to the repetition of the above process.

The most important material in the fixation of uranium from groundwater is carbonaceous matter. Also, limonite and clay, which are invariably found in uranium concentrations in Japan, play important roles.

The close relation between the distribution of the uranium deposits and granitic rocks has been mentioned. The concentration of uranium and thorium in the later stages of magmatic differentiation—thus in acidic rocks—is well known. Also, it has been shown experimentally that increase of bicarbonate in water increases the proportion of uranium dissolved from natural rocks (Szalay and Samsoni, 1969). Approximately half of the uranium of fresh granitic rocks in Japan can be leached by normal circulating water (K. Doi, unpub.). An analysis of groundwater of the Tono area shows that uranium content of groundwater in the granitic environment is higher by a factor of 10 compared with water in other rocks (Table 2). From these lines of evidence, it is concluded that uranium was leached from the basement granitic rocks and transported to the site of the deposits.

Uranium and Granitic Rocks

There were three major periods of acidic intrusive activity in Japan—late Paleozoic to early Mesozoic (180–220 m.y.), Cretaceous to Paleogene (60–120 m.y.), and Miocene (14–20 m.y.). Much older granitic bodies exist, but they are small. The activity was not as intensive, and the granites are of relatively small importance in relation to mineralization.

The Triassic to Jurassic granites are present west of the Hida Mountains and sporadically in the "Inner Zone" of Southwest Japan. The acidic bodies in the Hida area are called "Funatsu granite," and their composition is mainly granodiorite to adamellite. The high uranium content in the granodiorite is unusual.

The Cretaceous to Paleogene acidic intrusive bodies are by far the most important, and they are distributed extensively in the "Inner Zone" of Southwest Japan and in the "Outer Zone" of Northeast Japan. These granites played the most important roles in uranium mineralization, and the two large deposits of the Ningyo-toge and Tono areas are genetically related to these intrusive bodies.

The Miocene acidic intrusives are sparsely distributed in the "Inner Zone" of Northeast Japan and the "Outer Zone" of Southwest Japan. The uranium content in the rocks of the "Outer Zone" of Southwest Japan is not low, and the two small uranium deposits at Ashizuri and southern Kyushu (Taramizu) are derived from these bodies.

It is clearly demonstrated that the uranium content of the granitic rocks in Japan increases with the acidity (Kanaya and Ishihara, 1974); and from various geological considerations, it has been proposed that the target for future prospecting be the sedimentary formations overlying the Funatsu granites in the Hida Mountains (Ishihara and Kanaya, 1974).

Concluding Remarks

With the increasing trend in mining lower grade materials and the rapid increase of demand for energy, it is inevitable that we will be investigating the feasibility of lower grade resources. As we do not have significant bodies of alkaline rocks and other uraniferous material, one of the major resources of this category in Japan is the uraniferous glauconite and apatite which contain 0.01–0.08% U_3O_8. These have been overlooked in the past because of their lower uranium content, but the nature of their occurrence shows possibilities of large reserves.

REFERENCES CITED

Ando, A., T. Hamachi, and Y. Shimazaki, 1969, Distribution of minor elements in the Toge' and Yotsugi uranium deposits, Ningyo-toge area: Japan Geol. Survey Rept., no. 232, p. 233–267.

Ishihara, S., and H. Kanaya, 1974, Problems related to domestic uranium exploration, *in* Symposium on raw materials development: Atomic Energy Soc. Japan Preprint, p. 26–27.

Kanaya, H., and S. Ishihara, 1974, Uranium, thorium, and potassium content of Japanese granitic rocks—a summary up to 1972, *in* The natural radiation environment II; V. 2: U.S. Energy Resources Development Adm. Conf. Rept., CONF-720805-P2, p. 517–533.

Katayama, N., K. Kubo, and S. Hirono, 1974, Genesis of uranium deposits of the Tono mine, *in* Formation of uranium ore deposits: Internat. Atomic Energy Agency Proc. Series, no. STI/PUB/374, p. 437–452.

Muto, T., 1962, Paragenesis of the minerals of the Ningyo-toge mine: Mineralog. Jour., v. 3, p. 195–222.

——— et al, 1959, Ningyoite, a new uranous phosphate from Japan: Am. Mineralogist, v. 44, p. 633–650.

Patterson, J. A., 1970, Character of the United States uranium resources, *in* Uranium exploration geology: Internat. Atomic Energy Agency Panel Proc., p. 117–127.

Robertson, D. S., 1970, Uranium—its geological occurrence as a guide to exploration, *in* Uranium exploration geology: Internat. Atomic Energy Agency Panel Proc., p. 267–284.

Szalay, S., and Z. Samsoni, 1969, Investigation of the leaching of uranium from crushed magmatic rock: Geochem. Internat., v. 6, p. 613.

National Exploration Program for Base-Metal Deposits in Japan—Summary[1]

YOSHIHIRO SEKINE[2]

INTRODUCTION

A considerable discrepancy exists between annual demand and domestic production of base-metal ores in Japan. This gap has been filled with imports of raw metals and ores. The demand-supply relations and changing economic conditions have exerted a serious influence on the metal-mining industry. In recent years, domestic metal mines have been in a critical situation owing to depletion of ore reserves, increasing cost of mining, increasing expenditure for prevention of mining pollution, constraints by conservation of the natural environment, decreasing skilled manpower, and fluctuating metal prices. As a consequence, many metal mines have been closed, although new kuroko (black ore) deposits have contributed to the supply of domestic ores.

NATIONAL EXPLORATION PROGRAM

The most important factor for the stable supply of raw minerals to industry is the discovery of new ore deposits and, hence, the ever-increasing ore reserves. With this point in mind, the Government incentive policy to the mining industry has been directed toward promotion of effective and efficient exploration activities to be done by the mining company. The Government mining policy is most important in this respect in Japan, where private mining enterprises generally have little financial capability and are not apt to venture risky exploration programs.

Small-Scale Mines

For many years the Government of Japan has provided a governmental exploration subsidy to small-scale mining companies and concession holders for prospecting works. The subsidy, equivalent to one half of expenses for prospecting works, is allocated to about 100 minor mines upon application through the Provincial Bureau of the Ministry of International Trade and Industry (MITI). If new ore bodies are discovered and a profit made by exploiting them, recipients of the subsidy are required to pay back a regulated amount of money, but not more than the amount received.

New Mining Policy

In the early 1960s, because it was anticipated that even major mining companies and large mines would meet with critical difficulties under free trading conditions, the Government started to plan and execute a new policy for stabilizing the mining industry. The fundamental purpose of the policy is to promote exploration activities by furnishing major mining companies with governmental exploration loans and by beginning national exploration projects.

The national exploration program for copper, lead, and zinc mineral resources (Fig. 1) is an integrated exploration system in three stages: (1) regional geologic survey, (2) detailed geologic exploration, and (3) company prospecting financed by the governmental exploration loan. For the purpose of conducting this long-term exploration program, the Metal Mining Agency of Japan was established in 1963 as a governmental corporation.

REGIONAL GEOLOGIC SURVEY PROJECT

Most of the deposits now exploited in Japan were developed from the outcrop area downward. At present, localities of more than 6,000 old prospects and closed mines have been inventoried, and it seems probable that very few undiscovered outcrops remain. Accordingly, exploration for new ore deposits must be to greater depths—200 m or more below the surface—as proved by recent discoveries of kuroko deposits.

The regional geologic survey project is conducted in large areas by means of combined use of all methods of geological, geochemical, and geophysical prospecting and by deep structural drilling. Most field surveys and works are carried out by contract with consulting and drilling companies. Geologic maps made in the project are published at a scale of 1:20,000. Airborne magnetic, regional gravity, and refractive seismic surveys are commonly used; and ground magnetic, induced-polarization and specific-resistivity or electrical-sounding charged-potential methods and geochemical exploration are utilized in more

[1]Manuscript received, October 21, 1974.

[2]Manager, Overseas Department, Metal Mining Agency of Japan, Tokyo 105, Japan.

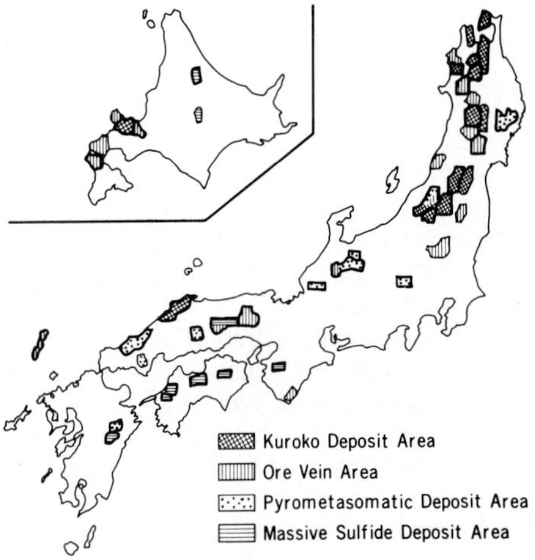

FIG. 1—Districts of national exploration projects for base-metal deposits. Hokkaido shown in inset.

restricted cases. Structural core drillings, though expensive, are indispensable for collection of geologic data at great depth.

Although the results of the project are published, and all information and data of the district are available to the public, the direct conclusion of the project is to select target areas within the district for the second stage of the program.

Detailed Geologic Exploration Project

The main scope of the second-stage project is to disclose detailed geologic features favorable to mineralization at selected target areas, and to find deep mineralized zones. Systematic or regularly spaced core drilling has been the principal type of work. Spacing of the exploratory drillholes ranges from about 400 to 1,000 m, depending on geologic conditions. Drillhole depth also varies depending on the geologic structures expected. Since 1969, a combined gravity–magnetic–induced-polarization survey has been used in several areas to effect better location of the exploratory drillholes. In 1970, exploratory tunneling with underground drilling was introduced as a substantial part of the project, and in 1973 a borehole induced-polarization method was used in some areas.

Information obtained in the project is published and available for use in follow-up prospecting works of mining companies.

Results of Exploration Projects

During 10 years of the national exploration program, substantial results have been achieved in the increase in payable ore reserves of base metals and the discovery of many new ore deposits, amounting to 93 million MT of commercial-grade copper ore and 29 million MT of lead-zinc ores. Some mines were opened in virgin areas, and many new orebodies have been developed and exploited in working mines.

Besides major discoveries in the Hokuroku and Nakatatsu districts, many mineralized zones and ore showings in drillholes and exploratory tunnels have been found in other districts. Many of these may contribute to finding new orebodies, such as massive sulfide deposits in the Shimokawa district, kuroko deposits in the Kunitomi district, pyrometasomatic copper deposits near the Kamaishi Mine in the Tono district, lead-zinc-copper veins near the Toyoha Mine in the Jozankei district, and copper-lead-zinc-tin-tungsten veins near the Akenobe Mine in the Bantan district.

Metallogenic Systems of Pacifides—Summary[1]

M. I. ITSIKSON[2]

INTRODUCTION

The idea of an independent Pacific ore belt was suggested by Smirnov (1946). He outlined within this belt two contrasting metallogenic zones—an inner, oceanic, essentially copper-bearing zone, and an outer, essentially stannic zone trending toward the continent. Smirnov's ideas were important for determining the trends of geologic exploration in the Soviet sector of the Pacific mobile belt (Pacifides).

As a result of integrated geological and geophysical research, extensive new data have been obtained which indicate the heterogeneity of the tectonosphere of the Pacific segment of the earth. Distributed in regular zones with respect to the ocean basin are volcanic-plutonic belts (Itsikson, 1973).

Within the Pacifides, the tectonosphere includes three types: (1) oceanic type, confined to the ocean basin and practically devoid of a granite-metamorphic layer; (2) continental type, with a well-developed granite-metamorphic layer and thicker crust characteristic of the continents framing the ocean; and (3) transitional type of mosaic structure, with great variations in crustal thickness. The distribution of the transitional-type tectonosphere is asymmetric with respect to the ocean basin and is limited by the marginal seas east of Asia and by some islands.

There is no transitional-type tectonosphere along the American continent; the oceanic plate is adjacent to the continental plate.

In accord with this zonation of the tectonosphere, the following types of large-scale volcanic-plutonic areas and belts extend from the ocean basin toward the continent:

1. Deep-seated intrusions and chains of volcanic islands of the basaltic area of the basin of the Pacific Ocean. The average content of SiO_2 in the basalts of the volcanic islands is 50.6%.

2. Ocean-margin volcanic-plutonic belt. This belt represents a chain of Cenozoic to Holocene structures that frame the ocean basin. The belt is confined to the boundary between the oceanic and transitional zones, and follows the highly mobile, discontinuous, circum-oceanic Benioff zone. This belt, also called the "andesite line," is actually a marginal oceanic andesite ring which includes the Andean-Cordilleran andesites on the continent. The composition of magmatic products is mainly andesitic, but some basalts and, less commonly, acidic differentiates are present. The average content of silicic acid is about 55%.

3. Continental-margin volcanic-plutonic belts of two varieties—Eastern Asiatic and Western American. They differ in their metallogenic characteristics.

The Eastern Asiatic belt, or "rhyolite line," is traced along the coast of Asia from Chukotsk to the Annam Mountains in Vietnam.

The belt follows a large-scale lineament clearly marked by gravitational variations, by a chain of positive and negative magnetic anomalies, and by great disparity in crustal thickness. The lineament separates the transitional and continental blocks. The Eastern Asiatic volcanogenic belt is believed to be a continental-margin rift zone developed during the late Mesozoic and part of the Paleogene. The volcanic-plutonic products are rhyolites and, to a lesser degree, dacites and andesites. The average content of silicic acid is 63.5%.

The Western American continental-margin, volcanic-plutonic belt joins the ocean-margin belt of the andesite ring.

4. The internal continental volcanic-plutonic chains (e.g., Khingan-Okhotsk, Great Khingan, Arguri-Kerulen), related to the rifting during the tectonic-magmatic activation of eastern Asia during the Mesozoic and part of the Cenozoic. This belt occupies an extreme peripheral position. These chains are formed on a totally continental crust of great thickness. The composition of magmatic products is mainly rhyolitic. Some differentiated basalt-rhyolite formations are observed. The average content of silicic acid is 70%.

Thus, in conformity with the asymmetric zonal distribution of tectonospheric blocks, the petrochemical composition of volcanic-plutonic belts grades from highly alkalic tholeiites and basalts within the oceanic trough, through the ocean-margin andesite zone and the continental-margin andesite-rhyolite zone, to prevailing rhyolite

[1]Manuscript received, November 11, 1974.

[2]VSEGEI (All-Union Geological Research Institute), Leningrad, USSR.

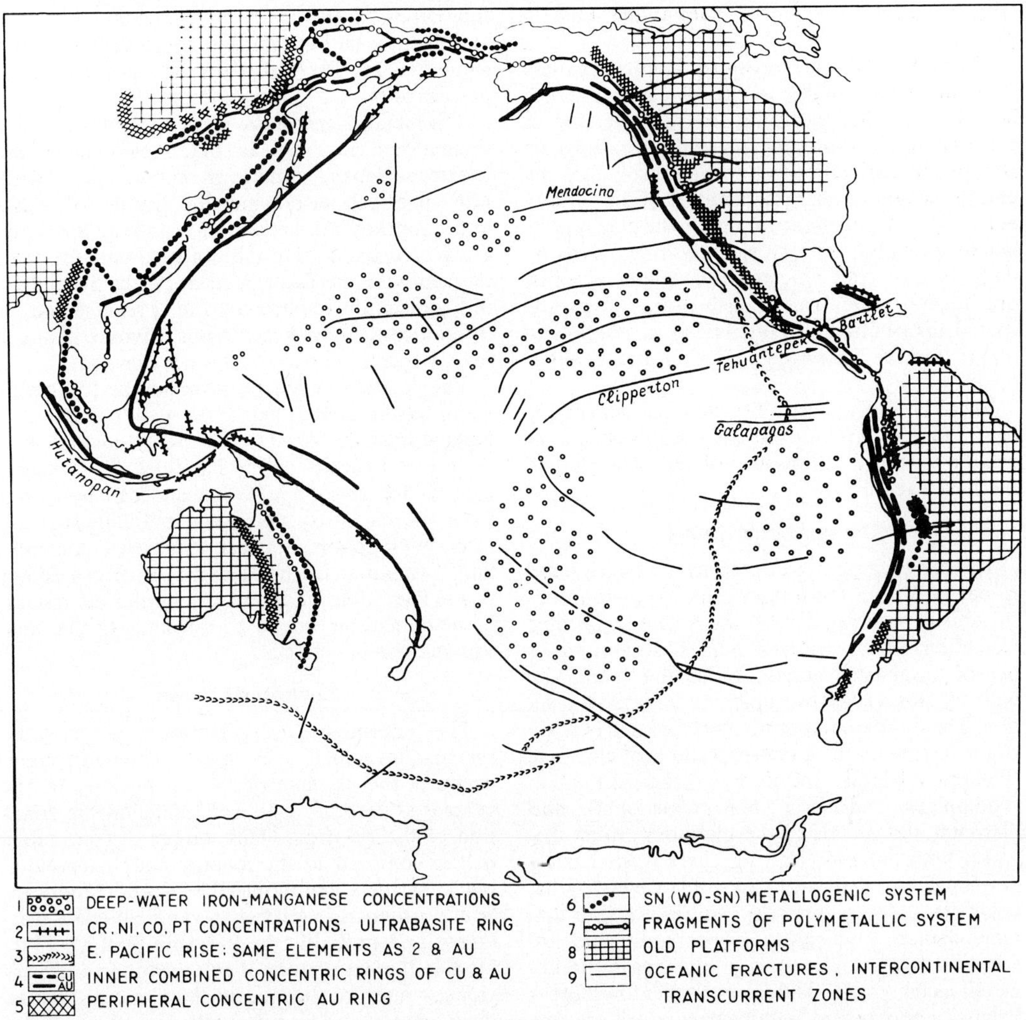

FIG. 1–Scheme of large-scale metallogenic zonation of Pacific segment. *1*. Main areas of distribution of deep-water iron-manganese concretions with copper, nickel, cobalt. *2*. Concentrations of chromium, nickel, cobalt, platinum, and platinum-bearing minerals in relation to ultrabasite ring of ocean frame. *3*. Hypothetical concentrations of same elements as *2* and gold in relation to East Pacific Rise. *4*. Inner combined concentric zones of "Great" copper and gold rings. *5*. Peripheral concentric auriferous ring. *6*. Stanniferous (tungsten-tin) metallogenic systems. *7*. Fragments of circumplatform polymetallic system. *8*. Old platforms. *9*. Oceanic fractures and intercontinental transcurrent zones.

(granitoid) magmas in the internal continental volcanic-plutonic chains.

Metallogenic Zones

In accordance with the outlined asymmetric quasi-concentric zonation of the tectonosphere and the corresponding changes in composition of volcanic-plutonic formations, there is also a geographic metallogenic zonation (Fig. 1). From the ocean basin toward the continent, four metallogenic zones are observed: (1) oceanic, (2) ocean margin, (3) continental margin, and (4) continental.

Oceanic Group

The group of oceanic metallogenic zones has been little studied. In the ultrabasite inclusions in basalts and andesite-basalts there are segregations of sulfides (pyrrhotite, pentlandite, chalcopyrite).

The accumulations of deep-sea iron-manganese concretions with high contents of copper, nickel, cobalt, and other elements are characteristic of the oceanic metallogeny. The study of radiogenic isotopes in concretions shows unmistakably the endogenic source of these isotopes (Cherdyntsev et al, 1971). A polygenic origin for the concretions is most probable, with an essential role played by degasification and volcanization which supplied ore minerals. This role is also indicated by the spatial distribution of concretions in relation to major fracture zones (e.g., Clarion and Clipperton).

There is also a possibility that concentrations of Cr, Pt, Co, Ni, and probably Au exist in connection with the ultrabasites of the axial part of the East Pacific Rise.

Ocean-Margin Group

The ocean-margin group of metallogenic zones is located within the framework of the ocean basin and is associated with the volcanic-plutonic formations of the andesite ring. A discontinuous belt of ultrabasite massifs, called the "peridotite belt" by Hess (1948), occupies the innermost position. The dunite-harzburgite facies of this belt are characterized by thick concentrations of chromite (Philippine Islands, Indonesia), nickel and cobalt (Philippines, Indonesia, New Caledonia), and platinum and metals of the platinum group. Together with the andesite ring, there is intensively developed mineralization of gold, silver, pyrite, polymetallic ores, mercury, antimony, barium, molybdenum, and sulfur. Copper and gold are the most typical elements of the ocean-margin metallogenic group. The inner concentric "Great Pacific" copper- and gold-bearing rings are discussed in a succeeding section.

Continental-Margin Group

The next boundary in the direction of the continents is marked by a group of metallogenic systems related to the continental-margin volcanic-plutonic belts.

The recorded asymmetry between Australian-Asiatic and American parts of the Pacifides resulted in great differences in metallogeny of the corresponding systems on the two sides of the ocean.

The continental-margin group in the Eastern Asiatic volcanogenic belt is characterized by concentrations of gold and silver in epithermal deposits; by mercury, polymetallic ores (Pb-Zn-Ag), and tin in cassiterite-silicate and cassiterite-sulfide deposits; and by boron (datelite-danburite), aluminoquartzites (alunite, pyrophyllite), and some fluorite. There also are some porphyry copper deposits.

The metallogeny of the Western American continental-margin belt, as determined from the Mexican–Central American section, has a definite silver-copper-polymetallic profile. Mercury and antimony are present in economic amounts. Gold predominates in the southern part. In comparison with the Eastern Asiatic belt, tin is definitely of minor importance and is represented by insignificant, though numerous, rhyolite deposits (e.g., Central plateau of Mexico; Nevada).

The Australian-Asiatic stanniferous (tungsten-stanniferous) zone is one of the geographic metallogenic systems of the continental-margin and continental blocks of the Pacifides. This system extends for about 19,000 km and can be traced from the Chukotsk Peninsula to Indonesia and, after interruption, appears in eastern Australia and Tasmania. In the Western American sector of the Pacifides, the Bolivian stanniferous system, of much shorter extent, corresponds to the Australian-Asiatic system.

Continental Group

The enormous circumplatform polymetallic system, developed over thick continental crust, occupies the extreme peripheral position in the general system of large-scale metallogenic zonation of the Pacifides. This unique discontinuous belt is confined to the mobile, marginal, folded block systems bordering the North American, South American, Australian, and Siberian platforms, and probably the Indostan platform (Fig. 1). The typical feature of the location of ore regions is their localization in the zones of thrusts, thrust sheets, and fractures following the boundaries of the platform structures.

Large-Scale Metallogenic Systems

Besides the metallogenic groups which are confined to a definite type of tectonospheric block, or to the margins of such zones, there are trans-Pacific metallogenic systems of the Pacific. The "Great Pacific" copper- and gold-bearing rings are examples.

Copper-Bearing Ring

The "Great Pacific" copper-bearing ring is one of the largest metallogenic systems, having a diameter in equatorial cross section of about 16,000 km. It has distinct biconcentric zonation. The inner concentric rings are related to Mesozoic and Cenozoic geosynclinal "ophiolite" depressions.

and to island arcs framing the Pacific basin. Copper, copper-pyrite, and copper-polymetallic systems with gold and silver are typical. This association contains some porphyry copper deposits, as well as manifestations of copper mineralization in connection with ultrabasites and gabbroids. In the peripheral concentric members nearer to continents there are ore-bearing complexes of nongeosynclinal deposition. Among these a porphyry copper deposit with molybdenum is of predominant importance.

The peripheral concentric member is most clearly displayed in the Western American section of the ring, where a unique Andean-Cordilleran belt of copper, copper-molybdenum, and molybdenum deposits can be traced from the Yukon to the Chilean Andes. A hypothetical, Eastern Asiatic homologue of this belt is outlined only by isolated deposits in southeastern Asia. It is assumed to be connected with a continental-margin rift system within the Eastern Asiatic volcanic-plutonic belt.

Gold-Bearing Ring

The "Great Pacific" gold-bearing ring is another of the major metallogenic systems. The idea of large-scale distribution of gold within the Pacific ring was suggested by Emmons (1937) and later was developed by other research workers. Indeed, the chain of ore regions and areas of gold placers circle the Pacific basin and its continental frame over the vast extent from the Chukotsk Peninsula to the Fiji Islands and New Zealand, and from the Yukon River to Cape Horn.

The inner auriferous concentric members coincide with the ocean-margin andesite ring. Most typical are deposits of gold-sulfide, gold-barite, and gold-barite-sulfide; shallow gold-silver deposits; and some tellurium and selenium. These deposits form a single formational complex with the pyrites, copper-pyrites, polymetallic deposits, and deposits of similar type of the inner concentric copper-bearing ring. They are also localized in the Mesozoic and Cenozoic geosynclinal depressions, and are found within the island arcs. The deposits are closely related to basalt and andesite magmas and their differentiates.

Within the continental tectonosphere, and along the boundary between the continental and transitional blocks, the peripheral concentric members of the gold-bearing ring are located. They differ from the inner concentric groups in association (gold-quartz, gold-chalcedony-quartz, gold-sulfide-quartz), as well as in the character of regional geotectonic control. The auriferous zones of the peripheral concentric members are more autonomous and rarely are associated with the zones of copper or other sulfide mineralization. They also differ in the type of magmatism, which is represented by rocks of mainly acidic or intermediate (diorite) composition.

The relation of auriferous zones to the latest stanniferous ones (e.g., In'yali-Debin zone in northeast USSR; Tasmania) is typical for the Australian-Asiatic part of the peripheral concentric ring in some provinces, as it is typical for the American part of the auriferous ring. One of the main features of the peripheral concentric ring is a belt of near-surface (epithermal) gold ore and gold-silver deposits connected with the evolution of the Eastern Asiatic continental-margin volcanic-plutonic belt.

Finally, extensive areas of auriferous mineralization related to shields and platforms frame the Pacifides. These occupy the outermost position in the general system of the gold-bearing ring, beyond the boundaries of the Pacific segment. Biconcentricity and some eccentricity (with displacement to the east) with respect to the ocean basin are common features of the trans-Pacific copper and gold metallogenic systems. Moreover, the inner concentric rings, which practically coincide for the two systems, are developed in the region of the contact of the oceanic and transitional blocks and are associated with the andesite ring. These concentric rings are characterized by ore derived from the early "ophiolite" depressions and island arcs. The peripheral continental-margin and continental concentric rings of the two global metallogenic systems are related to the continental tectonosphere and the region of its contact with the transitional type. Metallogenic systems are associated with volcanic-plutonic belts of the continental margin (Eastern Asiatic belt or rhyolite line) and internal continent. The ore formations may be paleogeosynclinal or, more commonly, may represent early orogenic stages of the Mesozoic mobile belts (Mesozoides). However, they mainly originate behind island arcs beyond the limits of geosynclines.

Thus the large-scale Circum-Pacific metallogenic zonation shown schematically on Figure 1 is only partly a zonation in the distribution of individual metals. Metallogenic zonation in the Pacific is created mainly by regular zonal distribution of individual formation groups of deposits of the same metals. The main point is that the largest trans-Pacific ring structures (copper- and gold-bearing rings, and probably mercury) developed beyond the boundaries of definite tectonosphere blocks, and are characterized by associations of certain metallogenic elements of the "mantle" (Itsikson, 1964). Ore-generating levels where these elements become isolated probably are situated at such a great depth that they are not related to the tectonosphere-block boundaries.

Stanniferous Metallogenic System

Tin behaves in a characteristic manner in the general system of zonation. Stanniferous metallogenic systems are confined to the peripheral zones of the Pacific that have thick continental crust and well-developed granite layers. Stanniferous systems are not characteristic of the inner concentric rings with crust of transitional or oceanic type. Thus, tin, a typical metallogenic element of crustal processes, is distinctly correlated with a definite type of tectonosphere. In contrast to copper, gold, mercury, and other "mantle" elements, tin has "single-element" metallogenic zonation which characterizes heterogeneous initial distribution of metals in the upper shells of the bedded model of the Pacifides. The ring character of zonal large-scale metallogenic systems (copper and gold) is strongly distorted in the near-equatorial region, whereas, in the southern hemisphere, the systems are distinctly displaced to the east. It can be assumed that the displacement occurs along the transcurrent fault system which unites the seismically active intercontinental Hutanopan (Sumatra)-Clipperton-Tehuantepec-Bartlett zones. The Caracas-Maracaibo fracture zone parallels this system in the South American sector. In individual zones, the length of fractures attains about 1,100–1,200 km; the shift is to the east. The transcurrent Hutanopan fracture divides the Australian-Asiatic stanniferous system, and has displaced its southeastern Australian Tasmanian flank far to the southeast. Also, the Central American gold-silver metallogenic system evidently is displaced in the same direction between Costa Rica, Panama, and Colombia (Itsikson, 1973).

It is necessary to assume a constant process of adaptation of the position of large-scale metallogenic systems in relation to rotational and gravitational fields of the earth in order to explain the marked disturbance of the initial metallogenic symmetry. At the mobile intercontinental zones, which are complicated by major faulting (e.g., Hutanopan and Bartlett fault zones), the greatest degree of asymmetry occurs. Large-scale faulting may also have affected the metallogenic systems in more northern latitudes. For instance, of great interest is the Anyuy-Yukon zone, which may divide the largest tectonosphere blocks of the Arctides and the Pacifides.

References Cited

Cherdyntsev, V. V., N. B. Kadirov, and N. V. Novichkova, 1971, The origin of iron-manganese concretions of the Pacific Ocean from radio-isotope data: Geochemistry, no. 3, p. 339–353.

Emmons, W. H., 1937, Gold deposits of the world, with a section on prospecting: New York, McGraw-Hill, 562 p.

Hess, H. H., 1948, Major structural features of the western North Pacific: Geol. Soc. America Bull., v. 59, no. 5, p. 417–445.

Itsikson, M. I., 1964, Some features of geochemistry and metallogeny of mobile belts of Pacific margin of the USSR: Soviet Geology, no. 1, p. 6–23.

——— 1973, The connection of metallogenic provinces with volcanogenic belts, large geotectonic elements, and types of earth crust, *in* Problems of regional metallogeny: VSEGEI, n.s., v. 191, p. 152–185.

Smirnov, S. S., 1946, On the Pacific ore belt: Akad. Nauk SSSR Ser. Geol., no. 2, p. 13–28.

Some Characteristics of Acid Volcanism of Southern Far East—Summary[1]

E. V. BYKOVSKAYA and E. I. BLUMSTEIN[2]

In the geologic history of the southern Far East, the postgeosynclinal acid volcanism was unrivaled in area of distribution, intensity, and volume of magmatic products. For the most part, the acid volcanism was an expression of the tectonic-magmatic activation processes, which became apparent in the various Mesozoic-Cenozoic structures—Bureya stable massif, Okhotsk-Nippon orogens, and Kuril-Japanese island arc. Within the limits of these structures, the intracontinental volcanic zone of the Lesser Khingan Range, the marginal volcanic belt of Sikhote Alin', the volcanic belt of southwest Japan, and the volcanic arc of northeast Japan were formed.

These manifestations of volcanism resulted from arc–fault-block movement of the earth's crust and the development of extensive linear, deep and shallow faults. Acid volcanism is characterized by its explosive character, predominance of ignimbrite eruptions, and close association with granite intrusions (Bykovskaya and Rotman, 1961; Krasnogo, 1973).

The postgeosynclinal complexes of these structures (Table 1) reveal variations in their mineralogy, petrochemistry, scale, and forms of ignimbrite volcanic manifestations.

Within the limits of the Lesser Khingan Range, the ignimbrite volcanism was characterized by the predominant development of high-potassic granitelike ignimbrite liparites closely associated with small-size, subvolcanic granite-porphyry intrusions. The liparite granitelike ignimbrites formed mainly extrusive domes, ranging in diameter from 10 to 40 km. The domes are located in the center of blocks of granite-metamorphic basement (Table 1). Their formation commonly was preceded by the effusion of andesites and liparites (in a ratio of 1:2), which are concentrated in the peripheral parts of the blocks and in the stable massif as a whole. The volume of the acid volcanic products exceeded 2,000 km^2 and, most probably, granitelike ignimbrites composed two thirds of it. Granitelike ignimbrites contain phenocrysts (40–50%), quartz, ultrapotassic sanidine, oligoclase, and biotite (<1.5–2%). The groundmass consists of pumice lentils ("fiamme") and devitrified materials such as quartz and alkali feldspar. Granophyric crystallization is common in the groundmass.

Characteristic of the volcanic belts of southwest Japan and Sikhote Alin' are: (1) simultaneous manifestations in both areas in Late Cretaceous and Paleogene time; (2) vast areas and large volumes of igneous rocks of intermediate and acidic composition with predominance of the acidic rocks; (3) parallel development (at least in Sikhote Alin') of granite-liparite and diorite-andesite series; (4) prevalence of ignimbrites; and (5) close association of volcanic rocks and granites, forming "self-burying" intrusions (i.e., those which crystallize under the crust of their own volcanic products).

During Late Cretaceous–Paleogene time, in the volcanic belt of the southern Sikhote Alin', there existed the following succession of magmatic processes: (a) intense and widespread eruptions of dacite liparite ignimbrites; (b) granite batholith intrusions; (c) effusions of liparites, obsidian, and dacite tuff-lavas and eruptions of pyroclastic liparite flows.

During the Late Cretaceous, in the earlier stage of magmatic evolution, the sheet type of volcanism prevailed and resulted in formation of wide volcano-tectonic depressions built by the liparite ignimbrites of the "large-volume fields" (Smith, 1960a, b).

At the end of the Late Cretaceous–Paleogene, in the "post batholithic" stage, ignimbrite volcanism was of considerably smaller scale compared with the initial stage but was most varied in form and character of association with granite magmatism and the process of caldera formation.

In the Olga-Therney volcanic field of southern Sikhote Alin', a variety of the ignimbrite volcanism forms was established. These forms varied in time from: (1) liparite crystal-ignimbrites of the "large-volume fields" (20–30% phenocrysts) to (2) granitelike ignimbrites of domes (phenocrysts = 40–50%), and then to (3) ignimbrite plateaus and autonomous flows (phenocrysts <15%). The liparite granitelike ignimbrites are easily distinguished in the topography by slope, steep cupolas, and fault-line scarps.

[1]Manuscript received, November 11, 1974.

[2]VSEGEI (All-Union Geological Research Institute), Leningrad, USSR.

Table 1. Main Characteristics of Acid Volcanism of Southern Far East

	Lesser Khingan Range	*Eastern Sikhote Alin'*	*Southwest Japan*	*Northeast Japan*
Time of acid volcanism[1]	End of Early Cretaceous to Late Cretaceous (1) 116-80, (2) 80-75	Late Cretaceous - Paleogene (1) 77-65, (2) 60-40	Late Cretaceous - Paleogene 95-60	Late Neogene - Quaternary
Predominant type of eruption	Central	Central and areal faults	Faults	Central
Morphology of volcanic and extrusive bodies	Cupolas (domes), rarely flows	"Large-volume fields," cupolas, flows	Flows	Flows
Volume in km^3	2,000	42,000	53,000	100
Main rock type	Granitelike ignimbrites and liparite lavas	Ignimbrites, granitelike ignimbrites of liparite and of liparite-dacite	Lavas, liparite, and liparite-dacite ignimbrites	Pumice and dacite ignimbrites
Main characteristics of chemical composition[2]	Rich in SiO_2, K_2O; poor in Al_2O_3, CaO, MgO. Di = 93.21	Normal quantity SiO_2; almost equal contents of K_2O, Na_2O. Di = 85.83	Normal quantity SiO_2; insignificant predominance of K_2O over Na_2O. Di = 79.56	Poor in (K_2O + Na_2O), K_2O, and FeO; rich in CaO; comparatively rich in MgO. Di = 74.78
Mineralization	Gold-tin	Tin-polymetal (Pb, Zn), possibly gold-silver (epithermal)	Tin-polymetal(?), gold-silver (epithermal), possibly copper-molybdenum	Silver-copper
Basement: Age and Composition	Precambrian metamorphic rocks, Paleozoic granites	Paleozoic and Mesozoic volcanogenic-terrigenous rocks	Mesozoic terrigenous rocks and quartz diorite	Late Mesozoic and Cenozoic (Miocene) volcanogenic-terrigenous rocks
Basement: Structural Type	Stable massif	Geosynclinal-orogenic	Geosynclinal-orogenic	Island arc

[1] Age in millions of years (absolute age determined by K-Ar method).
[2] Di = differential index, represented by sum of the normative quartz, orthoclase, and albite (Thornton and Tuttle, 1960).

The ignimbrite volcanism, which took place in the volcanic arcs of northeast Japan from the end of the Pliocene to the beginning of the Quaternary, appears to represent a terrestrial eruption of moderate to acidic volcanic rock. In the Quaternary the front line of active volcanoes was slightly more oceanward.

We are led to the conclusion that the main characteristics of postgeosynclinal acid volcanism are the geological and petrochemical zoning of the ignimbrite volcanism from the ocean zone (Japan) to the interior of the Asian continent (Lesser Khingan Range). Supporting reasons are: (a) an increase in K_2O and SiO_2 and corresponding decrease in content of CaO, MgO, Al_2O_3 and in the ratio of Fe_2O_3 and FeO; (b) the growing importance of ignimbrites of the "large-volume fields" and extrusive domes; (c) an increase of the differential index from 74 to 93%; and (d) association of Ag-Cu mineralization with the sodic volcanic rocks (northeast Japan), Sn-Pb-Zn and Sn-W-Mo with potassic-sodic intrusive-effusive associations (eastern Sikhote Alin' and southwest Japan), and Au-Sn mineralization associated with granite porphyries and potassic granitelike ignimbrites of the domes (Lesser Khingan Range). These features could be accounted for by the depth of crust anatexis and by tectonic development of structural forms.

The thick (35–40 km) continental crust of the Lesser Khingan Range and the long period of quiescence which preceded activation of the stable massif caused the formation of peripheral magmatic foci and contributed to the ultra-acidic composition and the increased potassium ratio.

The great ignimbrite masses and large granite plutons of similar composition in the volcanic belts of eastern Sikhote Alin' and southwest Japan, and the predominant sheet type of volcanism, prove the shallow location of the magma-formation level within the earth's crust. On the basis of mineral thermometry and comparison of experimental data (Tuttle and Bowen, 1958; Wones and Eugster, 1965), the magmatic process is characterized by minimum temperatures of crystallization (780–900°C) and sudden changes of oxygen partial pressure and moderate H_2O content in the initial melting. Stability of the "granite-water" system is known to be achieved at temperatures of 600–700°C and water-vapor pressure of 4 kb (Tuttle and Bowen, 1958). If the Sikhote Alin' geothermal gradient of 30°/km and the similarity of the ignimbrite composition to the gross composition of terrigenous-volcanic and terrigenous rocks of the upper Paleozoic–lower Mesozoic are taken into consideration, we can presume that the primary acid magma was formed as the result of the melting of these rocks at the depth of 12–20 km.

The intermediate character of the acid volcanism of the marginal volcanic belts, together with the above-mentioned peculiarities, is the result of the position of the volcanic belts on the border of the continent, the intermediate types of crust, the closeness in time of the tectonic-magmatic activation, and the general uplift processes.

The characteristic features of the Neogene-Quaternary volcanism of northeast Japan could be explained by tectonic rebuilding. The magmatic activity originated at depth in the mantle (Krasnogo, 1973; Rotman, 1968). Dacite ignimbrites are, for the most part, andesite magma products, and their appearance is related to melting in the zones of influence of deep faults.

References Cited

Bykovskaya, E. V., and V. K. Rotman, 1961, Geologitcheskaya pozitziya ignimbritov razlitchnykh vulkanitcheskikh zon Dal'nego Vostoka: Akad. Nauk SSSR Lab. Vulkanologii Trudy, v. 20, p. 151–156.

Krasnogo, L. I., 1973, Geologiya severo-vostotchnoy Azii pod red, *in* Magmatizm, V. 3: Izd-vo "Nedra," p. 395.

Rotman, V. K., 1968, Andezitovye dugi i ikh mesto v tektono magmatitcheskom razvitii severo-zapadnoy tchasti Tikhogo okeana: 23rd Mezhd. Geol. Cong., p. 187–192.

Smith, R. L., 1960a, Ash flows: Geol. Soc. America Bull., v. 71, p. 795–842.

——— 1960b, Zones and zonal variations in welded ash flows: U.S. Geol. Survey Prof. Paper 354-F, p. 149–159.

Thornton, C. P., and O. F. Tuttle, 1960, Chemistry of igneous rocks—pt. 1, Differentiating index: Am. Jour. Sci., v. 258, no. 9, p. 664–684.

Tuttle, O. F., and N. L. Bowen, 1958, Origin of granite in the light of experimental studies: Geol. Soc. America Mem. 74, 153 p.

Wones, D. R., and H. P. Eugster, 1965, Stability of biotite—experiment, theory, and application: Am. Mineralogist, v. 50, no. 9, p. 1228–1272.

Geologic Framework of Metallic Mineral Deposits of Northeast USSR[1]

N. A. SHILO[2]

Abstract The structure of the northeast USSR is composed of the following: the Yana-Kolyma and Chukotsk miogeosynclinal systems and the Oloy-Alazey eugeosynclinal system of the Mesozoides, median and residual massifs of pre-late Precambrian age (Omolon, Okhotsk) and Paleozoic age (Yablonskiy, Yeropolskiy), the Anadyr'-Koryak and Olyutor-Kamchatka geosynclinal systems of Cenozoic age, and the Okhotsk-Chukotsk volcanogenic belt.

Widespread deposits of Au, Ag, Sn, W, and Hg are controlled by the structure and igneous activity.

Gold-bearing quartz veins of plutonic origin form the Yana-Kolyma gold belt with its alluvial and bedrock deposits. The analogous deposits of the Chukotsk system form a less consistent gold belt. The gold here is associated closely with tin and tungsten mineralization.

The Okhotsk-Chukotsk volcanogenic belt is a province of volcanic gold-silver deposits, some of which are associated with tin and mercury mineralization.

Tin deposits of different types (sulfide, quartz, silicate, skarn, and pegmatite) are less abundant in the Yana-Kolyma system but are well developed in the Chukotsk system.

Commercial deposits of mercury are present in the Mesozoides, in the region of Cenozoic folding, and within the Okhotsk-Chukotsk volcanogenic belt. The deposits form linear zones either in carbonate rocks rimming the Yana-Kolyma system or in terrigenous rocks of widely differing age. Mercury also is present in volcanic rocks of Cenozoic age, as well as in the Okhotsk-Chukotsk belt. Conditions for the formation of copper, lead, zinc, and other minerals in the region have not been studied adequately. According to available data, they are related to the structures of the massifs of ancient consolidation, to the Oloy-Alazey eugeosynclinal system, and to the Okhotsk-Chukotsk volcanogenic belt. Platinum and chromium are located within ultramafic belts of the Olyutor-Kamchatka and Anadyr'-Koryak systems.

The major stage of endogenic ore deposition in the northeast USSR was at the end of the Jurassic and in the Cretaceous. In the Anadyr'-Kamchatka system and in Kamchatka, the main phase extended into the Paleogene, Neogene, and Quaternary. The most intensive period of placer deposition was in Pliocene-Pleistocene and Holocene times. The placers are related to relief-forming processes connected with tectonic activity. The sources of ore material for the placers were the Mesozoic-Cenozoic mineralized zones. In many regions, oxide weathering products also were formed as the result of peneplanation.

STRUCTURAL HISTORY

In the northern regions of the USSR Far East, geologic features related to commercial mineral deposits include the Mesozoides[3] of the Pacific margins and the subsequent Cenozoic geosynclines and volcanic belts. The latter are special earth-crust structures which may be the surface expressions of Benioff zones.

Our ideas about the general structure of this region have been clarified significantly. In the place of the so-called "middle Kolyma massif," which had been classified as a platform in the central part of the Verkhoyansk-Chukotsk Mesozoic folded area, there is a fully evolved eugeosynclinal system which has been named the "Oloy-Alazey system." This system is considered to be a new structural element of the Mesozoides in the northeast USSR. Thus, the Yana-Kolyma and Chukotsk miogeosynclinal systems, plus the Oloy-Alazey eugeosynclinal system, are part of the region's structure and are considered by many workers to be part of the Pacific Mesozoides. They include massifs consolidated during pre-Sinian time and several Paleozoic massifs (Yablonskiy, Yeropolskiy, and others). The eastern part of the northeastern USSR is occupied by the Anadyr'-Koryak and Olyutor-Kamchatka geosynclinal zones of Cenozoic age, which are separated from the Mesozoides by the Okhotsk-Chukotsk volcanic belt.

Completion of the structural framework of the northeast USSR occurred after the concluding stages of the Mesozoic; Cenozoic folded areas developed, and at the final stages there was orogenic eruptive activity in the Okhotsk-Chukotsk belt.

During the orogenic stage of Mesozoic geosynclinal development, late-geosynclinal and postgeosynclinal structures developed—principally marginal and residual troughs, riftogenic (taphrogenic) zones, and superimposed (successor) basins.

Late-geosynclinal structures of the Pacific Mesozoides (marginal troughs and inner superimposed basins) are characterized by molasse, including both allochthonous and autochthonous components—conglomerates and coarse-grained sandstones, tuffaceous volcanic rocks, marlstones, and limestones. In miogeosynclines, deposits of residual troughs are represented mainly by terrigenous flysch facies. In eugeosynclinal systems these residual troughs are filled mainly with siliceous volcanic rocks. They inherited the structural framework of the Mesozoic geosynclinal period, and commonly seem to accrete to the

[1]Manuscript received, November 6, 1974.

[2]Northeast USSR Scientific Research Institute, Magadan, USSR.

[3]Mesozoides—a fold system in which the age of the youngest rocks and the folding is Mesozoic.

underlying section of the geosynclinal complex. So-called riftogenic zones, which became independent structures quite recently, possess some other features: they have sutural (fault) limits and are as wide as 200 km and as long as 800–1,000 km. Some of them were simultaneously filled with siliceous volcanic, flysch, and molasse formations, which later were deformed into distinct, highly compressed, anticlinal structures.

Postgeosynclinal structures of the Mesozoides are mainly represented by superimposed basins of different ages. They commonly contain thick series of coal deposits. They are developed either within the central part of the fold system (Arkagala basin, etc) or within the margins of the fold system on the coast of the Sea of Okhotsk and along the Arctic shoreline. Some of the basins are filled with coarse-grained sedimentary rocks and contain placers of heavy metals. In the Cenozoic Koryak-Kamchatka region, the late-geosynclinal and postgeosynclinal structures are Neogene-Quaternary basins filled with unconsolidated continental sediments as thick as hundreds of meters. In places these sediments include coal facies and, elsewhere, gold and cassiterite placers are present in the basin margins.

The Okhotsk-Chukotsk volcanic belt, which stretches more than 3,000 km along a great lineament from the Bering Sea in the northeast to the upper reaches of the Maya River in the southwest, is part of the East Asiatic system of geologic structures. These structures border eastern Asia and probably delineate the intermediate continent-ocean zone. Southward, the Okhotsk-Chukotsk volcanic belt continues as the Sikhote Alin', the Honshū-Korean, and the East Chinese belts.

The Okhotsk-Chukotsk volcanic belt has been studied very thoroughly. Longitudinal and transverse structural zones permit the identification of the following major tectonic elements: outer and inner zones, as well as northwestern and southwestern flanks. The 3,000-km-long volcanic belt separates the Mesozoides (inner) and the Cenozoic folded belt (outer). Volcanogenic nappes of the outer zone discordantly overlie the folded structures of the inner Mesozoides. Many of the volcanic fields may be traced toward the inner parts of the Mesozoides along great abyssal faults which strike discordantly with respect to the general course of the volcanic belt. The inner zone of the belt is adjacent to the structures of the Cenozoic Anadyr'-Koryak system and shares a common extension with it. Here the fractures, which are well defined by the surface structures and geophysical data, serve as the limit of the belt.

The border between inner and outer zones of the Okhotsk-Chukotsk belt consists of contiguous, echelon faults and flexures, which in places intersect at right angles. The border may be recognized locally by straight-line sectors of river valleys or by the presence of young Neogene-Quaternary superimposed basins. The inner zone can be divided into "inherited" and Neogene subzones. The outer zone is divided into the following sectors: Central Chukotsk, Anadyr', Penzhina, and Okhotsk. These sectors are transverse zones of the volcanic belt. In each sector a special type of volcanic series and definitive ratios of volcanic to nonvolcanic formations are present; the sectors also are defined by different structures of the basement of the outer zone.

Recent data from the Northeast Institute at Magadan show that the Okhotsk-Chukotsk volcanic belt was formed as a special structure of northeast Asia during a short time-span (late Albian through early Cenomanian). During that time a mass of about 1,055,000 km^3 of volcanic deposits accumulated on the surface of the area (excluding the area of the Okhotsk Sea).

Magmatism

Among the volcanic rocks of the Okhotsk-Chukotsk belt, several groups of formations have different definitive (volcanic to nonvolcanic) ratios, structural character, and petrographic features. Andesitic formations are most abundant (41%), followed by high-alumina basalt and andesitic basalt (26%), ignimbrites (25%), liparite-andesite basalts and/or liparite basalts (6%), and trachybasalt (2%).

Andesites and bipyroxene-andesitic basalts occur mainly in the outer zone and along the flanks of the volcanic belt. High-alumina basalts and andesitic basalts, most characteristic of the Albian, are found in the inner zone; in the outer zone, the latter association occurs only in the Anadyr' sector. Within the inner zone, andesitic rocks make up a thick series of lavas and pyroclastic deposits which cover many kilometers; in them, the alumina content is greater than 17% and sodium is more abundant than potassium. Ignimbrites of late Albian age are present in the Central Chukotsk and Penzhina sectors, and Cenomanian ignimbrites are present in the inner zone of the Central Chukotsk and Anadyr' sectors. Most of the liparite basalt and/or liparite-andesitic basalt formations are on the flanks of the belt and in the outer zone, where they compose the volcanic section. Trachybasalt is common within the inner zone of the northern part of the belt. In many other parts, hyperalkalic basalts (analcite-bearing, etc) exist.

Different tectonic zones of the northeast USSR are characterized by specific kinds of magmatic rocks. In the zones under consideration, igneous rocks of different compositions occur together or separately; these include granitic, batholithic massifs (plutonic facies); minor intrusions (principally dikes of different compositions—hypabyssal facies); and volcanic rocks (effusive facies).

Magmatic manifestations in all three facies are typical for the region, from the earliest stages, when the Mesozoides and the Okhotsk-Chukotsk volcanic belt developed, to the end of the Quaternary, when the zone of volcanism shifted eastward. Increased magmatic activity is recorded during the early to middle Paleozoic, late Paleozoic to Early Triassic, and Late Cretaceous and Cenozoic. The most intense magmatic activity took place from Jurassic through Cretaceous time (i.e., during the epoch of large-scale formation of granite). At that time, within the Yana-Kolyma and Chukotsk systems, large belts of batholiths were formed along with associated swarms of dikes which cover many thousands of kilometers. On a scale of magmatic intensity, this epoch is rivaled only by the Aptian-Albian, when there were tremendous eruptions which formed the Okhotsk-Chukotsk volcanic belt.

The formation of batholiths in the final stage of folding in the Mesozoic geosynclines was accompanied by renewal of movement along earlier deep-seated faults, and also by the development of troughs and depressions superimposed on the geosynclinal substratum. In the Yana-Kolyma system, the main product of magmatic activity was great batholithic intrusions; they are composed of high-alumina, biotite- and two-mica granites of the Kolyma complex, as well as some other granitic series. Most of plutons parallel the fold axes, and they may have been controlled by the deep-seated faults. They are complex intrusions, with diorites in the peripheral parts and granodiorites, granites, and alaskites toward the center. Absolute ages of the plutons are 105–145 m.y.

Whereas the Yana-Kolyma system is mostly characterized by granitic batholiths and dikes, the Chukotsk system is characterized by the presence of only biotite-hornblende granodiorites and subalkalic granites intermediate between granodiorite and granite. These plutons are 86–132 m.y. old.

Intrusive magmatism in the Okhotsk-Chukotsk volcanic belt is remarkable for some distinct features. Intrusive massifs of its outer zone and flanks, which intrude consolidated structures of the Mesozoides and pre-Sinian basement, commonly occur in transverse rows which extend beyond the borders of the thick volcanic series. The rocks are gabbro, gabbro-diorite, diorite, granodiorite, adamellite, granite, and leucogranite. They compose the Okhotsk complex, which was formed during the end of Albian and the beginning of Cenomanian times. Gabbro-tonalite-plagiogranite developed within the outer zone of the belt.

Within the bounds of the Okhotsk-Chukotsk belt, the Kavralyan complex has been identified, mainly in the zone of the Anadyr' deep-seated fault. The complex includes polyphase intrusions of fine-grained syenodiorite and diorite and medium-grained and porphyritic granodiorite and quartz monzonite. Massifs of both the Okhotsk and Kavralyan complexes intrude thick series of volcanic rocks.

In the Anadyr'-Koryak geosynclinal system, siliceous volcanics, spilite-diabase complexes, and ultrabasite belts formed during Early Cretaceous time. In the Olyutor-Kamchatka zone, spilite, basalt, andesite basalt, and siliceous volcanic complexes appeared, together with ultrabasites, during Late Cretaceous time.

Mineral Deposits

The major features of the geologic structure of the northeast USSR occupy distinct geologic positions in the northwestern part of the Pacific margin, and it is believed that the positions, specifically of the features of the Mesozoides and the Cenozoic folded systems, played a decisive role in the distribution of mineral deposits there. The Asiatic Mesozoic structures differ from those of the North American Cordilleras in that the orientation of the tectonic zones is irregular and does not coincide with that of the Pacific. The asymmetric structure of the Mesozoides of the Pacific margin is repeated in the manner of distribution of Cenozoic geosynclinal systems. The Asiatic structures are characterized by faults which indicate large-scale extension, whereas thrusting processes (compression) are indicated in the Cordilleras. It is natural that the distinctive character of tectonic development in the northeast USSR most directly affects the distribution of magmatic and hydrothermal activities.

These factors produced the major qualitative and spatial differences in the metallogeny of the northeast area, where a spectrum of ore mineralization includes such elements as Au, Ag, Sn, W, and Hg. Although they generally duplicate the known list of the minerals of commercial deposits in the American Pacific margin, they differ in their original geologic position.

In the northeast USSR, plutonic and volcanic gold deposits are of major commercial importance, although minerals of metamorphic origin also may be present. Associated placers are noncommercial in the volcanic belt because practically all the formations in the belt are poor source rocks for placers. Most deposits are of the gold-quartz type, including dikes, quartz veins, stringer zones, and stockworks. Depending on their structural position in different Mesozoic and Cenozoic geosynclinal systems, the ore deposits are grouped in elongated zones making up either huge gold-bearing belts or extensive fields. Among these major zones is the Yana-Kolyma gold-bearing belt within the Yana-Kolyma miogeosynclinal system, which includes belts of batholithic intrusions of granitoids and dikes. This gold-bearing belt may be traced along the general course of the batholiths and folded structures of the Yana-Kolyma system over a distance of more than 1,000 km from the Okhotsk-Kolyma watershed to the basin of the Yana River, which discharges into the Arctic Ocean. The Yana-Kolyma gold-bearing belt is characterized by extensive alluvial-valley and bench placers, which spatially and genetically are related to the gold-bearing dikes, quartz veins, and stringer zones and stockworks in the flysch series of the geosynclinal complex. In the Yana-Kolyma Mesozoic zone, gold-antimony deposits occur locally, sporadically with gold-tourmaline (gold–rare metals) deposits.

Plutonic gold-ore deposits in the Chukotsk gold-bearing belt do not have the linear distribution which characterizes the Yana-Kolyma belt; they must be considered as separate fields associated with local folds of the Chukotsk Mesozoides. Gold mineralization is localized in linear zones that make up the mosaic structure of the belt. Among the gold deposits in the Chukotsk belt, quartz lodes are most important. Stringer zones and stockworks are of minor importance, a fact which probably explains the rather low incidence of placers. In addition to alluvial placers, gold placers of littoral origin are present.

Still less concentrated localization of gold mineralization is found in structures of the Cenozoic folded region, where quartz-lode deposits occur in separate, isolated gold-bearing zones; but alluvial-valley and bench placers controlled by faults are the most important deposits of gold there.

Within the limits of the Oloy-Alazey eugeosynclinal system, there are composite deposits of Cu, Pb, Zn, etc, with significant admixtures of Au and Ag. Ore mineralization forms isolated fields; the principles of distribution of the fields have not been investigated thoroughly, because the greater part of the zone is overlain by a thick cover of Neogene-Quaternary deposits and only its eastern flank is accessible for geologic mapping.

The Okhotsk-Chukotsk volcanic belt has practically no placers; those which do occur in some valleys are of limited commercial significance, owing to the limited source rocks in the volcanic formations.

The investigations of the Okhotsk-Chukotsk volcanic belt disclosed two kinds of deposit: gold-silver, which includes quartz-rhodochrosite and quartz-adularia types; and gold-sulfide (gold-polymetallic). Both are represented by different mineralogic types which formed at a rather wide temperature range in a subsurface environment.

The optimal zone of ore generation in volcanic tectonic-magmatic complexes is well correlated with adulariazation and formation of adularia-hydromicaceous quartzites and propylites. These conditions directly affect the formation of gold-silver deposits. Gold-sulfide characteristically forms in hypabyssal zones of epidote-albite propylitization. The formation of gold-silver mineralization and gold-sulfide mineralization within the Okhotsk-Chukotsk volcanic belt—under obvious conditions of ore-generating, metasomatic, textural-structural and, perhaps, thermodymanic and geochemical environments—occurred in strict conformance with these complex and well-established processes.

Volcanic and magmatic activities in the Okhotsk-Chukotsk belt were accompanied by metasomatic reworking of volcanic and volcanic-sedimentary rocks; hydrothermal metamorphism also affected a thick zone of the basement. The data show that two classes of hydrothermal alteration are distinct here—one regional in character and the other strictly related to ore. Each is characterized by many stages, diversity of facies, and a complex relationship with mineralization. In regionally metasomatized volcanic and intrusive rocks, metasomatites of hypabyssal facies are distinguished. Sulfide mineralization in and related to hypabyssal metasomatites generally contains gold-sulfide deposits; gold–rare metals deposits are connected with it less commonly. Gold-silver mineralization generally is confined to the fields of moderate-temperature propylitization, and associated with epidote, albite, chlorite, and hydromicas. Ore minerals include pyrite, chalcopyrite, galena, sphalerite, gray ores, and gold. Subvolcanic facies of hydrothermal alteration tend to be related to volcanic structures or to faults controlling them. They are represented by quartzite, argillites, propylites, propylitic rocks, and orthoclase metasomatites. Gold-silver mineralization is most common under the conditions of low-temperature propylitization and development of both adularia and hydromicaceous facies (both almost sodium-free).

A close relation exists between paragenesis of metasomatites and gold-silver mineralization, which contains abundant pyrargyrite, stephanite, and other sulfosalts, and commonly argentite and galena. These minerals belong to near-alkalic complexes. Gold-silver deposits include lode, stringer-metasomatic, and stockwork types, and are closely related to rocks of the folded basement of the volcanic belt, to small volcanic-intrusive bodies, and to lava flows and pyroclastic rocks. The distribution of these structural-morphologic types of deposits is not uniform. The deposits are of different commercial importance. The lode deposits are most promising, especially those represented by stringer-metasomatic bodies and stockworks.

The deposits are related to major ore-controlling faults, which generally dip steeply. Along the faults are large zones of hydrothermally altered volcanic rocks, which form extensive ore-bearing zones, including ore fields. Veins commonly are grouped into complex, nonlinear fields which are confined to volcanic collapse structures. These structures commonly are controlled by faults that cut the ore-bearing volcanic structures. Ore-bearing zones either extend northwesterly across the general trend of the volcanic belt or northeasterly along the trend of the structure.

Miogeosynclinal and eugeosynclinal structures of Mesozoic and Cenozoic ages are characterized by gold-quartz deposits containing rare sulfides (pyrite, arsenopyrite and, less commonly, sphalerite, galena, chalcopyrite, and jamesonite) and major high-grade gold segregations. In the zones of subaerial volcanism (Okhotsk-Chukotsk volcanic belt), gold occurs with silver, which in certain deposits is the dominant mineral. The gold-silver deposits contain a diverse association of ore minerals, including numerous sulfides and sulfosalts. Gold-silver-tin associations are common, where cassiterite is either an accessory mineral in gold-silver deposits or it is the major mineral of commercial tin deposits in which gold and silver are minor elements.

The northeastern region of the USSR may become a leading tin-producing region of the Pacific basin. Numerous occurrences of the metal have been identified in the region, and many of them form rather large deposits. They are spatially related to the three structural zones in the region—the Yana-Kolyma and Chukotsk systems and the Okhotsk-Chukotsk volcanic belt.

Within the limits of the Yana-Kolyma system, which is the richest in tin, deposits of cassiterite-quartz and cassiterite-silicate are widely distributed; deposits of tin-bearing skarn and cassiterite-sulfide also are common. The principal feature of the tin-bearing deposits of this zone is the presence of boron-aluminosilicates and chlorites; tungstenite also is common. The deposits of this zone generally are controlled by the faults which cut intrusions of Lower Cretaceous granitic bodies of complex composition.

Although the Okhotsk folded miogeosynclinal system is not as well mineralized as the Yana-Kolyma system, the frequency of ore deposits is high. In this district, deposits of cassiterite-quartz are much more abundant than cassiterite-silicate deposits. Cassiterite-sulfide deposits are uncommon. The background level of boron is much lower than in the Yana-Kolyma system, resulting in fewer boron-aluminosilicates in tin-ore deposits; these minerals primarily are associated with tungstenite, but tin also is a major element.

Ores in the Okhotsk-Chukotsk volcanic belt are characterized by the dominance of cassiterite-silicate deposits, although cassiterite-sulfide and cassiterite-quartz deposits also are important. Generally, deposits in the province have anomalously high backgrounds of boron, fluorine, magnesium, and iron—a phenomenon reflected in the character of mineralization and mineral associations (tourmaline, chlorite, fluorite, K-feldspar, etc).

Tin mineralization in the Okhotsk-Chukotsk volcanic belt is controlled by zones of hydrothermal, regional metamorphism where metasomatism prevails in volcanic structures along extensive fault zones. Thus, the general character of ore-controlling factors, mentioned above in the case of gold-silver mineralization, was repeated during tin mineralization. Reverse zonation is highly characteristic of deposits in the province; high-temperature minerals associated with cassiterite are present in the upper levels, but give way in the lower levels to lower temperature associations wherein cassiterite diminishes in quantity. The close relation of metasomatism to the formation of tin deposits probably determined the sizes of cassiterite segregations, which are distributed in the deposits as very thin accretions, thick lodes with indistinct contacts, or low-grade disseminations in the wall rock.

In many parts of the northeastern territory, deposits of native tin are associated with eluvial and alluvial cassiterite placers; cassiterite is locally present in gold-bearing placers as well. The valley and bench placers are Quaternary.

Mercury is one of the leading commercial metals in the northeastern territory. Mercury ores are extensively developed in different structural zones of the Mesozoides, in the Cenozoic folded region, and in the Okhotsk-Chukotsk volcanic belt. Mercury deposits form extensive zones that are to a large extent controlled by tectonic features. Lithologic controls also are very important. The mercury deposits of the northeast USSR form elongated zones. Most of the deposits are confined to sandstone, siltstone, and schist; limestone, dolomite, siliceous rocks, quartzite, and effusive rocks are of minor importance.

Based on the tectonic scheme of the territory and its geologic features, Babkin (1969) identified the following mercury-bearing regions in the northeast USSR: Chaun-Chukotsk, Anyuy, Koryak-Anadyr', and upper Kolyma. In these regions, 85–90% of the deposits are in areas of Triassic and Cretaceous quartzitic schist; the rest are in fields of felsic and intermediate effusive rocks.

Most of the mercury deposits are in zones of long-existing faults. Of major importance are tectonic marginal faults on the border between rigid structures and folded systems, major long-existing fault zones developed within folded geosynclinal structures, and major faults within the limits of the volcanic belt. Many mercury-bearing zones are elongated parallel with zones of gold and tin mineralization, suggesting a geochemical zonation probably associated with abyssal magmatic processes. Nevertheless, all mercury deposits compose a distinct class; the most important mineral associations are quartz-dolomite-dickite-cinnabar, quartz-calcite-cinnabar, and quartz-dickite-cinnabar. Mercury ores also are accompanied by pyrite, marcasite, native arsenic, realgar, galena, sphalerite, chalcopyrite, fluorite, and other minerals.

The principles and geology of occurrences of deposits of copper, zinc, and other metals have not been studied as thoroughly as they have been for gold, silver, mercury, and tin. However, such deposits are associated with the structures of ancient massifs (Omolon massif), the structures of the Oloy-Alazey eugeosynclinal systems, and the Okhotsk-Chukotsk volcanic belt. Platinum, chromium, and other metals are restricted to ultrabasite belts of the Olyutor-Kamchatka and the Anadyr'-Koryak systems.

On the basis of this discussion, it should be apparent that the major stage of endogenic ore formation in the northeastern USSR took place from the end of the Jurassic through Cretaceous time. In the Anadyr'-Kamchatka system, and on the Kamchatka Peninsula, major metallogenesis occurred during the Paleogene, Neogene, and Quaternary. Most placer formation occurred during Plio-Pleistocene and Holocene times and was related to relief-forming processes of modern tectonism. The sources of placer ores were Mesozoic-Cenozoic ore-bearing zones and, in many regions, oxidized zones (gossans) that were formed during peneplanation.

References Cited

Babkin, P. V., 1969, Mercury zones of the northeastern USSR and Kamchatka: Akad. Nauk SSSR Doklady, v. 181, no. 6, p. 1445–1448.

Lindgren, W., 1935, Waters, magmatic and meteoric: Econ. Geology, v. 30, no. 5, p. 463–477.

Ring Megastructures of the Pacific—Summary[1]

V. V. SOLOVYEV[2]

Introduction

By the application of the "method of direction anomalies" it is possible to distinguish central-type megastructures that are marked by geomorphologic lineaments, geophysical anomalies, and, in places, the concentric distribution of magmatic bodies of different ages. These long-developing structures result from the impulse discharge of tensions from separate energy-generating centers located in lower zones of the lithosphere and in the upper mantle. Central megastructures develop mainly in crust of continental type. Where the energy-generating center is located at the boundary of continental- and oceanic-type crust, the external concentric zone of the oceanic part of the megastructure forms an island arc.

Development of Central-Type Structures

Only well-defined, neotectonic, central-type structures can be interpreted by the application of usual geomorphologic methods. They appear as isolated structures everywhere except in young volcanic areas. As geomorphologic data are obviously insufficient to evaluate the geotectonic role of central-type megastructures, the writer (Solovyev, 1972, 1973) has worked out a special technique called "a method of direction anomalies." This technique establishes the position not only of recent central-type structures, but also of similar forms that have developed throughout the earth's history.

The theoretical basis for the method of direction anomalies was the fact that evidence of tectonic nonuniformities is preserved through long periods of geologic time in the form of geomorphologic lineaments that are found in the present topographic surface. Surface relief changes rapidly, whereas lineaments are stable in time. Owing to this property of lineaments, it is possible to interpret even Precambrian structures in aerial photographs and to recognize buried forms. Use of the method of direction anomalies enables delineation of central-type structures among the complex combination of heterogeneous lineaments. The patterns of lineaments are compiled from topographic maps and aerial photographs. Central structures, arcs, and ring elements are distinguished in these patterns by use of radial concentric symmetry. Further analysis of these structures can then be carried out from geologic, geophysical, metallogenic, and other data.

Central-type structures are classified according to genesis and morphology. The characteristic radial-concentric symmetry of the framework of all central-type structures is indicative of their environment of origin. They probably occur as the result of directed-impulse discharge of energy from deep-seated energy-generating centers. Depending on the mode of energy tension, the structures can be subdivided into several large genetic groups: intrusive (plutons), volcanic, subvolcanic, complex magmatic, and orthoconfocal. The orthoconfocal group includes structural forms which do not display a direct relation to magmatic bodies. They are bounded by ring fractures and zones of high-angle jointing that radiate from the energy-generating centers. These zones commonly are characterized by dikes, hydrothermal alteration of rocks, and individual minor intrusions. Volcanic pipes are a variety of orthoconfocal structures.

Morphologically, three large groups of central-type structure are distinguished on the basis of the characteristics of the transverse profile: (1) dome (with a positive profile), (2) ring (with a negative profile), and (3) dome-ring (combining elements of both). This subdivision reflects the general tectonic tension of the earth's crust—maximum tension for domes and minimum for ring forms. In addition, this classification might relate information on cryptomagmatic bodies. A combined morphostructural and geophysical analysis indicates that cryptoplutonic granitic bodies, such as stocks, commonly have domal forms. Lens-shaped cryptoplutonic bodies of granodiorite and diorite commonly conform to ring shapes. Dome-ring forms are the result of complicated magmatic complexes that were subject to repeated changes in tectonic movements during their development. It is apparent that a morphologic classification of the central-type structures gives a generalized indication of their origin.

[1]Manuscript received, November 6, 1974.

[2]VSEGEI (All-Union Geological Research Institute), Leningrad, USSR.

A definite pattern in the localization of different ore deposits is established empirically on the basis of spatial relation of different morphologic categories and on the origin of the central-type structures.

In particular, mineralization is widespread on the outer edges of these structures. In some cases (apical parts of cryptoplutons, vent facies of volcanoes, volcanic pipes) the interior areas of central forms are prospective for mineral deposits.

It should be noted that this systematic distribution of minerals enables one to proceed from a study of the exposed part of structures to an analysis of the entire structure. This permits full evaluation of the role of geotectonics in the development of all central-type structures. As a result, several general conclusions have been made regarding spatial-time patterns of these structures.

1. These structures are widely distributed, and are probably the main geotectonic entity, not only of the earth, but of other planets in the solar system. The ring structures are especially distinct on the surfaces of the Moon, Mars, and Venus, because of the lack of exogenic processes.

2. The mechanism of formation of central-type structures has not been studied sufficiently. The principle of a cause-effect relation between the redistribution of interior energy of the planet and the occurrence of central structures seems to be almost indisputable.

Energy flow in the lithosphere is accomplished by means of the formation of zones of radial-concentric fractures and fissures that are used by magma as passageways.

3. Dimensions of structures vary from units of kilometers to thousands of kilometers depending on the depth of energy-generating centers. These centers may be present both at the boundaries of different layers of the lithosphere and within the upper mantle and probably in deeper parts of the earth.

4. A diversity of ages of central structures is apparent—Precambrian to Cenozoic. Those structures formed during Precambrian times are predominantly gneiss and granite-gneiss domes; later structures are more complex.

5. The majority of the central-type structures are long-lived and stable in position. The larger the structure, the longer it will endure.

Successive periods of magmatism that conform to concentric zones of large structures occurred during several eras. The successive tectonomagmatism may involve the entire structure or only part of a large structure. Concentric and radial components of central structures revealed by the method of direction anomalies are projections in the surface of the lithosphere of lines of maximum tension related to the interior structural heterogeneities. This explains the high ore content of concentric and radial zones of the structures. Their general contours reflect a spatial position of the "spheres of influence" of individual abyssal energy-generating chambers.

Central-Type Megastructures of Pacific Basin

Large central-type megastructures reaching many hundreds and thousands of kilometers in diameter are of special interest. Smaller structural forms are generally produced by individual plutonic or volcanic bodies. Megastructures, however, join the systems of different and commonly different-age magmatic complexes subordinate to a single central geometry.

The interpretation of the contours of central-type megastructures by the method of direction anomalies enables one to establish for the first time a definite community of geotectonic complexes discrete in both time and space. The external concentric zones of megastructures, which are commonly characterized by geophysical anomalies, folded zones, or volcanic belts, are the most mobile. The interior areas of megastructures are tectonically more passive and commonly have a platform or subplatform regime.

The spatial-time development of tectonomagmatic processes within megastructures has the character of a spiral. Beginning in external areas, these processes are gradually drawn to the interior zones of megastructures. The displays of the next (new) tectonomagmatic cycle are established again in the external sphere of the megastructure. The greatest activity is displayed successively in individual sector blocks reflecting the energy effect of the wave of activation. The latter runs around all the parts of the megastructure like a radar ray. Of course, this process takes place on a scale of geologic time. The East Asian megastructure (Fig. 1) might be an example. Its location is marked by granitic intrusions of late Precambrian–early Paleozoic age found in external concentric zones (mainly in the Baykal sector). During Paleozoic and a greater part of Mesozoic time, magmatism was drawn in an antidromal direction to the interior zones of the megastructure. The megacycle was completed near the end of Early Cretaceous time. The beginning of the next megacycle was in the second half of the Cretaceous. The basic sphere of magmatism again moved into the external zones of the megastructure, but the Sikhote Alin' sector was affected this time. Moving gradually counterclockwise, the

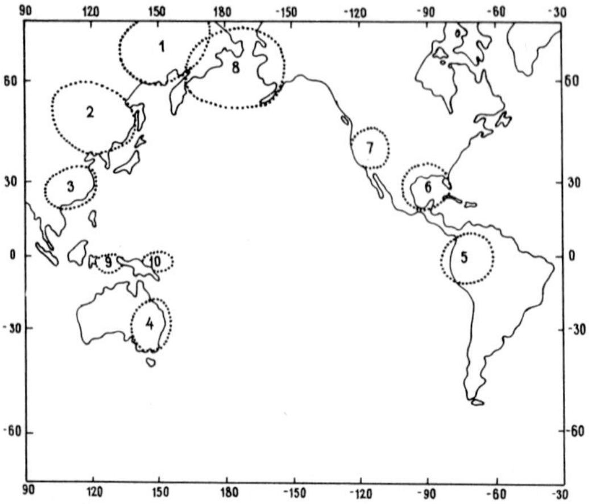

FIG. 1—Scheme of distribution of central-type megastructures: *1*, Yana-Kolyma; *2*, East Asiatic; *3*, South Chinese; *4*, Great Artesian basin; *5*, Upper Amazon; *6*, Gulf of Mexico; *7*, Great Basin; *8*, Aleutian-Chukotsk; *9*, Banda Sea; *10*, Bismarck Sea.

wave of activation during the Late Cretaceous, Paleogene, and Neogene caused the appearance of the series of intrusive and volcanic bodies in the north and south parts of Sikhote Alin' and farther inland in the USSR.

Large, long-developing ring megastructures of a wide age range are characteristic of the continental frame of the Pacific. Along the Asian continent the main megastructures are the Yana-Kolyma, East Asian, and South Chinese. The first is of Paleozoic age and is delineated by folded and volcanic belts. The East Asian megastructure is of late Proterozoic–early Paleozoic age and is characterized by the distribution of different-age magmatic complexes conformable to concentric elements of the framework. These magmatic complexes fix the time sequence of the energy impulses, which continued at intervals to Cenozoic times. The South Chinese megastructure is very old. In the eastern half it is marked by magmatic bodies mainly of Paleozoic and Mesozoic ages, and in the western half, by folded complexes. The average dimension of these megastructures is about 2,000 km. The chambers generating them apparently were located in the upper mantle and had a stable position throughout the Phanerozoic.

The ring megastructure of the Great Artesian basin is the most representative for Australia. Its parameters are similar to those cited above.

The megastructures of the American coast are younger. The Upper Amazon megastructure, the one most clearly defined for South America, most probably was formed in the late Paleozoic.

The maximum energy impulse that determined the conformable bend of folded complexes occurred in the Mesozoic.

A special case is the Aleutian-Chukotsk ring megastructure, the generating abyssal center of which is located near the boundary of areas with oceanic and continental crust. The south part of this structure has the form of an island arc, and the north part (continental) is marked by extrusives, subvolcanic structures, and arc fractures. The latter are traced along sector blocks both in Chukotsk and Alaska. This megaform is of interest because it shows the place of island arcs in the margins of ring structures developed in an oceanic-crust environment.

Ring megastructures are not characteristic of the interior parts of the Pacific Ocean. However, near the island of New Guinea two small megastructures of Cenozoic age are present. The boundaries of the Banda Sea coincide with one of them, and those of Bismarck Sea with the other.

Summary

The data cited give only general information on central-type megastructures of the Pacific basin. Each of the megastructures is complicated by several generations of low-order central forms which reflect a long, progressive and directed development of the earth's crust in the sphere of influence of energy flows of individual abyssal energy-generating chambers.

The consideration of this information might be useful both in general geologic investigations and in metallogenic prospecting, for the main ore regions of the Pacific margins are located mainly within the ring megastructures.

References Cited

Solovyev, V. V., 1972, Confocal paleomorphostructures as reflection of global regularities of tectogenesis, *in* Interorior geodynamics, V. 1: Leningrad, VSEGEI, p. 40–44.

——— 1973, Morphostructures of a central type in connection with some aspects of global tectonics and metallogeny, *in* Metallogeny and new global tectonics: Leningrad, VSEGEI, p. 148–152.

Evolutionary Series of Antimony Deposits of Eastern USSR—Summary[1]

V. I. BERGER[2]

INTRODUCTION

Antimony deposits are widely distributed in Mesozoic folded structures of the eastern USSR, in blocks of Paleozoic sedimentary rocks, and in ancient metamorphic complexes. The increase in intensity of antimony mineralization from the Pacific Ocean toward the interior Asiatic continent correlates with the increase in relative thickness of the granite-metamorphic layer of the earth's crust. The distribution of important ore concentrations of antimony in miogeosynclinal fold systems of the Mesozoides and in blocks of crystalline rocks bears out this relation. In the Koryak-Kamchatka area and the island of Sakhalin, representing fragments of the Nipponese circum-oceanic eugeosyncline, antimony is an "accessory" mineral and is subordinate to mercury. A similar distribution of antimony mineralization is found symmetrically on the other side of the Pacific in the western United States (White, 1962).

Three main types of antimony deposits are distinguished: (1) stratiform deposits of cinnabar-antimonite-fluorite; (2) lode deposits of gold-antimonite, and (3) lode deposits of antimonite and antimonite-ferberite. They are classified according to mineral composition, structural-morphologic features, age, and tectonic condition.

STRATIFORM DEPOSITS OF CINNABAR-FLUORITE-ANTIMONITE

This group is represented by a few occurrences in the continental interior at the margins of the Siberian platform. The minerals are present in early Paleozoic siliceous-terrigenous carbonate rocks present in near-geosynclinal and pericratonic troughs in Transbaykal along the eastern margin of the Aldan shield. The antimony-mercury mineralization of this type is distributed for 350 km within a 200-m thickness of dolomites at the base of the Early Cambrian Lena Stage.

The stratified distribution of mineralization is exceptionally constant, despite the isolation of blocks of Early Cambrian terrigenous carbonate rocks among Proterozoic metamorphic formations and granites.

Steeply dipping sheets of streakily impregnated ores occur in dolomites that have been subjected to intensive jasperization. Black, bituminiferous, micro-oolitic, siliceous dolomites are stratified in zones of jasperization and are almost completely replaced by microgranular quartz. Stable mineral associations characteristic of mineralization are cinnabar-fluorite-antimonite and antimonite-fluorite with barite generally present. Several generations of quartz, calcite, and ore minerals are noted, and there are distinct indications of repeated crushing, brecciation, recrystallization, and redistribution of the ore masses.

At the Ogney ore deposit, mineralization is superposed over paleokarst formations. Horizontal chimney deposits of mineralized breccias are found in dolomites of the Lena Stage. They bear paleokarst indications and are filled with slump breccias with quartz-barite-calcite cement containing sparse impregnations of cinnabar. The irregular distribution of fragments—concentrated on one side of pipe cavities—makes possible their use as a peculiar "paleoplumb." Simple reconstructions indicate that the formation of the pre-ore paleokarst and its cementation with minerals of the early association of jasperoid mineralization took place in unfolded, horizontal or gently dipping dolomites not later than the Middle Cambrian. Apparently, this date marks the lower age boundary of antimony-mercury jasperoid mineralization of northwestern Transbaykal.

At the Kelyan deposits, the late cinnabar-hydromicaceous association is superposed over the earlier jasperoid mineralization and differs from it in the form of manifestation (a net of small veins) and in composition (the presence of potassic hydromica). Late cinnabar-hydromicaceous streaks, ore-bearing jasperoids, and unaltered dolomites are dissected by complex lodelike bodies of diagenetic sandstone that filled the cavities in the post-ore paleokarst. The K-Ar age of potassic hydromica, according to two determinations, is 245 ± 5 and 240 ± 5 m.y. In the sandstones, miospores have been found which are constant components of Permian spore-pollen complexes (Berger and Murina, 1972). These formations define the upper boundary of mineralization as Permian.

[1]Manuscript received, November 11, 1974.

[2]VSEGEI (All-Union Geological Research Institute), Leningrad, USSR.

On the basis of the facts mentioned, the stratiform deposits of antimony-mercury jasperoid mineralization of Transbaykal are Paleozoic. The period of mineralization and subsequent transformation extended from Cambrian to Permian time. Such duration of the formative process and mineralization is generally characteristic of stratiform deposits (Smirnov, 1970). The Paleozoic cinnabar-fluorite-antimonite jasperoid mineralization of a stratiform type is the earliest manifestation of antimony-mercury mineralization in the eastern USSR.

Lode Deposits of Gold-Antimonite

Gold-antimonite deposits are one of the types of antimony mineralization in the Soviet sector of the Pacific mobile belt. They are distributed in regional gold-bearing zones of the Mesozoic folded systems. They are located on the flanks of areas of placer gold, forming zonal series with gold-quartz and gold-sulfide-quartz deposits.

Gold-antimonite deposits are distributed most widely within the Yana-Kolyma belt, which is composed of Triassic and Jurassic terrigenous formations pierced by late Mesozoic batholiths. The deposits have similar composition and structure along the extent (>1,000 km) of the belt. In the regional plan, they are controlled by extensive zones of crushing and of schist formation extending along the boundaries of blocks of different mobility (slightly faulted, intensively folded, amagmatic, and intensively granitized blocks). Regional fractures caused by zones of crushing extend along trough boundaries and control the distribution of gold-antimonite.

Gold-antimonite deposits are related to a lode type. In general, subconformable and conformable bodies predominate. Ladder veins are distributed in dikes. Ore bodies are confined to local zones of schist formation and mylonitization along upthrust shifts. Mineralization is accompanied by alteration of wall rocks to beresite and replacement by sericite, pyrite, ankerite, and albite. Parallel with gold and antimonite, berthierite is typomorphic in the ore composition. As a rule, two successive mineral associations are distinguished: early quartz-arsenopyrite and late quartz-antimonite (berthierite); gold is present in both associations. In some deposits the gold distribution is positively correlated with antimony content; this relation is characteristic of the richest antimony mineralization. In other deposits, gold is concentrated in sulfides of the early association and displays a stable positive correlation with arsenic.

In gold-antimonite deposits traced to a depth of several hundred meters, vertical zoning of mineralization is present, and there is a gradual increase with depth in the relative content of berthierite in ores and in the role of the early pyrite-arsenopyrite association.

A fineness of gold exceeding 950 is one of the most important diagnostic indications of mineralization of this type. The pure gold (950–995 fine) forms local aureoles in gold-bearing placers near such deposits and might be a good indication of gold-antimonite deposits (Bychok, 1974). However, the high-quality feature of gold is preserved irrespective of its relation with antimony in minerals of late ore association or with arsenic in early association.

A low silver ratio in ores (Au:Ag$\geq$10:1) is consistent with the insignificant contamination of gold by admixtures. The mercury content of the Yana-Kolyma belt, if not forming independent concentrations in gold-antimony deposits, is also very low (5–10 g/ton).

On the whole, the cited indications sufficiently characterize this type of antimony deposit; however, supplementary characteristics are observed in structures of ore aggregates. Among primary structures which are preserved only in relic form are macrocrystalline, columnar, drusy, and replacement structures. Primary structures are generally hidden by intensive dynamo-metamorphism and mylonitization and ore rolling typical of gold-antimony deposits. The character of "pre-ore," "intra-ore," and "early post-ore" deformation indicates that the development of the structure of deposits was predominantly under conditions of compression (Berger, 1973).

Gold-antimonite deposits form a single lateral series with gold-quartz and gold-sulfide-quartz. The age of the latter is 115–138 m.y. (Late Jurassic–Early Cretaceous; Zagruzina, 1970). Muscovite and sericite from ore bodies were used to determine K-Ar ages for three gold-antimonite deposits that were 900 km apart in the Yana-Kolyma belt. Their ages were 128, 124±4, and 118 ±2 m.y. The obtained figures are close to the age of the gold-quartz deposits and are conformable with determinations for Kolyma granitic bodies of about 120 m.y. (Nenashev, 1965). Geologic data and age determinations indicate that the gold-antimonite deposits are close in age to the gold-quartz ones, with which they form a single metallogenic complex. The time of formation of the complex was somewhat extended, but it was within the "inversion" stage of the development of the Verkhoyansk-Kolyma miogeosyncline—that is, mineralization apparently followed folding and was practically synchronous with polyphase granite magmatism.

Lode Deposits of Antimonite and Antimonite-Ferberite

This group includes antimony deposits in belts of subaerial volcanism of Cretaceous-Paleogene and Neogene ages. Mineralization is widely distributed along the western flank of the Eastern Asiatic volcanogenic belt and within the intracontinental volcanogenic zones of Transbaykal, Sikhote Alin', and Chukotsk. In regional ore-bearing zones, antimony mineralization is accompanied by mercury, but deposits are isolated. In several intracontinental volcanogenic zones, lateral zonal series of deposits have had successive rejuvenation of mineralization from epithermal gold-silver to antimony and mercury.

Antimonite-ferberite or, more rarely, cinnabar-antimonite-ferberite deposits appear in the regions where ore-bearing volcanogenic zones are superposed over blocks of metamorphic or sedimentary rocks that are pierced by tungsten-bearing granitic complexes. These deposits are distributed in Transbaykal (Shcheglov, 1959). Here, the Barun-Shiveya, Novo-Ivanovskoye, and other deposits form a chain in the section where the zone of Cretaceous epithermal mineralization is superposed over the central tin-tungsten belt.

The distribution of antimony and antimony-tungsten deposits is controlled by regional fractures. Mineralization is concentrated mainly in the folded or metamorphic basement of volcanic rocks (Berger et al, 1974).

In western Chukotsk (the Anyuy zone), the localization of antimony manifestations has been established in the margins of volcano-dome structures with a granite "core."

Although morphologic forms are varied (lodes, stockworks, sheet deposits, chimney deposits), antimonite and antimonite-ferberite deposits of volcanogenic zones have some common characteristics in composition (observed in 21 deposits).

1. Mineralization is accompanied by intensive argillization of wall rock that is replaced by dickite, mixed-layer montmorillonite-hydromicaceous aggregates, opal-like quartz, dolomite, etc.

2. A peculiar rhythm of ore deposition of at least two stages is established. In the early association, antimonite commonly forms cryptogranular colloform separations in chalcedonylike, porcelaneous quartz saturated with inclusions of other sulfides—pyrite, marcasite, arsenopyrite, etc. The second stage is identified by druse and ridge aggregates of quartz with macrocrystalline antimonite, ferberite, and cinnabar, and by the absence of admixtures of other sulfides. The development of mineralization of the second stage takes place partly at the expense of recrystallization and redistribution of earlier mineralization.

Chalcedony–argillized antimony deposits of volcanogenic zones are clearly individualized according to composition and to conditions of distribution. They form an independent group of the youngest antimony manifestations, being of Paleogene or Neogene age. These deposits are related to late- or post-orogenic stages of development of the mobile systems of the eastern USSR.

Conclusions

The three types of antimony deposits differ in composition, structure, and geologic conditions of distribution. They are related to different stages in the development of mobile systems of the eastern USSR.

1. Stratiform deposits of cinnabar-fluorite-antimonite jasperoid were formed before folding occurred in early Paleozoic siliceous-terrigenous carbonate complexes. The subsequent transformations of mineralization by superposed processes are fixed geologically and by K-Ar age determinations. Mineralization is preserved in isolated "relic" blocks.

2. Gold-antimony-quartz lode deposits distributed in gold-bearing belts of the Mesozoides bear distinct indications of synkinematic formation. They are almost synchronous with granite batholiths and are related to the "inversion" stage of the development of mobile systems of the Mesozoides.

3. Antimonite and antimonite-ferberite chalcedony-argillized deposits are related to the zones of subaerial volcanism belonging to late- and post-orogenic stages, or to the late Mesozoic–Cenozoic tectono-magmatic activation of ancient structures.

The deposits of the third type are located in zones commonly superposed over blocks and belts with antimony mineralization of the first and second types. In such sections there are distinct indications of the formation of later deposits at the expense of the regeneration of early antimony mineralization.

References Cited

Berger, V. I., 1973, Evolution of antimony and mercury mineralization of mobile areas, *in* Metallogeny and new global tectonics: Leningrad, VSEGEI, p. 110–115.

——— and G. A. Murina, 1972, New data on the age of mercury mineralization of the Baykal belt: Akad.

Nauk SSSR Doklady, v. 203, no. 3, p. 647–649.
——— N. S. Mudrogina, and V. G. Kuzmin, 1974, Mercury and antimony mineralization of Phanerozoic volcanogenic belts (abs.): Barcelona, 1st Internat. Mercury Cong. Abs., p. 24–25.
Bychok, B.G., 1974. Features of matter composition, the regularities of distribution and the rational methods of search of gold-antimony deposits of Eastern Yakutia, *in* Geological conference of Yakutia on gold, theses of reports: Yakutsk, p.112–115.
Nenashev, N. I., 1965, Meso-Cenozoic magmatism and ore formation of Eastern Yakutia: Moscow, "Nauka," 168 p.
Sidorenko, Z. V., V. I. Berger, and N. I. Popovich, 1972, Methods of the compilation of the prognostic-metallogenic map on mercury and antimony: Vses. Nauchno-Issled. Geol. Inst. Trudy, n. s., v. 176, p. 62–76.
Smirnov, V. I., 1970, Factor of time in the formation of stratiform ore deposits: Geology Ore Deposits, no. 6, p. 3–15.
Shcheglov, A. D., 1959, On some features of the formation of mercury-antimony-tungsten deposits of Transbaykal: Zapiski VMO, ch. 83, no. 1, p. 48–59.
White, D. E., compiler, 1962, Antimony in the United States, exclusive of Alaska and Hawaii: U.S. Geol. Survey Mineral Inv. Resource Map MR-20.
Zagruzina, I. A., 1970, Age of gold mineralization of the North-East: Geology Ore Deposits, no. 1, p. 31–36.

Meimechites of Kamchatka and Kimberlite Problems of Pacific Mobile Belt—Summary[1]

V. K. ROTMAN and B. A. MARKOVSKY[2]

Ultramafic volcanic rocks have been found in Kamchatka in the East Kamchatka mountain range. They are closely associated in space and time with the initial host volcanic rocks of Late Cretaceous–Paleogene age. The ultramafic volcanic rocks form dikes, sills, and diatremes. Their fragments also occur in strata of volcanic breccias and tuffs. Species among the ultramafic volcanic rocks approximate meimechites and picrites in mineralogic, chemical, and geochemical features. Inclusions of garnets of widely varying composition, as well as pyrope crystals, are typical. All ultramafic volcanic rocks include nodules of holocrystalline ultramafic rocks with verites prevailing. The ultramafic volcanic rocks also are present in the West Pacific on Kalimantan (Borneo) and Malaita (the Solomon Islands). They relate to the eugeosynclinal belt of the Late Cretaceous–Paleogene.

Thus, ultramafic volcanic rocks are widespread in the Pacific mobile belt, and it seems feasible that true kimberlites will be found there.

[1]From Circum-Pacific Energy and Mineral Resources Conference program booklet.

[2]VSEGEI (All-Union Geological Research Institute), Leningrad, USSR.

Structural Classification of North Pacific Shelves—Summary[1]

G. S. GANESHIN, V. V. SOLOVYEV, and YU. F. CHEMEKOV[2]

The structural classification of shelves includes three taxons. The first-order taxon comprises shelves as a whole complex of shallow submarine plains that frame the land. Second-order taxons are (1) orthoshelves, (2) parashelves, and (3) hemishelves.

Orthoshelves are submarine continuations of continental platforms or median masses of continental crust submerged below sea level. They are characterized by flat relief, insignificant variations in depth, considerable width, and minor gradients toward the ocean basin. These shelves developed during a quiet tectonic regime. Commercial deposits of oil, gas, and iron ore may exist on orthoshelves.

Parashelves form within folded areas in continental crust. These are narrow shelves with a significant gradient toward the continental slope. Mineralization is typical of folded belts (e.g., ore deposits of various metals, placers of gold, tin, etc).

Hemishelves are narrow bands of coastal shoal along intrageanticlinal ridges and island arcs of recent geosynclinal belts, volcanic ridges, islands, archipelagoes, atolls, etc. Mineralization is typical of the metallogenic zone of the Pacific mobile belt—Hg, Pb, Zn, Ti, Au, Ag, S, Cu, etc. These shelves form under conditions of intense tectonic activity.

Third-order taxons occur (1) within orthoshelves—those in ancient plates, young plates, or crystalline shields; (2) within parashelves—those of the Caledonides, Hercynides, or other folded areas; and (3) within hemishelves—those of young geosynclinal areas or ocean islands.

Each taxon of classification is characterized by definite geologic, tectonic, and geomorphologic structure, and by special endogenic and exogenic mineralization similar to that of the contiguous geostructural elements of land.

[1]Manuscript received, November 11, 1974.

[2]VSEGEI (All-Union Geological Research Institute), Leningrad, USSR.

Outline of Kuril-Kamchatka Arc—Summary[1]

V. K. ROTMAN[2]

The geologic history of the Kuril-Kamchatka arc since Late Cretaceous time is well known. Most of the older rocks are intensely metamorphosed. The development of the area is divided into two stages—a geosynclinal stage and a stage of arc development.

The geosyncline formed during latest Cretaceous time, when a thick pile of basaltic, volcaniclastic rocks was deposited in a deep-sea environment. Later, the axis of the eugeosyncline migrated gradually toward the ocean, and thick piles of alternating volcaniclastic and sedimentary rocks were deposited.

The alkali content (mostly K_2O) of magmatic rocks decreases from west to east. The occurrence of copper ore and of chromite ore is related to magmatic rocks high in potassium.

The second stage of tectonic evolution began in the early Miocene. This stage was characterized by the development of two island arcs—the Central Kamchatka arc and the south Kamchatka–Kuril arc. As a result of Neogene volcanism, some volcaniclastic sequences were formed; among them, two different associations can be distinguished. One association accumulated under shallow-water marine conditions, and the other was mainly subaerial. Both of the associations are composed of differentiated calc-alkalic rock suites in which andesites and basalts are predominant.

The Cu-Pb-Zn veins are related to the first association, whereas Au, Ag, Hg, and S are related to the second one. The volcanic zone of the Central Kamchatka arc is associated with Au, Hg, and S, whereas the volcanic zone of the South Kamchatka–Kuril arc is associated with Cu, Pb, and Zn. The Neogene volcanism was accompanied by intrusion of diorite. During the Neogene, the inner and outer belts and the Benioff zone of the central arc migrated toward the ocean; as a result, the single Kuril-Kamchatka volcanic arc and deep-sea trench were formed.

1Revised from Circum-Pacific Energy and Mineral Resources Conference program booklet.

2VSEGEI (All-Union Geological Research Institute), Leningrad, USSR.

Metal Provinces of Alaska—Summary[1]

ALLEN L. CLARK[2] and HENRY C. BERG[3]

Without detailed information concerning the age, controls, and genesis of most mineral deposits in Alaska, neither a metallogenic province map nor a metallogenic epoch map can be prepared. However, it is possible to map the distribution of kindred or diagnostic metals in large regions of Alaska. The definition of individual metal provinces was, therefore, prepared within the framework of the following concept: "A metal province is any area, ranging in size from an entire region to a single mining district, which is characterized by a distinct association and/or anomalous concentration of a metal or metals."

By use of this concept, large areas of Alaska were identified as metal provinces on the basis of known lode and placer deposits, geochemical sampling, extrapolation of geologic environments known to be favorable for metalliferous deposits, and the subjective evaluation by geologists who are experts on the various regions.

The major metal provinces of southeastern Alaska, proceeding inland from the Pacific Ocean to the Canadian border, are: Ni-Cr-Cu-Fe-Ti, Cu-Pb-Zn-Au-Ag, Mo-Cu-Au, Fe-Ti-Cr, and Mo-W-Cu-Pb-Zn-Au-Ag.

Metal provinces of the southern and central portions of Alaska, including the Alaska Range, are generally well known and conform to major tectonic elements related to plate tectonics. From the Gulf of Alaska to the Alaska Range, the metal provinces are: Cu-Au, Cr-Ni-Cu, Cu-Ag, Cu-Ag-Pt-Cr-Ni, Cu-Au-Ag, Mo-Cu, and Pb-Ag. Metal provinces in central Alaska are broad and ill-defined but are primarily typified by the metal assemblages Sn-W-Au-Pb-Zn and Sb-Au-Hg.

Only two metal provinces are defined in the northern portion of Alaska, primarily in the Brooks Range, and these are characterized by Au-Sb-W and Cu-Pb-Zn.

The Seward Peninsula area is perhaps the most completely studied area of Alaska. Within the region, metal provinces characterized by U-Th-Re, Sn-Au-F-Be, Au, and Sn-W-Au are defined.

[1]Manuscript received, September 26, 1975.

[2]Office of Resource Analysis, U.S. Geological Survey, Reston, Virginia 22092.

[3]Branch of Alaskan Geology, U.S. Geological Survey, Menlo Park, California 94025.

Metallogenic Provinces of Northeast Pacific[1]

RICHARD H. JAHNS[2]

Abstract Metal provinces of contrasting sizes and shapes in the Cordilleran region of western North America include deposits of many ages and appear to be no more than vaguely related to major elements of crustal tectonics. When considered in terms of respective structural and petrologic associations, apparent ages, and inferred genesis, however, the known deposits can be assigned to metallogenic provinces with a geologically systematic pattern. Five principal kinds of metal concentrations are especially useful in this connection: (1) relatively massive sulfide deposits associated with thick sections of subaqueous volcanic rocks; (2) stratabound deposits in marine sedimentary rocks; (3) stratabound deposits in continental sedimentary rocks; (4) deposits in various host rocks of continental orogens; and (5) deposits associated with major volcanic accumulations of continental affinities.

The volcanogenic sulfide associations, which provide a complex and long-term clue to crustal concentration processes, include (1) Fe-Cu-Zn-Au-Ag deposits of Precambrian age that may well reflect repeated contributions from a primitive mantle, (2) Fe-Cu-Pb-Zn-Ag deposits of younger Precambrian to Mesozoic ages in less mafic volcanic rocks and associated eugeosynclinal strata, and (3) post-Paleozoic Fe-Cu-Au deposits of the ophiolitic type that evidently represent mantle contributions along divergent boundaries between crustal plates. The mantle also appears to have been the principal source of Fe, Cu, Mn, and other metallic accumulations in pelagic sediments of deep ocean basins during Cenozoic time.

In marked contrast are other deposits that bespeak early separation into the earth's sialic crust of metals such as Mo, W, Sn, U, and V, and continuing differentiation in this direction for Ag, Pb, Zn, and other metals. Unlike those of more direct mantle derivation, these deposits evidently have required recycling of metals through various combinations of sedimentation, metamorphism, crustal melting, fluid transport, and new additions from the mantle to explain their present levels of concentration. Thus current models of metallization along zones of continental rifting, seafloor spreading, island-arc development, and subduction of oceanic crust can account directly for the formation of some deposits, but they must include at least partly related processes of concentration within the continental crust to explain all of the recognized metallogenic provinces. The copper province of Arizona and adjacent areas is perhaps the best example of such complicated interplay over a very long period of geologic time.

BACKGROUND SKETCH

Since the turn of the century, when Spurr (1903) noted that regional concentrations and combinations of ore metals can be usefully distinguished from one another, many investigators have applied the concept of metallogenic provinces to the Western Cordillera of North America. Starting with known occurrences of metalliferous deposits, they have variously identified such provinces in terms of inferred genesis of the deposits, kinds of associated rocks, structural or tectonic setting, age of mineralization, levels of metal concentration, distribution of trace elements, and other factors. They also have variously referred to the provinces as metalliferous, metal, metallographic, minerogenetic, metallogenic, and metallogenetic. Few clear distinctions among these terms have been suggested, and here it seems desirable to retain only two of them, "metal" for provinces defined solely on the basis of metal distribution, and "metallogenic" for provinces defined from additional criteria. It also seems desirable to follow Spurr (1903) and many others (e.g., Ferguson, 1944; Turneaure, 1955; Landwehr, 1967, 1968; Noble, 1970; Roberts et al, 1971; Clark, 1972) in recognizing that a single province may include deposits representing one or more ages or epochs of mineralization.

Even a limited sampling of past efforts to define and interpret metallogenic provinces reveals a wide spectrum of approach and emphasis. Coverage has ranged from individual mineral districts, belts, and regions (e.g., Billingsley and Locke, 1935; Ferguson, 1944; Field et al, 1974) to the entire Western Cordillera (e.g., Sainsbury et al, 1970; Clark 1972), and scope has ranged from essentially objective delineation of mineral arcs or belts (e.g., Landwehr, 1967, 1968; Sainsbury et al, 1970) to discussions and analyses of the entire subject (e.g., Turneaure, 1955; Stoll, 1965). Numerous correlations between zones of metallization and various observed or inferred structural features have been proposed (e.g., Schmitt, 1959; Wisser, 1960; Tweto and Sims, 1963; Roberts, 1966; several papers in Titley and Hicks, 1966; Jerome and Cook, 1967; Rehrig and Heidrick, 1972; Clark, 1972), and empirically based suggestions of this kind have been considerably refined during the half century since Spurr (1923, p. 459–463) so imaginatively advanced the notion of an intercontinental "Great Silver Channel." Especially wide attention has been given the tectonic "crossroads" concept of ore localization, as developed by Billingsley and others (Billingsley and Locke, 1935, 1941; Locke et al, 1940) and later expanded by Mayo (1958) and Schmitt (1966). In more recent years the much broader concept of

[1]Manuscript received, November 11, 1974.

[2]Professor of Geology and Applied Earth Sciences, Stanford University, Stanford, California 94305.

global plate tectonics has provided an inviting framework for analyzing the distribution of metalliferous deposits, and some promising ideas already have been expressed in this context (e.g., Sillitoe, 1972, 1973; Hutchinson, 1973; Field et al, 1974; Guild, 1974).

Metallogenic provinces commonly have been related, with or without genetic claims or conclusions, to stocks and batholiths, to belts of porphyries and other hypabyssal intrusives, and to volcanic centers or piles; and in recent years increasing emphasis has been placed on genetic associations between certain kinds of massive sulfide deposits and ancient sections of volcanic rocks (e.g., Anderson, 1969; Gilmour, 1971; Hutchinson, 1973). Delineation of provinces from quantitative data began with studies of trace elements in sphalerite from Western Canada (Warren and Thompson, 1945), and the geochemical approach later was used by Burnham (1959) in a comprehensive examination of trace-element levels in chalcopyrite and sphalerite from 172 mining districts in the southwestern United States and northern Mexico. In discussing geologic features of the same southwestern region, Anderson (1966) suggested that the association of copper and molybdenum in most porphyry copper deposits might be related to major chemical trends—that is, to crustal areas in which the Na_2O/K_2O weight ratio is equal to or slightly greater than unity. A different quantitative approach, involving lead-isotope data, was applied by Slawson and Austin (1962) in correlating a mineral belt in New Mexico with a northwest-trending lineament.

Although many investigators have considered economic factors in defining metallogenic provinces, only Noble (1970) has made intensive use of quantitative data on production plus estimated reserves for this purpose. His data, obtained from the most thorough search yet made, were carefully analyzed and then translated into maps of metal provinces, chiefly for the western United States. These maps were complemented by a presentation of reasons for his fundamental dependence on commercial factors rather than other features of metal occurrence; statements concerning limitations of data sources; a thoughtful discussion of subjectivity and constraints in the outlining of districts, subprovinces, and provinces; and some derivative geologic inferences that cannot be readily dismissed.

Geologic Framework

For the present discussions of metal and metallogenic provinces, the Northeast Pacific can be identified primarily with the Cordilleran region of western North America, which extends southeastward for a distance of about 8,000 km through Alaska, Western Canada, western parts of the conterminous United States, and Mexico This region ranges in width from 500 to 1,300 km, encompasses the Pacific and Cordilleran fold belts, and is bordered on the east by regions of platform deposits. The Colorado Plateau, Southern Rocky Mountains, and Rio Grande region, all lying east of the Cordillera, are considered in the following discussions because of their relation to some of the metallogenic provinces.

Widely scattered exposures of Precambrian rocks, including plutonic, metasedimentary, and metavolcanic types as well as some thick sections of relatively unmetamorphosed sedimentary strata, record a highly complex ancient history for the Cordilleran region. Some of these rocks contain important mineral deposits of Precambrian age, but their interrelations in space and time are not yet fully understood. Much of Phanerozoic time is represented by extensive sedimentary and volcanic terranes of pre–Late Cretaceous age—generally miogeosynclinal in eastern parts of the region and eugeosynclinal farther west. These sections are hosts for mineral deposits in many areas. Rocks of latest Mesozoic and Cenozoic age, also widespread, can be variously identified with marine and nonmarine basins, volcanic piles of considerable thickness, and, in Mexico and northwestern parts of the Cordilleran region, with chains of volcanoes. Of particular importance in connection with known mineral deposits are numerous, generally composite plutons that together constitute an enormous volume of igneous material. They range in dimension from small plugs to batholiths, in composition from diorite to granite, and in age from Triassic to Miocene. They include the so-called porphyry intrusives, most of which are relatively small. Scattered igneous masses of basic to ultrabasic composition, and chiefly of Mesozoic age, also are hosts of some mineral deposits in the region.

Extreme structural complexity characterizes much of the Western Cordillera. Thrust, extensional, and transcurrent faults, folds of many styles and degrees of continuity, several elongate domains of chaotic dislocation, and numerous lithologic belts in which degree of metamorphism does not correspond simply to rock age are among the features that denote a highly involved crustal history. The Precambrian rocks provide a remarkable record of tectonic and other crustal changes, but it is so fragmentary and poorly preserved that it is only beginning to yield broadly applicable results from evolutionary analyses (see Engel et al, 1974). Sedimentation, diastrophism, and volcanism appear to have occurred throughout Phanerozoic time, albeit with some variations in expression; plutonism evidently was more episodic, reaching by far its greatest dimensions during parts of Jurassic and Cretaceous time. Most of the observed rocks and structural features of late Mesozoic and younger ages can be reasonably, and at present more confidently, interpreted in terms of the plate-tectonics concept, although even the late Cenozoic history is far from simple. Currently active plate boundaries include the offshore Aleutian and Middle America Trenches representing subduction, the offshore East Pacific, Gorda, Juan de Fuca, and Explorer rises representing seafloor spreading, and the offshore-onshore Fairweather and San Andreas fault zones representing transcurrent movements.

However brief and generalized, the foregoing summary suggests the great complexity of the regional geologic base on which the pattern of metal and metallogenic provinces is superimposed.

Metal Provinces

Metallogenic provinces can be defined in so many ways and in terms of so many different elements that it is best to begin this overview with the more simple metal provinces outlined by Noble (1970). A combination of plots for eight metals—Cu, Au, Pb and Zn together, Hg, Mo, Ag, and W—yields many blotchlike to elongate provinces, some with broadly irregular margins. One extensive "superprovince" corresponds grossly to the Great Basin in the western United States, and another extends southward and southeastward from Arizona and New Mexico through much of northern and central Mexico. Some marginal irregularities are present, mainly from definition of the provinces by superimposing maps for individual metals rather than simultaneously plotting data for all the metals considered. Inclusion of other metals such as Cr, Co, Mn, Ni, Pt, Sn, U, and V would add a few new provinces and expand some others, but it would not violate the general pattern.

The plotted metal provinces are separated by narrow to very extensive blank areas. Some of these gaps are ascribable to covers of sedimentary or volcanic rocks younger than the mineralization, as in the Great Valley of California and parts of the Colorado and Columbia Plateaus, and others to major occurrences of relatively barren plutonites, as in the Sierra Nevada and the Peninsular Ranges of southern California and Baja California. Many, however, reflect low levels of metallization in diverse terranes that are reasonable hosts and, hence, they might be likened to doubly-closed anticlines with good reservoir rocks but no petroleum.

The provinces delineated on the basis of individual metals are less extensive, but most of them occupy large fractions of the combined-metals provinces. This delineation reflects the occurrence of only one metal in some regions and the coexistence of two or more metals in many deposits elsewhere. Examples are mercury in the California Coast Ranges west of the Great Valley, copper and molybdenum in parts of Western Canada, and precious and base metals in Colorado and Washington. The more coastward provinces and superprovinces show some parallelism with the continental margin and with major tectonic features but, in most parts of the region, according to Noble (1970, p. 1617), any ". . . control by 'lines,' 'lineaments,' and 'cross-roads' that have at times been cited as tectonic controls of metallization find almost no expression on the metal province maps." Noble has called attention, however, to a raylike distribution of metal occurrences about the margins of the Colorado Plateau, where piles of Tertiary volcanic rocks are present, and to another possible "focus of metallization" beneath the Cenozoic volcanic pile that covers parts of Washington, Oregon, northern California, and Nevada. Similar foci might be present beneath volcanic accumulations in southern British Columbia and large parts of Mexico.

Noble's maps for individual metal provinces indicate strong abundance maxima for some of the metals. The points of greatest concentration tend to be more or less randomly dispersed throughout the region, and some of them fall outside the main areas of occurrence for the respective metals. Very broad trends in metal content can be recognized throughout major parts of the region, and a synthesis of available reports and maps fortifies Noble's suggested west-to-east sequences of Hg-Cu-Au-Ag-W-Pb-Mo for the United States, Fe-Cu-Mo-Zn-Pb for British Columbia, and Cu-Ag-Pb for Mexico. It should be emphasized that these trends are very broad and include a great deal of "noise."

The ages of mineralization within most of the single-metal provinces and within nearly all of the combined-metal provinces are highly diverse. Most areal variations are irregular, but in some regions there is a tendency for age to decrease progressively from east to west. If all metal occurrences are considered together, there also appears to be a broad decrease in age of mineral deposits southeastward through the region as a whole. The epochs of metallization most commonly represented are Precambrian (several), Late Triassic and Jurassic, Cretaceous, and early to middle Tertiary.

Metallogenic Provinces

From respective structural and petrologic associations, apparent ages, and inferred modes of origin, the known metal deposits can be assigned to metallogenic provinces with complex but geologically more systematic patterns. Thus Roberts et al (1971) have referred to two major provinces in Nevada exclusive of the Ely copper district—a western one characterized by deposits of Au, Ag, W, Hg, and Sb, and an eastern one characterized by deposits of Pb and Zn with minor Ag and Au. Both provinces include deposits of Cu, Mo, and W. Host rocks in the western province are chiefly eugeosynclinal units laid down on simatic oceanic crust, whereas those in the eastern province are mainly carbonate strata that were deposited on sialic crust. The distribution of trace elements in sphalerite and chalcopyrite from the southwestern United States and northern Mexico led Burnham (1959) to define eastern, central, and western metallogenic belts consistent in trend and position with a beltlike distribution of major ore deposits and tectonic features. He did not, however, regard metallization and tectonism as closely related in time, but instead noted that both tectonic features and metallogenic belts ". . . are the effects of a more fundamental cause, which perhaps is a combination of compositional heterogeneities and associated physical discontinuities in the deep-seated source regions of the ores."

Clark (1972) defined six igneous-tectonic provinces in the context of molybdenum deposits, drawing distinctions between eugeosynclinal and miogeosynclinal to shelf environments of mineralization. Field et al (1974) discussed a large and complex province of porphyry copper–molybdenum deposits in the mobile tectonic belt of predominantly eugeosynclinal rocks in the northwestern Cordillera. The older mineralized intrusives in this province apparently are associated with occurrences of stratiform volcanogenic sulfides. Kerr (1946), in his summary of tungsten deposits in the United States, described contact-metamorphic, pegmatitic, and quartz-vein occurrences in terms of central and eastern arcs with prevailing Laramide affinities and a western arc with younger Nevadian affinities. In accordance with Noble's generalizations (1970, p. 1616–1617), the richest tungsten concentrations are highly localized deposits scattered throughout the region (Bateman, 1965; Gray et al, 1968). Although tin is highly consistent in its association with granitic host rocks, it is difficult to define workable parameters for tin provinces (Schuiling, 1967; Sainsbury et al, 1970). An irregular belt of known occurrences, few of them with economic concentrations of metal, extends the length of the Cordilleran region, and within it are provinces that can be defined according to respective associations of tin with granites, pegmatites, and rhyolitic volcanic rocks. The density of sampling is poor, however, and there is little correspondence to the three main belts defined by Burnham (1959) on the basis of high concentrations of tin in hypogene sulfide minerals.

The known mercury deposits in the Cordilleran region are geologically young and occur chiefly in two areas of coastal California, in parts of central and northern Nevada, and in central and southwestern Oregon (Bailey and Phoenix, 1944; Brooks, 1963; Bailey and Everhart, 1964; D. White, 1967; Linn, 1968; Dickson and Tunell, 1968; Henderson, 1969; Noble, 1970). Many evidently are genetically associated with volcanic activity. Mercury is the only metal that forms deposits thus far yielding substantial economic returns in the part of North America that lies southwest of the San Andreas fault.

The copper provinces, which include by far the most important commercial metal deposits in the Western Cordillera, reflect a wide variety of ages and environments of mineralization (Jacobsen, 1975). The occurrences can be divided into strata-bound concentrations of several pre-Tertiary ages (Kinkel et al, 1956; Albers and Robertson, 1961; Baker and Clayton, 1968; Gilmour and Still, 1968; Anderson, 1969), and porphyry-type occurrences (with or without molybdenum) of Triassic and younger ages (papers *in* Titley and Hicks, 1966; Eidel et al, 1968; Hammer and Peterson, 1968; Hernon and Jones, 1968; Field et al, 1974; Lowell, 1974). Genetic distinctions among the strata-bound deposits recently have been suggested by Hutchinson (1973), who has made some interesting tectonogenetic inferences with respect to pyrite-sphalerite-chalcopyrite bodies found in differentiated volcanic rocks of mafic to felsic composition, pyrite-galena-sphalerite-chalcopyrite bodies in more felsic volcanic rocks of calc-alkalic composition, and pyrite-chalcopyrite bodies that occur in mafic volcanic rocks of ophiolitic affinities. He has correlated the time-tectonic-stratigraphic interrelations among these kinds of mineralization with postulated evolutionary processes of crustal development.

Origin of Mineral Deposits

The foregoing summary indicates a complex pattern of metallization imprinted upon an equally complex lithologic-tectonic base. Extreme ranges in age characterize much of the metallization: tin occurs in Precambrian pegmatites and Pliocene rhyolites, scheelite in Precambrian and Miocene contact-metamorphic deposits, and extensive copper deposits in Precambrian volcanic sequences and Tertiary porphyries. Gold deposits in Nevada and Idaho have been related to five principal intrusive metallogenic episodes (Roberts et al, 1971), copper-molybdenum deposits in the northwestern Cordillera to intrusive bodies with an age range of 18–217 m.y. (Field et al, 1974), and numerous copper deposits in the southwestern United States to intrusives dated in the range 53–75 m.y. (Moorbath et al, 1967; Livingston et al, 1968).

Is it possible to establish some genetic order from the existing data? Several bold attempts toward answering this challenging question have been made during the past decade (e.g., Stoll, 1965; Krauskopf, 1967, 1971; D. White, 1968; Gilmour, 1971; Sillitoe, 1972; Hutchinson, 1973), in effect superimposed upon or derived from earlier theories of ore genesis (for a summary, see Ridge, 1968, p. 1713–1834) and the results of geochemical studies pertinent to the composition and behavior of ore-forming fluids. Two contrasting but related avenues of attack can be identified in the form of additional questions: Whence have the ore-forming metals come? By what means were they transferred from their sources?

Five principal kinds of metal concentrations found in the Western Cordillera are especially useful in considering these questions: (1) relatively massive sulfide deposits associated with thick sections of subaqueous volcanic rocks; (2) strata-bound deposits in marine sedimentary rocks; (3) strata-bound deposits in continental sedimentary rocks; (4) deposits in various host rocks of continental orogens; and (5) deposits associated with major volcanic accumulations of continental affinities. The volcanogenic sulfide associations, which provide a complex and long-term clue to crustal concentration processes, include Fe-Cu-Zn-Au-Ag deposits of Precambrian age that may well reflect repeated contributions from a primitive mantle, Fe-Cu-Pb-Zn-Ag deposits of younger Precambrian to Mesozoic ages in less mafic volcanic rocks and associated eugeosynclinal strata, and post-Paleozoic Fe-Cu-Au deposits of the ophiolitic type that evidently represent mantle contributions along zones of crustal-plate divergence or seafloor spreading. If the evolutionary model suggested by Hutchinson (1973) is correct, the strata-bound Precambrian deposits of Arizona and those of Mesozoic age in California can be systematically related to early stages of mantle differentiation and later episodes of seafloor spreading, island-arc formation, and mantle subduction. Mantle sources for such deposits have considerable appeal, as there is much evidence for movement of rock- and ore-forming materials out of the subcrustal domains.

Impressive on this score is the evidence for special concentrations of metals such as Fe, Mn, Cu, Cr, Ni, and Pb along the crests of the East Pacific Rise and Mid-Atlantic Ridge, where there is little opportunity for contamination by materials from the continental crust (Bostrom and Peterson, 1966; Dymond et al, 1973), and evidence for concentrations of Cu, Ag, K, Li, Sb, Pb, As, B, Be, Bi, Ga, and Au in muds and brines of the Red Sea and Salton Sea areas (D. White et al, 1963; Miller et al, 1966; Doe et al, 1966; D. White, 1968; Bischoff, 1969; Blissenbach, 1972), where relations of seafloor spreading and continental rifting provide ample opportunity for mixing of upward-moving mantle materials with those of the continental crust. If deep-seated processes are effective in mineralizing igneous and associated sedimentary materials at and near sites of seafloor spreading, why are relatively rich products of such environments not more widely manifested? Why, for example, should there be a rich and extensive copper concentration of ophiolitic origin on Cyprus, but none like it in the Franciscan assemblage of the California Coast Ranges? Perhaps the answer lies in mantle inhomogeneities, perhaps in limitations of onshore exposure, and perhaps in different degrees of effectiveness in the metal-concentration processes. Some modification of the deposits or redistribution of their contents also might be expected. For those few volcanogenic deposits that contain what appear to be unusual varieties of metals, one can think in terms of a Cyprus-like ophiolite complex developing at a site like the present Red Sea or Gulf of California region.

The primary formation of substantial mineral deposits along loci of seafloor spreading, with or without subsequent mechanical movements of the metal-bearing materials (Jenks, 1971; Sillitoe, 1973), must be admitted as a sound possibility, and it also seems reasonable to assume that similar deposits could be regenerated in island-arc environments around mobile continental margins. Such mechanisms and environments, however, could hardly account for all the known kinds of metal deposits. Some bespeak early separation from the mantle into the earth's sialic crust of metals such as Mo, W, Sn, U, and V, and continuing differentiation in this direction during much of the earth's history for Ag, Pb, Zn, and other metals. Tin, for example, is consistently associated with granitic rocks, and especially with rhyolites, biotite granites, and pegmatites, and tantalum and niobium are characteristically found together in granitic pegmatites. Many important deposits of Cu, Ag, Au, Pb, Zn, Mo, and W are genetically associated with felsic rocks of intermediate to silicic composition, even though the associations of individual metals with individual rock types are far from consistent. Also, the most common occurrences of Nb, rare-earth elements, Th, and Zr in mafic to felsic rocks of alkaline affinities are typically continental. All this is not to say that such sialic elements were not ultimately derived from the mantle, but it does raise the question of when and how they entered the continental environment. The habitat of metals such as tin and tantalum, which is neither simply nor systematically related to any pattern of plate tectonics, suggests that their entry was early in the earth's history, and that present concentrations may well be results of repeated cycles of mineral crystallization, detrital transport, incorporation into newly-formed melts, etc. Certainly lithophile elements of this kind are hyperfusible constituents of magmas, and expectably they would be among the first materials to be dissolved in crustal melting. Other metals such as Cu, Fe, Pb, and Zn, which are found in both oceanic and continental associations, might well have been derived progressively from the mantle during very long periods of geologic time. The generation of calc-alkaline magmas along zones of subduction provides a viable mechanism for transferring such materials from oceanic source rocks into overlying continental environments.

Both crustal and mantle sources for the metals of Cordilleran ore deposits thus seem reasonable, each or both for different deposits; an ultimate source probably would lie in a primordial "protomantle." Crustal sources have been argued for the metals concentrated in brines of the Salton Sea region (Doe et al, 1966) and for metals in numerous ore deposits (Tweto and Sims, 1963; Petrascheck, 1969; Jensen, 1971; Lange and Cheney, 1971; Richards, 1971). Deeper sources, generally in the mantle, have been suggested for other occurrences (C. White, 1945; Skerl, 1957; Hillebrand, 1954; Stoll, 1965; Slawson, 1967; Moorbath et al, 1967; Roberts et al, 1971; Chrismas et al, 1969; Field et al, 1974), and mixed or multiple sources for the metals in still other deposits also have been suggested (Doe et al, 1968; Clark, 1972; Field et al, 1974). Much information of value has derived from isotopic studies of elements such as lead, strontium, and sulfur. Even though problems of isotopic modeling remain difficult to resolve without ambiguity (Cannon et al, 1961; Patterson and Tatsumoto, 1964; Slawson, 1967; Hart, 1969; Richards, 1971), the level of understanding has been raised considerably since the early arguments of Holmes (1937) and Graton (1938).

Most concentrations of metals prompt a fundamental question that has been raised in two different ways by Noble (1970) and Krauskopf (1971): (1) Why are there so many blank areas on well-prepared maps of metal provinces? (2) Why are there so many barren rocks in regions where similar rocks contain abundant concentrations of one or more metals? The most tenable response may well be that many deposits in the continental crust reflect the recycling of metals through various combinations of sedimentation, metamorphism, crustal melting, fluid transport, and new additions from the mantle, and that such recycling is necessary to achieve the present levels of concentration. This might not be the case for the so-called magmatic ore deposits, such as those of Cr, Ni, and Pt in basic and ultrabasic rocks, but a multistage origin seems all but inescapable for most metal concentrations.

If a deposit is closely associated in space and time with a "cogenetic" intrusive of intermediate to felsic composition, it seems reasonable to view the igneous activity as an agent of transport and subsequent dispersion by any of several mechanisms (Burnham, 1967; Roedder, 1971; Haffty and Noble, 1972; Holland, 1972; Ridge, 1968, p. 1713–1834); but it may be no more than an incidental source. Similarly, current models of metallization along zones of continental rifting, sea-floor spreading, island-arc development, and subduction of oceanic crust might account directly or indirectly for the formation of some deposits, but they must include at least partly related processes of concentration within the continental crust to explain all of the recognized metallogenic provinces. The copper province of Arizona and adjacent areas is perhaps the best example of such complicated interplay over a very long period of geologic time.

Future Search for Ores

Much of the data input for present maps of metallogenic provinces has derived from surface inspection in contexts ranging from blind search to careful geologic mapping. Much also has resulted from the conception and testing of ideas variously directed toward the significance of gossans; haloes of alteration; patterns of fractures, lineaments, and other structural features; geochemical trends in surface and subsurface drainage; regional abundances and abundance ratios of major and trace elements; geophysical trends; and geometric or tonal features on remote imagery. Remaining to be tested are numerous specific suggestions of a geochemical nature, ranging from trends in sulfur isotopes (Field and Moore, 1971) to the use of tracers such as magnetite for cassiterite (Theobald et al, 1967; Sainsbury and Hamilton, 1968) and chlorine and tin for metallization in igneous rocks (Ivanova, 1963; Stollery et al, 1971). It seems reasonable to assume that further interpretive scrutiny of metallogenic provinces in the broad framework of geochemical, petrologic, tectonic, and crustal evolutionary theory can yield real dividends for the future. Just as improved knowledge of the plumbing in a crystallizing intrusive mass should lead to sounder predictions of ore occurrences relative to present levels of exposure, so should increased understanding of plate motions, thermal plumes, and the megaplumbing in the earth's mantle-crust system lead to significant new mineral discoveries.

References Cited

Albers, J. P., and J. F. Robertson, 1961, Geology and ore deposits of East Shasta copper-zinc district, Shasta County, California: U.S. Geol. Survey Prof. Paper 338, 107 p.

Anderson, C. A., 1966, Areal geology of the Southwest, *in* S. R. Titley, and C. L. Hicks, eds., Geology of the porphyry copper deposits, southwestern North America: Tucson, Arizona, Univ. Arizona Press, p. 3–16.

——— 1969, Massive sulfide deposits and volcanism: Econ. Geology, v. 64, no. 2, p. 129–146.

Bailey, E. H., and D. L. Everhart, 1964, Geology and quicksilver deposits of the New Almaden district, Santa Clara County, California: U.S. Geol. Survey Prof. Paper 360, 206 p.

——— and D. A. Phoenix, 1944, Quicksilver deposits in Nevada: Univ. Nevada Bull., v. 38, no. 5, 206 p.

Baker, Arthur, III, and R. L. Clayton, 1968, Massive sulfide deposits of the Bagdad district, Yavapai County, Arizona, *in* J. D. Ridge, ed., Ore deposits of the United States, 1933–1967: New York, Am. Inst. Mining, Metall. and Petroleum Engineers, p. 1311–1327.

Bateman, P. C., 1965, Geology and tungsten mineralization of the Bishop district, California: U.S. Geol. Survey Prof. Paper 470, 208 p.

Billingsley, P., and A. Locke, 1935, Tectonic position of ore districts in the Rocky Mountain region: Am. Inst. Mining and Metall. Engineers Trans., v. 115, p. 59–68.

——— and ——— 1941, Structure of ore districts in the continental framework: Am. Inst. Mining and Metall. Engineers Trans., v. 144, p. 9–64.

Bischoff, J. L., 1969, Red Sea geothermal brine deposits: their mineralogy, chemistry and genesis, *in* E. T. Degens and D. A. Ross, eds., Hot brines and recent heavy metal deposits in the Red Sea: New York, Springer-Verlag, p. 368–401.

Blissenbach, E., 1972, Continental drift and metalliferous sediments, *in* Oceanology international 72, conference papers: London, BPS Exhibitions, p. 412–416.

Bostrom, K., and M. N. A. Peterson, 1966, Precipitates from hydrothermal exhalations on the East Pacific Rise: Econ. Geology, v. 61, no. 7, p. 1258–1265.

Brooks, H. C., 1963, Quicksilver in Oregon: Oregon Dept. Geology and Mineral Industries Bull. 55, 223 p.

Burnham, C. W., 1959, Metallogenic provinces of the southwestern United States and northern Mexico: New Mexico Bur. Mines and Mineral Resources Bull. 65, 76 p.

——— 1967, Hydrothermal fluids at the magmatic stage, *in* H. L. Barnes, ed., Geochemistry of hydrothermal ore deposits: New York, Holt, Reinhart and Winston, p. 34–76.

Cannon, R. S., Jr., et al, 1961, The data of lead isotope geology related to problems of ore genesis: Econ. Geology, v. 56, no. 1, p. 1–38.

Chrismas, L., et al, 1969, Rb/Sr, S, and O isotopic analyses indicating source and date of contact metasomatic copper deposits, Craigmont, British Columbia, Canada: Econ. Geology, v. 64, no. 5, p. 479–488.

Clark, K. F., 1972, Stockwork molybdenum deposits in the Western Cordillera of North America: Econ. Geology, v. 67, no. 6, p. 731–758.

Dickson, F. W., and George Tunell, 1968, Mercury and antimony deposits associated with active hot springs in the western United States, *in* J. D. Ridge, ed., Ore deposits of the United States, 1933–1967: New York, Am. Inst. Mining, Metall. and Petroleum Engineers, p. 1673–1701.

Doe, B. R., C. E. Hedge, and D. E. White, 1966, Preliminary investigation of the source of lead and strontium in deep geothermal brines underlying the Salton Sea geothermal area: Econ. Geology, v. 61, no. 3, p. 462–483.

——— et al, 1968, Lead and strontium isotope studies of the Boulder batholith, southwestern Montana: Econ. Geology, v. 63, no. 8, p. 884–906.

Dymond, J., et al, 1973, Origin of metalliferous sediments from the Pacific Ocean: Geol. Soc. America Bull., v. 84, no. 11, p. 3355–3372.

Eidel, J. J., J. E. Frost, and D. M. Clippinger, 1968, Copper-molybdenum mineralization at Mineral Park, Mohave County, Arizona, *in* J. D. Ridge, ed., Ore deposits of the United States, 1933–1967: New York, Am. Inst. Mining, Metall. and Petroleum Engineers, p. 1258–1281.

Engel, A. E. J., 1974, Crustal evolution and global tectonics: a petrogenic view: Geol. Soc. America Bull., v. 85, no. 6, p. 843–858.

Ferguson, H. G., 1944, The mining districts of Nevada: Nevada Univ. Bull., v. 38, no. 4, p. 77–108.

Field, C. W. and W. J. Moore, 1971, Sulfur isotope study of the "B'"limestone and Galena fissure ore deposits of the U.S. Mine, Bingham mining district, Utah: Econ. Geology, v. 66, no. 1, p. 48–62.

——— M. B. Jones, and W. R. Bruce, 1974, Porphyry copper–molybdenum deposits of the Pacific Northwest: Soc. Petroleum Engineers Trans., v. 255, p. 9–22.

Gilmour, Paul, 1971, Strata-bound massive pyritic sulfide deposits—a review: Econ. Geology, v. 66, no. 8, p. 1239–1244.

——— and A. R. Still, 1968, The geology of the Iron King mine, *in* J. D. Ridge, ed., Ore deposits of the United States, 1933–1967: New York, Am. Inst. Mining, Metall. and Petroleum Engineers, p. 1238–1257.

Graton, L. C., 1938, Ores: From magmas or deeper? A reply to Arthur Holmes: Econ. Geology, v. 33, no. 3, p. 251–286.

Gray, R. F., et al, 1968, Bishop tungsten district, California, *in* J. D. Ridge, ed., Ore deposits of the United States, 1933–1967: New York, Am. Inst. Mining, Metall. and Petroleum Engineers, p. 1531–1554.

Guild, P. W., 1974, Distribution of metallogenic provinces in relation to major earth features: Österreichische Akad. Wiss., v. 1, p. 10–24.

Haffty, Joseph, and D. C. Noble, 1972, Release and migration of molybdenum during the primary crystallization of peralkaline silicic volcanic rocks: Econ. Geology, v. 67, no. 6, p. 768–775.

Hammer, D. F., and D. W. Peterson, 1968, Geology of the Magma Mine area, Arizona, *in* J. D. Ridge, ed., Ore deposits of the United States, 1933–1967: New York, Am. Inst. Mining, Metall. and Petroleum Engineers, p. 1282–1310.

Hart, S. R., 1969, Isotope geochemistry of crust-mantle processes, *in* P. J. Hart, ed., The earth's crust and upper mantle: Am. Geophys. Union Geophys. Mon. 13, p. 58–62.

Henderson, F. B., III, 1969, Hydrothermal alteration and ore deposition in serpentinite-type mercury deposits: Econ. Geology, v. 64, no. 5, p. 489–499.

Hernon, R. M., and W. R. Jones, 1968, Ore deposits of the Central mining district, Grant County, New Mexico, *in* J. D. Ridge, ed., Ore deposits of the United States, 1933–1967: New York, Am. Inst. Mining, Metall. and Petroleum Engineers, p. 1211–1237.

Hillebrand, J. R., 1954, A suggested source of the thiophile elements of the Tertiary ore deposits of the Southwest: Econ. Geology, v. 49, no. 8, p. 863–876.

Holland, H. D., 1972, Granites, solutions, and base metal deposits: Econ. Geology, v. 67, no. 3, p. 281–301.

Holmes, Arthur, 1937, The origin of primary lead ores: Econ. Geology, v. 32, no. 6, p. 763–782.

Hutchinson, R. W., 1973, Volcanogenic sulfide deposits and their metallogenic significance: Econ. Geology, v. 68, no. 8, p. 1223–1246.

Ivanova, G. F., 1963, Content of tin, tungsten, and molybdenum in granites enclosing tin-tungsten deposits: Geochemistry, no. 5, p. 492–500.

Jacobsen, J. B. E., 1975, Copper deposits in time and space: Minerals Sci. Eng., v. 7, no. 4, p. 337–371.

Jenks, W. F., 1971, Tectonic transport of massive sulfide deposits in submarine volcanic and sedimentary host rocks: Econ. Geology, v. 66, no. 8, p. 1215–1224.

Jensen, M. L., 1971, Provenance of Cordilleran intrusives and associated metals: Econ. Geology, v. 66, no. 1, p. 34–42.

Jerome, S. E., and D. R. Cook, 1967, Relation of some metal mining districts in the western United States to regional tectonic environments and igneous activity: Nevada Bur. Mines Bull. 69, 35 p.

Kerr, P. F., 1946, Tungsten mineralization in the United States: Geol. Soc. America Mem. 15, 241 p.

Kinkel, A. R., Jr., W. E. Hall, and J. P. Albers, 1956, Geology and base-metal deposits of West Shasta copper-zinc district, Shasta County, California: U.S. Geol. Survey Prof. Paper 285, 156 p.

Krauskopf, K. B., 1967, Source rocks for metal-bearing fluids, *in* H. L. Barnes, ed., Geochemistry of hydrothermal ore deposits: New York, Holt, Rinehart and Winston, Inc., p. 1–33.

——— 1971, The source of ore metals: Geochim. et Cosmochim. Acta, v. 35, p. 643–659.

Landwehr, W. R., 1967, Belts of major mineralization in western United States: Econ. Geology, v. 62, no. 4, p. 494–501.

——— 1968, The genesis and distribution of major mineralization in western United States: Econ. Geology, v. 63, no. 8, p. 967–970.

Lange, I. M., and E. S. Cheney, 1971, Sulfur isotopic reconnaissance of Butte, Montana: Econ. Geology, v. 66, no. 1, p. 63–73.

Linn, R. K., 1968, New Idria mining district, *in* J. D. Ridge, ed., Ore deposits of the United States, 1933–1967: New York, Am Inst. Mining, Metall. and Petroleum Engineers, p. 1623–1649.

Livingston, D. E., R. L. Mauger, and P. E. Damon, 1968, Geochronology of the emplacement, enrichment, and preservation of Arizona porphyry copper deposits: Econ. Geology, v. 63, no. 1, p. 30–36.

Locke, A., P. Billingsley, and E. B. Mayo, 1940, Sierra Nevada tectonic patterns: Geol. Soc. America Bull., v. 51, no. 4, p. 513–539.

Lowell, J. D., 1974, Regional characteristics of porphyry copper deposits in the Southwest: Econ. Geology, v. 69, no. 5, p. 601–617.

Mayo, E. B., 1958, Lineament tectonics and some ore districts of the Southwest: Min. Engineering, v. 10, no. 11, p. 1169–1175.

Miller, A. R., et al, 1966, Hot brines and recent iron deposits in the deeps of the Red Sea: Geochim. et Cosmochim. Acta, v. 30, no. 3, p. 341–359.

Moorbath, S., P. M. Hurley, and H. W. Fairbairn, 1967, Evidence for the origin and age of some mineralized Laramide intrusives in the southwestern United States from strontium isotope and rubidium-strontium measurements: Econ. Geology, v. 62, no. 2, p. 228–236.

Noble, J. A., 1970, Metal provinces of the western United States: Geol. Soc. America Bull., v. 81, no. 6, p. 1607–1624.

Patterson, Claire, and M. Tatsumoto, 1964, The significance of lead isotopes in detrital feldspar with respect to chemical differentiation within the Earth's mantle: Geochim. et Cosmochim. Acta, v. 28, no. 1, p. 1–22.

Petrascheck, W. E., 1969, Ore metals from the crust or mantle: Econ. Geology, v. 64, no. 5, p. 576–578.

Rehrig, W. A., and T. L. Heidrick, 1972, Regional fracturing in Laramide stocks of Arizona and its relationship to porphyry copper mineralization: Econ. Geology, v. 67, no. 2, p. 198–213.

Richards, J. R., 1971, Major lead ore bodies—mantle origin?: Econ. Geology, v. 66, no. 3, p. 425–434.

Ridge, J. D., ed., 1968, Ore deposits of the United States, 1933–1967: New York, Am. Inst. Mining, Metall. and Petroleum Engineers, 1880 p.

Roberts, R. J., 1966, Metallogenic provinces and mineral belts in Nevada: Nevada Bur. Mines Rept. 13, pt. A, p. 47–72.

——— A. S. Radtke, and R. R. Coats, 1971, Gold-bearing deposits in north-central Nevada and southwestern Idaho: Econ. Geology, v. 66, no. 1, p. 14–33.

Roedder, Edwin, 1971, Fluid inclusion studies on the porphyry-type ore deposits at Bingham, Utah, Butte, Montana, and Climax, Colorado: Econ. Geology, v. 66, no. 1, p. 98–120.

Sainsbury, C. L., and J. C. Hamilton, 1968, The geology of lode tin deposits, *in* Technical conference on tin (London): Internat. Tin Council, p. 313–349.

——— R. R. Mulligan, and W. C. Smith, 1970, The circum-Pacific "tin belt" in North America, *in* W. Fox, ed., A second technical conference on tin (Bangkok): London, Internat. Tin Council, p. 125–147.

Schmitt, H. A., 1959, The copper province of the Southwest: Min. Engineering, v. 11, no. 6, p. 597–600.

——— 1966, The porphyry copper deposits in their regional setting, *in* S. R. Titley and C. L. Hicks, eds., Geology of the porphyry copper deposits, southwestern North America: Tucson, Arizona, Univ. Arizona Press, p. 17–33.

Schuiling, R. D., 1967, Tin belts on the continents around the Atlantic Ocean: Econ. Geology, v. 62, no. 4, p. 540–550.

Sillitoe, R. H., 1972, A plate tectonic model for the origin of porphyry copper deposits: Econ. Geology, v. 67, no. 2, p. 184–197.

——— 1973, Environments of formation of volcanogenic massive sulfide deposits: Econ. Geology, v. 68, no. 8, p. 1321–1325.

Skerl, A. C., 1957, The cosmic origin of metallogenetic provinces: Econ. Geology, v. 52, no. 3, p. 307–310.

Slawson, W. F., 1967, Common lead isotope abundances, *in* H. L. Barnes, ed., Geochemistry of hydrothermal ore deposits: New York, Holt, Rinehart and Winston, p. 77–108.

——— and C. F. Austin, 1962, A lead isotope study defines a geological structure: Econ. Geology, v. 57, no. 1, p. 21–29.

Spurr, J. E., 1903, A consideration of igneous rocks and their segregation or differentiation as related to the occurrence of ores: Am. Inst. Mining and Metall. Engineers Trans., v. 33, p. 288–340.

——— 1923, The ore magmas; a series of essays on ore deposition: New York, McGraw-Hill Book Co., 915 p.

Stoll, W. C., 1965, Metallogenic provinces of magmatic parentage: Mining Mag., v. 112, no. 5, p. 312–323; no. 6, p. 394–405.

Stollery, G., M. Botcsik, and H. D. Holland, 1971, Chlorine in intrusives: a possible prospecting tool: Econ. Geology, v. 66, no. 3, p. 361–367.

Theobald, P. K., Jr., W. C. Overstreet, and C. E. Thompson, 1967, Minor elements in alluvial magnetite from the Inner Piedmont belt, North and South Carolina: U.S. Geol. Survey Prof. Paper 554-A, p. A1–A34.

Titley, S. R., and C. L. Hicks, eds., 1966, Geology of the porphyry copper deposits, southwestern North America: Tucson, Arizona, Univ. Arizona Press, 287 p.

Turneaure, F. S., 1955, Metallogenetic provinces and epochs, *in* Pt. 1 *of* A. M. Bateman, ed., Economic Geology: 50th Anniv. Vol., p. 38–98.

Tweto, Ogden, and P. K. Sims, 1963, Precambrian ancestry of the Colorado mineral belt: Geol. Soc. America Bull., v. 74, no. 8, p. 991–1014.

Warren, H. V., and R. M. Thompson, 1945, Sphalerites from western Canada: Econ. Geology, v. 40, no. 5, p. 309–335.

White, C. H., 1945, The abyssal versus the magmatic theory of ore genesis: Econ. Geology, v. 40, no. 5, p. 336–344.

White, D. E., 1967, Mercury and base-metal deposits associated with thermal and mineral waters, *in* H. S. Barnes, ed., Geochemistry of hydrothermal ore deposits: New York, Holt, Rinehart and Winston, p. 576–631.

——— 1968, Environments of generation of some base-metal ore deposits: Econ. Geology, v. 63, no. 4, p. 301–335.

——— E. T. Anderson, and D. K. Grubbs, 1963, Geothermal brine well—Mile-deep drill hole may tap ore-bearing magmatic water and rocks undergoing metamorphism: Science, v. 139, no. 3558, p. 919–922.

Wisser, E. H., 1960, Relation of ore deposition to doming in the North American Cordillera: Geol. Soc. America Mem. 77, 117 p.

Environments of Canadian Cordillera Depositional Basins[1]

H. GABRIELSE[2]

Abstract The stratigraphy of the Canadian Cordillera can be interpreted in terms of stratigraphic assemblages that are unique in distribution, gross lithology, and lateral facies variations. Models of depositional basins in which these assemblages accumulated are essential in exploration for mineral deposits whose distribution is controlled by stratigraphy.

Proterozoic and lower Paleozoic strata in the Cordillera make up an assemblage of clastic and carbonate rocks with minor volcanic rocks. The assemblage appears to represent a continental terrace wedge built along the margin of an earlier Precambrian continent. All units show a distinct polarity of facies distribution and thickness relative to the source area. This assemblage contains most of the known stratiform mineral deposits of gypsum, iron, copper, zinc, and lead in the Cordillera.

The distinctive elements of an Upper Devonian and Lower Mississippian assemblage suggest, at least in the northern Cordillera, the presence of a foredeep and related source areas to the west and northwest. In the southern and eastern parts of the Cordillera, however, the rocks reflect a continuing shelf-platform environment linked to the craton. The mineral potential of these rocks has been considered low, but it needs further study in view of an important zinc-lead deposit in the eastern Selwyn basin.

Distinctive rocks of oceanic character, ranging in age from Mississippian to Middle Triassic, underlie parts of the Cordilleran intermontane belt. They contain asbestos, gold, and some copper in the northern Cordillera, and copper, lead, and zinc on Vancouver Island. However, in general, mineral discoveries, other than those in ultramafic rocks, have been few. In the eastern Cordillera, a shelf environment prevailed.

The association of copper with Late Triassic and Early Jurassic volcanic rocks is well known. The volcanic deposits, together with spatially and temporally associated plutons, are thought to outline a system of evolving island arcs probably roughly coincident with the mapped distribution of these rocks. Between the arcs and the craton, strata were deposited in a marginal basin with little or no evidence of the volcanism that occurred farther west.

The remaining stratigraphic units, ranging in age from Early Jurassic to Cenozoic, are described as successor-basin and foredeep assemblages whose distribution and lithology reflect a close relation to bounding uplifts of metamorphic and plutonic terranes. Because they are a late-stage phenomenon in the evolution of the Cordillera, these assemblages have potential for a variety of placer deposits. They also contain all of the known coal reserves of the region.

INTRODUCTION

A geologic map of the Canadian Cordillera, on any scale but the very smallest, shows a bewildering mosaic of map units representing a broad spectrum of ages and lithologies of rocks. Except for the easternmost or miogeosynclinal part of the Cordillera, the great variety and discontinuity of rock units are characteristic of eugeosynclinal terranes wherever studied in detail. Two factors have determined this complexity—the character, shape, and extent of depositional basins, and the disruption of rock units by subsequent deformation. The degree to which the latter has affected the present distribution of stratigraphic assemblages is only now becoming apparent. At our current level of knowledge, it is clear that meaningful palinspastic restorations of Cordilleran geology will prove to be a formidable task. The amount of crustal shortening in the eastern Cordillera (Bally et al, 1966; Ross, 1968; Price and Mountjoy, 1970; Campbell, 1973) and the scale of offsets related to intense block faulting and probable major transcurrent faulting in the rest of the region (Roddick, 1967; Gabrielse, 1971; Muller, 1971; Paterson, 1971; Tipper, 1971; Richards, 1974) are controversial problems. Nevertheless, considerable progress has been made in relating gross tectonostratigraphic assemblages to environments of deposition (Jeletzky and Tipper, 1968; Ziegler, 1969; Douglas et al, 1970; Monger and Ross, 1971; Monger et al, 1972; Wheeler and Gabrielse, 1972; Gabrielse, 1973; Eisbacher, 1974a, b, c), and the present distribution of these assemblages provides some insight into the evolution of Cordilleran depositional basins in regard to character, space, and time.

Basin analysis has long been a key tool in hydrocarbon exploration, but only recently has it become important for concepts applied in mineral exploration in the Canadian Cordillera. Discovery of widespread stratiform mineral deposits in the eastern Cordillera and the close association of various metals with distinctive lithologies elsewhere focus attention on the importance of studies broadly included under basin analysis. This is true regardless of whether certain types of mineral occurrences are considered to be of syngenetic or epigenetic origin. It is the purpose of this paper to interpret the main characteristics and evolution of basin environments that can be inferred from

[1]Manuscript received, October 21, 1974.

[2]Geological Survey of Canada, Vancouver 3, British Columbia.

The writer is indebted to many of his colleagues in the Geological Survey for critical and stimulating discussions concerning the analysis of depositional basins in the Cordillera. A. Sutherland Brown critically read the manuscript. G. R. Heather and G. R. Dumas provided invaluable service in typing and editing, and in preparation of diagrams.

the geology of the Canadian Cordillera and to highlight their significance for mineral exploration.

MODELS OF CORDILLERAN BASINS

Five fairly distinct regional depositional environments, or combinations of environments, are considered to represent the main types of basins that existed during the history of the region. Examples shown in the cross sections (Figs. 2–8) are from areas where the particular stratigraphic assemblages and their relationships, vertically and laterally, are best known or where they can be most readily reconstructed. The basins of deposition and their contained tectonostratigraphic assemblages can be viewed as first-order features of Cordilleran geology. Second-order, local environments have contributed to rock types that may be common to several of the basins described herein. When taken in a regional context, however, the gross stratigraphy of each type of basin is essentially unique. The contrasting assemblages of rocks thus defined comprise the building blocks of the Cordilleran orogen.

The five depositional environments are not easily related in time and space, and therein lies the greatest challenge in studying the geology of the region. Not until the Mesozoic can the relative position of tectonic elements be reconstructed with any degree of confidence (Monger et al, 1972). Emphasis is placed on the region west of the easternmost Cordillera; the latter remained miogeosynclinal until Late Jurassic time. Figure 1, showing the distribution of geologic provinces, combines depositional and tectonic elements that may not be temporally related.

Miogeosyncline

The oldest and most long-lived of the depositional basins in the Canadian Cordillera is represented by a thick prism of predominantly nonvolcanic sedimentary rocks that extends continuously along the full length of the eastern Cordillera. The preserved western limit is commonly a fault contact with either upper Paleozoic rocks of oceanic aspect or Upper Triassic rocks of island-arc aspect. This assemblage is believed to have composed a continental terrace wedge that developed along the western margin of the North American craton from Proterozoic through Middle Devonian time, an interval of possibly more than a billion years. One of the most important unconformities in the Cordillera subdivides the prism into two great sedimentary units—the lower, Purcell through Belt sequence and the upper, Windermere through Middle Devonian sequence (Gabrielse, 1973). Each sequence seems to encompass about the same general time span of, perhaps, 500–600 m.y.

In the southern Canadian Cordillera the strata are much disrupted by faults. The cross section shown in Figure 2 is based on data from the craton across the Mackenzie Mountains to the Selwyn basin, where a combination of excellent exposure and relatively simple structure provides unusual scope for analysis. Salient points illustrated in the restored cross section are as follows:

1. Crystalline shield-type basement at or near sea level at the east end of the section must have been at a depth greater than 40,000 ft (12,200 m) along the east flank of the Selwyn basin during Middle Devonian time.
2. Along and near the major flexure, which may in part be fault controlled, deep faults must have penetrated through the sialic crust because mafic volcanic rocks and diabase sills and dikes, although minor, are present in rocks of all ages except Silurian and Devonian.
3. Clastic sedimentary rocks, best exemplified by the Lower Cambrian, thicken westward as relatively clean sandstones and then grade fairly abruptly into equally thick sequences of siltstone and shale.
4. Carbonate rocks, mainly dolomite and laminated dolomite with evaporites of platformal facies to the east, grade and thicken westward into limestone, argillaceous limestone, cherty limestone and dolomite, and carbonate reefs. This Middle Cambrian through Middle Devonian assemblage reaches maximum thickness at what is interpreted to be a deep-basin margin and grades abruptly to calcareous and noncalcareous shale, siltstone, and chert to the west. Only a few miles west of the basin margin the graptolitic shale facies is less than one third the thickness of the thickest coeval carbonate succession, and within much of the basin is generally less than about one tenth the thickness. Locally, a few thousand feet of chert are present in the graptolitic facies (S. L. Blusson, personal commun., 1974).
5. Bedded chert is present in the Selwyn basin in all systems except possibly the Cambrian.
6. Calcareous shales near the basin margin generally are replaced farther within the basin by noncalcareous shales.
7. Strata of a particular facies and thickness may extend for great distances parallel with the margin of the depositional basin, but except on the platform few formations persist very far laterally without marked changes in lithology and/or thickness.
8. Continued basinward tilting of the sedimentary prism is revealed by a multiple wedge of unconformities that result in increasing gaps in the stratigraphic column to the east.

Two additional phenomena related to these rocks are noteworthy. First, the main exposures of extensive metamorphic complexes in the eastern part of the Cordillera are composed of strata representing distal parts of the terrace wedge. Second, typical mid-Cretaceous quartz -monzonite batholiths in the eastern Cordillera are essen-

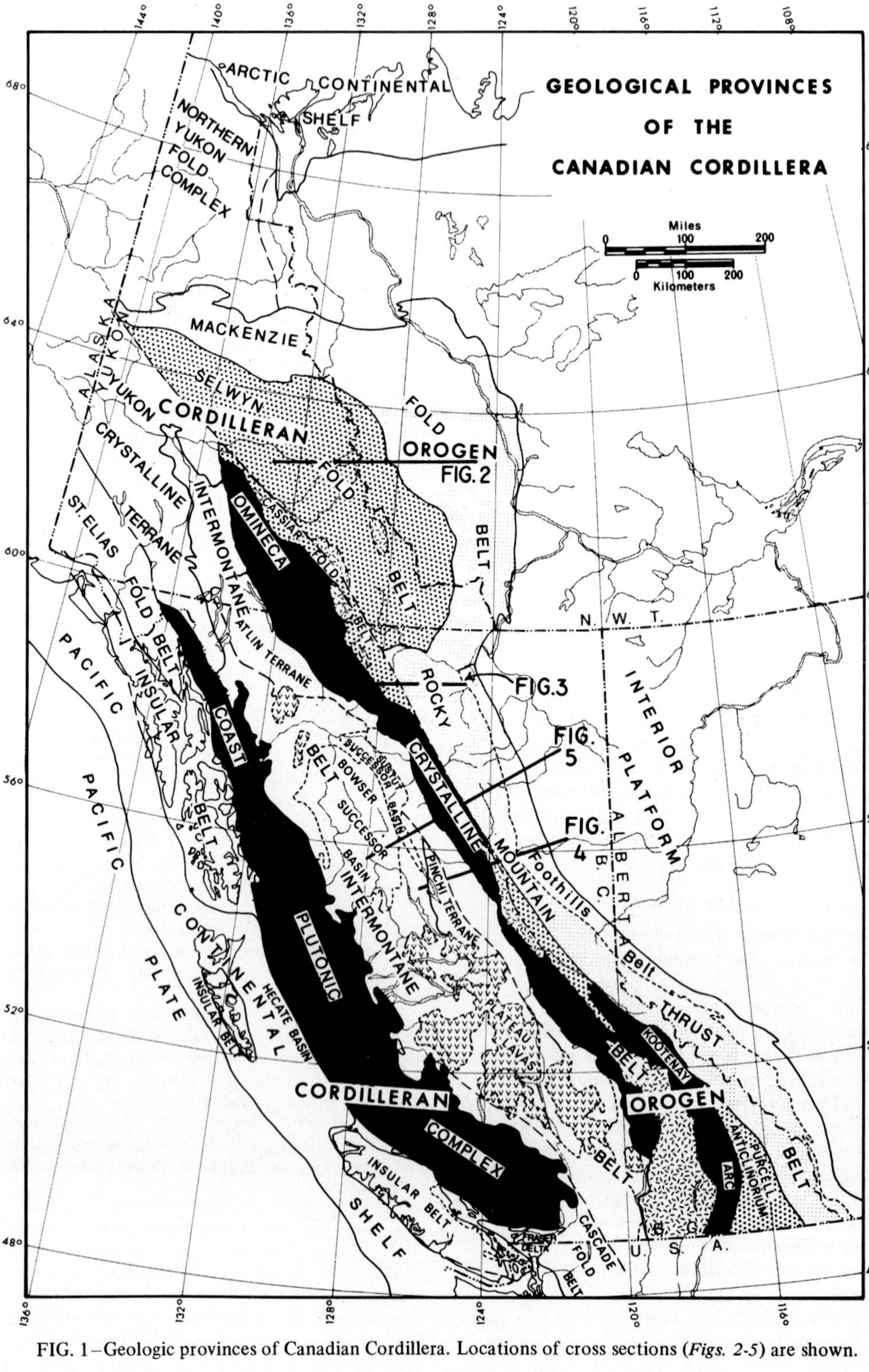

FIG. 1 – Geologic provinces of Canadian Cordillera. Locations of cross sections (*Figs. 2-5*) are shown.

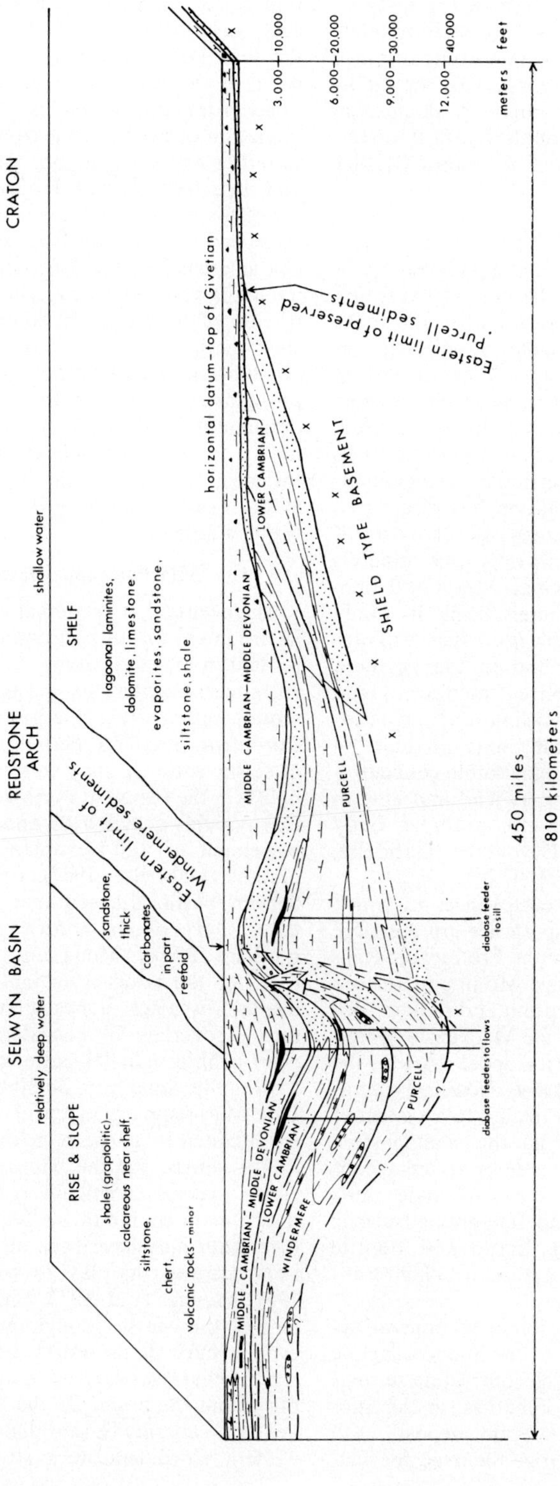

FIG. 2—Restored section of miogeosynclinal mid-Devonian and older rocks from craton to Selwyn basin. Location is shown in Figure 1.

tially confined to the outer part of the miogeosynclinal assemblage, although not directly related in space or time to the metamorphic terranes (Gabrielse and Reesor, in press). This may indicate that the westernmost limit of shield-type basement coincides approximately with the westernmost limit of miogeosynclinal rocks (Kistler and Peterman, 1973).

Mineral Deposits

Most of the large resources of a wide variety of mineral commodities, with the notable exception of Cu, Au, Hg, and Mo, are found in the rocks described above. Of particular importance are large, crudely stratiform deposits of Ag-Pb-Zn, such as the giant Sullivan mine in Proterozoic rocks of the southeastern Cordillera (Freeze, 1966) and the Faro mine in lower Paleozoic rocks of southeastern Yukon Territory (Tempelman-Kluit, 1972). Recent exploration has discovered many zinc and lead occurrences of broadly stratiform type in which predominantly zinc minerals form replacements of carbonate strata and fine-grained lead and zinc minerals occur in thinly bedded argillaceous assemblages. Much exploration in the Selwyn basin and in the northern Rocky Mountains has concentrated on facies boundaries in the miogeosynclinal units, particularly the transition from carbonate to shale. In addition, there seem to be remarkable concentrations of some minerals such as lead and zinc in formations of specific ages—for example, Early Cambrian and Middle Devonian (Gabrielse, 1969).

Other types of mineral occurrences in which stratigraphy is of prime importance are the large sedimentary iron deposits of the Proterozoic Rapitan Group in the Mackenzie Mountains (Green and Godwin, 1963), and gypsum and redbed copper in Proterozoic strata of the Mackenzie Mountains (Gabrielse et al, 1973). Copper veins in the northern Rocky Mountains also occur in Proterozoic strata but are spatially related to a system of diabase dikes. Tungsten in the southeastern Selwyn basin is found where mid-Cretaceous quartz monzonite plutons cut carbonate units within mixed carbonate and fine-grained clastic sequences (Blusson, 1968). Barite and fluorite have been found as veins and as roughly stratiform deposits.

The genesis of the many types of mineral occurrences within the realm of the miogeosyncline is a widely debated topic. The contrasting settings of the Faro and Sullivan orebodies, in the miogeosyncline, and the Pine Point deposit, well within the platform, emphasize the need for caution in selecting exploration targets. Beales and Jackson (1968) proposed a theory for the origin of the Pine Point deposit that could be applied to some of the lead-zinc occurrences in carbonate rocks of the miogeosyncline. The theory involves migration of metallic ions from shales, during dewatering and compaction, into carbonate rocks and their reaction with biogenically reduced sulfur of the carbonate-evaporite sequence. However, such a model is difficult to reconcile with the characteristics of the Faro and Sullivan orebodies, and a volcanogenic origin must be considered in view of their position some distance from the shelf margin or from known facies boundaries. Nonetheless, their empirical relations to stratigraphy illustrate the need for a comprehensive understanding of depositional environments. Fortunately, the miogeosynclinal assemblage is one of the most amenable to this kind of basin analysis in the Cordillera, and great promise is held for future studies.

Mid-Paleozoic Clastic Wedge

The events that accompanied this depositional environment are in part conjectural, but are important in that they terminated the miogeosynclinal environment in its western part and related sedimentation affected large areas of the northern Cordillera. In the northernmost Cordillera a flood of clastic rocks of great volume was derived from uplifts in the region of northwestern Yukon Territory possibly coeval with, and subsequent to, emplacement of mid-Devonian quartz monzonite plutons (Gabrielse, 1967). Farther south, in the Selwyn basin, the southern Mackenzie Mountains, and regions bordering the Tintina and northern Rocky Mountain trenches, carbonate rocks of the miogeosynclinal environment were overlain by black, noncalcareous shale, which, in turn, is overlain by considerable thicknesses of chert-pebble conglomerate and chert arenite. During this time, from possibly Late Devonian to Late Mississippian, much of the eastern Cordillera retained its miogeosynclinal character.

The source for the widespread chert-bearing clastic rocks is an enigma. It has been postulated that a main source in the Selwyn basin and farther south may have been allochthonous, oceanic(?) terranes that lay to the west (Gabrielse, 1967; Monger et al, 1972; Blusson, 1971; Fig. 3). One of the coarsest conglomerates known, however, occurs in the northeastern Selwyn basin, and its character suggests a nearby source, possibly within the basin. On the basis of present evidence, it can only be said that at least parts of the western Cordillera were strongly uplifted and

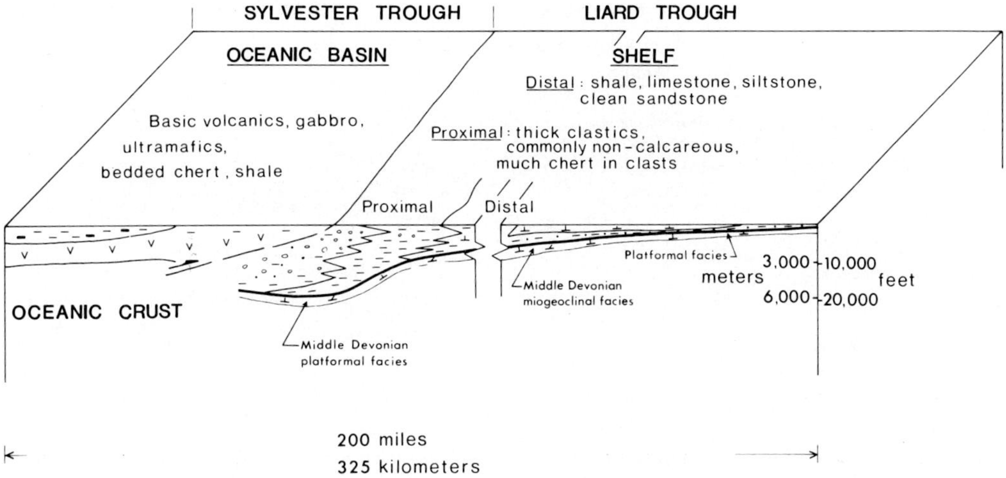

FIG. 3—Diagrammatic cross section of Late Devonian and Early Mississippian rocks showing hypothetical relation of oceanic suite to clastic wedge. Location is shown in Figure 1.

eroded at this time. Indeed, the accumulating evidence of deformation (Ross and Bames, 1972) and evidence of granitic intrusion in the south-central part of the region (A. V. Okulitch and R. K. Wanless, personal commun., 1974) indicate a major orogeny. Tentatively, the Upper Devonian and Lower Mississippian clastic rocks may be considered as a clastic wedge related to the uplift phase.

Mineral Deposits

Very few mineral deposits are known in the sedimentary rocks of the mid-Paleozoic clastic-wedge environment. A notable exception is a significant, roughly stratiform, zinc-lead deposit in the eastern Selwyn basin which contains much finely banded barite in a fine-grained, black non-calcareous slate. In places, pyritiferous black shales are sources for gossans with high zinc anomalies. A supposedly allochthonous, oceanic terrane of this age has asbestos, chromite, and nickel in ultramafic rocks. Volcanic rocks associated with the ultramafic rocks locally carry lode gold veins. These veins are clearly the source of some placer gold in northern British Columbia. Generally, sequences in the Canadian Cordillera that have much bedded chert or significant clastic chert are poor prospecting targets.

Oceanic Basin

One of the most distinctive assemblages of rocks in the western Cordillera comprises bedded chert, basic volcanic rocks, ultramafic rocks. and thick local units of carbonate rock; it is the Cache Creek Group (Monger, 1972; Danner, 1964). The group is generally thousands of feet thick and is known to range in age from Mississippian or older to mid-Triassic. In a few areas, limestones seem to contain a fairly complete record of deposition for several geologic periods (Monger, 1972). They appear to be coeval with nearby bedded chert. Most of the large alpine-type ultramafic bodies of the Cordillera occur in the assemblage. The Cache Creek Group has been shown by Monger and Ross (1971) to contain a Tethyan fusulinid fauna in contrast to the North American fusulinid faunas of bounding terranes. Relationships with coeval miogeosynclinal strata are unknown, as depicted in Figure 4. Nowhere has a distinct base to the group been recognized.

The general aspect of the Cache Creek Group suggests deposition on oceanic crust. Shallow-water limestones are thought to have been built on volcanic platforms near sea level, and carbonate deposition kept pace with subsidence.

Mineral Deposits

Except for asbestos in ultramafic rocks, few mineral deposits have been discovered in this assemblage. Possibly the paucity of acidic volcanic rocks is critical. The Western Mines copper-zinc-lead deposit on Vancouver Island (Northcote and Muller, 1972) and recently discovered copper prospects in the southeastern part of the Atlin terrane occur in relatively acidic volcanic rocks. The basalts certainly do not contain significant quantities of megascopic copper minerals and, in this respect, are similar to basalts of the Devonian-

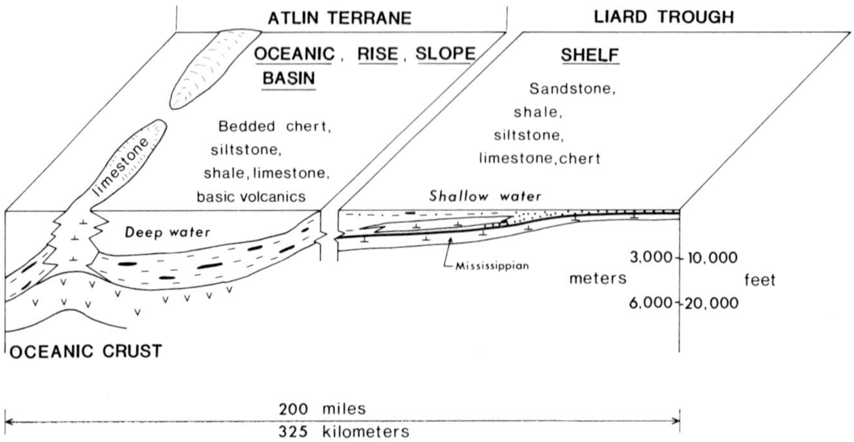

FIG. 4—Diagrammatic cross section showing inferred depositional environments of Pennsylvanian and Permian rocks. Location is shown in Figure 1.

Mississippian oceanic suite. In marked contrast are the copper-bearing, more acidic Upper Triassic volcanic rocks. A knowledge of the background amounts of metals in these rocks would be useful for concepts of metallogeny. Ultramafic and volcanic rocks are apparently the source rocks for placer gold deposits in the Atlin terrane and possibly elsewhere. Nonetheless, this suite of rocks appears to have been little explored for lode gold.

Island Arc

Two belts of Upper Triassic to Lower Jurassic volcanic rocks and related, relatively minor sedimentary rocks extend along much of the length of the Cordillera. One belt underlies a large part of Vancouver Island and trends northward through the Queen Charlotte Islands and southeastern Alaska (Muller, 1972; Sutherland Brown, 1966). The other belt can be traced from south-central British Columbia into southwestern Yukon Territory. Remarkably similar lithologies are present along the belt comprising great thicknesses of andesitic volcanic rocks with local cores of subvolcanic granodiorite plutons (Gabrielse and Reesor, in press; Fig. 5). Ultrabasic bodies of the Alaskan type are also believed to be part of the volcanic suite (Findlay, 1969; Irvine, 1974). Zoned ultrabasic to syenitic stocks are numerous, a good example being Copper Mountain in the southern Cordillera (Preto, 1972). Limestone, probably representing limestone reefs fringing volcanic islands, is fairly common. The thickest volcanic piles may be centered near the large plutons, and from these centers various units lense and overlap along trend. The lateral variations in the assemblage are less well known, mainly because of erosion or cover by younger strata. In places, volcanic rocks grade eastward into volcanogenic strata, which in turn grade into black shale. The latter probably interfingers with sedimentary rocks derived from the craton. This sedimentary assemblage can be regarded as outlining a marginal basin between the craton and a bordering island arc.

A fairly well-defined evolution of the volcanic suite is evident in the region depicted in the cross section (Fig. 5). The lower part of the sequence includes characteristic augite-andesite porphyry breccias, local pillow basalts, coarse-bladed feldspar porphyry, and considerable thicknesses of volcaniclastic rocks. Much of this sequence is marine and is overlain, possibly unconformably, by an upper sequence of dacitic, feldspar-porphyry ash flows which probably is in part of nonmarine origin.

These rocks are believed to represent an island-arc assemblage, and their original distribution may not have been much different from that observed today. On the basis of this hypothesis, it could be postulated that the large area occupied by Bowser basin (Fig. 1) is probably not floored by Upper Triassic and Lower Jurassic volcanic rocks. Instead, one might expect to find volcaniclastic and related sedimentary rocks which are derived from the arc and are considerably thinner than the central arc assemblage.

Mineral Deposits

The rocks of the island-arc assemblage host most of the large copper resources of the Canadian Cordillera. So far, mines have been developed mainly in the southern part of the region, but fu-

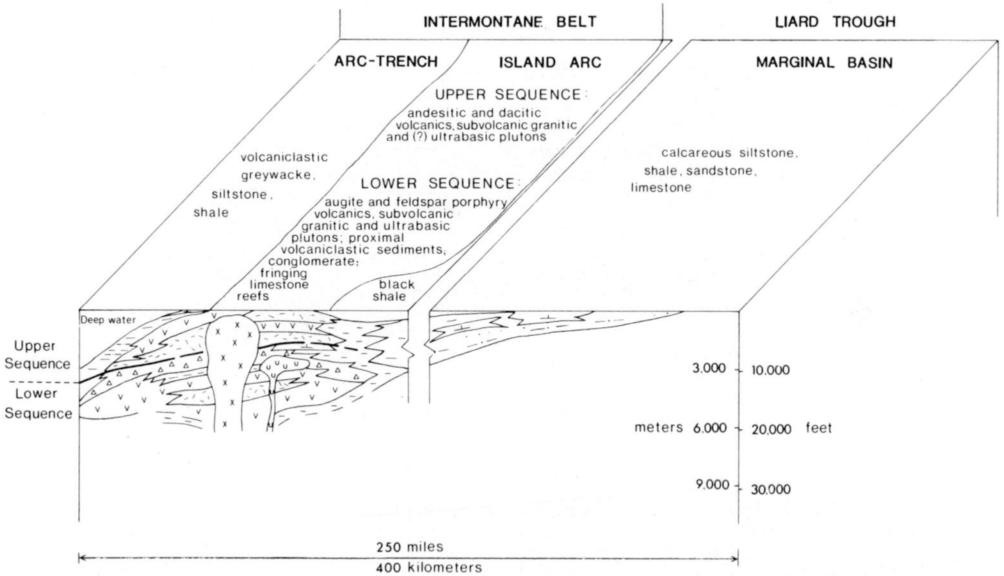

FIG. 5—Diagrammatic cross section of inferred volcanic arc, arc-trench, and marginal-basin sequences. Location is shown in Figure 1.

ture production undoubtedly will include important contributions from the central and northern parts as well. The most important districts seem to be in or near the larger granodiorite to syenite plutons, which possibly represent volcanic centers.

The ultimate source for the copper in the island-arc environment is a fundamental problem. It is evident that the volcanic rocks have a generally high background of copper, as do many of the subvolcanic granodiorite and syenite plutons. There is no doubt about the close temporal and spatial relation between copper mineralization and the arc rocks (White et al, 1968; Sutherland Brown et al, 1971). Presumably, copper could be concentrated from nearby host rocks. Alternatively, the metal is linked to the host rock because it is the product of the same general processes that accompanied island-arc development.

In the light of widely held current concepts that relate mineral deposits in a wide variety of environments to volcanism, it is interesting that strata of the marginal basin seem not to be enriched in copper. This suggests that transport of the metal from the source area has been limited and essentially restricted to the arc environment. Another characteristic of the island arc is the general lack of lead and zinc.

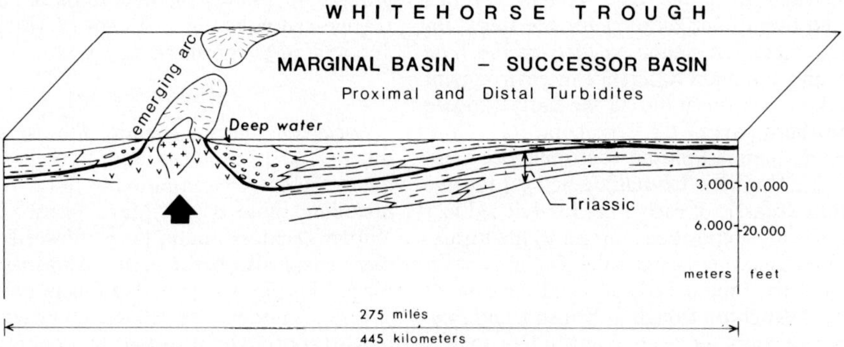

FIG. 6—Tectonic setting of Early Jurassic successor basin in northern Cordillera.

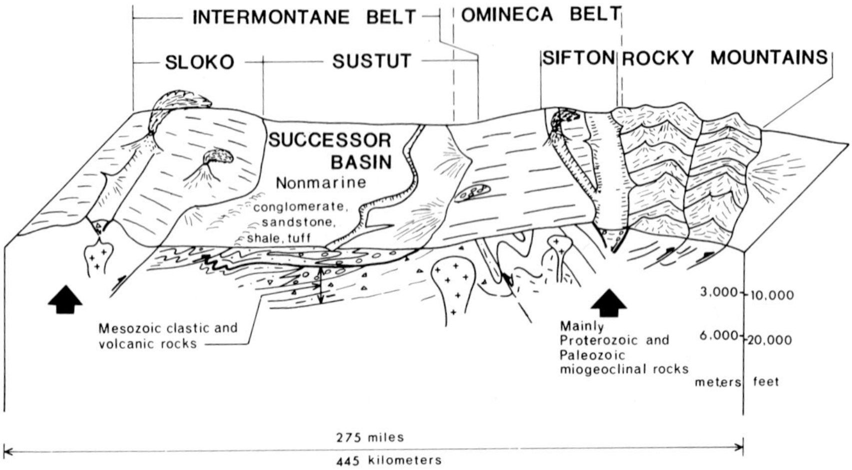

FIG. 7–Tectonic setting of Late Jurassic Bowser basin in northern Cordillera.

Successor Basins

These basins contain a record of the late-stage tectonic development of the Cordilleran orogeny. The cross sections in Figures 6–8 show the evolution of the north-central Cordillera following the island-arc stage. This region includes parts of three distinct basins—the Early to Middle Jurassic Whitehorse trough, the Late Jurassic to Early Cretaceous Bowser basin, and the Late Cretaceous to early Tertiary Sustut basin and early Tertiary Sloko and Sifton basins.

Lower Jurassic proximal and distal turbidites, the former including much coarse conglomerate derived from Upper Triassic rocks, appear closely related to uplift of the earlier emplaced island arc and related granodiorite terranes (Wheeler, 1961; Souther, 1971). Near the craton, fine-grained clastic strata were deposited in a marginal basin, perhaps inherited from the Late Triassic. In this model, the basin configuration for the Early Jurassic is depicted as similar to that for the Late Triassic, and the main difference in environment is the lack of volcanism during the Early Jurassic in the northern part of the Cordillera.

Regional metamorphism, deformation, and uplift of the Omineca Crystalline belt during the late Middle Jurassic or early Late Jurassic led to a restriction of the depositional area and the formation of three separate basins—the Tantalus basin in southwestern Yukon Territory, and the Bowser basin and Tyaughton trough in British Columbia. Important tectonic elements contributing to this segmentation were the Stikine and Skeena arches lying north and south of Bowser basin, respectively. Facies changes in the Bowser basin reveal a dominant source within the Omineca belt from which prograding deltas gradually effected a change from marine to nonmarine conditions (Eisbacher, 1974a). During its early history at least, the Bowser basin appears to have extended across the site of the Coast Mountains. The nature and timing of uplift that ended the marine environment to the west are not known in detail (Grove, 1971; Souther, 1972).

Renewed uplift of the Omineca Crystalline belt, possibly coeval with emplacement of quartz monzonite batholiths during Middle and Late Cretaceous time, resulted in the deposition of coarse, clastic nonmarine sediments represented in the Sustut basin by fluvial, deltaic, and lacustrine deposits. Widespread block faulting, particularly during the early Tertiary, was accompanied by acid volcanism, emplacement of high-level granitic intrusions, and deposition of coarse nonmarine sediments in a system of grabens along and west of the northern Rocky Mountain trench.

Mineral Deposits

Successor basins contain the western Cordillera's coal resources and have potential for paleo-placer concentrations. Silver-rich veins, including those of the Mayo district in central Yukon Territory and in the southwestern Bowser basin, may also occur in successor-basin assemblages. Porphyry copper deposits in north-central British Columbia are related to Eocene biotite-feldspar porphyry intrusions that appear to be localized along high-angle block faults (Carter, 1972; Richards, 1974). To date, most of the

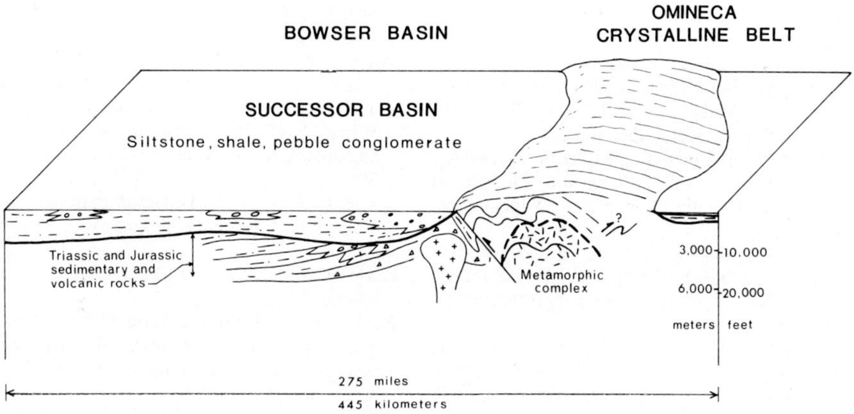

FIG. 8—Tectonic setting of early Tertiary Sloko, Sustut, and Sifton basins in northern Cordillera.

important copper occurrences seem to have been in areas where the intrusions cut Lower Jurassic volcanic rocks. In and near Bowser basin, the age of mineralization is closely related to the time of widespread block faulting and acid volcanism.

Conclusions

Depositional basins of the western Canadian Cordillera are characterized by distinctive assemblages of rocks which are hosts to well-defined types of mineral occurrences. Two contrasting basins, the Proterozoic through mid-Mesozoic miogeosyncline of the eastern Cordillera and the Upper Triassic–Lower Jurassic volcanic arc of the Intermontane belt, are most significant for mineral resources. A thorough understanding of the relation of mineral occurrences to tectonic elements and to various rock facies should contribute greatly to concepts regarding mineralization.

References Cited

Bally, A. W., P. L. Gordy, and G. S. Stewart, 1966, Structure, seismic data, and orogenic evolution of southern Canadian Rocky Mountains: Canadian Petroleum Geology Bull., v. 14, no. 3, p. 337–381.

Beales, F. W., and S. A. Jackson, 1968, Pine Point—a stratigraphical approach: Canadian Inst. Mining and Metallurgy Trans., v. 71, p. 193–204.

Blusson, S. L., 1968, Geology and tungsten deposits near the headwaters of Flat River, Yukon Territory and southwestern District of Mackenzie: Canada Geol. Survey Paper 67-22, 77 p.

——— 1971, Sekwi Mountain map-area, Yukon Territory and District of Mackenzie: Canada Geol. Survey Paper 71-22, 17 p.

Campbell, R. B., 1973, Structural cross-section and tectonic model of the southeastern Canadian Cordillera: Canadian Jour. Earth Sci., v. 10, no. 11, p. 1607–1620.

Carter, N. C., 1972, Granisle, *in* Copper and molybdenum deposits of the western Cordillera: 24th Internat. Geol. Cong., Guidebook for Field Excursions A09-C09, p. 27–36.

Danner, W. R., 1964, Permian—western Cordilleran region, *in* Geological history of Western Canada: Alberta Soc. Petroleum Geologists, p. 109–110.

Douglas, R. J. W., et al, 1970, Geology of Western Canada, *in* Geology and economic minerals of Canada: Canada Geol. Survey Econ. Geology Rept. 1, 5th ed., p. 365–488.

Eisbacher, G. H., 1974a, Deltaic sedimentation in the northeastern Bowser basin, British Columbia: Canada Geol. Survey Paper 73-33, 13 p.

——— 1974b, Evolution of successor basins in the Canadian Cordillera, *in* Modern and ancient geosynclinal sedimentation: Soc. Econ. Paleontologists and Mineralogists Spec. Pub. 19, p. 274–291.

——— 1974c, Sedimentary history and tectonic evolution of the Sustut and Sifton basins, north-central British Columbia: Canada Geol. Survey Paper 73-31, 57 p.

Findlay, C. R., 1969, Origin of the Tulameen ultramafic-gabbro complex, southern British Columbia: Canadian Jour. Earth Sci., v. 6, no. 3, p. 399–426.

Freeze, A. C., 1966, On the origin of the Sullivan orebody, Kimberley, B. C.: Canadian Inst. Mining and Metallurgy Spec. Vol., no. 8, p. 263–294.

Gabrielse, H., 1967, Tectonic evolution of the northern Canadian Cordillera: Canadian Jour. Earth Sci., v. 4, no. 2, p. 271–298.

——— 1969, Lower Cambrian strata and base metals: Western Miner, v. 42, no. 2, p. 22–28.

——— 1971, Sedimentary facies and northern Rocky Mountain trench (abs.), *in* Faults, fractures, lineaments and related mineralization in the Canadian Cordillera: Geol. Assoc. Canada Cordilleran Sec., Ann. Mtg. Prog. and Abs., p. 15.

——— 1973, Younger Precambrian of the Canadian Cordillera: Am. Jour. Sci., v. 272, no. 6, p. 521–536.

——— and J. E. Reesor, 1974, The nature and setting of Granitic plutons in the central and eastern parts of the Canadian Cordillera: Pacific Geology, v. 8, p. 109 –138.

——— S. L. Blusson, and J. A. Roddick, 1973, Geology of Flat River, Glacier Lake, and Wrigley Lake map-areas, District of Mackenzie and Yukon Territory: Canada Geol. Survey Mem. 366, 153 p.

Green, L. H., and C. I. Godwin, 1963, Mineral industry of Yukon Territory and southwestern District of Mackenzie 1962: Canada Geol. Survey Paper 63-38, p. 15–18.

Grove, E. W., 1971, Geology and mineral deposits of the Stewart area, northwestern British Columbia: British Columbia Dept. Mines and Petroleum Resources Bull. 58, 219 p.

Irvine, T. N., 1974, Ultramafic and gabbroic rocks in the Aiken Lake and McConnell Creek map-areas, British Columbia, *in* Report of Activities; part A, April to October 1973; petrology: Canada Geol. Survey Paper 74-1, p. 149–152.

Jeletzky, J. A., and H. W. Tipper, 1968, Upper Jurassic and Cretaceous rocks of Taseko Lakes map-area and their bearing on the geological history of southwestern British Columbia: Canada Geol. Survey Paper 67-54, 218 p.

Kistler, R. W., and Z. E. Peterman, 1973, Variations in Sr, Rb, K, Na, and initial Sr^{87}/Sr^{86} in Mesozoic granitic rocks and intruded wall rocks in central California: Geol. Soc. America Bull., v. 84, no. 11, p. 3489–3512.

Monger, J. W. H., 1972, Oceanic crust in the Canadian Cordillera: Canada Dept. Energy, Mines and Resources Earth Physics Br. Pub., v. 42, no. 3, p. 59–64.

——— and C. A. Ross, 1971, Distribution of fusulinaceans in the western Canadian Cordillera: Canadian Jour. Earth Sci., v. 8, p. 259–278.

——— J. G. Souther, and H. Gabrielse, 1972, Evolution of the Canadian Cordillera—a plate-tectonic model: Am. Jour. Sci., v. 272, p. 577–602.

Muller, J. E., 1971, Fault and fracture tectonics of the Insular belt (abs.), *in* Faults, fractures, lineaments and related mineralization in the Canadian Cordillera: Geol. Assoc. Canada Cordilleran Sec., Ann. Mtg. Prog. and Abs., p. 19–21.

——— 1972, The Insular belt, *in* Geology of the southern Canadian Cordillera: 24th Internat. Geol. Cong., Guidebook for Field Excursions A03-C03, p. 74–83.

Northcote, K. E., and J. E. Muller, 1972, Volcanism, plutonism and mineralization, Vancouver Island: Canadian Inst. Mining and Metallurgy Bull., v. 65, no. 726, p. 1–9.

Paterson, I. A., 1971, The Pinchi fault system in the Fort St. James area, B. C. (abs.), *in* Faults, fractures, lineaments and related mineralization in the Canadian Cordillera: Geol. Assoc. Canada Cordilleran Sec., Ann. Mtg. Prog. and Abs., p. 28–29.

Preto, V. A. G., 1972, Geology of Copper Mountain: British Columbia Dept. Mines and Petroleum Resources Bull. 59, 87 p.

Price, R. A., and E. W. Mountjoy, 1970, Geologic structure of the Canadian Rocky Mountains between Bow and Athabasca Rivers—a progress report, *in* J. O. Wheeler, ed., Structure of the southern Canadian Cordillera: Geol. Assoc. Canada Spec. Paper 6, p. 7–39.

Richards, T. A., 1974, Hazelton map-area (E½), British Columbia: Canada Geol. Survey Open File Rept. No. 215.

Roddick, J. A., 1967, Tintina Trench: Jour. Geology, v. 75, no. 1, p. 23–33.

Ross, J. V., 1968, Structural relations at the eastern margin of the Shuswap complex, near Revelstoke, southeastern British Columbia: Canadian Jour. Earth Sci., v. 10, no. 11, p. 831–850.

——— and W. C. Bames, 1972, Evidence for "Cariboοan orogeny" in the southern Okanagan region of British Columbia. Canadian Jour. Earth Sci., v. 9, no. 12, p. 1693–1702.

Souther, J. G., 1971, Geology and mineral deposits of Tulsequah map-area, British Columbia: Canada Geol. Survey Mem. 362, 84 p.

——— 1972, Telegraph Creek map-area, British Columbia: Canada Geol. Survey Paper 71-44, p. 13.

Sutherland Brown, A., 1966, Tectonic history of the Insular Belt of British Columbia, *in* H. C. Gunning, ed., Tectonic history and mineral deposits of the western Cordillera: Canadian Inst. Mining and Metallurgy Spec. Pub. No. 8, p. 83–100.

——— et al, 1971, Metallogeny of the Canadian Cordillera: Canadian Inst. Mining and Metallurgy Trans., v. 74, p. 121–145.

Tempelman-Kluit, D. J., 1972, Geology and origin of the Faro, Vangorda, and Swim concordant zinc-lead deposits, central Yukon Territory: Canada Geol. Survey Bull. 208, 73 p.

Tipper, H. W., 1971, The ages, magnitudes, and types of faults in the southern Coast Mountains and along their eastern flank (abs.), *in* Faults, fractures, lineaments and related mineralization in the Canadian Cordillera: Geol. Assoc. Canada Cordilleran Sec., Ann. Mtg. Prog. and Abs., p. 40.

Wheeler, J. O., 1961, Whitehorse map-area, Yukon Territory: Canada Geol. Survey Mem. 312, 156 p.

——— and H. Gabrielse, 1972, The Cordilleran structural province, *in* Variations in tectonic styles in Canada: Geol. Assoc. Canada Spec. Paper 11, p. 1–81.

White, W. H., J. E. Harakal, and N. C. Carter, 1968, Potassium-argon ages of some ore deposits in British Columbia: Canadian Inst. Mining and Metallurgy Bull.. v. 61, p. 1326–1334.

Ziegler, P. A., 1969, The development of sedimentary basins in Western and Arctic Canada: Alberta Soc. Petroleum Geologists Guidebook.

Metallic Mineral Resources of Canadian Cordillera[1]

A. SUTHERLAND BROWN[2]

Abstract The Canadian Cordillera has abundant minable deposits of Cu, Zn, Pb, and Mo. It has adequate Ni, W, Au, Ag, Hg, and Fe.

The Cordillera shows a pronounced regional zoning of metals in background abundances in rocks and an associated zoning in class of deposits and contained metals. These zonal changes occur sequentially across the strike of the five subparallel tectonic belts.

INTRODUCTION

The Canadian Cordillera is rich in minable deposits of Cu, Zn, Pb, and Mo, and has adequate Ni, W, Au, Ag, Hg, and Fe. The terrane has a long history of production, but in the last decade discovery, reserves, and production have escalated significantly. These deposits occur in a systematic arrangement that is related to the geologic pattern.

The Canadian Cordillera is a generally mountainous terrain that extends from just south of the 49th parallel to well north of the Arctic Circle—more than 1,500 mi (2,400 km) of the Circum-Pacific perimeter. The Cordillera is greater than 500 mi (800 km) in average width and has an area of about 585,000 sq mi (1,500,000 km^2). Politically, it consists of most of British Columbia and Yukon Territory and small parts of Alberta and Mackenzie Territory.

The mineral industry of the region has expanded rapidly in the last decade. The total value of metal production in 1973 was about C$946 million, in contrast to C$189 million in 1963. Production of major metals for 1973 totaled 778,000 MT. Copper accounted for 334,000 MT, zinc 252,000 MT, lead 188,000 MT, and molybdenum 14,000 MT.

GEOLOGIC FRAMEWORK

The Canadian Cordillera consists of two parallel orogens—the Columbian on the east or cratonic side, and the Pacific on the west or oceanic side (Wheeler and Gabrielse, 1972). The orogens are marked by two culminating core zones of deformation, metamorphism, and intrusion—the Omineca Crystalline belt on the east and the Coast Crystalline belt on the west. These core zones are separated by terrain characterized by plateaus—the Intermontane belt. Two other belts complete the Cordillera—the Eastern Marginal belt and the Insular belt. Hence, there are, in effect, five subparallel tectonic belts forming the Cordillera—from west to east, the Insular, Coast Crystalline, Intermontane, Omineca Crystalline, and Eastern Marginal (Fig. 1). The boundaries of these belts are fundamental tectonic interfaces—the continental margin, borders of core zones, and edges of disturbed belts. They are chiefly marked by major faults and commonly by lines of ultramafic plutons.

The tectonostratigraphic assemblages of the Cordillera have been described by Gabrielse (this volume). In effect, the Columbian orogen, or eastern two belts, consists of deformed miogeosynclinal rocks deposited as a great terrace wedge on the margin of the North American craton in Proterozoic and early Paleozoic times. In contrast, the Pacific orogen, or western three belts, consists of late Paleozoic and younger eugeosynclinal deposits of arc and oceanic nature which may be partly allochthonous.

Cycles of Sedimentation and Mineralization

Two separate cycles of mineralization, differing in character, are directly related in time and nature to the miogeosyncline and to the eugeosyncline (Sutherland Brown et al, 1971). The first cycle ranged from Proterozoic to Devonian. Mineralization consists largely of stratiform deposits which in many places appear to be syngenetic or to be of replacement origin and related to basin development. This cycle is therefore related to the building of the great terrace wedge on the edge of the continent. Many of the deposits are lead and zinc, and the primary source of metals has been the eroding craton. The second cycle extended from the late Paleozoic to the Holocene but was predominantly Mesozoic and early Tertiary. Mineralization chiefly affected the "new" crust west of the terrace wedge and was mainly epigenetic and related to volcanism or plutonism. Remobilization of first-cycle mineralization occurred during related tectonism within the Columbian orogen.

[1]Manuscript received, October 21, 1974.

[2]Deputy Chief Geologist, British Columbia Department of Mines and Petroleum Resources, Victoria, British Columbia.

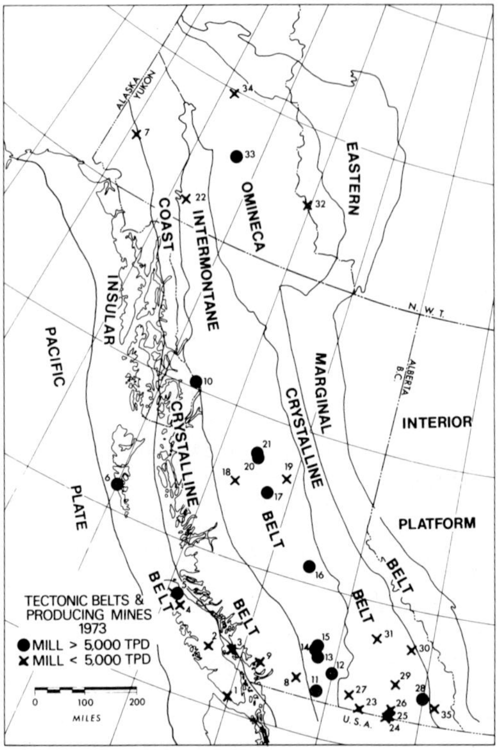

FIG. 1—Tectonic belts and producing mines in Canadian Cordillera.

Zonal Pattern of Distribution of Metals

The Canadian Cordillera shows a pronounced zoning of metals in background abundances in rocks and in mineral deposits, and a related zoning in genetic or morphologic classes of deposits (Sutherland Brown, 1969; Sutherland Brown et al, 1971). The zoning is clearly related to the patterns of geology and appears compatible with the concept of an accretionary origin of the continental crust (Sutherland Brown, 1974).

Background Abundances

Insufficient data on abundances in rocks of major ore metals (Cu, Zn, Mo, Pb) are available to provide a view of regional trends. However, sampling of the voluminous silt and soil geochemical data has revealed patterns in backgrounds of these metals so similar that they may be taken to reflect fairly accurately the regional background trends in rocks. A generally eastwardly decreasing gradient exists for background copper, one of lesser eastward slope for zinc, and a westwardly decreasing one for lead. Molybdenum rises to a central peak. This distribution has been related to a theoretical model dependent on time and degree of mantle and crustal processes in general, and to crustal accretion in particular (Sutherland Brown, 1974), in which high metal abundances and concentrations might be expected in the following order, from west to east—Ni, Fe, Zn, Cu, Mo, W, Sn, Ag, Hg, Pb, U, Th. A somewhat similar result was produced by Smith (1973) on the basis of a conventional plate-tectonic model.

The pattern of distribution of deposits of the various metals transverse to the belts matches the described pattern of background geochemistry to a considerable degree. This is shown substantially by considering Tables 1 and 2. Table 1 shows past production and reserves of all categories for major metals by tectonic belt. Table 2 is a matrix based on mine production in 1973.

The pattern of distribution of metal deposits, considered longitudinally, is characterized by an obvious irregularity with broad transverse and longitudinal regions in which metal deposits are rare (McCartney et al, 1974). Some of these areas of low metal endowment are caused by obvious geologic factors including thick cover rocks such as plateau lavas; deep glacial drift; or successor basins of thick clastic deposits, such as the Bowser basin. These features are concentrated in structural lows or transverse sags which may have innately low endowment (e.g., central British Columbia, where the low potential extends broadly beyond the cover rocks; cf. Noble, 1974). Some credence is given to this by the fact that transverse northeast-striking arches, such as the Stikine, Skeena, and a more diffuse one in southern British Columbia, are of innately high mineral potential.

Classes of Deposits and Their Zonal Pattern of Distribution

The classes of mineral deposits are zoned in a manner that conforms to the geologic pattern and zoning of metals. The writer groups metallic mineral deposits into six classes which are an amalgam of morphology and genesis. These classes are skarn, magmatic, massive sulfide, porphyry, vein, and stratiform. The stratiform class includes at least two types of deposits with such similarities that it is difficult to distinguish between them positively. These are deposits of syngenetic sedimentary character and replacement deposits that are strata-bound. A few deposits are of obscure nature because of metamorphism and deformation, and assignment is somewhat arbitrary.

The distribution of these classes is shown on Figure 2. The deposits are assembled by class and shown in their relative positions within the sub-

Table 1. Production and Reserves by Tectonic Belt

		Insular	*Coast*	*Intermontane*	*Omineca*	*Eastern*	*Total*
Copper[1]	P.[3]	285.5	1,022.5	1,239.0	351.0	16.5	2,914.5
	R.	2,147.5	1,727.0	18.793.5	1,538.5	216.0	24,422.5
	T.	2,433.0	2,749.5	20,032.5	1,889.5	232.5	27,337.0
Molybdenum[1]	P.	0.6	-	98.4	3.1	-	102.1
	R.	48.0	?	1,188.0	18.9	-	1,254.9
	T.	48.6	?	1,286.4	22.0	-	1,357.0
Zinc[1]	P.	166.6	208.9	18.8	10,359.6	89.4	10,843.3
	R.	160.7	53.4	?	5,913.7	107.5	6,235.3
	T.	327.3	262.3	18.8	16,273.3	196.9	17,078.6
Lead[1]	P.	19.5	60.4	15.5	8,878.5	56.7	9,030.6
	R.	24.4	23.5	?	8,579.5	103.5	8,730.9
	T.	43.9	83.9	15.5	17,458.0	160.2	17,761.5
Silver[2]	P.	5.0	76.7	9.6	500.3	1.0	592.6
	R.	10.9	18.2	250.0	250.8	20.0	549.9
	T.	15.9	94.9	259.6	751.1	21.0	1,142.5

[1]Thousands of short tons of metal.
[2]Millions of ounces.
[3]P. = Past production, R. = Current reserves of all categories, T. = Total.

parallel tectonic belts. Significant mines and major prospects include 159 deposits. The margins of the tectonic belts are preferred sites, having almost twice as many deposits as would occur in random distribution.

The zoning of classes of mineral deposits is shown clearly by Figure 2. The median position for each class is shown and is successively farther east from the westernmost median position (that of skarn deposits in the Insular belt) in the following order: skarn, magmatic, massive sulfide, porphyry, vein, and stratiform.

Skarn Deposits

Although the median for the skarn deposits is in the Insular belt, they are broadly distributed throughout the orogen and are second only to vein deposits in this regard. Skarn deposits are not known in the Eastern Marginal belt because it is practically devoid of plutons. They are scarce in the Coast Crystalline belt because of the density of plutons, high grade of metamorphism of remnant strata, and deep level of erosion. In the Intermontane belt they are more common than shown, but generally are included as a fringing segment of porphyry deposits.

The skarn deposits exhibit good metal zoning within the class itself. The trend is from Fe and Cu in the Insular belt, to Cu and Fe in the Coast, to Cu or Au in the Intermontane, to W and Cu or, rarely, Mo in the Omineca belt.

Most skarn deposits have a similar setting in which the host rocks are Triassic limestones. In the Insular belt they are situated at the contact between thick Upper Triassic basalt and the overlying limestone. The age of the genetically related intrusion ranges from Jurassic to Tertiary. Intrusions related to skarn deposits of the Omineca belt are mostly Cretaceous.

Magmatic Deposits

The median of the magmatic deposits is at the boundary between the Insular and Coast belts.

Table 2. 1973 Production, by Tectonic Belt

PRODUCTION VALUE 1973 $	INSULAR	COAST CRYSTALLINE	INTER-MONTANE	OMINECA	EASTERN MARGINAL
$>100 \times 10^6$	Cu		Cu	Zn, Pb	
>30 "	Fe	Cu	Mo	Ag	
>10 "	Zn,Au		Au	Cu,W	
>3 "	Ni,Ag	Ni,Ag	Ag		Cu
>1 "	Pb,Mo	Au	Zn,Hg	Au	
<1 "			Fe,Pb	Mo	Ag,Au

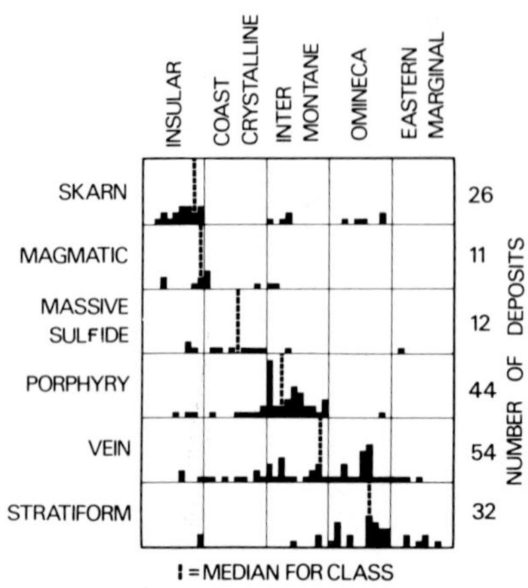

FIG. 2 – Relative transverse distribution of classes of mineral deposit by tectonic belt.

Known deposits are virtually confined to those two belts, particularly along their margins.

Both nickel-copper- and iron-bearing magmatic deposits are associated with small, zoned, unserpentinized ultramafic plutons similar to those of southeastern Alaska. These plutons are situated near major faults on the margins of the belts, and appear to range in age from late Paleozoic to Tertiary. Large alpine serpentinites, which are common in and on the borders of the Intermontane belt, have not been shown to contain any significant deposits other than asbestos and subeconomic chromite.

Massive Sulfide Deposits

The median of the massive sulfide deposits is located within the Coast Crystalline belt. Known deposits are distributed mainly in the two western belts. The deposits of the Insular and Coast belts range in age from Permian to Cretaceous. They are mostly associated with andesitic and dacitic volcanism. Many have been subjected to considerable metamorphism and deformation. They all have similar metal ratios: copper is the major base metal, and lead and zinc ratios are about 1:10. Precious metals are invariably important.

The two deposits in other belts shown on Figure 1 are different in several respects. The Sam Goosly deposit in the Intermontane belt is a finely disseminated deposit containing massive lenses; it is a copper-silver deposit with minor zinc. Its age is disputed but may be Jurassic or Eocene. The Hart River massive sulfide deposit in Yukon is on the border between the Eastern Marginal and Omineca belts. It is believed to be of Precambrian age and, hence, to be unrelated to the late Paleozoic and younger volcanism characteristic of the three western belts.

In general, Cordilleran massive sulfide deposits have characteristics transitional between Archean examples and Neogene kuroko deposits of Japan. This applies to the nature of iron sulfide, the metal ratios, the alteration types, and the presence or amount of barite or anhydrite. These transitions can be largely explained by evolutionary changes of ocean, atmosphere, and crust.

Porphyry Deposits

The median of the porphyry deposits, as well as their major distribution, is in the Intermontane belt. However, they are widely though more sparsely distributed in the two western belts, and there is also a single unusual representation in the Omineca belt—the Potato Hills tungsten deposit. Porphyry deposits have provided all the significant new copper and molybdenum production and the majority of the new reserves within the orogen.

A great diversity of age, petrology, contained metals, morphology, and setting exists for these deposits in the Canadian Cordillera. These facts must be considered in order to reveal the full significance of these deposits.

The ages range from Late Triassic to Miocene and are clustered in six groups. Porphyries associated with the deposits may belong to either of two magma series—the granite clan or the syenite clan. Porphyries of intermediate composition within the granite clan are associated with copper-molybdenum deposits in which these metals commonly have ratios comparable to their relative crustal abundances. Porphyries of very acidic character within the granite clan are associated with pure molybdenum deposits or, in the case of Potato Hills, a tungsten deposit. Porphyries of the syenite clan are associated with pure copper deposits.

Porphyry deposits may be divided in relation to their morphology into three subclasses—phallic, volcanic, and plutonic (Sutherland Brown, 1972). These subclasses are transitional but fairly discrete. Phallic porphyry deposits are associated with small, generally cylindrical plutons of moderately shallow depth that intruded unrelated host rocks in late- or postorogenic time. The plutons, which alter the host rocks to hornfels, commonly have associated breccia pipes or carapaces. Alteration and ore zones have cylindrical or concentric annular shell form developed in classical order.

Volcanic porphyry deposits are in direct contrast. During an orogenic period, they are intruded into consanguineous volcanic host rocks at a very shallow depth in the form of large dikelike to irregular plutons. They are characterized by major breccia pipes or linear zones and by close fault control of the pluton location, breccia bodies, the distribution of ore, and alteration. Most of the syenite-clan porphyry deposits belong to this subclass. Alteration in these deposits differs from the classic pattern.

Plutonic porphyries are relatively deeper and are associated with medium-sized, zoned plutons that may intrude related or unrelated hosts in orogenic or late-orogenic time. The intrusive rocks are chiefly phaneric with associated minor porphyry phases; breccias are absent. Orebodies tend to be large, equant shaped in plan, and of great vertical continuity. They are subtly controlled by intrusive phases or faults. Vein stockworks are regular and major veins are common. The alteration sequence is classical but variably developed.

No simple distribution pattern exists in the Canadian Cordillera (Carter, 1974) such as is reported in the Andes (Sillitoe, 1972), where there is a progressive eastward decrease in age and an increase in K content with increasing depth of generation. This complexity in the Canadian Cordillera applies if porphyries are considered as a group or in their subclasses. The volcanic porphyries are mostly Triassic and occur embedded within a narrow belt of coeval and consanguineous volcanic rocks in the Intermontane belt. However, at least one of Jurassic age is known in related volcanic rocks within the Insular belt. Known plutonic porphyries are also chiefly of Triassic age—although some are Jurassic—and generally occur in or near volcanic piles of similar age within the Intermontane belt. In contrast, phallic porphyries are all of Cretaceous or younger age, and their distribution extends across the three western belts, where they are generally somewhat divorced from volcanic piles of the same age. They occur in clusters but do not form a simple pattern of progressive change in chemistry or age.

Vein Deposits

Vein deposits exhibit the broadest distribution of any class, ranging across the whole orogen. However, they do have a concentration and their median lies within the Intermontane belt near the Omineca boundary. The vein deposits also exhibit a good transverse zoning in contained metals. Most of the veins are mined for their precious-metal content, but the associated metals reveal their affiliations. In the Insular belt, veins are mined for Cu and Au; in the Coast belt, for Au and Ag; in the Intermontane belt, for Ag, Ag-Pb-Zn, Ag-Cu-Zn, and Hg; in the Omineca belt, for Ag, Ag-Pb-Zn, and Au; and in the Eastern Marginal belt, for Ag-Pb-Zn and Cu. The age of many vein deposits is unknown, but many are Tertiary. Some appear to represent remobilized metals. This seems most evident in the Kootenay arc, where Ag-Pb-Zn veins abound in younger rocks in the same vicinity as the stratiform deposits of Proterozoic and early Paleozoic ages. The mobilization and upward movement in the section appear to result in a fractionation with enrichment in precious metals and lead relative to zinc.

Stratiform Deposits

Stratiform deposits have a median east of the vein median within the Omineca belt. These deposits are found across the whole width of the orogen, although none are known within the Coast Crystalline belt. The main concentration of deposits lies within the Omineca belt and consists of lead-zinc. In contrast, the stratiform deposits of other belts are chiefly copper rich.

The lead-zinc deposits of the Omineca belt include two general types (Fyles, 1967) which, because of deformation and metamorphism subsequent to their formation, may be difficult to distinguish. In one type (Shuswap), the ore beds are intercalated with other rocks (not necessarily carbonate rocks) on a fine to coarse scale, and are not associated with alteration; they may show sedimentary features and appear to be syngenetic deposits. These orebodies appear to be late Proterozoic or Eocambrian. The other type (Salmo) is strata-bound, but the orebodies may be irregular, generally occur in carbonate rocks, and commonly have associated alteration of dolomite and/or silica. They are largely Early Cambrian, but some are of varied early Paleozoic ages (Thompson and Panteleyev, 1974). The isotopic data for lead do not discriminate between them, but a common crustal source of lead is indicated (LeCouter, 1973). The Sullivan deposit, in fact, has characteristics that bridge the two types, and a complex origin combining syngenetic and epigenetic aspects has been proposed (Morris, 1972).

Stratiform deposits of the three western belts are copper deposits. Only one has been shown to be a significant deposit. Some have characteristics similar to Shuswap lead-zinc deposits, which suggests they are truly syngenetic, but the important deposit (Sustut) has more indications of deposition by copper-rich waters traveling through an aquifer. The known deposits are all Triassic.

Stratiform deposits of the Eastern Marginal belt include lead-zinc deposits formed, at least in part, near marked facies boundaries, particularly between carbonate rock and shale. There are also bedded copper deposits and a bedded hematite deposit of Proterozoic age. The major copper deposits have been proposed to be related to concentration by groundwater movement (Harrison and Domenico, 1973), whereas the iron deposit (Snake River) is believed to be a chemical precipitate from seawater (Dahlstrom, 1973).

Conclusion

The Canadian Cordillera has been very intensively explored for mineral deposits for more than a decade. Much of the basic geologic data is known, but interpretation of tectonics and metallogenesis is still in considerable flux. Plate-tectonic models are ill defined, and progress on metallogenic models is unlikely until there is greater resolution. Some form of accretionary origin of the Cordillera seems essential, first by accumulation of a great terrace wedge of craton-derived sediments, then by a complex history of arc formation, subduction, and possible rifting.

The metallogenic cycles fit such a model, as does the similarity of the patterns of metal-deposit distribution and metal backgrounds in rocks. These similarities cannot be accidental, and can be interpreted as showing that the mineral deposits owe their origin in the first instance to the general level of the background they inhabit. The metals of all classes of deposits are not distant travelers from the mantle by any more direct route than the rocks themselves. The deposits appear to be formed by concentration in one or more stages from metals available in background levels in nearby rocks. Nevertheless, the efficacy of plutonic processes in concentrating metals is attested to by the contrast in production and reserves between the Omineca Crystalline and Eastern Marginal belts.

References Cited

Carter, N. C., 1974, Geology and geochronology of porphyry copper and molybdenum deposits in west-central British Columbia: Unpub. PhD thesis, Univ. British Columbia.

Dahlstrom, C. D. A., 1973, The geology of the Snake River iron deposits, *in* Sedimentary geology and mineral deposits of the Canadian Cordillera: Geol. Assoc. Canada Cordilleran Sec.

Fyles, J. T., 1967, Geology of the Ainsworth-Kaslo area, British Columbia: British Columbia Dept. Mines and Petroleum Resources Bull. 53, 125 p.

Gabrielse, H., 1976, Environments of Cordilleran depositional basins: this volume.

Harrison, J. E., and J. A. Domenico, 1973, Association of copper mineralization in miogeosynclinal clastics of the Belt Supergroup, *in* Sedimentary geology and mineral deposits of the Canadian Cordillera: Geol. Assoc. Canada Cordilleran Sec.

Hollister, V. F., 1976, Evolution of the porphyry copper province of the northern Cordilleran orogen: this volume.

LeCouter, P. C., 1973, A study of lead isotopes from mineral deposits in south eastern British Columbia and the Anvil Range, Yukon Territory: Unpub. PhD. thesis, Univ. British Columbia.

Morris, H. C., 1972, Sullivan Mine: 24th Internat. Geol. Cong. Guidebook, Field Excursion A 24–C-24. p. 26–36.

McCartney, W. D., J. T. Fyles, and A. H. Matheson, 1974, Mineral capability maps for land-use planning in British Columbia: Western Miner, v. 47, no. 7, p. 9–16.

Noble, J. A., 1974, Metal provinces and metal finding in the western United States: Mineralium Deposita, v. 9, p. 1–25.

Sillitoe, R. H., 1972, A plate tectonic model for the origin of porphyry copper deposits: Econ. Geology, v. 67, p. 184–197.

Smith, D. M., Jr., 1973, Model for the distribution of selected metal provinces in a Cordilleran-type plate tectonic region and its comparison with northern Mexico: Asoc. Inc. Min. Metall. y Geol., Mexico (Preprint).

Sutherland Brown, A., 1969, Mineralization in British Columbia and the copper and molybdenum deposits: Canadian Inst. Mining and Metallurgy Bull., v. 62, p. 26–40.

——— 1972, Morphology and classification of porphyry deposits of the Canadian Cordillera (abs.): Vancouver, 9th Ann. West. Inter-Univ. Geol. Conf. Proc.

——— 1974, Aspects of metal abundance and mineral deposits in the Canadian Cordillera: Canadian Inst. Mining and Metallurgy Bull., v. 67, p. 48–55.

——— et al, 1971, Metallogeny of the Canadian Cordillera: Canadian Inst. Mining and Metallurgy Bull., v. 64, p. 37–61.

Thompson, R. I., and A. Panteleyev, 1974, Stratabound mineral deposits of the Canadian Cordillera: Unpub. ms.

Wheeler, J. O., and H. Gabrielse, 1972, The Cordilleran structural province: Geol. Assoc. Canada Spec. Paper No. 11, p. 1–81.

Porphyry Copper Province of Northern Cordilleran Orogen—Summary[1]

V. F. HOLLISTER[2]

INTRODUCTION

Porphyry copper deposits of the Cordilleran orogen north of the Columbia volcanic plateau and the Idaho batholith and east of the Alaska Range in Alaska may have characteristics distinct from those which occur elsewhere in the orogen. By use of aggressive procedures, porphyry copper exploration in British Columbia, the Yukon, and Alaska has matured in the past few years to the state where a substantial number of deposits are now known (Hollister et al, 1974). More deposits will be discovered, but enough have been described so that a statistically significant sample now exists. Valid summaries may be compiled illustrating the characteristics of the deposits, and these may be integrated into the regional geologic setting which has developed in the past few years (Fig. 1).

Two separate petrologic models exist for porphyry copper deposits in this area: (1) a silica-deficient or diorite model ("syenite suite" of Sutherland Brown et al, 1971) and (2) a silica-rich, quartz-bearing or quartz monzonite–quartz diorite model (commonly called a "quartz monzonite model"). The basic petrologic differences between the two indicate distinct evolutionary trends in alteration zoning and in habitat of mineralization. Several of the better known deposits of both models are listed in Table 1.

The quartz monzonite deposits are subdivided into a granitic-pluton type (after Field et al, 1973; called "plutonic" by Sutherland Brown, 1972) and a diapiric type (including the various types of piercements accompanying porphyry copper mineralization listed by Sutherland Brown [1969, 1972]).

Sillitoe and Sawkins (1971) divided porphyry deposits into breccia-pipe and stockwork models. Table 1 includes a column containing the appropriate designation as to stockwork model or breccia-pipe model.

GEOLOGY

No porphyry copper deposits have been dated older than 280 m.y. Therefore, it is assumed that geology most significant to porphyry copper development is Pennsylvanian or younger.

Comprehensive summaries of the tectonics of this area have been presented in the past by Brew et al (1966), Dercourt (1972), Monger et al (1972), and Wheeler et al (1972). Among the more significant alternatives proposed in this paper, which set it apart from the other studies, are the proposals that formations of the Nicola type and age fill a structural trough and thus are not a typical island-arc sequence; therefore, the allochthonous Paleozoic section moved as a result of tectonism, which included rifting. Monger et al (1972) preferred strike-slip movement as the mechanism forming the Paleozoic allochthon, but I favor involvement of rifting as well.

The major stratigraphic-structural units are, from west to east: (1) The allochthonous Paleozoic crust and associated pre–Late Jurassic island-arc volcanic sequences (Insular belt of Monger et al, 1972; Alexander terrane of Berg et al, 1972). Contiguous pre–Late Jurassic assemblages are included in unit 1 because the allochthon could have been mobile until deposition of the Late Jurassic rocks. (2) The volcanic-sedimentary trough that separates the Alaskan and coastal Paleozoic crust from the North American craton. For most of its length, unit 2 is occupied by the Takla–Hazelton–Cache Creek–Slide Mountain–Nicola or equivalent United States formations or by younger basinal rocks (partly coincident with Intermontane and Omineca belts of Sutherland Brown, 1969). East of these chiefly volcanic formations are (3) the Paleozoic rocks of the Rocky Mountain belt and the North American craton. Obliquely intruding both the allochthonous Paleozoic crust and the volcanic-sedimentary sequences that separate the North American craton from the coastal or allochthonous Paleozoic crust is (4) the Coast Range batholith. It is essentially younger than the three structural-stratigraphic units proposed.

[1]Manuscript received, October 21, 1974.

[2]Duval International Corporation, Vancouver 1, British Columbia.

H. Berg, of the U.S. Geological Survey, reviewed this paper and made many meaningful suggestions. A. Sutherland Brown, of the British Columbia Department of Mines, was equally generous with his criticisms, and R. Larson was helpful with the correlation of plate-tectonic events.

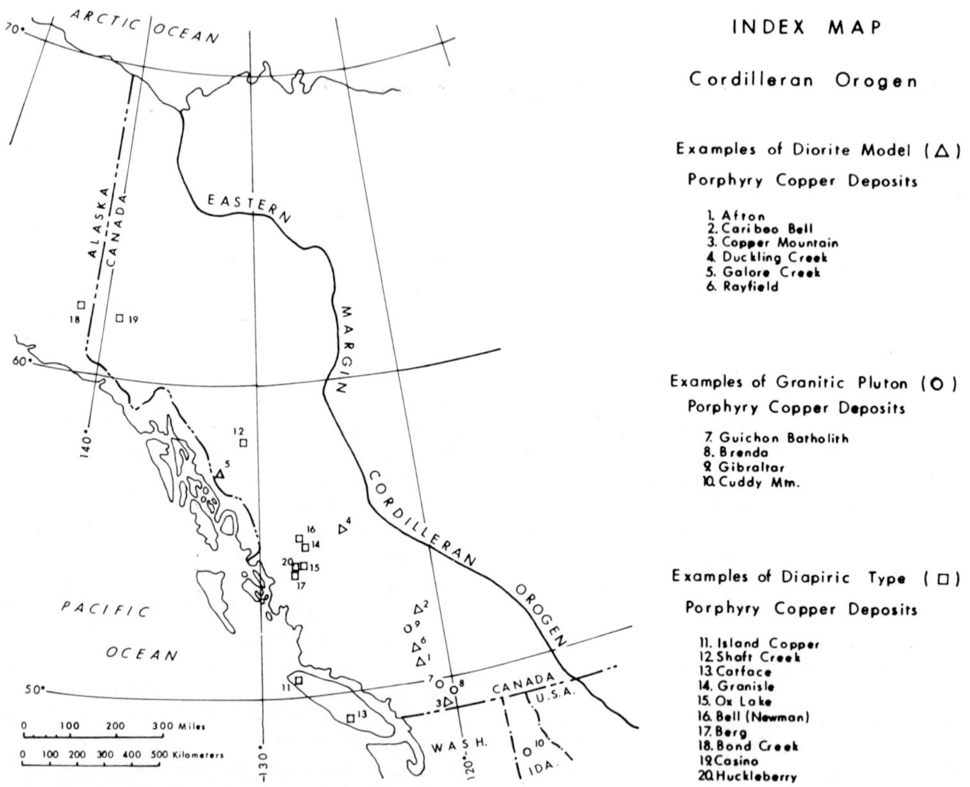

FIG. 1—Location of representative porphyry copper deposits in northern Cordilleran orogen. See Table 1 for names and data.

Diorite-type porphyry copper deposits are largely restricted to unit 2, whereas porphyry coppers of the diapiric quartz monzonite–quartz diorite type may occur in all four units. The granitic-pluton quartz monzonite–quartz diorite type occurs only in unit 2.

Monger et al (1972) suggested that Paleozoic rocks west of the outcropping Permian–Triassic–Lower Jurassic sedimentary and volcanic rocks—that is, west of the stratigraphically complex and areally extensive Takla-Hazelton-Nicola formations—are allochthonous with respect to the North American craton. They appear to have been moved to the northwest during late Paleozoic and early Mesozoic time. Berg et al (1972) implied that movement may have ceased by the Late Jurassic since strata of this age (their Gravina-Nutzotin belt) overlie allochthonous Paleozoic crust but do not appear to be rifted. Therefore, I have assumed that the allochthon may include pre–Late Jurassic rocks.

The allochthonous crust differs from strata of the same age which are part of the nearby North American craton in that it represents continental-margin accumulation (including volcanic deposits), whereas the cratonic strata represent a different geologic setting.

In spite of the presence of volcanic deposits on Vancouver Island contemporaneous with, and similar in some respects to, those of the Triassic and Early Jurassic Takla-Hazelton-Nicola formations, the interpretation presented here favors including Vancouver Island and the Insular belt with the allochthonous Paleozoic crust. This interpretation is supported by the existence in the Alaska allochthon of Permian and Triassic arc-trench units similar to those on Vancouver Island. Though these volcanic deposits may lie directly on oceanic crust on Vancouver Island (Stacey, 1973) and in parts of Alaska, they do appear to be spatially associated with the rifted Paleozoic crust, and appear themselves to be part of the pre–Late Jurassic allochthonous crustal block. The general interpretation is also supported by the subtle differences between the Nicola and the stratigraphically equivalent volcanic rocks of the

Table 1. Representative Porphyry Copper Deposits

Map No.	Name	Location	Intrusion Close to Ore	Intruded	Age in M.Y.	Alteration Sequence from Center	% Py in Pot Zone	Size of Py Halo	Structure
				DIORITE MODEL					
1	Afton	50°39′, 120°30′	Dio, Sye	Trias Vol	176	Prop	Rare	Erratic	Stockwork
2	Cariboo Bell	52°30′, 121°38′	Sye, Mon	Trias Vol	–	Pot-Phy-Prop	2%	1.0 × 1.1 km	Stockwork
3	Copper Mountain	49°21′, 121°34′	Sye	Trias Vol	193	Pot-Prop	3%	2.1 × 4.0 km	Stockwork
4	Duckling Creek	55°55′, 125°26′	Sye	Trias Vol	170	Pot-Prop	1%	2.0 × 3.0 km	Stockwork
5	Galore Creek	57°07′, 131°26′	Sye	Trias Vol	182	Pot-Phy-Prop	3%	4.0 × 5.0 km	Breccia
6	Rayfield River	51°15′, 121°04′	Sye, Dio	Trias Vol	–	Pot-Prop	1%	2.0 × 2.6 km	Stockwork
–	Poison Mountain	49°21′, 120°34′	Dio, Sye	Perm Sed	–	Pot-Prop	1%	1.0 × 1.7 km	Stockwork
				QUARTZ MONZONITE - QUARTZ DIORITE MODEL					
	Granitic-Pluton Type								
7	Guichon Batholith (Bethlehem, Lornex, Valley)	50°29′, 121°00′	Qtz Mon Por	Trias Vol	200	Pot-Phy-Arg-Prop	3-5%	Erratic	Bx and Stwk
8	Brenda	49°52′, 120°01′	Grdr	Trias Vol	168	Pot-Prop	1%	Erratic	Stockwork
9	Gibraltar	42°31′, 122°16′	Qtz Dio	Trias Vol	204	Pot-Phy-Prop	1%	Erratic	Stockwork
10	Cuddy Mtn.	45°10′, 116°15′	Grdr Por	Trias Vol	200	Pot-Phy-Prop	2%	Erratic	Bx and Stwk
	Diapiric Type								
11	Island Copper	50°36′, 127°28′	Qtz Por	Trias Vol	153	Pot-Phy-Arg-Prop	4%	1.6 × 2.2 km	Stockwork
12	Shaft Creek	57°19′, 130°50′	Qtz Dio Por	Trias Vol	182	Pot-Phy-Prop	3%	1.0 × 2 km	Stockwork
13	Catface	49°15′, 125°58′	Qtz Mon Por	Pal(?) Sed	48	Pot-Phy-Arg-Prop	3%	1.1 × 1.4 km	Stockwork
14	Granisle	54°58′, 126°19′	Qtz Mon Por	Trias Vol	51	Pot-Phy-Arg-Prop	4%	1.3 × 1.3 km	Breccia
15	Ox Lake	53°40′, 127°03′	Qtz Mon Por	Trias Vol	83	Pot-Phy-Prop	4%	1.3 × 1.3 km	Stockwork
16	Bell (Newman)	55°00′, 126°14′	Qtz Dio Por	Trias Vol	52	Pot-Phy-Arg-Prop	4%	1.5 × 2.1 km	Stockwork
17	Berg	53°47′, 127°28′	Qtz Mon Por	Trias Vol	50	Pot-Phy-Arg-Prop	4%	1.5 × 1.6 km	Stockwork
18	Bond Creek	62°15′, 142°46′	Qtz Dio Por	Trias Vol	109	Pot-Phy-Arg-Prop	5%	4.0 × 6.1 km	Breccia
19	Casino	62°45′, 138°48′	Qtz Por	Pal(?) Sed	70	Pot-Phy-Arg-Prop	6%	3.0 × 4.1 km	Breccia
20	Huckleberry	53°49′, 127°10′	Qtz Mon Por	Trias Vol	80	Pot-Phy-Arg-Prop	5%	1.3 × 1.3 km	Stockwork

Abbreviations

Pot: Potassic	Prop: Propylitic	Dio: Diorite	Trias: Triassic	Pal: Paleozoic	Bx: Breccia
Phy: Phyllic	Qtz: Quartz	Grdr: Granodiorite	Vol: Volcanics	Perm: Permian	Stwk: Stockwork
Arg: Argillic	Mon: Monzonite	Por: Porphyry	Sed: Sediments	Km: Kilometers	

same age on Vancouver Island. The former tend to be richer in alkalies.

Kanasewich's (1966) interpretations regarding crustal thicknesses are compatible with the concept that the Takla–Hazelton–Cache Creek–Nicola rocks occupy a rift zone between two blocks of Paleozoic crust.

The late Paleozoic (and possibly Early Triassic) Cache Creek–Slide Mountain rocks are believed to have partly filled a trough that resulted from movement of the Paleozoic crust (Monger et al, 1972), and I propose that the Takla-Hazelton-Nicola and equivalent marine Triassic-Jurassic volcanic rocks in Washington and Idaho completed the filling of the trough. If this is true, then volcanism, and probably rifting also, continued into the Early and Middle Jurassic. The crustal sequence in Alaska, possibly in the Coast Ranges of British Columbia, and also in the Washington Cascades suggests that the entire Paleozoic crust from the Cascades to Alaska's Alexander terrane is allochthonous, and that the volcanic deposits occupy a rift zone from Idaho to Alaska which separates allochthonous Paleozoic crust from the North American craton.

Cretaceous dates have been proposed for the Coast Range batholith (Roddick and Hutchison, 1972; Wanless, 1969). However, on the basis of dates currently available, part of this batholith appears to be Tertiary, with preliminary dates clustering in the 45–50-m.y. range.

Three periods of assumed rapid spreading—162–155 m.y., 110–85 m.y., and 50–45 m.y.—though apparently coincident with abundant formation of batholiths, do not appear to be particularly favorable for development of porphyry copper deposits. Few porphyry copper deposits in the area of interest have thus far been dated within the magnetic quiet zones. Since porphyry copper deposits uniformly exhibit extensional structural conditions, the compressive environment accompanying more rapid subduction would not encourage their formation. Lanphere and Reed (1973) concluded that the overall spacing, duration, and synchroneity of intrusive epochs in the Circum-Pacific batholiths are not yet well established. A negative correlation does seem to be emerging, however, between intrusion of batholiths and porphyry copper plutons in the Cordilleran orogen.

Monger et al (1972, p. 591) offered one explanation for the origin of the large transcurrent faults which are pronounced in British Columbia and Alaska, noting that the regional stress was transformed from compressive to strike slip in the Late Cretaceous. Wanless (1969) showed that some radiometric dates on batholithic plutons within the Coast Range batholith may be coincident with the strike-slip movement. The number of batholiths dated as younger than 85 m.y. appears to be smaller than for the Jurassic and early middle Cretaceous plutons. However, the number of dated porphyry copper deposits has apparently increased.

Sutherland Brown et al (1971) and Sutherland Brown (1972) noted that most of the porphyry copper deposits younger than 80 m.y. have many of the features of a diapir. The intrusion spatially close to ore is mostly quartz monzonite porphyry or a petrology close to that.

Diorite Model

Diorite-model porphyry copper deposits appear to be restricted largely to the area where Pennsylvanian to Middle Jurassic oceanic crust and successor formations (including andesitic and differentiated volcanic end products) occur between the rifted allochthonous Paleozoic crust and the North American craton (Fig. 1). Some diorite-model porphyry copper deposits have been found in the Pennsylvanian-Permian volcanic rocks of Alaska associated with the Paleozoic allochthon, but none of these deposits has yet proved to be commercial. Therefore, the diorite porphyry province at present is coincident with, and appears to be genetically related to, the Takla-Hazelton-Nicola volcanic rocks. Evidence suggests that rifting may have extended at least the length of the diorite porphyry copper province—from near the Idaho batholith northwest to the Pacific. Conceivably, therefore, diorite-model deposits could occur from Alaska to the Idaho batholith wherever upper Paleozoic–lower Mesozoic marine volcanic deposits are known to fill the trough.

Sutherland Brown et al (1971) and Field et al (1973) indicated that the diorite model actually includes syenites, monzonites, and other silica-deficient intrusions in addition to diorite. Most phases in the intrusive sequence are typically porphyritic.

The undersaturated intrusions other than diorite conceivably could be derived from a dioritic parent by potassic metasomatism or through processes of assimilation or differentiation or some combination of these. The scattered occurrence of magmatic garnet, nepheline, and other minerals not characteristic of calc-alkaline intrusions supports the hypothesis that they are a distinct magmatic series (Field et al, 1973). Similarly, the time relations (diorite appears early and syenite late in each porphyry copper–bearing magmatic sequence) suggest that the two are genetically associated in composite, zoned, differentiated intrusions.

The unifying feature of the silica-deficient intrusion which hosts porphyry copper mineralization, regardless of magmatic composition, is the fairly uniform response the rock has to potassic alteration. In detail, the mineral assemblages in the propylitic zone may vary widely from deposit to deposit but, if a potassic zone has developed, it is everywhere dominated by orthoclase-chlorite or orthoclase-biotite assemblages, or both. In only a few of these deposits is a phyllic or an argillic zone developed.

Mineralization tends to occur spatially associated with syenitic members of zoned diorite-monzonite-syenite intrusions. Only rarely is copper sulfide accompanied by molybdenum, but abnormally high values of gold are common. Sulfur accompanying the copper is commonly insufficient to consume the iron present in the host intrusive, and the altering mineralizing process leaves large amounts of magnetite accompanying the chalcopyrite. Magnetite may be as important a hydrothermal mineral in dissemination in diorite-model porphyries as is pyrite in the quartz monzonite–quartz diorite model. Mineralizing fluids appear to be silica-deficient, since the veinlets which contain the ore sulfides are usually free of quartz but may contain epidote, chlorite, calcite, or zeolite.

Stockworks are much more common in diorite-model porphyries than are breccias and, volumetrically, dissemination of the ore minerals is important in most intrusives.

Quartz Monzonite–Quartz Diorite Model

Granitic-Pluton Type

Granitic-pluton-type porphyry copper deposits are associated with large, usually zoned, quartz diorite–granodiorite–quartz monzonite intrusions; the porphyry deposit commonly occurs in the interior or core, spatially associated with the youngest phase. These plutons occur from the Cuddy Mountain–Peck Mountain area in Idaho to the Gibraltar deposit in British Columbia. They are all intrusive into the Triassic-Jurassic Takla-Hazelton-Nicola volcanic rocks or their equivalents in Idaho and Washington, and are coeval with their host. None of them have dates appreciably younger than their intruded rocks (see Table 1).

The granitic plutons which host porphyry copper deposits are large, upper mesozonal to epizonal, zoned multistage intrusions; they contain concentric phases with early, more mafic, nonporphyritic and sheared quartz diorite margins and later, more silicic, alkalic and younger porphyritic cores. The most common terminal phase, if magmatic, is close to quartz monzonite in composition and, if hydrothermal, is typical of porphyry copper mineralization; both tend to occupy the core area of the intrusive.

Northcote (1969), Ager et al (1972), and Hylands (1972) made an excellent case for magmatic differentiation in the Guichon batholith, and their petrographic interpretations are applicable to other plutons similar to the Guichon. Their strongest evidence is the gradational contacts that may be seen between any two adjacent petrographic zones. Acceptance of this hypothesis permits the speculation that mineralization may be the end product of differentiation of a calc-alkaline magma, and the evidence is continuous enough to be plausible. The most effective process of differentiation appears to have been the outward and upward movement of alkali and silica.

Table 1 lists the alteration zoning sequence from the center outward. All of the deposits cited have a well-developed potassic zone, and most have well-defined phyllic zones. An argillic zone is developed in a few Highland Valley ore deposits (Guichon batholith) but is missing in the other examples. The pyrite zones around the porphyry copper deposits are usually erratically developed, in that they do not extend far beyond the orebody, they do not ordinarily contain more than 3% pyrite, and the area containing pervasive disseminated pyrite includes large areas of rock barren of pyrite.

Copper may occur in both the potassic zone, which is characterized by orthoclase-biotite or orthoclase-chlorite mixtures or both, and the phyllic zone, which is essentially a quartz-sericite-pyrite mixture. Chalcopyrite is the dominant copper mineral, and copper sulfosalts are scarce in most deposits. All ore zones carry some molybdenum. Zinc and lead may be zonally arranged about the copper center.

Diapiric Type

The diapiric-type porphyry in the area of interest is a mineralized boss, dike zone, or small stock injected into its host rock. In porphyry copper deposits associated with small, injected intrusive bodies, the hydrothermal fluid may have been a lubricant, since contacts of the igneous rocks usually show a stronger hydrothermal metasomatism than do the central parts. The different structural types suggested by Sutherland Brown (1972) for porphyry copper deposits associated with the smaller intrusions all suggest the magma to be an igneous piercement, or diapir, intrusive into wall rock that yielded as the injection took place.

The diapiric type may occur anywhere in the Cordilleran orogen, but most commonly has been found east of the Coast Range batholith. Most intrude the Triassic-Jurassic volcanic rocks at the surface although they clearly postdate these rocks. Most of the diapiric-type porphyries are less than 83 m.y. old. The negative correlation with periods of batholith formation also suggests that each may occur under differing conditions of plate movement; the batholiths are more easily identified with rapid subduction, and the diapiric model is more closely associated with strike-slip tectonics.

The intrusive most commonly associated spatially with the diapir type of porphyry copper mineralization is quartz monzonite in composition. The quartz and orthoclase may be restricted to the groundmass, as at Granisle, or they may occur as prominent phenocrysts, as at Berg. If a compositional variation does occur, however, it commonly will be a quartz-rich intrusive close to quartz monzonite in composition. Most intrusions are mixtures of quartz, plagioclase, orthoclase, biotite, and hornblende. All known well-mineralized intrusions are porphyritic.

Other intrusive phases may exist in addition to the quartz monzonite, but their composition varies from intrusive center to intrusive center. The other igneous rocks usually preceded the quartz monzonite and the closely associated mineralization.

Alteration zones may not be as simply developed in the diapiric type as in the granitic pluton. Should the intrusive contact be the locus of a potassic zone, then the phyllic zone may occur both internally in the stock and externally away from it. However, if the center of alteration is well within the stock, then the phyllic, argillic, and propylitic zones will occur outside of it in concentric shells. This type of porphyry is typically a copper-molybdenum type with relatively low gold values.

CONCLUSIONS

Mineralization in the porphyry copper province of the northern Cordilleran orogen began in the late Paleozoic with what appears to have been a rifting environment. Diorite-model and granitic-pluton-type porphyry copper deposits formed in the Triassic and Early Jurassic as the trough was filled with volcanic debris. This was followed by a relatively sterile period when intrusive activity tended to be batholithic in nature. Beginning with the Late Cretaceous, batholithic intrusive activity declined and there was a resurgence of porphyry copper–related intrusions. These latter deposits tend to be related to smaller plutons which bear a tectonic resemblance to diapirs. They also coincide with the development of a strike-slip tectonic regime which was initiated at the end of the Cretaceous.

References Cited

Ager, C. A., W. J. McMillan, and T. J. Ulrych, 1972, Gravity, magnetics and geology of the Guichon Creek batholith: British Columbia Dept. Mines and Petroleum Resources Bull. 62.

Berg, H. C., D. L. Jones, and D. H. Richter, 1972, Gravina-Nutzotin belt: U.S. Geol. Survey Prof. Paper 800-D, p. D1–D24.

Brew, D. A., R. A. Loney, and L. J. Patrick-Muffler, 1966, Tectonic history of southeastern Alaska, *in* A symposium on the tectonic history and mineral deposits of the western Cordillera: Canadian Inst. Mining and Metallurgy Spec. Vol. 8, p. 149–170.

Dercourt, Jean, 1972, The Canadian Cordillera, the Hellenides, and the sea-floor spreading theory: Canadian Jour. Earth Sci., v. 9, p. 709.

Drummond, A. D., S. J. Tennant, and R. J. Young, 1972, The interrelationship of regional metamorphism, hydrothermal alteration, and mineralization at Gibraltar: Canadian Inst. Mining and Metallurgy Trans., v. 74.

Field, C. W., M. B. Jones, and W. R. Bruce, 1973, Porphyry copper-molybdenum deposits of the Pacific Northwest (abs.): Econ. Geology, v. 68, p. 137.

Hollister, V. F., S. A. Anzalone, and D. H. Richter, 1974, Porphyry copper belts of southern Alaska and contiguous Yukon Territory: Canadian Inst. Mining and Metallurgy Ann. Mtg. Preprint 62.

Hylands, J. J., 1972, Porphyry copper deposits of the Guichon Creek batholith, B.C.: 24th Internat. Geol. Cong., Sec. 4, p. 241.

Kanasewich, E. R., 1966, Deep crustal structure under the plains and Rocky Mountains: Canadian Jour. Earth Sci., v. 3, p. 937–945.

Lanphere, M. A., and B. L. Reed, 1973, Timing of Mesozoic and Cenozoic plutonic events in Circum-Pacific North America: Geol. Soc. America Bull., v. 84, p. 3773–3782.

Monger, J. W. H., J. G. Souther, and H. Gabrielse, 1972, Evolution of the Canadian Cordillera: a plate tectonic model: Am. Jour. Sci., v. 272, p. 577–602.

Northcote, K. E., 1969, Geology and geochronology of the Guichon Creek batholith: British Columbia Dept. Mines and Petroleum Resources Bull. 56.

Roddick, J. A., and W. W. Hutchison, 1972, Plutonic and associated rocks of the coast mountains of British Columbia: 24th Internat. Geol. Cong. Guidebook AC04.

Sillitoe, R. H., and F. J. Sawkins, 1971, Geologic, mineralogic, and fluid inclusion studies relating to origin of copper-bearing tourmaline breccia pipes, Chile: Econ. Geology, v. 66, p. 1028–1041.

Soregaroli, A. E., 1971, Geology of the Brenda copper-molybdenum deposit, B.C.: Canadian Inst. Mining and Metallurgy, 73d Ann. Mtg., Quebec.

Stacey, R. A., 1973, Gravity anomalies, crustal structure, and plate tectonics in the Canadian Cordillera: Canadian Jour. Earth Sci., v. 10, p. 615.

Sutherland Brown, A., 1969, Mineralization in British Columbia and the copper and molybdenum deposits: Canadian Inst. Mining and Metallurgy Trans., v. 72, p. 1–15.

——— 1972, Morphology and classification of porphyry deposits of the Canadian Cordillera (abs): Vancouver, 9th Ann. Western Inter-Univ. Geol. Conf. Proc.

——— et al, 1971, Metallogeny of the Canadian Cordillera: Canadian Inst. Mining and Metallurgy Trans., v. 74, p. 1–25.

Wanless, R. K., 1969, Isotopic age map of Canada: Canada Geol. Survey Map 1256A.

Wheeler, J. O., et al, 1972, The Cordilleran structural province, *in* Variations in tectonic styles in Canada: Geol. Assoc. Canada Spec. Paper No. 11.

Metallogenetic Chart of Mexico—Summary[1]

GUILLERMO P. SALAS[2]

INTRODUCTION

The *Metallogenetic Chart of Mexico* was prepared in order to comply with an international commitment subscribed by Mexico as a member of the North American subcommittee on the "Metallogenetic Chart of the World."

As author of the *Metallogenetic Chart of Mexico,* I have attempted to explain certain relationships between the diastrophic movements which cause orogenies and the metallogenetic provinces. Orogenic belts are used to subdivide the country broadly into metallogenetic provinces.

The metallogenetic epochs are very generally fixed as well, although much geochronologic work is needed to pinpoint the ages of metal formation.

METALLOGENETIC PROVINCES OF MEXICO

Figure 1 shows the *Metallogenetic Chart of Mexico.* It is presented in such a way that the symbols show the main minerals produced in an area, the age and environment of the deposit, the type of deposit, and its importance (shown by the dimensions of the circles). Other characteristics of the deposits are also shown.

In my attempt to subdivide the Mexican part of the North American continent into metallogenetic provinces, I have considered that areas affected by orogenies or diastrophism which have similar geologic history and physicochemical composition—even if the end mineral product is not nearly homogeneous—may be classified for exploration purposes as a metallogenetic province. The reason is that such events have given rise to genetic processes, irrespective of chemical composition of the products, at intervals that are coincident in time and space.

On that basis, I have subdivided Mexico into six such provinces, as follows.

1. The *Baja California province,* on the north, has an approximate surface area of 92,000 km^2, 70% of which is composed of one or more very large granodioritic batholiths, vast areas of metamorphic rocks of undetermined age, and restricted volcanic-neck areas. The southern part shows extensive andesitic and ignimbrite flows as well as silicic and mafic intrusive rocks. The most important ore deposits occur in the southern half of the peninsula. Copper occurs at Santa Rosalía in the marine Boleo Formation (lower Pliocene), and disseminated copper occurs in granodiorite at El Arco. Here the age of mineralization is 93 m.y. The Sierras of La Laguna and La Victoria, south and southeast from Puerto La Paz, in the extreme south part of the peninsula, are made up of batholithic granites, granodiorites, gabbros, and metamorphic rocks. Isotopic ages are now being checked. Considerable Ag, Pb, Zn, and As production has been obtained in the past at El Triunfo and San Antonio Districts in metamorphic rocks and faulted granodiorites.

The northern half of the peninsula has several mineralized zones. These generally occur at the contact of granitic rocks with metamorphic or sedimentary rocks. Metamorphic rocks throughout the peninsula are similar to the California Franciscan rocks.

2. The *Sierra Madre Occidental province* is a 266,000-km^2 area of volcanic rock intruded by granites and granodiorites and subordinate mafic rocks. Mineral deposits are present as veins in the intrusive and extrusive rocks. It is assumed that the intrusives were the source of the mineral deposits in the volcanic rocks. Some mineralization occurs in Cretaceous limestone as a result of contact metasomatism.

Perhaps the most important metallogenetic feature of this area is the cupriferous subprovince of the northeastern part of Sonora. This seems clearly to be a southern continuation of the Arizona-Utah copper province.

The Chihuahua part of this province has mostly precious and nonferrous metals. The so-called Sierra de Chihuahua has been a very important gold and silver producer in the past, and many mines, abandoned prematurely, are being reopened successfully.

Farther south in the State of Sinaloa, gold and silver districts occur along the western limits of

[1]Manuscript received, February 7, 1975.

[2]Director General, Consejo de Recursos Naturales no Renovables, Mexico 7, D.F., Mexico.

The writer appreciates the help of the late Genaro González-Reyna, and Rubén Pesquera and staff of the Consejo, without whose assistance the Chart would not have been possible.

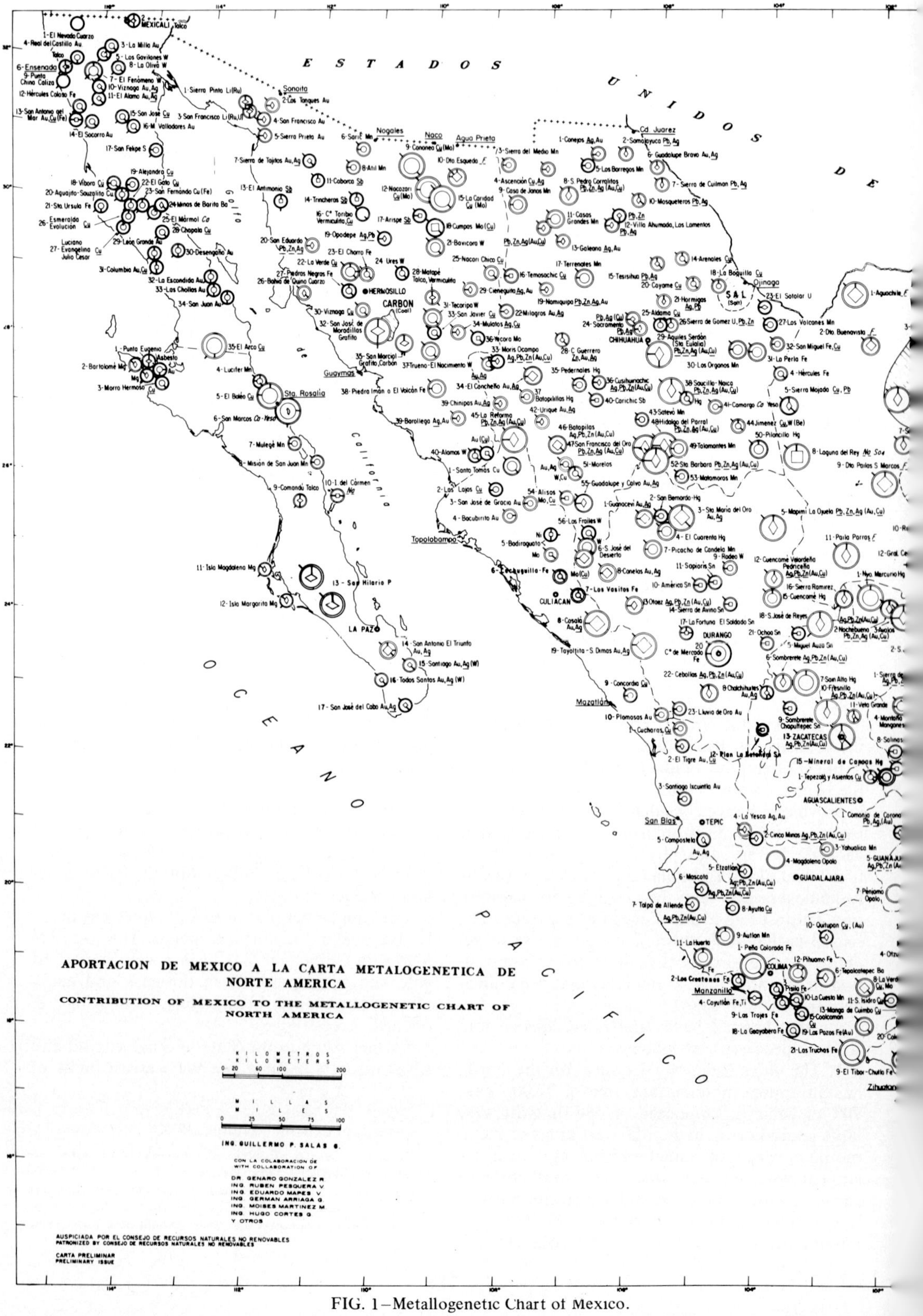

FIG. 1–Metallogenetic Chart of Mexico.

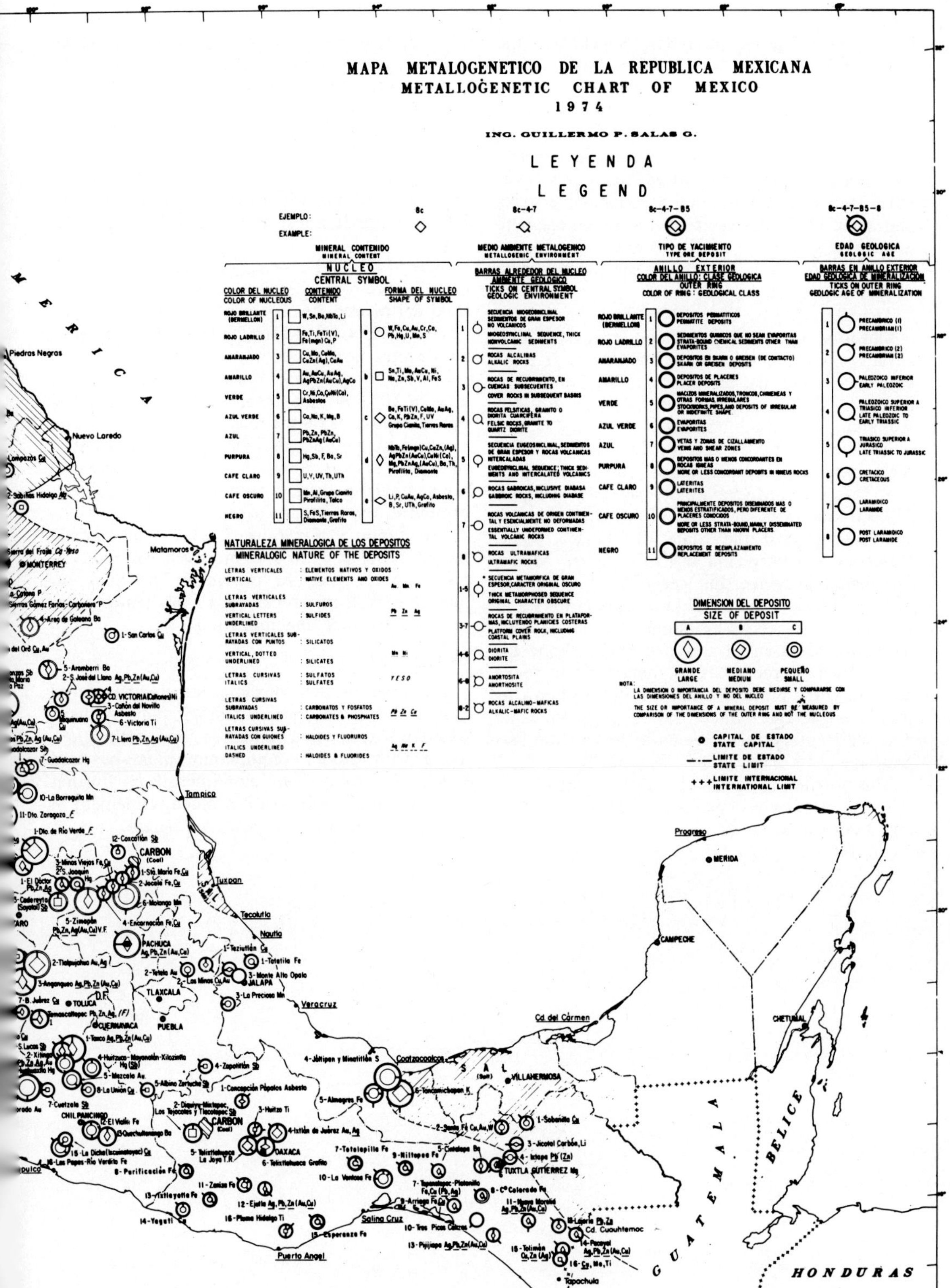
MAPA METALOGENETICO DE LA REPUBLICA MEXICANA
METALLOGENETIC CHART OF MEXICO
1974
ING. GUILLERMO P. SALAS G.
LEYENDA
LEGEND
EJEMPLO:
EXAMPLE:
MINERAL CONTENIDO
MINERAL CONTENT
NUCLEO
CENTRAL SYMBOL
COLOR DEL NUCLEO
COLOR OF NUCLEOUS
CONTENIDO
CONTENT
FORMA DEL NUCLEO
SHAPE OF SYMBOL
MEDIO AMBIENTE METALOGENICO
METALLOGENIC ENVIRONMENT
BARRAS ALREDEDOR DEL NUCLEO
AMBIENTE GEOLOGICO
TICKS ON CENTRAL SYMBOL
GEOLOGIC ENVIRONMENT
TIPO DE YACIMIENTO
TYPE ORE DEPOSIT
ANILLO EXTERIOR
COLOR DEL ANILLO: CLASE GEOLOGICA
OUTER RING
COLOR OF RING: GEOLOGICAL CLASS
EDAD GEOLOGICA
GEOLOGIC AGE
BARRAS EN ANILLO EXTERIOR
EDAD GEOLOGICA DE MINERALIZACION
TICKS ON OUTER RING
GEOLOGIC AGE OF MINERALIZATION
NATURALEZA MINERALOGICA DE LOS DEPOSITOS
MINERALOGIC NATURE OF THE DEPOSITS
DIMENSION DEL DEPOSITO
SIZE OF DEPOSIT
GRANDE
LARGE
MEDIANO
MEDIUM
PEQUEÑO
SMALL
CAPITAL DE ESTADO
STATE CAPITAL
LIMITE DE ESTADO
STATE LIMIT
LIMITE INTERNACIONAL
INTERNATIONAL LIMIT
GUATEMALA
BELICE
HONDURAS

the province. The Tayoltita–San Dimas Mine, the largest gold mine in Mexico, is typical. This is a large vein in a northwest-southeast-striking fault. Ground rock is andesite, and the dike is granodioritic.

3. The *Sierra Madre Oriental province,* a 379,000-km^2 area, occupies the great Mexican geosynclinal folded belt of Laramide age. Intrusive rocks and some volcanoes establish the metallogenetic processes, which are mostly contact metasomatism and vein filling. Metallogenesis appears to be of late Tertiary age.

The Sierra Madre Oriental province has mineral end products of the nonferrous type with high silver content predominant. Thus, the large mineral districts of the eastern part of Chihuahua State are all large-scale producers of Ag, Pb, Zn, Au, and Cu, in combinations.

Also important in this province are fluorite, mercury, antimony, manganese, and iron.

4. The *Sierra Madre del Sur province* occupies 114,000 km^2 from the State of Michoacan east-southeast toward the State of Oaxaca. Geologically, it is very similar to the Sierra Madre Occidental province, but seems to constitute a different block of generally lower topography and with more Cretaceous limestone. Limestone remnants overlie extensive intrusive rocks. Some are mineralized. Large areas of metamorphic rocks seem to carry mineralization of Precambrian or pre-Paleozoic age. Mineralization in the volcanic and sedimentary rocks seems to be of late Tertiary age.

This province is significant as an iron producer. Also important are deposits of copper, antimony and gold, and asbestos.

5. The *Eje Neo-Volcanico province,* a 190,000-km^2 volcanic zone or volcanic chain, crosses the continent from Bahia Banderas, in the vicinity of Puerto Vallarta on the Pacific coast, to the Sierra de San Andres Tuxtla on the Gulf of Mexico. Some authors have postulated that a large transverse fault crosses the continent as a continental expression of the Clarion fault zone. However, ERTS-1 data do not seem to support this postulation.

Silver and gold production have been significant in this province. One mine, the Pachuca y Real del Monte, has been in continuous production of silver and gold for the last 400 years. Mercury production also is important.

6. The *Mesa Central province* extends over an area of 105,000 km^2, from the northern edge of the volcanic axis on the south to northern Zacatecas and to part of the Durango on the Central plateau. Very vast flows of andesite are present in the southern part, and predominantly rhyolite is present in the northern part.

Two types of mineralization are characteristic in this province—tin and mercury. Antimony is also important in the province. The thick sequence of volcanic rocks shows low-temperature and low-pressure mineralization.

Conclusion

It is apparent that Mexico is largely unexplored from a modern economic-geologic point of view. Modern exploration methods and techniques, along with modern equipment, must be used much more extensively and intensively. Further study no doubt will result in much refinement of the broad classification proposed herein.

Principal Characteristics of Some Porphyry Copper Deposits in Sonora, Mexico—Summary[1]

JAVIER LOPEZ AVILA[2]

Introduction

Several porphyry copper deposits are located in the State of Sonora, Mexico. These deposits are similar to those of the copper province of the southwestern United States in Arizona and New Mexico. Physiographic, mineral, and geologic conditions can be traced from the southwestern United States southward and parallel with the Pacific Mexican coast. This paper describes the geologic characteristics, mineralization, and alteration of the main deposits in Sonora—Cananea, Pilares de Nacozari, and La Caridad (Fig. 1).

Cananea Mine

The Cananea mining district is in the north-central part of Sonora (Fig. 1), about 60 km south of the United States International Boundary.

The district has undergone sporadic mining activity since the 18th century. The Cananea mining district was organized in September 1899, when the Cananea Consolidated Copper Co. (now Compañia Minera Cananea, S.A. de C.V.) started operations. Since the discovery of the high-grade La Colorada orebody in 1926, the company has been operating continuously and has produced blister copper at an average annual rate of 27,000 tons. The district has produced more than 1 billion tons of Cu, a substantial amount of Mo, and minor amounts of Zn, Pb, Ag, and Au.

The most mineralized zone is in the central and southeastern part of the Cananea mountains, which consist of a series of parallel northwest-southeast-trending ranges. The entire mineralized area is about 3.5 km wide by 10 km long. The altitude above sea-level averages 1,800 m.

The oldest known rock at Cananea is the Capote Quartzite. This formation, which is 250 m thick, is well bedded and consists of quartz and sericite with abundant pyrite. About 800 m of limestone, stratigraphically placed between the Cambrian and the Carboniferous, overlies the quartzite.

Probably in late Mesozoic or early Tertiary time, the surface of deformed and eroded Paleozoic sedimentary rocks was covered discordantly by 5,000 m of volcanic deposits. Deep-seated granitic and porphyritic rocks intruded the sedimentary and volcanic sequence, following structural zones of weakness. The batholithic Cuitaca Granodiorite is found only in the northwestern part of the district; the El Torre Syenite, to the south; and the Cananea Granite, to the southwest. They are considered differentiates of the main intrusive. The porphyritic intrusive rocks, the Campana diabase and gabbro, occur in the form of dikes. La Colorada quartz porphyry commonly occurs as stocks or plugs. Characteristic distribution of these intrusives is along a general northwest-southeast-striking trend. Mineralizing solutions associated with the quartz porphyry intrusives produced the alteration of the country rock. Later fluids brought about the extensive copper mineralization of the district.

The structural pattern at the Cananea district is complex and obscures certain major dislocations that were probably pre-intrusive or contemporaneous with the abyssal intrusions that obliterated old zones of weakness. The northwest-striking Ricketts fault has an apparent offset of about 250 m and is sealed by a dike of Cananea Granite. It does not offset later and younger overlying volcanic rocks. The alignment of the quartz porphyry intrusives southeast of the Capote basin follows the general trend of the fault. Wide fracture and shear zones exist in the southwest and southeast parts of the district. They generally have steep dip and strike north-south.

Vein structures are absent at Cananea, but mineralized breccia pipes are characteristic. Ore deposits show marked variations.

The most important primary ore deposits have been associated with vertical breccia-pipe structures. The most typical are the Capote, Cananea-Duluth, and La Colorada. Other mineralized breccias are not true pipes. Perry (1961) described the Capote Breccia as a nearly vertical pipe, oval in cross section. The ore is erratically distributed throughout the entire structure in masses and lenses. On the upper levels, the orebody was about 90–120 m long and as much 50 m wide, but it is much smaller at the 486-m level. Rock frag-

[1]Manuscript received, October 21, 1974.

[2]Consejo de Recursos Naturales no Renovables, Mexico 7, D.F., Mexico.

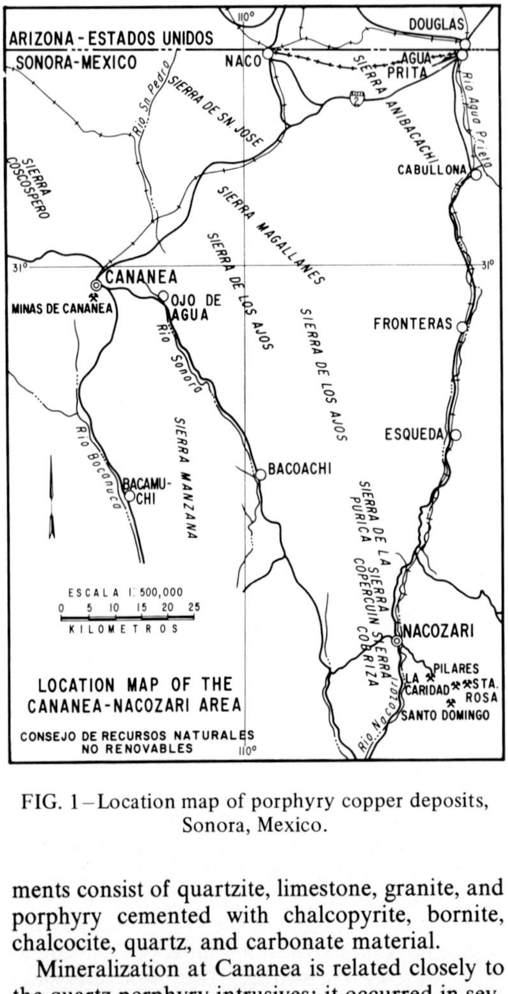

FIG. 1 – Location map of porphyry copper deposits, Sonora, Mexico.

ments consist of quartzite, limestone, granite, and porphyry cemented with chalcopyrite, bornite, chalcocite, quartz, and carbonate material.

Mineralization at Cananea is related closely to the quartz porphyry intrusives; it occurred in several surges. The deeper primary orebodies were localized in vertical pipelike breccias with pyrite and chalcopyrite, some bornite, quartz, sphalerite, and minor galena. Secondary sulfide deposits occur in economic amounts in the volcanic rocks and quartz porphyry. Primary copper and zinc mineralization is found in the sedimentary rocks. Orebodies also occur as replacement deposits in favorable limestone beds.

Hydrothermal alteration products consist of sericite, minor but significant amounts of kaolin. and abundant secondary quartz. Throughout the district the upper limestone beds are partly or wholly garnetized. The intense sericitization of the volcanic rocks and of the porphyry intrusives undoubtedly was directly due to the strong mineralizing activity.

Pilares de Nacozari

The Pilares Mine is located 11 km east of the town of Nacozari de Garcia, at an elevation of 1,500 m above sea level, in one of the north-trending ridges of the Sierra Madre Occidental.

The dominant rocks are andesitic lava flows, pyroclastic sandstones, and siliceous ignimbrites. These are cut by dikes, stocks, and batholiths. The volcanic rocks have been considered to be of early Tertiary age; they are represented in the area by seven units of andesite (the oldest), pyroclastic sandstones, ignimbrites, andesites, and latite at the top.

Volcanic activity was apparently followed by slight crustal warping. Folding was succeeded by igneous intrusions, when diorites, monzonites, quartz monzonites, and basaltic dikes were emplaced. More warping followed, accompanied by faulting. Volcanic rocks are approximately 900 m thick. They are early Tertiary in age, becoming younger eastward.

Several small bodies of intrusive rocks have broken into the extrusive section, especially in the Porvenir-Pilares area. These are a porphyritic augite-andesite, small dikes of latite, rhyolitic intrusives, and small dikes of quartz monzonitic porphyry.

In the vicinity of the mine, the volcanic rocks appear to form part of an asymmetric syncline which dips eastward, strikes northwestward, and has southeast pitch. Two main sets of faults are responsible for the shape of the orebody. They mark the zones of shattering that permitted the entrance of mineral-bearing solutions, and indicate the directions of movement that followed the period of dike intrusion. The first, or easterly, set of faults has an approximate N70°E strike; the second, or northwesterly, set has an approximate N40°E strike; they dip to the southeast.

The only volcanic rocks that have been cut by the mine workings are the latite and andesite breccias. The monzonite is the important intrusive from an economic standpoint, for it undoubtedly "brought in" the ore. The rock is dark, fine-grained holocrystalline, and of granitic texture, with the habit of a diorite rather than a monzonite. The intrusive tongues are slightly porphyritic and resemble dioritic dike rocks.

Although numerous contacts between monzonite-latite and monzonite-andesite are exposed, none of them shows well-marked alteration such as recrystallization due to the heat of the intrusive. All the alteration observed was produced by invading solutions that carried the ore. Alteration of each of the ore-bearing members was characteristic but special, because of the chemical differences.

Chalcopyrite is the main ore mineral. The lesser ore minerals are chalcocite, covellite, and bornite. Both tetrahedrite and native copper are scarce. Chalcocite and covellite are important as ore minerals only above the 200-m level. Bornite is scarce. Associated with the copper minerals are pyrite, sphalerite, scheelite, sericite, chlorite, calcite, quartz, specularite, epidote, gypsum, barite, and kaolin.

La Caridad

The La Caridad deposit is in the northeastern part of Sonora, about 15 km southeast of the town of Nacozari. Mining activity in the Nacozari district dates from the early 1900s, most notably with the operation of the old Pilares Mine. Renewed interest in the district began in 1964 with a joint exploration effort in northern Sonora, which resulted in the discovery of La Caridad deposit in 1967. Reserves are estimated to be 650 million MT of 0.65% copper.

The rocks in the area surrounding La Caridad are predominantly volcanic, consisting of a thick sequence of andestic to latitic rocks like those in the Pilares Mine area. A large area of altered premineral volcanic rocks is present around La Caridad, but no volcanic rocks occur within the deposit itself. Structurally, La Caridad seems to occur on a horst.

Detailed mapping and petrographic work has established five pre-mineral intrusive rock units within La Caridad deposit. They are represented from older to younger by diorite, granodiorite, quartz monzonite porphyry, aplite, and pegmatite.

La Caridad breccias generally consist of large blocks tightly fitted against one another, so that the main evidence for brecciation is the occurrence of small, angular cavities (2–5 cm across) that represent the void space left as these "megaclasts" were displaced with respect to one another.

Three major breccia classes are known: (1) mixed diorite, stockwork, and pebble breccias; (2) mixed porphyry breccia, and (3) contact and tourmaline breccias (Sell, 1970).

Little is known about faulting within La Caridad deposit. The entire deposit is highly fractured, and it is difficult to pick out any persistent linear zones that would indicate a distinct structure.

The mineralogy of La Caridad deposit consists mainly of pyrite and chalcopyrite partially replaced by chalcocite and covellite within a blanketlike zone of secondary enrichment. The enriched zone is overlain by an oxidized zone that is essentially devoid of secondary copper minerals. Widespread supergene mineralization is present and is a very important part of the deposit.

Guilbert and Echavarri (1970) established three fundamental types of alteration within La Caridad deposit: phyllic, argillic, and propylitic.

Conclusions

All the porphyry copper deposits known in Sonora are, in common, the result of quartz monzonite or quartz diorite intrusions into volcanic piles. Also, they occur in uplifted blocks that have permitted oscillation of the water table, which has caused supergene enrichment. Moreover, all deposits show vein or lode mineralization laterally or above the disseminated ore deposit.

References Cited

Guilbert, J. M., and Ariel Echavarri, 1970, Petrography and alteration at the La Caridad deposit, Nacozari, Sonora, Mexico: Unpub. ASARCO rept.

Perry, V. D., 1961, The significance of mineralized breccia pipes: Mining Eng., April, p. 367–376.

Sell, J. D., 1970, Field mapping La Caridad porphyry copper molybdenum deposit, Sonora, Mexico: Unpub. ASARCO rept.

Uranium and Its Projected Use in Nuclear Generation of Electricity in Mexico—Summary[1]

L. GÁLVEZ and C. VÉLEZ[2]

INTRODUCTION

Owing to the geographic, physical, economic, and human characteristics of Mexico, the problem related to its electrical needs must be resolved in order to improve its development.

Mexico is planning to make nuclear energy an efficient contributor to scientific, economic, and technologic development of the country through the generation of electricity from nuclear fuels. It will be necessary to incorporate current technology and develop new technology so that construction of power plants and the supply of nuclear fuels and other reactor materials can be facilitated.

In this paper we present a policy in relation to the exploration and exploitation of uranium ores, as well as the up-to-date determination of our national reserves.

In our estimate of nucleo-electrical demand of the country from 1978 until 1989, we consider that water-type reactors, possibly including heavy-water reactors (CANDU type), will be used. It is thought that during this period there will be no installation of breeder reactors.

Mexican uranium ore reserves are reported in Table 1; the IAEA classification is used.

Table 1. Uranium Ore Reserves

Location	*Tons (mineral)*	*Av. %*	*Tons (U_3O_8)*
PROVED RESERVES			
Cost lower than \$275 N. Cy./kg U_3O_8			
El Nopal, Chih.	140,000	0.250	350
La Coma, N. L.	490,000	0.205	1,000
Subtotal	630,000	0.215	1,350
Cost between \$275 and \$412 N. Cy./kg U_3O_8			
Los Amoles, Son.	1,000,000	0.047	470
La Preciosa, Dgo.	335,000	0.063	210
Subtotal	1,335,000	0.051	680
Total Proved Reserves	1,965,000	0.1033	2,030
ADDITIONAL RESERVES			
Cost lower than \$275 N. Cy./kg U_3O_8			
Las Margaritas, Chih.	2,000,000	0.103	2,060
TOTAL RESERVES	3,965,000	0.1032	4,090

Table 2. Projected Calendar for Generation of Electricity from Nuclear Plants

Year	*Installed Nuclear Power (Mw)*	*Accumulated Nuclear Power (Mw)*
1978	650	650
1979	650	1,300
1980	0	1,300
1981	0	1,300
1982	600	1,900
1983	1,400	3,300
1984	800	4,100
1985	2,400	6,500
1986	2,400	8,900
1987	1,000	9,900
1988	2,600	12,500
1989	3,000	15,500

DEMAND FOR NUCLEAR POWER

Based on the study by the IAEA ("Market Survey for Nuclear Power in Developing Countries," September 1973), we assume the calendar shown in Table 2 in relation to electricity generation from nuclear power.

A study carried out by the U.S. Atomic Energy Commission in December 1973 projected the quantity of 13,000 Mw of electricity to be generated by nuclear power for the year 1990, and we must consider this to be a corroboration of the more detailed study made by the IAEA. Both studies consider, for the sector of the economy related to electricity, an increase during the decade 1980–89, of about 2½ times present production (Table 3).

[1]Manuscript received, November 11, 1974.

[2]Instituto Nacional de Energía Nuclear de México, Mexico 18, D.F., Mexico.

Table 3. Projected Increase in Demand for Electricity

Year	*Production of electricity (Twh)*		*Maximum demand (Mw)*	
	IAEA	*USAEC*	*IAEA*	*USAEC*
1980	72.7	65.7	13,500	–
1990	178.9	165.2	33,200	–
2000	375.3	412.1	70,000	–

Nevertheless, the problems originated by the supply of hydrocarbons at the world level, combined with greater controls on environmental pollution, may determine a change in the economy of energy resources through a greater use of nuclear energy before 1990.

Uranium Demand and Services

In order to determine our needs in U_3O_8 (yellow cake) and fuel-cycle services, we assumed that light-water reactors (⅔ PWR, ⅓ BWR) will be used from 1980 on. However, if it is economically convenient to install heavy-water reactors of the CANDU type, there will be a consequent reduction in the requirements for U_3O_8, separative work units (SWU), and UF_6 conversion.

It has been estimated that, from 1978 to 1989, no high-temperature reactors (IITGR) or sodium- or helium-cooled fast-bender reactors (LMFBE, GCFBR) will be installed. The USAEC study did not consider installation of any of these reactors, which are supposed to be available on the market after 1986 and in operation by 1995.

The data in Table 4 have been used to integrate Mexico's needs for uranium and services, which are shown in Table 5.

To simplify the calculations, it was assumed that the U_3O_8 and the services for the initial load are needed 1 year before the considered capacity begins to operate, but that the needs concerning reloads are required during the year of operation. According to the hypothesis assumed in the following section, it will be supposed that, owing to the reprocessing of the irradiated fuel and the reuse of the uranium and plutonium, the needs of natural uranium and corresponding services for the reloads will decrease 20%, and the needs of enrichment 25%, beginning in 1987.

Uranium Supply and Services

The UF_6 corresponding to the first loads of units 1 and 2 of the Laguna Verde Plant has been supplied by URANEX Company. It is thought that all the uranium needed for reloads of those units, as well as for the future expansion of electricity produced through nuclear energy, will be supplied internally. The first uranium ore mill with a capacity of 125 tons of U_3O_8 (yellow cake) per year will be in service by 1975.

Permission to convert uranium concentrates to UF_6 for reloads of Laguna Verde units has been granted to COMURHEX Company until 1985. It

Table 4. Unit Data

	PWR	*BWR*
Initial load, t/100 Mw	8.00	13.23
Reloads, t/100 Mw-year	2.67	3.15
Enrichment of the initial load in %	2.51	2.00
SWU/kg of initial load (before 1980, tails of 0.2%)	–	2.19
Kg of nat. U of initial load (before 1980, tails of 0.2%)	–	3.52
SWU/kg initial load (after 1980, tails of 0.3%)	2.56	1.70
Kg of nat. U of initial load (after 1980, tails of 0.3%)	5.38	4.14
Enrichment of reloads in %	3.19	2.69
SWU/kg of reload (tails of 0.3%)	3.77	2.87
Kg of nat. U of reload (tails of 0.3%)	7.03	5.82

Table 5. Needs of Uranium and Services

Year	*U_3O_8 (Tons)*	*Conversion to UF_6 (Tons)*	*SWU (Tons)*	*Fuel Fabrication (Tons)*
1977	356	303	188	86
1978	356	303	188	86
1979	140	119	59	21
1980	281	239	118	41
1981	612	521	245	100
1982	1,054	896	414	177
1983	854	726	345	136
1984	2,045	1,738	820	331
1985	2,217	1,886	898	354
1986	1,972	1,677	835	286
1987	3,203	2,722	1,309	509
1988	3,538	3,007	1,433	577
1989	2,345	1,994	993	358
Total	18,973	16,131	7,845	3,062

is expected that, beginning in 1981, it will be possible to realize this operation in Mexico. The manufacture of fuel elements in Mexico is scheduled to begin by 1980.

Reprocessing of the irradiated fuel elements and the use of the obtained plutonium, if economically convenient, will be carried out after 1985, according to the present forecasts. Finally, we cannot expect that the enrichment of the uranium will be realized in Mexico before 1990.

Cost of Fuel

In the international market, it is estimated that the cost of nuclear fuel will remain constant and can be calculated at \$10 National Currency/10^6 kcal, which means the approximate equivalent of \$2.5/kwhr (Mexican) on January 1, 1973.

If national industrial participation in the nuclear fuel cycle increases, it might not produce an increase of the cost mentioned above. It can be assumed that uranium concentrates produced in Mexico in the near future will have a competitive cost in comparison with the international market.

According to this hypothesis and the percentages of the cost related to U_3O_8 (37%), conversion (5%), enrichment (43%), and fuel fabrication (15%), it is possible to determine the costs shown in Table 6.

Table 6. Fuel Cost

Year	*Generation Twh*	*Cost (Millions in National Currency)*	*National Participation (%)*	*Foreign Component (%)*
1978	2.3	57.5	0	100
1979	6.8	170.0	9	91
1980	9.1	227.5	18	82
1981	9.1	227.5	50	50
1982	11.2	280.0	55	45
1983	18.2	455.0	55	45
1984	25.9	647.5	55	45
1985	37.1	927.5	55	45
1986	53.9	1,347.5	55	45
1987	65.8	1,645.0	65	35
1988	78.4	1,960.0	65	35
1989	98.0	2,450.0	65	35

Mineral Resources of State of Sinaloa, Mexico—Summary[1]

D. A. CORDOBA and M. PATIÑO–A.[2]

The regional geologic mapping carried out during recent years in the State of Sinaloa by geologists from the Instituto de Geología brings new knowledge to the economic-geologic research of the state and provides help to the economic development of Mexico.

The five main groups of mineral deposits in the area can be divided, on the basis of their characteristics, as follows: (1) hydrothermal (epithermal, cavity filling); (2) metasomatic (contact metasomatism; acid igneous rocks—dissemination); (3) contact metamorphic (contact metamorphism); (4) magmatic or primary (dissemination in ultramafic or mafic rocks); and (5) other (deposits formed by mechanical concentration in clastic formations; residual and alluvial).

The important times for metallic concentration were Jurassic; Late Cretaceous; early, middle, and late Tertiary; and Pleistocene.

Most commonly mineralized are the volcanic rocks, chiefly those of intermediate composition. Similarly, other host rocks are the Albian-Cenomanian carbonate rocks. Mineralization is more abundant in andesites, though carbonate host rocks contain many mineral deposits in the area.

The mafic and ultramafic rocks show serpentinization at several localities and include nickel, platinum, and antimony minerals.

Clastic continental Tertiary formations contain radioactive minerals in the northern part of the state. Recent residual deposits commonly contain gold-placer deposits; Paleozoic host rocks contain gold, silver, zinc, and lead mineralization.

[1]From Circum-Pacific Energy and Mineral Resources Conference program booklet.

[2]Instituto de Geología, Universidad Nacional Autónoma de México, Mexico 20, D.F., Mexico.

Major Chemical Characteristics of Tertiary and Quaternary Volcanism of Baja California—Summary[1]

ALAIN DEMANT,[2] CLAUDE ROBIN,[3] CLAUDE BOBIER,[3] and DIEGO A. CORDOBA[4]

The volcanic series of Baja California are divided into two chemically different cycles.

A first calc-alkaline cycle is represented by deposits of pyroclastics and flows of andesites and dacites. This volcanism was most intense in the Miocene. The andesitic series of Sinaloa is comparable to series in Baja California, but began during the Oligocene; the imposing ignimbritic series of the Sierra Madre Occidental has no counterpart in Baja California.

Beginning in the Pliocene, an alkaline cycle was developed. This cycle is represented by large amounts of basalts and hawaiites and is partly of Holocene age. In fact, some eruptions have occurred within historic time (e.g., Tres Vírgenes, near St. Rosalia).

The chemical character of the lavas of Baja California is interpreted according to the pattern of geotectonic evolution which has been proposed for the western part of the Mexican continent, especially in relation to the Pliocene opening of the Gulf of California.

A compilation of recent chemical data concerning the volcanism of Mexico was processed by computers. The program defined the petrochemical characteristics. On the basis of these characteristics, a thesis is proposed on the petrogenic interpretation of the three large volcanic provinces of Mexico.

1. Generally alkaline volcanism (with tholeiitic manifestations) of eastern Mexico and the Gulf of Mexico. This volcanism is interpreted as the result of divergence of the North and South American continents and formation of the Gulf of Mexico. It began in Mesozoic time, was most active in the Tertiary, and has continued into the Quaternary.

2. Calc-alkaline volcanism of the Pacific Coast. This volcanism is related to Oligocene-Miocene subduction along the Farallon plate as it sank under the American plate. After the disappearance of this plate, strike-slip movements began along the San Andreas–Gulf of California fault system. In relation to this change, alkaline volcanism appeared after Pliocene time.

3. Mainly calc-alkaline volcanism along the Trans-Mexico axis, extending east-west between the Pacific Coast and the Gulf of Mexico. This volcanism is subdivided into (a) the calc-alkaline volcanism which began in early Miocene time along this belt and has continued to the present, and (b) the ignimbritic volcanism of the Sierra Madre Occidental on the north. The two types of volcanism occurred along different fracture systems and appear to be independent. The calc-alkaline volcanism may be related to the Circum-Pacific lineament and to movements of the Pacific plate under the American plate. It also can be interpreted as the southern part of the subduction system active along the Baja California coast until late Miocene time. The ignimbrites of the Sierra Madre Occidental may have developed in a volcano-tectonic rift zone above an active subduction area. Thus they would be related to the geodynamic phenomena which have affected the California area and would be contemporaneous with the Baja California andesitic volcanism, so that the end of the ignimbritic activity corresponds to the end of andesitic activity there.

[1]From Circum-Pacific Energy and Mineral Resources Conference program booklet.

[2]Instituto de Geología, Universidad Nacional Autónoma de México, Mexico 20, D.F., Mexico, and Laboratoire de Géologie Dynamique, Faculté des Sciences et Techniques St. Jerome, Marseille, France.

[3]Centre de Recherches Geodynamiques Sous-Marines 06230, Villefranche/Mer, France.

[4]Instituto de Geología, Univerisdad Nacional Autónoma de México, Mexico 20, D.F., Mexico.

Metallogenic Provinces of Southeastern Pacific Region[1]

GEORGE E. ERICKSEN[2]

Abstract Metalliferous mineral deposits of the southeastern Pacific region include: (1) hydrothermal, magmatic, and sedimentary deposits of the Andean region, one of the great mineral belts of the world; (2) scattered hydrothermal deposits in the Antarctic Peninsula; and (3) metal-enriched pelagic sediments and ferromanganese nodules on the seafloor. Andean metalliferous deposits, the chief topic of this report, are for the most part spatially and genetically related to calc-alkaline plutons, subvolcanic intrusions, and volcanic rocks emplaced during the Andean orogeny of Late Triassic to Quaternary age. They are components of a single metallogenic province superimposed on two or more pre-Andean metallogenic provinces of Paleozoic and Precambrian(?) age. Deposits in the Antarctic Peninsula are similar in age and origin to the deposits in the Andes and are considered to belong to the Andean province.

The Andean metallogenic province may be divided into several subprovinces, each parallel with the Andes and the continental margin and each having a dominant metal or suite of metals. The central Andes of Peru, northern Chile, and Bolivia, which contain the greatest concentration of exploitable deposits and the greatest variety of ore types, have as many as five linear, partly overlapping subprovinces. These subprovinces, from west to east (Pacific coast to the eastern Andean front), are characterized, respectively, by deposits of iron, copper, polymetallic base metals, tin, and gold.

Calc-alkaline plutonic and volcanic rocks of the Andes, which probably originated by partial melting in the Benioff zone, show a general, though nonuniform progression of decreasing age from west to east. Igneous rocks of Jurassic and Cretaceous age are most abundant near the coast, whereas those of Tertiary and Quaternary age dominate in the Andes. Locally, intrusive rocks and associated ore deposits of widely different ages are juxtaposed. Several sources of the metals are probable; they include oceanic crust and pelagic sediments in the Benioff zone, some of which may have been enriched previously in certain metals at the ancestral East Pacific Rise, and metal-rich zones in the overlying mantle and continental crust.

INTRODUCTION

This paper presents information about the distribution and environment of mineral deposits of three major subdivisions of the southeastern Pacific: the Andean region, marginal lands of western Antarctica (including the Antarctic Peninsula and Scotia Arc), and the ocean basins. The Andean region is treated more fully, whereas Antarctica and the seabed are but briefly summarized.

The Andean region, which is considered to include the coastal mountains and lowlands of westernmost South America as well as the Andean Highlands and low eastern foothills, can be divided into several physiographic provinces (Fig. 1), each of which tends to show distinctive geologic features and characteristic types of mineral deposits. The region constitutes the longest active orogenic belt in the world, stretching nearly 9,000 km from Venezuela to Tierra del Fuego. Mineral deposits are found throughout the Andes, but are by far the most numerous in the latitudes between northern Peru and central Chile. Within this region are thousands of mining districts and deposits, a concentration of exploitable mineral deposits equal to or greater than that in any comparable area.

The future of mining in the Andean region is bright, although many of the richest deposits have been mined out, and some metals such as silver, bismuth, and antimony now are produced largely as by-products of polymetallic base-metal and tin ores. Copper has the greatest potential for large future production; even though the region has been a major source of copper since the latter part of the 19th century, total production is only a fraction of the potentially exploitable resources. Moderately large production of lead and zinc, chiefly from Peru, and of iron ore from Peru and Chile probably will be sustained during the next several decades. An increase in gold production, particularly in the eastern Andes of Bolivia and Peru and the central Andes of Colombia, may be expected. Bolivia should continue to be a major source of tin and tungsten, although ore grades will continue to decrease in the next few decades.

The information presented and the conclusions drawn are based on many published and unpublished reports in addition to my own investigations of mineral deposits in the Andes, chiefly in Chile and Peru, which began in 1948. Some recent summary reports and maps on Andean metallogeny and metallogenic provinces are: Ahlfeld and Schneider-Scherbina (1964), Angelelli et al (1970), Bellido and de Montreuil (1972), Bellido

[1]Manuscript received, November 6, 1974; revised, December 18, 1975. Published by permission of the Director, U.S. Geological Survey.

[2]U.S. Geological Survey, National Center, Reston, Virginia 22092.

I express my thanks to my colleagues in the U.S. Geological Survey—Dennis P. Cox, Jacob E. Gair, and Philip W. Guild—who reviewed an early version of the manuscript, and to Gus H. Goudarzi, who made available his unpublished metal-deposit maps of South America.

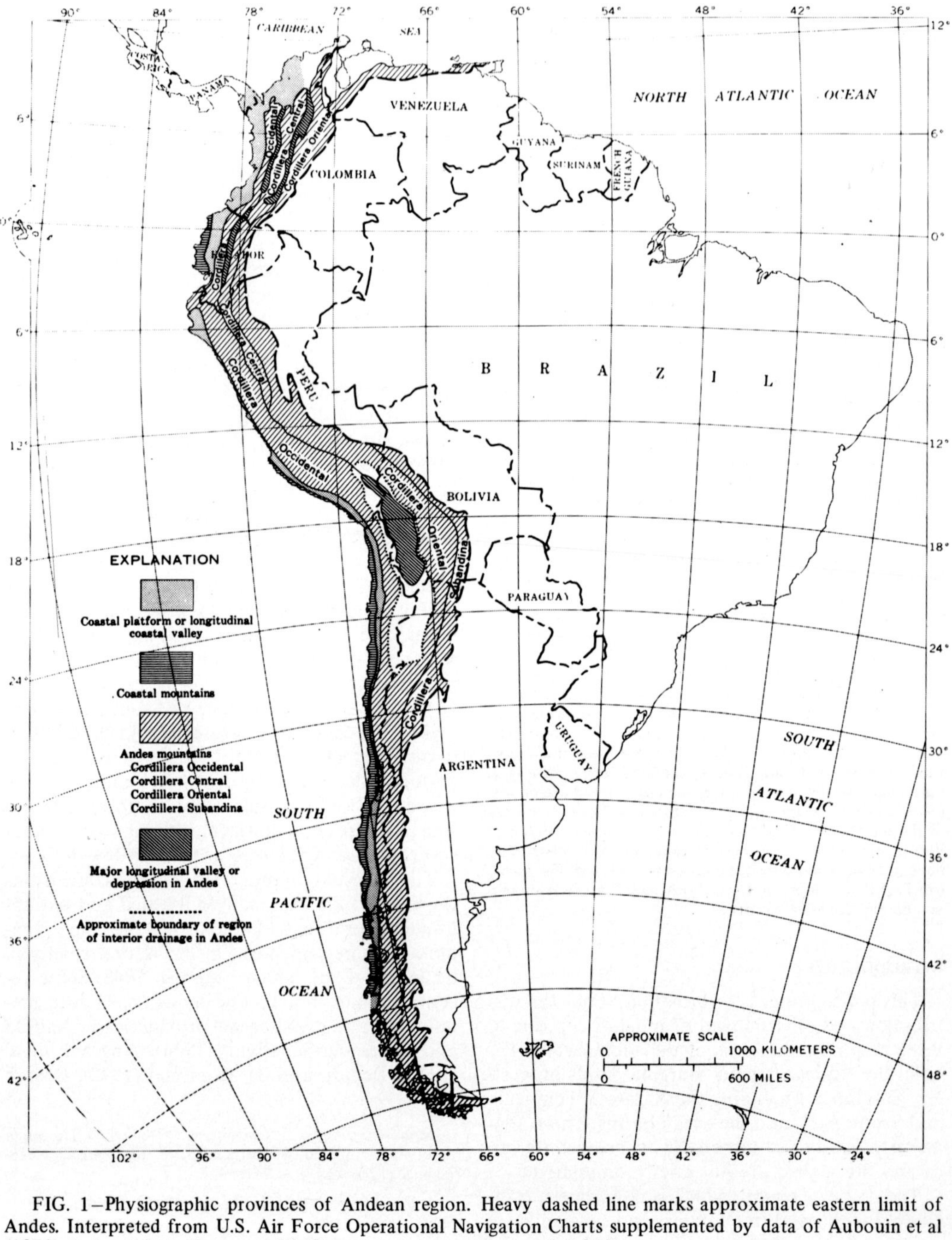

FIG. 1–Physiographic provinces of Andean region. Heavy dashed line marks approximate eastern limit of Andes. Interpreted from U.S. Air Force Operational Navigation Charts supplemented by data of Aubouin et al (1973), Irving (1971), Goossens (1969), and Bellido and Montreuil (1972).

et al (1972), Bellizia (1973), Bolivia Servicio Geol. (1968), de las Casas (1969), Goossens (1969, 1972, 1975), Hollister (1973), Kelly and Tureaure (1970), Petersen (1965, 1970), Ruiz and Ericksen (1962), Ruiz et al (1965), Sillitoe (1972), Turneaure (1960, 1971), and United Nations Development Programme (1968, 1970).

Tectonic and Geologic Setting

The mineral deposits of the southeastern Pacific reflect the nature of the local geologic environment at the time of deposition as well as regional or global processes such as orogeny, plutonism, and volcanism which are best explained by plate-tectonics theory and which may change in style and intensity with time. For example, the Andean metallogenic province formed (1) during the recent past, when plate configurations were similar to those of today, and (2) during Mesozoic time, when configurations were different (Herron, 1972; Larson and Pitman, 1972). In this report it is assumed that most Andean mineral deposits are genetically related to calc-alkaline magmas that were generated by partial fusion along the upper margin (Benioff zone) of a subducting oceanic plate. On the basis of distribution of earthquake foci, and present and Pliocene-Pleistocene plutonism and volcanism, calc-alkaline magmas appear to have formed on the Benioff zone at depths of 100–200 km and at distances on the order of 200–400 km from the axis of the Peru-Chile Trench (Fig. 2).

Structural controls of mineralization include local fault zones, favorable stratigraphic zones, and intrusions, as well as large regional structures such as longitudinal faults, folds, batholiths, and former geosynclines. Some of the Andean mineral deposits show affinities for certain rock types or geologic environments, suggesting that a major control of Andean mineralization was the gross lithology of the host rocks (Petersen, 1970, p. 888). Thus, copper deposits tend to be spatially related to the western part of the Mesozoic Andean geosyncline, where calc-alkaline igneous rocks and marine or continental clastic sedimentary rocks are the chief host rocks. Porphyry copper deposits in northern Chile occur chiefly in a narrow belt along the eastern margin of the geosyncline. Polymetallic base-metal deposits of Peru are associated with a marine sedimentary sequence of Mesozoic age having abundant limestone, as they are also in the eastern Andes of Colombia. Tin deposits of Bolivia are associated with plutons and subvolcanic intrusions in a thick sequence of marine clastic sedimentary rocks of middle Paleozoic age.

About 500 radiometric age dates for plutonic, volcanic, and metamorphic rocks of the Andes have been published since 1960. giving new insight into Andean orogenesis and metallogenic epochs. The radiometric ages indicate several stages of major plutonism in the Andean region prior to the Andean orogeny of Late Triassic to Quaternary age, as follows: late Precambrian, Late Ordovician and Early Silurian, Late Silurian and Early Devonian, Late Devonian and Early Carboniferous, and Late Permian (Clark et al, 1973; Corvalán and Munizaga, 1972; González Bonorino and Aguirre, 1970; Stewart et al, 1974). Younger plutonic rocks are chiefly of Jurassic to Pliocene age (Clark et al, 1973; Farrar et al, 1970; Quirt et al, 1971; Stewart et al, 1974). Intrusive rocks of Mesozoic and Tertiary age in the Andes south of Colombia tend to be oldest near the coast (generally Jurassic or Cretaceous) and youngest in the high Andes. For example, Stewart et al (1974) have shown that the batholith of coastal Peru was emplaced chiefly during the interval from Late Triassic to middle Oligocene time. The oldest rocks tend to occur in the southwestern part of any given segments of this batholith. Farrar et al (1970) have shown an eastward migration of magmatic activity with time in the region of Copiapó, northern Chile, whereby Jurassic plutons are at or near the coast and upper Eocene plutons are several tens of kilometers inland. In contrast, plutons of western Colombia tend to be youngest in the Western Cordillera (Fig. 1) and to become progressively older eastward.

Plutons and subvolcanic intrusions associated with metalliferous deposits appear to be chiefly Late Cretaceous (100 m.y.) to Pliocene (5 m.y.). Ages of intrusions associated with certain types of deposits tend to be within a characteristic time span. Thus, deposits of the copper province (Fig. 3) appear to be chiefly Cretaceous (100 m.y.) to Oligocene (30 m.y.) in age. Porphyry copper deposits, most of which are near the eastern margin of the copper province, are Paleocene (60 m.y.) to Pliocene (4 m.y.) in age (Quirt et al, 1971). The polymetallic base-metal deposits in the Andean Highlands of Peru, the youngest group of Andean metalliferous deposits, are chiefly of Miocene and Pliocene age (25–3 m.y.). Intrusive rocks in some of the most important mining districts here (Cerro de Pasco, Morococha, Yauricocha, and many districts in the Cordillera Blanca) are less than 10 m.y. old (Stewart et al, 1974), and the associated mineral deposits rank among the world's young-

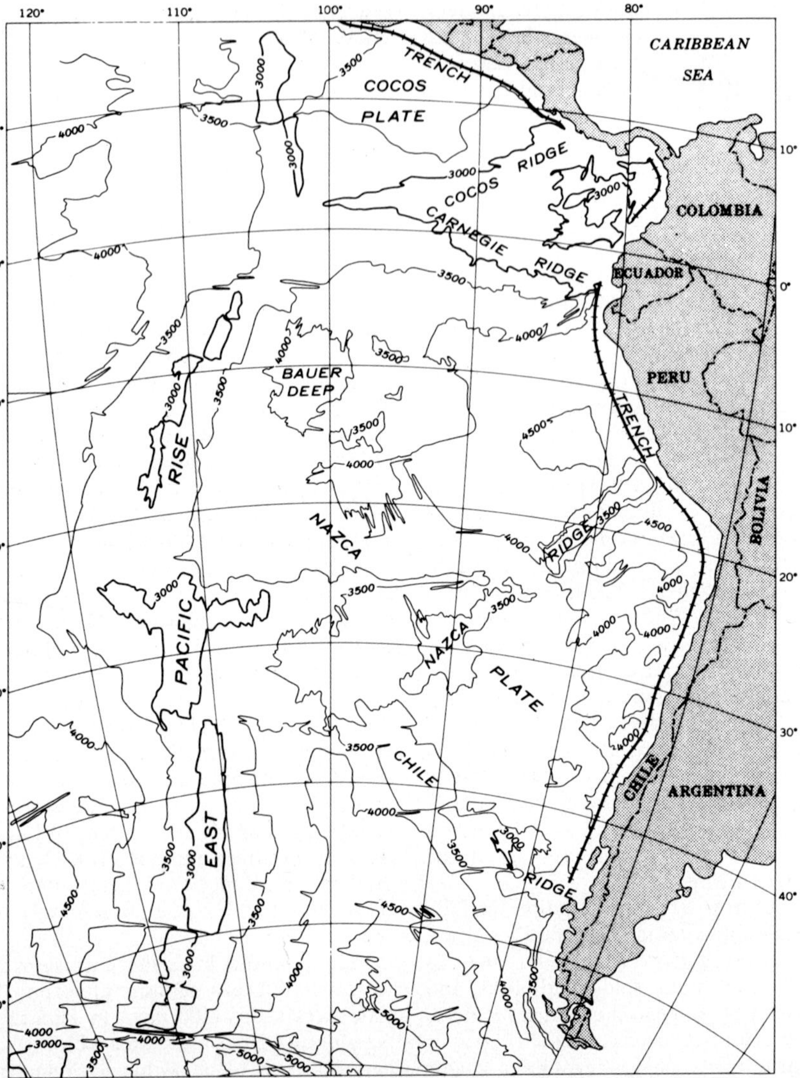

FIG. 2–Bathymetric map of southeastern Pacific. Isobaths in meters. From unpublished 1:10,000,000-scale map compiled by U.S. Geological Survey for Circum-Pacific Map Project and based on 1:6,500,000-scale maps published by Scripps Institution of Oceanography and Institute of Marine Resources, University of California.

est. Clark et al (1973) cited five different ages of tin mineralization in Bolivia: (1) Middle Triassic–Early Jurassic (210–180 m.y.), (2) Middle–Late Jurassic (160–145 m.y.), (3) Late Cretaceous–Paleocene (70–60 m.y.), (4) Oligocene (28–23 m.y.), and (5) Neogene (presumably <20 m.y.). Zentilli (1974) reported six pulses of intrusion and associated mineralization in the Copiapó region of northern Chile as follows: (1) Early Jurassic; (2) Middle–Late Jurassic; (3) Early–Late Cretaceous; (4) Paleocene; (5) Eocene–early Oligocene, and (6) early Miocene.

Mineral Deposits

Deposits of Andean Metallogenic Province

At least five linear subprovinces can be distinguished in the Andean metallogenic province (Fig. 3), each characterized by a metal or a suite of metals; locally, these subprovinces may be divided into mineral belts, each characterized by a dominant suite of minerals. Such mineral belts also are elongate and trend more or less parallel with the subprovinces. Although a certain type of deposit predominates in each subprovince, it is not necessarily the only type present, nor is it restricted to this subprovince.

By far the greatest number of major Andean mineral deposits are present in the region extending from northern Peru to central Chile; elsewhere, deposits tend to be scattered and small. Furthermore, the different types of deposits within this central Andean region tend to be clustered within given segments of the subprovinces (Fig. 3). For example, most polymetallic base-metal deposits occur in Peru, copper deposits in Chile, and tin deposits in Bolivia. More than 90% of the value of Andean mineral production has come from the deposits in these three countries. The only large production elsewhere in the Andes has been from gold deposits in Colombia and Ecuador and from a single lead-zinc deposit in northernmost Argentina which probably is of pre-Andean age.

Iron—The discontinuous iron province of the coastal region of northern Chile and southern Peru, which overlaps the copper province, contains several tens of deposits ranging from a few hundred thousand to about 100 million tons of high-grade iron ore (average grade of about 60% Fe). These deposits have been, and continue to be, the sources of nearly all the iron ore produced in the Andean region, which in the year 1971 was about 11 million MT in Chile and 9 million MT in Peru (U.S. Bureau Mines, 1974, p. 209, 640). Scattered iron deposits are found elsewhere in northern Chile and northwestern Argentina, in the eastern Andes of southern Peru, and in the eastern Andes of Colombia. Of these, only the deposits in Colombia are currently being exploited.

Nearly all the iron deposits are of either contact metasomatic or hydrothermal origin, and most are considered to be of Late Cretaceous age (Ruiz et al, 1965, Table 1; Bellido and de Montreuil, 1972). The iron deposits of coastal Chile consist of magnetite bodies and veins in sequences of andesite flows and breccias and continental clastic sedimentary rocks at or near the contacts of dioritic to granodioritic plutons (Ruiz et al, 1965, p. 147–148). Deposits in southern Peru are similar, but the host rocks in this region are schist, gneiss, and granite of probable Precambrian and early Paleozoic age (Bellido and de Montreuil, 1972, p. 40–41). Magnetite deposits at El Laco, northern Chile, which are interpreted as volcanic flows (Park, 1961; Ruiz et al, 1965, p. 246–247), are a unique type of iron deposit that has not been recognized elsewhere. The flows occur at five localities on the flanks of an eroded Quaternary volcano, probably at the sites of former craters (Sanchez, *in* Ruiz et al, 1965, p. 246–247).

Copper—A copper province can be traced nearly the full length of the Andes from northern Colombia to the southernmost tip of South America (Fig. 3). The deposits are most numerous in the western part of the Andes and the coastal region from central Peru to central Chile. A belt of copper deposits (not shown in Fig. 3) and intermingled polymetallic deposits occurs in southern Peru at the eastern side of the polymetallic base-metal province. The copper deposits in Chile and Peru have yielded most of the copper that has been produced in the Andes. Total copper production from Chile for the period 1601–1963 was estimated by Ruiz et al (1965, p. 172) to be about 18 million MT, most of which was produced in the present century from three porphyry copper deposits at Chuquicamata, Potrerillos, and El Teniente. Peruvian copper production is much less than that of Chile. Large-scale exploitation of a porphyry copper deposit (Toquepala) did not begin until the late 1950s. Production for the year 1971 from Chile was about 717,000 MT of fine copper, whereas that of Peru was 208,000 MT, which includes copper recovered from polymetallic ores (U.S. Bureau Mines, 1974, p. 209, 604).

Seven types of Andean copper deposits are recognized; in order of economic importance, they are as follows: porphyry copper, hydrothermal veins and replacements, contact metasomatic deposits, strata-bound deposits, redbed copper deposits, tourmaline breccia pipes, and at least one small fumarolic deposit. At least 40 porphyry copper deposits have been found in the Andes from central Chile to Colombia, and the economic potential of these deposits is far greater than that of all other Andean metalliferous deposits. Hydrothermal deposits, formed by fissure-filling and replacement, are the most widespread and numerous of all the copper deposits and have yielded important quantities of copper in Chile and Peru. In Chile, veins of this type also have yielded as by-products cobalt, mercury, and tungsten; these veins also contain the principal gold-bearing deposits. Contact metasomatic deposits are widespread but of less economic importance than the hydrothermal deposits. Strata-bound deposits occur chiefly in Cretaceous volcanic rocks and associated marine or lagoonal sedimentary rocks of northern Chile (Ruiz et al, 1965, Table 1). Scattered redbed copper deposits occur in a

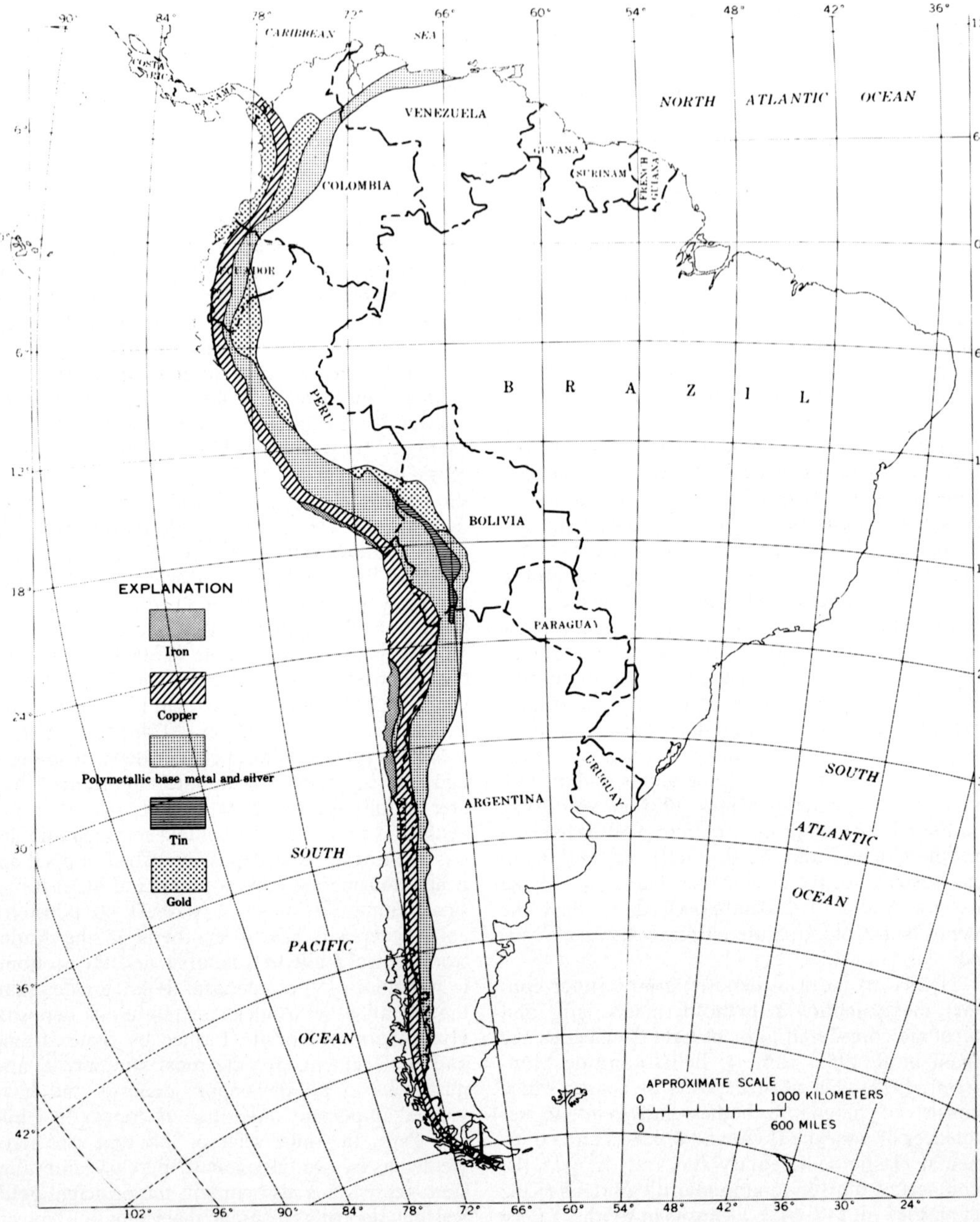

FIG. 3–Andean metallogenic province, showing five major subprovinces of (1) iron, (2) copper, (3) polymetallic base metal and silver, (4) tin, and (5) gold.

longitudinal belt in western Bolivia; only one deposit, Corocoro, has produced a significantly large amount of copper. Tourmaline breccia pipes ranging from a few meters to 1,200 m in diameter occur in a belt 2,000 km long, extending through northern Chile (Sillitoe and Sawkins, 1971, p. 1028). They have yielded relatively small amounts of copper. Small fumarolic copper veins, consisting of covellite associated with native sulfur, occur at the top of Volcan Aucanquilcha, a Holocene volcano near the Bolivian border in northern Chile.

Polymetallic base metals—The most numerous and widespread deposits of the Andean Highlands of Peru are hydrothermal polymetallic base-metal deposits that are characterized by a complex suite of sulfide and sulfosalt minerals of Pb, Zn, Cu, Ag, Sb, and As. Many deposits also contain recoverable amounts of other metals such as Bi, Cd, Au, Se, and Te. Similar deposits are moderately abundant in Bolivia, where many are peripheral with respect to tin deposits and districts. Polymetallic base-metal deposits are few and small north of Peru; to the south, they occur sporadically in Argentina and Chile.

In 1971, production of lead and zinc from Peruvian deposits was, respectively, 166,000 and 318,000 MT, whereas production from Bolivia was 20,600 and 45,000 MT (U.S. Bureau Mines, 1974). Copper production from these ores, though not reported, was considerably less than the lead production. During 1971, Peru produced nearly 40 million oz of silver, most of which was recovered from polymetallic base-metal ores.

Tin—The tin subprovince, which is characterized by widespread and abundant cassiterite, is almost wholly restricted to the eastern Bolivian Andes (Fig. 3). It contains hundreds of tin-bearing veins in a belt 25–80 km wide extending from northern to southern Bolivia, a distance of nearly 750 km (Turneaure, 1971; Kelly and Turneaure, 1970; Ahlfeld and Schneider-Scherbina, 1964). It extends across the borders into northwestern Argentina and into southern Peru, where only a few exploitable deposits are known. The tin deposits show a wide variation in chemical and mineralogic composition, and many are rich in silver and bismuth. In addition, three types of deposits can be recognized in the tin province: (1) tungsten deposits similar to the tin deposits but containing wolframite instead of cassiterite as the dominant mineral, (2) lead-zinc deposits having dominant galena and sphalerite and little or no tin, and (3) antimony deposits in which stibnite is the dominant mineral.

The grade of tin ore mined in Bolivia has decreased from an average of 12–15% during the early part of this century to 1–2%, with an upper range of about 3%, in the 1960s. Bolivian tin production in 1970 was about 26,000 MT, down from a maximum of about 42,000 MT in 1945 (U.S. Bureau Mines, 1974).

Gold—The gold that has been produced from the Andean region has come from three sources, all of which have been important economically: (1) gold-bearing quartz and quartz-pyrite veins, some of which also contain copper; (2) gold-bearing base-metal veins, where gold occurs in the primary ore and is enriched in the oxide capping; and (3) gold placers. Gold-bearing veins and placers are most numerous in the eastern Andes, where they occur sporadically from Ecuador to Bolivia; along the western flank of the Andes and the coastal region from Colombia to central Chile (the copper province shown in Fig. 3); and in the Cordillera Central of Colombia (Fig. 3). In 1971, gold production of Bolivia, Chile, Colombia, and Peru, which includes by-product gold from copper and polymetallic deposits, was about 360,000 oz (U.S. Bureau Mines, 1974). About half of this gold came from gold veins and placers in the Cordillera Central of Colombia (Fig. 1).

Other deposits—Andean metalliferous deposits of lesser importance can only be mentioned. Wolframite deposits in the polymetallic base-metal province of Peru have yielded significant quantities of tungsten. Sedimentary manganese deposits of Early Cretaceous age, which crop out in a belt between lats. 28 and 31°S, northern Chile, have been the principal source of manganese ore in the Andean region. Mercury has been produced from scattered deposits throughout the Andes, but by far the largest production has been from the Huancavelica district of south-central Peru. Platinum is recovered from placer deposits in southwestern Colombia. Small fissure-filling molybdenite veins have been exploited in Peru, but most molybdenum from the Andean region is a by-product of porphyry copper ores.

Pre-Andean Deposits

Metalliferous deposits of pre-Andean age (Fig. 4) include: (1) tin-tungsten deposits, beryllium-lithium pegmatites, sedimentary iron deposits, and certain lead-zinc deposits of northwestern Argentina, which formed during at least two stages of mineralization in early to middle Paleozoic time (Clark et al, 1973, p. 16; Angelelli et al, 1970); (2) gold deposits of Paleozoic age (Ruiz et al, 1965; González Bonorino and Aguirre, 1970) in

FIG. 4—Principal zones of metalliferous deposits of Precambrian and Paleozoic age in and near Andean region. Heavy dashed line marks approximate eastern limit of Andes.

coastal regions of southern Chile, and perhaps some of those in the eastern Andes from Argentina to Ecuador and the Cordillera Central of Colombia; and (3) metamorphic iron-formation and manganese deposits in southern Chile. Of these deposits, only the Aguilar lead-zinc deposit in northernmost Argentina (Fig. 4), which is of questionable early Paleozoic age, has yielded a significant quantity of metal. The Relun (Chile) and Zapla (Argentina) iron deposits (Fig. 4) may have moderately large tonnages of medium-grade iron ore (Ruiz et al, 1965, p. 244; Angelelli et al, 1970, p. 75). Other pre-Andean deposits apparently are of little or no economic importance.

The deposits in northwestern Argentina are the only ones of known or presumed pre-Andean age that have been exploited in the Andean region. In 1971, production of lead and zinc in Argentina was, respectively, 40,000 and 42,000 MT (U.S. Bureau Mines, 1974), nearly all of which came from the Aguilar deposit. On the basis of descriptions by Angelelli et al (1970), of the tin and tungsten deposits and the beryllium- and lithium-bearing pegmatites, total production of these four metals from several tens of mines in the areas shown in Figure 4 could not have been more than 30,000–40,000 MT of concentrates, of which tungsten concentrates made up more than half. Beryl, the only beryllium mineral that has been found, occurs widely in pegmatites in the areas shown in Figure 4; spodumene, the principal lithium-bearing mineral, occurs in fewer pegmatites, most of which also contain beryl (Angelelli et al, 1970, p. 39, 86).

Some of the gold veins in the eastern Andes of Peru and Bolivia and in the Cordillera Central of Colombia may be of Paleozoic or Precambrian age. However, because of the scarcity of radiometric age data for intrusions in these areas and the paucity of detailed information about the deposits, it is not possible to differentiate between veins formed during the Andean metallogenic epoch and those that might have formed earlier.

Mineral Occurrences in Antarctica

Within the Antarctic region, which in this report is considered to include the islands of the Scotia Arc, the Antarctic Peninsula, and the Pacific margin of the continent, only a few metallic occurrences are known (Fig. 5), none of which appears to be exploitable (Wright and Williams, 1974). Undoubtedly, future exploration in these areas will reveal other, perhaps exploitable deposits, but on the whole it seems unlikely that either numerous or very large deposits will be found in ice-free areas where they could be exploited under current technology and economic conditions. The paucity of deposits in the Antarctic Peninsula and in islands of the Scotia Arc may be compared with the mineralization pattern of the southern Andes, where relatively few and small deposits are known. Marginal lands of the Amundsen and Bellingshausen Seas (Fig. 5) are less well known and perhaps are more favorable for mineral deposits than the Antarctic Peninsula, but here, as elsewhere in Antarctica, exploration is too scanty to allow an evaluation of the mineral-resource potential. Furthermore, ice cover of about 95% will inhibit exploration for the foreseeable future. Even if deposits could be found beneath the ice cover, it is unlikely that they could be exploited unless the ice were thin and easily removed, or access could be gained by shafts and tunnels from nearby ice-free areas. Nevertheless, as pointed out by Wade (1974; this volume), the search for minerals in Antarctica has only begun, and the geologic similarities of this continent with other continents would indicate that it, too, should have important mineral resources.

Seafloor Deposits of Southeastern Pacific

Seafloor deposits include (1) ferromanganese nodules, which may contain, in addition to manganese and iron, significant quantities of copper, cobalt, and nickel; (2) metal-rich unconsolidated pelagic sediments; (3) phosphorite nodules and phosphatic sediments along continental margins; and (4) placer concentrations of heavy minerals at continental margins. Of these deposits, all except placer concentrations are known to exist on the seafloor in the southeastern Pacific, and even deposits of this type, particularly gold placers, may yet be found on the continental shelf of western South America.

Ferromanganese nodules and metal-rich pelagic sediments are widespread seafloor deposits in the southeastern Pacific, although potentially commercial deposits are yet to be found. Nodules, though widespread, are not as abundant as in some other areas (e.g., the North Pacific). Metal-rich pelagic sediments, which have been found on the East Pacific Rise and in the Bauer Deep (Fig. 2), contain anomalous amounts (on the order of a few hundred ppm) of Cu, Zn, Ni, and Pb (Bostrom and Peterson, 1966; Dasch et al, 1971, p. 176). The presence of metal-rich sediments on the East Pacific Rise suggests the possibility of underlying metal-rich volcanic rocks which, like the sediments, may contain metals deposited by hydrothermal solutions associated with volcanism

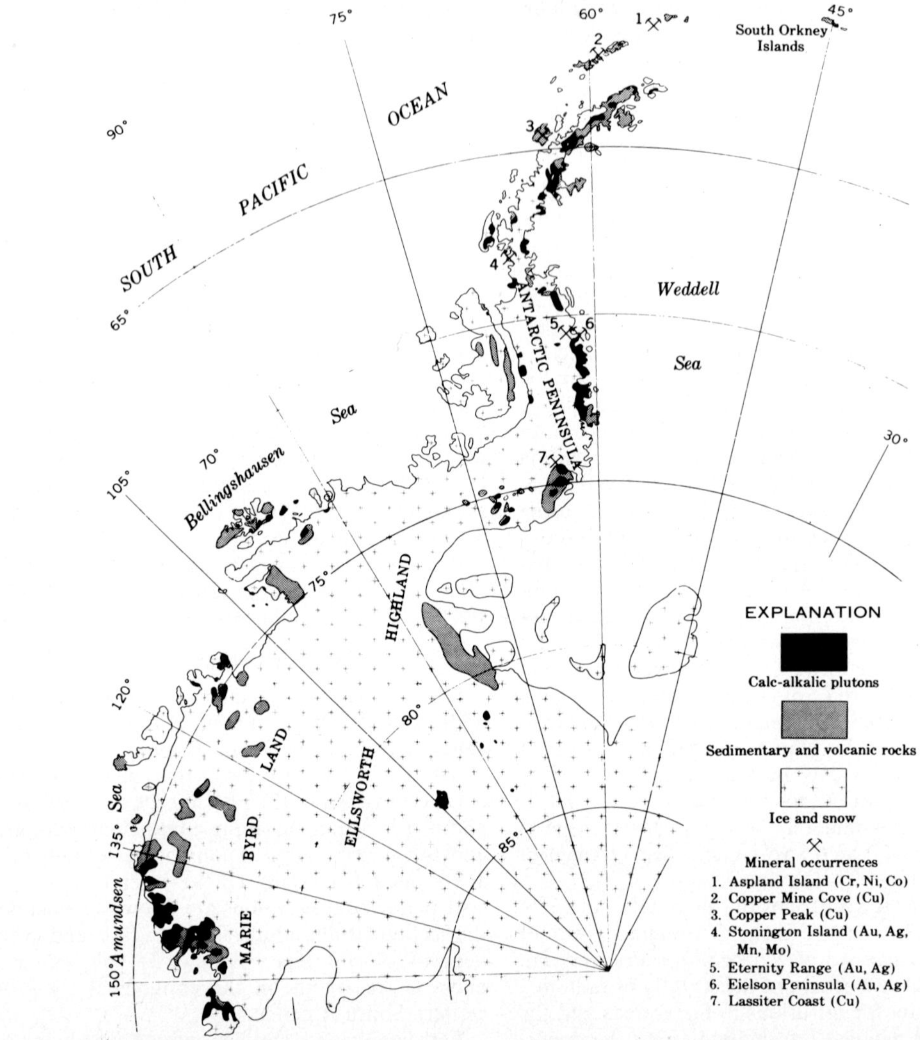

FIG. 5–Generalized geology and metallic mineral occurrences of Antarctic Peninsula and Pacific margin of Antarctica. Geology from Craddock (1970); mineral occurrences from Wright and Williams (1974).

at the ridge crest or along transform faults on its flanks.

CONCLUSIONS

The southern Pacific region—the Pacific basin and marginal lands of South America and Antarctica—shows a wide variety of metalliferous deposits of diverse origins, but those of greatest economic importance are hydrothermal deposits of the central Andean region. By far the most important deposits in terms of world resources are the porphyry copper deposits; other types of copper deposits, polymetallic base-metal and silver deposits, and tin deposits also are important sources of metals. Exploitable metallic mineral deposits have not been found in the Antarctic Peninsula or on the Pacific margin of Antarctica, and it seems unlikely that either numerous or large deposits will be found in ice-free parts of these regions. Ferromanganese nodules are widespread in the southeastern Pacific and may be exploited someday. However, nodules there are generally less abundant and have a lower recoverable metal content than nodules in extensive areas of the North Pacific Ocean.

The metals of Andean ore deposits were original constituents of calc-alkaline magmas that formed at the Benioff zone, augmented by metals that were mobilized or incorporated by the magmas as they rose through the overlying mantle and continental crust. Possible metal-rich source rocks are as follows: (1) oceanic crust and pelagic sediments in the Benioff zone (Sillitoe, 1972; Mitchell, 1973), which formed chiefly at or near the crest of the East Pacific Rise and its precursory spreading ridges; (2) longitudinal zones or layers in the South American lithospheric plate above the Benioff zone that are rich in certain metals such as tin (Clark et al, 1973); and (3) rocks in deeper parts of the continental crust in which metals may have been concentrated during an early stage of crustal development.

The age and geographic distribution of magmatism and mineral belts were probably influenced in part by position, inclination, and subduction rate of the oceanic lithospheric plate. Thus, the gradual eastward age migration of plutonism and associated mineralization was the result of one or more of the following: (1) slow eastward migration of the subduction zone relative to the continental margin, (2) generation of plutons at progressively deeper levels in the Benioff zone, and (3) slow decrease in inclination of the subducting slab. Changes in rates of subduction would influence these factors and could account for anomalies in the normal age-distribution pattern. It is not yet possible to evaluate the relative importance of each of these factors in the magmatic evolution in the Andes as a whole.

References Cited

Ahlfeld, F., and A. Schneider-Scherbina, 1964, Los yacimientos minerales y de hidrocarburos de Boliva: Bolivia Dept. Nac. Geología Bol. 5, 388 p.

Angelelli, V., et al, 1970, Descripción del mapa metalogenético de la República Argentina; minerales metaliferos: Argentina Dir. Nac. Geol. Minería Anales 15, 183 p.

Aubouin, J., et al, 1973, Esquisse paléogéographique et structurale des Andes méridionales: Rev. Géog. Physique et Géol. Dynamique, v. 15, p. 11–72.

Bellido B., E., and L. de Montreuil, 1972, Aspectos generales de al metalogenía del Peru: Peru Servicio Geología y Minería, Geol. Econ. No. 1, 149 p.

——— D. Girard, and J. Paredes, 1972, Mapa metalog énico del Peru: Peru Servicio Geología y Minería Map, scale 1:2,500,000.

Bellizia G.. A., 1973, Recursos minerales de Venezuela: Venezuela Ministerio Minas Hidrocarburos, map with text.

Bolivia Servicio Geológico, Division de Tecnología Minera, 1968, Investigaciones preliminares sobre tectónica y metalogénesis en las Cordilleras Real y Quimsa Cruz: Geobol (Bolivia Servicio Geol.), Bol. 9, 64 p.

Bostrom, K., and M. N. A. Peterson, 1966, Precipitates from hydrothermal exhalations on the East Pacific Rise: Econ. Geology, v. 61, no. 7, p. 1258–1265.

Clark, A. H., et al, 1973, Geochronological relationships of mineralization in the central Andean tin belt of Bolivia and Argentina (abs.): Caracas, Venezuela, Latinoamericano Geología Cong. Resúmenes, p. 16–17.

Corvalán D., J., and F. Munizaga S., 1972, Edades radiométricas de rocas intrusivas y metamórficas de la Hoja Valparaiso–San Antonio: Chile Inst. Inv. Geológicas Bol. 28, 40 p.

Craddock, C., compiler, 1970, Geologic map of Antarctica: Am. Geog. Soc., Antarctic Map Folio Ser., Folio 12, pl. 20, scale 1:10,000,000.

Dasch, E. J., J. R. Dymond, and G. R. Heath, 1971, Isotopic analysis of metalliferous sediment from the East Pacific Rise: Earth and Planetary Sci. Letters, v. 13, no. 1, p. 175–180.

de las Casas, F., 1969, Mapa metalogenético, Peru: Soc. Nac. Minería y Petróleo del Peru, scale 1:1,000,000.

Farrar, E., et al, 1970, K-Ar evidence for the post-Paleozoic migration of granitic intrusion foci in the Andes of northern Chile: Earth and Planetary Sci. Letters, v. 10, no. 1, p. 60–66.

Gonzalez Bonorino, F., and L. Aguirre, 1970, Metamorphic facies series of the crystalline basement of Chile: Geol. Rundschau, v. 59, no. 3, p. 979–994.

Goossens, P. J., 1969, Mineral index map, Republic of Ecuador: Ecuador Servicio Nac. Geología y Minería.

——— 1972, Metallogeny in the Ecuadorian Andes: Econ. Geology, v. 67, no. 4, p. 458–468.

——— 1975, Lithologic, geochemical, and metallogenic belts in the northern Andes and their structural relationships: Soc. Mining Engineers Preprints, 28 p.

Herron, E. M., 1972, Sea-floor spreading and the Cenozoic history of the east-central Pacific: Geol. Soc. America Bull., v. 83, no. 6, p. 1671–1692.

Hollister, V. F., 1973, Regional characteristics of porphyry copper deposits of South America: Mining Eng., v. 25, no. 8, p. 51–56.

Irving, E. M., 1971, La evolución estructural de los Andes mas septentrionales de Colombia: Colombia Servicio Geol. Nac. Bol. Geol., v. 19, no. 2, 89 p.

Kelly, W. C., and F. S. Turneaure, 1970, Mineralogy, paragenesis and geothermometry of the tin and tungsten deposits of the eastern Andes, Bolivia: Econ. Geology, v. 65, no. 6, p. 609–680.

Larson, R. L., and W. C. Pitman, III, 1972, World-wide correlation of Mesozoic magnetic anomalies, and its implications: Geol. Soc. America Bull., v. 83, no. 12, p. 3645–3662.

Mitchell, A. H. G., 1973, Metallogenic belts and angle of dip of Benioff zones: Nature; Phys. Sci., v. 245. no. 143, p. 49–52.

Park, C. F., 1961, A magnetite "flow" in northern Chile: Econ. Geology, v. 56, no. 2, p. 431–436.

Petersen, U., 1965, Regional geology and major ore deposits of central Peru: Econ. Geology, v. 60, no. 3, p. 407–476.

——— 1970, Metallogenic provinces in South America: Geol. Rundschau, v. 59, no. 3, p. 834–897.

Quirt, S., et al, 1971, Potassium-argon ages of porphyry copper deposits in northern and central Chile (abs.): Geol. Soc. America Abs. with Programs, v. 3, no. 6, p. 676–677.

Ruiz F., C., and G. E. Ericksen, 1962, Metallogenetic provinces of Chile, S. A.: Econ. Geology, v. 57, no. 1, p. 91–106.

——— et al, 1965, Geología y yacimientos metalíferos de Chile: Chile Inst. Invest. Geológicas, 305 p.

Sillitoe, R. H., 1972, Relation of metal provinces in western America to subduction of oceanic lithosphere: Geol. Soc America Bull., v. 83, no. 3, p. 813–818.

——— and F. J. Sawkins, 1971, Geologic, mineralogic, and fluid inclusion studies relating to the origin of copper-bearing tourmaline breccia pipes, Chile: Econ. Geology, v. 66, no. 7, p. 1028–1041.

Stewart, J. W., J. F. Evernden, and N. J. Snelling, 1974, Age determinations from Andean Peru—a reconnaissance survey: Geol. Soc. America Bull., v. 85, no. 7, p. 1107–1116.

Turneaure, F. S., 1960, A comparative study of major ore deposits of central Bolivia: Econ. Geology, v. 55, no. 2, p. 217–254; no. 3, p. 574–606.

——— 1971, The Bolivian tin-silver province: Econ. Geology, v. 66, no. 2, p. 215–225.

United Nations Development Programme, 1968, Mineral survey in the Andean Cordillera, Argentina: New York, United Nations, 176 p.

——— 1970, Investigation of porphyry copper type mineralization in the Provinces of Mendoza, Neuquén, and San Juan, Argentina: New York, United Nations, 335 p.

U.S. Bureau of Mines, 1974, Minerals yearbook, v. 3: Washington, D.C., U.S. Govt. Printing Office.

Wade, F. A., 1974, Antarctica: unprospected and unexploited continent (abs.): Circum-Pacific Energy and Mineral Resources Conference, Program and Abs., p. 42.

——— 1976, Antarctica: an unprospected, unexploited continent—summary: this volume.

Wright, N. A., and P. L. Williams, eds., 1974, Mineral resources of Antarctica: U.S. Geol. Survey Circ. 705, 29 p.

Zentilli, M., 1974, Geologic evolution and metallogenetic relationships in the Andes of northern Chile between 26° and 29° south: Kingston, Ontario, Queen's Univ., PhD thesis.

Metallogenesis in Southeast Pacific Ocean: Nazca Plate Project[1]

CYRUS W. FIELD,[2] JACK R. DYMOND,[3] JOHN B. CORLISS,[3] E. JULIUS DASCH,[2] G. ROSS HEATH,[3] RONALD G. SENECHAL,[2] and HERBERT H. VEEH[4]

Abstract The Nazca Plate Project is a detailed and systematic geological, geochemical, and geophysical investigation of a distinct and tectonically active lithospheric plate of the southeast Pacific Ocean basin. Major tectonic boundaries of the Nazca plate are defined by the East Pacific Rise on the west, along which new oceanic crust of basalt is being generated; the Peru-Chile Trench on the east, where older oceanic crust is being consumed by subduction beneath the overriding South American plate; and the Galapagos spreading center and the Chile Rise on the north and south, respectively. The motion of this plate relative to the Pacific plate is estimated to be eastward at more than 16 cm/yr, as documented by stratigraphic, structural, paleontologic, magnetic, and seismic evidence. Normal sediments of the deep-sea floor cover basaltic basement rocks over much of the plate, but are locally dominated by a metal-rich component on and near the East Pacific Rise and within the Bauer Deep. These metalliferous sediments, which are composed of crystalline ferromanganese hydroxyoxides (goethite, psilomelane, etc) and Fe-montmorillonite, contain anomalously high concentrations of Fe, Mn, Cu, Ni, Zn, and other transition metals. Elemental abundances and ratios, extraordinarily high rates of sedimentation, lead-isotope ratios, and statistical analysis of the data suggest that Fe, Mn, Cu, and possibly other metals were deposited by hydrothermal fluids emanating from sources of basaltic magmatism along the East Pacific Rise. However, isotopic abundances of O, Sr, S, and C and distribution patterns of the rare-earth elements imply that deposition of the metalliferous sediments occurred at normally low temperatures of the ocean floor and that the nonmetallic components were derived from and/or equilibrated with seawater.

INTRODUCTION

The Nazca Plate Project is a cooperative and continuing research endeavor between earth scientists at the Hawaii Institute of Geophysics, Oregon State University, and representatives of several South American countries, particularly Chile, Colombia, Ecuador, and Peru. It is sponsored by the National Science Foundation under the auspices of its International Decade of Ocean Exploration Program. Since the inception of this project in 1971, samples from piston cores and dredge hauls, together with cruise-track records obtained by personnel aboard the R/V *Kana Keoki* (Hawaii) and R/V *Yaquina* (Oregon) have added substantially to those originally available and subsequently provided by the Deep Sea Drilling Project. In 1974, Leg 34 of the Deep Sea Drilling Project drilled through the sedimentary column into the underlying basalt (Hart et al, 1974). Our research has been directed to the systematic application of a variety of geochemical, geological, and geophysical techniques to a detailed study of the Nazca plate, which is a well-defined, geologically distinct, and tectonically active portion of the earth's crust. We anticipate that results of the interdisciplinary research will materially enhance current knowledge relevant to (1) the evolution of a lithospheric plate from its origin to its destruction, and (2) the interactions and processes, such as magmatism, metallogeny, tectonism, sedimentation, and petroleum formation, that take place between the ocean basins and continental landmasses at, and marginal to, convergent plate boundaries.

The Nazca plate occupies approximately 16×10^6 km^2 of the southeast Pacific Ocean. Major bathymetric, geographic, and tectonic features of this plate are illustrated in Figure 1. Its eastern margin, off the west coast of South America, is defined by the Peru-Chile Trench, and the western margin by the East Pacific Rise. The northern boundary, subparallel with the equator, is defined by the Galapagos spreading center, and the southern boundary by fracture-zone offsets of the northwest-trending Chile Rise. The Galapagos Rise, an extinct center of seafloor spreading, trends north through the central part of the Nazca plate. With respect to adjoining lithospheric plates, it is bounded by the South American plate to the east, the Antarctic plate to the south, the Pacific plate to the west, and the Cocos plate

[1]Manuscript received, November 11, 1974.

[2]Department of Geology, Oregon State University, Corvallis, Oregon 97331.

[3]School of Oceanography, Oregon State University, Corvallis, Oregon 97331.

[4]Hawaii Institute of Geophysics, University of Hawaii, Honolulu, Hawaii; present address: University of South Australia, Adelaide, Australia.

Many of the data and interpretations cited herein have been developed and improved through the cooperation and assistance of our colleagues. In particular, we wish to thank R. J. Blakely, R. W. Couch, J. P. Dauphin, W. A. Edlund, M. Gemperle, D. M. Hussong, S. H. Johnson, L. D. Kulm, B. T. Malfait, T. C. Moore, G. E. Ness, R. A. Prince, D. K. Rea, and Tj. H. van Andel. L. D. Kulm of Oregon State University kindly read and edited an earlier version of this manuscript. We gratefully acknowledge the National Science Foundation for support of the Nazca Plate Project as a contribution to the International Decade of Ocean Exploration.

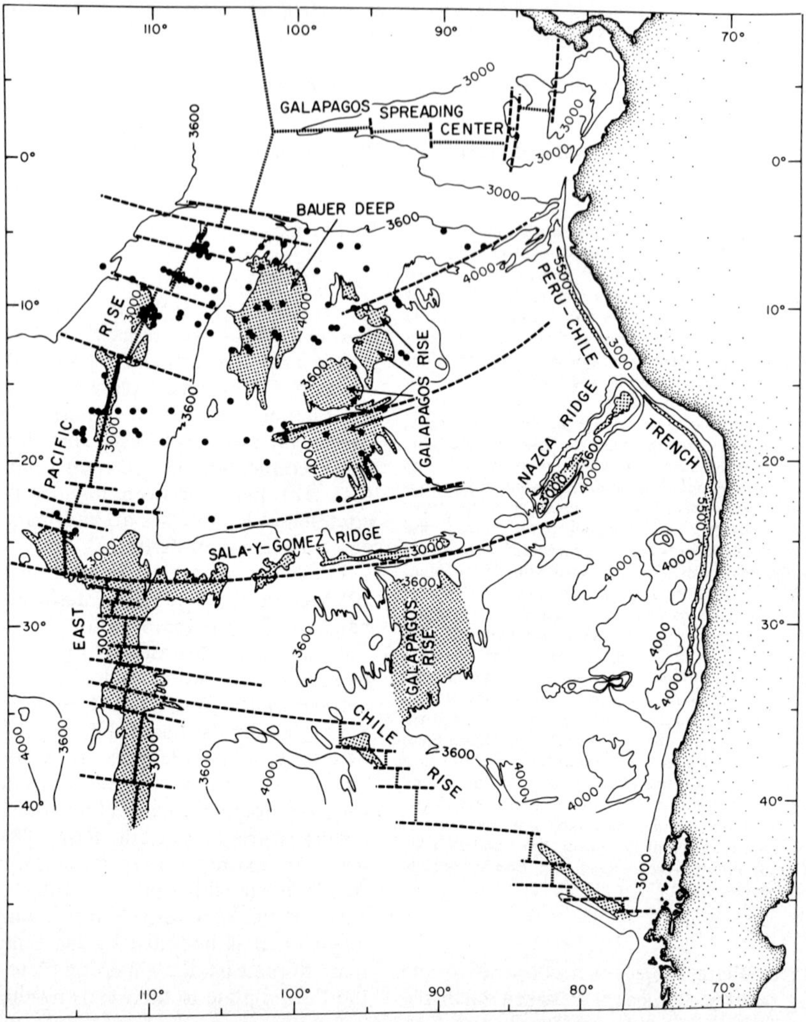

FIG. 1–Bathymetric and tectonic features of Nazca plate showing sample locations (dots), fracture zones (dashed lines), active rises (dotted lines), and important topographic ridges and basins (stippled areas).

to the north. In terms of general bathymetry, the central part of the lithospheric plate is formed of abyssal hills and basins with a thin veneer of sediment and scattered seamounts. Average depths to the seafloor range from less than 3,000 m along the East Pacific Rise crest to about 4,400 m in the basins, and to more than 5,500 m within the Peru-Chile Trench.

The Nazca plate is clearly representative of a dynamic system as defined by recently evolved concepts of the new global tectonics. In brief, new oceanic crust is being generated periodically by submarine basaltic volcanism along the axial crest of the East Pacific Rise. With time, that part of the crust forming the Nazca plate moves east from the rise crest by seafloor spreading and ultimately enters the Peru-Chile Trench, the submarine manifestation of an active subduction zone, where the oceanic crust is consumed by downthrusting along a steeply dipping Benioff zone beneath the overriding South American plate. Evidence for such dynamic behavior of the Nazca plate includes: (1) the distribution, age, and thickness of marine sediments that overlie the basaltic basement rocks of the plate (Hart et al, 1974); (2) the distribution of linear magnetic anomalies on the plate; (3) the pronounced seismicity of the East Pacific Rise, the Peru-Chile Trench, and the

Benioff zone beneath crustal rocks of the adjoining South American plate (Isacks et al, 1968); (4) deformational structures in sedimentary rocks along the east margin of the Peru-Chile Trench (Hussong et al, 1973; Kulm et al, 1973, 1974); and (5) geologically young to recent magmatism in the high Andes of South America.

Magnetic anomalies reported by Herron (1972), Rea et al (1973), and Rea and Malfait (1974) for the east-central Pacific demonstrate that the age of oceanic crust increases from west to east across the Nazca plate, and ranges from 3 m.y. and younger (Pleistocene-Holocene) along the East Pacific Rise crest to as much as 53 m.y. (Eocene) along the west margin of the Peru-Chile Trench. Moreover, paleontologic and stratigraphic evidence indicates that both the age of strata directly overlying the basaltic crust and the total thickness of this sedimentary column generally increase from west to east across the plate. Further evidence of dynamic motion is provided by Rea et al (1973), who estimated rapid movement (8.2 cm/yr) of the Nazca plate toward South America, and by Walker et al (1972) and Berg and Sutton (1974), who found a systematic correlation between seismicity and volcanism in the Andes and seismic activity of the East Pacific Rise. Also, paleontologic data reported by Kulm et al (1974) indicate a convergence rate of at least 2.8 cm/yr for the intersection of the Nazca Ridge (Fig. 1) with the Peru-Chile Trench.

Seismic reflection profiles and deep-sea drilling reveal that a veneer of sediment and sedimentary rock covers the underlying basaltic crust throughout much of the Nazca plate. Stratigraphic thicknesses range from variably thin or absent at and near the East Pacific Rise to as much as 150 m near the Peru-Chile Trench. Parts of this sequence contain anomalously high concentrations of transition metals, both locally at multiple levels and over regionally extensive areas such as the Bauer Deep (Fig. 1) submarine basin. This paper concerns these metalliferous sediments with respect to their chemical and mineralogic characteristics and origin and to speculative implications relevant to metallogeny on the adjacent South American continental plate.

METALLIFEROUS SEDIMENTS

Although metalliferous sediments were collected during the Challenger Expedition a century ago, it was not until the pioneering work of Bostrom and Peterson (1966, 1969) that the high concentration of transition metals in these sediments was accurately defined. Results of subsequent investigations suggest that metalliferous sediments are not uncommon components of seafloor sedimentation (Bostrom and Peterson, 1969; Bostrom et al, 1971; Cronan, 1972), particularly adjacent to active spreading centers such as the East Pacific Rise and the Red Sea (Degens and Ross, 1969). Unusually high concentrations of iron (~10–25%) and manganese (~2–8%) are characteristic of the metalliferous sediments. In addition, these sediments of the Nazca plate and elsewhere contain anomalous quantities of the base and precious metals. They include As (2–104 ppm), Co (15–240 ppm), Cu (46–1,240 ppm), Pb (45–145 ppm), Mo (0.5–310 ppm), Ni (30–820 ppm), Ag (2.7–5.1 ppm), and Zn (25–760 ppm), according to Bostrom and Peterson (1969) and Dymond et al (1973, and later unpub. analyses). Continental occurrences of these metals, particularly at such anomalously high concentrations and as a single polymetallic assemblage, would be of immediate interest to exploration geologists. Without question, the metalliferous sediments constitute an important natural resource of the future.

Lithology, Mineralogy, and Chemistry

Sediments of the deep-sea floor are normally fine grained and contain mineral phases that are representative of pelagic, detrital, and/or authigenic modes of deposition. The pelagic component may consist of two fractions: (1) a biogenic fraction dominated by siliceous and calcareous skeletal remains of planktonic microorganisms, plus phosphatic fish debris, and (2) a terrigenous fraction containing volcanic glass, clay minerals, and grains of quartz, feldspars, and ferromagnesian and other silicate minerals that in part are diagenetically altered. In contrast, the authigenic components are ferromanganese hydroxyoxides, zeolites, iron-rich smectite, and small amounts of other minerals including remobilized components of the biogenic and terrigenous fractions.

The metalliferous sediments are exceedingly fine grained, with particles ranging down to colloidal size. Colors typically vary from a distinctive orange-yellow in carbonate-rich facies to chocolate brown in carbonate-free deposits. They resemble pulverized gossan from the oxidized capping of a chalcopyrite-rich orebody in terms of color, texture (to some extent), and mineralogy. Although the sediments originally were considered to be amorphous, ultraslow-scan and differential X-ray diffraction analyses (Dasch et al, 1971) have revealed a variety of crystalline phases in them. On the basis of such analyses, Dymond et al (1973) reported goethite, Fe-montmorillonite (nontronite?), zeolite (phillipsite), phosphorite, psilomelane, δ-MnO_2 (todorokite?), and barite, in addition to calcite and silicate minerals, in metalliferous sediments of the eastern Pacific Ocean including the Nazca plate. Ferromanganese hydroxyoxides and Fe-montmorillonite predominate in the samples least contaminated by normal deep-sea sediments.

Chemical analyses of the metalliferous sediments are normally recalculated to carbonate-free values to eliminate dilution caused by the presence of pelagic carbonate (up to 90% $CaCO_3$ by weight). This transformation may enhance the actual concentrations of the elements by a factor of 10 or more over values in the original carbonate-rich samples. Analytical data reported by Bostrom and Peterson (1969) and Dymond et al (1973) clearly demonstrate that iron (5–34%), manganese (1–15%), alumina (1–16%), and silica (9–38%) are the dominant chemical constituents of the metalliferous sediments. Mineralogy is the chief determinant of bulk composition, although minor variations occur between different sieve-sized fractions of the sediment. This effect is well defined by analyses of separate mineral phases in a sediment sample (AII-54-14; see Dymond et al, 1973) from the Bauer Deep. According to Eklund (1974), the four mineral phases were chemically descernible as follows: Fe-smectite (SiO_2, 52.3%; Fe_2O_3, 29.7%; MgO, 6.1%; Al_2O_3, 2.4%; K_2O, 2.0%); todorokite (MnO_2, 74.9%; MgO, 3.1%; Fe_2O_3, 1.6%); iron hydroxyoxide, in part goethite (MnO_2, 36.2%; Fe_2O_3, 25.4%; SiO_2, 15.1%; MgO, 3.3%; CaO, 2.1%; Al_2O_3, 1.7%); and phosphatic fish debris (CaO, 45.2%; P_2O_5, 34.4%). In addition, various minor and trace elements are preferentially concentrated in the iron (Ni, Zn, Cu, Co), manganese (Cu, Ni), and phosphatic (La) mineral phases.

We have completed chemical analyses for a large suite of metalliferous sediments from the northwest part of the Nazca plate and East Pacific Rise. Locations of the core samples are shown in Figure 1, and the data are presented as ternary diagrams of the major elements (carbonate-free) in Figures 2 and 3. On the basis of compositional differences, most of the samples can be grouped into discrete or transitional populations that are correlative with regional and/or tectonic location. These sample groups are designated as separate populations, as are those of the ferromanganese nodules (Mero, 1965) and clay-rich Pacific sediments (Goldberg and Arrhenius, 1958), which are added for comparison. The ternary distribution of silicon-iron-manganese data (Fig. 2) shows that the ferromanganese nodules are compositionally distinct (high Mn) relative to all other groups that have nearly constant Fe:Mn ratios. Higher contents of silicon in the more northerly samples, particularly those from the northern Bauer Deep and East Pacific Rise crest, are ascribed to larger contributions of biogenic silica. We attribute the high and relatively invariant Fe:Mn ratios of the metalliferous sediments to a strong hydrothermal (Fe-rich) component. They contrast markedly with the low Fe:Mn ratios, typical of slow authigenic deposition, for ferromanganese nodules. However, the more iron-enriched members of this latter group may indicate a mixing of both authigenic and hydrothermal components. The ternary diagram for aluminum-manganese-iron (Fig. 3) illustrates another important compositional difference: metalliferous sediments and ferromanganese nodules are both depleted in aluminum relative to normal clay-rich pelagic sediments. High contents of iron (>20%) and manganese (>6%) are present in sediments of the Bauer Deep and southern parts of the East Pacific Rise. Q-factor analyses of these and other compositional variables suggest that 98% of the variability can be accounted for by three genetic end members. We interpret these to represent normal deep-sea sedimentation of clay minerals (factor 1; Ni and Al); hydrothermal additions (factor 2; Mn, Fe, and Cu); and biogenic deposition (factor 3; Si). In summary, the metalliferous sediments are readily distinguished from other sedimentary deposits of the deep-sea floor on the basis of major element characteristics such as iron enrichment, aluminum depletion, and high Fe:Mn ratios.

Minor and Trace Element Geochemistry

Anomalously high concentrations of minor elements, particularly the economically significant transition metals (Ag, Co, Cu, Mo, Ni, Pb, Zn), are a fundamental characteristic of the metalliferous sediments. The metals apparently do not form a critical component in the lattices of discrete mineral phases but, rather, they occur dispersed and absorbed in the ferromanganese hydroxyoxides (a scavenging effect well known to chemists and exploration geologists). Anomalous concentrations of nickel in core samples from the Nazca plate are regionally best defined by sediments of the Bauer Deep. The distribution of nickel is correlative with normal processes of deep-sea sedimentation (factor 1). Large concentrations of nickel are favored in products formed by slow authigenic deposition, and the unusually high enrichments displayed by the ferromanganese nodules (Mero, 1965) corroborate this effect. For the Nazca plate, the Fe:Ni ratios of metalliferous sediments are extremely variable relative

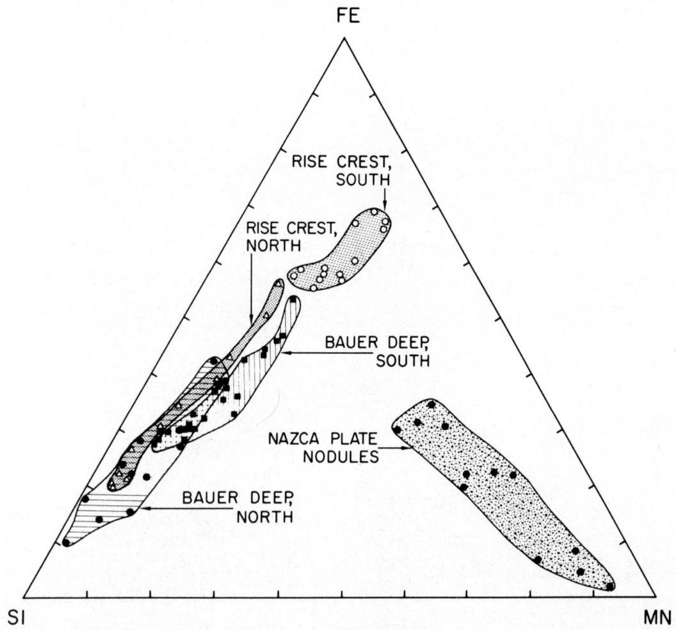

FIG. 2—Ternary diagram for silicon, iron, and manganese illustrating compositional differences between ferromanganese nodules (Mero, 1965) and regional occurrences of metalliferous sediments on Nazca plate.

to those of the nickel-enriched ferromanganese nodules (Fig. 4). Moreover, the distribution of Fe:Ni ratios in sediments appears to define two broad, yet distinct, regional populations in which samples of the Bauer Deep are enriched in nickel relative to those of the East Pacific Rise crest. This geochemical trend suggests an authigenic imprint to sedimentation in the Bauer Deep, and is supported by petrographic evidence that documents the ferromanganese phase as grain coatings and micronodules throughout these sediments. In Figure 5, an analogous relation is demonstrated between the concentrations of copper and nickel in the metalliferous sediments. Samples from the Bauer Deep, particularly the southern part, are markedly enriched in nickel, or they have larger Ni:Cu ratios than do those of the East Pacific Rise crest. This relation again points to an authigenic component to sedimentation in the Bauer Deep. However, the progressively diminishing concentrations of metals (total Cu-Ni) from south to north in both sample provinces (Fig. 5) is a function of increasing dilution from biogenic opaline silica.

Distribution patterns and abundances of the lithophile rare-earth elements in meteorites and various terrestrial rocks have received growing attention in recent years. They are particularly relevant to oceanographic investigations because one of the elements, cerium, may be significantly depleted in seawater relative to the other elements. Abundance patterns of the rare-earth elements (normalized with respect to their abundance in chondritic meteorites) in metalliferous sediments of the east Pacific Ocean are illustrated in Figure 6 (after Dymond et al, 1973). Samples A54-14 and V19-54 (Bender et al, 1971) are from the Bauer Deep and East Pacific Rise crest regions, respectively, of the Nazca plate. All the samples, regardless of their mineralogic composition, show the cerium depletion that is highly indicative of derivation and/or equilibration with seawater. Other correlations among the sample data, according to Dymond et al (1973), include a covariant relationship between total rare-earth element abundances and the amount of phosphatic fish debris, and an antipathetic relationship between Ce:Sm ratios and the iron content of the samples.

Radiogenic and Stable Isotope Geochemistry

Isotopic abundance data for a number of elements are widely applicable to geologic investigations concerned with provenance, genesis, and/or geothermometry of rocks and minerals. The available data for lead isotopes in metalliferous sediments are summarized in Dymond et al (1973). Comparisons with other Pb^{207}/Pb^{204} and Pb^{206}/Pb^{204} data show that the metalliferous sedi-

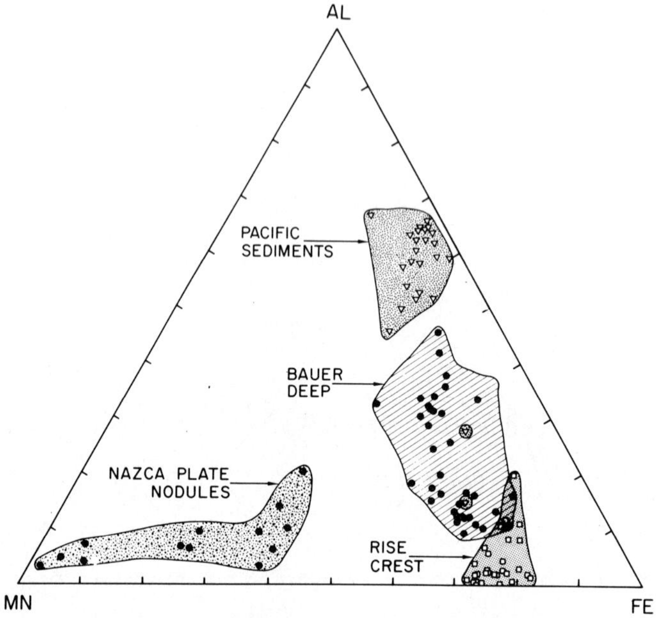

FIG. 3—Ternary diagram for manganese, aluminum, and iron illustrating compositional differences between normal deep-sea sediments (Goldberg and Arrhenius, 1958), ferromanganese nodules (Mero, 1965), and metalliferous sediments of East Pacific Rise and Bauer Deep regions of Nazca plate.

ments are isotopically similar to the oceanic subalkalic tholeiitic basalts (see Dasch et al, 1971), but differ appreciably from the ferromanganese nodules (Reynolds and Dasch, 1971). Although the isotopic data for the sediments imply a derivation from oceanic basalt or a mantle source, the interpretation does not uniquely demonstrate an origin by submarine magmatic or hydrothermal processes, as it does not preclude derivation via secondary processes such as high-temperature deuteric or low-temperature "weathering" types of alteration of a basaltic source of the deep-sea floor.

The present evidence from strontium-isotope abundance data of the metalliferous sediments is inconclusive with respect to origin (Bender et al, 1971; Dasch et al, 1971; Dymond et al, 1973). Ratios of Sr^{87}/Sr^{86} in the sediments closely bracket the ratio in modern-day seawater (Dymond et al, 1973). Although it is permissible to attribute the source of strontium to an ocean reservoir at the time of deposition, the strontium-isotope ratios of the sediments in fact are higher than those inferred for seawater at the early to middle Tertiary time of their formation. Alternative interpretations of these data include (1) the unlikely mixing of ancient (30–60 m.y.) Sr^{87}-depleted seawater with an Sr^{87}-enriched terrigenous component to form compositions similar to modern-day seawater, or (2) the more likely postdepositional equilibration of strontium isotopes between sedimentary mineral phases and modern deep-sea ocean water. Provided the latter interpretation is correct, strontium-isotope ratios are unlikely to contribute much useful information relating to the genesis or provenance of these sediments.

Concentrations of uranium in metalliferous sediments of the east Pacific Ocean, including the Nazca plate, range from 2 to 6 ppm and are somewhat higher than contents of 1–3 ppm reported for normal clay-rich pelagic sediments (Dymond et al, 1973). These weakly to moderately anomalous values may be related to the well-known scavenging effect of phosphatic fish debris for uranium, as well as for the rare-earth elements. Alternately, the contents may be controlled through coprecipitation by the amounts of iron being deposited, because large concentrations of up to 31.3 ppm uranium are contained in iron-rich deposits formed by the Red Sea geothermal brines (Ku, 1969). Although the activity ratios U^{234}/U^{238} indicate a closed system and isotopic equilibrium within the metalliferous hosts, the relatively old age of the sediments and the short half-life (0.25 m.y.) of U^{234} collectively render this technique of little value in determinations of

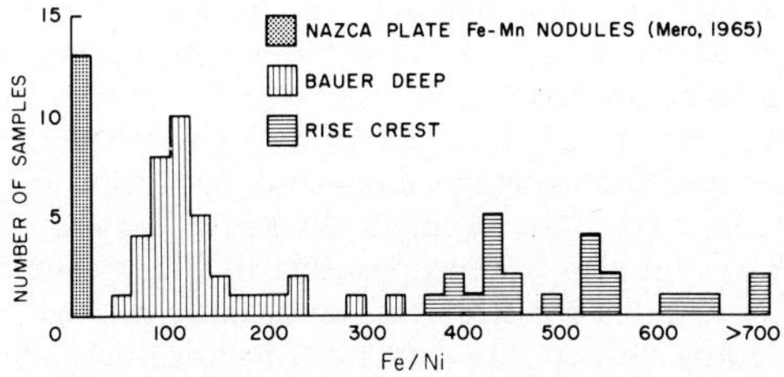

FIG. 4—Histogram of Fe:Ni ratios in samples of ferromanganese nodules (Mero, 1965) and metalliferous sediments of Bauer Deep and East Pacific Rise regions of Nazca plate.

uranium provenance. The Th^{230}-disequilibrium method, however, has been applied to measure rates of sedimentation (Bender et al, 1970, 1971). Data indicate that accumulation rates for iron and manganese are much higher in sediments of the East Pacific Rise (500×, Fe; 10–50×, Mn) and Bauer Deep (15–50×, Fe; 2–5×, Mn) than in typical authigenic deposits of these metals elsewhere on the deep-sea floor. Most investigators (Bostrom and Peterson, 1969; Bender et al, 1971; Dasch et al, 1971; Dymond et al, 1973) have attributed the origin of metal-rich sediments of the East Pacific Rise to submarine hydrothermal fluids emanating from a volcanic source near the rise crest.

According to Dymond et al (1973), the metalliferous sediments exhibit oxygen-isotope abundance variations ranging from +11.1 to +21.8 per mill that are correlative with sample mineralogy. This trend is illustrated in Figure 7 (after Dymond et al, 1973), where the δO^{18} per-mill values are plotted with respect to the combined percentage of iron plus manganese. The linear distribution of plotted data suggests a chemical-isotopic continuum between end members of the authigenic mineral fraction that consist of O^{18}-enriched silicates, phillipsite and montmorillonite, and the relatively O^{18}-depleted ferromanganese nodules. Because authigenic minerals have formed on the ocean floor at low temperatures and in isotopic equilibrium with seawater, the data for samples of metalliferous sediment that occur clustered along the tie line between ferromanganese nodules and phillipsite imply a similar environment of deposition (Dymond et al, 1973). X-ray diffractometer analyses support our contention that compositional variations among the metalliferous sediments are best attributed to contamination by authigenic silicates. In particular, the available δO^{18} data suggest that the oxide phases of the metalliferous sediments derived their oxygen through equilibration with seawater and under conditions lacking any significant isotopic imprint of temperatures or water of magmatic affinity.

The isotopic composition of sulfur in modern seawater is +21 per mill and has ranged from +10 to +33 per mill throughout all post-Precambrian time, as determined from analyses of evaporite sulfate in sedimentary rocks (Thode et al, 1961; Holser and Kaplan, 1966; Claypool et al, 1972). Sulfur is present in the metalliferous sediments at small concentrations not exceeding 2,500 ppm, and presumably as the mineral barite. Compositions exhibit a narrow range of δS^{34} values, from +17.8 to +20.8 per mill, according to six analyses reported by Dymond et al (1973). The isotopic data for sulfur in metalliferous sediments and other geologically important environments are summarized in Figure 8. Vertical lines at 0 and +21 per mill define the established isotopic compositions of magmatic and seawater sulfur, respectively. Relative to the data illustrated in Figure 8, sulfur in the metalliferous sediments is compositionally nearly identical to that in sulfates in the Red Sea geothermal brines (Kaplan et al, 1969) and in the hydrothermal replacement deposits of Morococha, Peru (Petersen, 1972a), for which seawater and/or evaporite sources of sulfur have been postulated. In contrast, sulfur of the sediments is significantly enriched in S^{34} relative to that in epigenetic sulfides of ocean-floor basalts, magmatic sulfide in continental basalt, sulfides from the Salton Sea geothermal brine (White, 1968), and sulfides and nearly all sulfates from hydrothermal deposits of North and South America. The isotopic data clearly suggest that essentially all sulfur in the metalliferous sediments has been derived from seawater.

Finally, recent summaries by Bender (1972) and Garlick (1972) indicate that the isotopic compositions of marine carbonates range from ap-

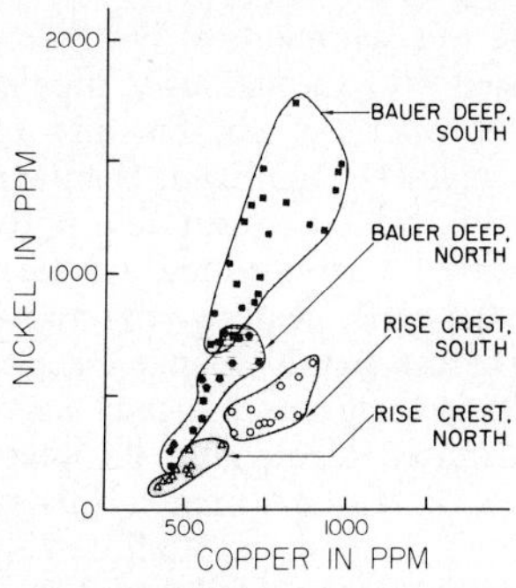

FIG. 5—Concentrations of copper and nickel (carbonate-free basis) in metalliferous sediments of East Pacific Rise and Bauer Deep regions of Nazca plate.

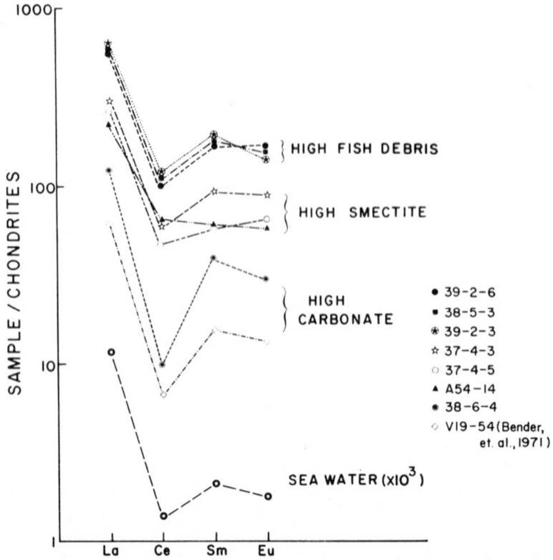

FIG. 6—Abundance patterns of rare-earth elements, normalized to their abundances in chondritic meteorites, in metalliferous sediments of Pacific Ocean including Nazca plate (after Dymond et al, 1973, Fig. 4).

proximately −3 to +3 per mill for carbon and from +26 to +36 per mill for oxygen. Analyses of the carbonate fraction of a single sample of metalliferous sediment from the northeast Pacific Ocean (Dymond et al, 1973, Table 1, sample 38-6-4) provide typically marine values for the carbon (+2.1 per mill) and oxygen (+30.4 per mill) isotopes. Moreover, the calculated isotopic temperature of deposition is 14°C, assuming equilibration of oxygen between carbonate and seawater. This isotopic-temperature estimate provides additional evidence for the low-temperature origin of the metalliferous sediments, and it is consistent with the derivation of carbonate from the skeletal remains of planktonic Foraminifera.

Genesis

Metalliferous sediments of the Nazca plate are characterized by anomalously high concentrations of Fe, Mn, Cu, Ni, Zn, and many other transition metals. Significant accumulations of these deposits, which constitute a potentially important future ocean resource, are localized principally on the geologically young and active East Pacific Rise and nearby Bauer Deep. Moreover, recent drilling suggests that they may be extensively distributed throughout the plate at depth within the sedimentary veneer (Hart et al, 1974). Chemical and mineralogic studies of these metal-rich sediments, combined with statistical analyses of the data, suggest that much of the iron, manganese, and copper originate as a hydrothermal component superimposed on the normal authigenic, biogenic, and terrigenous components of deep-ocean sediments. Evidence for the hydrothermal or volcanic exhalative origin of the metals in these sediments comes not only from their proximity to the magmatically active East Pacific Rise, but also from extraordinarily high accumulation rates for iron and manganese relative to authigenic deposits, and from lead-isotope compositions suggesting an ultimate derivation from ocean-floor basalts. Whether metallization was effected by juvenile fluids emanating from a submarine magmatic source, by convecting seawater heated locally at centers of magmatism, or possibly by more complex fluids related to either greenschist metamorphism or propylitic alteration cannot be documented from the available data (Dymond et al, 1973; Corliss, 1973–1974). Nevertheless, on the basis of mineralogic data and the bulk of the chemical evidence, we believe that the metalliferous sediments were deposited at low temperatures and in an oxygenated environment, and that most of their nonmetallic constituents were derived from or equilibrated with seawater. We anticipate that our continuing investigations of these and additional samples of the metal-rich sediments recently collected will provide further constraints to the existing genetic model.

Plate-Tectonic Implications

As noted in the introduction, evidence for the possible interaction of geologic materials and processes between ocean basin and continent at an active convergent plate boundary is an important objective of this project. In the context of rapidly evolving concepts of the new global tectonics, the discussion of the Nazca plate would be incomplete without reference to the adjacent continental plate of South America. Geologic evidence here and elsewhere amply documents the spatial and temporal coincidence of mobile tectonic belts, characterized by calc-alkaline magmatism, regional metamorphism, and structural deformation, to zones of lithospheric plate convergence that are predicted from plate-tectonic theory as a consequence of seafloor spreading. The Nazca and South American plates are particularly well suited for the study of ocean-continent interactions because they represent an active rise–convergent plate system that is manifested by contemporary volcanism along crests of both the East Pacific Rise and the Andean Cordillera, and by seismicity beneath the East Pacific Rise and the overriding continental plate of western South America.

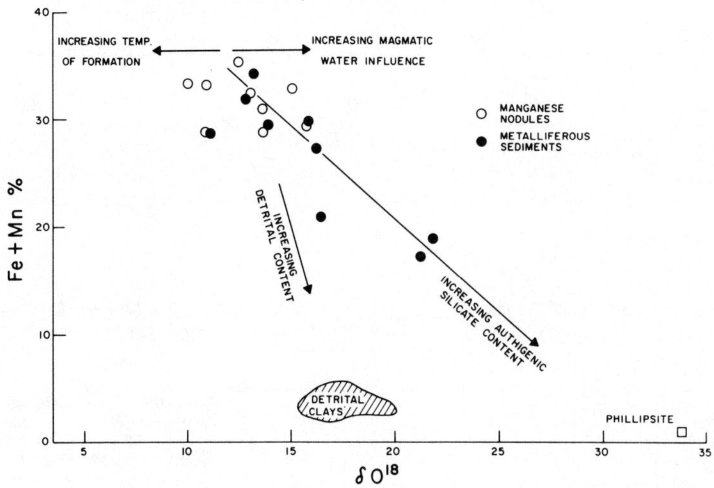

FIG. 7–Summary of oxygen-isotope data versus concentrations of iron plus manganese in samples of detrital clay minerals, authigenic phillipsite, ferromanganese nodules, and metalliferous sediments of Pacific Ocean including Nazca plate (after Dymond et al, 1973, Fig. 7).

Several major tectonic features of the adjoining Nazca and South American plates are illustrated in Figure 9. Magnetic anomalies reported by Herron (1972) demonstrate increasing ages of the basaltic oceanic crust, from its origin at the East Pacific Rise to its destruction by subduction at the Peru-Chile Trench, and document both the eastward motion and present dynamic behavior of the Nazca plate. Lithologies of the oceanic crust are dominated by submarine flows of basalt and related intrusions that are covered by a veneer of deep-sea sediments including the metal-rich varieties. In contrast, continental crust of the adjacent South American plate is dominated by calc-alkalic suites of igneous rock. They occur both as large batholithic plutons and as thick volcanic sequences that define a linear magmatic orogen largely coincident with the Andean Cordillera. Associated with this calc-alkalic suite of igneous rocks, and presumably related genetically to it, are large deposits of iron oxides and base metal- and precious-metal-bearing sulfides.

Plate-tectonic theory provides an attractive rationale for relating the geologic events on the continents to those related to seafloor spreading in the geographically separate ocean basins. For example, multiple lines of evidence now suggest that the orogenic evolution of the Cordillera of North and South America was closely related to seafloor spreading in the east Pacific Ocean, and to subduction of oceanic crust beneath the continental margins (Atwater, 1970; Burchfiel and Davis, 1972; Monger et al, 1972). Plate-tectonic theory is compatible with, and essential to, models for the geologic evolution of the Andean Cordillera as proposed by Hamilton (1969), James (1971), and Clark et al (1973). Moreover, it is fundamental to recent contributions by Petersen (1972b), Sawkins (1972), Sillitoe (1972a, b), Clark et al (1973), and Field et al (1974) to problems of ore genesis and regional aspects of metallogeny. In particular, Sillitoe (1972a, b) has suggested that the metallic constituents of Andean ore deposits were derived from sources within the mantle beneath the East Pacific Rise and accumulated in basalts and overlying sediments of the adjacent ocean floor. With seafloor spreading as the driving force of later subduction and partial fusion of oceanic crust, the metals were ultimately concentrated in hydrothermal deposits formed during the ascent and crystallization of calc-alkaline magmas that define the Andean Cordillera.

Our research efforts to date have been focused primarily on crustal lithologies of the Nazca plate, and particularly the uppermost sedimentary veneer that contains the metalliferous sediments. With a knowledge of the critical isotopic and chemical parameters of this regime, parallel investigations in collaboration with scientific colleagues from South America and elsewhere may reveal whether or not the continental rocks and ores bear vestiges of an oceanic parentage. In essence, forthcoming studies will consider topical questions such as possible contributions of oceanic crustal constituents to the calc-alkalic plutonic and volcanic rocks and associated mineral depos-

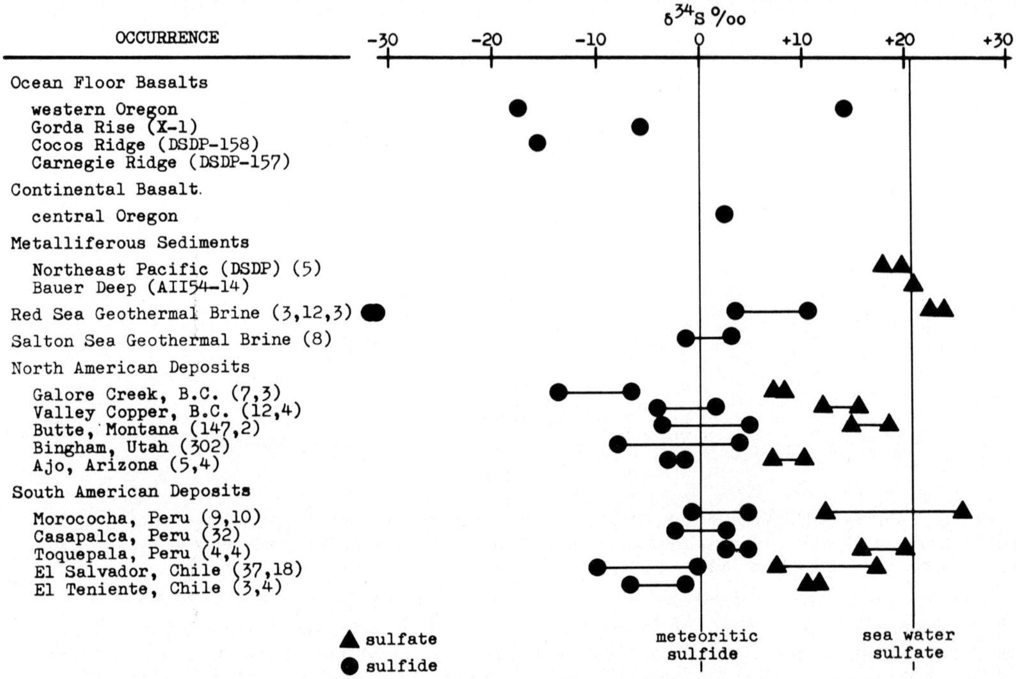

FIG. 8–Summary of sulfur-isotope data for sulfate and/or sulfide minerals in ocean-floor and continental basalts, metalliferous sediments, geothermal brines, and selected hydrothermal deposits of North and South America (see text for references; horizontal lines indicate range of isotopic values for each mineral group; sample number or number of samples analyzed is noted in parentheses to the right of locality).

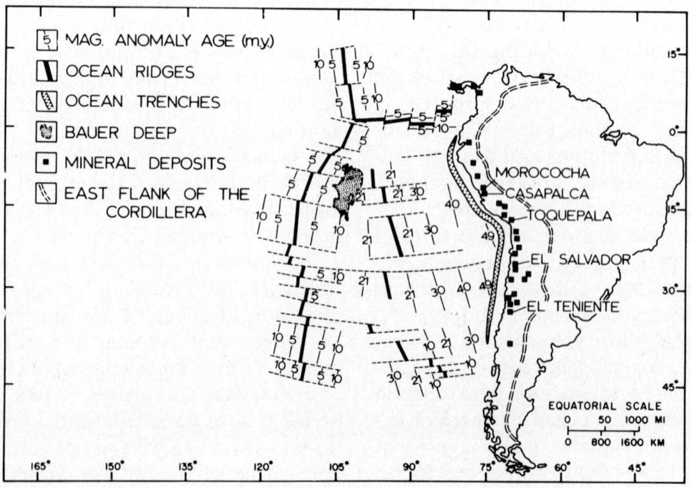

FIG. 9–Plate-tectonic summary of Nazca and South American plates showing ages of magnetic anomalies (Herron, 1972) for basaltic crust of ocean floor and distribution of selected hydrothermal mineral deposits of Andean Cordillera.

its of the Andean Cordillera, and of seawater to the saline brines localized in fluid inclusions (Corliss, 1973-1974) that effected hydrothermal transport and deposition of the ore and gangue minerals.

Continuing refinements to the interpretation of sulfur, lead, and strontium isotopes and other chemical data should enhance our ability to detect oceanic imprints in samples of rock and mineral from the continents. Results and conclusions derived from this systematic study of the dynamically active Nazca plate, and of its possible influence on the adjacent continental South American plate at a convergent boundary, are expected to provide important insight to the processes that characterized plate-tectonic regimes of the geologic past, and to well-documented spatial and temporal regularities in the distribution of igneous rocks and mineral deposits of the continents.

References Cited

Atwater, Tanya, 1970, Implications of plate tectonics for the Cenozoic tectonic evolution of western North America: Geol. Soc. America Bull., v. 81, p. 3513–3536.

Bender, M. L., 1972, Carbon isotope fractionation, *in* R. W. Fairbridge, ed., Encyclopedia of geochemistry and environmental sciences, Vol. IVA: New York, Van Nostrand Reinhold Co., p. 133–136.

——— T. L. Ku, and W. S. Broecker, 1970, Accumulation rates of manganese in pelagic sediments and nodules: Earth and Planetary Sci. Letters, v. 8, p. 143–148.

——— et al, 1971, Geochemistry of three cores from the East Pacific Rise: Earth and Planetary Sci. Letters, v. 12, p. 425–433.

Berg, E., and G. H. Sutton, 1974, Dynamic interaction of seismic and volcanic activity of the Nazca plate edges: Hawaii Inst. Geophysics Contr. 586, 17 p.

Bostrom, K., and M. N. A. Peterson, 1966, Precipitates from hydrothermal exhalations on the East Pacific Rise: Econ. Geology, v. 61, p. 1258–1265.

——— and ——— 1969, The origin of aluminum-poor ferromanganoan sediments in areas of high heat flow on the East Pacific Rise: Marine Geology, v. 7, p. 427–447.

——— B. Farquharson, and W. Eyl, 1971, Submarine hot springs as a source of active ridge sediments: Chem. Geology, v. 10, p. 189–203.

Burchfiel, B. C., and G. A. Davis, 1972, Structural framework and evolution of the southern part of the Cordilleran orogen, western United States: Am. Jour. Sci., v. 272, p. 97–118.

Clark, A. H., et al, 1973, The magmatic, tectonic, and metallogenetic evolution of the Central Andean orogen between latitudes 26° and 29° South (Abs., and circulated summary): Internat. Union Geodesy and Geophysics, Conf. Geodynamics Comptes Rendus, Lima, Peru, no. 18, pt. 1, p. 254 (CGD-44).

Claypool, G. E., et al, 1972, Sulfur and oxygen isotope geochemistry of evaporite sulfates (abs.): Geol. Soc. America Abs. with Programs, v. 4, no. 7, p. 473.

Corliss, J. B., 1973–74, The sea as alchemist: Oceanus, v. 17, Winter, p. 38–43.

Cronan, D. S., 1972, The Mid-Atlantic Ridge near 45° N, XVII; Al, As, Hg, and Mn in ferruginous sediments from the Median Valley: Canadian Jour. Earth Sci., v. 9, p. 319–323.

Dasch, E. J., J. R. Dymond, and G. R. Heath, 1971, Isotopic analysis of metalliferous sediment from the East Pacific Rise: Earth and Planetary Sci. Letters, v. 13, p. 175–180.

Degens, E. T., and D. A. Ross, eds., 1969, Hot brines and recent heavy metal deposits in the Red Sea: New York, Springer-Verlag, 600 p.

Dymond, Jack, et al, 1973, Origin of metalliferous sediments from the Pacific Ocean: Geol. Soc. America Bull., v. 84, p. 3355–3372.

Eklund, W. A., 1974, A microprobe study of metalliferous sediment components: Oregon State Univ., M.S. thesis, 77 p.

Field, C. W., M. B. Jones, and W. R. Bruce, 1974, Porphyry copper-molybdenum deposits of the Pacific Northwest: Am. Inst. Mining, Metall. and Petroleum Engineers Trans., v. 255, p. 9–22.

Garlick, G. D., 1972, Oxygen isotope geochemistry, *in* R. W. Fairbridge, ed., Encyclopedia of geochemistry and environmental sciences, Vol. IVA: New York, Van Nostrand Reinhold Co., p. 864–874.

Goldberg, E. D., and G. Arrhenius, 1958, Chemistry of Pacific pelagic sediments: Geochim. et Cosmochim. Acta, v. 13, p. 153–212.

Hamilton, W., 1969, The volcanic central Andes, a modern model for the Cretaceous batholiths and tectonics of western North America, *in* A. R. McBirney, ed., Proceedings of the Andesite Conference: Oregon Dept. Geology and Mineral Industries Bull. 65, p. 175–184.

Hart, S. R.. et al, 1974, Leg 34, oceanic basalt and the Nazca plate: Geotimes, v. 19, no. 4, p. 20–24.

Herron, E. M., 1972, Sea-floor spreading and the Cenozoic history of the east-central Pacific: Geol. Soc. America Bull., v. 83, p. 1671–1692.

Holser, W. T., and I. R. Kaplan, 1966, Isotope geochemistry of sedimentary sulfates: Chem. Geology, v. 1, p. 93–135.

Hussong, D. M., et al, 1973, Crustal structure of the eastern Nazca plate and Peruvian continental shelf from seismic refraction (abs.): Geol. Soc. America Abs. with Programs, v. 5, no. 1, p. 61.

Isacks, B., J. Oliver, and L. R. Sykes, 1968, Seismology and the new global tectonics: Jour. Geophys. Research, v. 73, p. 5855–5900.

James, D. E., 1971, Plate tectonic model for the evolution of the central Andes: Geol. Soc. America Bull., v. 82, p. 3325–3346.

Kaplan, I. R., R. E. Sweeney, and Arie Nissenbaum, 1969, Sulfur isotope studies on Red Sea geothermal brines and sediments, *in* E. T. Degens and D. A. Ross, eds., Hot brines and recent heavy metal deposits in the Red Sea: New York, Springer-Verlag, p. 474–498.

Ku, T. L., 1969, Uranium series isotopes in sediments from the Red Sea hot-brine area, *in* E. T. Degens and D. A. Ross, eds., Hot brines and recent heavy metal deposits in the Red Sea: New York, Springer-Verlag, p. 512–523.

Kulm, L. D., et al, 1973, Tholeiitic basalt ridge in the Peru Trench: Geology, v. 1, no. 1, p. 11–14.

——— et al, 1974, Transfer of Nazca Ridge pelagic sediments to the Peru continental margin: Geol. Soc. America Bull., v. 85, p. 769–780.

Mero, J. L., 1965, The mineral resources of the sea: New York, Elsevier, 312 p.

Monger, J. W. H., J. G. Souther, and H. Gabrielse, 1972, Evolution of the Canadian Cordillera, a plate tectonic model: Am. Jour. Sci., v. 272, p. 577–602.

Petersen, Ulrich, 1972a, Nuevas investigaciones de yacimientos Peruanos: Soc. Geol. Peru Bol., v. 42, p. 37–52.

——— 1972b, Geochemical and tectonic implications of South American metallogenic provinces: New York Acad. Sci. Annals, v. 196, art. 1, p. 1–38.

Rea, D. K., and B. T. Malfait, 1974, Geologic evolution of the northern Nazca plate: Geology, v. 2, no. 7, p. 317–320.

——— et al, 1973, New estimates of rapid sea-floor spreading rates and the identification of young magnetic anomalies on the East Pacific Rise, 6° and 11°S.: Earth and Planetary Sci. Letters, v. 19, p. 225–229.

Reynolds, P. H., and E. J. Dasch, 1971, Lead isotopes in marine manganese nodules and the ore-lead growth curve: Jour. Geophys. Research, v. 76, p. 5124–5129.

Sawkins, F. J., 1972, Sulfide ore deposits in relation to plate tectonics: Jour. Geology, v. 80, p. 377–397.

Sillitoe, R. H., 1972a, Relation of metal provinces in western America to subduction of oceanic lithosphere: Geol. Soc. America Bull., v. 83, p. 813–818.

——— 1972b, A plate tectonic model for the origin of porphyry copper deposits: Econ. Geology, v. 67, p. 184–197.

Thode, H. G., J. Monster, and H. B. Dunford, 1961, Sulphur isotope geochemistry: Geochim. et Cosmochim. Acta, v. 25, p. 159–174.

Walker, D. A., et al, 1972, Easter Island seismograph observations indicative of sea-floor spreading; plate-edge seismicity relationships and the prediction of earthquakes along the west coast of the western hemisphere: Hawaii Inst. Geophysics Rept. HIG-72-2, 25 p.

White, D. E., 1968, Environments of generation of some base-metal ore deposits: Econ. Geology, v. 63, p. 301–335.

Possibilities for Development of Mineral Resources of Colombia—Summary[1]

J. A. CRUZ[2]

Colombia's countrywide mineral exploration, started in 1964, is providing a basic geologic, structural, and geochemical framework within which further studies have been planned for developing mineral resources of the country.

It has been possible to recognize distribution patterns of the ore deposits and their relation to the tectonic framework of the Colombian Andes.

Metallogenic provinces occur in belts that are elongated along the three branches of the Andes. A Pb and Zn province is outlined along the Eastern Cordillera; Au, Ag, Mo, and Cu along the Central Cordillera; Hg along the boundary between the Central and Western Cordilleras; and Cu, Au, Pt, and Cr along the Western Cordillera.

The most important discoveries are the large geochemical anomalies reflecting copper mineralization in the Western Cordillera. One of these anomalies has been drilled and porphyry copper–type mineralization has been discovered.

[1]From Circum-Pacific Energy and Mineral Resources Conference program booklet.

[2]Ingeominas, Bogota, Colombia.

Mineral Resources of Ecuador: Their Development and Prospects for New Finds—Summary[1]

CARLOS F. MOSQUERA C.[2]

Early Exploration

Ecuador is an Andean country which, like other countries along the Pacific coast, should have important metallogenic provinces. Traces of mineral resources have been known since antiquity, but only recently has interest been aroused from the development of Ecuadorian mining prospects and recent petroleum discoveries.

Ecuador's mining history goes back to pre-colonial times when the Indians exploited the rich, gold-bearing veins and placers. There were sufficient quantities of precious metals to fill a room in order to pay, unsuccessfully, for the release of the Inca Atahualpa, who was taken prisoner by the Spaniards. This ransom must have been on the order of 20 million oz (about $3 billion at present prices) of gold. In addition, the ransom included jewels and works of art.

During the Spanish colonial period the major extraction of gold from mines and placers took place. In 1549, the Spaniards founded the town of Zaruma in the Portovelo mining district—the only area where there are working mines at present. At the same time, they went into the virgin Oriente jungle and built numerous villas at the sites of gold placers.

After independence, mining ceased completely, and the old mining prospects and explorations disappeared or were forgotten in the new era dedicated exclusively to farming activities.

In 1904, 80 years after the wars of independence, the Portovelo mining district began to be worked on a systematic basis. The South American Development Company (a North American company) exploited the mines with great success until 1950. Annual production was 60,000 oz of gold and 40,000 oz of silver. Until 1939 there was no recovery of copper, lead, or zinc, but the advent of World War II and the related rise in metal prices stimulated a new impulse of mining. In addition to precipitates of precious metals, concentrates of copper, lead, and zinc were produced.

The same year, another area that was exploited by the Spaniards—the Macuchi Mine—was worked by the Cotopaxi Exploration Company. In the war years, mining activities were supplemented with metallurgical treatment of the concentrates, yielding 3,500 tons annually of copper "blister."

In 1947 the mines closed at Macuchi, and in 1950 the American company withdrew from Portovelo. That year, operations of the Ecuadorian Compañía Industrial Minera (CIMA) were initiated. Stimulated by the increase in gold prices, the company continues to work the low-grade ore, but its reserves are nearing an end.

In recent years, areas with mineral indications that were explored, and at times exploited, by the Spaniards have been reopened by foreign and Ecuadorian enterprises. Of these, the following are the most important (Fig. 1).

The Toachi deposits (La Plata Mine) have similar geology and mineralogy to the Macuchi deposits. The principal ores are Au, Ag, Pb, and Cu. Toachi lies 100 km north of Macuchi, and is on the same, west flank of the Cordillera Occidental. At present, various foreign mining companies are interested in the area between the two mines, including the sector at Sigchos—an old Spanish gold prospect known to have copper.

The Molleturo deposits, 110 km north from Portovelo, on the same flank of the Cordillera, have polymetallic mineralization. Despite the possible importance of the deposits, lack of access roads owing to the rough topography has prevented significant exploration.

The Pilshum Mines are situated on Quaternary deposits of the Cordillera Real of the Andes, near the western boundary of the metamorphic rocks. At present, the government is hoping to stimulate interest in the evaluation of silver, lead, and barite of this deposit.

Recent Exploration

In 1965 the government concluded technical-assistance agreements with the United Nations, the French Institute of Petroleum, and the Japanese government in order to strengthen and rein-

[1]Manuscript received, October 21, 1974.

[2]Ministerio de Recursos Naturales y Energéticos, Dirección General de Geología y Minas, Quito, Ecuador.

The writer appreciates the help of J. Brian Kennerley and Roger Bristow C., who critically reviewed the manuscript.

force the recently created Servicio Nacional de Geología y Minas (today's Dirección General de Geología y Minas). Since 1969 the Dirección General de Geología Minas (DGGM) has received the technical cooperation of Great Britain in the geologic surveying of the country.

The cooperation of the United Nations lasted until February 1972, during which time intensive surveys of metallic and nonmetallic minerals were carried out as follows.

1. Coal investigations (Cuenca and Biblián-Loja areas). Extensive zones with coal seams led to the need to determine the total tonnage of coal by drilling. The investigations showed that, owing to the tectonic complexity, these lignites cannot be exploited economically by the usual mechanical method. The potential economic importance of the coal has diminished with the discovery of petroleum in the Ecuadorian Oriente. It is used at present only in the small regional industries.

2. Gold and base-metal sulfides (Portovelo-Zaruma area). The objective was to locate ore of commercial quality in the contiguous zones of the old mines. Geophysical exploration indicated the continuation of mineralization in the Viscaya zone, and this has permitted mining to continue until now.

3. Clays and travertine marble (Azuay-Cañar area). These operations covered essentially the same area as the coal investigations. The objective was to ensure both the quality and quantity of raw material needed by the ceramic and marble industries of the region. The results were favorable from the mining-geology point of view; there are large quantities and varieties of raw materials for the establishment of ceramic, refractory, and chemical industries, as well as sufficient limestone for cement plants, fertilizer, etc.

4. Glass sands and quartz (Santa Elena Peninsula). Efforts to locate raw material of good quality for the increased needs of the glass industries were not rewarded. Recently, however, the construction of new roads owing to the petroleum exploration in the Ecuadorian Oriente has led to the discovery along the foot of the Cordillera of great reserves of clean, white friable sandstones suitable for many industries.

5. Exploration and preliminary evaluation of metallic mineral deposits (Azuay-Cañar area). In this area, 14 prospects were evaluated on the basis of previously discovered geochemical anomalies of Cu, Mo, Zn, Pb, and Ag. The prospects with the more interesting results are the following:

Chaucha (Cu-Mo)—This was the first porphyritic copper-molybdenum deposit discovered in Ecuador. The preliminary studies, using geological, geochemical, and geophysical methods, gave good results in a central area of 3,500 ha. The government agreed to an exploration and exploitation program with the Overseas Mineral Resources Development Co. Ltd. of Japan, which proved 55 million tons of ore with 0.7% copper. A total of 83 diamond drillholes (7,983 m of perforation) and 497 m of exploration galleries were carried out. The prospect of increasing these reserve figures by enlarging the exploration area is good.

San Bartolome (Ag-Zn-Pb)—This prospect, lying 20 km southeast of Cuenca and accessible by road, is an area of mineralization of silver, zinc, and lead of economic possibilities. The detailed geologic work, with geochemical sampling, geophysical surveying, trenching, drilling (9 diamond drillholes with 1,330 m of perforation), and tunneling (300 m of exploration tunnels), led to an estimate of 23 million tons of ore with 2 oz Ag/ton. The mineralized area is 400 m long and 150 m wide and consists of numerous parallel and interlaced veinlets. Drilling showed that the mineralization reached a minimum depth of 200 m. The studies indicated the existence of a rich zone with 1 million tons of ore with a value of 10 oz Ag/ton.

Angas (Cu-Zn-Pb-Ag-Au)—This recently discovered polymetallic mineralized zone lies 40 km west of Cuenca, near Chaucha. Eight brecciated and silicified zones were located in an area of 6 km^2. Trenching of two of these revealed sulfurous mineralization which gave interesting values of Cu, Ag, Zn, Pb, and Au. Because of these promising results, the investigation is to continue, and the construction of 20 km of access road is programmed.

San Fernando (Ag)—In this zone, 45 km southwest of Cuenca, an intensely pyritized zone was discovered in association with silicified volcanic rocks. The zone has an area of 3 km^2, with sporadic values up to 3% zinc and 2 oz Ag/ST. Geological and geophysical surveying and diamond drilling have been carried out, but no concentrations of economic value have been found. The mineralogic association with extrusive volcanic deposits shows a similarity to deposits of other parts of the world where concentrations are of economic importance. In the Cordillera there are many mineralized zones analogous to those at San Fernando.

6. Investigation of placers of gold and platinum in northwestern Ecuador (near the Colombian border). Gold and platinum placers in this zone are gaining importance, and contracts with the government have been signed for the exploration

and exploitation of the placers. Because of administrative difficulties relative to the "Law of Mines and Gold Placers," it was not possible to realize the survey programmed by the United Nations for this zone. The government now is very much interested in promoting the gold mining of this region.

7. Investigation of iron ore and barite in the coastal region of Guayas and Manabí. In this region, diabasic rocks crop out in the core of the Coastal Cordillera; in localities near Pascuales, veins occur with iron ore. In this area of 8,315 km², airborne magnetic and radiometric surveys involving 19,420 km of flight lines were carried out, but the results were negative.

The barite of Pascuales is present in eluvial boulders which probably overlie siliceous veins containing barite. It was calculated that about 3,000 tons of barite-bearing boulders exist in this locality.

8. Systematic prospecting of the southern area of operation 5 (principally on the Western Cordillera as far as the border with Peru). This area of 8,300 km² was first covered by a geochemical survey of stream sediment samples at an average of one sample per 1.8 km². Because of inaccessibility of much of the area, a helicopter was employed for geochemical sampling, probably for the first time in the Andes. This support proved to be very efficient for an accelerated coverage. This work, together with regional geologic mapping, etc, led to the delineation of 15 zones of geochemical anomalies, most of which are located at contact zones between intrusives and volcanic rocks. Systematic evaluations of each were then made. The most important are the following.

Fierro Urco (Cu-Zn-Pb-Mo)—This porphyritic deposit covers an area of 93 km², but the zones with greater concentrations of copper and molybdenum have not yet been delineated. Near the mountain crest there is an old Spanish prospect over a lead-zinc vein which also carries silver and gold. This is considered as a fringe of lead mineralization related to the development of porphyritic copper-molybdenum ore of the zone. Although initial work has indicated the prospect is uneconomic, more exploration is needed.

Uritohuaser (Pb-Zn)—This geochemically anomalous zone led to the discovery of an important vein mineralization related to an intrusive-volcanic contact. The oxidized outcrop, cleared by trenching to each side of the stream bank, appeared to be 30–40 m wide and gave values in certain places of 25% zinc plus lead, and 3 oz Ag/ton. In other zones the average values are 7% and 1 oz/ton. At present, systematic soil sampling, geophysical surveying, and diamond drilling are being carried out. The mineralization may be 1 km long and up to 250 m high, indicating an important potential deposit.

Los Linderos (Cu-Mo)—This is a very arid zone near the border with Peru. Within the 5-km² anomalous zone is located an area of 0.75 km² with greater copper concentration associated with abundant malachite stains in a coarse quartz porphyry. Systematic geochemical sampling and geophysical surveys were carried out, followed by drilling of 10 shallow holes (50 m average). Sulfide mineralization with values of 0.20% copper and traces of molybdenum was encountered. It is proposed to build a road 20-km long to allow access to drilling rigs capable of reaching a depth of 300 m.

Río Playas (Cu-Mo)—This is also an arid zone, 40 km northwest from Los Linderos. The more important area is a thick hydrothermally altered zone of 10 km². Four coreholes totaling 653 m (total depths 112–227 m) were drilled prior to soil sampling and geophysical surveying. Values ranging from 0.1 to 0.5% copper were obtained. It is thought that these studies are not conclusive, and additional geophysical investigations and drilling to greater depths are proposed.

Saraguro (Sn-W)—This zone of mica schists, gneisses, and possible intrusives toward the east is situated 35 km north of Loja. During a regional geochemical reconnaissance, some panning concentrates were found to contain traces of cassiterite and scheelite—the first known occurrences in Ecuador.

Prospects for New Finds

As can be deduced from the foregoing, mining investigations have taken place principally in the southwest Andes. Here the Tertiary and Cretaceous rocks are intruded by granites and are not covered by more recent deposits. Also, the dry climate in some zones supports only sparse vegetation. A brief resume of the regional geologic picture of the country is one of north-south-trending provinces or belts. From west to east these are:

1. Coastal Cenozoic sedimentary belt—Tertiary and Quaternary strata, moderately to intensely folded and faulted.

2. Coastal Mesozoic belt—Mesozoic sedimentary and volcanic rocks, intensely folded and faulted along the central part of the coast (Chongón, Colonche Hills).

3. Mesozoic volcanic belt of the Western Andes—Mainly Mesozoic volcanic rocks, intensely folded and faulted along the Cordillera and cut by medium to large granodioritic batholiths.

4. Interandean Cenozoic belt (between Cordillera Occidental and Cordillera Real)—Tertiary and Quaternary lavas, tuffs, and pyroclastic units. Locally in the south, Tertiary freshwater sedimentary basins, moderately folded and faulted, occur along the Interandean valleys.

5. Eastern metamorphic belt—Mainly metamorphosed Mesozoic rocks, intensely folded and intruded by large granodioritic bodies in a belt that roughly coincides with the eastern slopes of the Cordillera Real.

6. Eastern Mesozoic belt—Mesozoic rocks, intensely folded and faulted on the anticlinal axis of the low Eastern Cordillera in the Amazon jungle.

7. Cenozoic sedimentary rocks of the Amazon basin—Mainly Tertiary strata with fans of Quaternary continental units in the west-central part toward the Cordillera Real.

The mineral deposits and the prospects outlined here are localized mainly in the Mesozoic volcanic belt of the Western Andes and, in part, in the Eastern metamorphic belt.

The prospecting work that is being carried out from the already-explored south area toward the central region is on the west flank of the Cordillera Occidental, but this work must overcome the problems of access and the dense vegetation. The belt being prospected is 120 km long, lying between 1 and 2°S, and is part of the area geologically surveyed by a technical-assistance geologic team of Great Britain and the DGGM.

The metamorphic belt of the Cordillera Real, because of its inaccessibility and high altitudes, presents serious difficulties to prospecting. At present, the DGGM is prospecting the sub-Andean valley of Valladolid, where it would seem there are important placers of gold and suspected mineralizations. In the high plateau of the southern Cordillera, the rich gold-bearing placers on the metamorphic schists which were exploited in ancient times are legendary. In general, much of the north-central Andes has a cover of Quaternary volcanic deposits mantling the older rocks but. until there are methods to probe through this thick mantle, the potential for mineralization remains unknown.

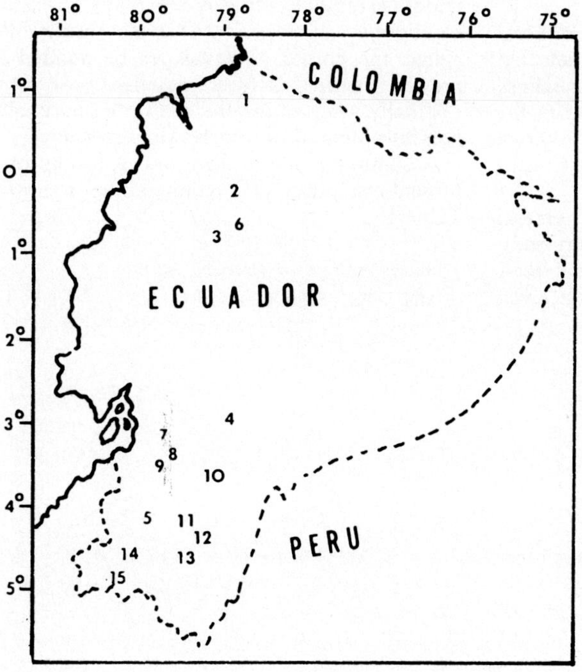

MINES

1. Playa de Oro (Esmeraldas)
2. La Plata
3. Macuchi
4. Pilshum
5. Portovelo

PROSPECTS

6. Sigchos
7. Molleturo
8. Angas
9. Chaucha
10. San Bartolome
11. Fierro Urco
12. Saraguro
13. Uritohuaser
14. Rio Playas
15. Los Linderos

FIG. 1–Location map of mines and prospects, Ecuador.

Mineral Resources Potential of Peru—Summary[1]

O. DAMIANI and C. CÁNEPA[2]

This paper is a preliminary, partial quantitative appraisal of the mineral resources of Peru.

Two criteria were established for a generalized classification of various mineral deposits: (1) To define a resource as a type of natural concentration of a chemical element in the earth's crust and to establish distinctions among resources on the basis of the amount of available data. (2) To categorize the mineral resources into two large groups based on whether or not profitability is taken into account; if it is, a further division is necessary to show whether such profitability is present profitability (at the time the appraisal is made) or future profitability (in the relatively near future).

Four categories are established under criterion (1). They are defined as follows:

Confirmed resources—mineral concentrations whose location, geometric characteristics, and chemical composition have been sufficiently defined by analysis and measurements and the geologically reasonable projections of these.

Indicated resources—concentrations of mineral substances whose occurrence has been detected by specific direct or indirect evidence, or whose existence within a given geologic unit and in the vicinity of mines or the vicinity of specific direct or indirect indications of mineralization is considered probable.

Inferred resources—concentrations of mineral substances about which no specific direct or indirect evidence has been detected, although their occurrence is considered possible within a given geologic unit and for a reasonable distance beyond the boundaries of the indicated resources.

Remote resources—occurrences of mineral substances with variable grades and within various geologic units which are assumed to exist around the areas with inferred resources. About such occurrences there are no concrete references.

Total resource potential—encompasses all four categories of resources, but its size is beyond our conventional assessment scale. In order to deal with more readily manageable (and meaningful) figures, it is necessary to define the concept of partial potential of mineral resources, the formulation of which is done by using the profitability criterion.

Criterion (2)—profitability—can be applied only to confirmed, indicated, and inferred resources. The partial potential of mineral resources is the sum of those parts of the confirmed, indicated, and inferred resources whose exploitation under currently existing and/or short-term future conditions can generate a certain profitability; hence, the partial potential can be divided into present potential or future potential.

Finally, *reserves* are the confirmed parts of the present potential of mineral resources.

A summary of the quantitative results of the present evaluation of 27 commodities is shown in Table 1.

[1]Manuscript received, December 26, 1974.

[2]Minero Peru, Lima, Peru.

Table 1. World Significance of Preliminary Evaluation of Peru's Principal Mineral Resources

Mineral	*Reserves of Peru (A)*	*World Reserves (B)*	*% (A/B)*	*Peru's Present Partial Potential (C)*	*Potential of World Recoverable Resources[1] (D)*	*% (C/D)*	*Production of Peru (1973)*	*Unit*
Antimony	111,600	3,629,000	3.08	440,020	8,709,100	5.05	1,300.00	MT
Arsenic	201,140	3,780,000	5.32	806,300	15,966,700	5.05	1,136.00	MT
Barite	1,814,400	181,440,000	1.00	7,257,600	471,744,000	1.54	226,800.00	MT
Bismuth	11,250	92,800	12.12	41,715	212,800	19.60	750.00	MT
Cadmium	31,800	516,710	6.15	112,210	1,538,400	7.29	117.30	MT
Coal[2]	124,936,000	643,885,200,000	0.02	1,030,936,000	8,618,000,000,000	0.01	NA	MT
Copper	25,463,700	335,664,000	7.59	102,777,100	698,600,000	14.71	218,815.00	MT
Diatomites	100,000,000	NA[4]	–	NA	NA	–	NA	MT
Tin	4,500	4,246,880	0.11	47,760	68,000,000	0.07	238.00	MT
Phosphates	550,000,000	4,649,400,000	11.83	707,500,000	17,000,000,000	4.16	NA	MT 30% P_2O_5
Gallium	660	110,000	0.60	2,320	240,000	0.97	NA	MT
Germanium	4,630	96,000	4.82	16,430	168,550	9.75	NA	MT
Iron	318,246,050	87,726,240,000	0.36	3,126,886,680	400,000,000,000	0.78	5,851,943.00	MT
Indium	585	9,500	6.16	2,065	28,300	7.30	3.72	MT
Mercury	E[3] 300	182,700	0.16	35,400	430,400	8.22	103.12	MT
Molybdenum	242,700	4,263,840	5.69	976,850	33,294,200	2.93	940.40	MT
Gold	174	31,100	0.56	1,950	150,000	1.30	3.00	MT
Silver	22,740	170,450	13.34	68,725	425,450	16.15	1,600.00	MT
Lead	3,214,000	130,636,800	2.46	11,917,800	550,000,000	2.17	199,164.00	MT
Rhenium	320	1,725	18.55	1,300	7,260	17.91	NA	MT
Selenium	11,500	151,050	7.61	46,400	314,350	14.76	7.70	MT
Thallium	39	1,260	3.10	140	18,150	0.77	NA	MT
Tellurium	2,280	33,570	6.79	9,160	107,390	8.53	20.39	MT
Tungsten	45,300	1,247,400	3.63	82,000	3,742,000	2.19	1,036.00	MT
Uranium	E U_3O_8 13,145	($10/lb) 1,022,400	1.29	18,345	NA	–	–	MT of U_3O_8
Vanadium	66,630	9,176,000	0.73	162,430	NA	–	–	MT
Zinc	7,318,300	118,843,200	6.16	25,808,300	353,840,000	7.29	348,365.00	MT

[1] For the potential of the recoverable resources of the world, those figures are taken which are estimated to correspond approximately to the present partial potential of Peru (C).

[2] Includes only bituminous coal plus anthracite.

[3] E = estimated.

[4] NA = not available.

Deposits of Bolivian Tin Belt—Summary[1]

S. RIVAS[2]

It is surprising that the 1,000-km-long Bolivian tin belt does not cross the national frontiers either north into Peru or south into Argentina. Lithologic and tectonic controls can explain this anomaly.

Three metallogenic cycles of different age can be distinguished. The oldest is in the northern part of the High Cordillera and is related to pegmatites and pneumatolytic veins in granitic batholiths of Triassic-Jurassic age. South of the latitude of La Paz, from the vicinity of Viloco to Colquiri, is the Miocene metallogenic cycle with its abundant tin and wolframite veins. South of Oruro, the lode deposits are related to small stocks and dikes of dacite and rhyolite of Pliocene age. The belt widens (100 km) and forms the famous polymetallic province, unique in the world because of the rare associations of tin, silver, and other minerals such as teallite, franckeite, aramayoite, andorite, and argyrodite (at Chocaya).

Inexplicably, the deposits are of shallow depth; the root zones of most are in the lower part of the hills, and the vertical difference in elevation of the zones is not more than 1,000 m (at Chorolque). There is zonal arrangement from tin-wolframite-bismuth, through lead-silver-zinc, to the outer zone of quartz-stibnite. In the southern province, the zonal arrangement is more local (at Tasna).

There are approximately 750 tin mines, 2,000 lead mines, 250 antimony mines, and about 30 mines for bismuth; most of them are small and shallow. The wealth of some of these mines is extraordinary. The total production from Llallagua (Catavi) is estimated as 600,000 tons of tin. There remains 1 million tons of proved and indicated reserves, 75% of which belongs to Corporacion Minera de Bolivia (COMIBOL).

The tin veins are exploited by selective underground methods despite high mining costs. The grade averages of the deposits have declined to limits which now can be considered marginal.

[1]From Circum-Pacific Energy and Mineral Resources Conference program booklet.

[2]Corporacion Minera de Bolivia (COMIBOL), La Paz, Bolivia.

Metallogenic Belts and Tectonic Evolution of Chilean Circum-Pacific Continental Border—Summary[1]

JOSÉ FRUTOS J.[2] and JORGE OYARZÚN M.[3]

A tectonic-metallogenic scheme shows the relations between the geologic evolution of the Chilean Circum-Pacific continental border from 17°S to 57°S and the genesis of the mineral deposits in the area. Thus the different metallogenic events can be analyzed from the viewpoint of palegeographic, magmatic, and tectonic-structural factors.

Mineral belts include deposits of the following metallogenic-tectonic events:

1. Paleozoic geosynclinal cycle: (a) early geosynclinal stage in early Paleozoic time—sedimentary-metamorphic iron and manganese deposits; (b) early to middle Paleozoic (Caledonian) diastrophism—nickel and chromium deposits.

2. Andean Mesozoic-Cenozoic geosynclinal cycle: (a) final marine-basin stage in Early Cretaceous time—sedimentary and volcanic manganese deposits; (b) mid-Cretaceous (Subhercynian) diastrophism—iron-apatite, copper, silver, and gold; (c) Paleocene to Miocene late orogenic phases—copper (porphyry type and exogenic) and silver; (d) Plio-Pleistocene taphrogenic phase—sulfur and effusive magnetite.

As a general conclusion of these relationships, it appears that Chilean metallogenesis is essentially controlled by the Andean orogenic episodes, to which most deposits are related.

As the Chilean continental border has been a mobile region since Precambrain time, it is likely that erosion during the successive tectonic cycles has erased most of the mineralization associated with its geologic development. In consequence, with the exception of Paleozoic iron, the age of economically important deposits is not older than Late Jurassic.

The metallogenesis of the Andean Mesozoic-Cenozoic geosynclinal cycle is controlled by relative position of former synclinal basins and the succeeding magmatic activity. Thus, eugeosynclinal facies are associated with the emplacement of copper deposits, and miogeosynclinal facies are related to polymetallic mineralization. Plutonic volcanism migrated to the east during the Andean cycle; therefore, younger deposits are coincident with the position of the present magmatic belt (iron, manganese, and sulfur associated with Andean volcanoes; porphyry copper deposits with the Andes south of lat. 31°S).

Faulting of continental, regional, and local magnitude has been a principal metallogenic factor because it controlled the emplacement of intrusives and extrusives, and thus mineralization. Faulting also controlled supergene enrichment and exogenic deposition through block uplift.

Volcanic or hybabyssal subcrustal calc-alkaline magmatism is related to most important copper, manganese, and sulfur mineralization. This fact may be related to the highly specialized copper-bearing metallogenic province of Chile and to the lack of stratiform polymettalic mineralization surrounding older crystalline massifs.

[1]Manuscript received, November 11, 1974.

[2]Departmento de Geologia, Universidad del Norte, Casilla 1280, Antofagasta, Chile.

[3]IRITUN, Universidad del Norte, Casilla 1280, Antofagasta, Chile.

INDEX

This index consists of two sections, which appear in the following order:

(1) Author Index and
(2) Keyword Index

The *author index* is arranged alphabetically according to each author's last name. For papers by more than one author, each author's name appears in the index in alphabetical order. The appearance of an author's name followed by the title of an article does not mean that he is the only author of that article. He may be one of two or more authors of the paper whose title follows his name. The author index does *not* show multiple authors in any single listing.

To locate a reference in the *keyword index*, the reader should begin by thinking of the significant words. Then he should look in the index for the keyword entry for each of those words. The reference codes will direct him to the pages.

The columns on the right-hand side of the keyword index give the page number and a code number (1 or 3) indicating the nature of the source. The code is:

(1) for phrase from title; and

(3) for phrase from abstract, text, table, figure, or figure caption.

The keyword for each entry is located at the left-hand side of the page. The (>) sign indicates the first word in each title or key phrase. The (<) sign indicates the end of the title or key phrase.

Author Index

Keyword Index